The Second Messenger Cyclic Di-GMP

The Second Messenger Cyclic Di-GMP

Edited by **Alan J. Wolfe** and **Karen L. Visick**

Loyola University Chicago Stritch School of Medicine
Maywood, Illinois

ASM PRESS *Washington, DC*

Address editorial correspondence to ASM Press, 1752 N St. NW, Washington, DC 20036-2904, USA

Send orders to ASM Press, P.O. Box 605, Herndon, VA 20172, USA
Phone: 800-546-2416; 703-661-1593
Fax: 703-661-1501
E-mail: books@asmusa.org
Online: estore.asm.org

Library of Congress Cataloging-in-Publication Data

The second messenger cyclic di-GMP / edited by Alan J. Wolfe and Karen L. Visick.
p. ; cm.
Includes bibliographical references and index.
ISBN 978-1-55581-499-1 (hard cover)
1. Cyclic guanylic acid. 2. Second messengers (Biochemistry) I. Wolfe, Alan J. II. Visick, Karen L. III. American Society for Microbiology.
[DNLM: 1. Cyclic GMP—physiology. 2. Bacterial Physiological Phenomena. 3. Cyclic GMP—analogs & derivatives. 4. Genome, Bacterial. 5. Signal Transduction—physiology. QU 58 S445 2010]
QP625.N89S43 2010
572.8'6293—dc22

2009036432

Current printing (last digit)
10 9 8 7 6 5 4 3 2 1

Cover illustration: A wrinkled colony formed by a mutant of *Vibrio cholerae.* The phenotype is associated with cyclic di-GMP production. (Courtesy of Nicholas J. Shikuma and Fitnat H. Yildiz.)

CONTENTS

CONTRIBUTORS

Sören Abel
Biozentrum, University of Basel, CH-4056 Basel, Switzerland

Phillip Aldridge
Centre for Bacterial Cell Biology and Institute for Cell and Molecular Biosciences, Newcastle University, Newcastle upon Tyne, NE2 4HH, United Kingdom

Dorit Amikam
P.O. Box 10356, Zichron Yakov 30900, Israel

Paul Babitzke
Department of Biochemistry and Molecular Biology, The Pennsylvania State University, University Park, PA 16802

Sinem Beyhan
Department of Microbiology and Environmental Toxicology, University of California, Santa Cruz, Santa Cruz, CA 95064

Alexander G. Bobrov
Department of Microbiology, Immunology, and Molecular Genetics, University of Kentucky, Lexington, KY 40536-0298

Ronald R. Breaker
Department of Molecular, Cellular and Developmental Biology, Department of Molecular Biophysics and Biochemistry, and Howard Hughes Medical Institute, Yale University, New Haven, CT 06520

Andrew Camilli
Howard Hughes Medical Institute and Department of Molecular Biology and Microbiology, Tufts University School of Medicine, Boston, MA 02111

J. Maxwell Dow
BIOMERIT Research Centre, Department of Microbiology, BioSciences Institute, University College Cork, Cork, Ireland

Thomas Ebensen
Department of Vaccinology and Applied Microbiology, Helmholtz Centre for Infection Research, 38124 Braunschweig, Germany

Michael Y. Galperin
National Center for Biotechnology Information, National Library of Medicine, National Institutes of Health, Bethesda, MD 20894

Mark Gomelsky
Department of Molecular Biology, University of Wyoming, Laramie, WY 82071

Nina Grantcharova
Department of Microbiology, Tumor and Cell Biology, Karolinska Institutet, SE-171 77 Stockholm, Sweden

Carlos A. Guzmán
Department of Vaccinology and Applied Microbiology, Helmholtz Centre for Infection Research, 38124 Braunschweig, Germany

Caroline S. Harwood
Department of Microbiology, The University of Washington, Seattle, WA 98195

Regine Hengge
Institut für Biologie—Mikrobiologie, Freie Universität Berlin, 14195 Berlin, Germany

John F. Hunt
Department of Biological Sciences, Columbia University, New York, NY 10027

Urs Jenal
Biozentrum, University of Basel, CH-4056 Basel, Switzerland

Kristina Jonas
Department of Microbiology, Tumor and Cell Biology, Karolinska Institutet, SE-171 77 Stockholm, Sweden

David K. R. Karaolis
Intragenics Research Institute, Havre de Grace, MD 21078, and Karagen Pharmaceuticals, Baltimore, MD 21210

Dorothy Kim
Department of Biological Sciences, Columbia University, New York, NY 10027

Agaristi Lamprokostopoulou
Department of Microbiology, Tumor and Cell Biology, Karolinska Institutet, SE-171 77 Stockholm, Sweden

Elaine R. Lee
Department of Molecular, Cellular and Developmental Biology, Yale University, New Haven, CT 06520

Rimma Libanova
Department of Vaccinology and Applied Microbiology, Helmholtz Centre for Infection Research, 38124 Braunschweig, Germany

Linda L. McCarter
Department of Microbiology, The University of Iowa, Iowa City, IA 52242

Yvonne McCarthy
BIOMERIT Research Centre, Department of Microbiology, BioSciences Institute, University College Cork, Cork, Ireland

Öjar Melefors
Department of Microbiology, Tumor and Cell Biology, Karolinska Institutet, SE-171 77 Stockholm, Sweden

Peter D. Newell
Department of Microbiology and Immunology, Dartmouth Medical School, Hanover, NH 03755

George A. O'Toole
Department of Microbiology and Immunology, Dartmouth Medical School, Hanover, NH 03755

Matthew Parsek
Department of Microbiology, University of Washington, School of Medicine, Seattle, WA 98195-7242

Robert D. Perry
Department of Microbiology, Immunology, and Molecular Genetics, University of Kentucky, Lexington, KY 40536-0298

Jason T. Pratt
Howard Hughes Medical Institute and Department of Molecular Biology and Microbiology, Tufts University School of Medicine, Boston, MA 02111

Tony Romeo
Department of Microbiology and Cell Science, University of Florida, Gainesville, FL 32611

Ute Römling
Department of Microbiology, Tumor and Cell Biology, Karolinska Institutet, SE-171 77 Stockholm, Sweden

Robert P. Ryan
BIOMERIT Research Centre, Department of Microbiology, BioSciences Institute, University College Cork, Cork, Ireland

Tilman Schirmer
Core Program Structural Biology and Biophysics, Biozentrum, University of Basel, CH-4056, Basel, Switzerland

Theodore J. Standiford
Department of Internal Medicine, Division of Pulmonary and Critical Care Medicine, University of Michigan Medical Center, Ann Arbor, MI 48109

Narasimhan Sudarsan
Howard Hughes Medical Institute, Yale University, New Haven, CT 06520

Rita Tamayo
Department of Microbiology and Immunology, The University of North Carolina at Chapel Hill, Chapel Hill, NC 27599

Karen L. Visick
Department of Microbiology and Immunology, Loyola University Chicago, Maywood, IL 60153

Christopher M. Waters
Department of Microbiology and Molecular Genetics, Michigan State University, East Lansing, MI 48824

Haim Weinhouse
Department of Biological Chemistry, The Alexander Silberman Institute for Life Sciences, Hebrew University of Jerusalem, Jerusalem 91904, Israel

Alan J. Wolfe
Department of Microbiology and Immunology, Loyola University Chicago, Maywood, IL 60153

Fitnat H. Yildiz
Department of Microbiology and Environmental Toxicology, University of California, Santa Cruz, Santa Cruz, CA 95064

PREFACE

In recent years, researchers have discovered a completely new type of signaling mechanism based on cyclic di-GMP [c-di-GMP, also known as bis-(3′,5′)-cyclic diguanylic acid or cyclic diguanylate]. This newly recognized second messenger is strongly associated with the transition between the motile and sessile states of many bacteria. This so-called "stick-or-swim" decision is fundamental to the formation and dispersion of biofilms, and the correlation between c-di-GMP and biofilm development has been found in many bacteria. This association appears to result from the ability of c-di-GMP to induce the synthesis of extracellular polysaccharides and to cause decreased motility. Additional phenotypes associated with c-di-GMP are also becoming apparent.

The near universality of this second messenger combined with the biomedical relevance and commercial importance of biofilms has generated a lot of interest in the emerging field that has arisen around c-di-GMP. Indeed, this interest is demonstrated in the numerous reviews that already exist on this topic. It is our hope that this book will provide an additional and more complete resource for existing members of this new field, a detailed state-of-the-field account for those on its fringes, and an overview for the uninitiated. As the field is sufficiently developed to benefit from such a book, we believe the timing is ideal.

This book examines the field of c-di-GMP signaling from a variety of angles. It begins with an introductory chapter that relates c-di-GMP to the more famous second messenger cAMP. It proceeds to recount the discovery of c-di-GMP, demonstrate the importance of bioinformatics in the development and continued evolution of the field, and describe the fundamental structure, function, regulation, and integration of those pathways. Other chapters describe the diversity of roles identified to date, including flagellar biogenesis and motility, extracellular polysaccharide biosynthesis, biofilm development, virulence, and innate host immunity. The book ends with a perspective on the future of the field.

We thank the authors of this book for putting together terrific chapters and also for serving as expert reviewers for other chapters. We also thank our outside reviewers, Ann Stock, John Kirby, Bob Belas, and Adam Driks. We particularly thank Greg Payne at ASM Press for rapidly and knowledgeably answering all of our many questions and providing us with advice. Finally, we also thank our senior production editor, Susan Birch.

Alan J. Wolfe
Karen L. Visick

I. SECOND MESSENGERS: AN OVERVIEW

The Second Messenger Cyclic Di-GMP
Edited by A. J. Wolfe and K. L. Visick

Chapter 1

Introduction to Second Messengers: Lessons from Cyclic AMP

ALAN J. WOLFE

THE DAWNING OF A NEW SECOND MESSENGER: c-di-GMP

At the turn of the 21st century, researchers linked the second messenger bis-(3′,5′) cyclic diguanylic acid (also known as cyclic di-GMP [c-di-GMP]) to proteins that contained either the GGDEF domain, the EAL domain, or both and showed that these proteins were ubiquitous. It is now accepted that proteins containing the GGDEF and EAL domains generally participate in c-di-GMP synthesis and degradation, respectively. The field evolved rapidly, beginning with a series of reports dating back to 1985, when Moshe Benziman and coworkers identified, isolated, and characterized c-di-GMP (reviewed in reference 21; see chapter 2). In 1995, Gregory Hecht and Austin Newton reported the existence of a novel domain, which they termed GGDEF (11). In 1998, Benziman and colleagues cloned and sequenced genes that encoded enzymes responsible for c-di-GMP synthesis (diguanylate cyclase) and degradation (phosphodiesterase [PDE]) (23). In 1999, Aldridge and Jenal provided evidence that the GGDEF domain functioned in signaling (1). In 2000, Römling and coworkers showed that Benziman's enzymes contained the GGDEF domain and demonstrated that a protein containing this domain played a role in a multicellular behavior (20). In 2001, two bioinformatic analyses revealed the ubiquity of GGDEF domain proteins (10, 16; see chapter 3), while Benziman and coworkers presented genetic evidence that linked the GGDEF domain to diguanylate cyclase activity (2). The first review that linked these reports appeared the following year (19). Although much remained to be learned, e.g., the biochemistry of c-di-GMP turnover, the structure of the enzymes that perform that turnover, the mechanisms by which c-di-GMP performs its functions, and the diverse roles played by c-di-GMP systems, a new field had been launched.

In many ways, the conceptual problems encountered in the c-di-GMP field—for example, how does a readily diffusible molecule permit specific signaling to occur?—parallel those that have been examined for years in the cAMP field. Thus, to help you fully appreciate the rapidly maturing c-di-GMP story, I will take you back in time to the discovery of cyclic 3′, 5′ AMP (cAMP).

THE ORIGINAL SECOND MESSENGER: THE DISCOVERY OF cAMP

In the late 1950s, Earl W. Sutherland was exploring the impact of the hormone epinephrine (also known as adrenaline) on liver tissue. His broader aim was to elucidate the biochemical mechanism by which epinephrine causes cells to degrade glycogen into sugar. Researchers had reported that epinephrine was produced by the adrenal gland and that this epinephrine traveled from the adrenal gland to the liver cells that store glycogen; however, the mechanism by which epinephrine elicited this effect remained unknown.

In 1958, with coworker T. W. Rall, Sutherland discovered that epinephrine stimulates the synthesis of cAMP. Further studies by Sutherland, Gilman, Rodbell, Krebs, and others over several decades subsequently provided us with a simple model for the action of this second messenger (Fig. 1A). The primary signal or first messenger (e.g., epinephrine) binds to the extracellular domain of an integral cytoplasmic membrane receptor. This causes a conformational change sensed by the cytoplasmic domain

Alan J. Wolfe • Department of Microbiology and Immunology, Loyola University Chicago, Maywood, IL 60153.

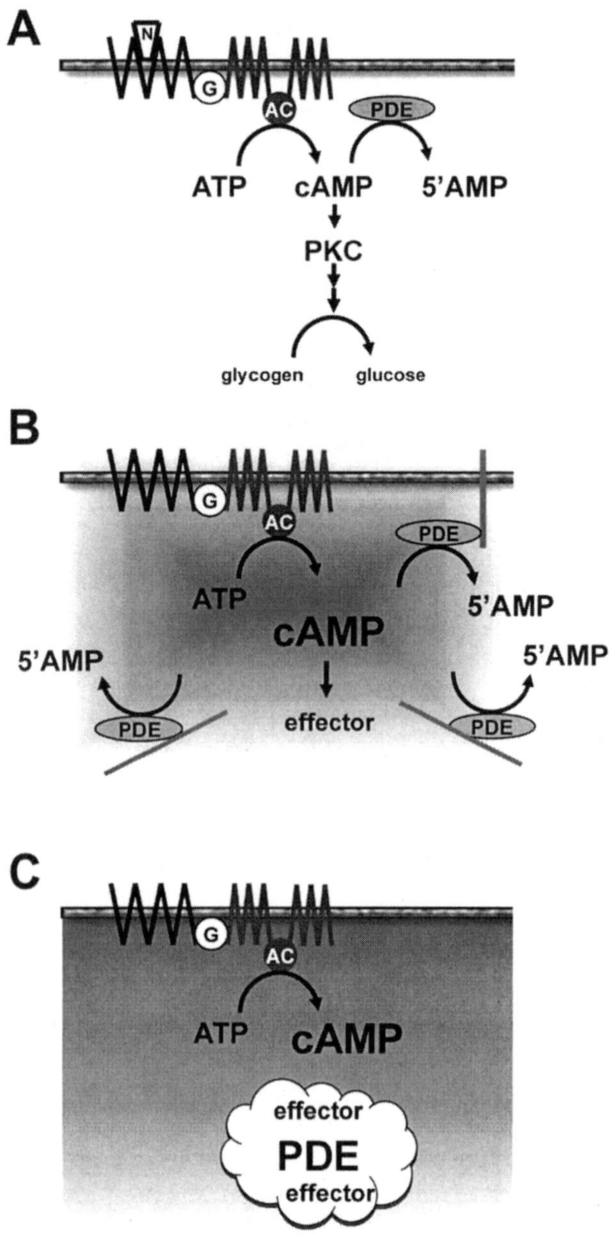

Figure 1. Models for cAMP-based signal transduction. (A) The simple model. An extracellular stimulus, e.g., norepinephrine (N), binds to and activates an integral membrane G-protein-associated receptor (black). The associated G protein (G) transduces that information to an integral membrane adenylyl cyclase (AC), which uses ATP to synthesize cAMP, which is degraded to 5′c-AMP by a soluble PDE. Thus, the balance between synthesis and degradation sets the concentration of cAMP, which binds to an effector, e.g., PKC. PKC activates the process by which glycogen becomes metabolized to glucose. (B) The barrier hypothesis. PDEs anchored to the membrane or to the cytoskeleton form an enzymatic barrier around an AC such that the cAMP synthesized by the associated AC remains localized. Thus, only effectors located in the vicinity of the AC become activated. (C) The sink hypothesis. PDE activity depletes its neighborhood of cAMP, ensuring that colocalized effectors do not become activated.

of the receptor, which interacts with a GTP-binding or G protein that conveys the primary hormone signal to adenylyl cyclase. Once activated, adenylyl cyclase converts ATP into cAMP, which is inactivated by a PDE. The action of these two opposing enzymes sets the concentration of this intracellular second messenger, which in turn activates an effector (e.g., protein kinase C [PKC]), which interacts with a downstream target to initiate a response (e.g., the degradation of glycogen). A similar scenario has been determined for cGMP. For an excellent history of cyclic nucleotide-based signal transduction, see reference 5.

THE cAMP NETWORK

Originally, adenylyl cyclase was thought to be a single entity. Today, the mammalian cAMP signal transduction network is known to include a dizzying array of G-protein-coupled receptors, a plethora of G-protein subunits, 10 adenylyl cyclases (9 membrane bound and 1 cytosolic), 11 PDE families, and multiple cAMP effectors, including PKA, PKC, diverse guanine exchange factors, and a variety of cyclic nucleotide-gated ion channels. It is now generally accepted that this complex network integrates numerous and disparate extracellular stimuli to control a wide range of diverse cellular processes in a remarkably stimulation-specific manner (4, 5, 8, 25).

The parallels between the mammalian cAMP network and the bacterial systems centered on c-di-GMP are striking. Both include multiple cyclases that can synthesize their respective second messenger and multiple PDEs that degrade it. These enzymes are or appear to be regulated by multiple sensory input domains. Both have evolved diverse second-messenger-binding effectors, and both regulate or appear to regulate a variety of cellular processes with a high degree of specificity. Finally, both networks revolve around a cyclic nucleotide that, in the absence of barriers, can freely diffuse throughout the cytoplasm. Note that the diffusion coefficient for cAMP in aqueous solution is in the hundreds of micrometers squared per second (7). Without some barrier, therefore, all cAMP effectors would possess an equal opportunity to become activated, with dire consequences for normal cell function (4, 5, 8).

COMPARTMENTALIZATION

How Can Specificity Be Achieved?

For the mammalian cAMP network, one answer to this question has become clear: compartmentalization. In these multicellular organisms, compartmentalization is achieved, in part, through cell-specific expression. This mechanism could be relevant to second-messenger-associated behaviors within multicellular communities of bacteria. Compartmentalization also can be achieved by organelle-specific expression, a mechanism perhaps relevant to bacterial species that possess cytoplasmic organelles. Finally, compartmentalization can be achieved via the localization of cAMP signaling pathways to microdomains within a given cell (reviewed in references 4, 5, and 25).

Why Would Cells Organize cAMP Signaling Complexes in Microdomains?

One obvious reason to compartmentalize proteins within microdomains is to bring related signaling components into close proximity. Other reasons could be to insulate signaling from the vagaries of cellular activity, to concentrate substrate and regulatory factors, and/or to protect the rest of the cell from the cAMP produced by any given pathway (4, 25).

How Are Microdomains Formed?

The mammalian plasma membrane is heterogeneous. This heterogeneity is produced, in part, by the concentration to specific locations of large amounts of cholesterol and sphingolipids. These regions of heterogeneity, called lipid rafts, possess properties distinct from the rest of the plasma membrane, including the propensity to attract certain types of proteins. Lipid raft domains also can contain caveolin, a scaffold protein that mediates the formation of invaginations, termed caveolae (14, 17, 22). Because of their propensity to recruit certain proteins and exclude others, it has been proposed that these rafts provide a mechanism for concentrating receptors, G proteins, adenylyl cyclases, effectors, and the rest of the machinery that comprise a cAMP signaling pathway (8, 25).

The concept that lipid rafts might concentrate the components of a particular signaling pathway and thus improve the efficiency of their interactions has been embraced by many but not all (3, 14, 18). However, evidence exists that the membrane-associated adenylyl cyclases do indeed localize to microdomains and that they form close associations with other components of their signaling pathways, including cAMP effectors and specific cAMP PDEs (25). For example, the development of probes that can monitor cAMP in living cells has revealed the existence of discrete gradients of cAMP that emanate from the plasma membrane following receptor stimulation. Given the large diffusion coefficient of cAMP, such gradients could only form if there existed some barrier to diffusion. These barriers could be physical, e.g., the plasma membrane or the underlying cytoskeleton. Alternatively, they could be enzymatic, e.g., via the action of PDEs that degrade the cAMP as it diffuses past them. Indeed, evidence supports the notion that PDEs play a central role in the formation of these gradients (4, 25).

How Are PDEs Compartmentalized?

To date, researchers have identified 11 distinct classes of PDEs, and within these classes, they have identified more than 20 different splice variants (5, 13, 25). Some PDEs specifically degrade cAMP, some degrade only cGMP, and some degrade both cyclic nucleotides; the substrate specificity depends on subtle structural differences in the catalytic domain (26). For example, PDE1 possesses dual-substrate specificity, while PDE4 specifically catalyzes the degradation of cAMP. Like adenylyl cyclases, PDEs are diversely regulated, responding to a wide range of stimuli. Unlike adenylyl cyclases, many PDEs are soluble (8, 25).

One way for a PDE to shape a cAMP gradient is to colocalize with an adenylyl cyclase. Colocalization of the mostly soluble PDEs is mediated primarily by scaffold (or anchor) proteins, e.g., A-kinase-anchoring proteins, β-arrestins, myomegalin, and the PKC-binding protein RACK1 (4, 25). Since many scaffold proteins can bind multiple signaling components, they function as building blocks to assemble macromolecular signaling complexes. Since they both bind multiple components and anchor to a specific cellular location, they provide spatial organization and, thus, specificity to signal routing (12, 15). For signaling pathways that rely on a diffusible second messenger, e.g., cAMP, such specificity is particularly critical (15).

How Do PDEs Facilitate the Formation of cAMP Gradients?

Two major hypotheses have been proposed to explain the ability of PDEs to shape cAMP gradients: the barrier hypothesis and the sink hypothesis (9, 24). The barrier hypothesis argues that PDEs anchored in the vicinity of an adenylyl cyclase form an enzymatic barrier that degrades cAMP as it passes by (Fig. 1B). The sink hypothesis argues that PDEs colocalize with specific cAMP effectors and deplete cAMP locally (Fig. 1C) (reviewed in reference 4).

Perhaps both hypotheses are correct. One subset of PDEs could function as a barrier, keeping cAMP from spreading from its source. Others might deplete cAMP from a discrete intracellular location, denying access to specific cAMP effectors anchored within that locale (6). In either case, the local concentration of cAMP will not faithfully reflect the overall or global cytosolic concentration (8).

What Other Mechanisms Impact Compartmentalization?

Certainly other mechanisms must contribute to the organization of the cAMP network. Most important, perhaps, microdomains are likely dynamic structures, assembling and disassembling depending on the physiological requirements of the cell (25). In response to its ever-changing environment (both external and internal), a cell may express a signaling component or not, localize that component or not, and activate it or not. The possibilities are endless. For example, an entire microdomain could assemble at a particular location, operate for a time, and then melt away to reassemble at a different time or a different place. Alternatively, a single component could be affected, altering the output of its associated pathway.

The evidence for such dynamics in mammals is relatively slim. It is generally accepted that most tissues and many cell types express more than one adenylyl cyclase species and that these enzymes are differentially expressed in different types of cells and at different stages of development. Furthermore, cyclases with different regulatory characteristics tend to be coexpressed, while those cyclases with similar properties tend not to. However, these patterns have not been confirmed at the single-cell level (25).

Thus, an opportunity arises for the new c-di-GMP field. Already, in evaluating single bacterial cells, the field has generated evidence for distinct—and dynamic—patterns of localization for enzymes involved in c-di-GMP synthesis. Discrete sets of synthetic and degradative enzymes appear to work in specific pathways. Of course, the problem of assaying small amounts of a signaling molecule is a prominent concern. However, with the insights garnered from the cAMP field and the tools already available to the c-di-GMP field—notably, the ability to genetically manipulate bacteria—rapid progress toward understanding the issues of specificity and bacterial compartmentalization might be expected.

In this book, you will learn about the history underlying the founding of the c-di-GMP field (chapter 2) and the bioinformatics that provided the critical thrust for its formation (chapter 3). You will be introduced to the biochemistry of the basic c-di-GMP module (chapters 4 and 5), learn how to detect c-di-GMP (chapter 6), and gain insights from crystal structures of c-di-GMP synthetic, degradative, and binding proteins (chapter 7). An overview of lifestyle choices, specifically the transition between motile and biofilm behaviors (chapter 8), will guide you as you become acquainted with c-di-GMP control in a variety of specific model organisms (chapters 9 to 19), including *Caulobacter crescentus, Salmonella enterica, Pseudomonas aeruginosa, Vibrio parahaemolyticus, Vibrio fischeri, Escherichia coli, Vibrio cholerae, Yersinia pestis,* and *Pseudomonas fluorescens.* Next, you will appreciate the importance of c-di-GMP in pathogenesis and host immunity (chapters 20 to 22). Finally, the concluding chapter will provide you with some perspective on the present state of the field and, through history, give you some insights into important future directions of the c-di-GMP field.

The diversity, complexity, and near-ubiquity of the c-di-GMP pathway components in bacteria, like that seen for the cAMP field, together are responsible, in large part, for capturing the imagination of the researchers in this newly established field. It is fertile ground both for the discovery of new signaling paradigms and for the design of therapies specific for some of the most tenacious bacterial infections. It has come of age exactly 25 years after the discovery of its central character, the fascinating second messenger c-di-GMP.

Acknowledgment. I thank my coeditor and long-term collaborator, Karen Visick, for all her help and encouragement.

REFERENCES

1. **Aldridge, P., and U. Jenal.** 1999. Cell cycle-dependent degradation of a flagellar motor component requires a novel-type response regulator. *Mol. Microbiol.* **32**:379–391.
2. **Ausmees, N., R. Mayer, H. Weinhouse, G. Volman, D. Amikam, M. Benziman, and M. Lindberg.** 2001. Genetic data indicate that proteins containing the GGDEF domain possess diguanylate cyclase activity. *FEMS Microbiol. Lett.* **204**:163–167.
3. **Babiychuk, E. B., and A. Draeger.** 2006. Biochemical characterization of detergent-resistant membranes: a systematic approach. *Biochem. J.* **397**:407–416.
4. **Baillie, G. S.** 2009. Compartmentalized signalling: spatial regulation of cAMP by the action of compartmentalized phosphodiesterases. *FEBS J.* **276**:1790–1799.
5. **Beavo, J. A., and L. L. Brunton.** 2002. Cyclic nucleotide research—still expanding after half a century. *Nat. Rev. Mol. Cell Biol.* **3**:710–718.
6. **Bender, A. T., and J. A. Beavo.** 2006. Cyclic nucleotide phosphodiesterases: molecular regulation to clinical use. *Pharmacol. Rev.* **58**:488–520.
7. **Bowen, W. J., and H. L. Martin.** 1964. The diffusion of adenosine triphosphate through aqueous solutions. *Arch. Biochem. Biophys.* **107**:30–36.
8. **Cooper, D. M. F.** 2003. Regulation and organization of adenylyl cyclases and cAMP. *Biochem. J.* **375**:517–529.

9. **Fischmeister, R., L. R. V. Castro, A. Abi-Gerges, F. Rochais, J. Jurevicius, J. Leroy, and G. Vandecasteele.** 2006. Compartmentation of cyclic nucleotide signaling in the heart: the role of cyclic nucleotide phosphodiesterases. *Circ. Res.* **99:**816–828.
10. **Galperin, M. Y., A. N. Nikolskaya, and E. V. Koonin.** 2001. Novel domains of the prokaryotic two-component signal transduction systems. *FEMS Microbiol. Lett.* **203:**11–21.
11. **Hecht, G. B., and A. Newton.** 1995. Identification of a novel response regulator required for the swarmer-to-stalked-cell transition in *Caulobacter crescentus. J. Bacteriol.* **177:**6223–6229.
12. **Jordan, J. D., E. M. Landau, and R. Iyengar.** 2000. Signaling networks: the origins of cellular multitasking. *Cell* **103:**193–200.
13. **Mehats, C., C. B. Andersen, M. Filopanti, S.-L. C. Jin, and M. Conti.** 2002. Cyclic nucleotide phosphodiesterases and their role in endocrine cell signaling. *Trends Endocrinol. Metab.* **13:**29–35.
14. **Munro, S.** 2003. Lipid rafts: elusive or illusive? *Cell* **115:**377–388.
15. **Pawson, T.** 2004. Specificity in signal transduction: from phosphotyrosine-SH2 domain interactions to complex cellular systems. *Cell* **116:**191–203.
16. **Pei, J., and N. V. Grishin.** 2001. GGDEF domain is homologous to adenylyl cyclase. *Proteins* **42:**210–216.
17. **Pike, L. J.** 2009. The challenge of lipid rafts. *J. Lipid Res.* **50:** S323–S328.
18. **Pike, L. J.** 2003. Lipid rafts: bringing order to chaos. *J. Lipid Res.* **44:**655–667.
19. **Römling, U.** 2002. Molecular biology of cellulose production in bacteria. *Res. Microbiol.* **153:**205–212.
20. **Römling, U., M. Rohde, A. Olsén, S. Normark, and J. Reinköster.** 2000. *AgfD,* the checkpoint of multicellular and aggregative behaviour in *Salmonella typhimurium* regulates at least two independent pathways. *Mol. Microbiol.* **36:**10–23.
21. **Ross, P., R. Mayer, and M. Benziman.** 1991. Cellulose biosynthesis and function in bacteria. *Microbiol. Mol. Biol. Rev.* **55:**35–58.
22. **Simons, K., and D. Toomre.** 2000. Lipid rafts and signal transduction. *Nat. Rev. Mol. Cell Biol.* **1:**31–39.
23. **Tal, R., H. C. Wong, R. Calhoon, D. Gelfand, A. L. Fear, G. Volman, R. Mayer, P. Ross, D. Amikam, H. Weinhouse, A. Cohen, S. Sapir, P. Ohana, and M. Benziman.** 1998. Three *cdg* operons control cellular turnover of cyclic di-GMP in *Acetobacter xylinum:* genetic organization and occurrence of conserved domains in isoenzymes. *J. Bacteriol.* **180:**4416–4425.
24. **Terrin, A., G. Di Benedetto, V. Pertegato, Y.-F. Cheung, G. S. Baillie, M. J. Lynch, N. Elvassore, A. Prinz, F. W. Herberg, M. D. Houslay, and M. Zaccolo.** 2006. PGE1 stimulation of HEK293 cells generates multiple contiguous domains with different [cAMP]: role of compartmentalized phosphodiesterases. *J. Cell Biol.* **175:**441–451.
25. **Willoughby, D., and D. M. F. Cooper.** 2007. Organization and Ca2+ regulation of adenylyl cyclases in cAMP microdomains. *Physiol. Rev.* **87:**965–1010.
26. **Zhang, K. Y. J., G. L. Card, Y. Suzuki, D. R. Artis, D. Fong, S. Gillette, D. Hsieh, J. Neiman, B. L. West, C. Zhang, M. V. Milburn, S.-H. Kim, J. Schlessinger, and G. Bollag.** 2004. A glutamine switch mechanism for nucleotide selectivity by phosphodiesterases. *Mol. Cell* **15:**279–286.

II. GENERAL PRINCIPLES: THE CORE

The Second Messenger Cyclic Di-GMP
Edited by A. J. Wolfe and K. L. Visick

Chapter 2

Moshe Benziman and the Discovery of Cyclic Di-GMP

DORIT AMIKAM, HAIM WEINHOUSE, AND MICHAEL Y. GALPERIN

The story of cyclic dimeric (3′→5′) GMP (c-di-GMP), the subject of this book, now recognized as a universal bacterial second messenger (27, 47, 48, 61, 62, 81), began in the early 1980s in the laboratory of Moshe Benziman (Fig. 1) at the Department of Biological Chemistry of The Hebrew University of Jerusalem, Israel. The discovery of this novel nucleotide in 1987 was a result of many years of studying the enzymatic mechanism and regulation of cellulose biosynthesis in the gram-negative bacterium *Acetobacter xylinum* (current name *Gluconacetobacter xylinus*) by Benziman and his students and colleagues (see references 29, 65, and 67). In the subsequent years and until his untimely death in 2003, Benziman and his colleagues published more than a dozen papers describing various aspects of c-di-GMP biochemistry (66, 89) and finally identifying the enzymes responsible for c-di-GMP synthesis and hydrolysis (79). Some details of the history of c-di-GMP discovery can be found in the 1991 review written by Benziman with his graduate students Peter Ross and Raphael Mayer (67) and in a book chapter by Deborah Delmer, who was at the time on a sabbatical in Benziman's lab (29). However, the full story remains to be told. The principal author of this chapter, Dorit Amikam (Fig. 2), spent the years 1979 and 1987 in Moshe's lab as a student and visiting scientist, later became Moshe's wife, and witnessed firsthand the fascinating story of the c-di-GMP discovery. Based on personal memories, recollections of Benziman's former students and scientists from his lab, and his published work, we try here to present a fuller picture of this great scientist and his work that led to the discovery of c-di-GMP.

Moshe Benziman was born in 1928 in Jerusalem, the city where his ancestors had lived for several generations and where he spent most of his life. His ancestors were among the first watchmakers in Palestine in the 19th century. His grandfather established the first modern winery in Palestine, which exported its products to Europe as early as the mid-19th century. Benziman's early years were tightly linked with the history of the formation of the State of Israel. As a 19-year-old-youth, Benziman interrupted his studies at the Hebrew University of Jerusalem to actively participate in Israel's War of Independence and was among the small number of defenders of the Old City of Jerusalem in 1948. When the Arab Legion captured the city, Benziman was taken prisoner of war and spent 9 months in a camp in Jordan before the armistice of 1949 allowed him to return home. After that, Benziman resumed his studies at the Hebrew University. He stayed at the Hebrew University for graduate studies in microbial biochemistry under the supervision of Andor Fodor and in 1957 received a Ph.D. for his thesis titled *Utilization of Amino Acids by Wine Yeasts.* After a postdoctorate from 1957 to 1959 with Irwin C. Gunsalus at the University of Illinois in Urbana, Benziman joined the newly formed Department of Biological Chemistry of the Hebrew University of Jerusalem, first as a lecturer, then a senior lecturer, and an associate professor. In 1977, he became a full professor in the same department and held this post for more than 20 years. Between the years 1973 and 1979, Benziman served as head of the department. In 1980, he spent time at R. Malcolm Brown's lab at the University of North Carolina, Chapel Hill. In 1984 and 1985, he served as President of the Israel Biochemical Society. In 1985, Benziman was awarded the Landau Prize for

Dorit Amikam • Molecular Oncology Laboratory, Rambam Medical Center, Haifa 31096, and Department of Biotechnology and Environmental Sciences, Tel-Hai Academic College, Tel-Hai 12210, Israel (retired). **Haim Weinhouse** • Department of Biological Chemistry, The Alexander Silberman Institute for Life Sciences, Hebrew University of Jerusalem, Jerusalem 91904, Israel. **Michael Y. Galperin** • National Center for Biotechnology Information, National Library of Medicine, National Institutes of Health, Bethesda, MD 20894.

Figure 1. Moshe Benziman in a 1996 photograph.

Science and Research, an annual award recognizing outstanding achievements by Israeli scientists. Figure 1 shows a photograph of Benziman taken circa 1996, when he became an Emeritus Professor. Moshe Benziman passed away 10 September 2003.

Figure 2. (A) Dorit Amikam as a visiting scientist in Moshe Benziman's laboratory in 1987. (B) Moshe Benziman with his favorite pipe.

THE PUZZLE OF CELLULOSE BIOSYNTHESIS IN PLANTS

In sheer amount, cellulose is one of the most abundant organic macromolecules on Earth and a key component of the planetary carbon cycle. A cellulose molecule is a homopolymer of D-glucose residues linked in β-1,4-glucosidic bonds. In plants and algae, individual cellulose molecules combine into fibrils, which appear to be aligned and form a quasicrystalline structure (67). Cellulose fibrils are insoluble in water and inelastic and, because of their molecular configuration, have a tensile strength comparable to that of steel (67). Cellulose fibrils are the primary architectural element in a plant cell wall, providing it with a mechanical support and a unique combination of chemical resilience and flexibility. These qualities make cellulose an indispensible raw material for such common products as paper, lumber, and cotton textiles. The processes of cellulose biosynthesis in various plant cells have been studied for many years, but there still remain unsolved puzzles. The nature of the immediate precursor of cellulose, UDP-glucose, has been identified, and the biochemical pathways leading to its formation have been established. Still, the mechanisms that account for the rapid and orderly assembly of cellulose molecules into a microfibril structure and govern its unidirectional elongation remain obscure.

The work of Benziman, his students, and his colleagues, who studied cellulose biosynthesis in a relatively simple bacterial model, made a major contribution toward a better understanding of cellulose biosynthesis. A detailed account of these studies and the current views on the mechanisms of cellulose biosynthesis can be found in a number of reviews (28, 30, 49, 70, 73, 83), in two excellent chapters by Alfred French and Deborah Delmer (29, 32) in the book *Discoveries in Plant Biology,* and in the recently published book *Cellulose: Molecular and Structural Biology,* edited by Malcolm Brown and Inder Saxena (25).

ACETOBACTER XYLINUM: A BACTERIAL MODEL OF CELLULOSE BIOSYNTHESIS

The complexity of the plant cell wall biogenesis makes studies of the cellulose biosynthesis in plants a very difficult task and prompted the search for a more convenient experimental model. The solution came from an unexpected source. The kombucha tea mushroom (actually, a syntrophic colony of acetobacteria and yeasts), used since the Qin Dynasty in China (around 250 BC), was known to produce a thick mat at its surface, although the nature of this mat, not to mention its chemical composition, remained a mystery. In 1886, Adrian Brown described a similar mat, produced by acetic acid bacteria, and showed that it consisted of cellulose (23, 24). The bacterium producing this mat was initially named *Bacterium xylinum* and was later referred to by a variety of names, including *Acetobacterium xylinum, Bacterium xylinodes,* and since 1925, *Acetobacter xylinum* (19). This last name remained the most popular one until 1983, when it was changed to *Acetobacter xylinus* (91). Finally, in 1997, the bacterium got its current name *Gluconacetobacter xylinus* (92). The existing rules dictate using the latter name, although at the time of the studies discussed in this chapter the organism was referred to as *A. xylinum* and that name was used in all of the cited publications. Since this bacterium is an obligate aerobe, the extracellular mat-like pellicle of cellulose fibrils (Fig. 3) that it produces appears to confer a selective advantage under static growth conditions. Although in the past several years biosynthesis of cellulose has been documented in a variety of bacteria (60), *G. xylinus* remains unrivaled in terms of both the sheer amount of the cellulose that it produces and its purity.

At the Hebrew University, studies of microbial cellulose production were started in the 1930s by M. Aschner (10) and continued by Shlomo Hestrin and his colleagues (11, 41, 43, 44, 55, 72). Two early papers by Aschner and Hestrin (11, 43) showed that *G. xylinus* produced a mesh of pure cellulose fibrils, forming a pellicle on the surface of media. The cellulose produced by cultures of *G. xylinus* resembled that from many plant and algal sources with respect to the crystalline unit structure and average microfibrillar width; it was also tough, of an exceptionally high purity, and deposited extracellularly (Fig. 3). These papers, published in *Nature,* paved the way to using *G. xylinus* as a convenient bacterial model for the study of cellulose formation. A later paper by Hestrin and Schramm showed that this process was also extremely efficient: the cellulose polymerization rate was estimated to be of the order of 1.5×10^8 glucose residues per cell per hour, which corresponds to an essentially irreversible incorporation into the growing β-1,4 glucan chains of more than 40,000 glucose molecules per second on the surface of a single cell (44). Furthermore, *G. xylinus* did not show any cellulase activity. These factors made *G. xylinus* a favorite model organism for the study of the mechanisms of cellulose biosynthesis.

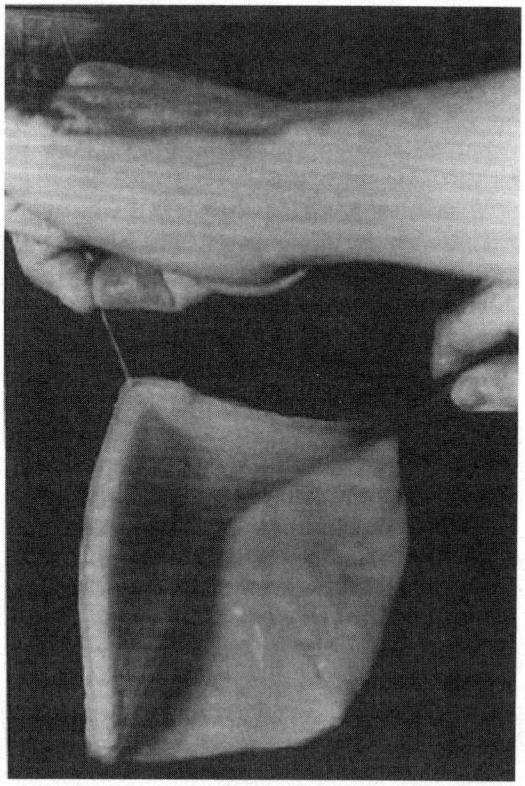

Figure 3. Cellulose pellicle formed by a static culture of *G. xylinus* after 24 h at 30°C (reprinted from reference 67 with permission).

STUDIES OF CARBOHYDRATE METABOLISM IN *ACETOBACTER XYLINUM*

When Benziman, a former student of Shlomo Hestrin, took over the lab, he shifted the emphasis from microbiological to biochemical studies. Previous studies showed that *G. xylinus* formed a cellulose pellicle when grown on fructose, glucose, glycerol, galactose, lactose, maltose, and sucrose but not on sorbose, mannose, cellobiose, erythritol, ethanol, or acetate (43, 72). Benziman and his group, including Haim Weinhouse, who worked with Benziman for more than 30 years, first as a student, then as a postdoc, and later as a scientist in the lab, embarked on a systematic study of the central metabolism of *G. xylinus,* characterizing the enzymes of hexose metabolism, pentose phosphate shunt, gluconeogenesis, and the tricarboxylic acid cycle and investigating the effects of various carbohydrates on cellulose formation (15–18, 77, 87, 88). The results obtained in those studies are still relevant for optimizing cellulose production by whole cells of *G. xylinus.*

The next series of experiments aimed at pinpointing the cellulose biosynthesis pathway and identifying the immediate precursor of cellulose. By following the in vivo incorporation of the radioactive label from [^{14}C]glucose into various sugar phosphates, nucleotide sugar phosphates, and finally, cellulose, Benziman and his colleagues were able to establish the following sequence: glucose → glucose 6-phosphate → glucose 1-phosphate → UDP-glucose → cellulose (78).

MAKING CELLULOSE SYNTHASE WORK IN VITRO

Despite the identification of UDP-glucose as the immediate precursor of cellulose (40) and the success in delineation of the UDP-glucose biosynthesis pathway (78), purification of the cellulose synthase posed an inscrutable problem. Although resting cells and particulate-membranous preparations of *G. xylinus* exhibited significant rates of cellulose synthesis (43, 44, 78), numerous attempts at creating a cell-free in vitro system for cellulose biosynthesis inexplicably failed to produce an active enzyme (see reference 29 for an excellent firsthand account). Part of the problem was the apparent association of cellulose synthase with the cell membrane. However, even isolated membrane preparations typically showed less than 1% of the *β*-1,4-D-glucan synthesis rate achieved in vivo (1).

In the early 1980s, the Benziman group focused its efforts on the isolation and characterization of an active preparation of the cellulose synthase from *G. xylinus.* Based on the earlier observations that (i) polyethylene glycol provided some activation of glucan synthase activities in plant systems and (ii) activation of yeast *β*-1,3-D-glucan synthase by GTP, Benziman and colleagues varied the concentrations of these compounds until these efforts finally paid off.

Dorit Amikam:

> It was one usual day while Moshe Benziman was sitting at his office, when Yehoshua Aloni, his Ph.D. student, then known as Shuky, knocked excitedly on his door. As Moshe answered "yes," in his deep serious voice, Shuky entered the room, which as always, smelled of a mixture of good tobacco aroma and good old books. "The machine is going crazy" Shuki said, laughing with excitement. Moshe took his pipe out of his lips and, as always, appearing restrained said semiquestioning, "really?" "Yes, the numbers are sky high!" "Well, let's go and see," said Moshe with one hand patting Shuki friendly on the shoulder and the other still holding his pipe. They both went down to the basement of the building, where the counter was, and watched the machine which indeed went crazy, and the numbers of ^{14}C incorporated into cellulose were of a dramatically high value, indicating that cellulose was produced at a very high quantity. This was almost too good to be real. Thus, they may have cracked the puzzle, the cellulose synthesis pathway, the ever-so-hard mechanism to crack with very hard work, a lot of tenacity, and volumes of creativity, vision, and wisdom. Moshe was still restrained and said, "Let's repeat this experiment."

This was the first step toward discovery of c-di-GMP, as Benziman and coworkers found out that the *G. xylinus* cellulose synthase (UDP-glucose:1,4-*β*-D-glucan 4-*β*-D-glucosyltransferase) showed marked and specific activation by GTP (1). This activation by GTP depended upon the presence of a protein factor which was heat labile, possessed high molecular weight, could be destroyed by protease treatment, and therefore appeared to be a regulatory protein subunit of the enzyme. This regulatory protein was easily dissociated from the membrane-bound enzyme and thus was not detected previously because it was lost during standard membrane preparation procedures. In the presence of polyethylene glycol 4000, the protein factor was preserved in membrane preparations, allowing the activation of cellulose synthase by GTP. This resulted in initial rates of 1,4-*β*-D-glucan synthesis that were 200 times greater than those previously reported for the purified enzyme and were comparable with the in vivo rates (1).

The discovery of the GTP-mediated activation of cellulose synthase provided the platform for the next important advance, solubilization of the high-activity enzyme system while preserving its capability to respond to the GTP-mediated regulatory mechanism (2). The catalytic and regulatory properties of the enzyme in the digitonin-solubilized state were remarkably similar to those observed for the membrane-bound form. As before, the GTP activation required the presence of a protein factor. In addition to GTP, Ca^{2+} ions were found to play a role in the regulation of cellulose biosynthesis in *G. xylinus,* a conclusion that had important implications for the future studies.

DISCOVERY OF THE c-di-GMP, THE ACTIVATOR OF CELLULOSE SYNTHASE

A careful examination of the activation of cellulose synthase by GTP revealed that this effect was absolutely specific for this compound. Aloni tried 10 μM and 1 mM concentrations of ADP, AMP, adenosine-2′-P, adenosine-3′-P, cyclic AMP, CDP, CMP, TTP, IMP, ADP-glucose, CDP-glucose, GDP-glucose, NAD, NADH, NADPH, flavinadenine dinucleotide, and NaF, but none of them had any effect on cellulose synthase activity. UTP and UDP were inhibitory, and only ATP and cGMP had a minor stimulatory effect (1). Remarkably, guanosine 5′-[γ-thio]triphosphate was even more effective than GTP, whereas GMP, guanosine 5′-[β-thio]diphosphate, and guanosine 5′-[β, γ-imino]triphosphate were completely inactive. This suggested that the actual activator could be some derivative of GTP.

The next series of experiments provided further evidence for the activator of cellulose synthase being a GTP derivative and allowed its preliminary characterization (63, 64). The regulatory protein, necessary for activation of cellulose synthase, was found to bind to an agarose-hexane-GTP column. However, following incubation with GTP, when this protein was washed away from the column, the activating factor remained bound to the column and could be eluted by GTP. This activating factor was found to be a low-molecular-mass, heat-stable compound that could be radioactively labeled by [8-^{3}H]GTP and [α-^{32}P]GTP but not by [γ-^{32}P]GTP. GDP, GMP, cGMP, diguanosine tetra- or pentaphosphate, guanosine 3′-diphosphate-*S*-diphosphate, and guanosine 3′-diphosphate-5′-triphosphate were all checked for their ability to stimulate cellulose synthase activity with negative results, indicating that the activator molecule was a previously unknown guanylate derivative (63, 64). Chemical analysis of this new compound indicated that it contained guanine, ribose, and phosphate in a ratio of 1:1:1 (63). Hydrolysis of this activator compound in mild alkali yielded two products comigrating with 2′-GMP and 3′-GMP, indicating that 3′-5′ or 2′-5′ phosphodiester bonds were being hydrolyzed. Furthermore, the compound was resistant to bacterial alkaline phosphatase but could be hydrolyzed by snake venom phosphodiesterase, yielding 5′-GMP as the only detectable product. This suggested that the activator had a cyclic nucleotide structure composed of GMP residues in 2′-5′ or 3′-5′ phosphodiester linkages (63). Further, cellulose synthase-containing membrane preparations were found to contain an enzyme, presumably a phosphodiesterase, which rapidly degraded the activator molecule into 5′-GMP. This phosphodiesterase activity was sensitive to Ca^{2+} ions, providing an explanation for the previously observed stimulation of cellulose biosynthesis by Ca^{2+} ions (63, 64).

The final series of experiments, culminating in the now-famous publication in *Nature* (65), included characterization of the new guanylate compound by mass spectrometry and its chemical synthesis. Chromatography on DEAE-Sephadex showed that the analyzed compound consisted of no more than two GMP moieties, whereas its sensitivity to the 3′-5′ phosphodiester-specific T1 endonuclease indicated that these two GMP moieties were linked by a 3′-5′ phosphodiester bond. Mass spectroscopic measurements, performed by Helmut Schwarz and Klaus Eckart at the Institute of Organic Chemistry, Technical University of Berlin, indicated that the activator compound had a molecular weight of 690, which corresponded to the molecular weight of a cyclic diguanylic acid. The final proof that the activator compound was indeed bis(3′→5′)-cyclic diguanylic acid (c-di-GMP) was obtained by chemical synthesis of this compound, performed by Jaques H. van Boom and his colleagues Erik de Vroom and Gijs van der Marel at the Gorlaeus Laboratories, University of Leiden, The Netherlands. The chemically synthesized c-di-GMP stimulated cellulose synthase activity and was indistinguishable from the native activator molecules in a variety of enzymatic and chemical tests, including nuclear magnetic resonance and high-pressure liquid chromatography (HPLC) (65).

The specificity of c-di-GMP-induced activation of cellulose synthase was studied using a series of 13 cyclic dimer and trimer nucleotide analogs that were synthesized by J. H. van Boom and colleagues. Of all these analogs, only deoxyribose and inosine variants of c-di-GMP were comparable to c-di-GMP in their ability to activate cellulose synthase. The same compounds also served as substrates to the Ca^{2+}-sensitive c-di-GMP-specific phosphodiesterase (66, 67). These data formed the basis of the proposed scheme of cellulose activation by c-di-GMP (Fig. 4) that was first published in 1987 and revised in 1991 (65, 67).

Identification of c-di-GMP as the direct activator of cellulose synthase led to the question about the exact mechanism of this activation. Several lines of evidence supported the idea that c-di-GMP func-

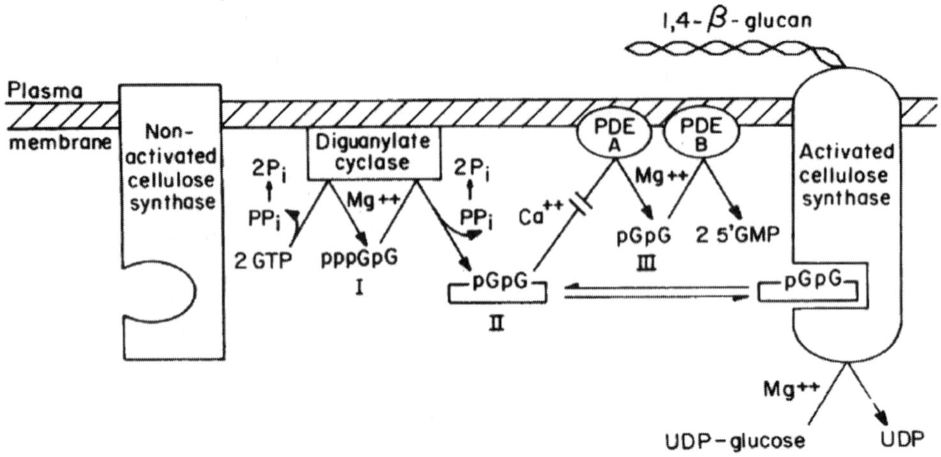

Figure 4. A model for regulation of cellulose biosynthesis in *G. xylinus* by c-di-GMP (reprinted from reference 67 with permission).

tioned as an allosteric regulator of cellulose synthase, binding to the enzyme in a reversible fashion at a regulatory site that was distinct from the catalytic—or, for that matter, substrate-binding—site(s). In vitro, activation of the cellulose synthase could be achieved at submicromolar levels of c-di-GMP with an activation constant (K_{act}) of 0.35 μM (65). In vivo, the cellular concentration of c-di-GMP determined in ^{32}P-labeled *G. xylinus* cells was estimated to be 5 to 15 μM, which is almost an order of magnitude higher than the affinity of its binding site on cellulose synthase (67). On the other hand, this value is 2 orders of magnitude lower than the cellular levels of GTP, which are in the millimolar range. Benziman and his colleagues resolved this conundrum by showing that at least part of the c-di-GMP pool resides in the membrane, tightly bound to a specific c-di-GMP-binding protein (K_d [dissociation constant] = 20 nM), which appeared to be an intrinsic component of the cellulose synthase complex (52, 89). It took another 10 years before Amikam and Galperin were able to identify this c-di-GMP-binding protein as the C-terminal domain of the catalytic subunit of the *G. xylinus* cellulose synthase (6).

The success in identification of c-di-GMP as the direct activator of cellulose synthase in *G. xylinus* attracted the interest of the Weyerhaeuser Co. of Tacoma, WA, one of the largest producers of cellulose fibers in the world, which provided Benziman's group with financing for further studies. The industrial sponsors were interested to see whether c-di-GMP could play the same role in other bacteria, plants, and algae. The first author of this chapter, Dorit Amikam, was able to demonstrate a c-di-GMP-responsive cellulose synthase in *Agrobacterium tumefaciens,* an alphaproteobacterial plant pathogen that synthesizes cellulose fibrils to promote its attachment to the surface of the plant cells (4). However, despite certain early indications that a similar mechanism could be functional in fungi and plants (7, 39, 52), attempts to unequivocally demonstrate c-di-GMP-stimulated cellulose biosynthesis in plant cells ultimately proved unsuccessful. Indeed, recent sequence comparisons of bacterial and plant cellulose synthases showed that the latter lack the c-di-GMP-binding PilZ domain (6).

MECHANISMS OF SYNTHESIS AND HYDROLYSIS OF c-di-GMP

The finding that the regulatory protein, necessary for activation of the *G. xylinus* cellulose synthase, was actually a diguanylate cyclase (c-di-GMP synthase, DGC) constituted a key step toward characterization of the enzymes involved in biosynthesis and hydrolysis of c-di-GMP. It was already clear that c-di-GMP synthesis proceeds from two molecules of GTP via the diguanosine tetraphosphate intermediate pppGpGp. The degradation of c-di-GMP was shown to occur in two distinct steps leading to 5′-GMP as the sole final product. c-di-GMP degradation was shown to be initiated by a Ca^{2+}-sensitive phosphodiesterase A (PDE-A) that cleaves a single 3′-5′ phosphodiester bond in the c-di-GMP structure, yielding an inactive linear dinucleotide pGpG, which is then converted to 5′-GMP by a second enzyme, a nonspecific PDE-B.

As mentioned above, DGC was shown to bind to an agarose-GTP column, which allowed its purification up to 2,000-fold (63). The c-di-GMP-specific PDE-A could also be purified on these columns, based on its sensitivity to Ca^{2+} ions. Purified enzymes were used to determine their partial amino acid sequences, prepare PCR primers to make probes to screen the library of *G. xylinus* genes, and isolate several clones expressing DGCs and PDE-A's. These clones corresponded to three separate operons, designated *cdg1, cdg2,* and *cdg3,* each containing a *pdeA* gene followed by a *dgc* gene and therefore encoding a PDE-A and a DGC (Fig. 5). Sequencing of the three *pdeA* and three *dgc* genes revealed that they were all very similar and their products shared a common sequence motif (79):

```
pdeA1:   GACAGCCTGACCGGCCTGCTCAACCGGGGCGCGCTG
         D  S  L  T  G  L  L  N  R  G  A  L
pdeA2:   --------------G-----------CCA--TCGCT
         D  S  L  T  G  L  L  N  R  T  S  L
pdeA3:   --------------------------CTCCT-GG--
         D  S  L  T  G  L  L  N  R  S  S  V
dgc1:    ---CCG-----G--G---T-------T--T-GCT-C
         D  P  L  T  G  L  F  N  R  G  A  L
dgc2:    --T-C--------------------C-----GCT-C
         D  T  L  T  G  L  L  N  R  G  G  F
Consensus:
         D  s  L  T  G  L  L  N  R  G  -  l
```

This work was performed by Benziman in collaboration with the Cetus Corporation of Emeryville, CA, in 1991 and resulted in a patent application, "Cyclic di-guanylate metabolic enzymes," filed with the U.S. Patent Office on 29 November 1991 on behalf of the Weyerhaeuser Co. (R. Tal, D. H. Gelfand, R. D. Calhoon, A. Ben-Bassat, M. Benziman, and H. C. Wong, U.S. patent application 07/800,218). An international patent application WO/1993/011294 was filed the next year (see http://www.wipo.int/pctdb/en/wo.jsp?IA=US1992008756&wo=1993011244), but the terms of Benziman's collaboration with Cetus Corp. and Weyerhaeuser Co. delayed open publication of these findings (79) for another 6 years, until the U.S. patent 5,759,828 was finally issued on 2 June 1998.

By that time, rapid growth of the nucleotide and protein sequence databases and progress in sequence analysis software (3) allowed far more sophisticated sequence comparisons, which led to the identification in DGC and PDE-A sequences of three separate protein domains (79) (Fig. 6). The N-terminal domains of DGCs and PDE-A's, respectively, showed similarity to two different types of nonenzymatic oxygen-sensing domains found in oxygen-responsive regulatory proteins from other bacteria (see below). The middle domain in all DGCs and PDE-A's was similar to the C-terminal output domain of the *Caulobacter crescentus* response regulator PleD (42), which was

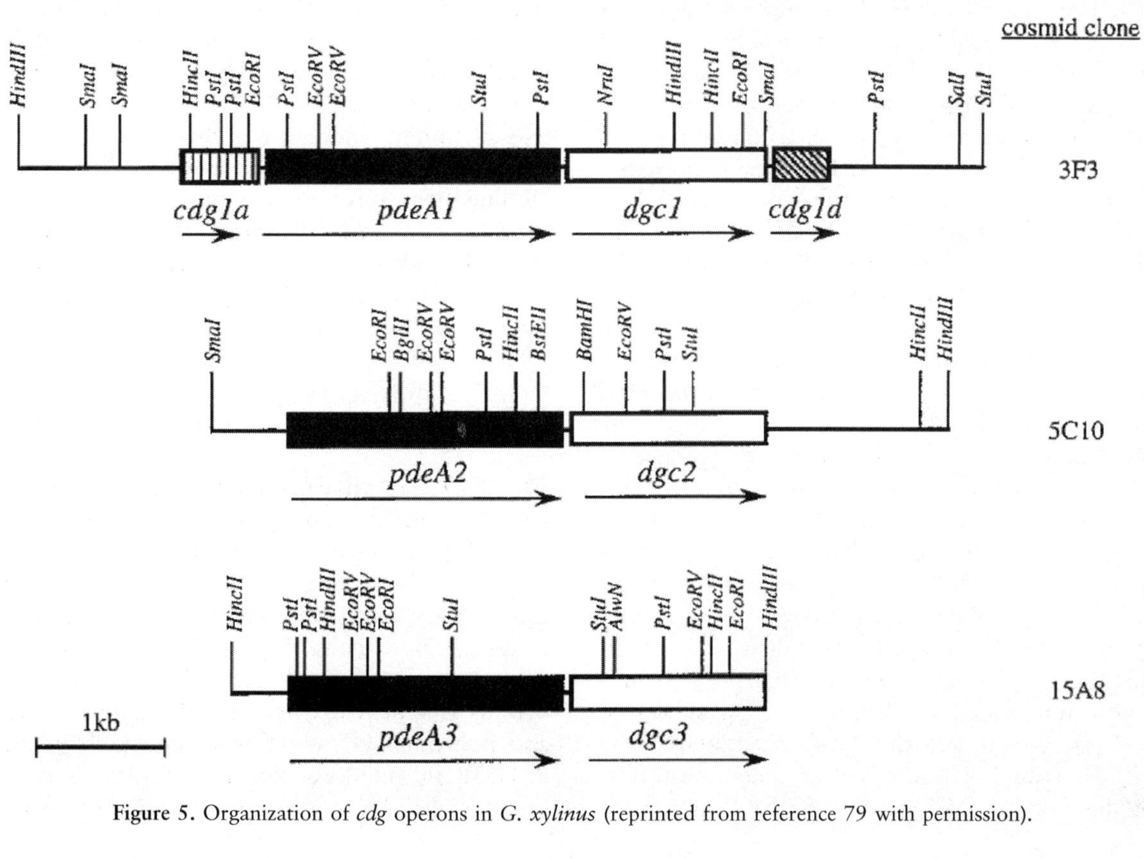

Figure 5. Organization of *cdg* operons in *G. xylinus* (reprinted from reference 79 with permission).

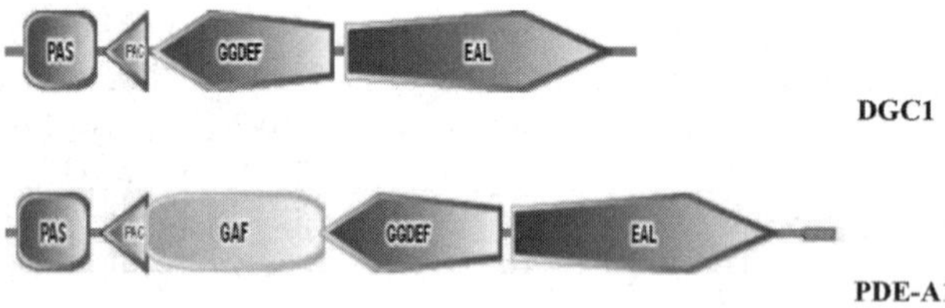

Figure 6. Domain architectures of DGC1 and PDE-A1. Graphical representations in the SMART database (51) of the domain structure of the *dgc1* and *pdeA1* gene products (GenBank accession no. AF052517; UniProt entries O87374 and O87373, respectively).

designated the GGDEF domain, based on its highly conserved Gly-Gly-Asp/Glu-Glu-Phe sequence motif. The C-terminal regions of DGCs and PDE-A's appeared to be new and were referred to as the EAL domain, based on the highly conserved Glu-Ala-Leu sequence motif near the start of this domain. The GGDEF and EAL domains were separated by a putative Q-linker, flexible segment known to connect functionally distinct domains in bacterial regulatory proteins (90). In contrast to the DGCs, the PDE-A's additionally contained an ~160-amino-acid spacer domain between the oxygen-binding and GGDEF domains (79). The current representation of the domain architectures of the six DGC and PDE-A sequences, determined by Benziman and coworkers (79; Tal et al., U.S. patent 5,759,828), is shown in Fig. 6.

The presence of the GGDEF and EAL domains in all six DGCs and PDE-A's sequenced by Tal and colleagues (79; Tal et al., U.S. patent application) made it impossible to decide whether synthesis of c-di-GMP was the property of either of these domains or required both of them. The key clues came from two sources, namely, genetic studies of cellulose biosynthesis in *Rhizobium leguminosarum* bv. *trifolii* and bioinformatics studies of the distribution of the GGDEF and EAL domains among completely sequenced bacterial genomes. A comparison of the number of genes encoding GGDEF, EAL, and the newly described HD-GYP domain in diverse bacterial genomes showed that the number of copies of the HD-GYP domain correlated with the prevalence of the GGDEF domain over the EAL domain (33). This observation suggested that the HD-GYP domain was also involved in cyclic diguanylate-mediated signaling and, owing to its similarity to the cAMP/cGMP phosphodiesterases, served as a diguanylate phosphodiesterase. Based on its complementary distribution with the EAL domain, it was suggested that the EAL domain, too, functioned as a diguanylate phosphodiesterase (33), which left the DGC function to the GGDEF domain (8). Additional arguments supporting the idea of the DGC activity of the GGDEF domain were provided by Pei and Grishin who demonstrated that this domain was evolutionarily related to the catalytic domain of eukaryotic adenylyl cyclase and concluded that it, too, catalyzed nucleotide cyclization, necessary for the formation of the c-di-GMP (58).

Meanwhile, Nora Ausmees and colleagues characterized the genes required for cellulose biosynthesis in the plant symbiont *Rhizobium leguminosarum* bv. trifolii and showed that this process is controlled by a PleD-like response regulator CelR2, which, like PleD, has a REC-REC-GGDEF domain architecture (12). Disruption of the *celR2* gene through transposon insertion or removal of the 249 bp from its 3′ end abolished cellulose biosynthesis. Given that cellulose synthase of *R. leguminosarum* was very similar to the c-di-GMP-responsive cellulose synthase of *Agrobacterium tumefaciens* (4), this suggested that the GGDEF domain alone, without the EAL counterpart, was sufficient for producing c-di-GMP. Indeed, expression of CelR2 and *G. xylinus* DGC1 in a cellulose-negative strain of *R. leguminosarum* resulted in the same cellulose-positive phenotype. A similar phenotype could be achieved by expressing in *R. leguminosarum* a GGDEF domain-encoding gene from *Escherichia coli* (13). These data showed that activation of cellulose synthase required the presence of an intact GGDEF domain, but not the EAL domain, indicating that the GGDEF domain alone was required for c-di-GMP synthesis. This 2001 paper happened to be the last one by Benziman, as his health was gradually deteriorating just as recognition of the importance of c-di-GMP in bacterial regulation was spreading among researchers worldwide.

Subsequent studies of c-di-GMP turnover, including the direct experimental proof that the GGDEF domain is responsible for the DGC activity (57, 68) and the EAL domain is responsible for the PDE-A activity (21, 71, 80), are described in detail in chapters 3, 4, and 9.

c-di-GMP AS A SIGNALING MOLECULE

Benziman's paper establishing a link between the GGDEF and EAL domains (79) and the turnover of c-di-GMP in *G. xylinus* paved the way for recognition of DGCs and c-di-GMP-specific phosphodiesterases as important components of the bacterial signal transduction machinery (47, 48, 61, 62, 81). The first evidence of a c-di-GMP-dependent system outside *G. xylinus* was provided in the 1989 paper by Amikam and Benziman (4), which showed involvement of c-di-GMP in cellulose biosynthesis in *Agrobacterium tumefaciens*. The presence of a GGDEF domain in the PleD response regulator of yet another alphaproteobacterium, *Caulobacter crescentus* (42), indicated that c-di-GMP is widespread in bacteria (at least alphaproteobacteria) and that it could be involved in signaling.

Additional evidence that c-di-GMP was part of a signal transduction cascade came from the analysis of the domain architectures of DGCs and PDE-A's. The N-terminal domains of all three DGCs were found to be similar to the flavin-binding oxygen-responsive PAS domains of the *E. coli* aerotactic sensor Aer (20) and *Klebsiella pneumoniae* regulatory protein NifL (31) and the redox-sensing domain of the *Arabidopsis thaliana* photosensor NPH1 (46). The N-terminal domains of all three PDE-A's were found to be similar to the heme-binding oxygen-sensing domain of the oxygen sensor FixL (53), a regulator of the nitrogen fixation *(nif)* operon in *Rhizobium meliloti,* and related FixL proteins from other bacteria (79). Thus, N-terminal domains of both DGCs and PDE-A's (representing two different variants of the widespread PAS domain) (82) were capable of responding to oxygen, in accordance with the early observations that cellulose synthesis benefits *G. xylinus* in static culture but is suppressed in well-aerated cultures (67). A joint paper from the groups of Benziman and Marie-Alda Gilles-Gonzalez (Ohio State University, Columbus) investigated the detailed mechanisms of PDE-A1 regulation and showed that the enzyme (i) is inactive in the apo form, (ii) can be activated by hemin, and (iii) has a higher activity in the deoxy form than in the oxygen-bound form (26).

For one of the authors of this review (M. Y. Galperin), the link between GGDEF and EAL domains and c-di-GMP offered a solution to a difficult question: what could be the functions of 29 *E. coli* proteins that encoded GGDEF and/or EAL domains (33, 34). Association with cytoplasmic signaling domains, such as PAS and GAF, as well as periplasmic and membrane-embedded ligand-binding domains, strongly suggested that GGDEF and EAL domains participated in some kind of signal transduction. However, the mechanism of this signaling and the nature of the transmitted signal(s) remained obscure. The link with c-di-GMP turnover indicated the presence in a variety of bacteria of an entirely novel signal transduction mechanism, mediated by c-di-GMP (33, 35, 36).

BENZIMAN'S LEGACY: PROKARYOTIC WORLD

In the last years of his life, Benziman enjoyed the discovery of the GGDEF and EAL domains in the growing number of sequenced bacterial genomes (33–35) and the gradual recognition of c-di-GMP as a signaling molecule in diverse bacteria (8, 9, 35, 54, 58, 60, 93). While he had always enjoyed respect, admiration, and friendship from his students and colleagues around the world, Benziman was proud that he had finally accomplished what he had set out to do 40 years earlier—solve the puzzle of cellulose biosynthesis in *G. xylinus.* The discovery of c-di-GMP was made possible by his determination, purposefulness, and dedication to the cause. The story of cellulose biosynthesis and c-di-GMP discovery offers many important lessons on the progress of science in general:

- A study of a supposedly arcane process of cellulose biosynthesis in an unusual bacterium turned out to have implications for the whole bacterial world (27, 47, 48, 61, 62, 81).
- A study that started primarily as an applied research led to fundamental discoveries, affecting our very understanding of bacterial signal transduction (35–38, 47, 48, 61, 62, 81).
- A study of a totally innocuous organism that can even be consumed by humans in the form of the kombucha tea made a major contribution toward the understanding of virulence mechanisms of *Vibrio cholerae, Yersinia pestis,* and other important human pathogens (22, 50, 81, 84, 85).
- A key reason for Benziman's success was the use of a convenient bacterial model. Although the c-di-GMP-mediated regulation does not seem to function in plants, the cellulose synthase genes proved to be homologous.

It is worth noting that the mechanisms of cellulose secretion and the assembly of the pellicle on the bacterial surface still remain obscure. The lack of clarity regarding these processes betrays a general lack of understanding of the mechanisms of assembly and functioning of the bacterial cell envelope, which could take many more years to resolve.

The studies of the role of c-di-GMP in various cellular processes, conducted all over the world and discussed in the subsequent chapters of this book, are the best contribution to the memory of Benziman and his lifelong work that led to the discovery of c-di-GMP.

BENZIMAN'S LEGACY: EUKARYOTIC WORLD

c-di-GMP, a Potential Cancer Therapy Compound?

The discovery of c-di-GMP presented a tempting platform to study molecular pathways possibly connected to this nucleotide, not only in other bacteria but also in eukaryotic systems. Cyclic nucleotides, such as cAMP and cGMP, have been well-known for their potential as mediators of diverse activation processes in different eukaryotic biological systems. Guanine nucleotides in general perform an important role in the regulation of eukaryotic processes. The unique structure and the natural occurrence of c-di-GMP prompted one of the authors of this review, Dorit Amikam, to examine whether this nucleotide has any effects on mammalian systems. The resulting published pioneer work of D. Amikam and colleagues (5, 75) showed, for the first time, that c-di-GMP affected various important cellular regulation mechanisms in human cancerous cell lines: the human acute lymphoblastic leukemia cell line, Molt 4; and the human T-cell leukemia cell line, Jurkat. Their innovative studies revealed that c-di-GMP exerted a marked increase of [^{3}H]thymidine incorporation into the studied cell lines which was in turn accompanied by a marked augmentation of DNA, interestingly enough not followed by a rise in cell proliferation. Thus, the authors proposed that cells treated with c-di-GMP enter the S phase uncoordinated with their overall replication rate. Indeed, when further characterizing the state of c-di-GMP-treated cells, Amikam and colleagues found that, strikingly, the percentage of cells in the S phase increased by 79% with the almost complete disappearance of G_2/M-phase cells, suggesting that c-di-GMP may cause a blockage of the cell cycle at the S phase, leading to impaired cell replication. This was further strengthened when light scatter analysis of c-di-GMP-treated cells showed the cells to be of a smaller size, a finding which is compatible with such cells being mostly in G_1 and S phases.

Taken together, it appears that even though c-di-GMP-treated cells exhibit elevated DNA synthesis while entering S phase, they could not undergo mitosis and overall cell replication was drastically impaired.

c-di GMP Binds to the Oncoprotein p21^{Ras}; a Possible Molecular Basis for the Inhibition of Proliferation in Human Cancerous Cell Lines

The GTP-binding, growth promoter oncoprotein p21^{Ras} has long been reported to interact with guanine nucleotides (74). Given the central role of the oncoprotein p21^{Ras} in transducing signals downstream, resulting in transformation and proliferation in various types of cell lines (69), taking into consideration the resemblance of the effects seen with c-di-GMP-treated cells and activated p21^{Ras}-harboring cells, and in view of the important fact that mutated forms of p21^{Ras} are found in 30% of all human cancers, D. Amikam and her group were intrigued by whether there was a possible interaction between the oncoprotein and the novel nucleotide. Innovative studies performed by Amikam and coworkers experimentally verified and ascertained c-di-GMP entrance into the lymphoblastoid cells, with the latter nucleotide being found to specifically bind to the p21^{Ras} oncoprotein, apparently in an irreversible fashion (5, 75). Since Ras is activated when bound to GTP, following transformation to the Ras-GDP inactive form, it has been proposed that the unusual effects exerted by c-di-GMP on the cell cycle of the cancerous cell lines described previously could be due to the irreversible binding of c-di-GMP to p21^{Ras}, which might result in a constitutive stable c-di-GMP-p21^{Ras} complex that would remain in the active conformation. The latter possibility was compatible with the observed, irreversible nature of the c-di-GMP effect on the synthesis of DNA (5). It was also supported by the notion that activated Ras plays an important role in the growth and cell size regulation of mammalian cells and is thereby able to cause growth arrest and inhibition of proliferation (14, 45, 59).

c-di-GMP Elevates the Expression of the Immune T-Cell Receptor CD4

Ras appears to play a role in inducing the expression of the T-cell receptors CD4 and CD8 and to take part in signaling the differentiation of T-lineage cells (56, 76, 86). Since c-di-GMP might activate Ras, a question of whether it affects the immune T-cell receptor CD4 in the CD4-bearing lymphoblastoid Jurkat cell line was raised. Interestingly enough, Amikam and coworkers showed for the first time that c-di-GMP exerts an effect on the immune system T-cell receptor CD4, with an observed marked elevated expression of the latter receptor, in c-di-GMP-treated cells (75). The c-di-GMP-induced upregulation of the CD4 receptor was specific and was not obtained with other guanine nucleotides.

The interesting effects of c-di-GMP are just beginning to be unveiled. c-di-GMP might offer a potentially powerful tool for studying signals involved in the control of the cell fate decision in lymphocytes, cell cycle-involved proteins, and Ras *trans*-signaled downstream proteins. Moreover, the striking aptitude of this unique nucleotide to significantly inhibit the proliferation of various human cancerous cell lines might advocate for its use in the development of new drug platform technologies and its playing a major role as an anticancer therapeutic agent. These studies further contribute to the scientific legacy of Benziman, the pioneer who discovered the novel nucleotide c-di-GMP.

Acknowledgments. This chapter is dedicated to Moshe Benziman of the Hebrew University of Jerusalem, beloved husband of Dorit Amikam. He left a scientific legacy as the pioneer who discovered the novel nucleotide c-di-GMP. We cherish his memory with love and admiration.

We thank Moshe Benziman's longtime colleagues Yoram Milner and Nathan De-Groot for their help and valuable comments.

M.Y.G. is supported by the Intramural Research Program of the National Institutes of Health, National Library of Medicine. M.Y.G.'s trip to Israel that made this chapter possible was sponsored in part by The Alexander Silberman Institute of Life Sciences, Hebrew University of Jerusalem.

REFERENCES

1. **Aloni, Y., D. P. Delmer, and M. Benziman.** 1982. Achievement of high rates of in vitro synthesis of 1,4-beta-D-glucan: activation by cooperative interaction of the *Acetobacter xylinum* enzyme system with GTP, polyethylene glycol, and a protein factor. *Proc. Natl. Acad. Sci. USA* **79**:6448–6452.
2. **Aloni, Y., R. Cohen, M. Benziman, and D. Delmer.** 1983. Solubilization of the UDP-glucose:1,4-beta-D-glucan 4-beta-D-glucosyltransferase (cellulose synthase) from *Acetobacter xylinum.* A comparison of regulatory properties with those of the membrane-bound form of the enzyme. *J. Biol. Chem.* **258**: 4419–4423.
3. **Altschul, S. F., T. L. Madden, A. A. Schaffer, J. Zhang, Z. Zheng, W. Miller, and D. J. Lipman.** 1997. Gapped BLAST and PSI-BLAST: a new generation of protein database search programs. *Nucleic Acids Res.* **25**:3389–3402.
4. **Amikam, D., and M. Benziman.** 1989. Cyclic diguanylic acid and cellulose synthesis in *Agrobacterium tumefaciens. J. Bacteriol.* **171**:6649–6655.
5. **Amikam, D., O. Steinberger, T. Shkolnik, and Z. Ben-Ishai.** 1995. The novel cyclic dinucleotide 3′-5′ cyclic diguanylic acid binds to p21ras and enhances DNA synthesis but not cell replication in the Molt 4 cell line. *Biochem. J.* **311**:921–927.
6. **Amikam, D., and M. Y. Galperin.** 2006. PilZ domain is part of the bacterial c-di-GMP binding protein. *Bioinformatics* **22**: 3–6.
7. **Amor, Y., R. Mayer, M. Benziman, and D. Delmer.** 1991. Evidence for a cyclic diguanylic acid-dependent cellulose synthase in plants. *Plant Cell* **3**:989–995.
8. **Anantharaman, V., and L. Aravind.** 2000. Cache—a signaling domain common to animal Ca^{2+}-channel subunits and a class of prokaryotic chemotaxis receptors. *Trends Biochem. Sci.* **25**: 535–537.
9. **Anantharaman, V., E. V. Koonin, and L. Aravind.** 2001. Regulatory potential, phyletic distribution and evolution of ancient, intracellular small-molecule-binding domains. *J. Mol. Biol.* **307**:1271–1292.
10. **Aschner, M.** 1937. Cultivation of cellulose-splitting bacteria on membranes of *Acetobacter xylinum. J. Bacteriol.* **33**:249–252.
11. **Aschner, M., and S. Hestrin.** 1946. Fibrillar structure of cellulose of bacterial and animal origin. *Nature* **157**:659.
12. **Ausmees, N., H. Jonsson, S. Hoglund, H. Ljunggren, and M. Lindberg.** 1999. Structural and putative regulatory genes involved in cellulose synthesis in *Rhizobium leguminosarum* bv. *trifolii. Microbiology* **145**:1253–1262.
13. **Ausmees, N., R. Mayer, H. Weinhouse, G. Volman, D. Amikam, M. Benziman, and M. Lindberg.** 2001. Genetic data indicate that proteins containing the GGDEF domain possess diguanylate cyclase activity. *FEMS Microbiol. Lett.* **204**:163–167.
14. **Bar-Sagi, D., and J. R. Feramisco.** 1985. Microinjection of the ras oncogene protein into PC12 cells induces morphological differentiation. *Cell* **42**:841–848.
15. **Benziman, M., and H. Burger-Rachamimov.** 1962. Synthesis of cellulose from pyruvate by succinate-grown cells of *Acetobacter xylinum. J. Bacteriol.* **84**:625–630.
16. **Benziman, M., and H. Goldhamer.** 1968. The role of ubiquinone in the respiratory chain of *Acetobacter xylinum. Biochem. J.* **108**:311–316.
17. **Benziman, M.** 1969. Role of phosphoenolpyruvate carboxylation in *Acetobacter xylinum. J. Bacteriol.* **98**:1005–1010.
18. **Benziman, M., and A. Mazover.** 1973. Nicotinamide adenine dinucleotide- and nicotinamide adenine dinucleotide phosphate-specific glucose 6-phosphate dehydrogenases of *Acetobacter xylinum* and their role in the regulation of the pentose cycle. *J. Biol. Chem.* **248**:1603–1608.
19. **Bergey, D. H., F. C. Harrison, R. S. Breed, B. W. Hammer, and F. M. Huntoon.** 1925. *Bergey's Manual of Determinative Bacteriology,* 2nd ed. The Williams and Wilkins Co., Baltimore, MD.
20. **Bibikov, S. I., L. A. Barnes, Y. Gitin, and J. S. Parkinson.** 2000. Domain organization and flavin adenine dinucleotide-binding determinants in the aerotaxis signal transducer Aer of *Escherichia coli. Proc. Natl. Acad. Sci. USA* **97**:5830–5835.
21. **Bobrov, A. G., O. Kirillina, and R. D. Perry.** 2005. The phosphodiesterase activity of the HmsP EAL domain is required for negative regulation of biofilm formation in *Yersinia pestis. FEMS Microbiol. Lett.* **247**:123–130.
22. **Bobrov, A. G., O. Kirillina, and R. D. Perry.** 2007. Regulation of biofilm formation in *Yersinia pestis. Adv. Exp. Med. Biol.* **603**:201–210.
23. **Brown, A. J.** 1886. On acetic ferment which forms cellulose. *J. Chem. Soc. Trans.* (London) **49**:432–439.
24. **Brown, A. J.** 1887. Note on the cellulose formed by *Bacterium xylinum. J. Chem. Soc. Trans.* (London) **49**:432–439.
25. **Brown, R. M., and I. M. Saxena (ed.).** 2007. *Cellulose: Molecular and Structural Biology.* Springer, New York, NY.
26. **Chang, A. L., J. R. Tuckerman, G. Gonzalez, R. Mayer, H. Weinhouse, G. Volman, D. Amikam, M. Benziman, and M. A. Gilles-Gonzalez.** 2001. Phosphodiesterase A1, a regulator of cellulose synthesis in *Acetobacter xylinum,* is a heme-based sensor. *Biochemistry* **40**:3420–3426.
27. **D'Argenio, D. A., and S. I. Miller.** 2004. Cyclic di-GMP as a bacterial second messenger. *Microbiology* **150**:2497–2502.
28. **Delmer, D. P.** 1999. Cellulose biosynthesis: exciting times for a difficult field of study. *Annu. Rev. Plant Physiol. Plant Mol. Biol.* **50**:245–276.
29. **Delmer, D. P.** 2000. Structure and biosynthesis of cellulose. Part II: biosynthesis, p. 199–216. *In* S.-D. Kung and S.-F. Yang (ed.), *Discoveries in Plant Biology,* vol. 3. World Scientific Publishing Co., Singapore, Singapore.
30. **Doblin, M. S., I. Kurek, D. Jacob-Wilk, and D. P. Delmer.** 2002. Cellulose biosynthesis in plants: from genes to rosettes. *Plant Cell Physiol.* **43**:1407–1420.
31. **Drummond, M. H., and J. C. Wootton.** 1987. Sequence of *nifL* from *Klebsiella pneumoniae:* mode of action and relationship to two families of regulatory proteins. *Mol. Microbiol.* **1**:37–44.
32. **French, A. D.** 2000. Structure and biosynthesis of cellulose. Part I: structure, p. 163–197. *In* S.-D. Kung and S.-F. Yang (ed.), *Discoveries in Plant Biology,* vol. 3. World Scientific Publishing Co., Singapore, Singapore.
33. **Galperin, M. Y., D. A. Natale, L. Aravind, and E. V. Koonin.** 1999. A specialized version of the HD hydrolase domain im-

plicated in signal transduction. *J. Mol. Microbiol. Biotechnol.* 1:303–305.

34. **Galperin, M. Y.** 2001. Conserved 'hypothetical' proteins: new hints and new puzzles. *Comp. Funct. Genomics* **2**:14–18.
35. **Galperin, M. Y., A. N. Nikolskaya, and E. V. Koonin.** 2001. Novel domains of the prokaryotic two-component signal transduction systems. *FEMS Microbiol. Lett.* **203**:11–21.
36. **Galperin, M. Y.** 2004. Bacterial signal transduction network in a genomic perspective. *Environ. Microbiol.* **6**:552–567.
37. **Galperin, M. Y.** 2005. A census of membrane-bound and intracellular signal transduction proteins in bacteria: bacterial IQ, extroverts and introverts. *BMC Microbiol.* **5**:35.
38. **Galperin, M. Y.** 2006. Structural classification of bacterial response regulators: diversity of output domains and domain combinations. *J. Bacteriol.* **188**:4169–4182.
39. **Girard, V., M. Fevre, R. Mayer, and M. Benziman.** 1991. Cyclic diguanylic acid stimulates 1,4-β-glucan synthase from *Saprolegnia monoica. FEMS Microbiol. Lett.* **82**:293–296.
40. **Glaser, L.** 1958. The synthesis of cellulose in cell-free extracts of *Acetobacter xylinum. J. Biol. Chem.* **232**:627–636.
41. **Gromet-Elhanan, Z., and S. Hestrin.** 1963. Synthesis of cellulose by *Acetobacter xylinum*. 6. Growth on citric acid-cycle intermediates. *J. Bacteriol.* **85**:284–292.
42. **Hecht, G. B., and A. Newton.** 1995. Identification of a novel response regulator required for the swarmer-to-stalked-cell transition in *Caulobacter crescentus. J. Bacteriol.* **177**:6223–6229.
43. **Hestrin, S., M. Aschner, and J. Mager.** 1947. Synthesis of cellulose by resting cells of *Acetobacter xylinum. Nature* **159**:64–65.
44. **Hestrin, S., and M. Schramm.** 1954. Synthesis of cellulose by *Acetobacter xylinum*. 2. Preparation of freeze-dried cells capable of polymerizing glucose to cellulose. *Biochem. J.* **58**: 345–352.
45. **Hirakawa, T., and H. E. Ruley.** 1988. Rescue of cells from ras oncogene-induced growth arrest by a second, complementing, oncogene. *Proc. Natl. Acad. Sci. USA* **85**:1519–1523.
46. **Huala, E., P. W. Oeller, E. Liscum, I. S. Han, E. Larsen, and W. R. Briggs.** 1997. *Arabidopsis* NPH1: a protein kinase with a putative redox-sensing domain. *Science* **278**:2120–2123.
47. **Jenal, U.** 2004. Cyclic di-guanosine-monophosphate comes of age: a novel secondary messenger involved in modulating cell surface structures in bacteria? *Curr. Opin. Microbiol.* **7**:185–191.
48. **Jenal, U., and J. Malone.** 2006. Mechanisms of cyclic-di-GMP signaling in bacteria. *Annu. Rev. Genet.* **40**:385–407.
49. **Joshi, C. P., and S. D. Mansfield.** 2007. The cellulose paradox—simple molecule, complex biosynthesis. *Curr. Opin. Plant Biol.* **10**:220–226.
50. **Kirillina, O., J. D. Fetherston, A. G. Bobrov, J. Abney, and R. D. Perry.** 2004. HmsP, a putative phosphodiesterase, and HmsT, a putative diguanylate cyclase, control Hms-dependent biofilm formation in *Yersinia pestis. Mol. Microbiol.* **54**:75–88.
51. **Letunic, I., T. Doerks, and P. Bork.** 2009. SMART 6: recent updates and new developments. *Nucleic Acids Res.* 37:D229–D232.
52. **Mayer, R., P. Ross, H. Weinhouse, D. Amikam, G. Volman, P. Ohana, R. D. Calhoon, H. C. Wong, A. W. Emerick, and M. Benziman.** 1991. Polypeptide composition of bacterial cyclic diguanylic acid-dependent cellulose synthase and the occurrence of immunologically crossreacting proteins in higher plants. *Proc. Natl. Acad. Sci. USA* 88:5472–5476.
53. **Monson, E. K., M. Weinstein, G. S. Ditta, and D. R. Helinski.** 1992. The FixL protein of *Rhizobium meliloti* can be separated into a heme-binding oxygen-sensing domain and a functional C-terminal kinase domain. *Proc. Natl. Acad. Sci. USA* 89:4280–4284.
54. **Mougel, C., and I. B. Zhulin.** 2001. CHASE: an extracellular sensing domain common to transmembrane receptors from prokaryotes, lower eukaryotes and plants. *Trends Biochem. Sci.* **26**:582–584.
55. **Ohad, I., I. O. Danon, and S. Hestrin.** 1962. Synthesis of cellulose by *Acetobacter xylinum*. 5. Ultrastructure of polymer. *J. Cell Biol.* **12**:31–46.
56. **Pastor, M. I., K. Reif, and D. Cantrell.** 1995. The regulation and function of p21^{ras} during T-cell activation and growth. *Immunol. Today* **16**:159–164.
57. **Paul, R., S. Weiser, N. C. Amiot, C. Chan, T. Schirmer, B. Giese, and U. Jenal.** 2004. Cell cycle-dependent dynamic localization of a bacterial response regulator with a novel diguanylate cyclase output domain. *Genes Dev.* **18**:715–727.
58. **Pei, J., and N. V. Grishin.** 2001. GGDEF domain is homologous to adenylyl cyclase. *Proteins* **42**:210–216.
59. **Ridley, A. J., H. F. Paterson, M. Noble, and H. Land.** 1988. Ras-mediated cell cycle arrest is altered by nuclear oncogenes to induce Schwann cell transformation. *EMBO J.* **7**:1635–1645.
60. **Römling, U.** 2002. Molecular biology of cellulose production in bacteria. *Res. Microbiol.* **153**:205–212.
61. **Römling, U., M. Gomelsky, and M. Y. Galperin.** 2005. C-di-GMP: the dawning of a novel bacterial signalling system. *Mol. Microbiol.* **57**:629–639.
62. **Römling, U., and D. Amikam.** 2006. Cyclic di-GMP as a second messenger. *Curr. Opin. Microbiol.* **9**:218–228.
63. **Ross, P., Y. Aloni, C. Weinhouse, D. Michaeli, P. Weinberger-Ohana, R. Mayer, and M. Benziman.** 1985. An unusual guanyl oligonucleotide regulates cellulose synthesis in *Acetobacter xylinum. FEBS Lett.* **186**:191–196
64. **Ross, P., Y. Aloni, H. Weinhouse, D. Michaeli, P. Weinberger-Ohana, R. Mayer, and M. Benziman.** 1986. Control of cellulose synthesis in *Acetobacter xylinum*. A unique guanyl oligonucleotide is the immediate activator of the cellulose synthase. *Carbohydr. Res.* **149**:101–117.
65. **Ross, P., H. Weinhouse, Y. Aloni, D. Michaeli, P. Weinberger-Ohana, R. Mayer, S. Braun, E. de Vroom, G. A. van der Marel, J. H. van Boom, and M. Benziman.** 1987. Regulation of cellulose synthesis in *Acetobacter xylinum* by cyclic diguanylic acid. *Nature* **325**:279–281.
66. **Ross, P., R. Mayer, H. Weinhouse, D. Amikam, Y. Huggirat, M. Benziman, E. de Vroom, A. Fidder, P. de Paus, and L. A. Sliedregt.** 1990. The cyclic diguanylic acid regulatory system of cellulose synthesis in *Acetobacter xylinum*. Chemical synthesis and biological activity of cyclic nucleotide dimer, trimer, and phosphothioate derivatives. *J. Biol. Chem.* **265**:18933–18943.
67. **Ross, P., R. Mayer, and M. Benziman.** 1991. Cellulose biosynthesis and function in bacteria. *Microbiol. Rev.* **55**:35–58.
68. **Ryjenkov, D. A., M. Tarutina, O. M. Moskvin, and M. Gomelsky.** 2005. Cyclic diguanylate is a ubiquitous signaling molecule in bacteria: insights into biochemistry of the GGDEF protein domain. *J. Bacteriol.* **187**:1792–1798.
69. **Satoh, T., M. Nakafuku, and Y. Kaziro.** 1992. Function of Ras as a molecular switch in signal transduction. *J. Biol. Chem.* **267**:24149–24152.
70. **Saxena, I. M., and R. M. Brown, Jr.** 2005. Cellulose biosynthesis: current views and evolving concepts. *Ann. Bot.* (London) **96**:9–21.
71. **Schmidt, A. J., D. A. Ryjenkov, and M. Gomelsky.** 2005. The ubiquitous protein domain EAL is a cyclic diguanylate-specific phosphodiesterase: enzymatically active and inactive EAL domains. *J. Bacteriol.* **187**:4774–4781.

72. **Schramm, M., Z. Grommet, and S. Hestrin.** 1957. Synthesis of cellulose by *Acetobacter xylinum:* 3. Substrates and inhibitors. *Biochem. J.* **67:**669–679.
73. **Somerville, C.** 2006. Cellulose synthesis in higher plants. *Annu. Rev. Cell Dev. Biol.* **22:**53–78.
74. **Stacey, D. W., and H. F. Kung.** 1984. Transformation of NIH 3T3 cells by microinjection of Ha-ras p21 protein. *Nature* **310:**508–511.
75. **Steinberger, O., Z. Lapidot, Z. Ben-Ishai, and D. Amikam.** 1999. Elevated expression of the CD4 receptor and cell cycle arrest are induced in Jurkat cells by treatment with the novel cyclic dinucleotide 3′,5′-cyclic diguanylic acid. *FEBS Lett.* **444:**125–129.
76. **Swat, W., Y. Shinkai, H. L. Cheng, L. Davidson, and F. W. Alt.** 1996. Activated Ras signals differentiation and expansion of CD4+8+ thymocytes. *Proc. Natl. Acad. Sci. USA* **93:**4683–4687.
77. **Swissa, M., and M. Benziman.** 1976. Factors affecting the activity of citrate synthase of *Acetobacter xylinum* and its possible regulatory role. *Biochem. J.* **153:**173–179.
78. **Swissa, M., Y. Aloni, H. Weinhouse, and M. Benziman.** 1980. Intermediary steps in *Acetobacter xylinum* cellulose synthesis: studies with whole cells and cell-free preparations of the wild type and a celluloseless mutant. *J. Bacteriol.* **143:**1142–1150.
79. **Tal, R., H. C. Wong, R. Calhoon, D. Gelfand, A. L. Fear, G. Volman, R. Mayer, P. Ross, D. Amikam, H. Weinhouse, A. Cohen, S. Sapir, P. Ohana, and M. Benziman.** 1998. Three *cdg* operons control cellular turnover of cyclic di-GMP in *Acetobacter xylinum:* genetic organization and occurrence of conserved domains in isoenzymes. *J. Bacteriol.* **180:**4416–4425.
80. **Tamayo, R., A. D. Tischler, and A. Camilli.** 2005. The EAL domain protein VieA is a cyclic diguanylate phosphodiesterase. *J. Biol. Chem.* **280:**33324–33330.
81. **Tamayo, R., J. T. Pratt, and A. Camilli.** 2007. Roles of cyclic diguanylate in the regulation of bacterial pathogenesis. *Annu. Rev. Microbiol.* **61:**131–148.
82. **Taylor, B. L., and I. B. Zhulin.** 1999. PAS domains: internal sensors of oxygen, redox potential, and light. *Microbiol. Mol. Biol. Rev.* **63:**479–506.
83. **Taylor, N. G.** 2008. Cellulose biosynthesis and deposition in higher plants. *New Phytol.* **178:**239–252.
84. **Tischler, A. D., and A. Camilli.** 2004. Cyclic diguanylate (c-di-GMP) regulates *Vibrio cholerae* biofilm formation. *Mol. Microbiol.* **53:**857–869.
85. **Tischler, A. D., and A. Camilli.** 2005. Cyclic diguanylate regulates *Vibrio cholerae* virulence gene expression. *Infect. Immun.* **73:**5873–5882.
86. **von Boehmer, H.** 1990. Developmental biology of T cells in T cell-receptor transgenic mice. *Annu. Rev. Immunol.* **8:**531–556.
87. **Weinhouse, H., and M. Benziman.** 1974. Regulation of hexose phosphate metabolism in *Acetobacter xylinum. Biochem. J.* **138:**537–542.
88. **Weinhouse, H.** 1977. *Regulation of Carbohydrate Metabolism in Acetobacter xylinum.* Ph.D. thesis. The Hebrew University of Jerusalem, Jerusalem, Israel.
89. **Weinhouse, H., S. Sapir, D. Amikam, Y. Shilo, G. Volman, P. Ohana, and M. Benziman.** 1997. c-di-GMP-binding protein, a new factor regulating cellulose synthesis in *Acetobacter xylinum. FEBS Lett.* **416:**207–211.
90. **Wootton, J. C., and M. H. Drummond.** 1989. The Q-linker: a class of interdomain sequences found in bacterial multidomain regulatory proteins. *Protein Eng.* **2:**535–543.
91. **Yamada, Y.** 1983. *Acetobacter xylinus* sp. nov., nom. rev., for the cellulose-forming and cellulose-less, acetate-oxidizing acetic acid bacteria with the Q-10 system. *J. Gen. Appl. Microbiol.* **29:**417–420.
92. **Yamada, Y., K. Hoshino, and T. Ishikawa.** 1997. The phylogeny of acetic acid bacteria based on the partial sequences of 16S ribosomal RNA: the elevation of the subgenus *Gluconoacetobacter* to the generic level. *Biosci. Biotechnol. Biochem.* **61:**1244–1251.
93. **Zhulin, I. B., A. N. Nikolskaya, and M. Y. Galperin.** 2003. Common extracellular sensory domains in transmembrane receptors for diverse signal transduction pathways in bacteria and archaea. *J. Bacteriol.* **185:**285–294.

The Second Messenger Cyclic Di-GMP
Edited by A. J. Wolfe and K. L. Visick

Chapter 3

Ubiquity of Cyclic Di-GMP Pathways: a Bioinformatic Analysis

MICHAEL Y. GALPERIN

The discovery of cyclic bis(3′→5′) di-GMP (c-di-GMP) in 1987, described in detail in the previous chapter, preceded the identification of enzymes involved in its turnover, diguanylate cyclase (c-di-GMP synthase, DGC) and c-di-GMP-specific phosphodiesterase (PDE-A). In turn, these enzymes were first described as conserved protein domains and only later assigned the respective enzymatic activities. This is a story where bioinformatic approaches played a surprisingly important role by providing valuable leads, subsequently verified by experimental studies. In this chapter, I show how the bioinformatic tools have contributed to the studies of the c-di-GMP-mediated signaling pathways—and continue to do so. A brief discussion of the roles of the GGDEF, EAL, and HD-GYP domains in c-di-GMP turnover and the role of the PilZ domain as a c-di-GMP adaptor protein is followed by an analysis of the phylogenetic distribution of these domains and a listing of the most common domain architectures that involve these four domains. These data are expected to stimulate experimental studies of the widespread c-di-GMP-related proteins that still remain poorly characterized.

PROTEIN DOMAINS INVOLVED IN c-di-GMP SIGNALING

The c-di-GMP signaling machinery involves a variety of membrane-bound and cytoplasmic proteins that serve as receptor, signal transduction, and signal output proteins. From the biochemical perspective, these proteins either synthesize, or hydrolyze, or bind c-di-GMP (Table 1). Accordingly, nearly all known c-di-GMP-related proteins include at least one of the four key protein domains identified to date, namely, the GGDEF domain, the EAL domain, the HD-GYP domain, and/or the PilZ domain (Table 1).

The GGDEF Domain, GGDEF Motif, and Its Enzymatic Activity

The first GGDEF domain was described in 1995 by Gregory Hecht and Austin Newton at Princeton University as the output domain of the unusual response regulator PleD that controls cell differentiation in the alphaproteobacterium *Caulobacter crescentus* (29). Analysis of the PleD sequence showed the presence of two N-terminal CheY-like response regulator receiver (REC) domains and a previously unknown 169-amino-acid C-terminal domain, which showed similarity to uncharacterized protein sequences from six different bacteria (29). The alignment of the seven sequences revealed four regions of particularly high sequence identity (now referred to as conserved sequence motifs). The third conserved sequence motif of this alignment, Arg-H-Gly-Gly-(Asp/Glu)-Glu-Phe (where H indicates a hydrophobic residue), corresponded to residues 366 to 372 in the *C. crescentus* PleD sequence and was used to give this sequence family a provisional designation, the GGDEF domain (29). This provisional name proved quite resilient and has since been assigned to homologous domains in more than 15,000 different proteins.

The year 1995 also marked the beginning of the genome era, the release of the first complete genomic sequences of two cellular life forms, the bacteria *Haemophilus influenzae* and *Mycoplasma genitalium.* Neither of these had any GGDEF domains (Table 2), but the complete genome sequence of the cyanobacterium *Synechocystis* sp. strain PCC 6803, released

Michael Y. Galperin • National Center for Biotechnology Information, National Library of Medicine, National Institutes of Health, Bethesda, MD 20894.

Table 1. General properties of c-di-GMP-related signaling domains

Domain name	Example (aa)[a]	Function	CDD entry[b] Name	CDD entry[b] No. of proteins	Pfam entry[c]	PDB entry[d]	Reference(s)
GGDEF	AAK24433 (291–450)	c-di-GMP synthesis	cl00291	9,923	PF00990	1w25	9, 29
EAL	AAC61686 (492–732)	c-di-GMP hydrolysis (class I)	cl00290	5,667	PF00563	2bas	66, 73
HD-GYP	CAB89843 (180–370)	c-di-GMP hydrolysis (class II)	COG2206	1,047			23, 61
PilZ	BAA36052 (110–230)	c-di-GMP binding	cl01260	1,455	PF07238	1ywu	2, 63
FleQ	AAG04486 (1–125)	c-di-GMP binding	cl05812	80	PF06490		31

[a]The protein entry in the NCBI Protein database (http://www.ncbi.nlm.nih.gov/) and the amino acids (aa) corresponding to the respective domain.
[b]The domain entry code in the NCBI Conserved Domain Database (CDD) (http://www.ncbi.nlm.nih.gov/Structure/cdd/cdd.shtml) (43) and the number of bacterial proteins containing the respective CDD entry in the NCBI Reference Sequence (RefSeq) (http://www.ncbi.nlm.nih.gov/RefSeq/) database (55) as of 31 December 2008.
[c]The domain entry code in the Protein Families (Pfam) (http://pfam.sanger.ac.uk) database (18).
[d]The entry code of a representative domain structure in the Protein Data Bank (PDB) (http://www.rcsb.org/pdb/) (30).

the next year, encoded 22 copies of the GGDEF domain (22, 23). The genome of *Escherichia coli* K-12, completed in 1997, encoded 19 GGDEF-containing proteins. None of them had been previously characterized, and as a result, most of them have been annotated simply "conserved hypothetical protein" (22). Seven of the GGDEF domain proteins from *Synechocystis* sp. were annotated "PleD-like protein," a questionable assignment as none of these proteins had the REC-REC-GGDEF domain architecture of PleD (one protein, Slr0687, had a REC-GAF-GGDEF domain structure and the other six did not even have a receiver domain). However, three *Synechocystis* sp. proteins that combined the REC and GGDEF domains were annotated as regulatory components of the sensory transduction system, an annotation that later proved to be correct. Continued microbial genome sequencing kept revealing large numbers of GGDEF-containing proteins encoded by diverse bacteria (Fig. 1), from 10 in *Clostridium acetobutylicum* and 11 in *Aquifex aeolicus* to 33 in *Pseudomonas aeruginosa,* 41 in *Vibrio cholerae,* and 52 in *Shewanella oneidensis* (Table 2) (20, 22, 23). The current record stands at 68 GGDEF-containing proteins encoded in the genome of *Shewanella amazonensis* with other representatives of *Shewanella* and *Vibrio* genera not far behind (21); see http://www.ncbi.nlm.nih.gov/Complete_Genomes/SignalCensus.html. The sheer abundance of the GGDEF-containing proteins begs the question, what roles do all these proteins perform (22, 23)?

The key to answering this question was the analysis of the domain architectures of various GGDEF-containing proteins that showed the GGDEF domain in association with cytoplasmic signaling domains, such as REC, PAS, and GAF (Fig. 1), suggesting that these proteins take part in signal transduction (23). In some proteins, the GGDEF domain was combined with periplasmic or membrane-embedded ligand-binding domains, indicating that these proteins could participate in transduction of environmental signals (22).

The identification of the GGDEF domain as a component of DGCs and c-di-GMP-specific phosphodiesterases, participating in c-di-GMP turnover, by Benziman and his colleagues (73) was a watershed event in at least three important aspects. This work provided the first evidence of an enzymatic function for this widespread protein domain and paved the way to the experimental demonstration that the GGDEF domain alone was responsible for the DGC activity (53, 62). Second, linking this widespread domain with c-di-GMP turnover provided evidence for the participation of c-di-GMP in a variety of signaling processes (20, 22, 23). Third, the presence of the GGDEF domain in DGCs and in c-di-GMP-specific phosphodiesterases, two kinds of enzymes with opposing activities, suggested that this domain had allosteric functions that regulate c-di-GMP turnover (11).

It is important to note here an important distinction between the GGDEF domain and the Gly-Gly-Asp-Glu-Phe sequence motif that gave it its name. The GGDEF domain is a protein that consists of ~170 amino acid residues, a separately folding structural unit that forms a well-defined globular structure. It can be found in a stand-alone form, although it is usually found as part of multidomain signaling proteins that additionally contain periplasmic, membrane-embedded, or cytoplasmic ligand-binding sensory domains. Some of these GGDEF domain proteins have the GGDEF sequence motif, some have the GGEEF motif, and others have all sorts of variant motifs that may differ from GGDEF in every single position. However, domains are classified on the ba-

Table 2. Distribution of genes encoding GGDEF, EAL, HD-GYP, and PilZ domain proteins in completely sequenced genomes of selected model organisms[a]

Organism	Phylum (class)	Total proteins	GGDEF	GGDEF+EAL	EAL	HD-GYP	PilZ
Agrobacterium tumefaciens	*Proteobacteria (Alphaproteobacteria)*	5,402	14	13	1	1	3
Anabaena sp. strain PCC 7120	*Cyanobacteria*	6,055	8	6	1	2	0
Aquifex aeolicus	*Aquificae*	1,560	7	4	3	1	5
Bacillus subtilis	*Firmicutes*	4,105	3	1	2	0	1
Borrelia burgdorferi	*Spirochaetes*	1,639	1	0	1	1	1
Campylobacter jejuni	*Proteobacteria (Epsilonproteobacteria)*	1,634	1	0	0	0	0
Caulobacter crescentus	*Proteobacteria (Alphaproteobacteria)*	3,737	4	7	3	0	2
Clostridium acetobutylicum	*Firmicutes*	3,848	7	3	1	9	5
Colwellia psychrerythraea	*Proteobacteria (Gammaproteobacteria)*	4,910	34	26	5	3	7
Cytophaga hutchinsonii	*Bacteroidetes*	3,785	0	0	0	0	0
Dechloromonas aromatica	*Proteobacteria (Betaproteobacteria)*	4,171	29	28	3	15	6
Deinococcus radiodurans	*Deinococcus-Thermus*	3,184	12	4	1	4	0
Desulfovibrio vulgaris	*Proteobacteria (Deltaproteobacteria)*	3,531	17	9	1	14	7
Escherichia coli K–12	*Proteobacteria (Gammaproteobacteria)*	4,242	12	7	10	0	2
Geobacter sulfurreducens	*Proteobacteria (Deltaproteobacteria)*	3,446	20	8	1	10	10
Haemophilus influenzae	*Proteobacteria (Gammaproteobacteria)*	1,711	0	0	0	0	0
Helicobacter pylori	*Proteobacteria (Epsilonproteobacteria)*	1,576	0	0	0	0	0
Lactobacillus casei	*Firmicutes*	2,771	0	0	0	0	0
Mycobacterium tuberculosis	*Actinobacteria*	3,927	0	1	1	0	0
Mycoplasma genitalium	*Firmicutes*	484	0	0	0	0	0
Myxococcus xanthus	*Proteobacteria (Deltaproteobacteria)*	7,331	18	0	2	6	23
Prochlorococcus marinus	*Cyanobacteria*	1,882	0	0	0	0	0
Pseudomonas aeruginosa	*Proteobacteria (Gammaproteobacteria)*	5,567	17	16	5	3	8
Ralstonia metallidurans	*Proteobacteria (Betaproteobacteria)*	6,319	18	13	10	2	5
Rhodobacter sphaeroides	*Proteobacteria (Alphaproteobacteria)*	4,126	9	7	0	0	1
Rhodospirillum rubrum	*Proteobacteria (Alphaproteobacteria)*	3,841	21	15	4	2	
Rickettsia prowazekii	*Proteobacteria (Alphaproteobacteria)*	835	1	0	1	0	0
Shewanella oneidensis	*Proteobacteria (Gammaproteobacteria)*	4,467	30	22	5	9	4
Streptococcus pneumoniae	*Firmicutes*	2,094	0	0	0	0	0
Streptomyces coelicolor	*Actinobacteria*	8,154	3	4	0	1	0
Synechococcus elongatus	*Cyanobacteria*	2,525	9	8	1	2	2
Synechocystis sp. strain PCC 6803	*Cyanobacteria*	3,567	13	9	4	2	0
Thermotoga maritima	*Thermotogae*	1,858	9	0	0	10	1
Thermus thermophilus	*Deinococcus-Thermus*	2,210	5	1	0	5	0
Vibrio cholerae	*Proteobacteria (Gammaproteobacteria)*	3,835	31	10	12	9	4
Yersinia pestis	*Proteobacteria (Gammaproteobacteria)*	4,067	2	3	4	0	1

[a]The data are taken from a web supplement to reference 2; see http://www.ncbi.nlm.nih.gov/Complete_Genomes/SigCensus/PilZphylo.html for a full list.

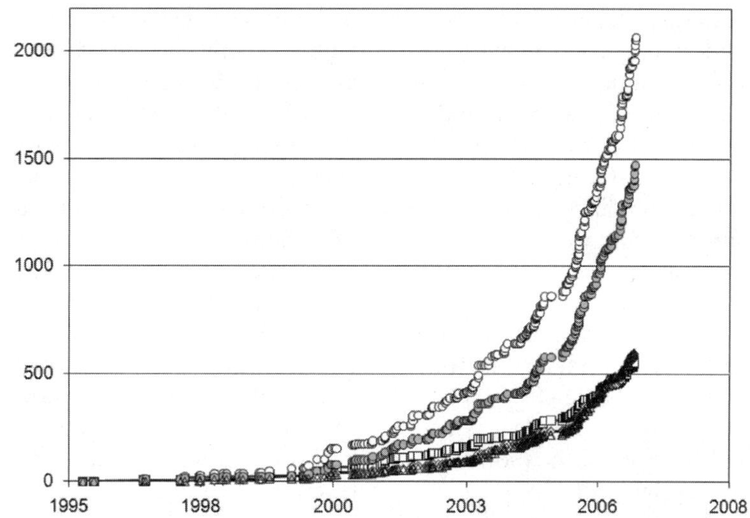

Figure 1. Growth of the number of c-di-GMP-related protein domains in the public protein databases. The numbers of proteins containing GGDEF domains but no EAL domains (open circles), both GGDEF and EAL domains (grey circles), only EAL domains (open squares), HD-GYP domains (diamonds), and PilZ domains (triangles) are indicated.

sis of predicted structure, and there is no evidence that alterations in the GGDEF motif alter the three-dimensional structure of the GGDEF domain. Thus, these variant domains should also be called GGDEF domains (and not GGEEF as some authors have done).

The distinction between the domain name and the actual sequence of the corresponding motif is also important for the assignment of the enzymatic activity. Several studies have convincingly demonstrated the DGC activity of GGDEF domains from a variety of diverse bacteria (42, 53, 62) and showed that this activity requires that all five residues of the GG(D/E)EF motif remain intact. The only allowable variation appeared to be the presence of either Asp or Glu in the middle position. All other changes in this sequence motif affected the enzymatic activity of the respective GGDEF domain but, in some cases, allowed retention of at least some activity (42). In addition, the DGC activity could be affected by amino acid changes outside the GG(D/E)EF motif. Thus, the presence of the GGDEF motif does not necessarily indicate the presence of the DGC activity, whereas the presence of certain substitutions does not necessarily render the domain inactive. This notion is extremely important for interpreting the annotation of GGDEF-containing proteins in newly sequenced bacterial genomes. When sequence comparisons identify the presence of a GGDEF domain in a given protein, it means only that this protein might have the DGC activity. A more reliable prediction is possible only through a detailed (often manual) analysis of the conservation of all the key residues, and even that might not be enough to make a decision.

The EAL Domain

The EAL domain was first described in 1986 in a joint paper from the groups of Nigel Brown at the University of Bristol, United Kingdom, and Simon Silver at the Washington University in St. Louis, St. Louis, MO, as an "unnamed open reading frame URF2" downstream of the mercury resistance determinants of plasmid R100 and transposon Tn*501* (8). This protein, whose association, if any, with mercury resistance still remains elusive, represents an example of a protein consisting of a single stand-alone full-length EAL domain. The next year, a very similar open reading frame was sequenced from *P. aeruginosa* and designated *toxR* because of its role as a positive regulator of the expression of the exotoxin gene (83). Subsequent work with purified ToxR (RegA) suggested that it directly interacts with DNA and RNA polymerase (57, 78, 79). In 1995, Tod Merkel and Scott Stibitz at the FDA (Bethesda, MD) identified a regulatory protein, named BvgR, that affected expression of virulence-related genes in *Bordetella pertussis* and noted its similarity to URF2 (48). Three years later, in 1998, they obtained the full-length sequence of BvgR and described a large family of proteins that shared with BvgR two regions of high sequence similarity (49). Just several months later, Benziman and colleagues identified EAL as a separate protein domain at the C termini of DGCs and c-di-

GMP-specific phosphodiesterases from *Gluconacetobacter xylinus* and assigned it its current name (73).

Again, since the domain name is based on its conserved sequence motif, it is important to note that not all EAL domains necessarily have the Glu-Ala-Leu motif. The EAL domain is an ~240-amino-acid protein that folds into a distinct globular structure. Like GGDEF, it is usually found as part of multidomain signaling proteins in combination with periplasmic, membrane-embedded or cytoplasmic ligand-binding sensory domains. Approximately half of all EAL domains are associated with GGDEF domains, in some organisms even more than that (Table 2). It is also important to note that although the Glu-Ala-Leu sequence motif appears to play a role in the c-di-GMP-specific phosphodiesterase activity of the EAL domain, the presence of this motif in a given protein does not guarantee that this protein would have any enzymatic activity (66).

The HD-GYP Domain

The HD-GYP domain is a member of a widespread superfamily of metal-dependent phosphohydrolases and is designated the HD superfamily after the conserved His-Asp doublet of predicted catalytic residues (4). Protein domains of this superfamily have been found in eukaryotic cGMP/cAMP phosphodiesterases, bacterial dGTP triphosphohydrolase, the ppGpp hydrolase SpoT, and a variety of other proteins. An HD-type domain was also found fused with the receiver domain in two-component response regulator-like proteins from several bacteria (4). A detailed analysis of those response regulators showed that they contained a special version of the HD domain that had an additional C-terminal subdomain with additional conserved residues, including a highly conserved Gly-Tyr-Pro motif (23). Owing to the presence of the highly conserved His and Asp residues implicated in metal binding and catalytic activity and this new GYP motif, the new domain (or, more likely, a combination of two independent domains) was named HD-GYP. While the classical HD domains displayed phosphodiesterase activity against a variety of substrates, the additional C-terminal subdomain was expected to confer stricter substrate specificity upon the HD-GYP domain. The existence of a domain fusion between HD-GYP and GGDEF in two distant bacteria, *Aquifex aeolicus* and *Deinococcus radiodurans,* as well as certain complementarity in the distribution of the HD-GYP and EAL domains, led to the suggestion that the HD-GYP domain could function as a c-di-GMP-specific phosphodiesterase (23). This suggestion proved to be correct (60, 61), as discussed in detail in chapter 5. Therefore, the EAL and HD-GYP domains comprise two different classes of phosphodiesterases that are highly specific for c-di-GMP and can be referred to as class I (the EAL type) and class II (the HD-GYP type) phosphodiesterases.

Again, it needs to be stressed that the mere presence of the eponymous residues in an HD-GYP domain does not necessarily mean that it is enzymatically active. Further, there is certain evidence that inactivated versions could have a regulatory role. For example, *Bacillus subtilis* YuxH protein contains a fusion of modified versions of the EAL and HD-GYP domains that both lack certain conserved residues. These amino acid substitutions are likely to impair the activity of the EAL domain, the HD-GYP domain, or both. However, this combination of supposedly inactivated domains is widespread among diverse bacteria: close homologs of YuxH are encoded in other bacilli and clostridia as well as in *Aquifex* and many beta-, gamma-, and deltaproteobacteria. Although the exact function(s) of YuxH and its homologs remains obscure, their wide phylogenetic distribution suggests that these proteins play an important role in signal transduction.

PilZ and Other c-di-GMP-Binding Domains

Despite numerous studies, the mechanism(s) of c-di-GMP-mediated signaling remains enigmatic. Part of the problem lies in the lack of information on the nature and the diversity of the c-di-GMP-binding proteins. As noted above, divergent and inactivated GGDEF and EAL domains could serve as c-di-GMP responsive modules. However, c-di-GMP binding by *G. xylinus* cellulose synthase (44, 82), as well as the effects of c-di-GMP on flagellar development in *E. coli* (67), indicated the existence of a dedicated c-di-GMP-binding protein, serving as a primary target of c-di-GMP action. A bioinformatics analysis (2) identified a conserved protein domain shared by several unrelated c-di-GMP-linked proteins, including cellulose synthases from *E. coli* and *G. xylinus, E. coli* protein YcgR, previously shown to affect flagellar expression, and the alginate biosynthesis protein Alg44 that is required for alginate biosynthesis in *P. aeruginosa* and *Azotobacter vinelandii* (41, 46). This domain was named the PilZ domain, after the short protein encoded by the *P. aeruginosa pilZ* gene (PA2960) that is required for the development of twitching motility in this organism and appears to act at some posttranslational stage (1).

Subsequent experimental studies unequivocally demonstrated c-di-GMP binding by PilZ domain-containing proteins from *E. coli, C. crescentus, V. cholerae,* and *P. aeruginosa* (12, 47, 54, 63). Upon binding the c-di-GMP molecule, the PilZ domain undergoes a significant conformational change, which appears to be an important part of its signaling mechanism (6). The PilZ domain is encoded in a variety of bacteria with a phylogenetic distribution that closely parallels those of the GGDEF and EAL domains (Table 2).

Despite the widespread distribution of the PilZ domain, its apparent absence in certain organisms that encode GGDEF and EAL domains (for example, in many cyanobacteria and members of the *Deinococcus-Thermus* group [Table 2]) shows that there must be alternative c-di-GMP-binding molecules. Recently, Jason Hickman and Caroline Harwood demonstrated c-di-GMP binding by the *P. aeruginosa* protein FleQ (31), a σ^{54}-interacting transcriptional regulator (see chapter 11). Its N-terminal domain (also referred to as FleQ) is closely related to the REC domain and appears to be a divergent variant of REC that has lost its phosphoacceptor function. FleQ domains have been found so far only in gammaproteobacteria. A study of the role of c-di-GMP in regulating PEL (pellicle) polysaccharide biosynthesis in *P. aeruginosa* identified yet another c-di-GMP-binding protein, PelD (40). Although PelD was assumed to be a novel c-di-GMP receptor, the predicted three-dimensional structure of its C-terminal fragment was similar to that of the GGDEF domain (40). Indeed, comparison of the PleD sequence against the Conserved Domain Database (43) showed that its C-terminal domain is a highly divergent GGDEF domain that had lost most of its conserved residues (including replacement of the GGDEF motif in its active site loop by RNDEG) but could have retained the c-di-GMP-binding I site (M. Y. Galperin, unpublished data). There are probably other highly diverged—and therefore overlooked—GGDEF and EAL domains that could serve as c-di-GMP-responsive elements in a variety of bacteria.

In addition to various c-di-GMP-binding proteins, a recent paper by Sudarsan and colleagues (71) described an RNA-based c-di-GMP receptor, a 110-nucleotide RNA molecule that was able to bind c-di-GMP with an extremely high affinity (equilibrium dissociation constant, K_D, ~1 nM). This RNA molecule (GEMM) was shown to function as a riboswitch, regulating expression of numerous genes in response to c-di-GMP binding (see chapter 15 for details). The current list of cellular receptors for c-di-GMP is probably still incomplete and can be expected to grow in the near future.

PHYLOGENETIC DISTRIBUTION OF THE c-di-GMP-RELATED DOMAINS

For reasons that still remain unclear, the c-di-GMP-mediated regulatory system appears to function only in bacteria. Searches of the public nucleotide or protein sequence databases occasionally find GGDEF, EAL, and even HD-GYP domains encoded in the newly sequenced (unfinished) eukaryotic genomes. However, so far, most of such sequences could be traced to the bacterial contamination of the DNA samples (e.g., transposon insertion into the source DNA). For example, the initial variant of the *Anopheles gambii* genome, deposited in GenBank, encoded at least eight proteins with GGDEF domains. Subsequent reanalysis and cleaning of low-quality genome fragments led to the gradual removal of these sequences, so the currently available revised version of the *A. gambii* genome does not contain a single GGDEF-encoding gene. At the time of this writing, the NCBI protein database listed 16 proteins from the starlet sea anemone *Nematostella vectensis* that contained GGDEF and/or EAL domains. While there seems to be no reason why this ancestral metazoan could not have a c-di-GMP turnover system, all of those GGDEF- and EAL-containing proteins look typically bacterial; some even have a fusion of the GGDEF domain with other specifically bacterial domains. It is very likely, therefore, that these sequences come from a bacterial contaminant (a parasite or a symbiont). Further, the current database lists two proteins from rice *Oryza sativa* subsp. *japonica* that carry GGDEF domains. Both proteins are extremely long (1,938 and 3,775 amino acid residues, respectively), carry a variety of domains, and are likely to arise from sequencing errors. Similarly, all EAL domains identified in eukaryotic genomes appear to come from bacteria. Again, it should be stressed that the coverage of the eukaryotic diversity by genome sequences is still very limited, and it is entirely possible that true c-di-GMP-mediated regulatory systems could still be found among some primitive—or poorly studied—eukaryotes. However, such a discovery would have to be proven by direct experiments and linked to a specific function of c-di-GMP in those eukaryotic cells, such as, for example, activation of the cellulose synthase. All eukaryotic cellulose synthases sequenced to date lack the c-di-GMP-binding PilZ domain and do not require c-di-GMP for their function.

Similar to the eukaryotes, c-di-GMP-related machinery seems to be missing in archaea (21). Of the 55 archaeal genomes completed by the end of 2008, only one, the genome of the uncultured methanogenic archaeon RC-I, encoded a single protein (locus tag RCIX2274) with full-length GGDEF and HD-GYP domains. Two more archaeal genomes encode truncated GGDEF domains but not EAL or PilZ domains. Again, there seems to be no reason that would prevent some archaeon from harboring a c-di-GMP signaling system, at least owing to a lateral gene transfer from bacteria. Nevertheless, none of the cultured representatives of archaea happen to encode such a system.

In the course of the characterization of the PilZ domain, we have compiled a listing of the GGDEF, EAL, HD-GYP, and PilZ domains encoded in the complete bacterial genomes that had been sequenced by the end of 2005 (2). As expected, the phylogenetic profiles of the enzymes involved in c-di-GMP turnover, the c-di-GMP synthase (the GGDEF domain) and the c-di-GMP-specific phosphodiesterase (the EAL and HD-GYP domains), and that of the postulated c-di-GMP-binding adaptor protein (the PilZ domain) proved to be very similar. An extension of that analysis, conducted at the end of 2007, largely confirmed the similarity of the phylogenetic profiles of the GGDEF, EAL and/or HD-GYP, and PilZ domains (Table 2; see http://www.ncbi.nlm.nih.gov/Complete_Genomes/c-di-GMP.html for the full listing). There appeared to be absolute correlation between the presence or absence of consensus GGDEF, EAL, and PilZ domains in the sequenced genomes of the representatives of such bacterial phyla as *Aquificae, Bacteroidetes, Chlamydiae, Deinococcus-Thermus, Spirochaetes,* and *Thermotogae,* as well as of *Betaproteobacteria, Gammaproteobacteria,* and *Deltaproteobacteria* (2). However, this correlation was not observed for many members of *Alphaproteobacteria, Actinobacteria, Cyanobacteria,* and *Firmicutes* (Table 2). For example, several intracellular bacterial parasites and endosymbionts, representing the family *Rickettsiaceae* (*Anaplasma, Ehrlichia, Neorickettsia, Orientia, Rickettsia,* and *Wolbachia* genera) were found to encode GGDEF (and sometimes EAL) domains but no PilZ domains. However, a recent work on *Anaplasma phagocytophilum* showed that its GGDEF domain is active as a DGC and plays an important role in regulating the intracellular infection by this organism (39). In these organisms, the GGDEF domains themselves could be involved in c-di-GMP-binding. They also could be harboring alternative c-di-GMP adaptors, such as the recently described c-di-GMP-binding riboswitch or just GGDEF or PilZ domains that have diverged beyond recognition by standard sequence comparison methods (2, 40, 71).

DOMAIN ARCHITECTURES OF THE c-di-GMP-RELATED PROTEINS

Bacterial genome sequencing revealed a large number of proteins containing GGDEF, EAL, HD-GYP, and PilZ domains (21, 22). Many of these proteins have complex domain architectures and combine these domains with a variety of other domains, including the response receiver (phosphoacceptor) REC domain, cytoplasmic ligand-binding domains PAS and GAF, and various periplasmic or membrane-embedded sensor domains (3, 20–22, 24, 50, 85). In other instances, apparently active GGDEF and/or EAL domains are fused with inactivated ones, with the latter serving as allosteric regulators of the former. The diversity of domain combinations extends to the genus and species levels, such that domain combinations seen in some strains may not be encoded in other strains from the same bacterial species. Nevertheless, some domain combinations clearly predominate and can be found across multiple species, sometimes even in different phyla. I describe here several of such common domain combinations.

GGDEF Domain Fusions

PleD (REC-REC-GGDEF)

Caulobacter crescentus response regulator PleD (29), the first experimentally characterized protein with the GGDEF domain, remains the best studied of all GGDEF-containing proteins (9, 52, 53, 81). Studies of a response regulator with the same REC-REC-GGDEF domain architecture from the plant symbiont *Rhizobium leguminosarum* provided genetic evidence that the GGDEF domain functions as a DGC (5). Later, the DGC activity of PleD was demonstrated biochemically (53), and it became the first GGDEF-containing protein to have a solved crystal structure and experimentally demonstrated allosteric properties (9, 11) (see chapters 4 and 7). Orthologs of PleD are found in most alphaproteobacterial genomes and a variety of other proteobacteria. In representatives of the order *Rickettsiales (Rickettsia, Anaplasma, Ehrlichia, Orientia,* and *Wolbachia),* PleD is the only GGDEF-containing protein encoded in the whole genome. A recent study showed that PleD is required for survival of *Anaplasma phagocytophilum* in the host cell (39).

WspR (REC-GGDEF)

The current protein databases contain more than 300 response regulators of the REC-GGDEF domain architecture. The best studied of them is the *Pseudomonas aeruginosa* response regulator WspR, which is responsible for the so-called wrinkly spreader phenotype, characterized by the production of cellulose fibers at the cell surface that result in biofilm formation at the air-liquid interface (14, 28, 32, 70). The *P. aeruginosa* WspR protein has been the subject of a comprehensive analysis of the DGC activity determinants (42). Its crystal structure has been solved (15) (see chapter 7).

AdrA (MASE2-GGDEF)

The AdrA (YaiC) protein was first characterized as a factor regulating cellulose production, which contributed to intercellular adhesion in *Salmonella enterica* serovar Typhimurium (59, 86) (see chapter 10). Its N-terminal region consists of six predicted membrane-spanning fragments with several conserved aromatic residues. Based on its association with the GGDEF domain (in AdrA and its homologs) and with an adenylate cyclase domain (in *P. aeruginosa* protein PA3217), this N-terminal region was recognized as a novel signaling domain and dubbed MASE2 (50). AdrA was shown to be one of the principal DGCs expressed in stationary cells of *S. enterica* serovar Typhimurium; its heterologous expression in plague-causing *Yersinia pestis* enhanced biofilm production in that organism (34, 68) (see also chapter 10). AdrA orthologs are found among gammaproteobacteria and in a handful of betaproteobacteria.

GReg (GCS-XXX-GGDEF)

Escherichia coli protein YddV is a globin-coupled sensor (GCS) that combines the GGDEF domain with an N-terminal globin sensor domain, which functions as a heme-containing oxygen-sensing signaling domain (19). Orthologs of YddV (globin regulators, or GReg) are found among alpha-, beta-, and gammaproteobacteria. Oxygen binding to GReg from the whooping cough pathogen *Bordetella pertussis* was shown to affect the DGC activity of the downstream GGDEF domain and regulate biofilm formation by this organism (80). These data suggest that *B. pertussis* could use GReg for monitoring oxygen levels in its environment and controlling colonization of human lungs.

EAL Domain Architectures

Stand-alone EAL

As noted above, URF2, ToxR (RegA), and BvgR proteins consisting of stand-alone EAL domains were the first EAL domains to be described and experimentally characterized (8, 48). *Escherichia coli* encodes two stand-alone EAL domains, products of the *ydiV* and *yhjH* genes, whose overexpression has been shown to improve motility and suppress biofilm formation in *E. coli, Pseudomonas putida,* and other organisms (13, 26, 37, 56). Stand-alone EAL domains are found in members of diverse bacterial phyla, including *Proteobacteria, Cyanobacteria,* and *Firmicutes.*

PvrR (REC-EAL)

The first experimentally characterized response regulator with the EAL output domain was described in a clinical isolate of *P. aeruginosa* and designated phenotypic variance regulator, PvrR, as it was found to affect the frequency of switching between the two cell phenotypes, which differed in colony morphology, susceptibility to antibiotics, and the ability to form biofilms (17). Overexpression of this protein resulted in a decreased intracellular level of c-di-GMP, reduced fimbriae expression, and affected colony morphology (45). Another protein with the same REC-EAL domain architecture was subsequently characterized in the laboratory strain *P. aeruginosa* PAO1 and designated RocR; it appeared to counteract the activation of fimbrial gene expression by the NarL family response regulator RocA1 (38). Phylogenetic distribution of PvrR/RocR orthologs appears to be limited to proteobacteria, cyanobacteria, and spirochaetes.

VieA (REC-EAL-HTH)

Response regulators of the VieA family, found so far exclusively in *Vibrio* spp., differ from PvrR-like proteins by the presence of an additional DNA-binding helix-turn-helix (HTH) domain at their C termini. The VieA protein from *V. cholerae* has been shown to function as a c-di-GMP phosphodiesterase, regulating biofilm formation and expression of virulence genes including the cholera toxin gene (74, 76, 77).

YcgF (BLUF-EAL)

Escherichia coli and several other enterobacteria encode an unusual photoreceptor YcgF that combines the EAL domain with the BLUF (*b*lue *l*ight sensor *u*sing *f*lavin adenine dinucleotide) domain (27). This protein has been shown to contain noncovalently bound flavin adenine dinucleotide and renamed blue-light-regulated phosphodiesterase, BlrP (27, 58). However, the name YcgF is still commonly used, particularly because *E. coli* YcgF apparently lacks the PDE-A activity. However, both YcgF homologs encoded in *Klebsiella pneumoniae* are active and their activities are increased by light (M. Gomelsky, personal communication). Functions of YcgF might include sensing whether the organism is outside or inside the host, as is the case with other blue-light receptors (72). Outside enterobacteria, YcgF orthologs are found in *Alteromonas macleodii* and *Bordetella avium*.

GGDEF and EAL Domain Fusions

DGC (PAS-GGDEF-EAL)

In their groundbreaking paper, Moshe Benziman and colleagues characterized six open reading frames, three of which functioned as DGCs and the other three functioned as c-di-GMP phosphodiesterases (73). All three DGCs had very similar sequences consisting of PAS, GGDEF, and EAL domains. As PAS domains are often involved in sensing oxygen and/or redox conditions inside the cell (75), it was proposed that the activity of these DGCs could be regulated by oxygen (73). Proteins with the same domain architecture have been shown to affect swarming motility and colony morphology in *Vibrio parahaemolyticus* and *Pseudomonas putida* (36, 64). In the first of these, ScrG, both GGDEF and EAL domains appear to be active; the full-length protein has a PDE-A activity, but its truncated variant that lacks the C-terminal EAL domain displays a DGC activity (36). Proteins with the PAS-GGDEF-EAL domain architecture are found in representatives of *Proteobacteria, Cyanobacteria, Chloroflexi, Planctomycetes,* and *Firmicutes.*

PDE-A (PAS-GAF-GGDEF-EAL, PAS-PAS-GGDEF-EAL)

The three c-di-GMP phosphodiesterases, characterized by Tal and colleagues (73), had the PAS-GAF-GGDEF-EAL domain architecture and were also proposed to be regulated by oxygen (73). The c-di-GMP phosphodiesterase activity of a related protein, *E. coli* YddU, which has the PAS-PAS-GGDEF-EAL domain architecture, was experimentally demonstrated to be regulated by oxygen (10, 16). This protein was therefore renamed direct oxygen sensor (DOS) and became a favorite model for studying mechanisms of oxygen sensing by PAS domains. Remarkably, DOS (YddU) and the above-mentioned GReg (YddV) proteins are encoded by adjacent genes that form an operon, allowing the *E. coli* cell to sense oxygen by two different mechanisms and accordingly regulate synthesis and hydrolysis of c-di-GMP. In a separate study, the DOS protein was reported to have a hydrolytic activity against 2′-O-(*N*-methylanthraniloyl)-cAMP, a fluorescent derivative of cAMP, but not against bis(*p*-nitrophenyl) phosphate (65), which is a standard substrate for phosphodiesterases, including PDE-A (7). In addition, Δ*dos* mutants were reported to have increased levels of cellular cAMP (84). However, a subsequent report failed to detect any activity of DOS against cAMP proper and noted that its activity against cAMP derivative was 1,000-fold lower than the activity of DOS against c-di-GMP (66). Therefore, the assumption that the DOS protein functions as a cAMP phosphodiesterase in vivo (65, 84) remains to be confirmed (see chapter 4).

REC-GGDEF-EAL

Response regulators combining GGDEF and EAL domains are more numerous than PvrR-type (REC-EAL) response regulators and have very similar phylogenetic distribution, being found primarily in proteobacteria (33). So far, none of these proteins has been experimentally characterized.

FimX (REC-PAS-GGDEF-EAL)

The *P. aeruginosa* FimX protein contains both GGDEF and EAL domains, but only EAL is enzymatically active (35). Instead, the divergent GGDEF domain of FimX appears to activate the c-di-GMP phosphodiesterase activity of the EAL domain, whereas the N-terminal REC domain serves as a polar localization signal (35). Proteins with this domain architecture have been found exclusively in gammaproteobacteria.

HD-GYP Domain Fusions

RpfG (REC-HD-GYP)

Although the HD-GYP domain can be found in a variety of domain combinations (23), the only experimentally characterized protein with the HD-GYP domain is the response regulator RpfG, which controls virulence and virulence factor production in the phytopathogen *Xanthomonas campestris* (61, 69) (see chapter 5). Proteins of the RpfG family are found in a variety of bacterial phyla. Many organisms encode both classes of c-di-GMP phosphodiesterases, the EAL and HD-GYP domains.

PilZ Domain Fusions

Cellulose synthase BcsA (Glycos_transf_2-PilZ)

As discussed in the previous chapter, the requirement for an allosteric activator of cellulose synthase in *G. xylinus* was how c-di-GMP was discovered in the first place. Catalytic subunits of many bacterial cellulose synthases contain a C-terminal PilZ domain that is missing in eukaryotic (and certain bacterial) enzymes and was proposed to be the unit responsible for c-di-GMP binding (2). This proposal has been experimentally confirmed, although the purified PilZ domain of BcsA showed much lower affinity for c-di-GMP than intact BcsA protein (63). Remarkably, cyanobacterial cellulose synthases can be of either type (*Nostoc punctiforme* even encodes both of them), suggesting that plant cellulose synthases originated from PilZ-less cyanobacterial ancestors. Cellulose synthases are found in representatives of diverse bacterial phyla, suggesting that cellulose biosynthesis is very common in the bacterial world.

YcgR (PilZN-PilZ)

The *ycgR* mutation was found to stimulate the motility of *E. coli*, indicating its regulatory role (37). In accordance with that, in *Burkholderia cenocepacia* and several related bacteria, the *ycgR* gene is part of the *flgH-flgI-flgJ-ycgR-flgK-flgL* flagellar operon, suggesting that its product might be part of the flagellar machinery. A recent study demonstrated c-di-GMP binding by the full-length YcgR protein as well as by its PilZ domain alone (63). Therefore, YcgR appears to be a central regulator controlling bacterial flagellar motility in response to the changes in cellular c-di-GMP levels (13, 25, 63).

Alg44 (PilZ-HlyD)

The Alg44 protein is required for alginate formation by *P. aeruginosa* and *Azotobacter vinelandii* (41, 46). Sequence analysis showed that these proteins contain an N-terminal PilZ domain fused to a domain that is somewhat similar to HlyD, the membrane component of a multidrug secretion system (2). The ability of Alg44 to bind c-di-GMP and its transmembrane topology have been confirmed (47, 51). It appears that Alg44 either directly participates in the secretion of alginate precursor or regulates this process. Orthologs of Alg44 are encoded in members of the family *Pseudomonadaceae* as well as in the recently sequenced genome of the epsilonproteobacterium *Nitratiruptor* sp.

CONCLUSIONS

The examples listed above show that certain domain architectures involving c-di-GMP-related domains, GGDEF, EAL, HD-GYP, and PilZ, are commonly found in a variety of diverse bacteria. There are certain indications that these conserved genome architectures carry out similar (or at least related) functions in all organisms that encode such proteins. This makes these widespread proteins attractive targets for experimental studies. On the other hand, less-widespread domain fusions often involve uncharacterized protein domains, which open new avenues for further studies of c-di-GMP-mediated signal transduction.

Despite the almost universal distribution of c-di-GMP-mediated signaling among diverse bacteria and the obvious importance of this system in regulating a variety of bacterial developmental processes, it remains poorly characterized. Experimental characterization of the most widespread combinations of c-di-GMP-related domains, including those described above, remains a promising venue of research that can be expected to provide much-needed insights into the functioning of this fascinating signaling system and its role in bacterial adaptation mechanisms.

Acknowledgments. This work was supported by the Intramural Research Program of the National Institutes of Health at the National Library of Medicine.

REFERENCES

1. **Alm, R. A., A. J. Bodero, P. D. Free, and J. S. Mattick.** 1996. Identification of a novel gene, *pilZ*, essential for type 4 fimbrial biogenesis in *Pseudomonas aeruginosa*. *J. Bacteriol.* **178:**46–53.
2. **Amikam, D., and M. Y. Galperin.** 2006. PilZ domain is part of the bacterial c-di-GMP binding protein. *Bioinformatics* **22:** 3–6.
3. **Anantharaman, V., and L. Aravind.** 2003. Application of comparative genomics in the identification and analysis of novel families of membrane-associated receptors in bacteria. *BMC Genomics* **4:**34.
4. **Aravind, L., and E. V. Koonin.** 1998. The HD domain defines a new superfamily of metal-dependent phosphohydrolases. *Trends Biochem. Sci.* **23:**469–472.
5. **Ausmees, N., H. Jonsson, S. Hoglund, H. Ljunggren, and M. Lindberg.** 1999. Structural and putative regulatory genes involved in cellulose synthesis in *Rhizobium leguminosarum* bv. *trifolii*. *Microbiology* **145:**1253–1262.
6. **Benach, J., S. S. Swaminathan, R. Tamayo, S. K. Handelman, E. Folta-Stogniew, J. E. Ramos, F. Forouhar, H. Neely, J. Seetharaman, A. Camilli, and J. F. Hunt.** 2007. The structural basis of cyclic diguanylate signal transduction by PilZ domains. *EMBO J.* **26:**5153–5166.
7. **Bobrov, A. G., O. Kirillina, and R. D. Perry.** 2005. The phosphodiesterase activity of the HmsP EAL domain is required for negative regulation of biofilm formation in *Yersinia pestis*. *FEMS Microbiol. Lett.* **247:**123–130.
8. **Brown, N. L., T. K. Misra, J. N. Winnie, A. Schmidt, M. Seiff, and S. Silver.** 1986. The nucleotide sequence of the mercuric resistance operons of plasmid R100 and transposon Tn501: further evidence for *mer* genes which enhance the activity of the mercuric ion detoxification system. *Mol. Gen. Genet.* **202:** 143–151.
9. **Chan, C., R. Paul, D. Samoray, N. C. Amiot, B. Giese, U. Jenal, and T. Schirmer.** 2004. Structural basis of activity and allosteric control of diguanylate cyclase. *Proc. Natl. Acad. Sci. USA* **101:**17084–17089.

10. **Chang, A. L., J. R. Tuckerman, G. Gonzalez, R. Mayer, H. Weinhouse, G. Volman, D. Amikam, M. Benziman, and M. A. Gilles-Gonzalez.** 2001. Phosphodiesterase A1, a regulator of cellulose synthesis in *Acetobacter xylinum,* is a heme-based sensor. *Biochemistry* **40:**3420–3426.
11. **Christen, B., M. Christen, R. Paul, F. Schmid, M. Folcher, P. Jenoe, M. Meuwly, and U. Jenal.** 2006. Allosteric control of cyclic di-GMP signaling. *J. Biol. Chem.* **281:**32015–32024.
12. **Christen, M., B. Christen, M. G. Allan, M. Folcher, P. Jeno, S. Grzesiek, and U. Jenal.** 2007. DgrA is a member of a new family of cyclic diguanosine monophosphate receptors and controls flagellar motor function in *Caulobacter crescentus. Proc. Natl. Acad. Sci. USA* **104:**4112–4117.
13. **Claret, L., S. Miquel, N. Vieille, D. A. Ryjenkov, M. Gomelsky, and A. Darfeuille-Michaud.** 2007. The flagellar sigma factor FliA regulates adhesion and invasion of Crohn disease-associated *Escherichia coli* via a cyclic dimeric GMP-dependent pathway. *J. Biol. Chem.* **282:**33275–33283.
14. **D'Argenio, D. A., M. W. Calfee, P. B. Rainey, and E. C. Pesci.** 2002. Autolysis and autoaggregation in *Pseudomonas aeruginosa* colony morphology mutants. *J. Bacteriol.* **184:**6481–6489.
15. **De, N., M. Pirruccello, P. V. Krasteva, N. Bae, R. V. Raghavan, and H. Sondermann.** 2008. Phosphorylation-independent regulation of the diguanylate cyclase WspR. *PLoS Biol.* **6:**e67.
16. **Delgado-Nixon, V. M., G. Gonzalez, and M. A. Gilles-Gonzalez.** 2000. Dos, a heme-binding PAS protein from *Escherichia coli,* is a direct oxygen sensor. *Biochemistry* **39:** 2685–2691.
17. **Drenkard, E., and F. M. Ausubel.** 2002. *Pseudomonas* biofilm formation and antibiotic resistance are linked to phenotypic variation. *Nature* **416:**740–743.
18. **Finn, R. D., J. Tate, J. Mistry, P. C. Coggill, S. J. Sammut, H. R. Hotz, G. Ceric, K. Forslund, S. R. Eddy, E. L. Sonnhammer, and A. Bateman.** 2008. The Pfam protein families database. *Nucleic Acids Res.* **36:**D281–D288.
19. **Freitas, T. A., S. Hou, and M. Alam.** 2003. The diversity of globin-coupled sensors. *FEBS Lett.* **552:**99–104.
20. **Galperin, M. Y.** 2004. Bacterial signal transduction network in a genomic perspective. *Environ. Microbiol.* **6:**552–567.
21. **Galperin, M. Y.** 2005. A census of membrane-bound and intracellular signal transduction proteins in bacteria: bacterial IQ, extroverts and introverts. *BMC Microbiol.* **5:**35.
22. **Galperin, M. Y., A. N. Nikolskaya, and E. V. Koonin.** 2001. Novel domains of the prokaryotic two-component signal transduction systems. *FEMS Microbiol. Lett.* **203:**11–21.
23. **Galperin, M. Y., D. A. Natale, L. Aravind, and E. V. Koonin.** 1999. A specialized version of the HD hydrolase domain implicated in signal transduction. *J. Mol. Microbiol. Biotechnol.* **1:**303–305.
24. **Galperin, M. Y., T. A. Gaidenko, A. Y. Mulkidjanian, M. Nakano, and C. W. Price.** 2001. MHYT, a new integral membrane sensor domain. *FEMS Microbiol. Lett.* **205:**17–23.
25. **Girgis, H. S., Y. Liu, W. S. Ryu, and S. Tavazoie.** 2007. A comprehensive genetic characterization of bacterial motility. *PLoS Genet.* **3:**1644–1660.
26. **Gjermansen, M., P. Ragas, and T. Tolker-Nielsen.** 2006. Proteins with GGDEF and EAL domains regulate *Pseudomonas putida* biofilm formation and dispersal. *FEMS Microbiol. Lett.* **265:**215–224.
27. **Gomelsky, M., and S. Kaplan.** 1998. AppA, a redox regulator of photosystem formation in *Rhodobacter sphaeroides* 2.4.1, is a flavoprotein. Identification of a novel FAD-binding domain. *J. Biol. Chem.* **273:**35319–35325.
28. **Goymer, P., S. G. Kahn, J. G. Malone, S. M. Gehrig, A. J. Spiers, and P. B. Rainey.** 2006. Adaptive divergence in experimental populations of *Pseudomonas fluorescens.* II. Role of the GGDEF regulator WspR in evolution and development of the wrinkly spreader phenotype. *Genetics* **173:**515–526.
29. **Hecht, G. B., and A. Newton.** 1995. Identification of a novel response regulator required for the swarmer-to-stalked-cell transition in *Caulobacter crescentus. J. Bacteriol.* **177:**6223–6229.
30. **Henrick, K., Z. Feng, W. F. Bluhm, D. Dimitropoulos, J. F. Doreleijers, S. Dutta, J. L. Flippen-Anderson, J. Ionides, C. Kamada, E. Krissinel, C. L. Lawson, J. L. Markley, H. Nakamura, R. Newman, Y. Shimizu, J. Swaminathan, S. Velankar, J. Ory, E. L. Ulrich, W. Vranken, J. Westbrook, R. Yamashita, H. Yang, J. Young, M. Yousufuddin, and H. M. Berman.** 2008. Remediation of the protein data bank archive. *Nucleic Acids Res.* **36:**D426–D433.
31. **Hickman, J. W., and C. S. Harwood.** 2008. Identification of FleQ from *Pseudomonas aeruginosa* as a c-di-GMP-responsive transcription factor. *Mol. Microbiol.* **69:**376–389.
32. **Hickman, J. W., D. F. Tifrea, and C. S. Harwood.** 2005. A chemosensory system that regulates biofilm formation through modulation of cyclic diguanylate levels. *Proc. Natl. Acad. Sci. USA* **102:**14422–14427.
33. **Jenal, U., and M. Y. Galperin.** 2009. Single-domain response regulators: molecular switches with emerging roles in cell organization and dynamics. *Curr. Opin. Microbiol.* **12:**152–160.
34. **Kader, A., R. Simm, U. Gerstel, M. Morr, and U. Römling.** 2006. Hierarchical involvement of various GGDEF domain proteins in rdar morphotype development of *Salmonella enterica* serovar Typhimurium. *Mol. Microbiol.* **60:**602–616.
35. **Kazmierczak, B. I., M. B. Lebron, and T. S. Murray.** 2006. Analysis of FimX, a phosphodiesterase that governs twitching motility in *Pseudomonas aeruginosa. Mol. Microbiol.* **60:**1026–1043.
36. **Kim, Y. K., and L. L. McCarter.** 2007. ScrG, a GGDEF-EAL protein, participates in regulating swarming and sticking in *Vibrio parahaemolyticus. J. Bacteriol.* **189:**4094–4107.
37. **Ko, M., and C. Park.** 2000. Two novel flagellar components and H-NS are involved in the motor function of *Escherichia coli. J. Mol. Biol.* **303:**371–382.
38. **Kulasekara, H. D., I. Ventre, B. R. Kulasekara, A. Lazdunski, A. Filloux, and S. Lory.** 2005. A novel two-component system controls the expression of *Pseudomonas aeruginosa* fimbrial *cup* genes. *Mol. Microbiol.* **55:**368–380.
39. **Lai, T. H., Y. Kumagai, M. Hyodo, Y. Hayakawa, and Y. Rikihisa.** 2008. *Anaplasma phagocytophilum* PleC histidine kinase and PleD diguanylate cyclase two-component system and role of cyclic di-GMP in host-cell Infection. *J. Bacteriol.* **191:** 693–700.
40. **Lee, V. T., J. M. Matewish, J. L. Kessler, M. Hyodo, Y. Hayakawa, and S. Lory.** 2007. A cyclic-di-GMP receptor required for bacterial exopolysaccharide production. *Mol. Microbiol.* **65:**1474–1484.
41. **Maharaj, R., T. B. May, S. K. Wang, and A. M. Chakrabarty.** 1993. Sequence of the *alg8* and *alg44* genes involved in the synthesis of alginate by *Pseudomonas aeruginosa. Gene* **136:** 267–269.
42. **Malone, J. G., R. Williams, M. Christen, U. Jenal, A. J. Spiers, and P. B. Rainey.** 2007. The structure-function relationship of WspR, a *Pseudomonas fluorescens* response regulator with a GGDEF output domain. *Microbiology* **153:**980–994.
43. **Marchler-Bauer, A., J. B. Anderson, F. Chitsaz, M. K. Derbyshire, C. DeWeese-Scott, J. H. Fong, L. Y. Geer, R. C. Geer, N. R. Gonzales, M. Gwadz, S. He, D. I. Hurwitz, J. D. Jackson, Z. Ke, C. J. Lanczycki, C. A. Liebert, C. Liu, F. Lu, S.

Lu, G. H. Marchler, M. Mullokandov, J. S. Song, A. Tasneem, N. Thanki, R. A. Yamashita, D. Zhang, N. Zhang, and S. H. Bryant. 2009. CDD: specific functional annotation with the Conserved Domain Database. *Nucleic Acids Res.* 37:D205–D210.

44. Mayer, R., P. Ross, H. Weinhouse, D. Amikam, G. Volman, P. Ohana, R. D. Calhoon, H. C. Wong, A. W. Emerick, and M. Benziman. 1991. Polypeptide composition of bacterial cyclic diguanylic acid-dependent cellulose synthase and the occurrence of immunologically crossreacting proteins in higher plants. *Proc. Natl. Acad. Sci. USA* 88:5472–5476.

45. Meissner, A., V. Wild, R. Simm, M. Rohde, C. Erck, F. Bredenbruch, M. Morr, U. Römling, and S. Haussler. 2007. *Pseudomonas aeruginosa cupA*-encoded fimbriae expression is regulated by a GGDEF and EAL domain-dependent modulation of the intracellular level of cyclic diguanylate. *Environ. Microbiol.* 9:2475–2485.

46. Mejia-Ruiz, H., J. Guzman, S. Moreno, G. Soberon-Chavez, and G. Espin. 1997. The *Azotobacter vinelandii alg8* and *alg44* genes are essential for alginate synthesis and can be transcribed from an *algD*-independent promoter. *Gene* 199:271–277.

47. Merighi, M., V. T. Lee, M. Hyodo, Y. Hayakawa, and S. Lory. 2007. The second messenger bis-(3′-5′)-cyclic-GMP and its PilZ domain-containing receptor Alg44 are required for alginate biosynthesis in *Pseudomonas aeruginosa*. *Mol. Microbiol.* 65:876–895.

48. Merkel, T. J., and S. Stibitz. 1995. Identification of a locus required for the regulation of *bvg*-repressed genes in *Bordetella pertussis*. *J. Bacteriol.* 177:2727–2736.

49. Merkel, T. J., C. Barros, and S. Stibitz. 1998. Characterization of the *bvgR* locus of *Bordetella pertussis*. *J. Bacteriol.* 180: 1682–1690.

50. Nikolskaya, A. N., A. Y. Mulkidjanian, I. B. Beech, and M. Y. Galperin. 2003. MASE1 and MASE2: two novel integral membrane sensory domains. *J. Mol. Microbiol. Biotechnol.* 5: 11–16.

51. Oglesby, L. L., S. Jain, and D. E. Ohman. 2008. Membrane topology and roles of *Pseudomonas aeruginosa* Alg8 and Alg44 in alginate polymerization. *Microbiology* 154:1605–1615.

52. Paul, R., S. Abel, P. Wassmann, A. Beck, H. Heerklotz, and U. Jenal. 2007. Activation of the diguanylate cyclase PleD by phosphorylation-mediated dimerization. *J. Biol. Chem.* 282: 29170–29177.

53. Paul, R., S. Weiser, N. C. Amiot, C. Chan, T. Schirmer, B. Giese, and U. Jenal. 2004. Cell cycle-dependent dynamic localization of a bacterial response regulator with a novel diguanylate cyclase output domain. *Genes Dev.* 18:715–727.

54. Pratt, J. T., R. Tamayo, A. D. Tischler, and A. Camilli. 2007. PilZ domain proteins bind cyclic diguanylate and regulate diverse processes in *Vibrio cholerae*. *J. Biol. Chem.* 282:12860–12870.

55. Pruitt, K. D., T. Tatusova, W. Klimke, and D. R. Maglott. 2009. NCBI Reference Sequences: current status, policy and new initiatives. *Nucleic Acids Res.* 37:D32–D36.

56. Rahman, M., R. Simm, A. Kader, E. Basseres, U. Römling, and R. Mollby. 2007. The role of c-di-GMP signaling in an *Aeromonas veronii* biovar *sobria* strain. *FEMS Microbiol. Lett.* 273:172–179.

57. Raivio, T. L., D. Hoffer, R. W. Prince, M. L. Vasil, and D. G. Storey. 1996. Linker insertion scanning of *regA*, an activator of exotoxin A production in *Pseudomonas aeruginosa*. *Mol. Microbiol.* 22:239–254.

58. Rajagopal, S., J. M. Key, E. B. Purcell, D. J. Boerema, and K. Moffat. 2004. Purification and initial characterization of a putative blue light-regulated phosphodiesterase from *Escherichia coli*. *Photochem. Photobiol.* 80:542–547.

59. Römling, U., M. Rohde, A. Olsen, S. Normark, and J. Reinkoster. 2000. AgfD, the checkpoint of multicellular and aggregative behaviour in *Salmonella typhimurium* regulates at least two independent pathways. *Mol. Microbiol.* 36:10–23.

60. Ryan, R. P., Y. Fouhy, J. F. Lucey, B. L. Jiang, Y. Q. He, J. X. Feng, J. L. Tang, and J. M. Dow. 2007. Cyclic di-GMP signalling in the virulence and environmental adaptation of *Xanthomonas campestris*. *Mol. Microbiol.* 63:429–442.

61. Ryan, R. P., Y. Fouhy, J. F. Lucey, L. C. Crossman, S. Spiro, Y. W. He, L. H. Zhang, S. Heeb, M. Camara, P. Williams, and J. M. Dow. 2006. Cell-cell signaling in *Xanthomonas campestris* involves an HD-GYP domain protein that functions in cyclic di-GMP turnover. *Proc. Natl. Acad. Sci. USA* 103:6712–6717.

62. Ryjenkov, D. A., M. Tarutina, O. M. Moskvin, and M. Gomelsky. 2005. Cyclic diguanylate is a ubiquitous signaling molecule in bacteria: insights into biochemistry of the GGDEF protein domain. *J. Bacteriol.* 187:1792–1798.

63. Ryjenkov, D. A., R. Simm, U. Römling, and M. Gomelsky. 2006. The PilZ domain is a receptor for the second messenger c-di-GMP: the PilZ domain protein YcgR controls motility in enterobacteria. *J. Biol. Chem.* 281:30310–30314.

64. Sarand, I., S. Osterberg, S. Holmqvist, P. Holmfeldt, E. Skarfstad, R. E. Parales, and V. Shingler. 2008. Metabolism-dependent taxis towards (methyl)phenols is coupled through the most abundant of three polar localized Aer-like proteins of *Pseudomonas putida*. *Environ. Microbiol.* 10:1320–1334.

65. Sasakura, Y., S. Hirata, S. Sugiyama, S. Suzuki, S. Taguchi, M. Watanabe, T. Matsui, I. Sagami, and T. Shimizu. 2002. Characterization of a direct oxygen sensor heme protein from *Escherichia coli*. Effects of the heme redox states and mutations at the heme-binding site on catalysis and structure. *J. Biol. Chem.* 277:23821–23827.

66. Schmidt, A. J., D. A. Ryjenkov, and M. Gomelsky. 2005. Ubiquitous protein domain EAL encodes cyclic diguanylate-specific phosphodiesterase: enzymatically active and inactive EAL domains. *J. Bacteriol.* 187:4774–4781.

67. Simm, R., M. Morr, A. Kader, M. Nimtz, and U. Römling. 2004. GGDEF and EAL domains inversely regulate cyclic di-GMP levels and transition from sessility to motility. *Mol. Microbiol.* 53:1123–1134.

68. Simm, R., J. D. Fetherston, A. Kader, U. Römling, and R. D. Perry. 2005. Phenotypic convergence mediated by GGDEF-domain-containing proteins. *J. Bacteriol.* 187:6816–6823.

69. Slater, H., A. Alvarez-Morales, C. E. Barber, M. J. Daniels, and J. M. Dow. 2000. A two-component system involving an HD-GYP domain protein links cell-cell signalling to pathogenicity gene expression in *Xanthomonas campestris*. *Mol. Microbiol.* 38:986–1003.

70. Spiers, A. J., J. Bohannon, S. M. Gehrig, and P. B. Rainey. 2003. Biofilm formation at the air-liquid interface by the *Pseudomonas fluorescens* SBW25 wrinkly spreader requires an acetylated form of cellulose. *Mol. Microbiol.* 50:15–27.

71. Sudarsan, N., E. R. Lee, Z. Weinberg, R. H. Moy, J. N. Kim, K. H. Link, and R. R. Breaker. 2008. Riboswitches in eubacteria sense the second messenger cyclic di-GMP. *Science* 321: 411–413.

72. Swartz, T. E., T. S. Tseng, M. A. Frederickson, G. Paris, D. J. Comerci, G. Rajashekara, J. G. Kim, M. B. Mudgett, G. A. Splitter, R. A. Ugalde, F. A. Goldbaum, W. R. Briggs, and R. A. Bogomolni. 2007. Blue-light-activated histidine kinases: two-component sensors in bacteria. *Science* 317:1090–1093.

73. Tal, R., H. C. Wong, R. Calhoon, D. Gelfand, A. L. Fear, G. Volman, R. Mayer, P. Ross, D. Amikam, H. Weinhouse, A. Cohen, S. Sapir, P. Ohana, and M. Benziman. 1998. Three *cdg* operons control cellular turnover of cyclic di-GMP in *Ace-*

tobacter xylinum: genetic organization and occurrence of conserved domains in isoenzymes. *J. Bacteriol.* **180:**4416–4425.

74. **Tamayo, R., A. D. Tischler, and A. Camilli.** 2005. The EAL domain protein VieA is a cyclic diguanylate phosphodiesterase. *J. Biol. Chem.* **280:**33324–33330.
75. **Taylor, B. L., and I. B. Zhulin.** 1999. PAS domains: internal sensors of oxygen, redox potential, and light. *Microbiol. Mol. Biol. Rev.* **63:**479–506.
76. **Tischler, A. D., and A. Camilli.** 2004. Cyclic diguanylate (c-di-GMP) regulates *Vibrio cholerae* biofilm formation. *Mol. Microbiol.* **53:**857–869.
77. **Tischler, A. D., and A. Camilli.** 2005. Cyclic diguanylate regulates *Vibrio cholerae* virulence gene expression. *Infect. Immun.* **73:**5873–5882.
78. **Walker, S. L., L. S. Hiremath, and D. R. Galloway.** 1995. ToxR (RegA) activates *Escherichia coli* RNA polymerase to initiate transcription of *Pseudomonas aeruginosa toxA. Gene* **154:**15–21.
79. **Walker, S. L., L. S. Hiremath, D. J. Wozniak, and D. R. Galloway.** 1994. ToxR (RegA)-mediated in vitro transcription of *Pseudomonas aeruginosa toxA. Gene* **150:**87–92.
80. **Wan, X., J. R. Tuckerman, J. A. Saito, T. A. Freitas, J. S. Newhouse, J. R. Denery, M. Y. Galperin, G. Gonzalez, M. A. Gilles-Gonzalez, and M. Alam.** 2009. Globins synthesize the second messenger bis-(3′-5′)-cyclic diguanosine monophosphate in bacteria. *J. Mol. Biol.* **388:**262–270.
81. **Wassmann, P., C. Chan, R. Paul, A. Beck, H. Heerklotz, U. Jenal, and T. Schirmer.** 2007. Structure of BeF3− -modified response regulator PleD: implications for diguanylate cyclase activation, catalysis, and feedback inhibition. *Structure* **15:** 915–927.
82. **Weinhouse, H., S. Sapir, D. Amikam, Y. Shilo, G. Volman, P. Ohana, and M. Benziman.** 1997. c-di-GMP-binding protein, a new factor regulating cellulose synthesis in *Acetobacter xylinum. FEBS Lett.* **416:**207–211.
83. **Wozniak, D. J., D. C. Cram, C. J. Daniels, and D. R. Galloway.** 1987. Nucleotide sequence and characterization of *toxR:* a gene involved in exotoxin A regulation in *Pseudomonas aeruginosa. Nucleic Acids Res.* **15:**2123–2135.
84. **Yoshimura-Suzuki, T., I. Sagami, N. Yokota, H. Kurokawa, and T. Shimizu.** 2005. DOS(Ec), a heme-regulated phosphodiesterase, plays an important role in the regulation of the cyclic AMP level in *Escherichia coli. J. Bacteriol.* **187:**6678–6682.
85. **Zhulin, I. B., A. N. Nikolskaya, and M. Y. Galperin.** 2003. Common extracellular sensory domains in transmembrane receptors for diverse signal transduction pathways in *Bacteria* and *Archaea. J. Bacteriol.* **185:**285–294.
86. **Zogaj, X., M. Nimtz, M. Rohde, W. Bokranz, and U. Römling.** 2001. The multicellular morphotypes of *Salmonella typhimurium* and *Escherichia coli* produce cellulose as the second component of the extracellular matrix. *Mol. Microbiol.* **39:**1452–1463.

The Second Messenger Cyclic Di-GMP
Edited by A. J. Wolfe and K. L. Visick

Chapter 4

The Core Pathway: Diguanylate Cyclases, Cyclic Di-GMP-Specific Phosphodiesterases, and Cyclic Di-GMP-Binding Proteins

Mark Gomelsky

Cyclic di-GMP (c-di-GMP) is a widely recognized second messenger that affects numerous processes in the *Bacteria*. In the *Proteobacteria*, it affects an important lifestyle choice, between single-cellular motility and the surface-attached, sessile mode of growth, which often results in formation of multicellular communities (biofilm). The c-di-GMP-dependent signaling pathways are not limited to regulation of this important transition and also affect bacterial cell cycle and cell fate, virulence, long-term survival, and other processes, most of which are described in this book. It took some time for the scientific community to appreciate the scope and significance of c-di-GMP as a second messenger affecting a broad spectrum of processes. However, following such acceptance, spectacular progress has been achieved, especially in the last few years. By now, the main pathways through which c-di-GMP is made and degraded have been elucidated and several types of c-di-GMP receptors have been identified. However, our knowledge about c-di-GMP targets is far from complete, while our understanding of the molecular mechanisms through which c-di-GMP works is almost nonexistent.

The discovery of c-di-GMP by Moshe Benziman and his colleagues at the Hebrew University of Jerusalem and their subsequent groundbreaking work are described by Dorit Amikam, Haim Weinhouse, and Michael Galperin in chapter 2 of this book. Moshe Benziman and his colleagues not only identified c-di-GMP and solved its structure but they also established the biochemical pathway for c-di-GMP synthesis and hydrolysis and purified the first diguanylate cyclase (DGC) and c-di-GMP-specific phosphodiesterase (PDE) (77, 79, 80). Critical for the understanding of the potential role of c-di-GMP as a signaling molecule was determination of sequences of these enzymes from *Gluconacetobacter xylinus* (formerly *Acetobacter xylinum*) and identification of the GGDEF and EAL domains in both enzymes (94). The fact that a DGC that catalyzes synthesis of c-di-GMP from two GTP molecules and a PDE that catalyzes phosphoester bond hydrolysis in c-di-GMP have the same arrangement of GGDEF and EAL domains presented an enzymatic conundrum that is discussed later in this chapter.

A much bigger puzzle emerged from the realization that GGDEF and EAL domains were among the most abundant domains of unknown functions present in diverse bacterial genomes. The bioinformatics analyses by Michael Galperin and colleagues revealed the potential for c-di-GMP to be a broadly distributed signaling molecule (26–28; see also chapter 3 of this book). These publications boosted awareness about c-di-GMP in the microbiology community. Another useful bioinformatics insight into the function of the GGDEF domain came from the study performed by Pei and Grishin (63). These researchers showed that the GGDEF domain bears significant structural similarity to adenylate cyclases, and thus, they linked the DGC activity to the GGDEF domain. For early reviews predicting c-di-GMP to be an important second messenger in the *Bacteria*, see references 21, 40, 74, and 82.

This chapter describes the studies that resulted in the experimental verification of the core signaling pathway (Fig. 1). It discusses advances in enzymology of c-di-GMP synthesis and hydrolysis, identification and characterization of c-di-GMP receptors, end targets, and molecular mechanisms, where these are

Mark Gomelsky • Department of Molecular Biology, University of Wyoming, Laramie, WY 82071.

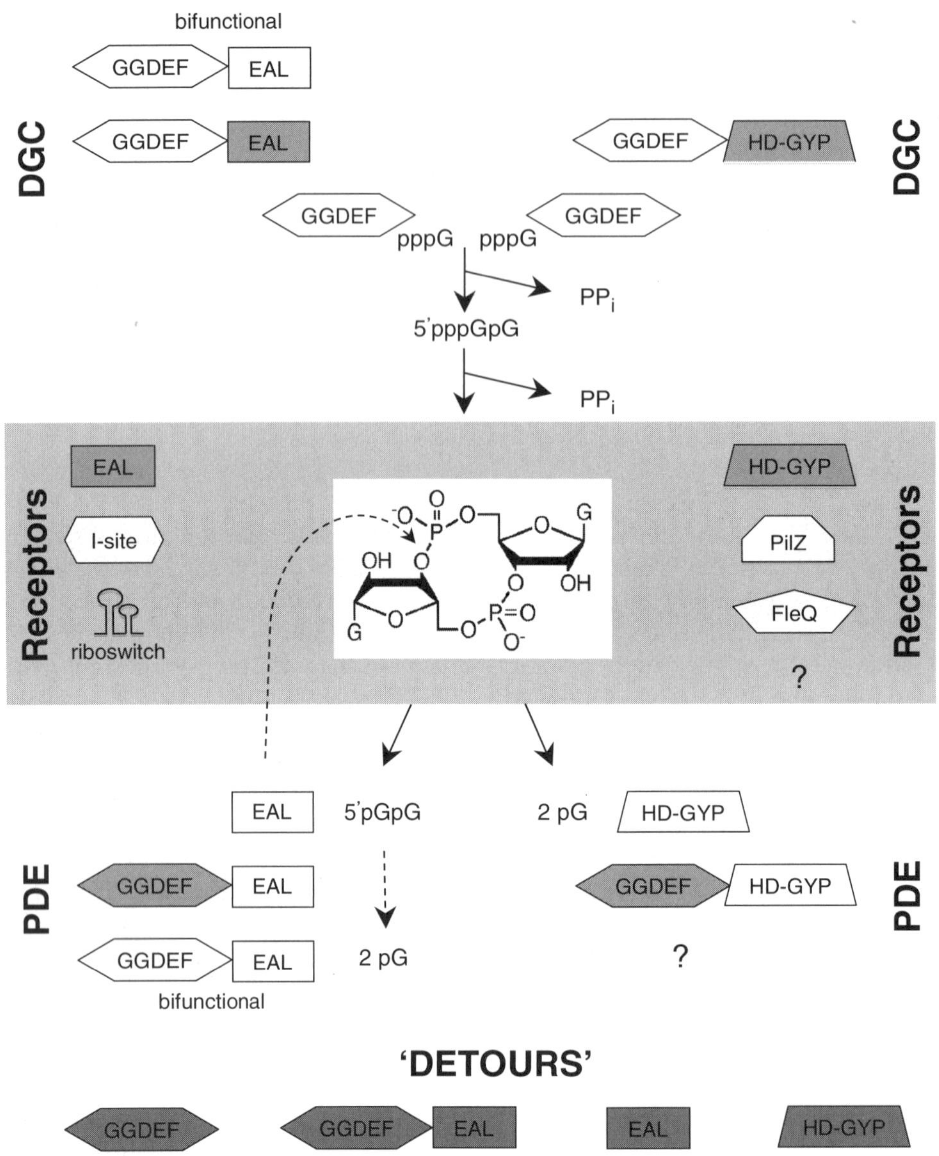

Figure 1. Core c-di-GMP signaling pathway and detours from the core. Shown are key protein domains and domain combinations involved in c-di-GMP signaling. Enzymatically active GGDEF, EAL, and HD-GYP domains are drawn on the white background; enzymatically inactive domains involved in substrate binding are shown as light grey, while domains that are no longer associated with c-di-GMP are shown as dark grey. G, guanine.

known. Where reasonable, I try to fill the gaps in knowledge with hypothetical scenarios to stimulate the reader's thinking and experimental work that could accelerate the progress of deciphering the intricacies of c-di-GMP signaling. At the end of the chapter, I present general principles of organization of c-di-GMP-dependent regulatory pathways in bacterial cells, pose questions, and describe challenges in the field. Note that this chapter is focused on generalities and has only minimal descriptions of phenomenology, which is extensively covered by other contributors to this book.

ENZYMOLOGY OF c-di-GMP SYNTHESIS AND DEGRADATION

Insights from Genetic Studies

The first line of experimental evidence that supported the prediction (63) that DGC activity resides in the GGDEF domain came from genetic studies. Ausmees et al. (5) tested whether a protein containing the GGDEF domain but no EAL could functionally replace a DGC from *G. xylinus*. The test system used in that study involved a *Rhizobium leguminosarum* mutant impaired in cellulose synthesis because of the mutation in the GGDEF domain gene. The researchers found that not only the DGC gene from *G. xylinus* but also GGDEF domain proteins from *Escherichia coli* and *Agrobacterium tumefaciens* could restore cellulose production. These data showed that the GGDEF domain proteins from proteobacterial species possess DGC activity and that EAL was not required. Yet another report, from Urs Jenal's group (1), demonstrated the interchangeability of GGDEF domains, i.e., the GGDEF domain protein WspR from *Pseudomonas fluorescens* could complement a mutation in the PleD protein from *Caulobacter crescentus,* which subsequently became an important model for understanding DGC activity and its regulation (15, 17, 61).

If GGDEF domains were to function as DGC, the predicted activity of the EAL domains would have to be c-di-GMP hydrolysis. However, it was unclear whether or not EAL domains are sufficient for PDE activity, or whether GGDEF domains are needed as well. Simm and colleagues (90) made a strong case for the sufficiency of the GGDEF domain for DGC activity and the sufficiency of EAL domains for c-di-GMP-specific PDE activity. These researchers overexpressed proteins containing either GGDEF or EAL domains, or combinations of those proteins, in *Salmonella enterica* serovar Typhimurium, *E. coli,* and *Pseudomonas aeruginosa.* They showed that overexpression of the GGDEF domain proteins resulted in similar phenotypes, i.e., increased biofilm formation, and that overexpression of the EAL domain proteins countered these phenotypes. This study made an important step forward by establishing a correlation between overexpression of the GGDEF/EAL proteins and intracellular c-di-GMP concentration. The same conclusions about apparent sufficiency of the EAL domain for PDE activity were reached by Tischler and Camilli (99) who showed that the *Vibrio cholerae* EAL domain protein VieA decreases the intracellular concentration of c-di-GMP, which is opposite to the effects of overexpressed DGCs.

In Vitro Activity of the GGDEF Domain Proteins

The remaining logical step in characterizing enzymatic activities of the GGDEF and EAL domains was direct demonstration of the DGC and PDE activities in vitro. This task was accomplished by several groups in 2004 and 2005 (34, 61, 84). Urs Jenal's group published the first report demonstrating DGC activity of the GGDEF domain in vitro. The researchers purified *C. crestentus* PleD and showed that it converts GTP to c-di-GMP, is highly specific to GTP as a substrate, and lacks c-di-GMP PDE activity (61) (Fig. 1).

Several important insights pertaining to the properties of enzymes involved in c-di-GMP metabolism were revealed in that study, i.e., highly specific spatiotemporal regulation of the DGC activity of PleD as well as activation of the DGC activity of PleD by phosphorylation by a cognate kinase. Spatiotemporal and stimulus-specific regulation proved to be crucial for maintaining specificity of responses and avoiding unwanted cross talk in the bacteria that possess vast networks of DGCs and PDEs. I briefly discuss this topic later in this chapter, and other authors of this book illustrate it using their particular bacterial systems.

Another important finding was made during subsequent analysis of PleD. This protein contains an RXXD motif, where the first Arg is located several residues upstream of the GGDEF motif. The RXXD site was shown to bind c-di-GMP and serve as an allosteric inhibitory site (I-site) of many DGCs (15, 17, 103; for more extensive coverage of the I-site, see chapter 7 by Dorothy Kim, John Hunt, and Tilman Schirmer and chapter 9 by Sören Abel and Urs Jenal).

While GGDEF domains from selected proteobacterial species appeared to possess DGC activity, it remained unclear whether this applies to all other GGDEF domains. In light of the similarity of the GGDEF domain to adenylate cyclases (63) and a reported cAMP-dependent PDE activity of the *E. coli* GGDEF-EAL domain protein Dos (86), the possibility that some GGDEF-EAL systems were c-di-GMP specific while others were cAMP specific remained real. Ryjenkov et al. (84) addressed this possibility by testing the enzymatic activities of six GGDEF domain proteins representing diverse phylogenetic branches of the *Bacteria,* i.e., *Thermotoga, Deinococcus-Thermus, Cyanobacteria, Spirochaetes,* and *Proteobacteria.* All six proteins that were chosen randomly were shown to possess DGC activity, with GTP as their exclusive substrate. These results essentially rejected the possibility of an alternative enzymatic activity for GGDEF domains and provided experimental evidence that the sphere of c-di-GMP influence encompasses all of the *Bacteria.* A subsequent study by Schmidt et al. showed that the cAMP-specific PDE activity of Dos was likely an artifact (89).

Ryjenkov et al. (84) demonstrated that purified GGDEF domains by themselves possess low-level DGC activity, which was much lower than that of the full-length proteins, and the prevalent activity of GGDEF domains was GTPase. This observation implied that GGDEF domains, when present at high concentrations, form homodimers and adopt enzymatically active conformation. That DGCs must form dimers was obvious from their similarity to adenylate/

guanylate cyclases, and this was observed experimentally. It also implied that the sensory domains linked to the GGDEF domain proteins are important for bringing GGDEF domains together in the enzymatically active conformation. The researchers stumbled upon a striking example of the importance of the regulatory domains in DGCs. One of the six purified proteins, *Borrelia burgdorferi* Rrp1, composed of a response regulator receiver domain, Response_reg (Pfam PF00072) (25) and GGDEF, was completely inactive in vitro. However, the DGC activity of Rrp1 could be revealed by phosphorylation using acetyl phosphate, which is often used in vitro as a phosphoryl group donor to purified response regulators (51). The DGC activity was proportional to the extent of Rrp1 phosphorylation (84).

Since functional DGCs are present in representatives from various branches of the phylogenetic tree of the *Bacteria,* including branches comprised mostly of thermophiles, one can predict that c-di-GMP originated early in bacterial evolution. It is peculiar that the *Archaea,* which lack GGDEF domain proteins, contain a related protein domain designated GGDN (PF05165). The GGDN family encodes a specific class of GTP cyclohydrolases (class III) which, together with GGDEF and guanylate cyclase domains, forms a nucleotide cyclase clan (CL0276). Based on the presence of the GGDN domains exclusively in the *Archaea* and GGDEF domains exclusively in the *Bacteria,* it is tempting to speculate that each family has evolved from a common ancestor nucleotide cyclase early after the split between the archael and bacterial lineages.

In Vitro Activity of the EAL Domain Proteins

The c-di-GMP PDE activity of the EAL domains in vitro was demonstrated shortly after demonstration of in vitro DGC activity. Bobrov et al. (13) used a nonspecific PDE substrate, bis(*p*-nitrophenyl) phosphate, to show that the purified EAL domain protein HmsP from *Yersinia pestis* possesses PDE activity (see chapter 18 by Robert Perry and Alex Bobrov). Schmidt et al. (89) tested several potential phosphoester-containing substrates using the *E. coli* EAL domain protein YahA and found that the PDE activity is highly c-di-GMP specific. Furthermore, the purified individual EAL domains from YahA and *E. coli* Dos were shown to carry all sequence determinants necessary for PDE activity and substrate specificity.

Schmidt et al. (89) showed that the EAL domain catalyzed c-di-GMP hydrolysis in two steps. In the first, fast reaction, c-di-GMP was converted to linear diguanylate, 5′pGpG, that had a 5′ as opposed to a 3′ phosphate. In the second step, pGpG was converted to two molecules of GMP. However, this later activity was 2 to 3 orders of magnitude less efficient than the c-di-GMP hydrolysis (Fig. 1). The strong preference for c-di-GMP was entirely consistent with the earlier observations made by Benziman and colleagues using the *G. xylinus* PDEs (80). Additional properties of YahA also conformed to those observed by the Benziman group, i.e., PDE activity was dependent on either Mn^{2+} or Mg^{2+} and was strongly inhibited by Ca^{2+}. Unlike a GGDEF domain, EAL did not require dimerization, as the enzymatically active EAL domain from YahA was monomeric in vitro (89). However, several EAL domain proteins purified later were found to form dimers, tetramers, or higher-order oligomers (6, 71, 97). At least for one of the tested PDEs, BlrP1 from *Klebsiella pneumoniae,* dimerization was found to be necessary for activation of the PDE activity by the environmental stimulus blue light (6).

In vitro studies characterizing c-di-GMP-specific PDE activity were performed on the EAL domain proteins from other sources, *C. crescentus* CC3396 (18) and *V. cholerae* VieA (95), and very similar conclusions regarding the nature of enzymatic activity were reached.

While in vitro EAL domains can completely hydrolyze c-di-GMP to GMP, the much slower rate of hydrolysis of pGpG raises doubts about the relevance of this activity in vivo. In this regard, it is instructive that Benziman and colleagues have always distinguished the c-di-GMP-specific PDE activity resulting in linear diguanylate, c-di-GMP → 5′ pGpG, from the secondary PDE activity, 5′ pGpG → 2 pG (2 GMP) (Fig. 1). The former activity was designated PDE-A, while the latter one was designated PDE-B (77, 80). In their partially purified membrane preparations used to characterize PDE activity, the PDE-A activity was strongly inhibited by Ca^{2+}, whereas PDE-B activity was not, thus suggesting that different enzymes were responsible for these two activities in vivo. One can readily envision that RNases present in bacterial cells would readily break pGpG down to mononucleotides. Consistent with this possibility, nucleotide extracts prepared from different kinds of bacteria, where c-di-GMP metabolism has been manipulated, do not show appreciable accumulation of pGpG. There is an interesting possibility that pGpG can play some regulatory role on its own. Given the lack of measurable accumulation of pGpG in bacteria, its role (if any) would have to be on a local, rather than global, scale.

In summary, genetic and biochemical assays have unambiguously assigned DGC activity to GGDEF domains and PDE (PDE-A) activity to EAL domains. Importantly, GGDEF and EAL domains were shown to be exclusively specific to c-di-GMP synthesis and hydrolysis, respectively. Experimental evidence of functional DGCs in diverse bacteria together with the evidence of diverse functions associated with the GGDEF/EAL domain proteins proved the cautious prediction made by Benziman and colleagues in 1998 that "if [GGDEF and EAL domains]...are specifically associated with c-di-GMP metabolism, the possibility arises that c-di-GMP has wider significance as a regulatory molecule" (94). By the mid-2000s, c-di-GMP had been firmly established as a ubiquitous bacterial second messenger.

ENZYMATIC CONUNDRUM OF THE GGDEF-EAL PROTEINS

If GGDEF acts as a DGC and EAL as a PDE, what is the function of proteins containing both domains? The proteins containing GGDEF and EAL domains arranged in GGDEF-EAL tandems are abundant and in some species reach up to one-third of all proteins containing either one of these domains (see Table 2 in chapter 3 for statistics). While the GGDEF-EAL domain architecture is by far the most common, proteins with the opposite domain order also exist, usually with an intervening domain (X) between EAL and GGDEF, i.e., EAL-X-GGDEF.

Theoretically, there are two scenarios by which GGDEF-EAL enzymes can work. In the first scenario, only one domain in a GGDEF-EAL tandem is enzymatically active, while the other one is not. There are numerous examples that show that in the course of evolution, enzymatic domains have often retired from catalysis to serve alternative functions, most commonly sensing of former substrates or products or engaging in new protein-protein or protein-polynucleotide interactions.

In the alternative scenario, both GGDEF and EAL domains in the GGDEF-EAL tandems may be enzymatically active, yet their activities may be differentially regulated by environmental or intracellular signals, so that under a given set of conditions one activity would prevail over another. A well-known example of bifunctional regulatory proteins is protein histidine kinases of the two-component regulatory systems (64). These kinases transfer phosphate to their cognate response regulators following activation by specific stimuli. In addition to having kinase domains, many histidine kinases contain phosphatase domains capable of dephosphorylating their cognate response regulators. The latter activity helps downregulate the level of phosphorylated response regulators following a burst in kinase activity and can prevent accumulation of phosphorylated response regulators by noncognate kinases.

Enzymatically Inactive Domains in GGDEF-EAL Tandems

The retirement scenario appears to explain the functions of most GGDEF-EAL domain enzymes (18, 89). Schmidt et al. (89) noticed that, based on sequence conservation, EAL domains can be divided into two categories. Those EAL domains that possessed PDE activity, as evidenced from genetic or biochemical data, contained several conserved regions, e.g., EXL, DDFGTG, KXD, and EGXE (89). The first of these regions corresponds to the EAL (Glu-Ala-Leu) signature motif that gave rise to the domain name. These regions were in general much more degenerate in the EAL domains from the GGDEF-EAL domain DGCs described by Benziman and colleagues (94). While at the time this observation was made the exact functions of conserved motifs were unknown, the lack of sequence conservation in the EAL domains from GGDEF-EAL DGCs pointed to the possibility that they are impaired in PDE activity. Indeed, the purified EAL domain from DgcA1, the GGDEF-EAL DGC from *G. xylinus* (94), is enzymatically inactive (our unpublished data).

The charged residues present in the conserved motifs identified by Schmidt et al. (89) were subsequently tested by site-directed mutagenesis, and most of them were shown to be important for PDE activity (71). The X-ray crystal structure of the BlrP1 protein revealed that some of the conserved residues coordinate catalytic metal ions, either Mg^{2+}or Mn^{2+}, others contact c-di-GMP, yet others stabilize the overall structure (6). The insights into PDE activity of the EAL domain derived from structural analyses are discussed in chapter 7.

What is the role of enzymatically inactive EAL domains in the GGDEF-EAL tandems? We suggest that most of these domains bind their former substrate, c-di-GMP, and function as sites for feedback inhibition of the DGC activity associated with the upstream GGDEF domains. Consistent with this assessment is our observation that the GGDEF domains from the GGDEF-EAL DGCs lack alternative inhibitory c-di-GMP binding sites, or I-sites. However, some degenerate EAL domains may be engaged in new kinds of interactions, which are independent of c-di-GMP.

Similar to EAL domains, GGDEF domains from the GGDEF-EAL tandems also belong to enzymatically active and inactive categories. An enzymatically inactive GGDEF domain may still be able to bind its former substrate, GTP. Christen et al. (18) identified precisely this scenario while characterizing CC3396, the GGDEF-EAL PDE from *C. crescentus*. In vitro, the PDE activity of CC3396 is enhanced by approximately 40-fold upon addition of GTP. The authors of the study showed by UV cross-linking that GTP binds to the former active site of the GGDEF domain, which contains a degenerate GEDEF motif. They mutagenized the GEDEF motif and identified a mutant with significantly increased basal level PDE activity that was essentially irresponsive to GTP. This study established that divergent GGDEF motifs of the GGDEF-EAL PDEs may act as sites for allosteric regulation by GTP.

The physiological significance of GTP-dependent regulation of PDE activity, however, is not entirely clear. Affinity of CC3396 to GTP was found to be in the micromolar range, which is about 2 to 3 orders of magnitude lower than intracellular GTP concentrations (18). It is therefore likely that under most circumstances, GTP serves a structural, but not regulatory, role. However, if GTP levels were to drop precipitously, e.g., as a result of deep starvation or a specific developmental stage, the GTP deficiency could potentially become inhibitory for the PDE activity (18). Alternatively, affinity of CC3396 to GTP in vivo may differ significantly from that in vitro, for example, as a result of interaction between the uncharacterized N terminus of CC3396 and GGDEF. Interestingly, GTP had an inhibitory, as opposed to a stimulatory, effect on the PDE activity of the *G. xylinus* GGDEF-EAL PDEs (78). Additional studies of the GGDEF-EAL PDEs are clearly warranted to better understand the role of GTP in these enzymes.

Bifunctional GGDEF-EAL Enzymes

The second scenario concerning GGDEF-EAL enzymes involves bifunctional enzymes, where opposing c-di-GMP synthase and hydrolase activities are differentially regulated. Several cases of potentially bifunctional GGDEF-EAL enzymes have been described. However, a detailed picture of a protein that functions as a DGC under one set of conditions and turns into a PDE as conditions change has yet to emerge. Importantly, most GGDEF-EAL proteins appear to have either one activity or another, i.e., they fall into the group described earlier, which suggests that bifunctional enzymes are likely to be rare. Below, I describe three reports of potentially bifunctional GGDEF-EAL enzymes.

Tarutina et al. (97) characterized the first potentially bifunctional GGDEF-EAL enzyme, *Rhodobacter sphaeroides* BphG1. This protein contains a photoreceptor module, PAS-GAF-PHY, typical of red/far-red light receptor bacteriophytochromes (72), upstream of the GGDEF-EAL domain tandem. The photoreceptor module of BphG1 binds the chromophore, biliverdin IXα, and properly responds to illumination. The sequence analysis of the GGDEF and EAL domains suggested that both domains may be enzymatically active; however, BphG1 possessed only PDE activity. Further, the PDE activity was essentially light independent. Unexpectedly, upon expression in *E. coli,* BphG1 underwent proteolytic cleavage precisely at the end of the GGDEF domain. Surprisingly, the N-terminal product of proteolysis, PAS-GAF-PHY-GGDEF, had DGC activity. Even more surprising, the DGC activity was strongly upregulated by light, which suggests (but does not prove) that the light-dependent regulation of DGC activity may be relevant in vivo. The authors of the study proposed that if proteolysis were to take place in the native host, then BphG1 could act as a bifunctional enzyme, i.e., as a constitutive PDE prior to proteolytic cleavage and as a light-dependent DGC following the cleavage. The fate of the EAL domain, which retained partial PDE activity, is unknown. One can envision that it would be physically separated from the DGC activity in order not to interfere with it. Tarutina et al. speculated that the EAL domain of BphG1, which has a tendency to dimerize, limits the mobility of the upstream GGDEF domains, which must come together in a very precise manner to achieve the DGC activity. Thus, the EAL domain in the full-length protein effectively locks BphG1 in the PDE conformation (97).

Yet another example of a putative bifunctional GGDEF-EAL enzyme, *Vibrio parahaemolyticus* ScrC, was described by Ferreira et al. (24). The GGDEF and EAL domains of ScrC appear to be intact, thus suggesting that they can be enzymatically active. In the presence of two interacting proteins, ScrA and ScrB, ScrC acts as a PDE. ScrB is a predicted periplasmic protein that may interact with SrcC through the transmembrane sensory domain. ScrA is predicted to be a cytoplasmic protein that may also interact with SrcC. In the absence of ScrA and ScrB, ScrC appears to act as a DGC. Furthermore, deletion or site-directed inactivation of the EAL domain of ScrC results in the protein that possesses DGC activity. Therefore, it appears that interactions with ScrA and ScrB can cause ScrC to switch from PDE activity to DGC activity (24; see chapter 12 by Linda McCarter). Both examples, i.e., BphG1 and SrcC, underscore the importance of protein-protein interactions in controlling the PDE-DGC switch of the potentially bifunctional GGDEF-EAL tandem enzymes.

Yet another example of a potentially bifunctional protein was presented by Kumar and Chatterji (47).

The *Mycobacterium smegmatis* GGDEF-EAL protein MSDGC-1, when mixed with GTP in vitro, produces c-di-GMP, which after some delay is partially hydrolyzed to pGpG. The MSDGC-1 gene knockout lacked intracellular c-di-GMP, thus suggesting that DGC activity is prevalent in vivo, at least under the tested conditions. Whether PDE activity of MSDGC-1 has any significance in vivo remains unknown. However, since the genome of *M. smegmatis* carries no identifiable PDE genes, besides MSDGC-1, it is likely that both DGC and PDE activities of MSDGC-1 may indeed be functional. Note that c-di-GMP is a very stable compound (92); therefore, a cellular PDE activity is needed to reset the sensitivity of its targets.

In summary, most proteins containing GGDEF-EAL domain tandems appear to have one activity, either PDE or DGC. It is often (but not always) possible to predict the activity of a given protein based on the primary sequence analysis, especially after the residues involved in substrate binding and catalysis were deciphered by structure-function analyses. The prevalent role of the enzymatically inactive domains appears to be substrate binding: GTP for the enzymatically inactive GGDEF domains and c-di-GMP for the enzymatically inactive EAL domains. While evidence of bifunctional GGDEF-EAL proteins is gradually accumulating, such proteins likely represent a minority among GGDEF-EAL proteins. Whether both activities play functional roles in vivo is as yet unclear, and how these potentially bifunctional enzymes switch from c-di-GMP synthesis to degradation remains poorly understood.

c-di-GMP HYDROLYSIS BY THE HD-GYP FAMILY...AND BEYOND

The HD-GYP domain was predicted to have c-di-GMP-specific PDE activity (26, 27). Ryan et al. (81) tested this hypothesis by using an HD-GYP domain protein RpfG from *Xanthomonas campestris*. As predicted, RpfG had c-di-GMP-specific PDE activity in vitro and could functionally replace an EAL domain PDE when expressed in a heterologous host. Interestingly, the primary product of c-di-GMP hydrolysis by RpfG was found to be GMP, not pGpG, the common product of PDE activity catalyzed by EAL domains (Fig. 1). pGpG was observed by the authors of that study only at high enzyme concentrations. Therefore, in contrast to EAL domain PDEs, HD-GYP domain PDEs may have similar affinities for c-di-GMP and pGpG. Thus, they would not release pGpG prior to the hydrolysis of the second phosphodiester bond. However, it is somewhat premature to draw conclusions about the HD-GYP-mediated c-di-GMP hydrolysis. More studies are needed to derive the common features regarding the HD-GYP domain PDEs (see chapter 5 by Robert Ryan and colleagues).

Some HD-GYP domains exist in tandems with GGDEF domains. One can envision that the GGDEF–HD-GYP tandem proteins would function similarly to the GGDEF-EAL domains. However, to our knowledge, this has not yet been tested. Interestingly, HD-GYP domains sometimes occur in tandem with EAL domains. These EAL–HD-GYP fusions may function in several different ways. The most likely scenario is that either the EAL domain or the HD-GYP domain is enzymatically inactive and instead serves as a c-di-GMP-binding site. Various combinations of GGDEF, EAL, and HD-GYP protein domains have arisen throughout evolution, showing that Mother Nature plays with protein domain combinations to ultimately select those that work—in either old or new ways.

Can protein domains unrelated to EAL or HD-GYP hydrolyze c-di-GMP? In our view, this possibility is viable because relatively few changes in the substrate-binding pocket of a preexistent class of phosphoesterases may be sufficient to accommodate c-di-GMP as a new substrate. In fact, the HD-GYP family has likely evolved as a c-di-GMP-specific branch of a large and diverse HD_PDEase superfamily (PF01966) (4). Members of this superfamily catalyze hydrolysis of diverse phosphoesters, including cyclic mononucleotides. Interestingly, analysis of protein domain architectures suggests that c-di-GMP-specific PDEs belong to two different protein domains of the HD_PDEase superfamily, i.e., HD (PF01966) and HDOD (PF08668). It is possible that c-di-GMP specialization evolved in the HD and HDOD branches independently, and the HD-GYP family may not encompass all c-di-GMP-specific PDEs from the HD_PDEase superfamily. To understand the basis for c-di-GMP specificity, future biochemical and structural analyses of the c-di-GMP binding members of the HDOD and HD domains would be very instructive.

While we should not rule out the possibility that new, as yet unrecognized, c-di-GMP PDE domains exist, whether within the HD_PDEase superfamily or outside, it is important to realize that according to protein domain architecture analysis, any new c-di-GMP-specific PDE domains would clearly have much smaller distribution than EAL or HD-GYP. Therefore, EAL and HD-GYP domain PDEs will remain the workhorses of c-di-GMP hydrolysis.

If more than one type of c-di-GMP-specific PDE has evolved, is it possible that DGCs unrelated to GGDEF domains have also evolved? We cannot know for sure and should remain open-minded. The pros for this possibility are that several different classes of adenylate/guanylate cyclases exist. Further, a protein with diadenylate cyclase (c-di-AMP synthase) activity was recently identified. The protein domain responsible for this activity, DisA_N (PF02457), has no similarity to GGDEF (76, 107). Therefore, a nonhomologous-to-GGDEF DGC activity may remain hidden among hundreds of domains of unknown function. The best argument against this possibility is that thus far there is no experimental or bioinformatics evidence of a non-GGDEF DGC.

In summary, at least two classes of c-di-GMP-specific PDEs have evolved, the EAL domain and the HD-GYP family. Additional domains with c-di-GMP-specific PDE and DGC activities may exist, however, probably with much more limited distribution. Therefore, the GGDEF-based DGCs and EAL- or HD-GYP-based PDEs remain the workhorses of c-di-GMP metabolism. However, the HD-GYP family may not fully represent c-di-GMP-specific PDEs from the HD_PDEase superfamily.

GGDEF, EAL, OR HD-GYP DOMAIN PROTEINS UNFAITHFUL TO c-di-GMP

Thus far, we have been discussing GGDEF, EAL, and HD-GYP as key domains involved in c-di-GMP synthesis and hydrolysis. Even when discussing proteins containing two domains with apparently opposing activities, e.g., GGDEF-EAL or GGDEF–HD-GYP, we focused on those where at least one domain is involved in c-di-GMP metabolism. Now, we will take a detour from the core pathway and consider proteins that contain familiar domains but lost their ties to c-di-GMP (Fig. 1).

The first GGDEF-EAL protein that had diverged from the mainstream of c-di-GMP signaling was characterized by Suzuki et al. (93). *E. coli* CsrD contains a GGDEF-EAL domain tandem where neither GGDEF nor EAL domains are enzymatically active. CsrD binds two small regulatory RNAs, designated CsrB and CsrC, and targets them for degradation. The function of these small RNAs is to counteract the effect of the carbon storage regulatory protein CsrA (93). The Csr regulatory network is described in detail by Tony Romeo and Paul Babitzke in chapter 14.

Another example of a traditional c-di-GMP-associated domain that has lost ties to c-di-GMP comes from staphylococci. The GGDEF domain protein GdpS found in *Staphylococcus epidermidis* and *Staphylococcus aureus* controls expression of the exopolysaccharide biosynthesis locus, *icaADBC*. GdpS contains a transmembrane domain linked to the GGDEF domain. However, when purified, neither GdpS nor its GGDEF domain showed DGC activity. Importantly, the integrity of the GGDEF domain was not required for regulation of the *icaADBC* expression in vivo and the GdpS protein containing a mutant GGF motif was just as active as the wild-type GdpS (37). How the GGDEF domain functions in the staphylococcal GdpS proteins remains a mystery. One possibility is that the GGDEF domain of GdpS has evolved into a protein interaction module that transmits the signal from the transmembrane domain to a downstream partner or, in the opposite orientation, from the cytoplasmic partner to the transmembrane output.

Tschowri et al. (100) described yet another example, this time of an EAL domain protein that has lost its ties with c-di-GMP in favor of acting via protein-protein interactions. YcgF from *E. coli* is a BLUF-EAL protein (68), where BLUF is a blue light sensory domain (31, 66). The EAL domain of YcgF lacks several conserved residues involved in c-di-GMP binding and hydrolysis (6). Accordingly, in vitro it does not bind or hydrolyze c-di-GMP. Instead, YcgF interacts with the transcriptional repressor YcgE in a light-dependent manner. The authors of this study put forth a model where the unusually acidic EAL domain of YcgF competes with DNA (representing the YcgE binding site) for binding by the YcgE repressor (see chapter 16 by Regine Hengge for further discussion).

It is noteworthy that some enterobacterial orthologs of YcgF that share the same domain organization and significant sequence similarity actually function as light-dependent PDEs. One such protein, *K. pneumoniae* BlrP1, has recently been described (6). The very different modes of action represented by YcgF and BlrP1 are but one illustration of the incredible flexibility in bacterial regulatory systems. It reminds us to be careful when extrapolating data even on orthologous proteins from closely related species. This warning must be particularly relevant to the c-di-GMP regulatory networks in those bacteria that contain numerous GGDEF, EAL, and HD-GYP proteins because large numbers of interchangeable modules provide greater opportunities for variability.

It is peculiar that the three described systems—CsrABCD, GdpS, and YcgF—regulate biofilm formation, a process traditionally associated with c-di-GMP signaling. They all engage typical c-di-GMP domains, yet they no longer involve c-di-GMP itself. An HD-GYP domain protein that is unfaithful to c-di-GMP has not yet been described but will likely be identified in the future.

In summary, the existence of enzymatically inactive proteins containing GGDEF, EAL, and HD-

GYP domains and combinations of these domains once again proves, to paraphrase Shakespeare, that not all that shines is gold. It should be taken as a warning against the blind belief in existing protein annotations that often equate enzymatic activity with the presence of a protein domain normally associated with that activity. The enzymatically inactive domains that used to be involved in c-di-GMP metabolism evolved to interact with new partners. Interestingly, in many instances these proteins remain engaged in the processes that traditionally belong to the sphere of influence of c-di-GMP. These alternative, c-di-GMP-independent mechanisms are regulatory gems in their own right. Their understanding enriches our knowledge about the structure and evolution of c-di-GMP pathways.

c-di-GMP-BINDING PROTEINS

Currently, our knowledge of the enzymology of c-di-GMP synthesis and hydrolysis surpasses our understanding of protein targets of c-di-GMP (c-di-GMP receptors), which in turn surpasses our understanding of the molecular mechanisms through which c-di-GMP works. Fortunately, this field is rapidly progressing. Several types of c-di-GMP receptors have been discovered, and molecular details of c-di-GMP action are beginning to emerge.

PilZ Domain c-di-GMP Receptors

Discovery of PilZ domains, the first identified c-di-GMP receptors

Benziman and colleagues described the first c-di-GMP target as subunits of the *G. xylinus* cellulose synthase complex (77, 79). They also showed that a c-di-GMP-binding protein was tightly associated with cellulose synthase. Mayer et al. (53) UV cross-linked radioactively labeled c-di-GMP to *G. xylinus* cell extract and analyzed labeled proteins. They reported that c-di-GMP was bound to the BcsB subunit of the cellulose synthase. However, a later report suggested that c-di-GMP was bound to a 200-kDa membrane-bound protein complex that copurified with cellulose synthase (105). Since the BcsB subunit shows no significant similarity to other proteins, the search for additional c-di-GMP receptors based on comparative genomics was not productive.

Michael Galperin took a fresh look at this issue. He analyzed sequences of both BcsB and BcsA subunits essential for cellulose synthase activity and found that an approximately 100-amino-acid-long C-terminal extension on BcsA forms a separate protein domain. Strikingly, the new domain was found downstream of some EAL or GGDEF-EAL proteins, suggesting its involvement with c-di-GMP. The distribution of this new domain in bacterial genomes correlated well with the presence of the protein domains involved in c-di-GMP metabolism. This prompted Amikam and Galperin to hypothesize that the C-terminal domain of BcsA is the long sought after c-di-GMP receptor (3). The domain was named PilZ (PF07238) after the PilZ protein required for pili formation in *P. aeruginosa* (2) that consists of the PilZ domain only (Fig. 1).

c-di-GMP binding to PilZ domains in vitro

Ryjenkov and colleagues (85) tested experimentally the hypothesis that PilZ domain is the c-di-GMP receptor. They overexpressed and purified two different PilZ domains, one from *G. xylinus* BcsA and another from the YcgR protein from *E. coli,* and showed, using an equilibrium dialysis technique, that both PilZ domains specifically bind c-di-GMP but not cyclic mononucleotides or trinucleotides. The full-length YcgR has a submicromolar dissociation constant, K_d, for c-di-GMP, which is within the working range of the c-di-GMP concentration in the cells, believed to be in the submicromolar to lower-micromolar range (61, 90, 105). Ryjenkov et al. also mutagenized several conserved residues in the PilZ domain consensus sequence deciphered by Amikam and Galperin, R(X3)R(X19–23)DXS(X2)G, and verified the importance of conserved residues for c-di-GMP binding in vitro. Further, they used the X-ray structure of the *V. cholerae* apoprotein PlzD (VCA0042), which has a domain organization similar to YcgR, to locate residues comprising the signature motif of the PilZ domain. These residues, while separated on the primary sequence, turned out to be in close proximity in the tertiary structure and, therefore, likely represented the c-di-GMP binding site. c-di-GMP was predicted to bind on the outside of the globular core of the PilZ domain at the interface between the PilZ domain and the N-terminal globular domain of PlzD (and by analogy the N-terminal domain of YcgR [PF07317]).

Ryjenkov et al. (85) also observed that the YcgR–c-di-GMP complex appeared smaller in size than YcgR alone, as judged by size exclusion chromatography, which was unexpected of a monomeric protein. This suggested that c-di-GMP binding resulted in a more compact protein, perhaps by bringing PilZ and the N-terminal domain of YcgR in closer proximity. Such a scenario would be consistent with the proposed c-di-GMP binding site located between two globular domains.

The ability of PilZ domains to bind c-di-GMP in vitro and the importance of conserved residues for c-di-GMP binding were soon demonstrated for a number of PilZ domain proteins from other sources: PlzC and PlzD from *V. cholerae* (65), DgrA and DgrB from *C. crescentus,* PA4608 from *P. aeruginosa,* YcgR from *Salmonella enterica* (18), and Alg44 from *P. aeruginosa* (55). All tested PilZ domain proteins bound c-di-GMP in the submicromolar (<50 to 850 nM) range, as determined by filter binding assays using radioactively labeled c-di-GMP, which is consistent with their ability to function under physiological conditions.

The X-ray structure of PlzD complexed with c-di-GMP was determined by Benach et al. (8). It showed that c-di-GMP was indeed bound by the conserved signature motif. The R(X3)R motif termed the c-di-GMP switch comprised the primary site for c-di-GMP binding. Upon binding, the c-di-GMP switch swings toward PilZ to complete the binding pocket between the N-terminal and PilZ domains, i.e., in agreement with the model proposed earlier (85). Essentially the same picture has emerged from the nuclear magnetic resonance structures of the PilZ domain protein, *P. aeruginosa* PA4608 (18, 69). An in-depth discussion of PilZ domain structures is presented in chapter 7.

Similar to GGDEF and EAL domains, PilZ domains can be classified as active or inactive. In this case, inactivity means conservation of the overall structure but lack of c-di-GMP binding. Ironically, the *P. aeruginosa* PilZ protein used to name the c-di-GMP receptor domain lacks conserved residues required for c-di-GMP binding and does not bind c-di-GMP in vitro (50). Similarly, some of the PilZ domain proteins from *V. cholerae* showed no detectable c-di-GMP binding in vitro (65).

Functions of the PilZ domain proteins

What processes do PilZ domains regulate? How do they function at the molecular level? We are beginning to learn answers to the first question but remain almost clueless about the latter. Below, I describe what we do know and discuss possible mechanisms.

(i) Regulation of flagellum-based motility. Some PilZ domain proteins from the *Proteobacteria* control flagellum function. Ko and Park (45) first observed that YcgR could inhibit flagellum-based motility of *E. coli,* long before the association of YcgR with c-di-GMP was revealed. This early observation was subsequently confirmed for *S. enterica* (85) and for other *E. coli* strains (20, 30). The function of YcgR is to block flagellum-based motility at high c-di-GMP levels, which normally correlate with the synthesis of exopolysaccharides and adhesive pili, i.e., sessile lifestyle. In the strains with increased c-di-GMP levels, inactivation of *ycgR* improves motility, whereas *ycgR* overexpression suppresses motility. The studies by Ko and Park (45) and Girgis et al. (30) suggest that the number of flagella and their appearance are not affected by the *ycgR* mutation; instead, flagellum rotation is defective.

Regulation of flagellum rotation is achieved by the so-called switch complex, FliGMN (for a recent review on flagellum-based motility, see reference 98). The well-known factor that controls the frequency of switching from counterclockwise (promotes motility) to clockwise (corresponds to cell tumbling) is CheY, the chemotaxis response regulator. When phosphorylated, CheY~P binds to FliM and increases the probability that the flagellar motor will rotate clockwise instead of counterclockwise. In enterobacteria, this brings about the dismantling of the flagellar bundle and further results in cell tumbling and reorientation in space.

How does YcgR affect flagellum rotation? Wolfe and Visick proposed that YcgR may directly interact with the FliGMN switch complex (108). Several lines of evidence (including coimmunoprecipitation, colocalization in vivo, two-hybrid analysis, and mutagenesis studies) appear to support this insightful prediction (our unpublished data; D. Blaire, unpublished data; R. Harshey, unpublished data). However, a detailed picture of YcgR-switch complex interactions remains to be elucidated. Interestingly, a mechanism termed molecular clutch has recently been described in *Bacillus subtilis,* where a putative glycosyl transferase involved in exopolysaccharide biosynthesis, EspE, directly interacts with the FliG subunit of the switch complex to disengage the flagellum from its proton-driven motor (10, 12). The *B. subtilis* molecular clutch blocks motility when expression of a polysaccharide biosynthesis gene cluster, including *espE,* is activated. While this reciprocal regulation of biofilms and motility in *B. subtilis* is reminiscent of c-di-GMP-regulated processes in the *Proteobacteria,* EspE does not appear to be a c-di-GMP-binding protein (our unpublished data).

PilZ domain proteins from *C. crescentus* (19) and *V. cholerae* (65) are also involved in motility regulation. Deletions of *dgrA* and/or *dgrB* in *C. crescentus* in the strain with increased c-di-GMP levels improved flagellum-based motility, whereas overexpression of *dgrA* or *dgrB* had an opposite effect. DgrA and DgrB apparently have overlapping function in controlling motility. The target of Dgr action has been suggested to be the membrane protein FliL. In *C. crescentus,* unlike in enterobacteria, FliL is essential for flagellum function and motility (41). The level of FliL was greatly reduced in the *dgrA-* or *dgrB-*overexpressing strains (19).

(ii) Regulation of glycosyl transferase activities. Another function of PilZ domains is regulation of glycosyl transferases involved in polysaccharide biosynthesis. The BcsA subunit of the bacterial cellulose synthase was of course the first identified PilZ domain protein (3, 85). BcsA is an inner membrane protein that uses UDP-glucose as its substrate (88). Together with the BcsB subunit, they are required for poly-β1,4-glucoside chain synthesis and translocation across the cytoplasmic membrane. We do not really know what is being activated upon the binding of c-di-GMP to the PilZ domain: proper BcsA/BcsB complex formation, polymerization reaction, translocation of the growing polymer across the membrane, or some combination of the above (Fig. 2). Interestingly, in some strains, BcsA and BcsB subunits are present on the same polypeptide chain (87) (Fig. 2).

The alginate biosynthesis system in *P. aeruginosa* is also regulated by a PilZ–c-di-GMP switch and in some ways resembles the cellulose synthase BcsBA. Alginate is a β-1,4 polysaccharide made of D-mannuronate and L-guluronate monomers responsible for the mucoid phenotype of *P. aeruginosa,* which is often found in isolates from patients with chronic pulmonary diseases (70). Alg44 is a c-di-GMP-binding PilZ domain protein that is linked to a periplasmic domain, which belongs to the membrane fusion protein family (55). Alg44 cooperates with glycosyl transferase Alg8 in alginate synthesis (52, 60). Both Alg8 and BcsA are type 2 glycosyl transferases. Therefore, the PilZ–c-di-GMP complex in both systems may function in a similar manner. However, all the work on deciphering the molecular basis of c-di-GMP regulation in these systems lies ahead. Several polysaccharide-producing systems regulated by c-di-GMP are presented in Fig. 2.

(iii) PilZ domains as universal c-di-GMP sensors. The functions of PilZ domains are not limited to regulation of flagellum-based motility and polysaccharide synthesis. Experimental data on PilZ domain mutants suggest that some PilZ domains are involved in regulation of protein secretion and bacterial virulence independent of polysaccharide production or motility (11, 54, 65; see also chapter 20 by Jason Pratt and colleagues). However, the mechanisms are unknown.

It is clear that the PilZ domain can perform a diverse range of functions. After all, PilZ may be viewed as a c-di-GMP responsive switch that undergoes large conformational changes upon c-di-GMP binding (8). These changes can affect interactions with various downstream partners. In the *Proteobacteria,* the key physiological role of c-di-GMP is con-

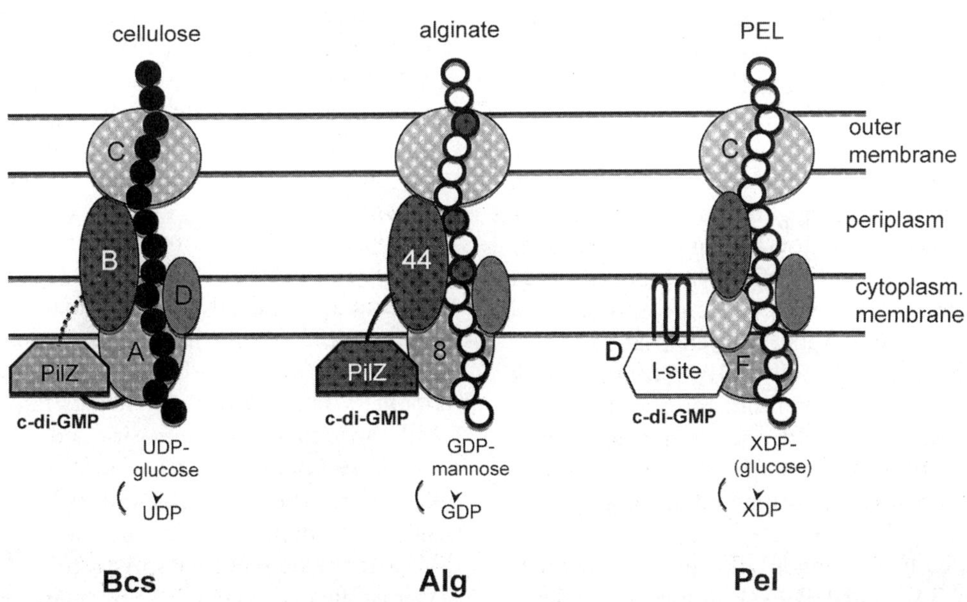

Figure 2. c-di-GMP-binding domains in exopolysaccharide biosynthetic and translocation machines. Bcs, bacterial cellulose synthase present in *G. xylinus* and enterobacteria; Alg, alginate synthase; Pel, glucose-based PEL polysaccharide. Both Alg and Pel are from pseudomonads. cytoplasm., cytoplasmic.

trolling the transition from the motile to the surface-attached, sessile lifestyle (see chapter 8 by Mathew Parsek and Paul Aldridge and chapters 9 to 19). Therefore, some c-di-GMP-binding PilZ proteins from this bacterial branch are involved in this transition. It is as yet unclear that PilZ domains, or c-di-GMP for that matter, play the same physiological role in bacteria from phylogenetic branches distant from the *Proteobacteria*. The data on the role of c-di-GMP outside the *Proteobacteria* are scarce. Existing reports on *M. smegmatis (Actinobacteria)* (47), *Anabaena* sp. *(Cyanobacteria)* (58), *B. burgdorferi (Spirochaetes)* (73), and obligatory intracellular *Anaplasma phagocytophilum (Rickettsiales)* (48) suggest functions unrelated to the motile-sessile transition.

PilZ domains are ubiquitous. While *E. coli* has only two PilZ domain proteins, YcgR and BcsA, other bacteria have many more. *Bdellovibrio* has 15 and is the champion among currently sequenced bacteria. Among PilZ domain proteins, the vast majority are stand-alone PilZ domains whose functions are difficult to predict. However, multidomain proteins exist and include PilZ domain fusions to methyl-accepting chemotaxis proteins, σ^{54}-type transcription regulators, helix-turn-helix domains often involved in DNA binding, protein Ser/Thr kinases, heat shock-like domains, and adenylate/guanylate cyclases. Therefore, in the future, we will likely see some creative ways by which bacteria have utilized PilZ domains as universal c-di-GMP-specific molecular switches.

Non-PilZ Domain c-di-GMP Receptors

The number of GGDEF, EAL, and HD-GYP domain proteins involved in c-di-GMP metabolism in many bacterial species far exceeds the number of PilZ domain proteins. This imbalance implies that non-PilZ receptors of c-di-GMP must exist. The genetic data confirm this prediction. For example, deletion of all PilZ domain-encoding genes in *V. cholerae* does not abolish the effect of c-di-GMP on colony rugosity (11). Furthermore, some organisms lack PilZ domain proteins altogether despite the presence of functional c-di-GMP enzymes (see Table 2 in chapter 3).

What do non-PilZ c-di-GMP receptors look like? One can envision several possibilities. From the earlier discussion, one could predict that an inhibitory site, I-site, of a DGC that binds c-di-GMP specifically and with high affinity could serve as a c-di-GMP receptor in the absence of the intact GGDEF domain. Other obvious candidates involve enzymatically inactive EAL and HD-GYP domains that have lost enzymatic activity but retain c-di-GMP binding. It is instructive to look at the issue of c-di-GMP binding from a broader perspective. Diverse protein sites have evolved to function in nucleotide binding. Given that c-di-GMP-binding proteins (I-site, PilZ, EAL) which are currently characterized require only a handful, or fewer, of strategically positioned amino acid residues, one can envision that different types of c-di-GMP-binding sites may exist.

c-di-GMP receptors containing I-sites

Below, I describe two examples of c-di-GMP-binding receptors based on I-sites, one potentially unrelated to the GGDEF domain and another one clearly involving a degenerate GGDEF domain.

The PelD protein from *P. aeruginosa* represents the first published c-di-GMP receptor that utilizes the I-site but has no DGC activity (49). PelD is a membrane-bound protein encoded by the *pel* operon along with other genes involved in pellicle formation. Pellicle, a biofilm formed at the liquid-air interface, is composed of a glucose-based exopolysaccharide (83). PelD has a characteristic RXXD motif and a second Arg (R) residue located some 35 amino acids downstream, which is typical of I-sites. Importantly, while the GGDEF motif is not recognizable in PelD, the predicted secondary structure around the I-site is similar to that of a GGDEF domain. It is therefore possible that, despite the lack of recognizable similarity to GGDEF, PelD may have evolved from an I-site containing GGDEF domain.

How PelD functions remains unknown. One wonders whether c-di-GMP-binding PelD interacts with and activates glycosyl transferase, PelF, in the manner that is reminiscent of BcsBA and Alg8-Alg44. It is noteworthy that just like PelD is a non-PilZ c-di-GMP receptor, PelF is a type 4 glycosyl transferase, not type 2 as are BcsA and Alg8. However, it is possible that mechanistic aspects of c-di-GMP regulation are similar in these nonhomologous systems (Fig. 2).

PopA represents another example of the I-site c-di-GMP receptor whose precise spatiotemporal localization is required for a normal cell cycle in *C. crescentus* (23). PopA has the same domain architecture as PleD, i.e., REC-REC-GGDEF, yet its GGDEF domain is enzymatically inactive. PopA binds c-di-GMP through the I-site positioned upstream of the degenerate GGDEF motif. PopA is localized to cell poles during the cell cycle, and its localization to the new pole is dependent on the intact I-site and c-di-GMP. At the new pole, PopA recruits the master regulator of the cell cycle CtrA via the intermediary partner RcdA. After being sequestered to the new pole, CtrA is degraded by the cell-pole-localized ClpXP protease. This opens the way for unrepressed initiation of chromosome replication (67). This system is described in more detail in chapter 9.

Yet another candidate for a c-di-GMP receptor that uses the I-site is *V. cholerae* CdgG, which controls colony rugosity (11; see also chapter 17 by Sinem Beyhan and Fitnat Yildiz). Like PopA, it has a degenerate GGDEF domain but an intact I-site, which is essential for protein function.

Inactive EAL and HD-GYP domains as c-di-GMP receptors

While enzymatically inactive EAL and HD-GYP domains are natural candidates for c-di-GMP-binding receptors, this pool of potential receptors has not yet been adequately assessed (Fig. 2). At the time of this writing, only one EAL-based c-di-GMP receptor had been characterized in vivo and in vitro. The LapD protein is necessary for secretion of the outer membrane adhesin LapA, which in turn is absolutely required for stable surface attachment and biofilm formation in *P. fluorescens* (59).

LapD contains a GGDEF-EAL tandem, yet both GGDEF and EAL domains appear to be degenerate based on the sequence analysis. Newell and colleagues showed that LapD lacks DGC or PDE activities but binds c-di-GMP specifically with a reasonable K_d of 5.5 μM (59). Using a genetic approach targeting conserved residues near the binding pocket, these researchers showed that c-di-GMP binds to the degenerate EAL in vitro, most likely in its substrate-binding pocket. How LapD functions is not completely clear. However, the authors of the study made a strong case for the inside out signaling, where c-di-GMP binding to the EAL domain affects the output periplasmic domain, which is separated from the GGDEF-EAL tandem via a HAMP and transmembrane domains (see chapter 19 by Pete Newell and George O'Toole for more details).

It is intriguing to contemplate that glycosyl transferases may use a similar mode of activation, where cytoplasmic c-di-GMP levels control translocation of the polysaccharide chains across the cytoplasmic membrane and periplasm (Fig. 2). Are we witnessing the emergence of c-di-GMP as a key regulator of translocation of diverse macromolecules to the cell surface? While speculative, this hypothesis is worth pursuing because export of biomolecules is critical for bacterial virulence and biofilms. Elucidation of common molecular mechanisms of c-di-GMP action and appreciation of the scope of c-di-GMP involvement may have profound implications for our understanding and eventual control of these phenomena.

Is regulation of pili formation by PilZ proteins consistent with the proposed hypothesis? The phenotype of the *pilZ* mutation in *P. aeruginosa,* i.e., blocked export of pili subunits and their accumulation in the cytoplasm (2), is fully consistent with this hypothesis. The problem is that *P. aeruginosa* PilZ protein does contain the c-di-GMP switch and does not bind c-di-GMP in vitro (50). The question remains as to whether PilZ proteins from other species bind c-di-GMP and regulate pili subunit secretion in a c-di-GMP-dependent manner. What other secretion systems may be regulated by c-di-GMP-binding proteins?

Bacillus subtilis YkuI represents another candidate c-di-GMP receptor that is based on the enzymatically inactive EAL domain. YkuI has been characterized structurally but not yet functionally. YkuI binds c-di-GMP but lacks PDE activity (56; see also chapter 7). The overall structure of the EAL domain of YkuI complexed with c-di-GMP is quite similar to the structure of the active PDE, *K. pneumoniae* BlrP1. However, YkuI binds only one Mg^{2+} ion, whereas two ions appear to be required for c-di-GMP hydrolysis (6). YkuI has a C-terminal PAS-like domain. It remains to be determined whether the PAS-like domain functions as a signal input to affect c-di-GMP binding by the EAL domain or as an output whose conformation depends on c-di-GMP binding by EAL.

Unpredictable c-di-GMP receptors

The c-di-GMP receptors described so far were predictable, except for PelD. The discovery that the transcriptional regulator FleQ from *P. aeruginosa* is a c-di-GMP-binding protein was clearly unpredictable (Fig. 1). This discovery opened a new page in c-di-GMP signaling because FleQ does not fall into any of the known classes of c-di-GMP receptors (33). Hickman and colleagues showed that FleQ is a key transcriptional regulator coordinating the choice between motile and sessile lifestyles in *P. aeruginosa*. In the absence of c-di-GMP, FleQ, in partnership with FleN, activates transcription of flagellum biosynthesis genes and represses transcription of the *pel* and *psl* operons involved in exopolysaccharide biosynthesis. In vivo, FleQ-based repression is relieved by increased c-di-GMP levels. In vitro FleQ binding to c-di-GMP decreases its affinity to the *pel* operon promoter, as demonstrated by electrophoretic mobility shift assays. The FleQN system and the upstream signaling cascade involved in controlling its activity are described by Carrie Harwood in chapter 11.

FleQ proves that, despite significant progress achieved in the last few years, our knowledge of potential c-di-GMP receptors is far from complete. It is therefore reasonable to anticipate identification of new kinds of c-di-GMP-binding proteins.

c-di-GMP-specific riboswitches

While discovery of every new type of c-di-GMP receptor is very exciting, the discovery of a c-di-GMP-specific riboswitch (92) was an unanticipated bonus (Fig. 1). Riboswitches are noncoding segments of mRNA that adopt specific secondary structures. When cognate small molecules bind to these structures, their stability changes, which in turn can drastically affect mRNA stability and/or conformation and, therefore, affect mRNA level or translation efficiency (7). Sudarsan and colleagues found that c-di-GMP binds specifically to a particular class of riboswitches, GEMM, that had been identified earlier but lacked known ligands (92). In some bacteria, the c-di-GMP-specific riboswitch was found in the upstream regions of genes encoding DGCs and PDEs as well as genes that are traditionally controlled by c-di-GMP, including virulence genes and genes for biogenesis of pili and flagella. However, predicted c-di-GMP riboswitches appear to regulate many new targets as well (see chapter 15 by Elaine Lee and colleagues).

In summary, several types of c-di-GMP receptors have evolved. One class involves PilZ domains that catch c-di-GMP via an extended N-terminal c-di-GMP switch containing the R(X3)R motif and subsequently bring the switch close to the PilZ domain core. The second class involves I-sites, RXXD(X30–35)R, located in degenerate GGDEF domains or, possibly, in the unrelated sequences. The third class involves enzymatically inactive EAL or HD-GYP domains that retain c-di-GMP-binding sites. (The latter has not yet been experimentally verified.) *P. aeruginosa* FleQ binds c-di-GMP via an uncharacterized site, which indicates that the full range of c-di-GMP receptors has yet to be discovered. Finally, riboswitches represent a nonprotein class of c-di-GMP receptors that are present in a variety of bacteria.

REGULATORY NIGHTMARE OR FLEXIBLE ORDER?

Maintenance of the intracellular pool of c-di-GMP to which dozens of DGCs and dozens of PDEs are contributing, as is the case in many bacteria, must be a regulatory nightmare. How do cells control numerous c-di-GMP signaling systems? The more we learn about c-di-GMP-dependent signaling, the more we realize that c-di-GMP regulatory systems are arranged in a hierarchical manner, that they are well coordinated, that they serve specific targets, and that they are fine-tuned to environmental and intracellular cues. The first three features represent hardwired (i.e., genetically predetermined) order, while the last one allows for flexibility. While we do not yet fully understand the organization of all c-di-GMP-dependent pathways in any given bacterium, a number of studies listed below appear to confirm this general view (Fig. 3).

The intracellular level of c-di-GMP is likely to match a particular bacterial lifestyle or physiological state by setting activities of c-di-GMP-binding transcription factors and receptors. Under any given condition, only one or a few DGCs and PDEs appear to contribute the most to establishing this background

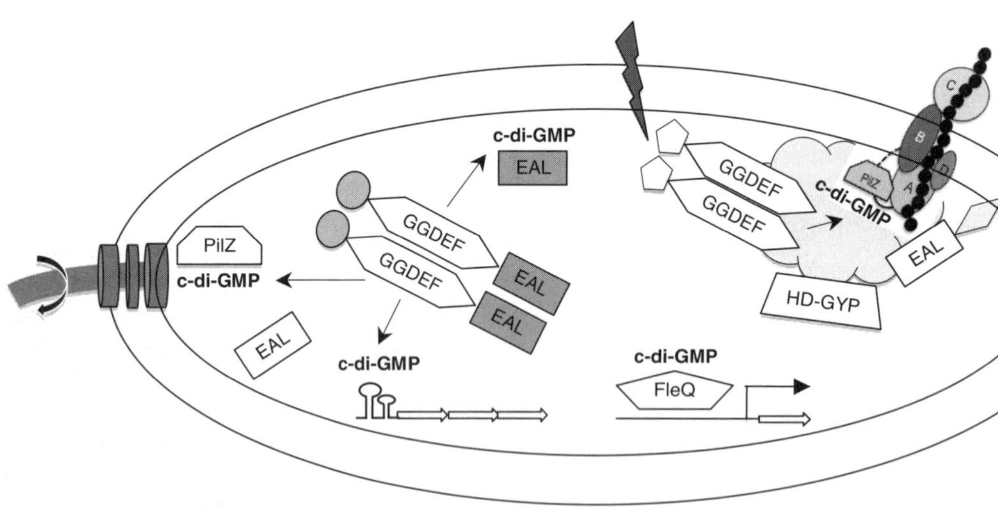

Figure 3. General principles of c-di-GMP regulation. Various DGCs, c-di-GMP PDEs, and c-di-GMP receptors are shown. A cloud of c-di-GMP generated by a DGC that is activated by an environmental signal is shown to affect a specific target, exopolysaccharide synthase.

level. For example, in *G. xylinus,* the DGC1 cyclase contributes ~85% of total DGC activity and PDEA1 contributes ~85% total PDE activity (94). CC3396 contributes >80% of total PDE activity in *C. crescentus* (18). Dependent on environmental conditions, the leading DGC or PDE may change (42). Furthermore, c-di-GMP networks vary significantly even among *E. coli* strains, as nicely illustrated by a study by Monteiro et al. (57).

While there may be general purpose DGCs and PDEs, many, if not most, c-di-GMP metabolizing systems are likely to be (i) specific to their targets and localized in close proximity to them and (ii) highly regulated by environmental signals. Several studies revealed that intracellular levels of c-di-GMP do not always explain the observed phenotypes (29, 35, 46, 91). Target specificity makes c-di-GMP signaling in bacteria similar to cAMP and cGMP signaling in eukaryotes. Despite the general lack of membrane-bound organelles that serve for spatial separation in eukaryotes, bacteria contain myriads of complex protein assemblages that perform specialized functions, which may serve as c-di-GMP targets. Since localized control is in many ways better understood in eukaryotes, it may be instructive to take cues from these systems (9, 14, 101). If we do so, the following picture would emerge.

A specific DGC produces a cloud of c-di-GMP in its immediate vicinity in response to a specific environmental stimulus that regulates its activity, expression, and/or localization. A spike in local c-di-GMP level is sensed by a c-di-GMP receptor that generates a specific outcome. A dedicated PDE controls the amplitude and duration of the c-di-GMP spike around its target or/and ensures that the c-di-GMP cloud is not spread to the unintended targets (Fig. 3).

More and more evidence is accumulating to substantiate this general view of c-di-GMP signaling pathways. Below, I present selected examples to illustrate key regulatory mechanisms involved. The mechanisms that organisms use and the extent to which they use them are highly species specific, as discussed by the contributors to this book.

Target-Specific c-di-GMP Signaling Modules

Benziman and colleagues were the first to observe the local, target-specific control of c-di-GMP levels. They found that >80% of cellular c-di-GMP in *G. xylinus* was bound to the membrane protein fraction containing cellulose synthase (105). Knowing that cellulose synthase by itself binds c-di-GMP, it is safe to assume that >80% of c-di-GMP is bound to cellulose synthase. Further, most cellular c-di-GMP PDE activity was associated with the partially purified membrane fraction enriched in cellulose synthase. This suggests that a cellulose synthase-specific PDE tightly guards its target.

This early assessment of c-di-GMP distribution showing that most c-di-GMP is protein bound should serve as a warning against overreliance on total cellular c-di-GMP levels as a measure of functionality of a given c-di-GMP signaling system. The total c-di-GMP contains all protein-bound and potentially unbound c-di-GMP and may not adequately represent c-di-GMP saturation of a particular c-di-GMP receptor.

The *C. crescentus* DGC, PleD, provides yet another excellent example of target specificity (61, 62). This phosphorylation-dependent DGC is specifically recruited to assist in flagellum ejection from a stalk cell just prior to cell division. The function of the PleD is described in detail in chapter 9.

Spatiotemporal Regulation

PleD also serves as an example of precise spatiotemporal control. It is localized to cell poles in a cell cycle-dependent manner. Huitema et al. (39) suggested that PleD may be a partner to the EAL domain protein TipF, which in itself is a remarkable example of spatiotemporal specificity. TipF is a membrane-bound protein whose localization is cell cycle dependent. TipF localizes to the sites of cell constriction during cell division and subsequently moves to the cell poles (39).

The precise spatiotemporal regulation of the c-di-GMP receptor PopA (23) was described earlier. Note that a high level of spatiotemporal regulation is by no means limited to *C. crescentus,* which is renowned for its complex developmental program. Huang et al. (38) and Kazmierczak et al. (44) showed polar localization of the GGDEF-EAL PDE, FimX from *P. aeruginosa,* involved in twitching motility regulation, while Güvener and Harwood (32) presented a compelling example of the differential subcellular location of the DGC, WspR, from the same bacterium (see also chapter 11).

Regulation at the Level of Gene Expression

This type of regulation is very prominent in bacteria; therefore, it is not surprising that transcriptional activation or repression of genes encoding DGCs and PDEs and c-di-GMP receptors play significant roles in c-di-GMP regulation. An early example of such regulation was described by Römling et al. (75). In *S. enterica,* AdrA, the key DGC responsible for the biofilm lifestyle (42), is under strict control of the transcriptional factor AgfD. Additional levels of transcriptional control of DGC and PDE genes have been described for this bacterium (see chapter 10 by Ute Römling and colleagues) as well as for *E. coli* (reference 104 and chapter 16), *V. cholerae* (chapter 17), *Vibrio parahaemolyticus* (chapter 12), and *P. aeruginosa* (chapter 11).

Regulation of DGC and PDE Activities by Environmental Signals

The notion that activities of the enzymes involved in c-di-GMP metabolism are highly regulated by environmental stimuli has been appreciated for a long time because most GGDEF and EAL domain proteins have discrete sensory domains (26–28; see also chapter 3). The first DGCs and PDEs described by Tal and colleagues (94) are no exceptions. The main c-di-GMP PDE from *G. xylinus,* PDEA1, contains a heme-binding PAS domain that acts as an oxygen sensor (16) and controls PDE activity in an oxygen-dependent manner (96). The *E. coli* ortholog of PDEA1, Dos, works in a similar fashion (22). The heme-binding PAS domain is but one of the known oxygen-sensing modules that are hooked to DGCs and PDEs, e.g., oxygen regulates the activity of a *Bordetella pertussis* DGC via a globin domain (102).

A number of DGCs and PDEs are regulated by light. The bacteriophytochrome BphG from *R. sphaeroides,* whose DGC activity is enhanced by red light (97), was described earlier in this chapter. A similar GGDEF-EAL phytochrome was found in cyanobacteria (106). Recently, Barends et al. (6) described the structure and mechanism of activation by blue light of the PDE BlrP1 from *K. pneumoniae.*

Numerous other signals, e.g., polyamines (43), phosphate (see chapter 19), growth on surfaces (see chapter 11), and aminoglycoside antibiotics (36), affect activities of c-di-GMP signaling enzymes.

In summary, the c-di-GMP signaling systems resemble cyclic nucleotide regulation in eukaryotes. These systems are arranged in a hierarchical manner. At any given time, only a subset of genes encoding DGCs, PDEs, and their receptors is expressed. The products of these systems are fine-tuned to environmental and intracellular cues and are often controlled with an amazing spatiotemporal resolution.

CLOSING THOUGHTS AND PERSPECTIVES

In the span of only a few years, the field of c-di-GMP signaling has spectacularly progressed from infancy to adolescence. The core c-di-GMP signaling pathway has been elucidated. We understand the enzymology of c-di-GMP synthesis and hydrolysis reasonably well. We know several kinds of c-di-GMP receptors. However, most of the end targets of c-di-GMP remain unknown, as are the detailed molecular mechanisms through which c-di-GMP operates.

The enzymology of c-di-GMP metabolism involves a GGDEF domain containing DGCs and EAL and HD-GYP domains containing PDEs. While this picture is likely to remain largely unchanged, there appears to be room for expansion, particularly for novel c-di-GMP PDEs. Some of these may come from the HD_PDEase superfamily and involve HD and HDOD domains that lack the GYP motif, while others may still emerge from unrelated phosphoesterases that evolved to use c-di-GMP as their substrate.

We have learned from structure-function studies how GGDEF and EAL domains bind c-di-GMP. We have also learned in great detail how EAL domains function as PDEs but still lack mechanistic details on the mechanism of the DGC reaction. We have yet to learn how HD-GYP PDEs operate.

The conundrum of enzymes containing GGDEF-EAL domain tandems has for the most part been resolved. Most of these appear to have one enzymatic activity and use the second domain as a sensor. The GGDEF-EAL PDEs use substrate-binding sites of their enzymatically inactive GGDEF domains to sense intracellular GTP. One can predict that the GGDEF-EAL DGCs use their enzymatically inactive EAL domains for c-di-GMP binding as a mechanism of feedback inhibition (akin to the I-sites present in some DGCs); however, this has yet to be tested experimentally. One can also envision that inactive EAL domains in some GGDEF-EAL proteins may have evolved to perform different functions, e.g., interactions with other proteins. The truly bifunctional GGDEF-EAL proteins remain somewhat of a mystery. What seems to be clear is that proteins that have physiological roles as both DGCs and PDEs are rare. How exactly these proteins switch from c-di-GMP synthesis to hydrolysis and back remains unclear, but it appears that such switches may involve interactions with additional proteins. Based on the framework of GGDEF-EAL tandems, one can predict how the GGDEF–HD-GYP and other domain tandems would operate. However, no proteins containing non-GGDEF-EAL tandems have yet been investigated.

Those GGDEF-EAL proteins where none of the domains is enzymatically active represent a very peculiar set. Some function as c-di-GMP receptors (*P. fluorescens* LapD). Others turned away from c-di-GMP signaling to perform new functions, e.g., binding small regulatory RNAs (*E. coli* CsrD) or yet unknown partners (*S. epidermidis* GdpS).

Several c-di-GMP receptor classes have been identified—PilZ domain, I-site, enzymatically inactive EAL domain. The enzymatically inactive HD-GYP domain is a readily predictable candidate for a new c-di-GMP receptor class that will likely emerge. The fact that some c-di-GMP-binding proteins do not belong to these classes (e.g., *P. aeruginosa* FleQ) is exciting and suggests that new, unpredictable c-di-GMP receptors are yet to be discovered.

Some of the most intriguing questions in the c-di-GMP field concern mechanisms employed by c-di-GMP receptors and the nature of their targets. Only a few end targets of c-di-GMP signaling are currently known. Among these are glycosyl transferases. What exactly does c-di-GMP binding to the PilZ domain affect—polysaccharide synthesis, translocation across the cytoplasmic membrane, or both? Is there a common theme in how c-di-GMP affects glycosyl transferases of different kinds (BcsBA, Alg44-Alg8, PelD-PelF)? How do non-PilZ c-di-GMP receptors affect secretion of outer membrane-localized adhesins (LapD-LapA)? Is c-di-GMP emerging as a regulator of protein and polysaccharide translocation to the cell surface, i.e., inside out signaling?

How do PilZ domain proteins affect flagellar rotation? What transcriptional factors, other than FleQ, bind c-di-GMP? What is the connection, if any, between c-di-GMP and small regulatory RNAs? How do c-di-GMP-specific riboswitches fit into the overall c-di-GMP signaling system? How does c-di-GMP-dependent regulation work in the systems unrelated to motile-sessile switches?

Despite our general understanding of the core c-di-GMP signaling pathways and some interesting detours from the core, numerous important questions remain. The quest for answers to these questions will keep the c-di-GMP field flourishing for years to come. This important bacterial second messenger is just coming of age.

Acknowledgments. I thank my laboratory coworkers and collaborators for their hard work and invaluable intellectual contributions. I am grateful to the book editors and to the anonymous reviewer of this chapter for editorial improvements and suggestions. I apologize to those researchers whose c-di-GMP studies may have been inadvertently overlooked or mentioned only in passing.

The c-di-GMP work in my laboratory is supported by the National Science Foundation (MCB 0645876) and the U.S. Department of Agriculture Cooperative State Research Education Extension Service (AD-417) via the University of Wyoming Agricultural Experimental Station.

REFERENCES

1. **Aldridge, P., R. Paul, P. Goymer, P. Rainey, and U. Jenal.** 2003. Role of the GGDEF regulator PleD in polar development of *Caulobacter crescentus*. *Mol. Microbiol.* **47:**1695–1708.
2. **Alm, R. A., A. J. Bodero, P. D. Free, and J. S. Mattick.** 1996. Identification of a novel gene, *pilZ*, essential for type 4 fimbrial biogenesis in *Pseudomonas aeruginosa*. *J. Bacteriol.* **178:**46–53.
3. **Amikam, D., and M. Y. Galperin.** 2006. PilZ domain is part of the bacterial c-di-GMP binding protein. *Bioinformatics* **22:**3–6.
4. **Aravind, L., and E. V. Koonin.** 1998. The HD domain defines a new superfamily of metal-dependent phosphohydrolases. *Trends Biochem. Sci.* **23:**469–472.
5. **Ausmees, N., R. Mayer, H. Weinhouse, G. Volman, D. Amikam, M. Benziman, and M. Lindberg.** 2001. Genetic data indicate that proteins containing the GGDEF domain possess diguanylate cyclase activity. *FEMS Microbiol. Lett.* **204:**163–167.
6. **Barends, T. R. M., E. Hartmann, J. Griese, T. Beitlich, N. V. Kirienko, D. A. Ryjenkov, J. Reinstein, R. L. Shoeman, M. Gomelsky, and I. Schlichting.** 2009. Structure and mechanism of a bacterial light-regulated cyclic nucleotide phosphodiesterase. *Nature* **459:**1015–1018.
7. **Barrick, J. E., and R. R. Breaker.** 2007. The distributions, mechanisms, and structures of metabolite-binding riboswitches. *Genome Biol.* **8:**R239.
8. **Benach, J., S. S. Swaminathan, R. Tamayo, S. K. Handelman, E. Folta-Stogniew, J. E. Ramos, F. Forouhar, H. Neely, J. Seetharaman, A. Camilli, and J. F. Hunt.** 2007. The structural basis of cyclic diguanylate signal transduction by PilZ domains. *EMBO J.* **26:**5153–5166.
9. **Bender, A. T., and J. A. Beavo.** 2006. Cyclic nucleotide phosphodiesterases: molecular regulation to clinical use. *Pharmacol. Rev.* **58:**488–520.
10. **Berry, R. M., and J. P. Armitage.** 2008. Microbiology. How bacteria change gear. *Science* **320:**1599–1600.
11. **Beyhan, S., L. S. Odell, and F. H. Yildiz.** 2008. Identification and characterization of cyclic diguanylate signaling systems controlling rugosity in *Vibrio cholerae*. *J. Bacteriol.* **190:**7392–7405.
12. **Blair, K. M., L. Turner, J. T. Winkelman, H. C. Berg, and D. B. Kearns.** 2008. A molecular clutch disables flagella in the *Bacillus subtilis* biofilm. *Science* **320:**1636–1638.
13. **Bobrov, A. G., O. Kirillina, and R. D. Perry.** 2005. The phosphodiesterase activity of the HmsP EAL domain is required for negative regulation of biofilm formation in *Yersinia pestis*. *FEMS Microbiol. Lett.* **247:**123–130.
14. **Brunton, L. L.** 2003. PDE4: arrested at the border. *Sci. STKE* **204:**pe44.
15. **Chan, C., R. Paul, D. Samoray, N. C. Amiot, B. Giese, U. Jenal, and T. Schirmer.** 2004. Structural basis of activity and allosteric control of diguanylate cyclase. *Proc. Natl. Acad. Sci. USA* **101:**17084–17089.
16. **Chang, A. L., J. R. Tuckerman, G. Gonzalez, R. Mayer, H. Weinhouse, G. Volman, D. Amikam, M. Benziman, and M. A. Gilles-Gonzalez.** 2001. Phosphodiesterase A1, a regulator of cellulose synthesis in *Acetobacter xylinum,* is a heme-based sensor. *Biochemistry* **40:**3420–3426.
17. **Christen, B., M. Christen, R. Paul, F. Schmid, M. Folcher, P. Jenoe, M. Meuwly, and U. Jenal.** 2006. Allosteric control of cyclic di-GMP signaling. *J. Biol. Chem.* **281:**32015–32024.
18. **Christen, M., B. Christen, M. Folcher, A. Schauerte, and U. Jenal.** 2005. Identification and characterization of a cyclic di-GMP-specific phosphodiesterase and its allosteric control by GTP. *J. Biol. Chem.* **280:**30829–30837.
19. **Christen, M., B. Christen, M. G. Allan, M. Folcher, P. Jenö, S. Grzesiek, and U. Jenal.** 2007. DgrA is a member of a new

family of cyclic diguanosine monophosphate receptors and controls flagellar motor function in *Caulobacter crescentus*. *Proc. Natl. Acad. Sci. USA* **104**:4112–4117. (Erratum, **104**: 7729.)

20. **Claret, L., S. Miquel, N. Vieille, D. A. Ryjenkov, M. Gomelsky, and A. Darfeuille-Michaud.** 2007. The flagellar sigma factor FliA regulates adhesion and invasion of Crohn disease-associated *Escherichia coli* via a cyclic dimeric GMP-dependent pathway. *J. Biol. Chem.* **282**:33275–33283.
21. **D'Argenio D. A., and S. I. Miller.** 2004. Cyclic di-GMP as a bacterial second messenger. *Microbiology* **150**:2497–2502.
22. **Delgado-Nixon, V. M., G. Gonzalez, and M. A. Gilles-Gonzalez.** 2000. Dos, a heme-binding PAS protein from *Escherichia coli,* is a direct oxygen sensor. *Biochemistry* **39**: 2685–2691.
23. **Duerig, A., S. Abel, M. Folcher, M. Nicollier, T. Schwede, N. Amiot, B. Giese, and U. Jenal.** 2009. Second messenger-mediated spatiotemporal control of protein degradation regulates bacterial cell cycle progression. *Genes Dev.* **23**:93–104.
24. **Ferreira, R. B., L. C. Antunes, E. P. Greenberg, and L. L. McCarter.** 2008. *Vibrio parahaemolyticus* ScrC modulates cyclic dimeric GMP regulation of gene expression relevant to growth on surfaces. *J. Bacteriol.* **190**:851–860.
25. **Finn, R. D., J. Tate, J. Mistry, P. C. Coggill, S. J. Sammut, H. R. Hotz, G. Ceric, K. Forslund, S. R. Eddy, E. L. Sonnhammer, and A. Bateman.** 2008. The Pfam protein families database. *Nucleic Acids Res.* **36**:D281–D288.
26. **Galperin, M. Y., D. A. Natale, L. Aravind, and E. V. Koonin.** 1999. A specialized version of the HD hydrolase domain implicated in signal transduction. *J. Mol. Microbiol. Biotechnol.* **1**:303–305.
27. **Galperin, M. Y., A. N. Nikolskaya, and E. V. Koonin.** 2001. Novel domains of the prokaryotic two-component signal transduction systems. *FEMS Microbiol. Lett.* **203**:11–21.
28. **Galperin, M. Y.** 2004. Bacterial signal transduction network in a genomic perspective. *Environ. Microbiol.* **6**:552–567.
29. **García, B., C. Latasa, C. Solano, F. García-del Portillo, C. Gamazo, and I. Lasa.** 2004. Role of the GGDEF protein family in *Salmonella* cellulose biosynthesis and biofilm formation. *Mol. Microbiol.* **54**:264–277.
30. **Girgis, H. S., Y. Liu, W. S. Ryu, and S. Tavazoie.** 2007. A comprehensive genetic characterization of bacterial motility. *PLoS Genet.* **3**:1644–1660.
31. **Gomelsky, M., and G. Klug.** 2002. BLUF: a novel FAD-binding domain involved in sensory transduction in microorganisms. *Trends Biochem. Sci.* **27**:497–500.
32. **Güvener, Z. T., and C. S. Harwood.** 2007. Subcellular location characteristics of the *Pseudomonas aeruginosa* GGDEF protein, WspR, indicate that it produces cyclic-di-GMP in response to growth on surfaces. *Mol. Microbiol.* **66**:1459–1473.
33. **Hickman, J. W., and C. S. Harwood.** 2008. Identification of FleQ from *Pseudomonas aeruginosa* as a c-di-GMP-responsive transcription factor. *Mol. Microbiol.* **69**:376–389.
34. **Hickman, J. W., D. F. Tifrea, and C. S. Harwood.** 2005. A chemosensory system that regulates biofilm formation through modulation of cyclic diguanylate levels. *Proc. Natl. Acad. Sci. USA* **102**:14422–14427.
35. **Hisert, K. B., M. MacCoss, M. U. Shiloh, K. H. Darwin, S. Singh, R. A. Jones, S. Ehrt, Z. Zhang, B. L. Gaffney, S. Gandotra, D. W. Holden, D. Murray, and C. Nathan.** 2005. A glutamate-alanine-leucine (EAL) domain protein of *Salmonella* controls bacterial survival in mice, antioxidant defence and killing of macrophages: role of cyclic diGMP. *Mol. Microbiol.* **56**:1234–1245.
36. **Hoffman, L. R., D. A. D'Argenio, M. J. MacCoss, Z. Zhang, R. A. Jones, and S. I. Miller.** 2005. Aminoglycoside antibiotics induce bacterial biofilm formation. *Nature* **436**:1171–1175.
37. **Holland, L. M., S. T. O'Donnell, D. A. Ryjenkov, L. Gomelsky, S. R. Slater, P. D. Fey, M. Gomelsky, and J. P. O'Gara.** 2008. A staphylococcal GGDEF domain protein regulates biofilm formation independently of c-di-GMP. *J. Bacteriol.* **190**: 5178–5189.
38. **Huang, B., C. Whitchurch, and J. S. Mattick.** 2003. FimX, a multidomain protein connecting environmental signals to twitching motility in *Pseudomonas aeruginosa*. *J. Bacteriol.* **185**:7068–7076.
39. **Huitema, E., S. Pritchard, D. Matteson, S. K. Radhakrishnan, and P. H. Viollier.** 2006. Bacterial birth scar proteins mark future flagellum assembly site. *Cell* **124**:1025–1037.
40. **Jenal, U.** 2004. Cyclic di-guanosine-monophosphate comes of age: a novel secondary messenger involved in modulating cell surface structures in bacteria? *Curr. Opin. Microbiol.* **7**:185–191.
41. **Jenal, U., J. White, and L. Shapiro.** 1994. *Caulobacter* flagellar function, but not assembly, requires FliL, a non-polarly localized membrane protein present in all cell types. *J. Mol. Biol.* **243**:227–244. (Erratum, **248**:883, 1995.)
42. **Kader, A., R. Simm, U. Gerstel, M. Morr, and U. Römling.** 2006. Hierarchical involvement of various GGDEF domain proteins in rdar morphotype development of *Salmonella enterica* serovar Typhimurium. *Mol. Microbiol.* **60**:602–616.
43. **Karatan, E., T. R. Duncan, and P. I. Watnick.** 2005. NspS, a predicted polyamine sensor, mediates activation of *Vibrio cholerae* biofilm formation by norspermidine. *J. Bacteriol.* **187**: 7434–7443.
44. **Kazmierczak, B. I., M. B. Lebron, and T. S. Murray.** 2006. Analysis of FimX, a phosphodiesterase that governs twitching motility in *Pseudomonas aeruginosa*. *Mol. Microbiol.* **60**:1026–1043.
45. **Ko, M., and C. Park.** 2000. Two novel flagellar components and H-NS are involved in the motor function of *Escherichia coli*. *J. Mol. Biol.* **303**:371–382.
46. **Kulasakara, H., V. Lee, A. Brencic, N. Liberati, J. Urbach, S. Miyata, D. G. Lee, A. N. Neely, M. Hyodo, Y. Hayakawa, F. M. Ausubel, and S. Lory.** 2006. Analysis of *Pseudomonas aeruginosa* diguanylate cyclases and phosphodiesterases reveals a role for bis-(3′-5′)-cyclic-GMP in virulence. *Proc. Natl. Acad. Sci. USA* **103**:2839–2844.
47. **Kumar, M., and D. Chatterji.** 2008. Cyclic di-GMP: a second messenger required for long-term survival, but not for biofilm formation, in *Mycobacterium smegmatis*. *Microbiology* **154**: 2942–2955.
48. **Lai, T. H., Y. Kumagai, M. Hyodo, Y. Hayakawa, and Y. Rikihisa.** 2009. The *Anaplasma phagocytophilum* PleC histidine kinase and PleD diguanylate cyclase two-component system and role of cyclic di-GMP in host cell infection. *J. Bacteriol.* **191**:693–700.
49. **Lee, V. T., J. M. Matewish, J. L. Kessler, M. Hyodo, Y. Hayakawa, and S. Lory.** 2007. A cyclic-di-GMP receptor required for bacterial exopolysaccharide production. *Mol. Microbiol.* **65**:1474–1484.
50. **Li, T. N., K. H. Chin, J. H. Liu, A. H. Wang, and S. H. Chou.** 2008. XC1028 from *Xanthomonas campestris* adopts a PilZ domain-like structure without a c-di-GMP switch. *Proteins* **75**: 282–288.
51. **Lukat, G. S., W. R. McCleary, A. M. Stock, and J. B. Stock.** 1992. Phosphorylation of bacterial response regulator proteins by low molecular weight phospho-donors. *Proc. Natl. Acad. Sci. USA* **89**:718–722.
52. **Maharaj, R., T. B. May, S. K. Wang, and A. M. Chakrabarty.** 1993. Sequence of the *alg8* and *alg44* genes involved in the synthesis of alginate by *Pseudomonas aeruginosa*. *Gene* **136**: 267–269.

53. **Mayer, R., P. Ross, H. Weinhouse, D. Amikam, G. Volman, P. Ohana, R. D. Calhoon, H. C. Wong, A. W. Emerick, and M. Benziman.** 1991. Polypeptide composition of bacterial cyclic diguanylic acid-dependent cellulose synthase and the occurrence of immunologically crossreacting proteins in higher plants. *Proc. Natl. Acad. Sci. USA* **88:**5472–5476.
54. **McCarthy, Y., R. P. Ryan, K. O'Donovan, Y. Q. He, B. L. Jiang, J. X. Feng, J. L. Tang, and J. M. Dow.** 2008. The role of PilZ domain proteins in the virulence of *Xanthomonas campestris* pv. *campestris*. *Mol. Plant Pathol.* **9:**819–824.
55. **Merighi, M., V. T. Lee, M. Hyodo, Y. Hayakawa, and S. Lory.** 2007. The second messenger bis-(3′-5′)-cyclic-GMP and its PilZ domain-containing receptor Alg44 are required for alginate biosynthesis in *Pseudomonas aeruginosa*. *Mol. Microbiol.* **65:**876–895.
56. **Minasov, G., S. Padavattan, L. Shuvalova, J. S. Brunzelle, D. J. Miller, A. Baslé, C. Massa, F. R. Collart, T. Schirmer, and W. F. Anderson.** 24 February 2009. Crystal structures of YkuI and its complex with second messenger c-di-GMP suggests catalytic mechanism of phosphodiester bond cleavage by EAL domains. *J. Biol. Chem.* **284:**13174–13184. [Epub ahead of print.]
57. **Monteiro, C., I. Saxena, X. Wang, A. Kader, W. Bokranz, R. Simm, D. Nobles, M. Chromek, A. Brauner, R. M. Brown, Jr., and U. Römling.** 2009. Characterization of cellulose production in *Escherichia coli* Nissle 1917 and its biological consequences. *Environ. Microbiol.* **11:**1105–1116.
58. **Neunuebel, M. R., and J. W. Golden.** 2008. The *Anabaena* sp. strain PCC 7120 gene all2874 encodes a diguanylate cyclase and is required for normal heterocyst development under high-light growth conditions. *J. Bacteriol.* **190:**6829–6836.
59. **Newell, P. D., R. D. Monds, and G. A. O'Toole.** 2009. LapD is a bis-(3′,5′)-cyclic dimeric GMP-binding protein that regulates surface attachment by *Pseudomonas fluorescens* Pf0-1. *Proc. Natl. Acad. Sci. USA* **106:**3461–3466.
60. **Oglesby, L. L., S. Jain, and D. E. Ohman.** 2008. Membrane topology and roles of *Pseudomonas aeruginosa* Alg8 and Alg44 in alginate polymerization. *Microbiology* **154:**1605–1615.
61. **Paul, R., S. Weiser, N. C. Amiot, C. Chan, T. Schirmer, B. Giese, and U. Jenal.** 2004. Cell cycle-dependent dynamic localization of a bacterial response regulator with a novel diguanylate cyclase output domain. *Genes Dev.* **18:**715–727.
62. **Paul, R., S. Abel, P. Wassmann, A. Beck, H. Heerklotz, and U. Jenal.** 2007. Activation of the diguanylate cyclase PleD by phosphorylation-mediated dimerization. *J. Biol. Chem.* **282:** 29170–29177.
63. **Pei, J., and N. Grishin.** 2001. GGDEF domain is homologous to adenylyl cyclase. *Proteins* **42:**210–216.
64. **Perego, M., and J. A. Hoch.** 1996. Protein aspartate phosphatases control the output of two-component signal transduction systems. *Trends Genet.* **12:**97–101.
65. **Pratt, J. T., R. Tamayo, A. D. Tischler, and A. Camilli.** 2007. PilZ domain proteins bind cyclic diguanylate and regulate diverse processes in *Vibrio cholerae*. *J. Biol. Chem.* **282:**12860–12870.
66. **Purcell, E. B., and S. Crosson.** 2008. Photoregulation in prokaryotes. *Curr. Opin. Microbiol.* **11:**168–178.
67. **Quon, K. C., B. Yang, I. J. Domian, L. Shapiro, and G. T. Marczynski.** 1998. Negative control of bacterial DNA replication by a cell cycle regulatory protein that binds at the chromosome origin. *Proc. Natl. Acad. Sci. USA* **95:**120–125.
68. **Rajagopal, S., J. M. Key, E. B. Purcell, D. J. Boerema, and K. Moffat.** 2004. Purification and initial characterization of a putative blue light-regulated phosphodiesterase from *Escherichia coli*. *Photochem. Photobiol.* **80:**542–547.
69. **Ramelot, T. A., A. Yee, J. R. Cort, A. Semesi, C. H. Arrowsmith, and M. A. Kennedy.** 2007. NMR structure and binding studies confirm that PA4608 from *Pseudomonas aeruginosa* is a PilZ domain and a c-di-GMP binding protein. *Proteins* **66:** 266–271.
70. **Ramsey, D. M., and D. J. Wozniak.** 2005. Understanding the control of *Pseudomonas aeruginosa* alginate synthesis and the prospects for management of chronic infections in cystic fibrosis. *Mol. Microbiol.* **56:**309–322.
71. **Rao, F., Y. Yang, Y. Qi, and Z. X. Liang.** 2008. Catalytic mechanism of cyclic di-GMP-specific phosphodiesterase: a study of the EAL domain-containing RocR from *Pseudomonas aeruginosa*. *J. Bacteriol.* **190:**3622–3631.
72. **Rockwell, N. C., Y. S. Su, and J. C. Lagarias.** 2006. Phytochrome structure and signaling mechanisms. *Annu. Rev. Plant Biol.* **57:**837–858.
73. **Rogers, E. A., D. Terekhova, H. M. Zhang, K. M. Hovis, I. Schwartz, and R. T. Marconi.** 2009. Rrp1, a cyclic-di-GMP-producing response regulator, is an important regulator of *Borrelia burgdorferi* core cellular functions. *Mol. Microbiol.* **71:**1551–1573.
74. **Römling, U., M. Gomelsky, and M. Y. Galperin.** 2005. C-di-GMP: the dawning of a novel bacterial signalling system. *Mol. Microbiol.* **57:**629–639.
75. **Römling, U., M. Rohde, A. Olsén, S. Normark, and J. Reinköster.** 2000. AgfD, the checkpoint of multicellular and aggregative behaviour in *Salmonella typhimurium* regulates at least two independent pathways. *Mol. Microbiol.* **36:**10–23.
76. **Römling, U.** 2008. Great times for small molecules: c-di-AMP, a second messenger candidate in Bacteria and Archaea. *Sci. Signal.* **1:**pe39.
77. **Ross, P., H. Weinhouse, Y. Aloni, D. Michaeli, P. Weinberger-Ohana, R. Mayer, S. Braun, E. de Vroom, G. A. van der Marel, J. H. van Boom, and M. Benziman.** 1987. Regulation of cellulose synthesis in *Acetobacter xylinum* by cyclic diguanilic acid. *Nature* **325:**279–281.
78. **Ross, P., R. Mayer, H. Weinhouse, D. Amikam, Y. Huggirat, M. Benziman, E. de Vroom, A. Fidder, P. de Paus, L. A. Sliedregt, G. A. van der Marel, and J. H. van Boom.** 1990. The cyclic diguanylic acid regulatory system of cellulose synthesis in *Acetobacter xylinum*. Chemical synthesis and biological activity of cyclic nucleotide dimer, trimer, and phosphothioate derivatives. *J. Biol. Chem.* **265:**18933–18943.
79. **Ross, P., Y. Aloni, C. Weinhouse, D. Michaeli, P. Weinberger-Ohana, R. Meyer, and M. Benziman.** 1985. An unusual guanyl oligonucleotide regulates cellulose synthesis in *Acetobacter xylinum*. *FEBS Lett.* **186:**191–196.
80. **Ross, P., Y. Aloni, H. Weinhouse, D. Michaeli, P. Weinberger-Ohana, R. Mayer, and M. Benziman.** 1986. Control of cellulose synthesis in *Acetobacter xylinum*. A unique guanyl oligonucleotide is the immediate activator of the cellulose synthase. *Carbohydr. Res.* **149:**101–117.
81. **Ryan, R. P., Y. Fouhy, J. F. Lucey, L. C. Crossman, S. Spiro, Y. W. He, L. H. Zhang, S. Heeb, M. Camara, P. Williams, and J. M. Dow.** 2006. Cell-cell signaling in *Xanthomonas campestris* involves an HD-GYP domain protein that functions in cyclic di-GMP turnover. *Proc. Natl. Acad. Sci. USA* **103:**6712–6717.
82. **Ryan, R. P., Y. Fouhy, J. F. Lucey, and J. M. Dow.** 2006. Cyclic di-GMP signaling in bacteria: recent advances and new puzzles. *J. Bacteriol.* **188:**8327–8334.
83. **Ryder, C., M. Byrd, and D. J. Wozniak.** 2007. Role of polysaccharides in *Pseudomonas aeruginosa* biofilm development. *Curr. Opin. Microbiol.* **10:**644–648.
84. **Ryjenkov, D. A., M. Tarutina, O. M. Moskvin, and M. Gomelsky.** 2005. Cyclic diguanylate is a ubiquitous signaling mol-

ecule in bacteria: insights into biochemistry of the GGDEF protein domain. *J. Bacteriol.* **187**:1792–1798.

85. **Ryjenkov, D. A., R. Simm, U. Römling, and M. Gomelsky.** 2006. The PilZ domain is a receptor for the second messenger c-di-GMP: the PilZ domain protein YcgR controls motility in enterobacteria. *J. Biol. Chem.* **281**:30310–30314.
86. **Sasakura, Y., S. Hirata, S. Sugiyama, S. Suzuki, S. Taguchi, M. Watanabe, T. Matsui, I. Sagami, and T. Shimizu.** 2002. Characterization of a direct oxygen sensor heme protein from *Escherichia coli*. Effects of the heme redox states and mutations at the heme-binding site on catalysis and structure. *J. Biol. Chem.* **277**:23821–23827.
87. **Saxena, I. M., K. Kudlicka, K. Okuda, and R. M. Brown, Jr.** 1994. Characterization of genes in the cellulose-synthesizing operon (*acs* operon) of *Acetobacter xylinum:* implications for cellulose crystallization. *J. Bacteriol.* **176**:5735–5752.
88. **Saxena, I. M., R. M. Brown, Jr., and T. Dandekar.** 2001. Structure-function characterization of cellulose synthase: relationship to other glycosyltransferases. *Phytochemistry* **57**: 1135–1148.
89. **Schmidt, A. J., D. A. Ryjenkov, and M. Gomelsky.** 2005. Ubiquitous protein domain EAL encodes cyclic diguanylate-specific phosphodiesterase: enzymatically active and inactive EAL domains. *J. Bacteriol.* **187**:4774–4781.
90. **Simm, R., M. Morr, A. Kader, M. Nimtz, and U. Romling.** 2004. GGDEF and EAL domains inversely regulate cyclic di-GMP levels and transition from sessility to motility. *Mol. Microbiol.* **53**:1123–1134.
91. **Simm, R., A. Lusch, A. Kader, M. Andersson, and U. Römling.** 2007. Role of EAL-containing proteins in multicellular behavior of *Salmonella enterica* serovar Typhimurium. *J. Bacteriol.* **189**:3613–3623.
92. **Sudarsan, N., E. R. Lee, Z. Weinberg, R. H. Moy, J. N. Kim, K. H. Link, and R. R. Breaker.** 2008. Riboswitches in eubacteria sense the second messenger cyclic di-GMP. *Science* **321**: 411–413.
93. **Suzuki, K., P. Babitzke, S. R. Kushner, and T. Romeo.** 2006. Identification of a novel regulatory protein (CsrD) that targets the global regulatory RNAs CsrB and CsrC for degradation by RNaseE. *Genes Dev.* **20**:2605–2617.
94. **Tal, R., H. C. Wong, R. Calhoon, D. Gelfand, A. L. Fear, G. Volman, R. Mayer, P. Ross, D. Amikam, H. Weinhouse, A. Cohen, S. Sapir, P. Ohana, and M. Benziman.** 1998. Three *cdg* operons control cellular turnover of cyclic di-GMP in *Acetobacter xylinum:* genetic organization and occurrence of conserved domains in isoenzymes. *J. Bacteriol.* **180**:4416–4425.
95. **Tamayo, R., A. D. Tischler, and A. Camilli.** 2005. The EAL domain protein VieA is a cyclic diguanylate phosphodiesterase. *J. Biol. Chem.* **280**:33324–33330.
96. **Tanaka, A., H. Takahashi, and T. Shimizu.** 2007. Critical role of the heme axial ligand, Met95, in locking catalysis of the phosphodiesterase from *Escherichia coli* (Ec DOS) toward cyclic diGMP. *J. Biol. Chem.* **282**:21301–21307.
97. **Tarutina, M., D. A. Ryjenkov, and M. Gomelsky.** 2006. An unorthodox bacteriophytochrome from *Rhodobacter sphaeroides* involved in turnover of the second messenger c-di-GMP. *J. Biol. Chem.* **281**:34751–34758.
98. **Terashima, H., S. Kojima, and M. Homma.** 2008. Flagellar motility in bacteria structure and function of flagellar motor. *Int. Rev. Cell. Mol. Biol.* **270**:39–85.
99. **Tischler, A. D., and A. Camilli.** 2004. Cyclic diguanylate (c-di-GMP) regulates *Vibrio cholerae* biofilm formation. *Mol. Microbiol.* **53**:857–869.
100. **Tschowri, N., S. Busse, and R. Hengge.** 2009. The BLUF-EAL protein YcgF acts as a direct anti-repressor in a blue-light response of *Escherichia coli*. *Genes Dev.* **23**:522–534.
101. **Vandecasteele, G., F. Rochais, A. Abi-Gerges, and R. Fischmeister.** 2006. Functional localization of cAMP signalling in cardiac myocytes. *Biochem. Soc. Trans.* **34**:484–488.
102. **Wan, X., J. R. Tuckerman, J. A. Saito, T. A. Freitas, J. S. Newhouse, J. R. Denery, M. Y. Galperin, G. Gonzalez, M. A. Gilles-Gonzalez, and M. Alam.** 2009. Globins synthesize the second messenger bis-(3′-5′)-cyclic diguanosine monophosphate in bacteria. *J. Mol. Biol.* **388**:262–270.
103. **Wassmann, P., C. Chan, R. Paul, A. Beck, H. Heerklotz, U. Jenal, and T. Schirmer.** 2007. Structure of BeF_3-modified response regulator PleD: implications for diguanylate cyclase activation, catalysis, and feedback inhibition. *Structure* **15**: 915–927.
104. **Weber, H., C. Pesavento, A. Possling, G. Tischendorf, and R. Hengge.** 2006. Cyclic-di-GMP-mediated signaling within the sigma network of *Escherichia coli*. *Mol. Microbiol.* **62**: 1014–1034.
105. **Weinhouse, H., S. Sapir, D. Amikam, Y. Shilo, G. Volman, P. Ohana, and M. Benziman.** 1997. c-di-GMP-binding protein, a new factor regulating cellulose synthesis in *Acetobacter xylinum*. *FEBS Lett.* **416**:207–211.
106. **Wilde, A., B. Fiedler, and T. Borner.** 2002. The cyanobacterial phytochrome Cph2 inhibits phototaxis towards blue light. *Mol. Microbiol.* **44**:981–988.
107. **Witte, G., S. Hartung, K. Büttner, and K. P. Hopfner.** 2008. Structural biochemistry of a bacterial checkpoint protein reveals diadenylate cyclase activity regulated by DNA recombination intermediates. *Mol. Cell* **30**:167–178.
108. **Wolfe, A. J., and K. L. Visick.** 2008. Get the message out: cyclic-di-GMP regulates multiple levels of flagellum-based motility. *J. Bacteriol.* **190**:463–475.

The Second Messenger Cyclic Di-GMP
Edited by A. J. Wolfe and K. L. Visick

Chapter 5

The HD-GYP Domain and Cyclic Di-GMP Signaling

ROBERT P. RYAN, YVONNE MCCARTHY, AND J. MAXWELL DOW

Cyclic di-GMP (c-di-GMP) was originally described in 1987 as an allosteric regulator of cellulose synthesis in *Acetobacter xylinum* (now *Gluconacetobacter xylinus*) by the laboratory of Moshe Benziman (32). Subsequent work from the same laboratory showed that c-di-GMP synthesis and degradation in *G. xylinus* are controlled by enzymes containing two protein domains that were named GGDEF and EAL after conserved amino acid motifs (40). The enzymatic activity of the GGDEF domain as a diguanylate cyclase (DGC) catalyzing c-di-GMP synthesis and the EAL domain as a phosphodiesterase (PDE) catalyzing hydrolysis of c-di-GMP were determined by 2005 (7, 30, 36, 37, 41). In the same period, whole-genome sequencing revealed an abundance of GGDEF and EAL domain-containing proteins across the majority of bacterial species (both gram positive and gram negative), and functional analysis demonstrated that proteins with GGDEF and/or EAL domains regulate diverse functions in a wide range of organisms.

The possible occurrence of other classes of enzyme capable of c-di-GMP degradation was first indicated by comparative genome analysis (14, 15). Studies by Galperin and colleagues showed that a number of bacterial genomes encode proteins with the GGDEF domain but no EAL domain (14, 15). A role for the HD-GYP domain in c-di-GMP hydrolysis was then proposed based on examination of the distribution and numbers of GGDEF, EAL, and HD-GYP domains encoded by different bacterial genomes, coupled with the known activities of other members of the HD superfamily of enzymes as metal-dependent hydrolases. In addition, association of the HD-GYP domain with a CheY-like two-component receiver domain in many bacterial proteomes indicated a role in signaling (14, 15).

Experimental studies of HD-GYP domain proteins have focused largely on RpfG, a two-component regulatory protein from the plant pathogen *Xanthomonas campestris* pv. campestris *(Xcc)*. RpfG and cognate sensor RpfC act to positively control the synthesis of virulence factors and virulence to plants and negatively control biofilm formation (10, 39). Recent studies have demonstrated that the HD-GYP domain of RpfG is indeed active in c-di-GMP degradation (33) and have begun to elucidate the role of the signature and other conserved motifs in domain function. Here we review our current understanding of the HD-GYP domain in c-di-GMP signaling. In particular, we discuss (i) the identification of the HD-GYP domain and bioinformatic prediction as a novel c-di-GMP PDE, (ii) the experimental evidence implicating the HD-GYP domain in c-di-GMP degradation, (iii) the biological role of the HD-GYP domain protein RpfG in signal transduction in *Xcc*, (iv) regulatory interplay between RpfG and other c-di-GMP signaling proteins in *Xcc*, and (v) emerging information on the role of HD-GYP domain proteins in other bacteria.

BIOINFORMATIC ANALYSIS OF THE STRUCTURE, FUNCTION, AND DISTRIBUTION OF THE HD-GYP DOMAIN

The HD-GYP Domain Was Initially Implicated in c-di-GMP Hydrolysis by Bioinformatics

As outlined above, the proposition that the HD-GYP is a second PDE involved in c-di-GMP signaling came from bioinformatic studies. HD-GYP is a subgroup of the HD superfamily of metal-dependent

Robert P. Ryan, Yvonne McCarthy, and J. Maxwell Dow • BIOMERIT Research Centre, Department of Microbiology, BioSciences Institute, University College Cork, Cork, Ireland.

phosphohydrolases, named after the principal conserved residues implicated in metal binding and catalysis (3). The HD superfamily includes the bacterial nucleotide hydrolyzing enzymes ppGpp(p) hydrolase (SpoT) and dGTP triphosphohydrolase and the RNA-modifying enzyme tRNA nucleotidyl transferase. The HD residues are the presumed catalytic dyad and are known to be required for the multiple enzymatic activities of tRNA nucleotidyl transferase (46). Analysis of predicted proteomes from diverse bacteria indicated that a number of response regulator-like proteins had a version of the HD domain fused to a CheY domain. This subfamily had many additional highly conserved residues, including a conserved GYP motif, and was hence named HD-GYP (14, 15). The fusion with the CheY domain implicated the HD-GYP domain in signal transduction. A role for HD-GYP in c-di-GMP hydrolysis was proposed based on an examination of the distribution and numbers of GGDEF, EAL, and HD-GYP domains encoded by different bacterial genomes, where several genomes (e.g., those of *Thermotoga maritima* and *Treponema pallidum*) encode proteins with the GGDEF and HD-GYP domains but no EAL domain (14–16).

Conserved Residues within the HD-GYP Domain

A consensus sequence of the HD-GYP domain determined using 211 proteins is shown in Fig. 1. These alignments indicate that the signature HD and GYP motifs may be more usefully considered as part of the larger motifs HDXGK and HHEXXDGXGYP, respectively. In addition, there are a number of other well-conserved charged and hydrophobic residues. The proposed role of the HD dyad in enzymatic activity of the HD-GYP domain against c-di-GMP (3,

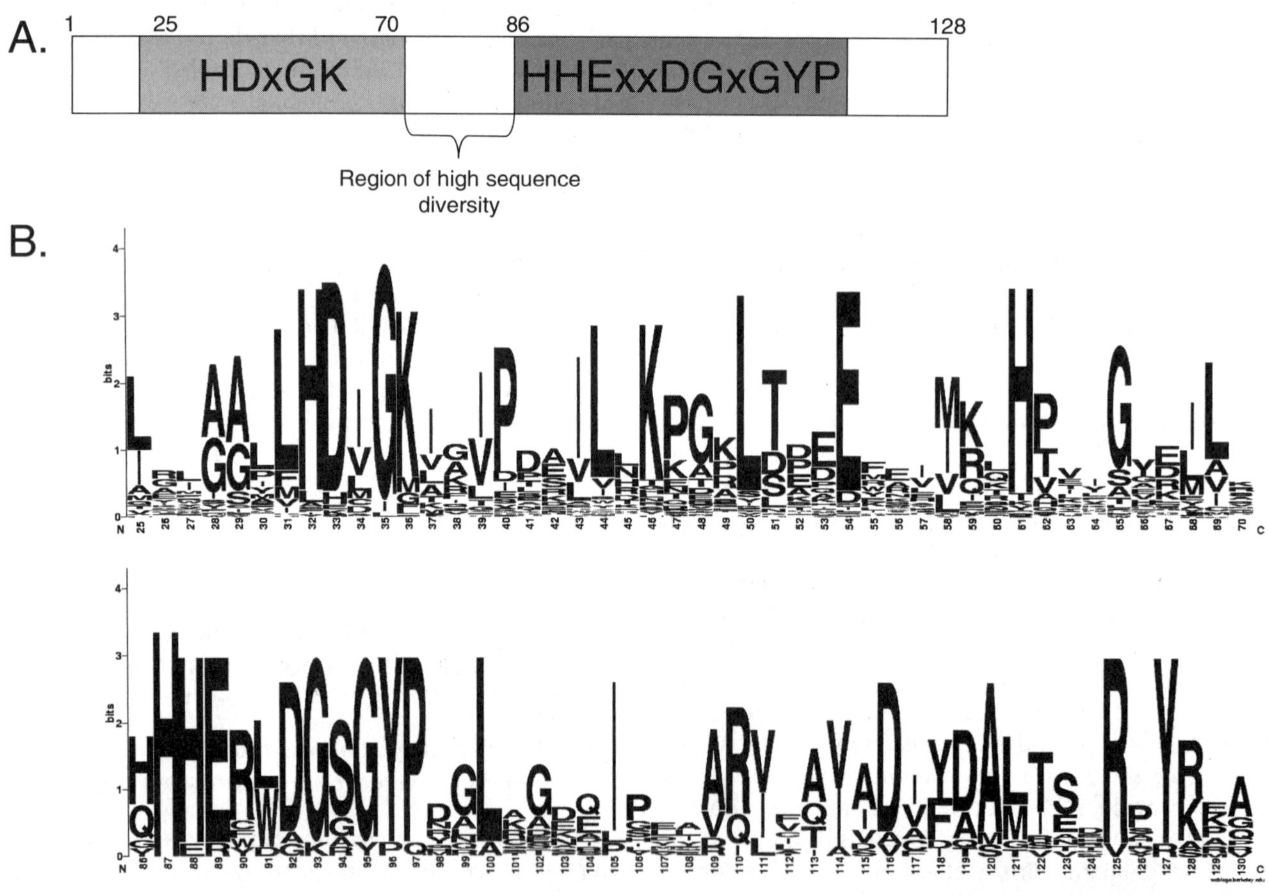

Figure 1. Conservation of amino acid residues and motifs within the HD-GYP domain. (A) Schematic representation of the HD-GYP domain indicating the separation of the regions containing the HD dyad and the GYP motif by a region of high sequence diversity. (B) Alignments of 211 HD-GYP domains encoded by complete microbial genomes (http://www.ncbi.nlm.nih.gov/Complete_Genomes/RRcensus.html) were used to establish conserved residues, and the output was drawn using the WebLogo program (http://weblogo.berkeley.edu). The letters in each position represent amino acid residues found in that position; the height of each letter reflects the fraction of sequences with the corresponding amino acid residue in that position. The residue numbering is from *Xanthomonas campestris* RpfG (XC_2335) as designated by the SMART nrdb database (http://smart.embl-heidelberg.de).

14) has now been confirmed experimentally through studies of the HD-GYP domain of RpfG from *Xcc* (33). These findings will be discussed in detail below. These H and D residues are not conserved in all HD-GYP domains, however. Variations include HT, HG, HN, HE, HQ, and YN, the last of which appears to be restricted to the pseudomonads. Recent work suggests that the YN-GYP variant domain found in PA2572 from *Pseudomonas aeruginosa* has no activity against c-di-GMP (35), but whether the other variants have enzymatic activity is not known. Enzymatically inactive HD-GYP domains could conceivably act in a regulatory capacity as seen or proposed for variant GGDEF and EAL domains (7, 37). We will return to this point later.

The role of other conserved residues in the HD-GYP domain is as yet unknown. It has been proposed that other conserved H and D residues (i.e., those not in the HD dyad) have a role in metal binding in these metal-dependent enzymes and that the GYP motif has a role in determining substrate specificity (14, 15), but currently, there is no experimental evidence. Interestingly, in a class of HD-GYP proteins in *Pseudomonas* spp., the two signature motifs are separated by ~230 amino acids, whereas in the canonical domain, this separation is 56 amino acids. Since the role of the GYP motif is as yet unknown, the significance of this difference is unclear.

Distribution of HD-GYP Domain Proteins in Bacterial Proteomes

GGDEF and EAL domains are highly abundant and widely distributed in bacteria, although they are not found in archaea (16). The HD-GYP domain is similarly distributed, although less abundant, with over 590 HD-GYP domains in over 140 genomes (http://www.ncbi.nlm.nih.gov/Complete_Genomes/SignalCensus.html). Most bacterial genomes encode 0 to 3 HD-GYP domain proteins, although they can be more highly represented. For example, HD-GYP domain proteins are absent from the proteomes of *Escherichia coli* and *Bacillus subtilis* but present in *Xylella fastidiosa* (1 protein), *Xanthomonas* spp. (3 proteins), different *Vibrio* spp. (5 to 13 proteins), *Shewenella* spp. (6 to 8 proteins), and various deltaproteobacteria (5 to 14 proteins). The domain is found not only in association with the CheY-like response receiver regulatory domain but also in association with other signaling domains such as HAMP and GAF, with uncharacterized protein domains and as a stand-alone domain (Fig. 2). The HAMP domain, which is found in a number of bacterial sensor and chemotaxis proteins and in eukaryotic histidine kinases, has an unspecified role in signal transduction, whereas the GAF domain has a role in the binding of cyclic nucleotides and other small molecules. The subgroup of HD-GYP domain proteins from *Pseudomonas* spp. in which the HD and GYP motifs are more widely separated has N-terminal HAMP and GAF domains.

In one of the three HD-GYP domain proteins found in *Xanthomonas* spp., the HD-GYP domain is in the central region and is flanked by uncharacterized domains (Fig. 2). This domain organization (including the same uncharacterized domains) is also found in proteins from unrelated bacteria such as *Agrobacterium tumefaciens* and *Ralstonia solanacearum* (11). The association of the HD-GYP domain with other signaling domains suggests that the activities of HD-GYP domain proteins are responsive to environmental cues, as is seen in proteins with GGDEF and/or EAL domains. The HD-GYP domain is also found in association with domains implicated in c-di-GMP signaling, including the PilZ domain which is involved in c-di-GMP binding (1; see also chapter 4) and the GGDEF domain. *Erwinia carotovora* susbsp. *atroseptica* has one HD-GYP domain protein (ECA 3548) in which the domain has the variant sequences HT-MGK and is attached to a PilZ domain. Fusions between the HD-GYP and GGDEF domains occur predominantly in *Clostridium* spp. (Fig. 2).

EXPERIMENTAL EVIDENCE OF THE FUNCTION OF THE HD-GYP DOMAIN AS A c-di-GMP PDE

Experimental evidence for the role of the HD-GYP domain in c-di-GMP degradation was provided through studies of the two-component regulator RpfG in the plant pathogen *Xcc*. RpfG, which has an HD-GYP domain fused to a CheY-like domain, acts to positively regulate synthesis of extracellular enzymes and extracellular polysaccharide (which are virulence factors), to positively regulate virulence, and to negatively regulate biofilm formation (10, 39) (Fig. 3). Indirect evidence for a role for RpfG in c-di-GMP signaling was provided by examination of the effects of expression of heterologous PDEs and a DGC in wild-type and *rpfG* mutant *Xcc* strains (33). Expression of genes encoding EAL domain proteins with established PDE activity from *Pseudomonas aeruginosa* in the *Xcc rpfG* mutant restored the synthesis of extracellular enzyme and extracellular polysaccharide (EPS) to near wild-type levels and blocked biofilm formation. In contrast, expression in wild-type *Xcc* of *wspR19*, which encodes a GGDEF domain protein with DGC activity, caused a reduc-

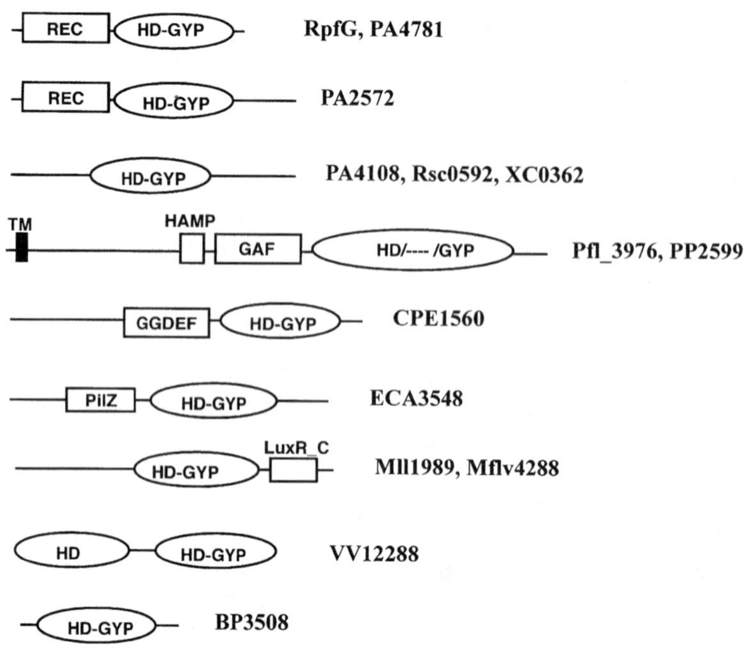

Figure 2. Domain architectures of proteins with an HD-GYP domain. The HD-GYP domain can occur singly but is found more often in association with other signaling and regulatory domains, commonly with REC, the receiver domain of two-component regulators. RpfG and PA4781, which are described in the text, have this domain organization. Some proteins with REC and HD-GYP domains such as PA2572 have an additional C-terminal domain that has not been characterized, but has a regulatory role. HD-GYP is also found in association with PilZ and GGDEF domains, with HD domains, with domains related to the C-terminal DNA binding domain of LuxR family regulators (LuxR_C), with HAMP and GAF domains, and with uncharacterized flanking domains. In some proteins, the HDX GK and HHEXXDGXGYP motifs are more widely separated than in the canonical sequence; this is indicated by HD—GYP. TM indicates the proposed transmembrane helix. Organism abbreviations: PA, *Pseudomonas aeruginosa;* XC, *Xanthomonas campestris;* Rsc, *Ralstonia solanacearum;* Pfl, *Pseudomonas fluorescens;* PP, *Pseudomonas putida;* CPE, *Clostridium perfringens;* ECA, *Erwinia carotovora;* Mll, *Mesorhizobium loti;* Mflv, *Mycobacterium gilvum;* VV, *Vibrio vulnificus;* BP, *Bordetella pertussis.*

tion in extracellular enzyme and EPS synthesis and promoted biofilm formation, thus giving a phenocopy of the *rpfG* mutant (33). Direct evidence for a role for RpfG in c-di-GMP hydrolysis was provided by studies of the isolated HD-GYP domain. Expression of the HD-GYP domain alone restored extracellular enzyme and EPS synthesis to the *rpfG* mutant, although expression of the CheY-like domain had no effect. Biochemical studies of the purified His_6-tagged protein demonstrated that the isolated HD-GYP domain could hydrolyze c-di-GMP to GMP. This reaction proceeded via an intermediate tentatively identified as the linear nucleotide pGpG and depended upon Mn^{2+} for which Mg^{2+} could not substitute. Mutation of the HD residues in the presumed catalytic dyad abolished both the enzymatic activity against c-di-GMP and the regulatory activity in controlling extracellular enzyme and EPS synthesis (33). Furthermore, the cellular levels of c-di-GMP were elevated in the *rpfG* mutant compared to the wild type (35), consistent with the role of RpfG as a PDE. Taken together, these findings establish the HD-GYP domain as a novel c-di-GMP PDE.

More recent work has extended these studies to provide evidence that the HD-GYP domain from other bacteria has PDE activity. Expression of isolated HD-GYP domains from PA4781 and PA4108 of *Pseudomonas aeruginosa* in the *rpfG* mutant of *Xcc* is able to restore the production of extracellular enzymes to wild-type levels (35). These effects are associated with a reduction of the (elevated) level of c-di-GMP seen in the *rpfG* mutant. Furthermore, mutation of PA4108 and PA4781 in *P. aeruginosa* results in an increase in c-di-GMP level. A partial restoration of extracellular enzyme synthesis in the *rpfG* strain of *Xcc* is also seen upon expression of the isolated HD-GYP domains of proteins from diverse organisms such as *Treponema denticola* (TDE2659), *Vibrio cholerae* (VC2497), and *Clostridium perfringens* (CPE1560) (R. P. Ryan, Y. McCarthy, and J. M. Dow, unpublished data), although in these cases, links to c-di-GMP levels were not examined. Overexpression of another *V. cholerae* protein, VCA0681, reduces c-di-GMP levels in *V. cholerae* (17). These findings indicate that the HD-GYP domain from a range of bacteria has PDE activity, although at the

Figure 3. The HD-GYP domain protein RpfG controls multiple functions in *Xanthomonas campestris*. Mutation of *rpfG* leads to reduced virulence in plants (top left), reduced production of extracellular enzymes such as protease (top right), reduced motility (bottom right), and increased formation of aggregates in certain media (bottom left).

time of this writing, only the HD-GYP domain of RpfG has been studied biochemically.

THE HD-GYP DOMAIN REGULATOR RpfG OF *Xcc* LINKS CELL-CELL SIGNALING TO c-di-GMP TURNOVER

RpfG acts together with the sensor histidine kinase RpfC to positively regulate virulence and the synthesis of extracellular enzymes and to negatively regulate biofilm formation in *Xcc* (9, 10, 19, 33). RpfC is a complex sensor kinase comprising a membrane-associated sensory input domain and a histidine kinase domain fused to a CheY-like receiver domain and a C-terminal histidine phosphotransfer domain. The *rpfG* and *rpfC* genes are transcribed as the *rpfGHC* operon, although no function has yet been ascribed to RpfH, which is related in amino acid sequence to the input domain of RpfC. Extracellular enzyme synthesis and biofilm dispersal in *Xcc* are also regulated by cell-cell signaling, mediated by the diffusible signal molecule DSF (4, 33). DSF has been characterized as the unsaturated fatty acid *cis*-11-methyl-2-dodecenoic acid (44). The synthesis of the DSF signal is partially dependent on RpfB, a long-chain fatty acyl coenzyme A ligase, and fully dependent on RpfF, a protein with some similarity to enoyl coenzyme A hydratase (4). The *rpfB* and *rpfF* genes are contiguous and adjacent to the *rpfGHC* operon, which is convergently transcribed (4, 39).

The addition of DSF can restore extracellular enzyme synthesis to *rpfF* mutants but not to strains with mutations in *rpfG* or *rpfC*. Mutations of *rpfC, rpfG,* or *rpfF* all lead to the formation of biofilms by *Xcc* in certain media in which the wild type grows in a dispersed planktonic fashion (9, 10). The addition of DSF causes dispersal of biofilms produced by the *rpfF* mutant but not by the *rpfG* or *rpfC* mutants. These findings are consistent with a role for RpfC/RpfG in perception and transduction of the DSF signal (for recent reviews, see references 12 and 20).

Further support for this model has come from experiments in which the RpfC/RpfG two-component system has been reconstructed in *Pseudomonas aeruginosa* and shown to confer responsiveness to exogenously added DSF, as seen through effects on swarming motility (33). It has been proposed that recognition of DSF by RpfC leads to autophosphorylation of RpfC, followed by phosphotransfer to RpfG, which activates RpfG for c-di-GMP hydrolysis (33) (Fig. 4). Accordingly, the level of c-di-GMP in an *rpfF* mutant is higher than in the wild type but is reduced upon addition of DSF (Ryan et al., unpublished). In addition, mutation of those H and D residues of RpfC predicted to be involved in phosphorelay abrogates synthesis of extracellular enzymes even in the presence of exogenous DSF (19). Expression of either the isolated HD-GYP domain or full-length RpfG can restore enzyme synthesis to the *rpfG* mutant. However, only the HD-GYP domain can restore enzyme synthesis to the *rpfGHC* mutant, which lacks the RpfC sensor component. These findings are consistent with the model that, in the absence of phosphorylation, the CheY-like receiver domain acts to inhibit the PDE activity of RpfG; conformational changes that occur upon phosphorylation relieve this effect. However, the influence of phosphorylation of RpfG on the enzymatic activity against c-di-GMP has not been specifically examined.

The phenotypes of *rpf* mutants indicate that c-di-GMP regulates a wide range of functions in *Xcc,* as has been shown in other bacteria. High levels of c-di-GMP promote biofilm formation but inhibit the synthesis of virulence determinants in *Xcc*. This is consistent with a theme emerging from a number of studies with a range of human and animal pathogens that high levels of c-di-GMP promote biofilm formation and sessility but low levels promote motility and the synthesis of virulence factors (25, 31, 38). However, the regulatory influence of c-di-GMP in

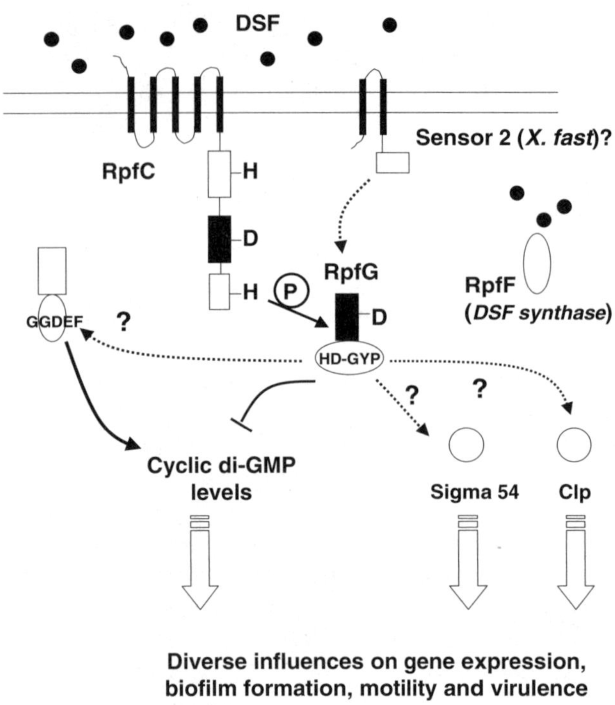

Figure 4. Model for role of the HD-GYP domain regulator RpfG in transduction of the DSF cell-cell signal in xanthomonads. The synthesis of the DSF signal requires RpfF, whereas RpfC is a complex sensor kinase responsible for DSF perception. Structural changes that occur upon DSF binding to the RpfC sensory input domain trigger autophosphorylation of RpfC at a histidine (H) residue and phosphorelay to RpfG (solid arrow) via aspartate (D) and histidine (H) residues of the receiver domain (black) and histidine phosphotransfer domain (white) of RpfC, respectively. Phosphorylation of RpfG activates c-di-GMP degradation by the HD-GYP domain. RpfG may also physically interact with GGDEF domain proteins to influence c-di-GMP concentrations and with other regulators, modulating their activity in transcription (dotted lines). In *Xcc,* RpfG influences expression of the CRP-like protein Clp (19) by an unknown mechanism, leading to activation of a downstream signaling cascade involving other transcriptional regulators, Zur and FhrR (not shown). In *Xylella fastidiosa (X. fast),* a second sensor may phosphorylate RpfG in response to DSF. Although shown as a membrane-bound protein, this sensor may equally be located in the cytoplasm.

Xcc is probably not restricted to these phenotypes. Transcriptome profiling of wild-type *Xcc* and an *rpfF* strain with and without exogenous DSF has revealed that DSF signaling has a widespread influence on *Xcc* gene expression, including effects on genes implicated in multidrug resistance, motility, chemotaxis, the tricarboxylic acid cycle, and iron uptake as well as the effects on genes encoding extracellular enzymes and the synthesis of EPS that had been established previously (18). It is not clear that all of these effects of DSF are exerted through RpfG and alterations of c-di-GMP levels, however, since a comparative transcriptome analysis of the role of RpfG has yet to be reported. Furthermore, the possibility that other perceptions, system(s), or signal transduction pathways for DSF occur in *Xcc* cannot currently be excluded (see below).

The Rpf/DSF System Regulates Virulence in Diverse Xanthomonads

Genome sequencing has revealed that the Rpf/DSF regulatory system occurs in other *Xanthomonas* spp., in the related plant pathogen *Xylella fastidiosa,* and in the nosocomial pathogen *Stenotrophomonas maltophilia.* A role for DSF signaling in virulence has now been described for a number of these xanthomonads (5, 6, 13, 29, 42). By extension of the work in *Xcc,* DSF perception through RpfC may be linked to the degradation of c-di-GMP by RpfG in these related bacteria. Differences between organisms in the regulatory functions of the Rpf/DSF system are apparent. These may be explained in part by different regulatory actions of c-di-GMP in different bacteria. In addition, it should be pointed out that both the number of c-di-GMP signaling systems and the nature of the proteins involved differ significantly between different xanthomonads. As some of these other signaling systems may potentially be involved in regulation of the same phenotypes as the Rpf/DSF system, the importance of the Rpf/DSF system to the regulation of specific phenotypes may differ between organisms.

The above considerations do not, however, explain the marked contrast seen between the positive regulatory influence of DSF on virulence seen in most bacteria and the negative influence seen in *Xylella fastidiosa,* the causal agent of Pierce's disease of grapes. In *Xylella fastidiosa,* mutation of *rpfF* and *rpfC* has opposite effects on virulence and on expression of genes encoding adhesins or involved in biofilm formation but similar effects on expression of a subset of virulence genes including *pglA,* which encodes an extracellular polygalacturonase (6). To explain these findings, Chatterjee and colleagues propose the existence of two distinct pathways for DSF perception in *Xylella fastidiosa* (6). In the first pathway, in which RpfC acts as the sensor, DSF negatively regulates expression of *pglA* independently of RpfG and activates genes involved in biofilm formation through RpfG. The second proposed pathway involves a separate sensor and RpfG to positively regulate expression of genes linked to biofilm formation (6, 12) (Fig. 4). The model proposes some specific actions of RpfG that can be tested experimentally by analysis of the phenotypes of an *rpfG* strain and related double mutants.

REGULATORY INTERPLAY BETWEEN RpfC/RpfG AND OTHER c-di-GMP SIGNALING SYSTEMS IN *Xcc*

The number of proteins with a potential role in c-di-GMP signaling encoded by bacterial genomes increases with genome size but in a nonlinear fashion (15). For example, the genome of *Xylella fastidiosa* (~2.7 Mb) encodes only 5 proteins with a potential role in c-di-GMP signaling, whereas the genome of *Xcc* at ~5 Mb encodes 37 proteins. These large numbers in the larger genomes indicate that there must be considerable complexity in the organization of c-di-GMP signaling within a single organism, which is poorly understood.

One model of c-di-GMP-dependent regulation is that specific signaling systems modulate the levels of c-di-GMP in a highly localized fashion, thus affecting the activity of colocalizing proteins. Regulatory influences exerted via effects on discrete pools of c-di-GMP would allow the cell to have distinct responses to different environmental cues. The specific subcellular localization, such as that of TipF and PleD in *Caulobacter crescentus* (23, 30) and FimX in *Pseudomonas aeruginosa* (22, 27), may be suggestive of the occurrence of such dedicated, localized, signaling systems. Two analyses of the role of c-di-GMP signaling proteins in biofilm formation in *Pseudomonas aeruginosa* indicate apparently anomalous roles of particular signaling systems; biofilm formation is enhanced by mutation of certain genes encoding GGDEF domain DGCs and prevented by mutation of other genes encoding EAL domain PDEs (21, 28). These effects have been ascribed to highly localized effects on discrete pools of c-di-GMP.

In contrast to effects associated with individual signaling proteins, a number of systems may impinge on the same pool of c-di-GMP. This nodal arrangement would allow integration of information about different aspects of the environment with consequent effects on specific bacterial functions. It has been proposed that both specific and nodal regulation of c-di-GMP occur in the same cell. For example, studies of *Salmonella* have revealed a hierarchical arrangement within a c-di-GMP signaling network that acts in the regulation of cellulose synthesis (25). Two GGDEF-EAL domain proteins (STM3388 and STM2123) additively contribute to the expression of the transcriptional regulator CsgD. This protein regulates transcription of *adrA,* which encodes a GGDEF domain protein with DGC activity that is directly implicated in the regulation of cellulose synthesis but not in the expression of *csgD* (26).

The genome of *Xcc* encodes 34 proteins with GGDEF and/or EAL domains and 3 HD-GYP domain proteins, including RpfG. The functional organization of the RpfG/RpfC signaling system with respect to the other c-di-GMP signaling systems was addressed through a comprehensive mutational analysis of the role of all 37 proteins in RpfG-regulated phenotypes. The regulation of virulence to plants, the production of extracellular enzymes under a range of growth conditions in vitro, motility, and biofilm formation were examined in each case (34).

Genes with significant roles in virulence to plants included those encoding proteins whose probable function is as DGCs as well as others (including RpfG) implicated as PDEs. Furthermore, RpfG positively controlled the expression of a subset of these genes, although it is not known if this is a direct or indirect effect. A partially overlapping set of elements controlled the production of extracellular enzymes in vitro, although only RpfG had a role under all conditions tested. Other signaling proteins implicated in virulence control the synthesis of extracellular enzymes in minimal medium with glutamine as a sole carbon source (XC1582 and XC1841) or under conditions of low oxygen tension (XC2324). All three of these proteins have both GGDEF and EAL domains, although single-amino-acid variations in the signature motifs are seen in XC1582 (GSDEF-EVL), XC1841 (GGDEF-EVL), and XC2324 (AGDEF-EAL). The conditions under which the regulatory effects of these proteins are seen may be significant for a vascular pathogen such as *Xcc.* Glutamine is the most abundant amino acid found in the guttation fluid of *Arabidopsis thaliana,* which is a host for *Xcc.* Both the presence of glutamine and low oxygen tension are therefore environmental conditions likely to be encountered by the pathogen that is largely restricted to plant xylem elements during disease. The molecular basis of the effects of amino acid cues or low oxygen tension on the synthesis of extracellular enzymes and the signals or cues recognized by the other virulence-related systems remain obscure. Other GGDEF-EAL domain proteins influenced the motility of *Xcc* but had no apparent effect on extracellular enzyme synthesis. Specifically, loss of XC2161 (APMRF-EAL) and XC2226 (SDDEF-EAL) caused reduced and increased motility, respectively, compared to the wild type. Mutation of *rpfG* also led to reduced motility. However, with the exception of RpfG, none of the signaling proteins affecting enzyme synthesis had an effect on motility.

These findings are largely consistent with the notion that a number of c-di-GMP signaling systems in *Xcc* are organized in a network that acts to integrate information from cell-cell signaling with other environmental inputs to modulate extracellular enzyme synthesis, whereas other signaling systems are dedicated to specific cellular tasks, such as control of motility. There are, however, several observations that indicate a more complex arrangement of systems impinging on the same overall process. For example, mutation of genes encoding certain GGDEF-EAL, EAL, and GGDEF domain proteins all had an effect on extracellular enzyme levels; there was no simple relationship between the predicted activities against c-di-GMP and the effect of mutation of the cognate gene on extracellular enzyme synthesis. This is difficult to reconcile with the concept that all components that affect enzyme synthesis influence a single pool of c-di-GMP or a single process.

The mutational analysis also indicated that RpfG is unique in its ability to influence both motility and extracellular enzyme synthesis. A trivial explanation is that mutation of *rpfG* leads to such a strong increase in the level of c-di-GMP that it swamps several different pools of the nucleotide in the cell. Another possibility is that activated RpfG is able to localize to a range of cellular sites, thus influencing a number of pools of c-di-GMP within the cell. In this context, recent findings from yeast two-hybrid analysis that reveal a physical interaction between the HD-GYP domain of RpfG and a subset of GGDEF domain proteins are particularly intriguing (2). Although the biological relevance of such interactions has yet to be tested, these observations suggest a scenario whereby the HD-GYP domain may modulate the activity of GGDEF domain proteins that are not associated with an EAL domain (Fig. 4). These interactions between HD-GYP and GGDEF domains were revealed using proteins from *Xanthomonas axonopodis* pv. citri, which is related to *Xcc* and has all but one of the c-di-GMP signaling proteins found in *Xcc.* The mutational analysis of *Xcc* did not, however, reveal any correlation between the effects of mutation of individual genes encoding GGDEF targets and mutation of *rpfG,* although the effects of multiple mutations have not been examined (34).

Overall, these observations raise a number of questions about the detailed interactions and regulatory circuits within c-di-GMP signaling in *Xanthomonas* spp. that remain to be resolved. Can evidence of the occurrence of interdomain interactions within bacteria be obtained? What are the detailed regulatory roles of the target GGDEF domain proteins? Do all three HD-GYP domain proteins in *Xcc* interact with the same subset of GGDEF domain proteins, or do these interactions exhibit some specificity? What residues in GGDEF and HD-GYP domains determine the protein-protein interaction, and what consequences does interaction have for the enzymatic activities of the respective domains?

Does RpfG Exert Regulatory Effects That Are Independent of c-di-GMP?

The yeast two-hybrid analysis discussed above also revealed interactions of the HD-GYP domain of RpfG of *X. axonopodis* pv. citri with the sigma factor σ^{54}, NtrBC (both of which are involved in nitrogen-regulated transcription) and other regulatory proteins (2). It is not known whether such interactions occur in *Xcc* (rather than in *Saccharomyces cerevisiae*) and if they play a regulatory role. A precedent for such a regulatory mechanism involving a c-di-GMP signaling protein comes from studies of the EAL domain protein YcgF of *Escherichia coli* (43). YcgF acts as a direct antirepressor of a blue-light-induced stress response, although it does not have PDE activity and does not bind c-di-GMP (43).

ROLE OF HD-GYP DOMAIN PROTEINS IN OTHER BACTERIA

Studies of the biological role of HD-GYP domain proteins have been largely restricted to work on *Xcc* and related organisms. However, very recent work has addressed the role of these proteins in the unrelated human pathogens *Pseudomonas aeruginosa* and *Vibrio cholerae* (17, 35). The genome of the human pathogen *P. aeruginosa* PAO1 encodes two proteins (PA4108 and PA4781) with an HD-GYP domain and a third protein, PA2572, which contains a variant domain with the sequence YNIGK in place of HDMGK seen in RpfG. Heterologous expression of the isolated HD-GYP domains of PA4108 and PA4781 in the *rpfG* mutant of *Xcc* reduced c-di-GMP levels (35). Furthermore, mutation of *PA4108* and *PA4781* led to an increase in the level of c-di-GMP in *P. aeruginosa.* These findings are consistent with the predicted PDE activity of the encoded proteins. In contrast, expression of the variant HD-GYP domain of PA2572 in *Xcc* or mutation of *PA2572* in *P. aeruginosa* had no effect on c-di-GMP levels, indicating that the YN-GYP variant domain was enzymatically inactive.

Mutation of *PA4108, PA4781,* and *PA2572* had distinct effects on biofilm formation and the architecture of *P. aeruginosa.* In addition, all three proteins contributed to the virulence of *P. aeruginosa* to larvae of the greater wax moth *Galleria mellonella,* although they had differing effects on production of *P. aeruginosa* virulence factors and on swarming motility. Mutation of either *PA4108* or *PA4781* led to a reduction in swarming, whereas mutation of *PA2572* had no effect. However, PA2572 had a dominant negative influence on swarming that was cryptic and was revealed only after removal of an uncharacterized C-terminal domain (35). Complementation of the *PA4108* or *PA4781* mutant with the cognate cloned wild-type gene restored motility and biofilm formation to the wild type. Mutated alleles expressing variant proteins with alanine substitution in the HD dyad were, however, unable to complement the strains (Ryan et al., unpublished), indicating that the regulatory effects of PA4108 and PA4781 on motility and biofilm formation required their enzymatic activity against c-di-GMP. The effects on virulence are interesting in light of previous work that showed that expression of all three genes is substantially increased during coculture of *P. aeruginosa* with human epithelial cells (8, 24) and that *PA2572* and *PA4781* are induced by mucopurulent fluid from cystic fibrosis patients (45).

These findings illustrate two aspects of HD-GYP domain proteins that are consonant with our understanding of the diverse roles of GGDEF and EAL domain proteins. The first is that different HD-GYP domain proteins appear to have distinct regulatory roles. The second is that although domains with sequences that diverge from the consensus may have lost enzymatic activity, they may still play a regulatory role. For example, the GGDEF-EAL domain protein (CC3396) from *Caulobacter crescentus* has an altered GGDEF domain (actually GEDEF) that lacks the ability to synthesize c-di-GMP. However, this domain is still able to bind GTP, leading to activation of the attached EAL domain in c-di-GMP hydrolysis (7). As well as intramolecular effects, enzymatically inactive domains could regulate cellular processes through intermolecular interactions with other proteins. As mentioned above, the EAL domain protein YcgF of *Escherichia coli* acts as a direct antirepressor of a blue-light-induced stress response but does not have PDE activity or bind c-di-GMP (43). How the regulatory action of PA2572 is achieved is as yet unknown. By analogy with work on the protein-protein interactions of the HD-GYP domain of RpfG and the role of inactive domains discussed above, plausible mechanisms include interaction with GGDEF domain proteins to modulate their activity in c-di-GMP synthesis or interactions with other regulatory proteins.

Work with *Vibrio cholerae* has addressed the role that HD-GYP domain proteins may have in signal transduction mechanisms associated with quorum sensing (QS) in the El Tor biotype (17). Using *lux* fusions, a number of genes whose expression is upregulated either in response to exogenous autoinducers or in QS mutants that are locked in a high-cell-density mode were identified. It was shown that expression of four genes (*vca0210, vca0681, vca0931,* and *vc2340*) that encode HD-GYP domain proteins in *V. cholerae* El Tor was activated by QS. Each of the encoded proteins has the conserved HD and GYP motifs as well as several other conserved residues, suggesting that they are active as PDEs. In *V. cholerae* El Tor, QS negatively regulates the production of exopolysaccharide and biofilm formation. Accordingly, overexpression of VCA0681 in *V. cholerae* El Tor led to reduced cellular levels of c-di-GMP, decreased expression of the *vps* gene cluster that directs EPS synthesis, and reduced biofilm formation (17). These regulatory effects are associated with the PDE activity of VCA0681, since a variant with a D-to-A substitution within the HD catalytic dyad had little effect on c-di-GMP levels and did not influence either *vps* expression or biofilm formation. In contrast to overexpression, mutation of individual genes encoding HD-GYP domain proteins did not lead to any alteration in *vps* expression or biofilm phenotype, suggesting either a possible redundancy in the action of these proteins (17) or that the natural targets of regulation are distinct from those involved in biofilm formation. Thus far, no investigations have been made of the role of HD-GYP domain proteins in gram-positive bacteria and spirochetes or in organisms such as *Thermotoga maritima* that have GGDEF domain proteins but lack EAL domain proteins. The work on HD-GYP domain proteins is currently limited to studies of a handful of proteins from a few bacteria. Nevertheless, the available data suggest that these proteins will prove to have as diverse a range of cellular functions as is seen with GGDEF and EAL domain proteins.

WHY HAVE A SECOND c-di-GMP PDE?

The occurrence of a second PDE unrelated to the EAL domain immediately raises the question of why such an alternative activity arose. Differences between the activity of the EAL and HD-GYP domains are seen in the relative activities against pGpG and in requirements for divalent metal ions. Biochemical studies of intact EAL domain proteins as well as isolated EAL domains have shown that, in the majority of cases, the major product of the enzymatic action on c-di-GMP is the linear nucleotide pGpG. This product is converted to GMP at a much slower rate. This activity of the EAL domain is absolutely dependent on the presence of Mg^{2+} or Mn^{2+} (37). In contrast, enzyme assays using the HD-GYP domain of RpfG reveal only a transient appearance of pGpG, and the major product is GMP (33). Furthermore, the enzymatic activity of the HD-GYP domain depends upon Mn^{2+} and is not supported by Mg^{2+}. The significance, if any, of these differences is unclear. It has been suggested that pGpG may have a regulatory role in c-di-GMP signaling or may itself act as a signal molecule (31). In this case, the HD-GYP domain could serve to regulate pGpG levels, although this task may be performed by other, nonspecific nucleases. The physical interaction between the HD-GYP domain of RpfG and a subset of GGDEF domain proteins revealed by yeast two-hybrid analysis (2) may suggest that the HD-GYP domain evolved to modulate the activity of GGDEF domain proteins that are not associated with an EAL domain. This may allow additional regulation of specific cellular processes in response to environmental cues.

CONCLUDING REMARKS

Although c-di-GMP was discovered in 1987, the appreciation of the importance of the molecule in bacterial signaling has occurred only relatively recently. This understanding has been strongly promoted by the increased interest in microbial biofilm formation and by microbial genome sequencing. These studies demonstrated functions for a number of proteins with GGDEF and/or EAL domains and indicated the wide distribution of these domains in the bacterial kingdom. It is now known that c-di-GMP acts to regulate diverse bacterial properties that include not only aggregative behavior, adhesion, and biofilm formation but also developmental transitions and the virulence of human and plant pathogens. Although identification of the role of the HD-GYP domain in c-di-GMP degradation increases our understanding of c-di-GMP signaling, our knowledge of this domain is still very limited. A systems biology approach, integrating information from biochemistry and genetics with cell biology and physiology, may provide answers to a number of the following questions within the next few years. What is the role of GYP and other conserved motifs and residues? Does the HD-GYP domain have other cellular functions, for example, in protein-protein interactions with GGDEF domain proteins or regulators? Do different HD-GYP domain proteins interact with different GGDEF domain proteins or show specific cellular localizations that depend upon activation? What are the cues that activate or inactivate the signaling proteins?

Acknowledgments. We acknowledge support from the Science Foundation of Ireland through Principal Investigator Awards (03/IN3/B373 and 07/IN.1/B955 to J.M.D.), from Enterprise Ireland (IC/2006/61), and from EMBO for a short-term fellowship to R.P.R.

REFERENCES

1. **Amikam, D., and M. Y. Galperin.** 2006. PilZ is part of the bacterial c-di-GMP binding protein. *Bioinformatics* **22:**3–6.
2. **Andrade, M. O., M. C. Alegria, C. R. Guzzo, C. Docena, M. C. Rosa, C. H. Ramos, and C. S. Farah.** 2006. The HD-GYP domain of RpfG mediates a direct linkage between the Rpf quorum-sensing pathway and a subset of diguanylate cyclase proteins in the phytopathogen *Xanthomonas axonopodis* pv. *citri. Mol. Microbiol.* **62:**537–551.
3. **Aravind, L., and E. V. Koonin.** 1998. The HD domain defines a new superfamily of metal-dependent phosphohydrolases. *Trends Biochem. Sci.* **23:**469–472.
4. **Barber, C. E., J. L. Tang, J. X. Feng, M. Q. Pan, T. J. Wilson, H. Slater, J. M. Dow, P. Williams, and M. J. Daniels.** 1997. A novel regulatory system required for pathogenicity of *Xanthomonas campestris* is mediated by a small diffusible signal molecule. *Mol. Microbiol.* **24:**555–566.
5. **Chatterjee, S., and R. V. Sonti.** 2002. *rpfF* mutants of *Xanthomonas oryzae* pv. *oryzae* are deficient for virulence and growth under low iron conditions. *Mol. Plant-Microbe Interact.* **15:**463–471.
6. **Chatterjee, S., C. Wistrom, and S. E. Lindow.** 2008. A cell-cell signaling sensor is required for virulence and insect transmission of *Xylella fastidiosa. Proc. Natl. Acad. Sci. USA* **105:** 2670–2675.
7. **Christen, M., B. Christen, M. Folcher, A. Schauerte, and U. Jenal.** 2005. Identification and characterization of a cyclic di-GMP-specific phosphodiesterase and its allosteric control by GTP. *J. Biol. Chem.* **280:**30829–30837.
8. **Chugani, S., and E. P. Greenberg.** 2007. The influence of human respiratory epithelia on *Pseudomonas aeruginosa* gene expression. *Microb. Pathog.* **42:**29–35.
9. **Crossman, L., and J. M. Dow.** 2004. Biofilm formation and dispersal in *Xanthomonas campestris. Microbes Infect.* **6:**623–629.
10. **Dow, J. M., L. Crossman, K. Findlay, Y. Q. He, J. X. Feng, and J. L. Tang.** 2003. Biofilm dispersal in *Xanthomonas campestris* is controlled by cell-cell signaling and is required for full virulence to plants. *Proc. Natl. Acad. Sci. USA* **100:**10995–11000.
11. **Dow, J. M., Y. Fouhy, J. F. Lucey, and R. P. Ryan.** 2006. The HD-GYP domain, cyclic di-GMP signaling, and bacterial virulence to plants. *Mol. Plant-Microbe Interact.* **19:**1378–1384.
12. **Dow, M.** 2008. Diversification of the function of cell-to-cell signaling in regulation of virulence within plant pathogenic xanthomonads. *Sci. Signal.* **1:**pe23.
13. **Fouhy, Y., K. Scanlon, K. Schouest, C. Spillane, L. Crossman, M. B. Avison, R. P. Ryan, and J. M. Dow.** 2007. Diffusible signal factor-dependent cell-cell signaling and virulence in the nosocomial pathogen *Stenotrophomonas maltophilia. J. Bacteriol.* 189:4964–4968.
14. **Galperin, M. Y., D. A. Natale, L. Aravind, and E. V. Koonin.** 1999. A specialized version of the HD hydrolase domain implicated in signal transduction. *J. Mol. Microbiol. Biotechnol.* **1:**303–305.
15. **Galperin, M. Y., A. N. Nikolskaya, and E. V. Koonin.** 2001. Novel domains of the prokaryotic two-component signal transduction systems. *FEMS Microbiol. Lett.* **203:**11–21.
16. **Galperin, M. Y.** 2005. A census of membrane-bound and intracellular signal transduction proteins in bacteria: bacterial IQ, extroverts and introverts. *BMC Microbiol.* **5:**35.
17. **Hammer, B. K., and B. L. Bassler.** 2008. Distinct sensory pathways in *Vibrio cholerae* El Tor and classical biotypes mod-

ulate cyclic dimeric GMP levels to control biofilm formation. *J. Bacteriol.* **191**:169–177.
18. **He, Y. W., M. Xu, K. Lin, Y. J. Ng, C. M. Wen, L. H. Wang, Z. D. Liu, H. B. Zhang, Y. H. Dong, J. M. Dow, and L. H. Zhang.** 2006. Genome scale analysis of diffusible signal factor regulon in *Xanthomonas campestris* pv. *campestris:* identification of novel cell-cell communication-dependent genes and functions. *Mol. Microbiol.* **59**:610–622.
19. **He, Y. W., C. Wang, L. Zhou, H. Song, J. M. Dow, and L. H. Zhang.** 2006. Dual signaling functions of the hybrid sensor kinase RpfC of *Xanthomonas campestris* involve either phosphorelay or receiver domain-protein interaction. *J. Biol. Chem.* **281**:33414–33421.
20. **He, Y. W., and L. H. Zhang.** 2008. Quorum sensing and virulence regulation in *Xanthomonas campestris. FEMS Microbiol. Rev.* **32**:842–857.
21. **Hoffman, L. R., D. A. D'Argenio, M. J. MacCoss, Z. Zhang, R. A. Jones, and S. I. Miller.** 2005. Aminoglycoside antibiotics induce bacterial biofilm formation. *Nature* **436**:1171–1175.
22. **Huang, B., C. B. Whitchurch, and J. S. Mattick.** 2003. FimX, a multidomain protein connecting environmental signals to twitching motility in *Pseudomonas aeruginosa. J. Bacteriol.* **185**:7068–7076.
23. **Huitema, E., S. Pritchard, D. Matteson, S. K. Radhakrishnan, and P. H. Viollier.** 2006. Bacterial birth scar proteins mark future flagellum assembly site. *Cell* **124**:1025–1037.
24. **Ichikawa, J. K., A. Norris, M. G. Bangera, G. K. Geiss, A. B. van 't Wout, R. E. Bumgarner, and S. Lory.** 2000. Interaction of *Pseudomonas aeruginosa* with epithelial cells: identification of differentially regulated genes by expression microarray analysis of human cDNAs. *Proc. Natl. Acad. Sci. USA* **97**: 9659–9664.
25. **Jenal, U., and J. Malone.** 2006. Mechanisms of cyclic-di-GMP signaling in bacteria. *Annu. Rev. Genet.* **40**:385–407.
26. **Kader, A., R. Simm, U. Gerstel, M. Morr, and U. Römling.** 2006. Hierarchical involvement of various GGDEF domain proteins in rdar morphotype development of *Salmonella enterica* serovar Typhimurium. *Mol. Microbiol.* **60**:602–616.
27. **Kazmierczak, B. I., M. B. Lebron, and T. S. Murray.** 2006. Analysis of FimX, a phosphodiesterase that governs twitching motility in *Pseudomonas aeruginosa. Mol. Microbiol.* **60**:1026–1043.
28. **Kulesekara, H., V. Lee, A. Brencic, N. Liberati, J. Urbach, S. Miyata, D. G. Lee, A. N. Neely, M. Hyodo, Y. Hayakawa, F. M. Ausubel, and S. Lory.** 2006. Analysis of *Pseudomonas aeruginosa* diguanylate cyclases and phosphodiesterases reveals a role for bis-(3′-5′)-cyclic-GMP in virulence. *Proc. Natl. Acad. Sci. USA* **103**:2839–2844.
29. **Newman, K. L., R. P. Almeida, A. H. Purcell, and S. E. Lindow.** 2004. Cell-cell signaling controls *Xylella fastidiosa* interactions with both insects and plants. *Proc. Natl. Acad. Sci. USA* **101**:1737–1742.
30. **Paul, R., S. Weiser, N. C. Amiot, C. Chan, T. Schirmer, B. Giese, and U. Jenal.** 2004. Cell cycle-dependent dynamic localization of a bacterial response regulator with a novel diguanylate cyclase output domain. *Genes Dev.* **18**:715–727.
31. **Römling, U., M. Gomelsky, and M. Y. Galperin.** 2005. C-di-GMP: the dawning of a novel bacterial signalling system. *Mol. Microbiol.* **57**:629–639.
32. **Ross, P., H. Weinhouse, Y. Aloni, D. Michaeli, P. Weinberger-Ohana, R. Mayer, S. Braun, E. de Vroom, G. A. van der Marel, J. H. van Boom, and M. Benziman.** 1987. Regulation of cellulose synthesis in *Acetobacter xylinum* by cyclic diguanylate. *Nature* **325**:279–281.
33. **Ryan, R. P., Y. Fouhy, J. F. Lucey, L. C. Crossman, S. Spiro, Y. W. He, L. H. Zhang, S. Heeb, M. Camara, P. Williams, and J. M. Dow.** 2006. Cell-cell signaling in *Xanthomonas campestris* involves an HD-GYP domain protein that functions in cyclic di-GMP turnover. *Proc. Natl. Acad. Sci. USA* **103**:6712–6717.
34. **Ryan, R. P., Y. Fouhy, J. F. Lucey, B. L. Jiang, Y. Q. He, J. X. Feng, J. L. Tang, and J. M. Dow.** 2007. Cyclic di-GMP signalling in the virulence and environmental adaptation of *Xanthomonas campestris. Mol. Microbiol.* **63**:429–442.
35. **Ryan, R. P., J. Lucey, K. O'Donovan, Y. McCarthy, L. Yang, T. Tolker-Nielsen, and J. M. Dow.** 2008. HD-GYP domain proteins regulate biofilm formation and virulence in *Pseudomonas aeruginosa. Environ. Microbiol.* **11**:1126–1136.
36. **Ryjenkov, D. A., M. Tarutina, O. M. Moskvin, and M. Gomelsky.** 2005. Cyclic diguanylate is a ubiquitous signaling molecule in bacteria: insights into biochemistry of the GGDEF protein domain. *J. Bacteriol.* **187**:1792–1798.
37. **Schmidt, A. J., D. A. Ryjenkov, and M. Gomelsky.** 2005. Ubiquitous protein domain EAL encodes cyclic diguanylate-specific phosphodiesterase: enzymatically active and inactive EAL domains. *J. Bacteriol.* **187**:4774–4781.
38. **Simm, R., M. Morr, A. Kader, M. Nimtz, and U. Römling.** 2004. GGDEF and EAL domains inversely regulate cyclic di-GMP levels and transition from sessility to motility. *Mol. Microbiol.* **53**:1123–1134.
39. **Slater, H., A. Alvarez-Morales, C. E. Barber, M. J. Daniels, and J. M. Dow.** 2000. A two-component system involving an HD-GYP domain protein links cell-cell signalling to pathogenicity gene expression in *Xanthomonas campestris. Mol. Microbiol.* **38**:986–1003.
40. **Tal, R., H. C. Wong, R. Calhoon, D. Gelfand, A. L. Fear, G. Volman, R. Mayer, P. Ross, D. Amikam, H. Weinhouse, A. Cohen, S. Sapir, P. Ohana, and M. Benziman.** 1998. Three *cdg* operons control cellular turnover of cyclic di-GMP in *Acetobacter xylinum:* genetic organization and occurrence of conserved domains in isoenzymes. *J. Bacteriol.* **180**:4416–4425.
41. **Tamayo, R., A. D. Tischler, and A. Camilli.** 2005. The EAL domain protein VieA is a cyclic diguanylate phosphodiesterase. *J. Biol. Chem.* **280**:33324–33330.
42. **Tang, J. L., J. X. Feng, Q. Q. Li, H. X. Wen, D. L. Zhou, T. J. Wilson, J. M. Dow, Q. S. Ma, and M. J. Daniels.** 1996. Cloning and characterization of the *rpfC* gene of *Xanthomonas oryzae* pv. *oryzae:* involvement in exopolysaccharide production and virulence to rice. *Mol. Plant-Microbe Interact.* **9**:664–666.
43. **Tschowri, N., and R. Hengge.** 2008. The EAL protein YcgF acts as an anti-repressor in a blue-light stress response in *Escherichia coli,* abstr. BP-57. *XIIth Int. Cong. Bacteriol. Appl. Microbiol.,* Istanbul, Turkey.
44. **Wang, L. H., Y. He, Y. Gao, J. E. Wu, Y. H. Dong, C. He, S. X. Wang, L. X. Weng, J. L. Xu, L. Tay, R. X. Fang, and L. H. Zhang.** 2004. A bacterial cell-cell communication signal with cross-kingdom structural analogues. *Mol. Microbiol.* **51**: 903–912.
45. **Wolfgang, M. C., J. Jyot, A. L. Goodman, R. Ramphal, and S. Lory.** 2004. *Pseudomonas aeruginosa* regulates flagellin expression as part of a global response to airway fluid from cystic fibrosis patients. *Proc. Natl. Acad. Sci. USA* **101**:6664–6668.
46. **Yakunin, A. F., M. Proudfoot, E. Kuznetsova, A. Savchenko, G. Brown, C. H. Arrowsmith, and A. M. Edwards.** 2004. The HD domain of the *Escherichia coli* tRNA nucleotidyltransferase has 2′,3′-cyclic phosphodiesterase, 2′-nucleotidase, and phosphatase activities. *J. Biol. Chem.* **279**:36819–36827.

The Second Messenger Cyclic Di-GMP
Edited by A. J. Wolfe and K. L. Visick

Chapter 6

Methods for Cyclic Di-GMP Detection

CHRISTOPHER M. WATERS

The first decade of the 21st century has seen an explosion of interest in the bacterial second messenger signal cyclic di-GMP (c-di-GMP). The study of c-di-GMP during this period has been fueled by the development of methods to visualize and quantify this molecule. These methods, which were first developed to examine production of c-di-GMP in vitro, were later adapted to quantify the intracellular concentration of c-di-GMP in a number of different bacterial species. This chapter will review the history of the development of methods to detect c-di-GMP, summarize the currently available technologies, and highlight the new challenges that must be overcome to deepen our understanding of this molecule and its actions.

INITIAL DISCOVERIES: c-di-GMP IN A TEST TUBE

As with every aspect of c-di-GMP research, the initial development of methods to detect this molecule arose in the laboratory of Moshe Benziman. Benziman found that the cellulose synthase enzymes of *Gluconacetobacter xylinum* could be activated by incubating GTP with an unidentified protein factor from membrane extracts (see chapter 2). However, after a 20-minute incubation, the stimulatory activity of the extracts remained even after removal of the protein component. This finding suggested that the protein component contained an enzyme that converted GTP into a signaling molecule. In a key experiment that unraveled the structure of c-di-GMP, Benziman incubated the membrane fraction with two types of radioactive GTP: [α-^{32}P]GTP and [γ-^{32}P]GTP. Benziman knew the active signal was a molecule that absorbed light at 254 nm, the wavelength of light absorbed by guanine, suggesting that the guanine moieties from GTP were incorporated into the signaling molecule. Importantly, incubation of the membrane protein fraction with [α-^{32}P]GTP, but not [γ-^{32}P]GTP, resulted in comigration of radioactivity with the active signal and absorption at 254 nm. This led Benziman to conclude that the signaling molecule contained the α but not γ phosphate from the GTP substrate (4). Besides this important conclusion, another major advance resulting from this experiment was that c-di-GMP could now be radioactively tagged using [α-^{32}P]GTP, providing a sensitive method to track the molecule.

Benziman also developed methods to separate c-di-GMP from the other components of the in vitro synthesis reaction, which would later be useful in quantification of c-di-GMP from living bacteria. These methods relied on two separation techniques, liquid chromatography (LC) and thin-layer chromatography (TLC). Both of these techniques separate molecules based on their differential interaction with a solid substrate phase and a mobile liquid phase. For both techniques, Benziman purified radiolabeled c-di-GMP to a single peak (LC) or single spot (TLC) using the stimulation of cellulose synthesis activity and absorbance at 254 nm as a way to track the active molecule (4).

The final elucidation of the chemical structure of c-di-GMP was determined by Benziman using the powerful technique of high-performance liquid chromatography (HPLC). HPLC is similar to LC and TLC except that the liquid phase is eluted over the solid phase at high pressure, resulting in better separation. Using HPLC, Benziman was able to resolve c-di-GMP from its precursor, GTP, the intermediate synthesis product, pppGpGp, and the c-di-GMP degradation product pGpG. These molecules were then examined

Christopher M. Waters • Department of Microbiology and Molecular Genetics, Michigan State University, East Lansing, MI 48824.

by mass spectrometry (MS). MS is a technique that allows determination of the specific mass of molecules based on the movement of their ionization products through an electric field. MS can be done in either the positive-ion mode or negative-ion mode in which molecules are converted into positive or negative ions, respectively. MS allows researchers to determine the mass/charge *(m/z)* ratio for any given molecule. As the charge is typically only 1 unit, the *m/z* value is essentially the molecular mass of the molecule with an extra hydrogen atom (positive-ion mode) or missing a hydrogen atom (negative-ion mode). Benziman examined the active compound purified by HPLC using MS analysis and found it had an *m/z* of 689, the expected *m/z* of two linked GMP molecules in negative-ion mode (5). These experiments were critical in determining the structure of c-di-GMP, and the methods Benziman developed to examine this active fragment in vitro would later provide the tools necessary to quantify the concentration of c-di-GMP in bacterial cells.

THE REAL WORLD: DETERMINING THE CONCENTRATION OF c-di-GMP IN LIVING CELLS

The initial studies of c-di-GMP analyzed the in vitro-produced active molecule generated from mixing GTP with membrane fractions of *G. xylinum.* These experiments laid the foundation for the study of c-di-GMP; however, the next advancement in the field was the analysis of c-di-GMP from growing bacteria. The development of techniques to accomplish this task was essential for determining the intracellular concentration of c-di-GMP in different species and studying how the concentration of this molecule changes in response to different environments and genetic mutations. The first determination of the in vivo levels of c-di-GMP was done in the Benziman laboratory. This approach utilized the in vitro cellulose synthesis reaction (see chapter 2). The reaction was standardized by measuring cellulose synthesis over a concentration range of chemically synthesized c-di-GMP. Samples of *G. xylinum* were boiled for 3 minutes to denature protein (c-di-GMP is heat stable) and extracted with perchloric acid, and the denatured protein was removed by centrifugation. This sample was applied to the in vitro cellulose synthesis reaction mixture, and the amount of cellulose synthesis was determined. By comparing the rate of cellulose synthesis stimulated by the nucleotide extracts to the known standards, the intracellular concentration of c-di-GMP was estimated to be 5 to 10 μM (12) in *G. xylinum.*

Although the cellulose synthesis reaction was sufficient to estimate the in vivo concentration of c-di-GMP, it possesses some limitations. First, this reaction requires numerous components, including purified cellulose synthesis proteins, the development of the synthesis reaction, and quantification of the incorporation of radioactive UDP-[^{14}C]glucose into 1,4-β-D-glucan. More important, the reaction measures the concentration of c-di-GMP indirectly. Although there is no evidence to suggest that other molecules besides c-di-GMP from *G. xylinum* possess any activity in this assay, it remains possible that unidentified molecules could interfere with genuine c-di-GMP, leading to an incorrect estimate of the levels of c-di-GMP. For these reasons, quantification of c-di-GMP using the cellulose synthesis assay as a standard was never widely used. Instead, two other approaches, two-dimensional TLC (2D-TLC) and HPLC-MS-MS, have emerged as the dominant methods to quantify the intracellular concentration of c-di-GMP.

HITTING THE SPOT: MEASURING c-di-GMP BY 2D-TLC

In vitro-synthesized c-di-GMP could be separated by Benziman using TLC on a polyethyleneimine-cellulose plate with a variety of different mobile liquid phases producing reproducible R_f values (the distance the molecule travels relative to the solvent front). Radiolabeled c-di-GMP was easily tracked by exposing the TLC plates to film. c-di-GMP could be radioactively labeled in vivo via the addition of radiolabeled inorganic phosphate, $^{32}P_i$, to cultures of bacteria grown in phosphate-limiting media. Unfortunately, this technique also radioactively labels every other cellular nucleotide. Therefore, 1D-TLC is not sufficient to resolve c-di-GMP from every nucleotide that is labeled with ^{32}P. To circumvent this problem, the Benziman laboratory adopted a technique developed by Bruce Ames in 1982 that utilized 2D-TLC to resolve $^{32}P_i$-labeled nucleotides of the cell (2). 2D-TLC works as follows. First, the nucleotide extract is resolved in the vertical dimension along the edge of a large square plate. For the Ames technique, this first dimension resolves nucleotides based on the number of phosphate groups. After drying, the plate is then turned 90° counterclockwise, and the nucleotides are resolved vertically with a different liquid phase. This second dimension separates the nucleotides based on the content of their base groups. Performing this technique with different combinations of liquid mobile phases allowed Ames to resolve and identify all of the known cellular nucleotides, although c-di-GMP had not yet been discovered at that time.

The Benziman group modified this technique for the separation of c-di-GMP from the other radiolabeled nucleotides in *Agrobacterium tumefaciens* cell extracts on polyethyleneimine-cellulose plates using the liquid phases they had previously developed to resolve in vitro-produced c-di-GMP. Specifically, the first mobile phase consisted of 0.75 M Tris-0.45 N HCl-0.5 M LiCl, while the second mobile phase was 1.5 M potassium dihydrogen phosphate. Nucleotides were extracted with cold formic acid. Chemically synthesized and enzymatically produced [^{32}P]c-di-GMP were also separated under the same 2D-TLC conditions. A radiolabeled spot from extracts of *A. tumefaciens* was found to comigrate identically to the control samples. This spot was recovered from the plate and shown to be genuine c-di-GMP via further 1D-TLC analysis (1).

The laboratory of Andrew Camilli rekindled the 2D-TLC technique to determine the intracellular levels of c-di-GMP from different strains of *Vibrio cholerae* (Fig. 1; see chapter 20). The technique was essentially the same as that developed by Ames and Benziman, although a different primary mobile phase was used (ammonium bicarbonate instead of HCl). Using 2D-TLC, Tischler and Camilli were able to observe differences in the intensities of the c-di-GMP spot in the wild-type strain (Fig. 1A), an EAL mutant strain (Fig. 1B), and a GGDEF-overexpressing strain (Fig. 1C). The identity of the c-di-GMP spot was confirmed by removing the resolved spot from the plate and showing that the recovered molecule was resistant to the activity of alkaline phosphatase, which removes free phosphate groups, but sensitive to snake venom phosphodiesterase, which can cleave diesterfied phosphate groups (9). The technique has been developed to allow normalization of levels of c-di-GMP by expressing the relative intensity of the c-di-GMP spot to a control spot. This modification of the technique allows quantitative estimates of c-di-GMP from different samples. A number of different controls spots have been used for this purpose including GTP, GDP (Fig. 1), and the total $^{32}P_i$ signal on the blot.

The 2D-TLC approach to determine the intracellular concentration of c-di-GMP has been adopted by a number of researchers because the experimental protocol is relatively simple and can be performed in any standard laboratory. However, the technique is labor intensive, as each sample must be extracted, separated in two dimensions, dried, exposed, and quantified. None of the steps in this process are automated. Additionally, the 2D-TLC approach requires the use of radioactivity. Another limitation of this approach is that it requires the use of phosphate-limiting media, restricting the environmental conditions that can be studied. Finally, the sensitivity of the 2D-TLC method is not ideal, and oftentimes samples with low concentrations of c-di-GMP will have no visible c-di-GMP spot. For example, no c-di-GMP spot could be observed from wild-type *Vibrio cholerae* using this technique (Fig. 1A). Nevertheless, the 2D-TLC method has been a workhorse for many laboratories, and it has provided much insight into the

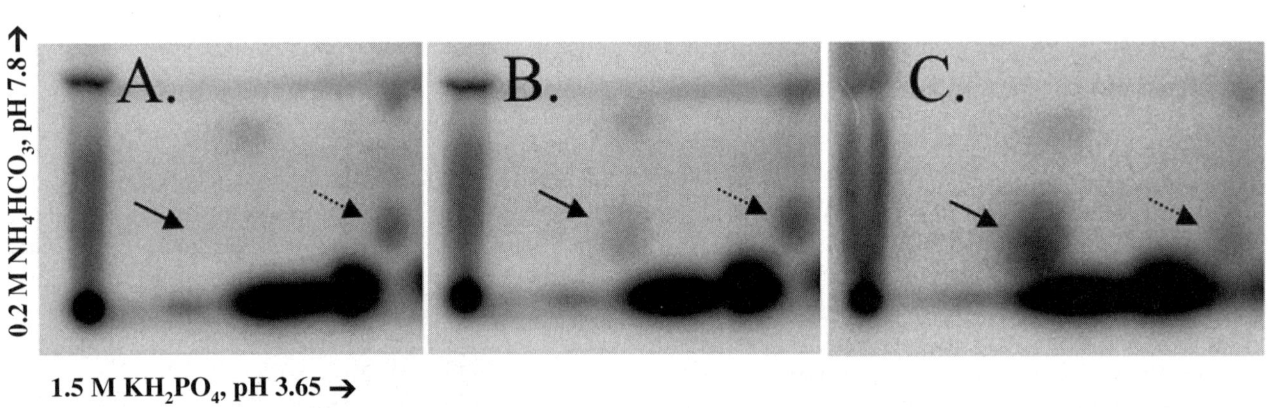

Figure 1. Measurement of c-di-GMP by 2D-TLC. Nucleotides from *V. cholerae* labeled with $^{32}P_i$ were first resolved using 0.2 M NH_4HCO_3, pH 7.8. After drying, the plate was turned counterclockwise and exposed to 1.5 M KH_2PO_4, pH 3.65. Radioactive molecules were detected by exposing the plate to film. The c-di-GMP spot is indicated by the solid arrow, while the GDP spot is shown by the dotted arrow. In panel A, the extract from wild-type *V. cholerae* is shown. This condition is known to be low in c-di-GMP due to the phosphodiesterase activity of the EAL type protein, VieA, and thus, no visible c-di-GMP is observed. In panel B, VieA has been rendered inactive through site-directed mutagenesis, and a c-di-GMP spot is now visible. In panel C, a GGDEF protein, VCA0956, is overexpressed, and a large spot of c-di-GMP can be observed. Reprinted from *Infection and Immunity* (10) with permission.

intracellular levels of c-di-GMP from a number of different bacterial species.

WISDOM OF THE MASSES: THE HPLC-MS-MS METHOD

The second major approach used to quantify the intracellular levels of c-di-GMP combines HPLC separation followed by MS quantification, both of which were used by the Benziman group to determine the chemical structure of in vitro-synthesized c-di-GMP. In one approach developed by the Römling group (see chapter 10), these techniques are used independently. First, cultures were fixed with 0.19% formaldehyde, and nucleotides were extracted from bacterial cultures by boiling at 100°C for 5 minutes or by formic acid extraction as described above. The extract was separated using reverse-phase HPLC, and the fractions that had an absorbance at 254 nm were analyzed by matrix-assisted laser desorption ionization–time of flight MS analysis to identify those containing an *m*/*z* of 689 (the predicted *m*/*z* of c-di-GMP). These positive fractions were further analyzed using electrospray ionization-MS.

For this technique, sample molecules are fragmented (generally by inert gasses), and the resulting fragmentation products are determined. The fragments generated from any given molecule can be predicted based on its chemical structure. Therefore, analysis of the fragmentation products yields a high degree of confidence that the correct molecule is being examined. For the above experiment, the predicted fragmentation products of c-di-GMP have *m*/*z* values of 540.03 and 248.06. Both of these fragments were observed in HPLC fractions that had molecules with an *m*/*z* of 689, confirming that these fractions contained c-di-GMP. Ultimately, the concentration was calculated by measuring the absorbance of the c-di-GMP-containing fraction at 254 nm. The method was standardized by adding known quantities of synthetic c-di-GMP to the bacterial extracts, and the limit of detection for c-di-GMP was found to be 1 fmol (6).

A recent detailed description of this method utilized simultaneous detection of a c-di-AMP internal standard to more accurately produce a standard curve for the quantification of c-di-GMP (7). Comparison of different concentrations of c-di-GMP to c-di-AMP at concentrations of 0.1 μM and 1 μM generated a linear standard curve for c-di-GMP in the range of 5 to 200 nM. This approach was used to show that different bacteria show a wide range of c-di-GMP concentrations from 0.1 pmol/mg of cells in *Salmonella enterica* serovar Typhimurium and *Escherichia coli,* 0.6 pmol/mg of cells in *Pseudomonas aeruginosa,* and 2.0 pmol/mg of cells in *V. cholerae.* Using this information, the authors calculated that *S. enterica* serovar Typhimurium contains between 80 and 250 molecules of c-di-GMP per cell.

New MS instrumentation has combined the HPLC separation component with the MS analysis component by linking these instruments in series. These instruments are termed HPLC-MS. The advantage of these systems is that metabolites resolved by HPLC can be immediately analyzed by MS, negating the need to measure the absorbance of HPLC fractions at 254 nm. This technique also allows a real-time quantification of the relative molecules of interest. Multiple MS analyzers can be linked in series. These instruments are called HPLC-MS-MS, or HPLC-tandem MS, and generally function as follows (Fig. 2A). After nucleotide extraction and resolution by HPLC, the first MS isolates molecules with a specific *m*/*z* of interest. These molecules can then be fragmented, and the second MS can analyze the amounts of the specific fragmentation products. Isolation of only the subset of molecules with a specific *m*/*z* reduces the number of fragments produced that are not derived from the molecule of interest. This increases the sensitivity of the technique by essentially decreasing the noise.

A recent paper from the Bassler laboratory used such a setup to determine the intracellular levels of c-di-GMP from *V. cholerae* (11). In this case, a reverse-phase HPLC was connected to a triple quadrupole MS (Fig. 2A). As described above, the first quadrupole filters ions for a specific *m*/*z* value. In this case, it was 689, the *m*/*z* of c-di-GMP analyzed in negative-ion mode. The second quadrupole fragments the molecule, and these fragments are measured by the third quadrupole. c-di-GMP is predicted to fragment into three detectable products with *m*/*z* values of 540, 344, and 150 (Fig. 2B). Analysis of acetonitril-formic acid extracts from *V. cholerae* revealed a molecule with an *m*/*z* of 689 resolved on the HPLC that formed the expected fragmentation products. This molecule eluted from the HPLC at the identical time as chemically synthetic c-di-GMP (Fig. 3), confirming that the molecule from the extracts is genuine c-di-GMP. Analysis of extracted nucleotides from different *V. cholerae* strains that had mutations in the quorum-sensing signaling circuit revealed a 20-fold difference in the levels of c-di-GMP. Using synthetic c-di-GMP to generate a standard curve, this technique was found to have a sensitivity of 5 ng/ml and linear detection from 5 ng/ml to 10 μg/ml. Therefore, the HPLC-MS-MS technique is quite sensitive and can measure c-di-GMP to a concentration

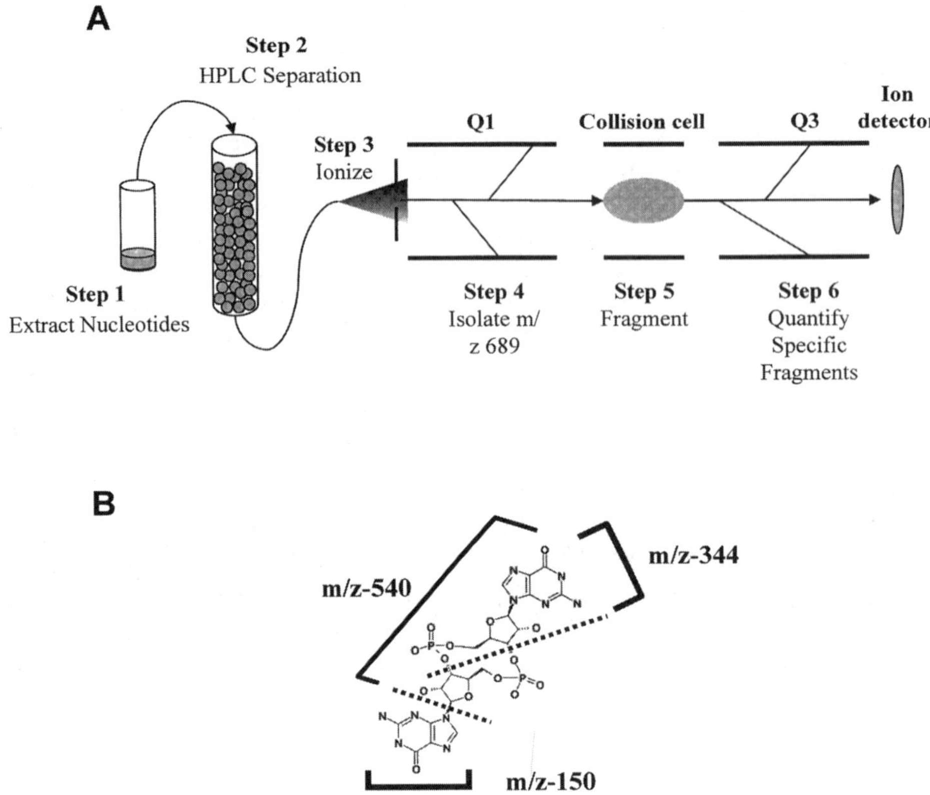

Figure 2. HPLC-MS-MS. (A) This cartoon depicts an HPLC-MS-MS triple quadrupole instrument. Nucleotide extracts are separated based on polarity via HPLC and analyzed in real time. The first quadrupole (Q1) isolates molecules of a specific *m/z*, the molecules are fragmented in the collision cell, and the fragmentation products are analyzed in the third quadrupole (Q3). (B) c-di-GMP is predicted to fragment into three specific products with *m/z* values of 540, 344, and 150.

as low as 7.2 nM. Analysis of extracted nucleotides from a wild-type *V. cholerae* strain compared with this standard curve indicated that the intracellular concentration of c-di-GMP is in the low micromolar range, similar to the value observed in *G. xylinum* by Benziman.

One additional molecule with an *m/z* of 689 also contained the *m/z* 344 and 150 predicted fragments (Fig. 3). This molecule has a slightly different retention time on the HPLC (26 min versus 27.5 min). Using an Orbitrap mass spectrometer with a mass accuracy of <2 ppm, it was found that the *m/z* for the molecule at 26 min corresponded to 691.1356, while the *m/z* for the molecule at 27.5 min was 691.1011. The predicted weight of protonated c-di-GMP is 691.1021, confirming that the molecule at 27.5 min is c-di-GMP, while the molecule at 26 min is an unknown interfering compound. Nevertheless, this example highlights the challenge of quantifying the in vivo concentrations of c-di-GMP because cell extracts may contain unidentified metabolites that closely resemble the *m/z* and fragmentation patterns of c-di-GMP. Therefore, each technique must be standardized with synthetic c-di-GMP to ensure that only genuine c-di-GMP, and not closely related interfering compounds, are measured.

The HPLC-MS-MS approach has many advantages over the 2D-TLC method. The MS component specifically measures c-di-GMP based on its unique *m/z* and ignores most interfering compounds, improving the signal-to-noise ratio and sensitivity of the assay. Also, the HPLC-MS-MS approach is more quantitative than 2D-TLC because the MS provides numeric determination of specific ion fragments, while calculation of the intensity of the 2D-TLC spot is more prone to experimental interpretation, mainly due to the decision on what to include in the spot. This technique is also not limited to a specific medium type, allowing the researcher to study c-di-GMP in any environmental condition. The HPLC-MS-MS approach is much less time intensive than the 2D-TLC approach because much of the process is automated. The researcher only needs to perform the extraction (Fig. 2A), and the HPLC-MS-MS will automatically load, resolve, and quantify each sample (Fig. 2A). Nevertheless, the disadvantage to this tech-

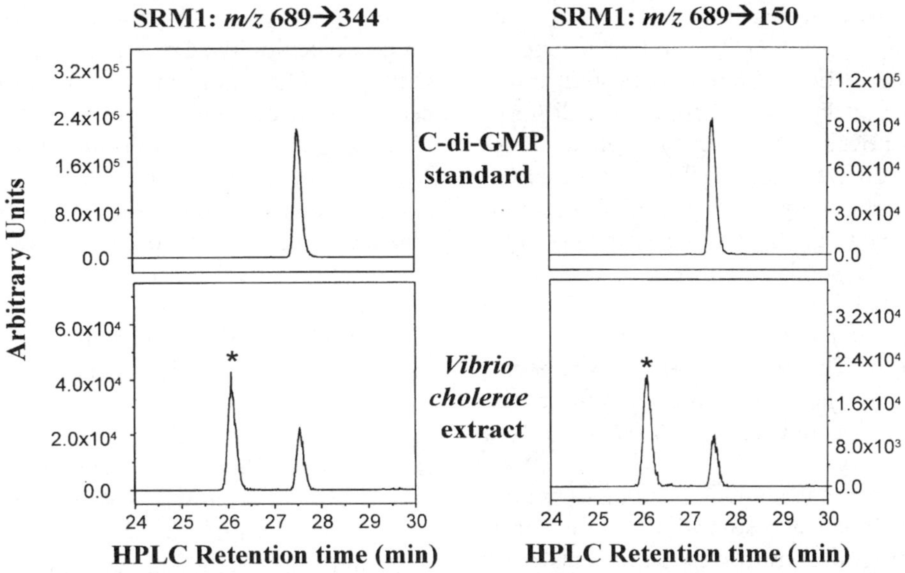

Figure 3. c-di-GMP measured by HPLC-MS-MS. A chemically synthesized c-di-GMP standard (top two boxes) is compared next to nucleotide extracts from *V. cholerae* (bottom two boxes). The left panels show fragmentation products identified by selected reaction monitoring (SRM) analysis with an m/z of 344 originating from ions with an m/z of 689. Likewise, the right panels show fragmentation products with an m/z of 150. The x axis indicates the time that each molecule was retained on the HPLC. Notice that the c-di-GMP peak from the *V. cholerae* extracts elutes from the HPLC at the identical time as the standard. The relative amount of c-di-GMP can be quantified by determining the area under the peak. The molecule in the starred peak also generates both fragmentation products, but it elutes from the HPLC at a different time than the chemical standard. Orbitrap MS analysis showed that this molecule is not c-di-GMP. Reprinted from *Journal of Bacteriology* (11) with permission.

nique is the requirement for a well-equipped MS facility.

FUTURE DIRECTIONS: ANALYSIS OF c-di-GMP AT CELLULAR RESOLUTION

The techniques described above have provided researchers with the tools to determine the intracellular concentrations of c-di-GMP. This work has greatly expanded the field from Benziman's original in vitro experiments on cellular extracts of *G. xylinum* to the in vivo study of c-di-GMP from a variety of different bacterial populations. These techniques have begun to allow researchers to probe how different environmental inputs, genes, and signaling pathways alter the levels of c-di-GMP and correlate how the intracellular concentration of c-di-GMP affects downstream behaviors. There is no doubt that the 2D-TLC and HPLC-MS-MS methods will remain important techniques for future c-di-GMP studies.

However, for our understanding of c-di-GMP to continue to advance, new methods that probe c-di-GMP in vivo at even finer resolution must be developed. Because the techniques described above examine nucleotides extracted from bacterial populations, both the 2D-TLC and HPLC-MS-MS methods can provide only the average concentration of c-di-GMP in the population. Any subtleties that exist within populations of bacteria cannot currently be resolved. For example, if mutation of a gene increases the overall levels of c-di-GMP in a population as determined by the 2D-TLC or HPLC-MS-MS techniques, it cannot be determined if this result is due to a moderate increase in c-di-GMP in each member of the population or a dramatic increase of c-di-GMP in only a subset of the population. Additionally, understanding c-di-GMP changes in multispecies populations is not possible, as the current methods are unable to distinguish the specific species origin of the c-di-GMP from nucleotide extracts. To address questions like these, a new method to probe the concentration of c-di-GMP at the single-cell level must be developed.

The most promising technology to accomplish this task may utilize a recently described c-di-GMP riboswitch (8). This riboswitch is a 100-bp nucleotide sequence found in the 5′ untranslated region of certain mRNAs in multiple bacterial species. This riboswitch forms a specific secondary and tertiary structure that directly binds to c-di-GMP to control expression of the downstream open reading frame through an unknown mechanism. Based on the position of these motifs, c-di-GMP binding to the riboswitch can either activate or repress protein translation. Control of protein production by c-di-GMP riboswitches is a novel mechanism through which c-di-GMP controls different bacterial behaviors. However, from a methodology perspective, this riboswitch is quite exciting, as it can be harnessed to provide readout for single-cell levels of c-di-GMP. Specifically, this riboswitch acts as a direct link between the intracellular c-di-GMP concentration and protein production. Therefore, fusion of a reporter protein, such as green fluorescent protein *(gfp)*, to the c-di-GMP riboswitch could report c-di-GMP levels in single cells.

The development of this new technology would open many new doors of research. First, the intracellular c-di-GMP concentration in individuals from a given population could be profiled. This information could lead to a much greater understanding of c-di-GMP signaling and the response of bacterial populations to changes in c-di-GMP levels. Additionally, with a single-cell reporter, c-di-GMP could be studied in mixed populations by insertion of the reporter element into one species. This technology should also increase the number of conditions that are accessible for study. Currently, a minimum number of bacteria, typically 1 million, is required to produce a detectable signal using the 2D-TLC and HPLC-MS-MS methods. With a single-cell technique, the levels of intracellular c-di-GMP from populations of bacteria at very low cell densities where it is not possible to harvest 1 million bacteria could theoretically be examined. This attribute would allow researchers to study the levels of c-di-GMP in populations outside the test tube, such as environmental samples where the density of bacteria may not be great.

Although analysis of c-di-GMP at the single-cell level would be a major breakthrough, detection at even a great resolution is needed. Thus far, experiments have suggested that the concentration of c-di-GMP is not uniform throughout the bacterial cell. For example, a gradient of c-di-GMP is predicted to exist from the stalked to swarmer cell poles in *Caulobacter cresentus* (see chapter 9). Also, a GGDEF protein, WspR, in *Pseudomonas aeruginosa* localizes upon activation (see chapter 11). These findings have led researchers to propose that c-di-GMP produced by different diguanylate cyclases may be spatially segregated in the intracellular cytoplasm. Currently, no technology exists to test these predictions.

Although examining the intracellular levels of c-di-GMP at such a fine resolution appears daunting, precedence for such experiments comes from the use of fluorescence-based sensor proteins to measure small molecules like calcium, phosphate, glucose, and glutamate (3). These molecular sensors use fluorescent resonance energy transfer (FRET) to report binding to their target molecules. FRET is the process whereby the emission from one fluorescent protein excites the chromophore of another fluorescent protein when they are in close spatial proximity (Fig. 4). The amount of FRET signal is inherently correlated with the distance between the two fluorescent protein partners. Therefore, movement of these partner fluorescent proteins relative to one another can be measured by a change in the FRET signal. The general strategy for these small-molecule fluorescent sensors is to fuse two fluorescent proteins capable of FRET to each end of a known protein that binds the target small molecule. Binding of the small molecule to the internal receptors then alters the conformation of the fluorescent proteins relative to one another, leading to an observable change in the FRET signal (Fig. 4). These small-molecule fluorescent sensors have led to much new advancement in neurobiology and allowed researchers to view chemical signaling from living cells in real time.

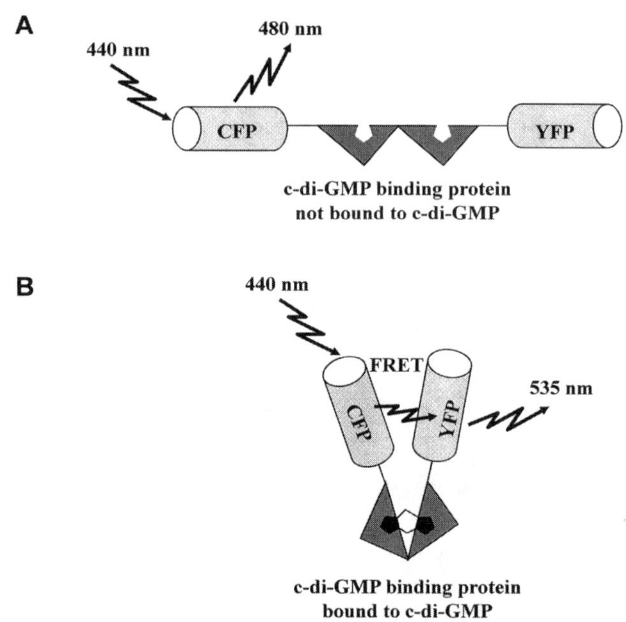

Figure 4. Putative fluorescence-based c-di-GMP sensor. The typical architecture of fluorescence-based sensors is the attachment of two compatible FRET-based protein partners to each end of a binding domain (CFP, cyan fluorescent protein; YFP, yellow fluorescent protein). In panel A, c-di-GMP is not bound to the receptor component of the sensor protein, resulting in no FRET signal because of the large distance between CFP and YFP. In panel B, binding of c-di-GMP to the receptor component decreases the distance of the FRET protein pairs, resulting in an increased FRET signal.

It is conceivable that such a small-molecule fluorescent sensor could be developed to bind to and report the levels of c-di-GMP in localized regions of the cell. One hurdle to construct this sensor is the identification of compatible c-di-GMP binding proteins that can function as the receptor component of the hybrid proteins. Currently, a few proteins that bind c-di-GMP have been identified, including those containing PilZ domains, the RXXD inhibition motif, or even GGDEF and EAL domains themselves (see chapter 4 for more details of known c-di-GMP binding proteins). The identification of additional c-di-GMP binding proteins will provide a greater pool of potential binding partners to construct fluorescence-based sensor proteins. Once these sensors have been constructed, another challenge will be to direct the sensors to specific regions of the cell to gauge localized c-di-GMP levels. Localization of these sensors to the cell poles could conceivably be accomplished by fusing them to known polarly localized proteins, but targeting these sensors to other regions of the cell will pose a greater challenge. Most of the research using small-molecule fluorescence-based sensor proteins has been done in eukaryotic systems. For a fluorescence-based c-di-GMP sensor, it will be technically challenging to resolve the FRET signal from these sensors in the much smaller volume of bacterial cells.

SUMMARY

The methods to measure and quantify c-di-GMP from living cells have sprung forth from the early work of the Benziman laboratory. Two major approaches, 2D-TLC and HPLC-MS-MS, currently dominate the approaches to quantify the intracellular levels of c-di-GMP, and it is likely that these two techniques will remain important tools in the toolboxes of researchers in this field. However, to gain a further understanding of this signaling molecule, new technologies measuring c-di-GMP at the single-cell level and in localized regions of the cells must be developed. The development of these new technologies will ensure the continuing renaissance in the study of c-di-GMP signaling.

REFERENCES

1. **Amikam, D., and M. Benziman.** 1989. Cyclic diguanylic acid and cellulose synthesis in *Agrobacterium tumefaciens*. *J. Bacteriol.* 171:6649–6655.
2. **Bochner, B. R., and B. N. Ames.** 1982. Complete analysis of cellular nucleotides by two-dimensional thin layer chromatography. *J. Biol. Chem.* **257**:9759–9769.
3. **Medintz, I. L.** 2006. Recent progress in developing FRET-based intracellular sensors for the detection of small molecule nutrients and ligands. *Trends Biotechnol.* **24**:539–542.
4. **Ross, P., Y. Aloni, H. Weinhouse, D. Michaeli, P. Weinberger-Ohana, R. Mayer, and M. Benziman.** 1986. Control of cellulose synthesis in *Acetobacter xylinum*. A unique guanyl oligonucleotide is the immediate activator of the cellulose synthase. *Carbohydr. Res.* **149**:101–117.
5. **Ross, P., H. Weinhouse, Y. Aloni, D. Michaeli, P. Weinberger-Ohana, R. Mayer, S. Braun, E. de Vroom, G. A. van der Marel, J. H. van Boom, and M. Benziman.** 1987. Regulation of cellulose synthesis in *Acetobacter xylinum* by cyclic diguanylic acid. *Nature* **325**:279–281.
6. **Simm, R., J. D. Fetherston, A. Kader, U. Romling, and R. D. Perry.** 2005. Phenotypic convergence mediated by GGDEF-domain-containing proteins. *J. Bacteriol.* **187**:6816–6823.
7. **Simm, R., M. Morr, U. Remminghorst, M. Andersson, and U. Romling.** 2009. Quantitative determination of cyclic diguanosine monophosphate concentrations in nucleotide extracts of bacteria by matrix-assisted laser desorption/ionization-time-of-flight mass spectrometry. *Anal. Biochem.* **386**:53–58.
8. **Sudarsan, N., E. R. Lee, Z. Weinberg, R. H. Moy, J. N. Kim, K. H. Link, and R. R. Breaker.** 2008. Riboswitches in eubacteria sense the second messenger cyclic di-GMP. *Science* **321**: 411–413.
9. **Tischler, A. D., and A. Camilli.** 2004. Cyclic diguanylate (c-di-GMP) regulates *Vibrio cholerae* biofilm formation. *Mol. Microbiol.* 53:857–869.
10. **Tischler, A. D., and A. Camilli.** 2005. Cyclic diguanylate regulates *Vibrio cholerae* virulence gene expression. *Infect. Immun.* 73:5873–5882.
11. **Waters, C. M., W. Lu, J. D. Rabinowitz, and B. L. Bassler.** 2008. Quorum sensing controls biofilm formation in *Vibrio cholerae* through modulation of cyclic di-GMP levels and repression of *vpsT*. *J. Bacteriol.* **190**:2527–2536.
12. **Weinhouse, H., S. Sapir, D. Amikam, Y. Shilo, G. Volman, P. Ohana, and M. Benziman.** 1997. c-di-GMP-binding protein, a new factor regulating cellulose synthesis in *Acetobacter xylinum*. *FEBS Lett.* **416**:207–211.

The Second Messenger Cyclic Di-GMP
Edited by A. J. Wolfe and K. L. Visick

Chapter 7

Making, Breaking, and Sensing of Cyclic Di-GMP: Structural, Thermodynamic, and Evolutionary Principles

DOROTHY KIM, JOHN F. HUNT, AND TILMAN SCHIRMER

Structural biology, particularly protein crystallography, has contributed fundamentally to our understanding of signal transduction. While a tremendous body of structural information is available for eukaryotic signaling pathways, less is known about the structural basis of signaling in prokaryotes. Best studied are the bacterial two-component systems, consisting of sensory histidine kinase (35) and response regulator (22, 39, 57), sensory rhodopsin (24), and mononucleotide cyclases (56).

Rather quickly after the realization of the importance of cyclic di-GMP (c-di-GMP) signaling in bacteria, structure-function studies started to reveal mechanistic aspects of this network on the atomic and molecular levels. For the major players involved in c-di-GMP signaling, the GGDEF, EAL, and PilZ domains, three-dimensional structures are now known, and these provide profound and detailed insight into their roles in metabolizing and recognizing the second messenger. Moreover, structural principles are emerging (also reviewed in reference 53) that govern signal transduction from the regulatory or sensory input domain to the catalytic GGDEF and EAL domains, and similar structural mechanisms have been proposed to be involved in signaling to downstream targets via c-di-GMP binding to PilZ domains.

c-di-GMP, A SELF-ASSOCIATING MESSENGER MOLECULE

Nucleotides can form 3′-5′ phosphodiester bonds, resulting in the formation of cyclic mononucleotides or oligonucleotides, such as cAMP and RNA or DNA, respectively. A special case is the formation of two symmetric intermolecular phosphodiester bonds as in c-di-GMP (Fig. 1A). Only in 1985, this compound was discovered as the activating factor for cellulose synthase in *Gluconacetobacter xylinus* (48), but it has now received ample recognition as an important second messenger, as outlined in this book.

The large size (688 daltons) and distinctly amphiphilic chemical structure of c-di-GMP (Fig. 1C) are advantageous properties for a small molecule mediating signal transduction. This combination of characteristics enables c-di-GMP to have high binding energy and to be recognized with high specificity. Importantly, high binding energy increases the potential of the molecule to induce large allosteric conformational changes in receptor proteins while maintaining high binding affinity. Such changes typically require a significant amount of the ligand-binding energy to be devoted to inducing the conformational change, which results in weak ligand-binding affinity unless the intrinsic interaction energy of the ligand is very high. The proven ability of c-di-GMP to induce large conformational changes at submicromolar concentrations (8, 14, 44, 51) indicates that it can have very strong interaction energy with cognate receptors.

The first structure of an isolated cyclic dinucleotide was that of c-di-dAMP determined by nuclear magnetic resonance (NMR) (9) and X-ray crystallography (20). These studies were followed by crystal structure analyses of c-di-GMP (18, 25, 34). In all these structures, and in the complexes with proteins discussed further below, the 12-member macrocycle formed by the ribose and phosphate moieties is seen in virtually the same conformation (20). Thus, the

Dorothy Kim and John F. Hunt • Department of Biological Sciences, Columbia University, New York, NY 10027. **Tilman Schirmer** • Core Program Structural Biology and Biophysics, Biozentrum, University of Basel, CH-4056 Basel, Switzerland.

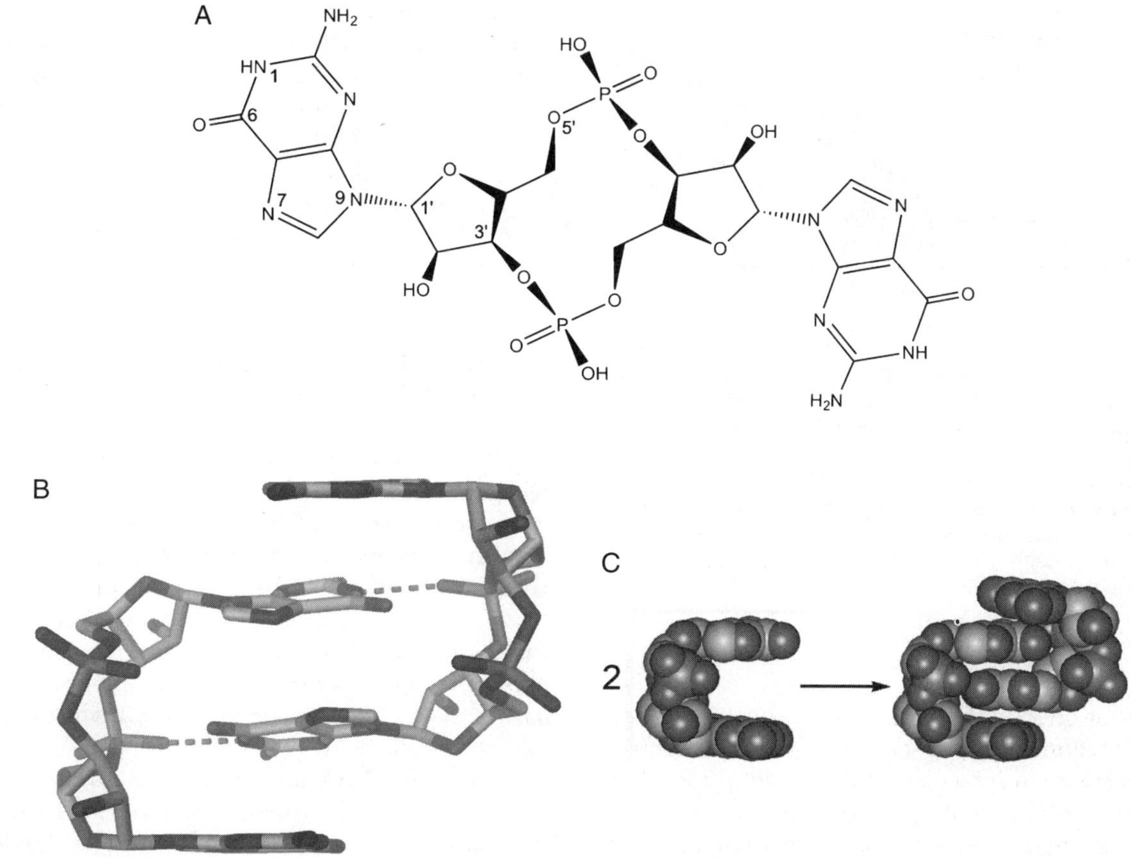

Figure 1. Cyclic di-GMP. (A) Chemical structure. (B) Crystal structure (18) showing a dimer with intercalated bases. The same arrangement has been found in complex with PleD (PDB codes 1w25 and 2v0n) and WspR (3bre). (C) Molecular surfaces of the c-di-GMP monomer and dimer, shaded as in panel B.

backbone of the twofold symmetric molecule appears to be rigid, and the only remaining conformational freedom of the molecule corresponds to rotation around the glycosidic C-1′–N-9 bond (Fig. 1A) that defines the orientation of the base with respect to the macrocycle. The limited flexibility of c-di-GMP is advantageous for the energetics of binding to cognate protein sites.

In all structural studies of isolated dinucleotides, the compound is found in a homodimeric form with intercalated bases (Fig. 1B). Only for c-di-dAMP was a monomer found in addition to the dimer (20). Association with Mg^{2+} or Co^{2+} is observed in some structures, where these cations are found on the twofold symmetry axis coordinated by the N-7 atoms of the two central bases, but their presence does not significantly change the structure of the dimer (18, 34). Most important, this homodimeric form of c-di-GMP has a dramatically different molecular contour from the monomer (Fig. 1C), which means that the monomeric versus dimeric forms are likely to interact with different receptors.

The concentration of dimer versus monomer at any total c-di-GMP concentration will depend on its dissociation constant for dimerization (K_d), but unfortunately, this value has not yet been defined experimentally even though it could have important consequences for the physiology of c-di-GMP signaling. Solution NMR studies have hitherto been carried out only at rather high (>15 mM) concentrations (9, 65), indicating the presence of oligomers. UV spectra at millimolar to submillimolar concentrations show a hyperchromic shift, i.e., increase of absorbance, with increasing temperature (64). This shift probably reflects destacking of the intercalated guanine bases during dissociation of the c-di-GMP dimer and is most pronounced in the presence of Mg^{2+} (65). While the exact K_d for c-di-GMP dimer dissociation has not yet been determined, this observation suggests that it is in the micromolar range, in which case the oligomeric state of c-di-GMP may vary across its physiological concentration range (Fig. 2). In this case, there will be nonlinear activation of receptors recognizing dimeric c-di-GMP as the total c-di-GMP

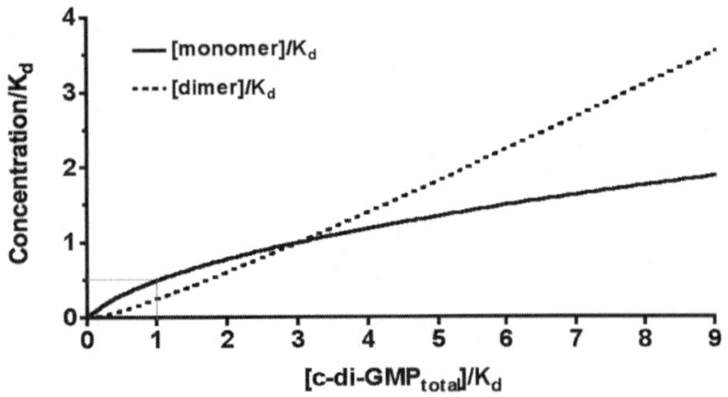

Figure 2. Concentration of c-di-GMP monomer and dimer as a function of total c-di-GMP concentration, expressed as a ratio relative to the K_d for dimerization. Note that K_d is not yet known, although it is likely to be in the micromolar range at physiological salt concentration. Given a reversible dimerization reaction (2 monomers ↔ dimer) and defining K_d = [monomer]^2/[dimer], the monomer concentration is given by [monomer]/K_d = $(-1 + (1 + 8[\text{c-di-GMP}_{total}]/K_d)^{1/2})/4$, while the dimer concentration is given by [dimer]/K_d = [monomer]^2/K_d^2.

concentration rises (i.e., a quadratic increase in activation of dimer-binding receptors in response to a linear increase in total c-di-GMP concentration). Furthermore, differential dimerization of c-di-GMP across its physiological concentration range creates the opportunity for bimodal responses, with different receptors being activated at high concentrations where the dimer predominates compared to low concentrations where the monomer predominates (Fig. 2). The structural results reviewed below provide examples of protein binding sites that seem to be specific for either the monomeric or dimeric states of c-di-GMP, making such bimodal responses likely under some physiological conditions.

Spectroscopic studies (65) indicate that an additional structurally distinct c-di-GMP oligomer forms at submillimolar concentrations in the presence of 100 mM potassium (K^+) ions (although other cations were absent in these studies). Circular dichroism spectroscopy suggests the formation a "G-quartet" structure, which involves coordination of a central potassium ion by four coplanar guanine bases. In the same study (65), NMR spectroscopy has been used to characterize related structures formed at very high c-di-GMP concentrations, but no related structure has been observed yet bound to a protein receptor. Therefore, while there is strong evidence for formation of an alternative oligomeric polymorph at physiological concentrations of c-di-GMP and K^+, it is unclear whether there will be a significant concentration of this species in the more complex ionic environment of the cytosol, where other ions compete with K^+ for interaction with c-di-GMP.

Nonetheless, it is clear that the oligomeric interactions of c-di-GMP are likely to influence its physiological properties. Future work should focus on accurate characterization of the dissociation constant for the well-established, base-intercalated dimer under physiological conditions. Furthermore, researchers should be alert to the possibility of additional cation-dependent conformational variations, particularly in the presence of K^+ ions.

STRUCTURAL PARADIGMS FOR c-di-GMP BINDING AND RECEPTOR ACTIVATION

The proteins involved in c-di-GMP formation, recognition, and degradation are drawn primarily from large gene families which have multiple representatives encoded by most bacterial genomes. c-di-GMP synthesis is catalyzed by diguanylate cyclases (DGCs) which are often called GGDEF proteins based on a stereotyped amino acid sequence motif present in their active sites. GGDEF proteins also have a separate allosteric regulatory site called the inhibitory site (I-site) (12, 62) that binds c-di-GMP to mediate feedback inhibition of its synthesis.

Some GGDEF proteins, such as PelD from *Pseudomonas aeruginosa* (32) (note that PelD should not be confused with PleD, which is a GGDEF DGC; see below) and PopA from *Caulobacter crescentus* (17), appear to have lost catalytic function altogether and act exclusively as c-di-GMP receptors via the action of their I-sites. This functional divergence among GGDEF domains is readily rationalized by the accumulated structural results on c-di-GMP-binding proteins, as explained in the following paragraphs. Domains from at least two additional structural families function as c-di-GMP receptors. The so-called PilZ domain has been proven to bind c-di-GMP and mediate signal transduction. Like the GGDEF proteins, this domain has very broad phylogenetic distribution, and multiple representatives are found in most bacterial genomes. The FleQ proteins, which have a substantially more restricted phylogenetic distribution, are not homologous to PilZ domains but have been shown to act as c-di-GMP receptors (26). c-di-GMP can also bind specifically and with high affinity to certain mRNA riboswitches upstream of some open reading frames for DGCs, phosphodiesterases (PDEs), and proteins that are regulated by c-di-GMP (58), thus probably constituting a means to control gene expression as riboswitches.

Finally, degradation of c-di-GMP is mediated by two families of specific PDEs called EAL and HD-GYP proteins based on a stereotyped amino acid sequence motif present in their active sites, and these proteins once again have a broad and frequently redundant phylogenetic distribution. Multiple structures are available for proteins from all of these families with the exception of HD-GYP PDEs and FleQ receptors, which remain structurally uncharacterized.

While GGDEF and PilZ proteins have unrelated backbone folds, both proteins respond to c-di-GMP binding using an equivalent structural mechanism in which the ligand controls the geometry of interdomain interactions by binding in an interdomain interface (8, 12, 16, 62) (Fig. 3). The fact that the I-site in GGDEF proteins regulates their own catalytic activity using a structural mechanism similar to that used by PilZ receptor proteins explains the evolutionary adaptation of some GGDEF proteins to also function as receptors, in some cases coupled to loss of catalytic activity (17, 32). The energetically preferred interdomain interaction geometry in the absence of c-di-GMP differs from that with bound ligand, so its binding produces substantial changes in the surface topography of the protein. The interdomain interaction geometry stabilized by c-di-GMP binding presumably has high affinity for specific downstream effector proteins. In this manner, c-di-GMP binding to receptors can gate their physical interaction with effector proteins and thereby change the physiological activity of the effectors. Elucidating the structural details of such interactions represents a central goal for future research on the structural biology of c-di-GMP signaling. However, progress in this area is impeded by the fact that most effector pathways have not yet been elucidated.

Structural studies have revealed a remarkable similarity in the local stereochemistry of the c-di-GMP binding sites in the GGDEF (12, 16, 62) and PilZ (8) proteins, in spite of the fact that the protein backbone structures are different even in the binding sites. In both proteins, the side chains of arginine residues make key atomic contacts to both the guanine base and the cyclic phosphodiester backbone of c-di-GMP. Their interaction geometry both with c-di-GMP and with each other is very similar in these otherwise structurally unrelated binding sites. Most remarkably, stereochemically similar arginine interactions are observed with both monomeric and dimeric c-di-GMP ligands in structures that seem to be specific for the different oligomeric species. In the PilZ domains, these arginines are located in a compact seven-residue c-di-GMP switch sequence that changes its backbone conformation upon c-di-GMP binding (8), thereby mediating changes in interdomain interactions that likely gate interaction with effector proteins.

The ability of simple arginine-based sequence motifs to make key structural interactions with c-di-GMP may explain the evolutionary selection of this molecule to control diverse physiological pathways that respond to environmental conditions (as out-

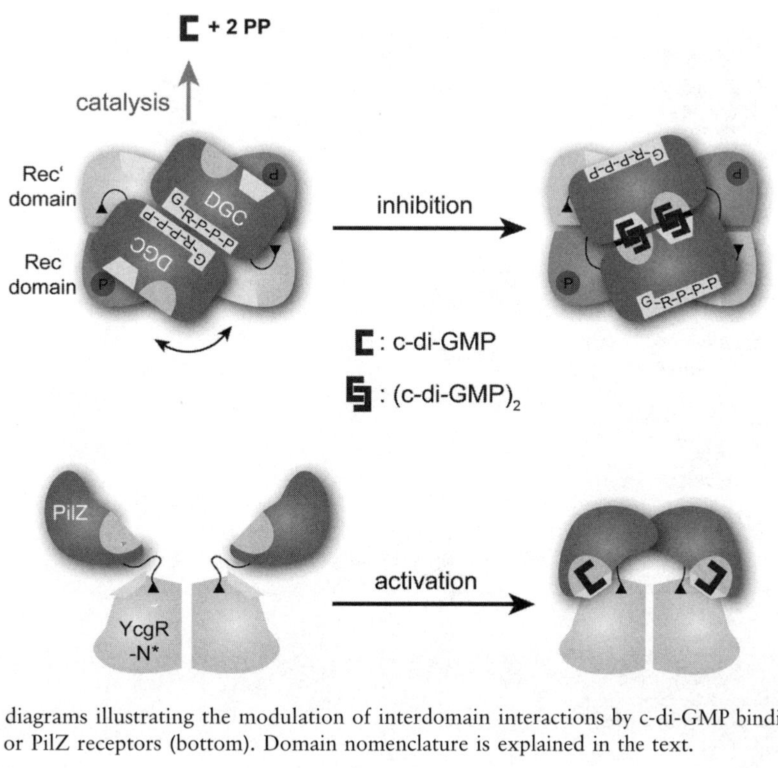

Figure 3. Schematic diagrams illustrating the modulation of interdomain interactions by c-di-GMP binding to representative DGC proteins (top) or PilZ receptors (bottom). Domain nomenclature is explained in the text.

lined elsewhere in this book). Because such responses must vary in organisms adapted to different environmental niches, it should be advantageous for the related signal transduction systems to undergo facile evolutionary adaptation. Great evolutionary flexibility is afforded by the ability of simple sequence motifs to mediate not just high-affinity c-di-GMP binding but also c-di-GMP-dependent conformational changes. These properties enable diversification of c-di-GMP-responsive proteins to proceed either via fusion of c-di-GMP-binding domains to alternative partner domains (Table 1) or via sequence variations within receptors with conserved domain organization (both represented in Table 1). The latter process is facilitated by the fact that there are only very weak sequence constraints associated with maintenance of a high-affinity c-di-GMP-dependent conformational switch. Therefore, the relative simplicity of c-di-GMP-responsive sequence motifs contributes to both the modular organization of the enzyme families involved in c-di-GMP signaling and their facile evolutionary diversification. In this context, the ease of recognizing and responding to c-di-GMP seems likely to explain its evolutionary selection to control environmental response systems in bacteria, which need to be highly adaptable.

Table 1. Structures of representative c-di-GMP-related proteins

Protein	Organism	Domain organization	Main ligand(s)	PDB code	Source or reference
DGCs					
GSU0542	*Geobacter sulfurreducens*	?-GGDEF	*apo*	3ezu	JCSG[a]
PleD	*C. crescentus*	Rec-Rec'-GGDEF	c-di-GMP	1w25	12
			BeF_3, c-di-GMP, GTP-αS	2v0n	62
WspR	*P. aeruginosa*	Rec-GGDEF	c-di-GMP	2bre	16
PDEs					
Blrp1	*B. subtilis*	BLUF-EAL	c-di-GMP, Mn^{2+}, FMN[b]	3gg0[c]	7
TBD1265	*Thiobacillus denitrificans*	PBP-TM-GGDEF-EAL	*apo*	2r6o[d]	MCSG[e]
YkuI	*B. subtilis*	EAL-PAS	*apo*	2bas	37
			c-di-GMP, Ca^{2+}	2w27	37
PilZ domain proteins (c-di-GMP receptors)					
PA4608	*P. aeruginosa*	PilZ	*apo*	1ywu[f]	45
PilZ (PA2960), XC1028	*P. aeruginosa*, *X. campestris*	PilZ⁻ (No c-di-GMP switch)	*apo*	3dsg[g]	33
PlzD (VCA0042)	*V. cholerae*	YcgR-N*-PilZ	*apo*	1yln	MCSG
			c-di-GMP	2rde	8
PP4397	*P. putida*	YcgR-N*-PilZ	*apo*	2gjg	JCSG
Alg44	*P. aeruginosa*	PilZ-?-?			
Cellulose synthase	*E. coli*	?-CelS-GlyT2-CelS-PilZ			
Cellulose synthase	*Rhodobacter sphaeroides*	MdoH-CelS-GlyT2-CelS-PilZ			
Glucosyl transferase (*S. enterica*), cellulose synthase (*T. elongatus*)	*S. enterica*, *T. elongatus*	MdoH-GlyT2-CelS-PilZ			
PA2989	*P. aeruginosa*	?-PilZ			
PilZ3	*Rhodopseudomonas palustris* BisB18	PilZ-PilZ			
PlzC (VCA2344)	*V. cholerae*	?-?-PilZ-?			
YcgR	*E. coli*	YcgR-N-PilZ			

[a] JCSG, Joint Center for Structural Genomics.
[b] FMN, flavin mononucleotide.
[c] Additional structures were determined at a different pH, without metal ions, and complexed with Ca^{2+}.
[d] Crystal structure comprises only the EAL domain.
[e] MCSG, Midwest Center for Structural Genomics.
[f] NMR structure. All other structures were determined by X-ray crystallography.
[g] Crystal structure is of protein XC1028.

DGCs

c-di-GMP is synthesized from GTP by the formation of two 3′-5′ intermolecular phosphodiester bonds:

$$2\ \text{GTP} \rightarrow \text{c-di-GMP} + 2\ \text{PP}_i$$

The reaction is catalyzed by DGC, a protein that typically occurs as part of multidomain proteins. The architecture of the DGC domain bears similarity with adenylyl cyclase as predicted by profile-based sequence comparisons (43) and verified by the crystal structure of the DGC PleD from *C. crescentus* (12) (Table 1). The similarity extends to the triphosphate binding subsite, while the guanine binding subsite is unique in that it is completely provided by the protomer and not, as in adenylyl cyclase, by residues of the other subunit of the heterodimer. The presence of two Mg^{2+} ions in the substrate analog complex of PleD (Table 1) (62) suggested a catalytic mechanism similar to that of mononucleotide cyclases and type I DNA polymerase. A detailed comparison with these enzymes is given in reference 12 and extended in the excellent review by Sinha and Sprang (56).

The DGC domain carries a GGDEF signature motif that locates to a β-hairpin of the central β-sheet that constitutes part of the active site and is one of the defining features of the domain (Pfam PF00990). It was the ubiquity of the so-called GGDEF domain in prokaryotic genomes (21) that indirectly suggested a fundamental role for c-di-GMP. The DGC function of the GGDEF domain was first revealed by genetic complementation (4) and later confirmed in vitro (42, 52).

Activation of DGCs

The reaction catalyzed by DGCs is quite special in that two identical substrates are processed to form a twofold symmetric product. Not surprisingly, an enzymatic domain has evolved that binds one GTP substrate molecule and performs the reaction as a homodimer (40). It may be speculated that the additional domains or extensions found in many DGCs constitute dimerization modules. Indeed, a GGDEF protein structure with an N-terminal dimerization domain has been deposited recently into the Protein Data Bank (PDB; accession code 3EZU). Recent data also demonstrate that YdeH, a GGDEF protein from *Escherichia coli*, is dimeric in solution (10). Signal-dependent dimerization would be an attractive and simple means to regulate DGC activity. Upon dimerization, the reaction type for the encounter of substrate-loaded DGC domains would be changed from bi- to unimolecular, and thus, the efficiency would be dramatically increased. Support for this notion is provided by the structure analysis of PleD (Fig. 4).

PleD is a response regulator protein containing a tandem pair of receiver (Rec) domains at its N terminus followed by a GGDEF domain (i.e., Rec-Rec′-GGDEF in Table 1). Only the first receiver domain (Rec) carries an active aspartate (Asp53) that, for activation of the enzyme, gets phosphorylated by cognate histidine kinases (41, 42). The crystal structure of native PleD shows a loose dimer with the intersubunit contacts mediated by the Rec-Rec′ stem domains (Fig. 4A) (12). To obtain the structure of activated PleD, phosphorylation was mimicked by BeF_3^- modification (62). Upon activation, the stem is considerably tightened, which increases the intersubunit area by more than 50% due to a Rec-Rec′ domain rearrangement within each protomer (Fig. 4B). A corresponding decrease in the dimer dissociation constant K_d (from 100 μM to about 3 μM) was observed in solution (62). Interestingly, in the two crystal structures, there is a distinct arrangement of the GGDEF domain with respect to the stem. Indeed, apart from the linker, there are virtually no contacts between the stem and the catalytic domains.

In both crystal structures of PleD, the two GGDEF domains of the dimer are found in a nonproductive configuration (see below). However, considering the flexibility of the GGDEF-Rec′ linker, rearrangement of the two GGDEF domains appears possible, which could result in an antiparallel arrangement of the two enzyme-bound GTP molecules ready for intermolecular attack of the 3′ hydroxyl on the α-phosphorus of the juxtaposed second substrate molecule (Fig. 4C). The events leading to dimerization and catalysis are schematically summarized in the top half of Fig. 5.

The WspR DGC from *P. aeruginosa* has a simpler Rec-GGDEF domain organization (Table 1). Its crystal structure (16) shows a tetramer which again has dimeric stems, formed here by the Rec domain and its symmetry mate (Fig. 6). The Rec dimer in this structure, although not modified, is essentially in the activated conformation as observed in other response regulator structures (5, 6, 22). This is intriguing but with precedence. Nonactivated PhoP(Rec) crystallizes in the active conformation, which probably is induced by the high protein concentration needed for crystallization as suggested before (5). Thus, at low concentration, the canonical nonmodified conformation would be predicted for WspR in solution. To our knowledge, it has not yet been investigated whether WspR changes its oligomeric state upon

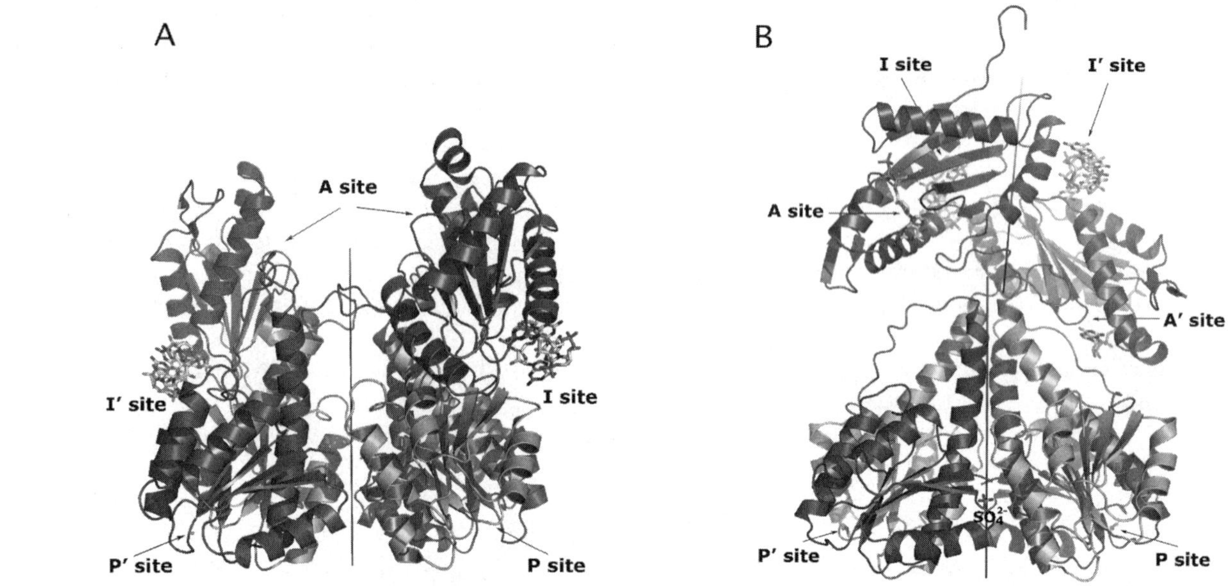

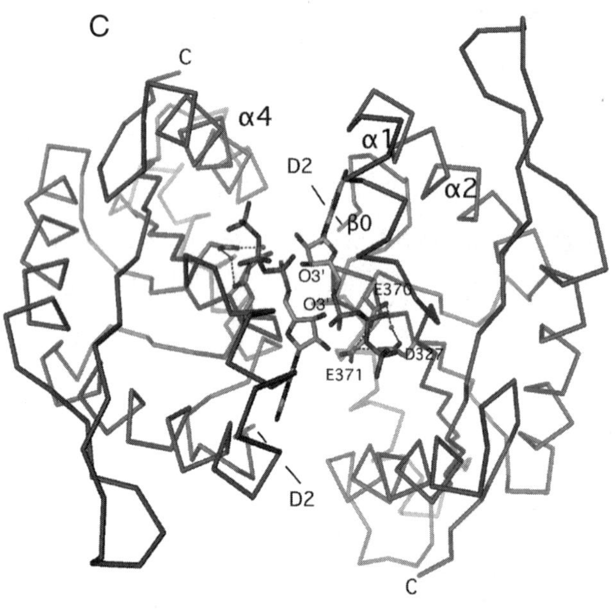

Figure 4. Structure of PleD from *C. crescentus.* (A) Native PleD (12). Loose dimer contacts are mediated by the Rec and Rec′ domains to form the stem at the bottom. The active sites (A-site) of the two GGDEF domains (top) are well separated. (B) BeF3$^-$-activated PleD (62). Note that the stem is tightened up due to the modification at the P-sites (Asp53). The GGDEF domains are in a different orientation with respect to the stem compared to native PleD (panel A). (C) Model of the catalytically competent GGDEF dimer constellation with GTP bound to each domain. View along molecular twofold with the stem (not shown) beneath the projection plane. Figure reprinted from reference 62 with permission.

phosphorylation like PleD, although such a mechanism seems likely.

The requirement of DGCs to operate as dimers to catalyze the condensation of two GTP molecules to form the c-di-GMP product provides a simple and modular way to control its activity. For the control of catalytic activity, any accessory domains that homodimerize in a signal-dependent manner may be used, as long as the mobility of the GGDEF domain with respect to the dimeric stem is not impeded. From the existing structural data, it can be inferred that phosphorylation-dependent dimerization mediated by Rec domains most likely represents a general mechanism for the regulation of DGCs with Rec domains. Regulation of transcription by orthodox response regulators appears to work analogously, because dimerization of these is required for efficient binding to the operator sequence (as reviewed in reference 22).

Feedback Inhibition of DGCs

Strong noncompetitive product inhibition with a K_i of about 1 μM has been reported for DGCs PleD and DgcA from *C. crescentus* (12, 13). The physiological significance of this regulation may relate to

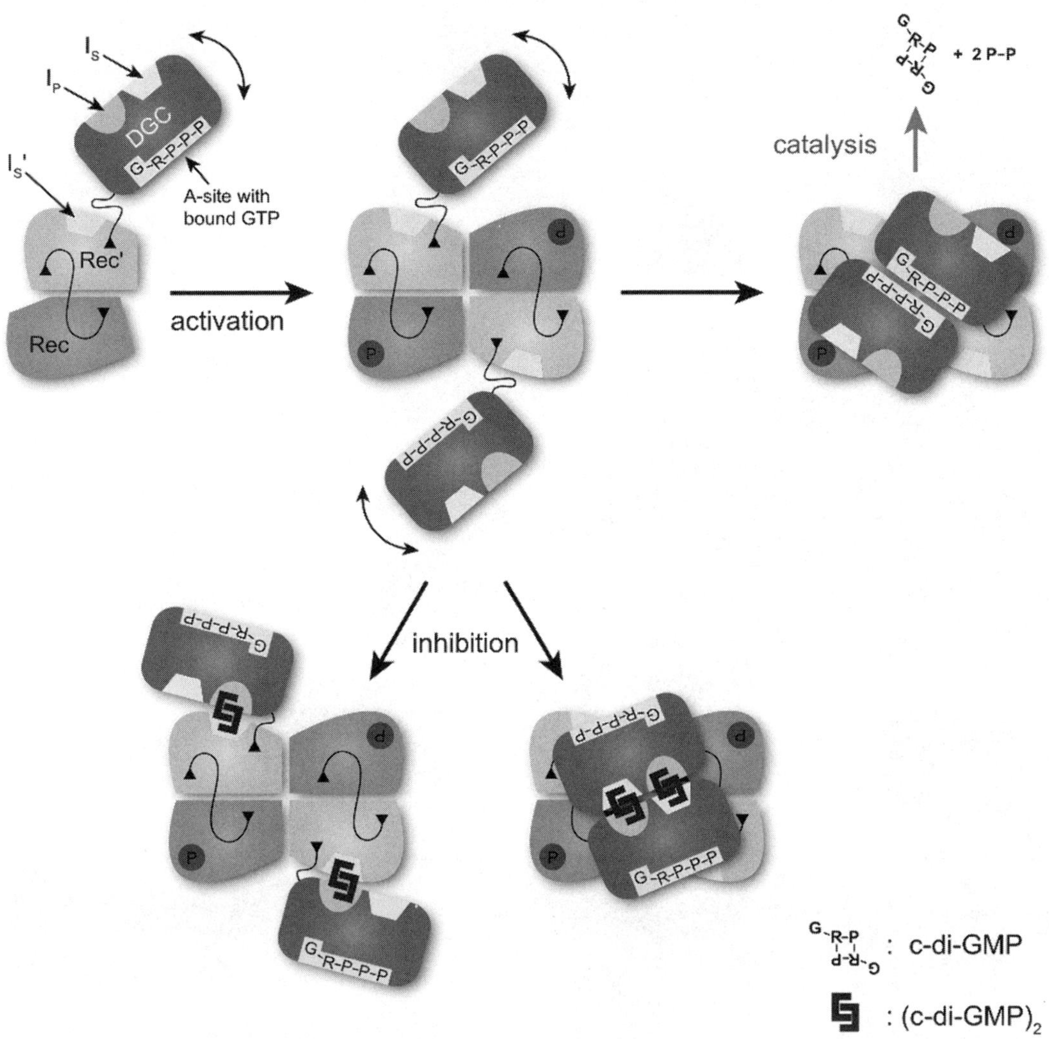

Figure 5. Schematic scheme of PleD catalysis and regulation, adapted from reference 62. (Upper row) The three-domain (Rec-Rec'-GGDEF) protein is monomeric in solution (left). Dimerization via the Rec-Rec' stem is induced upon phosphorylation of the Rec domains, allowing formation of the catalytically competent GGDEF dimer (Fig. 4C) to catalyze formation of two symmetric phosphodiester bonds between two GTP (G-R-P-P-P) molecules (right). (Lower row) Binding of (c-di-GMP)$_2$ to PleD causes either GGDEF-Rec' (left, 1w25) or GGDEF-GGDEF cross-linking (right, 2v0n) and prevents encounter of the active sites, shown here with bound GTP.

setting an upper limit for the concentration of c-di-GMP in the cell and to prevent futile consumption of GTP. This is underscored by the toxic effect of overexpression of DGCs that have lost feedback control (13).

Both PleD crystal structures revealed dimeric c-di-GMP bound to an allosteric binding site formed by residues of the β3-α3 loop (RXXD motif) and Arg390 (called primary inhibition site Ip) of the GGDEF domain. Simultaneously, the ligand was interacting with one or two arginines on an adjacent domain (bottom of Fig. 5). In the structure with PDB code 2V0N, it interacts with Arg313 in the neighboring GGDEF domain, while in the structure with PDB code 1W25, it interacts with Arg148 and Arg178 in the Rec' domain. These observations immediately suggested that the inhibitory action of c-di-GMP is due to this domain cross-linking, which would effectively prevent the encounter of the two active half-sites of the dimer. This inhibition by immobilization mechanism has been corroborated by site-directed mutagenesis, although residual product inhibition (with a 60-fold-increased K_i compared to the wild type) was still observed in a triple mutant that had the residues of the two secondary inhibition sites truncated (62). This residual inhibition may be due to direct intradomain communication between the Ip and active sites as suggested by a recent mo-

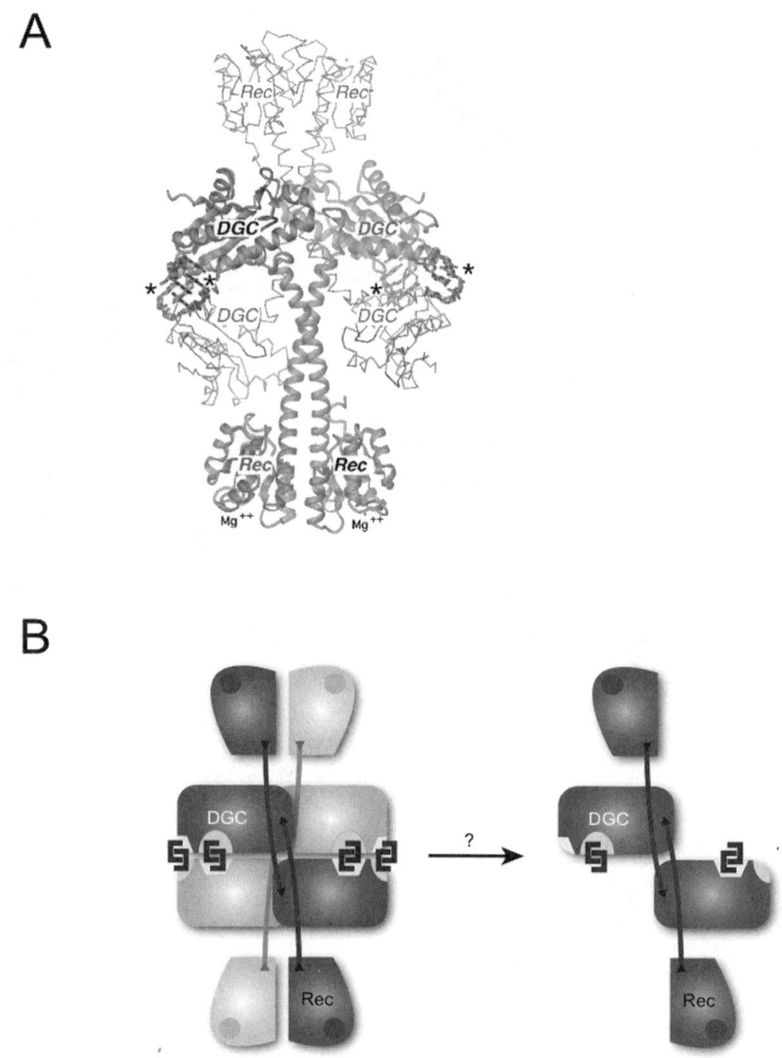

Figure 6. Crystal structure of WspR from *P. aeruginosa* (16) (3bre). (A) Two dimers (ribbon and thin line representation, respectively) are related by a twofold symmetry axis that runs approximately along the viewing direction. The common twofold axis of the dimers is oriented vertically. Each protomer is composed of a Rec and a GGDEF domain. The Rec domains maintain the intradimeric contact, whereas the two dimers are joined together by four $(c\text{-di-GMP})_2$ molecules (shown in full and marked with asterisks), each cross-linking an Ip-site with Arg198 of an adjacent dimer. (B) Schematic representation of the WspR tetramer and the elongated dimer proposed to represent the inhibited state (16).

lecular dynamics study (54). Mutation of the Ip-site in DgcA completely abolished product inhibition of that enzyme (13).

Direct feedback inhibition seems to be operational in the majority of DGCs judged from the conservation of the Ip-site residues. A large proportion of these DGCs seem to operate via DGC-DGC cross-linking, as observed for activated PleD. This inference derives from the fact that 64% of the DGC sequences with the RXXD motif show an arginine at position 313 (as in PleD) or one turn of the helix $\alpha0$ further C-terminal (positions 316 and 317) (62).

The crystal structure of WspR (16) has been determined in the presence of c-di-GMP. In the structure, the ligand is bound as a dimer to the canonical Ip-sites of the tetramer. From there, the ligands bridge to residue Arg198 on helix $\alpha0$ of an adjacent dimer, effectively cross-linking the two dimers of the tetramer (Fig. 6B; due to the 222-symmetry, this cross-link occurs four times). Interestingly, Arg198 is homologous to Arg313 of PleD. Therefore, by using homologous residues, the intradimeric cross-link observed for activated PleD has changed to an interdimeric cross-link in the crystal structure of WspR. In solution, WspR is observed primarily in the dimeric form irrespective of the presence or absence of c-di-GMP (16). Thus, the observed tetramerization seems not to be the mechanism of inhibition for WspR.

Rather, based on the occurrence of dimers with distinct mobility in size-exclusion chromatography, it was hypothesized that two types of dimers may exist: the canonical Rec-mediated dimer as well as an elongated dimer with *trans*-associated subunits (Fig. 6B) that would represent catalytically incompetent WspR originating from a transient tetramer. We feel that further work is needed to test this hypothesis, especially because the thermal stability of the proposed elongated dimer appears weak as judged from the model.

c-di-GMP-SPECIFIC PDEs

c-di-GMP is hydrolyzed by specific PDEs to yield the linear dinucleotide pGpG. This reaction is catalyzed by EAL (15, 47, 55, 59) and HD-GYP (50) domains that often occur in combination with accessory domains that presumably regulate the enzymatic activity. Here, we focus on the EAL proteins that are most abundant and structurally best characterized.

The first EAL crystal structures have been determined by structural genomics initiatives (Table 1). Subsequently, structures in complex with c-di-GMP and divalent cations were determined for YkuI from *Bacillus subtilis* (37) and BlrP1 from *Klebsiella pneumoniae* (7) (Fig. 7). The fold of the EAL domain (Pfam PF00563) is that of a classical TIM barrel. However, the first helix after strand $\beta1$ is missing, and accordingly, $\beta1$ runs antiparallel to the remaining strands. The substrate is bound to a shallow groove located at the C-terminal side of the central β-barrel. Both guanine bases are interacting with the protein via interactions that include stacking with aromatic residues and H bonds to the backbone. The

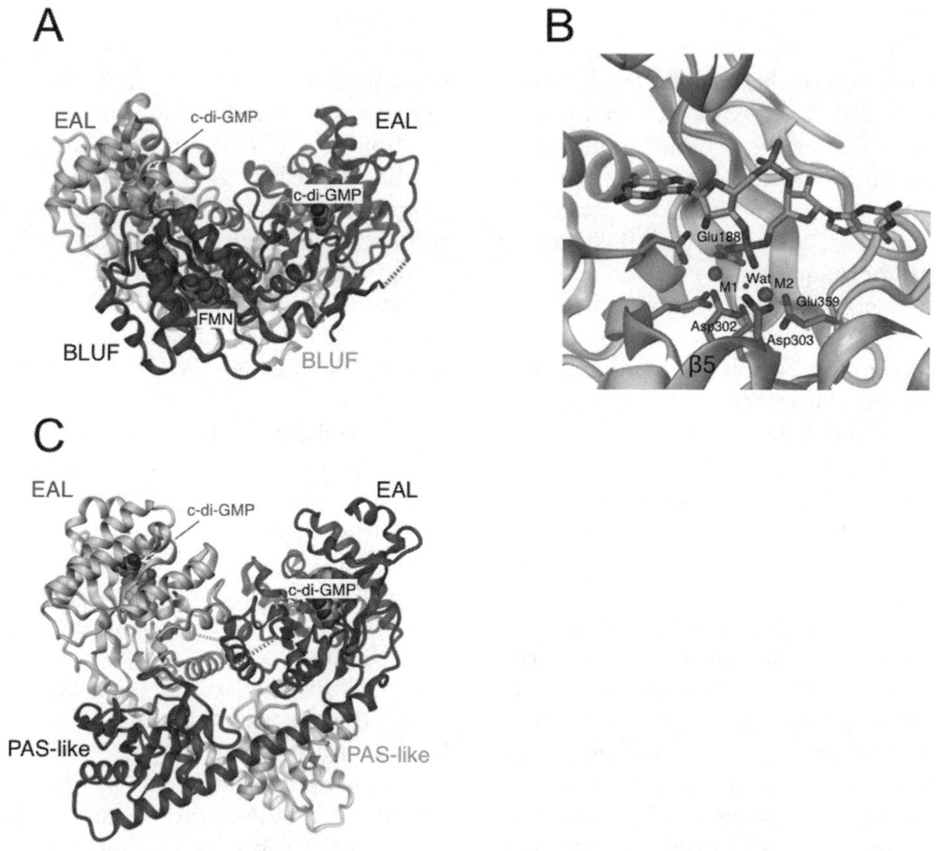

Figure 7. Crystal structures of EAL proteins in complex with c-di-GMP. (A) Ribbon diagram of the BLUF-EAL PDE BlrP1 from *K. pneumoniae* (7) (3gfz). The two chains of the homodimer are drawn in light and dark grey. The molecular twofold symmetry axis is oriented vertically. The c-di-GMP molecules bound to the EAL domain and the flavin mononucleotide (FMN) chromophore bound to the BLUF domain are drawn as van der Waals spheres. (B) Detailed view of the active site with bound c-di-GMP, Mn^{2+} ions (metals M1 and M2), and active site residues shown in full. Also shown is the water (Wat) molecule (small sphere) that is coordinated by both metal ions and well positioned for nucleophilic attack on the phosphorus in line with the scissile phosphodiester bond (approximately vertical). (C) Ribbon diagram of the EAL-PAS protein YkuI from *B. subtilis* (37) (2w27). Representation is as described for panel A.

latter contacts in particular should make the ligand binding base specific.

Figure 7B shows that, in the BlrP1 complex structure, the manganese ions are sandwiched between protein and substrate, where they are coordinated by several carboxylic residues and an asparagine, all of which project from the β-barrel backbone, as well as by an oxygen atom of one of the two substrate phosphates. While the alanine of the EAL signature motif is part of the hydrophobic core of the protein, the glutamate is one of the metal ligands and the leucine is part of the substrate binding site, with the substrate macrocycle positioned directly on top of it.

Catalysis by PDEs

In the structure of the c-di-GMP/Mn^{2+} complex of BlrP1, a water molecule appears to be appropriately positioned for a nucleophilic attack on the phosphorus in line with the scissile P-O-3′ ester bond (7) (Fig. 7B). This activity (PDE-A activity) would yield the linear 5′-pGpG dinucleotide consistent with in vitro characterization of the product produced by YahA from *E. coli* (55). The water molecule is proposed (7) to be activated by the two metal ions which define its precise position and promote deprotonation. Lys323 is in close proximity and suggested to act as a proton acceptor, consistent with the fact that the enzyme shows optimum activity at high pH. The EAL domain bears no similarity with the PDEs specific for cAMP or cGMP (reviewed in reference 28), whose conserved catalytic core is composed of three α-helical subdomains. However, these enzymes are also suggested to operate via a two-metal assisted mechanism. In the c-di-GMP/Ca^{2+} complex of YkuI, only one metal ion is observed, which may be attributed to the presence of an asparagine instead of the Asp303 of BlrP1 that is coordinating the second metal (Fig. 7B). Under standard conditions, YkuI is catalytically inactive (37), but this is probably not entirely due to this substitution (see below). A corresponding Asp → Asn mutation in RocR from *Pseudomonas aeruginosa* reduced the activity only about 30-fold, whereas all other metal coordinating groups and the conserved lysine turned out to be indispensable (47). Following the suggestion of Rao et al. (47) and based on the structure of YkuI in complex with c-di-GMP/Ca^{2+}, Minasov et al. (37) proposed a different catalytic mechanism involving only one cation, with the equivalent of glutamate 359 of BlrP1 acting as a general base to deprotonate the attacking water molecule. There are other inactive EAL proteins (see Fig. 7 in reference 47), such as YcgF from *E. coli*, that are lacking the equivalent of this glutamate (Fig. 7B). Whether some of these inactive domains constitute c-di-GMP receptors remains to be investigated. However, YcgF does not bind c-di-GMP but is involved in direct interaction with the repressor YcgE (61).

Further processing of the pGpG product to yield two molecules of GMP (PDE-B) would require rebinding of the linear dinucleotide to position the phosphodiester in the active site. This, however, may cause steric problems at the distal phosphate binding site that is built to accommodate a phosphodiester and not a terminal phosphate. In fact, PDE-B activity is probably not physiologically relevant (15, 49, 55, 59) and the linear dinucleotide may be cleaved by other enzymes in vivo.

Regulation of PDEs

EAL proteins show a large variety of associated regulatory or sensory domains that putatively enable the control of PDE activity by chemical or physical input signals. The structures of the two known full-length EAL proteins YkuI (37) and BlrP1 (7) give some ideas into possible mechanisms of signal transduction. Despite the different nature and sequence location of the accessory domains (i.e., EAL-PAS in YkuI and BLUF-EAL in BlrP1), the three-dimensional domain organization of the two dimeric proteins is surprisingly similar (Fig. 7A and C) with a small conserved EAL-EAL interface and *trans* association of the accessory domain with the EAL domain of the other subunit. The same interface is observed in the structure of the isolated EAL domain of tbd_1265 (PDB code 2r6o).

In YkuI, the C-terminal domain with a PAS-like fold is connected to the EAL domain by a long helix. PAS-like domains frequently constitute sensors involved in signaling pathways (21). They can harbor chromophores or heme cofactors for light and oxygen sensing, respectively, or they can recognize various other small molecules, such as cGMP. The natural ligand of YkuI has not yet been elucidated. BlrP1 has an N-terminal BLUF (blue light using flavin adenine dinucleotide) domain with a noncovalently bound chromophore.

Strikingly, both accessory domains contact the EAL domain of the other subunit at the rear side, i.e., opposite the substrate binding site, so that modulation of catalytic activity by occlusion can be excluded. Based on the known light-induced changes in the isolated BLUF domain (29), Barends et al. (7) suggest a detailed path for the transduction of the signal to the EAL active site and raise the possibility of an accompanying change in quaternary structure, as has also been discussed for YkuI (37). Indeed, a signal-induced change in the dimeric arrangement that would alter the EAL-EAL interface and, in particular, the $\beta5$-$\alpha5$ dimerization loop, may be transmitted to the nearby $\beta5$ strand that carries catalytically indispensable aspartates (D302 and D303 of BrlP1) (see reference 37 for a description of the canonical naming of secondary structure elements in EAL domains). In this context, it is interesting that D152 of YkuI (structurally equivalent to D302 of BlrP1) has been found shifted out of the active site by 1.7 Å, which may be the reason for the catalytic inactivity of YkuI. Additional structural data will be needed to determine whether this structural difference is coupled to the observed difference in the dimerization loop conformation and whether this difference in loop conformation is controlled by the regulatory domain. However, recently reported site-directed mutagenesis experiments have provided evidence that the $\beta5$-$\alpha5$ dimerization loop influences catalytic activity (46).

Summarizing these considerations, EAL domain enzyme activity could be regulated by a signal-induced change in interdomain packing geometry, which could produce a concomitant change in the quaternary structure (domain arrangement) that alters the position of the metal-coordinating aspartates of strand $\beta5$. Such a mechanism could be employed by various EAL proteins with accessory sensory or regulatory domains and, in fact, would be a particularly attractive mechanism to control the enzyme in response to a signal-induced change in the oligomerization state. Recently, light-induced dimerization has been demonstrated for the BlrP1 homolog YcgF, an EAL protein with a BLUF photosensory domain (38), which appears, however, to be an EAL protein with a corrupted active site (61) (see above).

GGDEF-EAL COMPOSITE PROTEINS

A substantial fraction of GGDEF or EAL proteins contains both domains arranged in tandem in GGDEF-EAL succession. The existence of these composite (or tandem) proteins raises an important question concerning their enzymatic identity and regulation. Emerging evidence argues for variable biochemical functions within this group. Although some members are clearly bifunctional, others are either DGCs or PDEs (11, 15, 19, 23, 30, 31, 60, 63). The best-understood example is the *C. crescentus* PDE PdeA. Its GGDEF domain shows a slightly altered active site signature motif (GEDEF) and lacks DGC activity. Intriguingly, GTP binding to this domain strongly stimulates PDE activity, which demonstrates that the GGDEF domain has adopted a regulatory role, the physiological significance of which is still unclear.

Only a few composite GGDEF-EAL proteins have been reported that are likely to retain both cyclase and PDE activities (19, 31, 60). It is unclear whether or how such proteins can coordinate their antagonistic enzymatic activities. The RelA/SpoT homolog from *Streptococcus dysgalactiae* subsp. *equisimilis* is a well-characterized bifunctional enzyme with two opposing activities, namely, formation and hydrolysis of the intracellular signaling alarmone ppGpp. Its crystal structure showed that the two activities are reciprocally coupled by a switch-like conformational antagonism (27). If the bifunctional GGDEF-EAL proteins are regulated by a similar mechanism, tight interdomain interaction may be expected, or the activities may be sensitive to the oligomeric state of the protein, which could provide a more general mechanism for their differential regulation.

c-di-GMP RECEPTORS CONTAINING PilZ DOMAINS

Identification of PilZ Domains as c-di-GMP Sensors

Elucidation of the role of PilZ domains (Pfam PF07238) in c-di-GMP signaling can be traced to a bioinformatics analysis published by Amikam and Galperin in 2006 (3). Their analysis was based on the seminal biochemical work of Benziman and coworkers, who had shown that c-di-GMP is an allosteric regulator of cellulose synthetase from the fruit-degrading eubacterium *Gluconacetobacter xylinus* (48). Amikam and Galperin performed sequence profiling analyses to identify domains homologous to those in cellulose synthetase (3). These analyses revealed significant similarity between the C-terminal domain in cellulose synthetase, a domain of previously uncharacterized molecular function, and the PilZ protein from *Pseudomonas aeruginosa* (also called PA2960). This 118-residue single-domain protein was one of the few with unassigned molecular function in the *pil* operon, which is responsible for type IV pilus biosynthesis and assembly as well as regulation of twitching motility (2). Mutations in the pilZ gene produce normal amounts of pilin but are unable to correctly assemble pili into functional units (2). Amikam and Galperin found that PilZ domain homologs are broadly distributed in eubacteria in many proteins involved in processes likely to be regulated by c-di-GMP. Based on this analysis combining bioinformatics with insightful biological reasoning, they proposed that PilZ domains generally function as c-di-GMP sensors (3). This prediction was subsequently verified in biochemical (14, 36, 44, 51) and structural studies (8, 45) of a variety of PilZ domain-containing proteins. Ironically, even though most PilZ domains probably bind c-di-GMP, the PilZ protein from *Pseudomonas aeruginosa* from which the domain family derives its name has recently been demonstrated not to bind c-di-GMP (33) (see below).

The phylogenetic analyses of Amikam and Galperin demonstrated that the PilZ domain family is one of the most abundant encoded by bacterial genomes but that it is absent from archaeal and eukaryotic genomes (3). While it sometimes occurs on its own in single-domain proteins (like PilZ and PA4608 from *P. aeruginosa*), the PilZ domain is much more frequently found in larger multidomain proteins with extremely diverse domain organizations and biochemical functions (Table 1). Many of the associated processes are known to be regulated by c-di-GMP, in addition to some of the encoded enzymatic activities (like cellulose synthetase). A particularly common organization matches that in the two-domain protein YcgR, which was recently implicated in regulating motility in enterobacteria (51). YcgR has an N-terminal domain of equal size fused to a C-terminal PilZ domain. Surprisingly, this domain has the same fold as the PilZ domain even though it does not share any detectable sequence homology (8). The occurrence of the PilZ domain in a very large number of different protein families is consistent with its role as a global physiological regulator controlling a wide variety of physiological processes. The tendency of these families to be restricted to specific phylogenetic lineages likely reflects regulation of different physiological processes by c-di-GMP in species adapted to different environmental niches.

The overall level of sequence conservation in PilZ domains is extremely low, so sophisticated sequence profiling techniques are required to identify family members (3). Two short but functionally important regions contain the most highly conserved residues in the sequence profile, which also includes a number of modestly conserved hydrophobic residues spread throughout the domain. The first highly conserved region is a seven-residue arginine-rich segment at the N terminus of the domain that was designated the c-di-GMP switch based on crystallographic studies showing that it acts as a c-di-GMP sensor (8) (see below). This segment has the consensus sequence (q/e)rRXXXR. The second two arginine residues in this sequence are invariant and essential for c-di-GMP binding (14, 36, 44, 51), while the first has weaker conservation and may contribute to recognizing the ligand in some PilZ domains. A second conserved sequence motif with consensus dΦSXXG (where Φ is any hydrophobic amino acid) occurs ~20 to 30 residues later and also participates in c-di-GMP binding (8).

Overview of PilZ Domain Protein Structures

Atomic structures have been determined for four different proteins containing PilZ domains. These domains all share a very similar fold in spite of their low degree of sequence similarity (PDB no. 2GJG and 1YLN) (8, 33, 45). The core of the PilZ domain comprises a six-stranded antiparallel β-barrel, which has a variable number of α-helices bound to its surface in different family members (typically two). The dΦSXXG motif is located on the surface of the β-barrel, where it makes key interactions with one guanine base in c-di-GMP. The seven-residue c-di-GMP switch is located at the N terminus of the PilZ domain in front of its first β-strand, but this site is proximal to the dΦSXXG motif in the three-dimensional structure of the domain. This geometry enables the arginine-rich switch sequence to wrap around c-di-GMP molecules interacting with the dΦSXXG motif on the surface of the β-barrel to form the complete c-di-GMP binding site. The change in the conformation of the switch upon c-di-GMP binding combined with the presence of the ligand itself creates a new surface that seems likely to interact with effector proteins, at least in single-domain PilZ proteins.

The fact that the c-di-GMP switch is located at the N terminus of the PilZ domain enables c-di-GMP-dependent conformational changes in its conformation to directly alter the packing geometry of the PilZ domain relative to any preceding domains in the receptor protein. Such c-di-GMP-dependent changes in interdomain interactions represent a second mechanism to produce new interaction surfaces upon c-di-GMP binding, and this mechanism seems likely to play a central role in signal transduction by multidomain receptors containing PilZ domains. The fact that the PilZ domain is almost always located at the C terminus of multidomain receptors (Table 1) is likely to be explained by the fact that this domain organization enables the N-terminal c-di-GMP switch in the PilZ domain to control its packing interactions with N-terminally attached domains. The exact separation between the c-di-GMP switch and the N-terminal β-strand of the PilZ domain varies in different family members (3, 8), as does the conformation of this β-strand in some family members (33), which enables the switch to effect distinct changes in interdomain interactions in different PilZ domain-containing receptors (as explained in greater detail in the discussion of the PP4397 protein below). This structural plasticity provides great flexibility in adapting multidomain PilZ proteins to control different effector proteins and physiological processes.

Recently, a crystal structure was solved for the XC1028 protein from *Xanthomonas campestris,* which shares 65% sequence identity and 70% similarity with the PilZ protein from *Pseudomonas aeruginosa* (PA2960). As explained above, this single-domain protein was the source of the name for the PilZ domain family. Ironically, neither the PilZ protein nor the XC1028 protein binds c-di-GMP in biochemical assays, and the structure of XC1028 does not contain either c-di-GMP binding motif described above (33) (although its backbone fold is otherwise homologous to that of c-di-GMP-binding PilZ domains except for the absence of the N-terminal β-strand). Therefore, the namesake protein for the PilZ domain family does not bind c-di-GMP. These observations demonstrate that the PilZ domain fold must be capable of performing some biochemical function that does not involve c-di-GMP binding. Based on the likely contribution of the PilZ domain to mediating interprotein interactions in c-di-GMP receptors, the PilZ domain proteins that do not have the ability to bind c-di-GMP seem likely to mediate interprotein interactions. This observation rationalizes the most common domain organization of c-di-GMP receptor proteins, in which a c-di-GMP-binding PilZ domain is fused at the C terminus of a structurally homologous domain that lacks c-di-GMP-binding motifs (called either YcgR-N or YcgR-N* domains in Table 1). In such two-domain c-di-GMP receptors, both domains are likely to participate in mediating interactions with effector proteins even though only the C-terminal domain directly binds c-di-GMP.

NMR Structure of the Single-Domain PilZ Protein PA4608

NMR spectroscopy was used to determine the structure of PA4608 from *P. aeruginosa,* which provided the first view of PilZ domain architecture (45). Titration of c-di-GMP into the NMR sample resulted in a shift in most protein resonances, consistent with the hypothesis that PA4608 is a c-di-GMP-binding protein, but the binding site could not be localized from the NMR data. Protein backbone resonance could not be observed for the invariant arginine residues in the N-terminal c-di-GMP switch (Arg9 and Arg13 in PA4608), indicating that this region is flexible in the absence of c-di-GMP (45).

Crystallographic and Thermodynamic Studies of the Two-Domain Protein PlzD

PlzD, which is the product of the VCA0042 gene, is one of five PilZ domain-containing proteins identified in *Vibrio cholerae.* This organism depends on c-di-GMP signaling during pathogenesis, and either PlzD or PlzB (another PilZ domain protein) is required for efficient infection of mice (44). In addition to a C-terminal PilZ domain, PlzD has an N-terminal domain of similar size that was called a YcgR-N* domain based on the fact that sequence profiling techniques detect significant similarity to the N-terminal domain of the YcgR protein from *E. coli* (3, 8). PlzD and YcgR share one of the most commonly observed domain organizations observed for PilZ domain-containing proteins (Table 1), with an N-terminal YcgR-like domain fused to a C-terminal PilZ domain. The crystal structure of PlzD (Fig. 8) was determined both with and without bound c-di-GMP (PDB no. 1YLN) (8), providing the first insight into the mechanistic details of c-di-GMP signal transduction by PilZ domain-containing proteins.

The YcgR-N* and PilZ domains in PlzD are structurally homologous to one another, which was unexpected prior to determination of the crystal structure because these domains do not have significant sequence similarity (~5% identity). They share an equivalent backbone fold in which a pair of three-stranded antiparallel β-sheets abutted by an α-helix interact to form the six-stranded β-barrel that constitutes the core of the domain. Hydrodynamic studies show that both *apo* and c-di-GMP-bound PlzD are dimers, while the crystal structures show that the major intersubunit contacts occur between the YcgR-N* domains and that these are preserved upon c-di-GMP binding (8).

The crystal structure of PlzD complexed with c-di-GMP shows a single c-di-GMP molecule bound to each subunit in the physiological dimer (8). No divalent cation is present in the c-di-GMP-binding site, even though Mg^{2+} ions were included in the crystallization solution. Isothermal titration calorimetry experiments on PlzD gave data that could be fit equally well assuming binding of either a monomer or a dimer of c-di-GMP per subunit in solution (due to uncertainty in the quantitation of c-di-GMP concentration) (8). However, no evidence of electron density for a second c-di-GMP molecule per subunit could be found in the crystal structure, and modeling a c-di-GMP dimer into the observed protein conformation produces severe steric clashes between the second c-di-GMP molecule and three β-strands from the YcgR-N* domain in the same subunit (data not shown). Given these observations, it seems unlikely that each PlzD subunit could bind a c-di-GMP dimer even in solution. In contrast, the single PilZ domain protein PA4608 was observed to bind a dimer of c-di-GMP in solution NMR experiments (14). Therefore, other PilZ domains may also bind dimeric c-di-

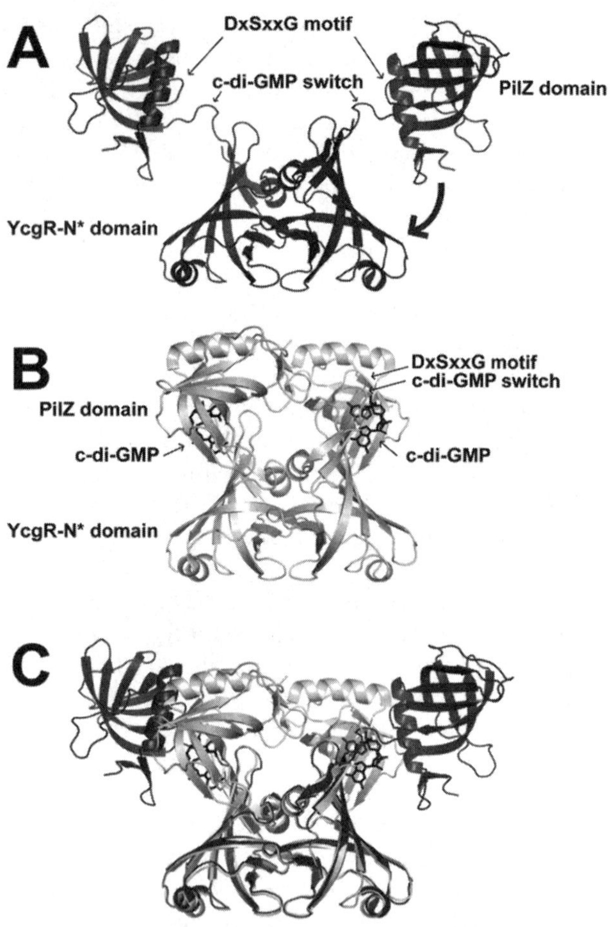

Figure 8. Crystallographic studies show a large conformation change in PlzD upon c-di-GMP binding (PDB code 2rde). The PlzD protein from *V. cholerae* has an N-terminal YcgR-N* domain fused to a C-terminal PilZ domain (PDB no. 1YLN) (8). Although there is no significant sequence similarity between them, both domains form structurally homologous six-stranded β-barrels. (A) The *apo* structure (i.e., in the absence of c-di-GMP, PDB code 1YLN) shows no contacts between the YcgR-N* and PilZ domains in each protomer because the c-di-GMP switch that links them adopts an extended conformation. (B) c-di-GMP binding produces a 123° rotation of each PilZ domain toward the YcgR-N* in the same protomer. The c-di-GMP molecule is sandwiched between the two domains and held into place by contacts from the c-di-GMP switch as well as conserved residues in the β-barrel core of the PilZ domain. (C) Superposition of the *apo* (dark) and c-di-GMP-bound (light) structures based on least-squares alignments of the YcgR-N* domains.

GMP. Notably, PlzD is lacking a partially conserved arginine residue that is present in the c-di-GMP switch in PA4608 and most PilZ domains. Given the importance of arginine residues in binding c-di-GMP (see below), this sequence variation is conceivably related to differences in affinity for monomeric versus dimeric c-di-GMP.

The most striking difference observed in comparing the *apo* structure of PlzD to the c-di-GMP-bound structure is a major change in interdomain interactions within each protomer. In the *apo* structure, the dimer is found in an extended configuration in which the C-terminal PilZ domain and N-terminal YcgR-N* domain do not directly contact one another (Fig. 8A). The ligand-bound structure shows a 123° rotation of the PilZ domain toward the YcgR-N* domain that results in encapsulation of a single c-di-GMP molecule in the interface between the two domains (Fig. 8B and C) (8). The YcgR-N* domain dimer interface is essentially unchanged during this conformational change, which results in a much more compact dimer structure stabilized by new contacts between the PilZ domains of the two protomers (Fig. 8B). The binding of c-di-GMP produces at most minor perturbations in domain structure except in the c-di-GMP switch (residues 134 to 140), which links the end of the N-terminal YcgR domain to the start of the C-terminal PilZ domain. This segment, which corresponds to the dynamic protein segment in the NMR structure of PA4608, undergoes a major backbone conformational change. Its structural plasticity couples formation of tight c-di-GMP contacts to a substantial change in interdomain packing geometry within each protomer (Fig. 8C). The invariant arginine residues R136 and R140 (corresponding to R9 and R13 in PA4608) both directly contact c-di-GMP in the bound structure (Fig. 9A). These interactions would not be possible in the *apo* structure given the stereochemical configuration of the c-di-GMP switch in its open conformation (Fig. 9B). Additional contacts to c-di-GMP are made by the β-barrel in the PilZ domain, especially the dΦSXXG motif spanning β-strands 2 to 3.

Although the crystal structure of PlzD in the absence of the c-di-GMP ligand shows a single ordered conformation of the c-di-GMP switch, this conformation may be stabilized by crystal lattice packing constraints. The c-di-GMP switch seems likely to be dynamic in *apo* in PlzD in solution, as previously observed for the equivalent segment in the NMR structure of PA4608. Isothermal titration calorimetry showed that a substantial entropy loss is coupled to c-di-GMP binding to PlzD, suggesting that some protein segment undergoes a disorder-to-order transition (8). Given the lack of significant conformational changes outside the c-di-GMP switch, this observation suggests that the switch is likely to be dynamic in the *apo* state and become ordered upon binding the c-di-GMP ligand. This entropy loss upon binding creates a strong temperature-dependent change in binding affinity, which could be used to modulate c-di-GMP-dependent signaling pathways as a function of growth temperature (8).

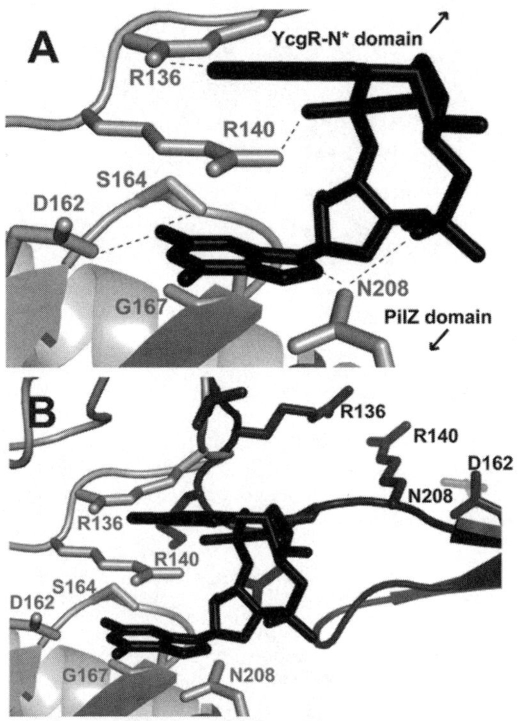

Figure 9. A cooperative hydrogen-bonding network stabilizes c-di-GMP in the binding pocket of PlzD (PDB code 2rde). The c-di-GMP binding pocket in PlzD is formed by the c-di-GMP switch (residues 134 to 140) and several β-strands containing residues conserved in PilZ domains (residues 162 to 170 and 219 to 221) (8). A single c-di-GMP molecule (black) binds in a *cis* conformation with its two guanine bases stacked parallel to one another. (A) The invariant arginine residues in the c-di-GMP switch make critical interactions with the ligand. Arg136 contacts the guanine base, while Arg140 hydrogen bonds with one of the phosphate groups and simultaneously makes π-π interactions with one base. The other guanine base is stabilized by hydrogen bonds to residues Asp162 and Ser164 in the dΦSXXG motif, while the Cα of Gly167 in this motif contacts its surface. This base and a backbone phosphate group are contacted by Asn208, a residue present in VCA0042 and a minor fraction of other PilZ domains (8). (B) Comparison of the *apo* (dark grey) and c-di-GMP-bound (light grey) structures shows the conformational change in the c-di-GMP switch, which moves the core of the PilZ domain and the other key elements of the binding pocket away from the YcgR-N* domain. The conformation of the switch in the *apo* state is clearly incompatible with c-di-GMP binding.

Similarity in the Stereochemistry of c-di-GMP Binding in PleD/WspR and PlzD

The I-sites of the DGCs PleD of *C. crescentus* and WspR of *P. aeruginosa* share considerable stereochemical similarity with the c-di-GMP binding site of PlzD (Fig. 10). In both instances, key arginine residues in the binding pocket serve to stabilize the c-di-GMP dimer. The PleD pocket contains dimeric c-di-GMP which is interdigitated by Arg359 of the RXXD Ip-site motif, and Arg313′ of the second GGDEF domain (replaced by Arg178 of the Rec′ domain in the nonactivated structure 1w25). Similarly, c-di-GMP binding to WspR is mediated by Arg242 of the conserved Ip-site and Arg198′ of the GGDEF domain of another molecule (see above). This binding stereochemistry is replicated in PlzD even though it only binds a monomer of c-di-GMP. In this case, the essential arginine residues come from the c-di-GMP switch. The side chains of Arg136 and Arg140 in the switch stack on top of one another, and their guanidino groups lock the c-di-GMP molecule in place (Fig. 10A). Arg359 and Arg313′ in PleD interact with each other in an equivalent geometry, and their guanidino groups make essentially the same interactions with c-di-GMP (Fig. 10B), although R313′ of PleD is additionally interacting with one of the guanine bases of the second c-di-GMP molecule of the intercalated dimer. It is noteworthy that R390 of PleD, which has no analog in PlzD, interacts with the other guanine base of the second c-di-GMP.

Structural Divergence near the c-di-GMP Switch in the Two-Domain Protein PP4397

Protein PP4397 from *Pseudomonas putida* KT2440, which contains an N-terminal YcgR-N* domain fused to a C-terminal PilZ domain, is found in an operon involved in flagellar biosynthesis. A homologous operon is found in *P. aeruginosa*. The crystal structure of PP4397 has been determined in the absence of c-di-GMP (PDB no. 2GJG). Although the individual YcgR-N* and PilZ domains in PP4397 each have only 11% sequence identity to the corresponding domain in PlzD, they both have equivalent structures including essentially identical backbone topologies. The arginines in the c-di-GMP switch that are critical for ligand binding to PlzD (44) are both conserved in PP4397, which is expected to bind c-di-GMP using this protein segment. The crystal structure of PP4397 reveals a dimer associating through its PilZ domains (not shown), rather than the YcgR-N* domain interaction observed in PlzD (8).

While the *apo* structure of PP4397 is in an extended conformation crudely reminiscent of the PlzD *apo* structure, a surprising difference is observed in the region of the c-di-GMP switch. Alignment of the N-terminal YcgR-N* domains in PP4397 and PlzD shows that the c-di-GMP switch is located at the equivalent structural site relative to this domain, but it is displaced relative to the PilZ domain, even though it is the most conserved sequence feature in the PilZ domain. In PP4397, the switch and the three residues immediately following it form an α-helix that together with an ensuing loop push the PilZ domain away from the YcgR-N* domain compared to

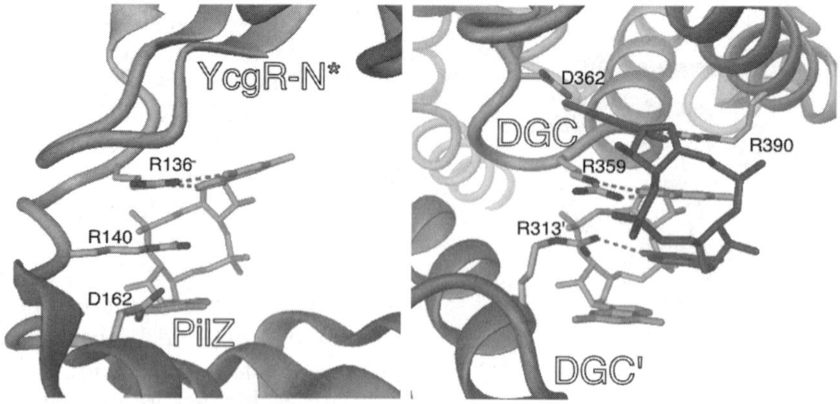

Figure 10. Binding mode of monomeric c-di-GMP to PlzD (left) (PDB code 2rde) and of a self-intercalated c-di-GMP dimer to the allosteric inhibition site of PleD (right) (PDB code 2v0n). In each case, two arginine side chains (PlzD, R136 and R140; PleD, R313′ and R359) extend at right angles and are involved in ligand binding. H-bonds between the guanidino groups and the O-6, N-7 (non-Watson-Crick) edge of the guanyl bases are shown as dashed lines.

PlzD (Fig. 11). Structural superposition of the PilZ domains indicates that an eight-residue insertion occurs between the N-terminal c-di-GMP switch and the core of the PilZ domain in PP4397 compared to PlzD. This insertion plays a role in producing both the altered conformation of the switch and its relative displacement from the PilZ domain observed in the *apo* structure of PP4397 (Fig. 11). The two invariant arginine residues in the c-di-GMP switch of PP4397 (Arg123 and Arg127) sit on the same face of the α-helix, meaning that the conformation of the c-di-GMP switch observed in the *apo* structure would have to change substantially to adopt the c-di-GMP-bound switch conformation observed in PlzD. Additionally, the DXSXXG motif, which is conserved in both PP4397 and PlzD and makes substantial contacts to the c-di-GMP ligand in the PlzD structure, is considerably more remote from the c-di-GMP switch in PP4397 than PlzD. These intriguing local structural variations, combined with the differences in dimer structure, suggest that there will be significant differences between PP4397 and PlzD in their conformational responses to c-di-GMP binding in spite of the fact that they possess homologous c-di-GMP switches and PilZ domain structures.

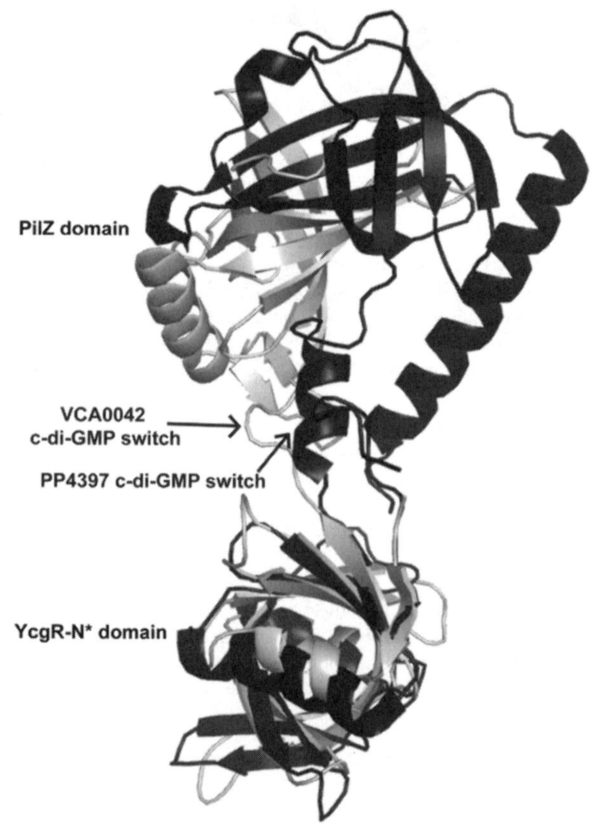

Figure 11. Structural divergence near the c-di-GMP switch in PP4397. Superposition of protomers from the *apo* crystal structures of PP4397 (black, PDB code 2rde) and PlzD (gray, PDB code 2GJG) based on least-squares alignment of their N-terminal YcgR-N* domains. The PP4397 protein contains an N-terminal YcgR-N* domain and a C-terminal PilZ domain, both of which show strong structural homology to the corresponding domains in PlzD. However, PP4397 has an eight-residue insertion between the core of the PilZ domain and the c-di-GMP switch. The interdomain geometry in PP4397 is dramatically different, and its c-di-GMP switch adopts a different conformation in the absence of c-di-GMP (i.e., an α-helix in PP4397 versus an unstructured loop in PlzD).

CONCLUDING REMARKS

c-di-GMP is an exquisite multivalent molecule that can immobilize disordered protein segments or change the relative orientation of protein domains. These large structural changes probably constitute the output signal that triggers events downstream in the signaling pathway. However, the manner in which input signals sensed by the various accessory domains are communicated to the cyclase and esterase domains to control cellular c-di-GMP levels is an important question that needs to be addressed in future research. With the elucidation of the activation mechanism of Rec-domain-containing DGCs, the first step has been taken toward achieving this goal. Future work should also focus on elucidating the structure of c-di-GMP receptor/effector complexes as well as understanding the physiological significance of the binding of c-di-GMP to different receptors in possibly different oligomeric states.

REFERENCES

1. Reference deleted.
2. **Alm, R. A., A. J. Bodero, P. D. Free, and J. S. Mattick.** 1996. Identification of a novel gene, *pilZ*, essential for type 4 fimbrial biogenesis in *Pseudomonas aeruginosa*. *J. Bacteriol.* **178:**46–53.
3. **Amikam, D., and M. Y. Galperin.** 2006. PilZ domain is part of the bacterial c-di-GMP binding protein. *Bioinformatics* **22:** 3–6.
4. **Ausmees, N., R. Mayer, H. Weinhouse, G. Volman, D. Amikam, M. Benziman, and M. Lindberg.** 2001. Genetic data indicate that proteins containing the GGDEF domain possess diguanylate cyclase activity. *FEMS Microbiol. Lett.* **204:**163–167.
5. **Bachhawat, P., and A. M. Stock.** 2007. Crystal structures of the receiver domain of the response regulator PhoP from *Escherichia coli* in the absence and presence of the phosphoryl analog beryllofluoride. *J. Bacteriol.* **189:**5987–5995.
6. **Bachhawat, P., G. V. Swapna, G. T. Montelione, and A. M. Stock.** 2005. Mechanism of activation for transcription factor PhoB suggested by different modes of dimerization in the inactive and active states. *Structure* **13:**1353–1363.
7. **Barends, T., E. Hartmann, J. Griese, T. Beitlich, N. Kirienko, D. Ryjenkov, J. Reinstein, R. Shoeman, M. Gomelsky, and I. Schlichting.** 2009. Structure and mechanism of a bacterial light-regulated cyclic nucleotide phosphodiesterase. *Nature* **459:**1015–1018.
8. **Benach, J., S. S. Swaminathan, R. Tamayo, S. K. Handelman, E. Folta-Stogniew, J. E. Ramos, F. Forouhar, H. Neely, J. Seetharaman, A. Camilli, and J. F. Hunt.** 2007. The structural basis of cyclic diguanylate signal transduction by PilZ domains. *EMBO J.* **26:**5153–5166.
9. **Blommers, M. J., C. A. Haasnoot, J. A. Walters, G. A. van der Marel, J. H. van Boom, and C. W. Hilbers.** 1988. Solution structure of the 3′-5′ cyclic dinucleotide d(pApA). A combined NMR, UV melting, and molecular mechanics study. *Biochemistry* **27:**8361–8369.
10. **Boehm, A., S. Steiner, F. Zaehringer, A. Casanova, F. Hamburger, D. Ritz, W. Keck, M. Ackermann, T. Schirmer, and U. Jenal.** 2009. Second messenger signalling governs Escherichia coli biofilm induction upon ribosomal stress. *Mol. Microbiol.* **72:**1500–1516.
11. **Boles, B. R., and L. L. McCarter.** 2002. *Vibrio parahaemolyticus scrABC,* a novel operon affecting swarming and capsular polysaccharide regulation. *J. Bacteriol.* **184:**5946–5954.
12. **Chan, C., R. Paul, D. Samoray, N. C. Amiot, B. Giese, U. Jenal, and T. Schirmer.** 2004. Structural basis of activity and allosteric control of diguanylate cyclase. *Proc. Natl. Acad. Sci. USA* **101:**17084–17089.
13. **Christen, B., M. Christen, R. Paul, F. Schmid, M. Folcher, P. Jenoe, M. Meuwly, and U. Jenal.** 2006. Allosteric control of cyclic di-GMP signaling. *J. Biol. Chem.* **281:**32015–32024.
14. **Christen, M., B. Christen, M. G. Allan, M. Folcher, P. Jenö, S. Grzesiek, and U. Jenal.** 2007. DgrA is a member of a new family of cyclic diguanosine monophosphate receptors and controls flagellar motor function in Caulobacter crescentus. *Proc. Natl. Acad. Sci. USA* **104:**4112–4117.
15. **Christen, M., B. Christen, M. Folcher, A. Schauerte, and U. Jenal.** 2005. Identification and characterization of a cyclic di-GMP-specific phosphodiesterase and its allosteric control by GTP. *J. Biol. Chem.* **280:**30829–30837.
16. **De, N., M. Pirruccello, P. V. Krasteva, N. Bae, R. V. Raghavan, and H. Sondermann.** 2008. Phosphorylation-independent regulation of the diguanylate cyclase WspR. *PLoS Biol.* **6:**e67.
17. **Duerig, A., S. Abel, M. Folcher, M. Nicollier, T. Schwede, N. Amiot, B. Giese, and U. Jenal.** 2009. Second messenger-mediated spatiotemporal control of protein degradation regulates bacterial cell cycle progression. *Genes Dev.* **23:**93–104.
18. **Egli, M., R. V. Gessner, L. D. Williams, G. J. Quigley, G. A. van der Marel, J. H. van Boom, A. Rich, and C. A. Frederick.** 1990. Atomic-resolution structure of the cellulose synthase regulator cyclic diguanylic acid. *Proc. Natl. Acad. Sci. USA* **87:** 3235–3239.
19. **Ferreira, R. B., L. C. Antunes, E. P. Greenberg, and L. L. McCarter.** 2008. Vibrio *parahaemolyticus* ScrC modulates cyclic dimeric GMP regulation of gene expression relevant to growth on surfaces. *J. Bacteriol.* **190:**851–860.
20. **Frederick, C. A., M. Coll, G. A. van der Marel, J. H. van Boom, and A. H. Wang.** 1988. Molecular structure of cyclic deoxydiadenylic acid at atomic resolution. *Biochemistry* **27:** 8350–8361.
21. **Galperin, M. Y., A. N. Nikolskaya, and E. V. Koonin.** 2001. Novel domains of the prokaryotic two-component signal transduction systems. *FEMS Microbiol. Lett.* **203:**11–21.
22. **Gao, R., T. R. Mack, and A. M. Stock.** 2007. Bacterial response regulators: versatile regulatory strategies from common domains. *Trends Biochem. Sci.* **32:**225–234.
23. **Garcia, B., C. Latasa, C. Solano, F. Garcia-del Portillo, C. Gamazo, and I. Lasa.** 2004. Role of the GGDEF protein family in Salmonella cellulose biosynthesis and biofilm formation. *Mol. Microbiol.* **54:**264–277.
24. **Gordeliy, V. I., J. Labahn, R. Moukhametzianov, R. Efremov, J. Granzin, R. Schlesinger, G. Büldt, T. Savopol, A. J. Scheidig, J. P. Klare, and M. Engelhard.** 2002. Molecular basis of transmembrane signalling by sensory rhodopsin II-transducer complex. *Nature* **419:**484–487.
25. **Guan, Y., Y. G. Gao, Y. C. Liaw, H. Robinson, and A. H. Wang.** 1993. Molecular structure of cyclic diguanylic acid at 1 A resolution of two crystal forms: self-association, interactions with metal ion/planar dyes and modeling studies. *J. Biomol. Struct. Dyn.* **11:**253–276.
26. **Hickman, J. W., and C. S. Harwood.** 2008. Identification of FleQ from Pseudomonas aeruginosa as a c-di-GMP-responsive transcription factor. *Mol. Microbiol.* **69:**376–389.
27. **Hogg, T., U. Mechold, H. Malke, M. Cashel, and R. Hilgenfeld.** 2004. Conformational antagonism between opposing active sites in a bifunctional RelA/SpoT homolog modulates (p)ppGpp metabolism during the stringent response [corrected]. *Cell* **117:**57–68.
28. **Jeon, Y. H., Y. S. Heo, C. M. Kim, Y. L. Hyun, T. G. Lee, S. Ro, and J. M. Cho.** 2005. Phosphodiesterase: overview of protein structures, potential therapeutic applications and recent progress in drug development. *Cell. Mol. Life Sci.* **62:** 1198–1220.

29. **Jung, A., J. Reinstein, T. Domratcheva, R. L. Shoeman, and I. Schlichting.** 2006. Crystal structures of the AppA BLUF domain photoreceptor provide insights into blue light-mediated signal transduction. *J. Mol. Biol.* **362:**717–732.
30. **Kuchma, S. L., K. M. Brothers, J. H. Merritt, N. T. Liberati, F. M. Ausubel, and G. A. O'Toole.** 2007. BifA, a cyclic di-GMP phosphodiesterase, inversely regulates biofilm formation and swarming motility by *Pseudomonas aeruginosa* PA14. *J. Bacteriol.* **189:**8165–8178.
31. **Kumar, M., and D. Chatterji.** 2008. Cyclic di-GMP: a second messenger required for long-term survival, but not for biofilm formation, in Mycobacterium smegmatis. *Microbiology* **154:** 2942–2955.
32. **Lee, V. T., J. M. Matewish, J. L. Kessler, M. Hyodo, Y. Hayakawa, and S. Lory.** 2007. A cyclic-di-GMP receptor required for bacterial exopolysaccharide production. *Mol. Microbiol.* **65:**1474–1484.
33. **Li, T. N., K. H. Chin, J. H. Liu, A. H. Wang, and S. H. Chou.** 2009. XC1028 from Xanthomonas campestris adopts a PilZ domain-like structure without a c-di-GMP switch. *Proteins* **75:** 282–288.
34. **Liaw, Y. C., Y. G. Gao, H. Robinson, G. M. Sheldrick, L. A. Sliedregt, G. A. van der Marel, J. H. van Boom, and A. H. Wang.** 1990. Cyclic diguanylic acid behaves as a host molecule for planar intercalators. *FEBS Lett.* **264:**223–227.
35. **Marina, A., C. D. Waldburger, and W. A. Hendrickson.** 2005. Structure of the entire cytoplasmic portion of a sensor histidine-kinase protein. *EMBO J.* **24:**4247–4259.
36. **Merighi, M., V. T. Lee, M. Hyodo, Y. Hayakawa, and S. Lory.** 2007. The second messenger bis-(3′-5′)-cyclic-GMP and its PilZ domain-containing receptor Alg44 are required for alginate biosynthesis in Pseudomonas aeruginosa. *Mol. Microbiol.* **65:**876–895.
37. **Minasov, G., S. Padavattan, L. Shuvalova, J. S. Brunzelle, D. J. Miller, A. Baslé, C. Massa, F. R. Collart, T. Schirmer, and W. F. Anderson.** 2009. Crystal structures of YkuI and its complex with second messenger c-di-GMP suggests catalytic mechanism of phosphodiester bond cleavage by EAL domains. *J. Biol. Chem.* **284:**13174–13184.
38. **Nakasone, Y., T. A. Ono, A. Ishii, S. Masuda, and M. Terazima.** 2007. Transient dimerization and conformational change of a BLUF protein: YcgF. *J. Am. Chem. Soc.* **129:**7028–7035.
39. **Neiditch, M. B., M. J. Federle, A. J. Pompeani, R. C. Kelly, D. L. Swem, P. D. Jeffrey, B. L. Bassler, and F. M. Hughson.** 2006. Ligand-induced asymmetry in histidine sensor kinase complex regulates quorum sensing. *Cell* **126:**1095–1108.
40. **Paul, R., S. Abel, P. Wassmann, A. Beck, H. Heerklotz, and U. Jenal.** 2007. Activation of the diguanylate cyclase PleD by phosphorylation-mediated dimerization. *J. Biol. Chem.* **282:** 29170–29177.
41. **Paul, R., T. Jaeger, S. Abel, I. Wiederkehr, M. Folcher, E. G. Biondi, M. T. Laub, and U. Jenal.** 2008. Allosteric regulation of histidine kinases by their cognate response regulator determines cell fate. *Cell* **133:**452–461.
42. **Paul, R., S. Weiser, N. C. Amiot, C. Chan, T. Schirmer, B. Giese, and U. Jenal.** 2004. Cell cycle-dependent dynamic localization of a bacterial response regulator with a novel diguanylate cyclase output domain. *Genes Dev.* **18:**715–727.
43. **Pei, J., and N. V. Grishin.** 2001. GGDEF domain is homologous to adenylyl cyclase. *Proteins* **42:**210–216.
44. **Pratt, J. T., R. Tamayo, A. D. Tischler, and A. Camilli.** 2007. PilZ domain proteins bind cyclic diguanylate and regulate diverse processes in Vibrio cholerae. *J. Biol. Chem.* **282:**12860–12870.
45. **Ramelot, T. A., A. Yee, J. R. Cort, A. Semesi, C. H. Arrowsmith, and M. A. Kennedy.** 2007. NMR structure and binding studies confirm that PA4608 from Pseudomonas aeruginosa is a PilZ domain and a c-di-GMP binding protein. *Proteins* **66:** 266–271.
46. **Rao, F., Y. Qi, H. H. Chong, M. Kotaka, B. Li, J. Li, J. Lescar, K. Tang, and Z. X. Liang.** 2009. The functional role of a conserved loop in EAL domain-based cyclic di-GMP-specific phosphodiesterase. *J. Bacteriol.* **191:**4722–4731.
47. **Rao, F., Y. Yang, Y. Qi, and Z. X. Liang.** 2008. Catalytic mechanism of cyclic di-GMP-specific phosphodiesterase: a study of the EAL domain-containing RocR from *Pseudomonas aeruginosa*. *J. Bacteriol.* **190:**3622–3631.
48. **Ross, P., Y. Aloni, C. Weinhouse, D. Michaeli, P. Weinberger-Ohana, R. Meyer, and M. Benziman.** 1985. An unusual guanyl oligonucleotide regulates cellulose synthesis in Acetobacter xylinum. *FEBS Lett.* **186:**191–196.
49. **Ross, P., R. Mayer, and M. Benziman.** 1991. Cellulose biosynthesis and function in bacteria. *Microbiol. Rev.* **55:**35–58.
50. **Ryan, R. P., Y. Fouhy, J. F. Lucey, L. C. Crossman, S. Spiro, Y. W. He, L. H. Zhang, S. Heeb, M. Camara, P. Williams, and J. M. Dow.** 2006. Cell-cell signaling in Xanthomonas campestris involves an HD-GYP domain protein that functions in cyclic di-GMP turnover. *Proc. Natl. Acad. Sci. USA* **103:** 6712–6717.
51. **Ryjenkov, D. A., R. Simm, U. Römling, and M. Gomelsky.** 2006. The PilZ domain is a receptor for the second messenger c-di-GMP: the PilZ domain protein YcgR controls motility in enterobacteria. *J. Biol. Chem.* **281:**30310–30314.
52. **Ryjenkov, D. A., M. Tarutina, O. V. Moskvin, and M. Gomelsky.** 2005. Cyclic diguanylate is a ubiquitous signaling molecule in bacteria: insights into biochemistry of the GGDEF protein domain. *J. Bacteriol.* **187:**1792–1798.
53. **Schirmer, T., and U. Jenal.** 2009. Structural and mechanistic determinants of cyclic di-GMP signaling. *Nat. Rev. Microbiol.* **7:**724–735.
54. **Schmid, F. F., and M. Meuwly.** 2007. All-atom simulations of structures and energetics of c-di-GMP-bound and free PleD. *J. Mol. Biol.* **374:**1270–1285.
55. **Schmidt, A. J., D. A. Ryjenkov, and M. Gomelsky.** 2005. The ubiquitous protein domain EAL is a cyclic diguanylate-specific phosphodiesterase: enzymatically active and inactive EAL domains. *J. Bacteriol.* **187:**4774–4781.
56. **Sinha, S. C., and S. R. Sprang.** 2007. Structures, mechanism, regulation and evolution of class III nucleotidyl cyclases. *Rev. Physiol. Biochem. Pharmacol.* **157:**105–140.
57. **Stock, A. M., V. L. Robinson, and P. N. Goudreau.** 2000. Two-component signal transduction. *Annu. Rev. Biochem.* **69:** 183–215.
58. **Sudarsan, N., E. R. Lee, Z. Weinberg, R. H. Moy, J. N. Kim, K. H. Link, and R. R. Breaker.** 2008. Riboswitches in eubacteria sense the second messenger cyclic di-GMP. *Science* **321:** 411–413.
59. **Tamayo, R., A. D. Tischler, and A. Camilli.** 2005. The EAL domain protein VieA is a cyclic diguanylate phosphodiesterase. *J. Biol. Chem.* **280:**33324–33330.
60. **Tarutina, M., D. A. Ryjenkov, and M. Gomelsky.** 2006. An unorthodox bacteriophytochrome from Rhodobacter sphaeroides involved in turnover of the second messenger c-di-GMP. *J. Biol. Chem.* **281:**34751–34758.
61. **Tschowri, N., S. Busse, and R. Hengge.** 2009. The BLUF-EAL protein YcgF acts as a direct anti-repressor in a blue-light response of Escherichia coli. *Genes Dev.* **23:**522–534.
62. **Wassmann, P., C. Chan, R. Paul, A. Beck, H. Heerklotz, U. Jenal, and T. Schirmer.** 2007. Structure of $BeF3^{-}$-modified response regulator PleD: implications for diguanylate cyclase

activation, catalysis, and feedback inhibition. *Structure* **15**: 915–927.

63. **Weber, H., C. Pesavento, A. Possling, G. Tischendorf, and R. Hengge.** 2006. Cyclic-di-GMP-mediated signalling within the sigma network of Escherichia coli. *Mol. Microbiol.* **62**:1014–1034.

64. **Zhang, Z., B. L. Gaffney, and R. A. Jones.** 2004. c-di-GMP displays a monovalent metal ion-dependent polymorphism. *J. Am. Chem. Soc.* **126**:16700–16701.

65. **Zhang, Z., S. Kim, B. L. Gaffney, and R. A. Jones.** 2006. Polymorphism of the signaling molecule c-di-GMP. *J. Am. Chem. Soc.* **128**:7015–7024.

III. CYCLIC DI-GMP AND LIFESTYLE CHOICES

The Second Messenger Cyclic Di-GMP
Edited by A. J. Wolfe and K. L. Visick

Chapter 8

Choosing the Right Lifestyle: Regulation of Developmental Pathways by Cyclic Di-GMP

MATTHEW PARSEK AND PHILLIP ALDRIDGE

Over the last decade, the perception that bacterial life simply involves individual cells growing and dividing symmetrically has been drastically altered. We now accept the concept and importance of temporal and spatial coordination for the core macromolecular units required to achieve growth in a single cell. More so, we have also been sensitized to the fact that many bacterial species can adopt different lifestyles (Fig. 1).

One lifestyle involves growth on a surface as a sessile community. Biofilms are surface-associated bacterial communities that are important in many different environments (19). Biofilm communities are generally embedded within an extracellular matrix. The matrix composition is complex, consisting of a mixture of exopolysaccharides (EPS), proteins, and nucleic acids. This matrix serves as scaffolding, holding bacteria together and to the surface.

In contrast, the planktonic lifestyle is associated with growth of individual bacteria primarily in liquid environments. A significant proportion of bacterial species that utilize the planktonic lifestyle possess the ability to move (57). Bacterial movement is defined as either swimming, swarming, or twitching. Swimming describes bacterial motility through an aqueous environment, whereas swarming or twitching motility relates to movement across a surface. Specific environmental signals dictate the direction of movement toward chemoattractants and away from chemorepellents (86).

The ability to grow in liquid as well as on surfaces and the ability to direct the transition between these two lifestyles play a significant role during host-bacterium interactions. We have come to appreciate that these lifestyles involve a series of developmental events, with evidence of cross-regulation between these developmental pathways. For the coordinated transition between environmentally dictated lifestyles, cyclic di-GMP (c-di-GMP) has emerged as a central player (Fig. 1). In this chapter, our goal is to provide you with an understanding of the general developmental concepts associated with c-di-GMP regulation that will be discussed in detail in the following chapters. We will provide you with a broad understanding of the concepts of planktonic and biofilm lifestyles and the modes utilized by bacteria to achieve directed movement. Thus, in later chapters, this foundation will allow those authors to focus more on the importance of c-di-GMP than explain in detail the underlying specific pathways being discussed.

THE PROS AND CONS OF SESSILITY VERSUS MOTILITY

The ability of motile bacteria to swim in an aqueous environment is a tremendous advantage contributing to the fitness of the bacterium. Motile bacteria are able to respond to both chemoattractants and repellents, facilitating the acquisition of nutrients and helping avoid stressful environments. Motile cells are also able to actively seek out and colonize new environments. Why then would a motile bacterium wish to give up such benefits by attaching to a surface?

There are several advantages for adopting a biofilm lifestyle that are not mutually exclusive and that

Matthew Parsek • Department of Microbiology, University of Washington, School of Medicine, Seattle, WA 98195-7242. **Phillip Aldridge** • Centre for Bacterial Cell Biology and Institute for Cell and Molecular Biosciences, Newcastle University, Newcastle upon Tyne, NE2 4HH, United Kingdom.

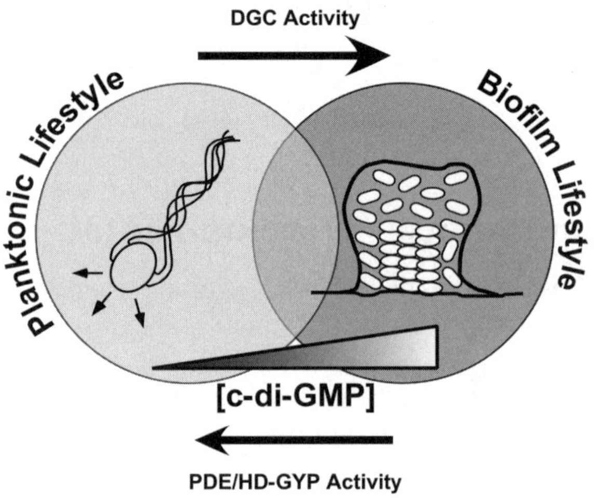

Figure 1. Influence of c-di-GMP in controlling the transition from the planktonic to biofilm lifestyle and vice versa. DGC enzymes make c-di-GMP, while its breakdown is mediated by either PDE or HD-GYP activity. The observation that many mutants in the genes that encode DGC and PDE of HD-GYP enzymes result in pleiotropic or incomplete phenotypes suggests that cells use c-di-GMP concentrations ([c-di-GMP]) as a means to measure when it is optimal to fully commit to either lifestyle. As a result, there is a level of c-di-GMP where both planktonic and biofilm traits can still be functional (overlapping circles). Only once the thresholds of this intersection have been passed will a cell commit to either lifestyle.

depend upon the local environment (19, 20, 107, 129). A characteristic of biofilm communities is their ability to tolerate different types of antimicrobial stress. Depending upon the antimicrobial agent, embedded biofilm bacteria can be up to 1,000 times more resistant to antimicrobial stress than planktonic bacteria of the same species (35, 127). The presence of slow-growing subpopulations in the interior of a biofilm community, reduced penetration of antimicrobials into the biofilm, and patterns of biofilm-specific gene expression have all been suggested to contribute to biofilm antimicrobial tolerance (87, 107, 126, 127). In addition to antimicrobial tolerance, biofilm communities can also tolerate physical stressors. For example, biofilms retain moisture and can provide a measure of tolerance against desiccation (15). Therefore, in a stressful environment, entering a biofilm state may protect the bacterium.

Biofilm formation can also provide nutritional advantages. Aquatic surfaces are known to accumulate nutrients (19). In some cases, surfaces themselves can serve as a nutritional source, such as the shells of aquatic invertebrates that are colonized by chitin-utilizing bacteria (6, 66, 113, 119). In multispecies biofilms, some bacterial species can position themselves in proximity to other species that produce nutritional substrates as a by-product of their own metabolism. For example, nitrifying bacteria in wastewater biofilms proliferate in close spatial proximity to one another (6, 66, 119). Ammonia-oxidizing bacteria produce nitrite, a substrate of nitrite-oxidizing bacteria. Not surprisingly, nitrite-oxidizing bacteria are seen to position themselves next to ammonia oxidation in wastewater biofilms. Thus, biofilm formation may be a means to immobilize oneself in a favorable nutritional environment. This is also an important aspect of oral microbial communities, which form biofilms on the enamel and gingival tissue present in the oral cavity (72).

The efficiency of horizontal gene transfer can also be greatly improved in biofilms. Being spatially stabilized has been shown to facilitate conjugal plasmid transfer. This has been demonstrated for the transfer of both antibiotic resistance and catabolism plasmids (95, 98). The extracellular environment of many biofilm systems is also rich in DNA (143). It has been hypothesized that this DNA may participate in transformation events, although this has not been definitively demonstrated in the laboratory.

In summary, there are distinct advantages to both free-swimming and biofilm lifestyles. In general, free-swimming bacteria may be better suited to identifying new environments, permitting colonization, and proliferation. In contrast, biofilms are better at allowing bacteria exposed to suboptimal or stressful conditions to survive. There are probably multiple environmental stimuli that would provide input into making the decision to stick or swim, and what these stimuli are has been the focus of much biofilm-related research.

We will now begin our discussion of these two lifestyles by defining the regulatory principles associated with c-di-GMP important for the transition between these lifestyles. We will then discuss the biofilm lifestyle and the transition to this lifestyle from a planktonic state. This will then be followed by a discussion of the major mechanisms by which bacteria achieve movement.

THE REGULATORY MECHANISMS OF c-di-GMP

Intracellular c-di-GMP levels are controlled by a few general enzymatic activities responsible for its synthesis and its breakdown (58). c-di-GMP is synthesized from two GTP molecules by a protein domain with diguanylate cyclase (DGC) activity (see chapters 4 and 5). Two unrelated protein domains with phosphodiesterase (PDE) activity catalyze its breakdown: the EAL domain that breaks c-di-GMP into the linear form pGpG and the HD-GYP domain that breaks it down into two GMP molecules (58) (Fig. 1). These enzymatic domains (DGC, EAL, and HD-GYP) can be found associated with a wide range of well-characterized signaling or sensory proteins including two-component response regulators (34; chapter 3). This suggests that, in many species, a wide range of environmental stimuli modulate c-di-GMP levels. The studies described in the following chapters will highlight that these sensory proteins are responsible for modulating the temporal and spatial activity of DGC and PDE domains. Considering the developmental role of c-di-GMP, this association is of no surprise. What is intriguing is the immense diversity found within bacterial genomes when the arrangements of these domains and proteins are compared (34).

The diversity of proteins harboring DGC and PDE domains is reflected by the growing list of cellular processes regulated by these proteins. c-di-GMP has been associated with the regulation of transcription, translation, and the activity of macromolecular protein complexes such as the flagellum and type IV pili (58). The regulatory activities of c-di-GMP include (i) allosteric regulation of an enzyme activity through a direct interaction (115), (ii) regulation of a cellular process through interaction of proteins containing PilZ domains (116), (iii) direct interaction with transcription factors (47), and (iv) direct interaction with RNA molecules (131). The finding of Sudarsan et al. that c-di-GMP directly interacts with riboswitches (RNA) offers an added level of regulatory control at the level of translation (131). The wide-ranging ability of c-di-GMP to interact with a number of cellular components and its ability to regulate different functions at multiple levels make it difficult to define the mechanisms by which c-di-GMP regulation works. A central concept of c-di-GMP regulation is that its accumulation favors biofilm formation and inhibits motility and the planktonic lifestyle (Fig. 1) (144). Alteration of the intracellular c-di-GMP levels or deletion of specific c-di-GMP catalytic enzymes in many cases does not lead to a complete loss or gain of a given phenotype. This is partly due to the redundancy of enzymes capable of synthesizing or degrading c-di-GMP (58). Therefore, understanding c-di-GMP regulatory effects is not as simple as measuring the effect of modulating cytoplasmic c-di-GMP levels. Importantly, c-di-GMP synthesis/degradation can be spatially compartmentalized (58). This suggests that a major regulatory role of c-di-GMP is to modulate the amplitude of a given response to a particular stimulus rather than acting as an all-or-nothing switch. This is very similar to the mechanisms involved in modulating the response of the chemotaxis machinery. During the perception of attractants and repellents, the amplitude of the signal is modulated through the modification of the methyl-accepting chemoreceptors by a methylase (137). This suggests that c-di-GMP offers the cell the choice to continually assess whether the decision to pass through a transition from one lifestyle to another is optimal or not. As a result, a cell will only commit fully to a change once a threshold level of c-di-GMP input has been reached (Fig. 1).

One final concept to consider with respect to c-di-GMP regulation is at which stage in a process it acts. The lifestyle choices we will describe, once initiated, are extremely robust. For example, once a biofilm is established, it will exist in a state of flux for a considerable length of time and there are not many factors that can completely destroy it (129). However, at the regulatory level, c-di-GMP depletion is one of the few ways to remove a biofilm (133). From this standpoint, c-di-GMP can be considered a master regulator of biofilm formation in many species.

The motors that drive assisted movement also exhibit this property. The assembly pathways of the main driving forces are robust. Very few external inputs are known to affect the assembly pathway once it is started. The one input that can "sweep a bacterium off its feet" by inhibiting motility is c-di-GMP (144). It is important to note at this point that the majority of other systems that modulate motility often do so through an indirect or direct regulation of global inputs fed into these systems through their associated master regulators (125). As a result, what is occurring in these cases is a change in the rate of initiation of assembly rather than a direct inhibition of the assembly pathway or function. Importantly c-di-GMP possesses the ability to directly inhibit or promote function and/or assembly of motility systems (144). The result of this flexibility is that pathways can be shut down at both ends: functional components can be inhibited while the assembly pathway can also be stopped.

THE COMMUNAL LIFESTYLE AND c-di-GMP

Many bacteria in natural and clinical environments have a tendency to stick to surfaces. These surface-associated communities called biofilms have been an intense focus of microbiological research in the last 5 to 10 years. Except for some infections, most biofilms contain multiple species. Biofilms usually consist of cells in close proximity and at high cell density, resulting in a high potential for interspecies cooperation and competition.

What Is the Definition of a Biofilm?

How one defines a biofilm is frequently the source of debate. Some claim that simple association with a surface is the defining feature of biofilm cells, while others have argued that cells firmly fixed to a surface and encased in an extracellular matrix are representative of the true biofilm state. Is, therefore, a monolayer of individual cells on a surface a biofilm? A colony on an agar plate? Currently, biofilms are usually defined on the basis of qualitative properties, such as the formation of aggregates or the acquisition of antimicrobial tolerance. Hampering our ability to definitively identify bacteria in the biofilm state is the lack of universal, biofilm-specific molecular markers. For the purpose of this chapter, we will define biofilms as surface-associated communities embedded in a self-produced matrix.

The Practical Consequences of Biofilm Formation

Depending upon the species, biofilm formation can have several consequences in industrial and clinical environments. The CDC has estimated that upwards of 65% of all bacterial infections are caused by biofilm bacteria. In these situations, usually opportunistic pathogens colonize dead tissue or indwelling medical devices (20). Once a biofilm infection is established, it is difficult to eradicate with standard antimicrobial therapy. *Pseudomonas aeruginosa* causes biofilm infections in the airways of people suffering from cystic fibrosis (CF) (41, 50), where it persists for years, despite aggressive antibiotic therapy. In industry, bacterial biofilms foul surfaces important in many industrial processes.

Thus, there is strong general interest in preventing biofilm formation. While some strategies are geared toward producing surfaces that are recalcitrant to microbial colonization (such as the silver-coated venous catheters used in the medical industry), researchers are also interested in manipulating microbial populations in an effort to control biofilm formation. One promising avenue has been the use of the transition metal gallium, which serves as an iron analog (64). Iron limitation is a known environmental signal that dissuades some species of bacteria from initiating biofilm development (122). At concentrations subinhibitory to growth, gallium is able to prevent biofilm accumulation by *P. aeruginosa*. However, researchers are also interested in broader strategies that may be effective against a wider range of bacterial species. c-di-GMP control potentially represents such a therapeutic target of biofilm formation.

Biofilm Formation as a Developmental Program

Biofilm formation has been likened to a developmental process consisting of several steps common to many different species (103, 120). These steps include attachment to a surface, the formation of cellular aggregates called microcolonies, and the active departure of cells from a biofilm in a process called detachment or dispersion. Intense research over the past decade has focused on the genetic factors that contribute to these different steps in biofilm formation. Some of these functions are discussed in the next section. Interestingly, c-di-GMP signaling is now known to control different aspects of, if not all of, these developmental steps (Fig. 2).

Identification of the genetic determinants that contribute to *P. aeruginosa* biofilm formation on abiotic surfaces has demonstrated a number of important points. Importantly, the biofilm culturing conditions (e.g., incubation in static or flow environments) influences the requirement of a particular factor (8, 69, 114, 121). As a result, many factors that influence biofilm formation are not essential but modulate the developmental process in some way. For example, the requirement of the single polar flagellum for initial adherence to a surface depends upon the carbon source on which *P. aeruginosa* is grown (69). The second important point is that biofilm formation is a complex process: it is not simply bacterial growth on a surface. Biofilm development can involve subpopulations within a community exhibiting different behaviors that contribute to the final biofilm structure and its persistence.

Several groups have demonstrated that *P. aeruginosa* is capable of forming two general types of biofilms in the laboratory (28, 67) (Fig. 2). A relatively confluent, uniform community of bacteria is called a flat biofilm (Fig. 2). In contrast, a structured biofilm consists of cell aggregates or mushrooms separated by channels or spaces. Both biofilm types exhibit the hallmark characteristic of biofilms—enhanced tolerance to antimicrobial treatment relative to planktonic cells (75, 132).

The formation of both flat and structured biofilms is thought to proceed through discrete steps or stages: initial attachment, the formation of mature biofilms, and ultimately, dispersion, where a subpopulation of the community swims away from the mature biofilm, reinitiating the developmental cycle (Fig. 2). However, there are developmental features specific to either the flat or structured biofilms. For example, the formation of a flat biofilm is characterized by a significant degree of surface motility (121). Ultimately, bacteria multiply and continue to move on the surface, forming a confluent mat of cells.

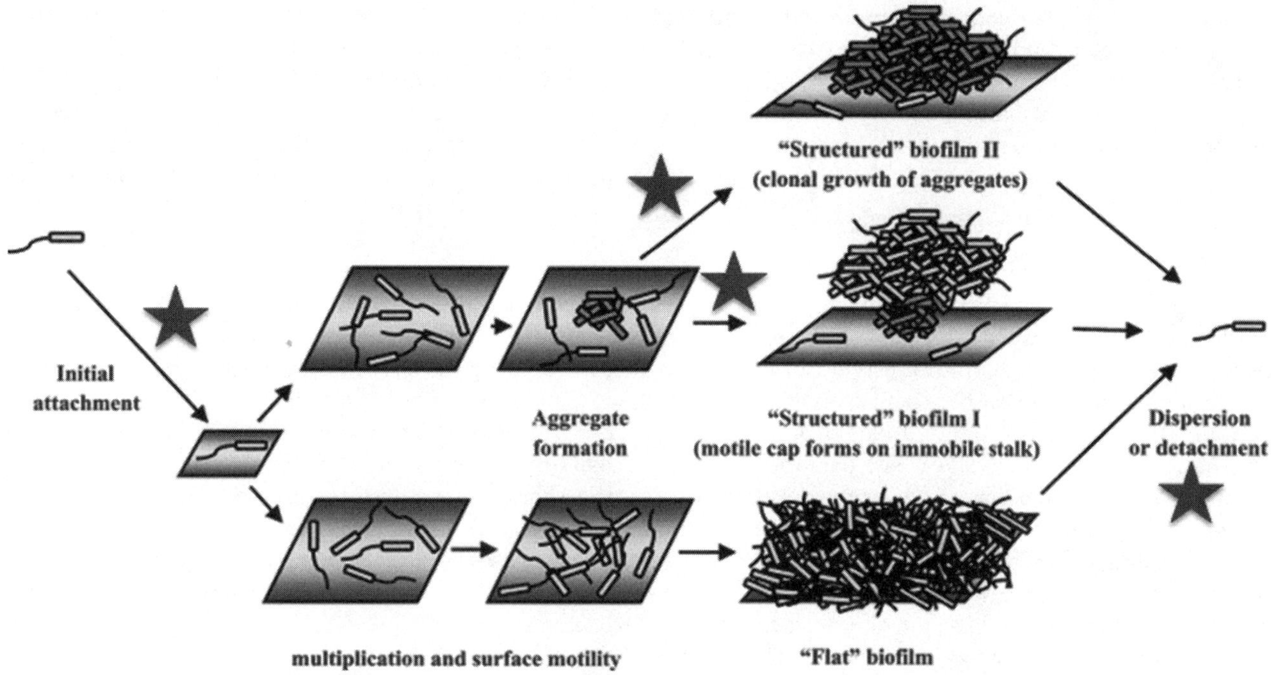

Figure 2. Diagram showing the different stages of *P. aeruginosa* biofilm formation that are influenced by c-di-GMP (indicated by the stars). Elevated c-di-GMP promotes initial attachment, which involves adherence of free-swimming cells to a surface. In the case of a flat biofilm, cells continue to multiply and move on the surface, forming a confluent, flat mat of cells. In the case of structured biofilms, we present two alternative routes to their formation. The first we call structured biofilm I. Here, dark gray cells represent the immobile stalks of structured biofilms, while light gray cells represent the motile subpopulation that produce the cap. The second we call structured biofilm II. In both cases, the structured biofilms are characterized by EPS production (either Pel, Psl, or alginate, depending upon the strain). EPS production is promoted by elevated c-di-GMP. In this case, small cell aggregates grow clonally, forming large cell aggregates consisting of cells primarily derived from cells in the small cell aggregates. This is indicated by the large aggregate of dark gray cells in the figure, indicating that the cells in the aggregate are progeny derived from the initial small aggregates. Finally, cells can actively leave the biofilm to reinitiate the cycle in a process called dispersion or detachment. c-di-GMP has also been implicated in this step.

The development of structured biofilms initially involves the formation of cell aggregates on a surface. Eventually, a motile subpopulation of the biofilm migrates to the surface of a nonmotile subpopulation called a stalk, forming the cap of a mushroom-like structure (68). Lequette and Greenberg suggested that the stalk population may be producing the surfactant rhamnolipid, which is thought to grease the surface of the stalk, facilitating movement of the motile subpopulation onto it (78). Finally, rhamnolipid production is required to maintain the open channels and spaces between the cell aggregates of structured biofilms (27).

Dispersion is an event that has been best studied in pseudomonads. In this case, cells actively leave a biofilm in response to prevailing environmental conditions. This process is thought to involve both an active release from the biofilm matrix and an induction of swimming motility. c-di-GMP signaling is important for this process (discussed later in the motility section). Gjermansen and Tolker-Nielsen demonstrated that carbon starvation of *Pseudomonas putida* biofilms induced rapid dispersion (38). This process required a GGDEF/EAL domain-containing protein encoded by the *pp0165* gene (37). This group has now shown that induction of c-di-GMP PDE activity is key for dispersion in this organism. Morgan et al. identified an MCP-like protein in *P. aeruginosa,* BdlA, that is required for dispersion in response to the nutrient environment in *P. aeruginosa.* They went on to link BdlA to modulating c-di-GMP levels in the cell in response to the nutritional environment, although the direct molecular nature of this link is unclear (96).

Determinants That Influence Biofilm Formation in Gram-Negative Species

A few classes of determinants and conditions are known to impact biofilm formation in several gram-negative bacterial species. Many of these have also been linked to c-di-GMP signaling, either directly or indirectly. We will summarize a few of these findings and discuss their known relationship to c-di-GMP signaling. When considering these factors, it is important to note that many play a role in modulating biofilm formation rather than being essential.

Surface Appendages

Several surface features of different bacterial species have been shown to contribute to biofilm formation. During the initial stages of biofilm formation, cells converting to the biofilm lifestyle can alter their attachment from a reversible to an irreversible interaction. This change is brought about by a number of alternative surface appendages. For example, the flagellum (to be described later) has been shown to contribute to initial reversible adherence to a surface for *Escherichia coli, P. aeruginosa, Vibrio cholerae,* and a number of other motile species (104, 111, 141). The flagellum is thought to increase the surface area of the bacterium, maximizing bacterium surface interactions. Some species, such as *P. aeruginosa* and *E. coli,* have been shown to attach to abiotic surfaces at the tip of the flagellum.

Different fimbrae and pili have also been shown to contribute to surface attachment. Type IV pili in *P. aeruginosa,* curli in *E. coli,* and the mannose-sensitive pilus in *V. cholerae* are all important adherence factors (69, 112, 142). Surface-associated adhesins can also assist in bacterium-surface interactions. The large adhesin, LapA, of *Pseudomonas fluorescens* is important for the transition to the irreversible state, where the long axis of the rod-shaped cell becomes affixed to the surface (49).

These surface appendages also contribute to postattachment steps during biofilm formation. One example is the three-chaperone usher pathway *(cup)* fimbriae in *P. aeruginosa.* Three gene clusters designated *cupA-C* encode these appendages. The *cupA* and *cupC* fimbriae contribute to biofilm maturation (134). Furthermore, expression of the *cupC* fimbriae is under control of the c-di-GMP PDE, RocR (73).

EPS

EPS have the potential to impact bacterial biofilm development at multiple levels. Secreted polysaccharides represent the most abundant and extensively studied component of the extracellular biofilm matrix. Depending upon the species and environmental conditions, the amount and types of EPS produced can vary extensively. Some bacteria can synthesize more than one type of EPS during biofilm formation. For example *P. aeruginosa* can produce at least three (alginate, Pel, and Psl) (31, 32).

EPS can either positively or negatively influence attachment depending upon the EPS type and the organism. For example, in 1973, Fletcher and Floodgate (29) reported an acidic polysaccharide associated with a marine pseudomonad that enhanced attachment to surfaces. Similarly, different EPS types have been reported to promote adherence in a variety of other species such as the VPS of *Vibrio cholerae* and PIA production by certain staphylococci (84, 146).

There are also examples where EPS has been shown to inhibit attachment, as in the case of colanic acid of uropathogenic *Escherichia coli.* Colanic acid is a polymer of glucose, galactose, fucose, and glucuronic acid. Interestingly, although colanic acid impairs initial attachment, its production is required for the characteristic architecture seen in mature *E. coli* biofilms (22). Biofilm-derived EPS from a mucoid isolate of *P. aeruginosa* deposited on a glass surface has also been shown to inhibit attachment of planktonic bacteria (40). EPS production can also impact biofilm development downstream of attachment. *P. aeruginosa* produces mucoid colony variants, which are characterized as overproducers of the EPS alginate (Fig. 3). When comparing nonmucoid to mucoid *P. aeruginosa* strains, mucoid strains produced robust biofilms with more extensive architecture (46, 99). The mucoid biofilms were also demonstrated to show heightened resistance to tobramycin compared to the nonmucoid biofilms, suggesting that alginate overproduction by biofilm cells in the CF lung may play a protective role.

Two other EPSs, Pel and Psl, have also been linked to biofilm development in nonmucoid strains of *P. aeruginosa.* Psl is important for both attachment to surfaces and maintenance of a mature biofilm, while Pel appears to maintain biofilm biomass (31, 56, 83, 136). Finally, in *P. aeruginosa,* a type of lipid-linked EPS, rhamnolipid, was also demonstrated to maintain heterogeneous biofilm architecture (27). Rhamnolipid is thought to function as a surfactant, keeping the channels that run through the depths of the biofilm cell free. The result is that nutrients can penetrate the biofilm and waste products can leave.

Vibrio cholerae, the causative agent of cholera, can be isolated as two different colonial variants: smooth and rugose. Rugose strains produce wrinkled, smaller colonies on solid medium and are characterized by enhanced EPS production. The EPS of rugose *V. cholerae* O1 El Tor, termed EPS^{ETr}, is composed of mostly glucose and galactose and is encoded by the *vps* locus (146). In the laboratory, rugose variants can be isolated from stationary-phase batch liquid cultures of the smooth variant. The rugose isolates revert to the smooth colony phenotype with high frequency, suggesting a phase variation mechanism. The

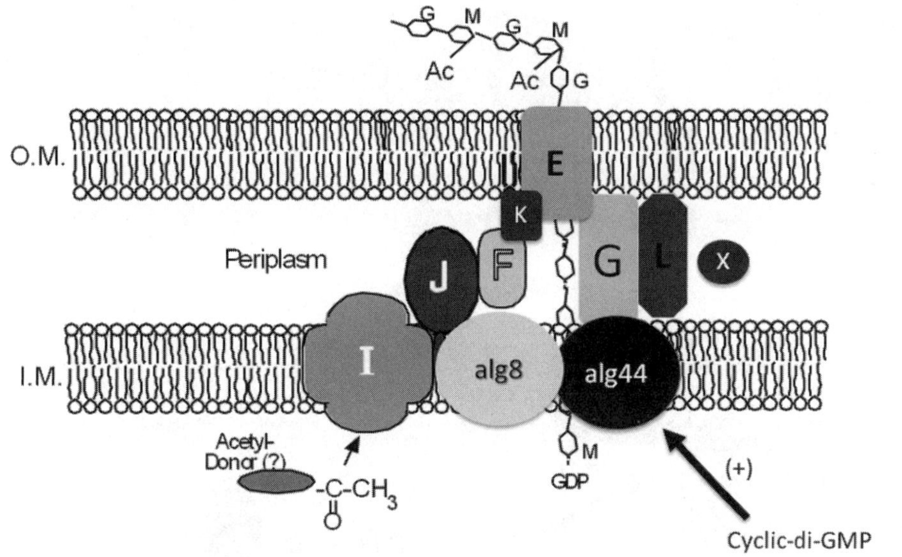

Figure 3. Schematic diagram depicting the alginate biosynthetic machinery. c-di-GMP binds the PilZ domain of Alg44, acting as a positive allosteric activator. Alg8 and Alg44, inner membrane (I.M.) proteins; K (AlgK), periplasmic lipoprotein with lipid anchor in the outer membrane (O.M.) and a tetratricopeptide repeat scaffold protein; E (AlgE), outer membrane porin-like protein that interacts with AlgK; G (AlgG), periplasmic protein that has a right-handed beta-helical domain and contains the epimerase active site; L (AlgL), a periplasmic protein likely involved in alginate processing; F (AlgF), periplasmic protein; J (AlgJ), periplasmic protein anchored in the inner membrane by a noncleaved signal peptide; I (AlgI), inner membrane protein with seven membrane-spanning domains; X (AlgX), periplasmic protein (function not known yet).

rugose variant produces thicker, more structured biofilms than the smooth variant. This EPS also provides rugose strains with added protection against chlorine. In fact, when purified EPS^{ETr} was added to cultures of smooth *V. cholerae* variants, survival against chlorine treatment was greatly enhanced. The rugose phenotype was hypothesized to help *V. cholerae* form protective biofilms in the environment.

Importantly, the production of EPS during biofilm development is a common target for c-di-GMP signaling. In almost every case studied, c-di-GMP has been linked, in some context, to control of EPS abundance during biofilm development. Historically, this mode of c-di-GMP regulation led to its discovery through its allosteric control of cellulose synthesis in *Gluconacetobacter xylinus* (previously *Acetobacter xylinum*) by Benziman and coworkers (115; chapter 2). *G. xylinus* utilizes the cellulose it produces to form a pellicle. Pellicles are in essence a biofilm of bacteria able to float at a liquid-surface interface. There is evidence that c-di-GMP signaling can regulate alginate production at the allosteric level. A PilZ domain-containing protein (see later), Alg44, was shown to positively regulate alginate production (92) (Fig. 3). Positive allostery also regulates the Pel biosynthetic machinery, with PelD being the target for activation (77) (Fig. 4). There is also evidence that c-di-GMP positively regulates the Psl enzyme activity (48, 67).

Bomchil et al. identified a gene in a *V. cholerae* El Tor strain designated *mbaA* (for *m*aintenance of *b*iofilm *a*rchitecture) that was necessary for formation of biofilms with characteristic mushroom and tower structures. Mutant biofilms exhibited densely packed, homogeneous biofilms with little or no architecture (12). The MbaA primary amino acid indicates that it is a transcriptional regulator harboring a DGC motif. Scanning electron microscopy analysis of *mbaA* mutant biofilms shows increased production of some biopolymer, suggesting that MbaA regulates EPS production.

The Nutritional Environment

Several environmental factors have been shown to influence biofilm formation. The nutritional environment is crucial in this regard. Klausen et al. demonstrated that *P. aeruginosa* grown on glucose as the sole carbon source forms structured biofilms, while glutamate- and succinate-grown biofilms were flat (68). Another example of the nutritional environment controlling biofilm formation is the important role played by catabolite repression during biofilm formation by *E. coli.* The cAMP receptor protein

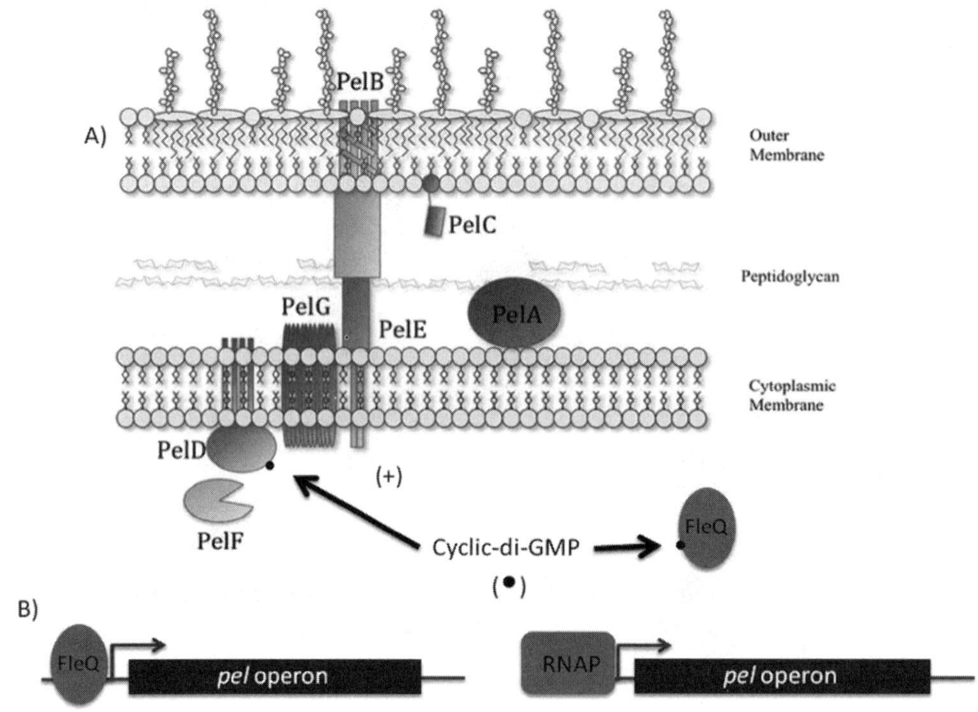

Figure 4. c-di-GMP controls Pel production at two levels. (A) Schematic depicting the predicted localization of the Pel biosynthetic machinery. The localization is primarily based upon bioinformatics, except in the case of PelC, which has been shown experimentally to localize to the outer membrane. c-di-GMP acts as an allosteric activator of PelD. The Pel proteins are predicted to have the following functions. PelA is a large soluble, periplasmic protein with no clear function; it has a large amount of disorder but may adopt a TIM barrel structure. PelB is an outer membrane protein with a large N-terminal TPR domain followed by a C-terminal porin domain. PelC is an outer membrane lipoprotein that may adopt a TolB-like structure. PelD is a cytoplasmic membrane protein with four transmembrane helices (TMs) that has a c-di-GMP binding domain. PelE is a cytoplasmic membrane protein with two TMs followed by a periplasmic region (residues 90 to 320) carrying four or five TPR motifs (residues 155 to 310). Residues 90 to 150 are predicted to carry loops and helices; residues ~115 to 127 may be disordered. PelF is a cytoplasmic glycosyltransferase. PelG is a cytoplasmic membrane protein with 12 TMs that is a member of the polysaccharide transporter family and also resembles Na^+/H^+ antiporters. (B) c-di-GMP also binds to the transcriptional repressor FleQ. When c-di-GMP levels are low, FleQ represses *pel* expression (left). At elevated levels, c-di-GMP binds FleQ, relieving repression of *pel* expression (right).

controls catabolite repression in *E. coli* by assessing relative levels of cAMP. When glucose was the primary carbon source, Jackson et al. (55) observed a decrease in biofilm formation.

Iron is another important nutritional parameter for some species. Iron depletion is a signal for *P. aeruginosa* to inhibit biofilm formation. Kaneko et al. demonstrated that low levels of iron did not impact planktonic growth and reduced biofilm biomass accumulation and increased twitching motility. In agreement, gallium, a cation analog of iron, has been shown to specifically disrupt *P. aeruginosa* biofilms (64).

c-di-GMP—a Molecular Metric for Assessing the Biofilm State?

As alluded to above, a question that has plagued biofilm microbiologists is "how do you define a biofilm?" There are several different definitions, and almost all involve some qualitative criteria, such as microscopic verification that aggregates are present or the observation that the community shows elevated resistance to antimicrobial stressors (129). Unfortunately, the absence of consistent molecular markers has hampered scientists in their attempt to rigorously quantify whether microbes in complex systems are in the biofilm state. This is a key consideration, as several economically important infections have been tentatively proposed to involve biofilms using qualitative criteria. Molecular confirmation of the biofilm lifestyle might alter a therapeutic approach to some infections or even alter our perception of the pathogenesis of some pathogens.

Assessing c-di-GMP levels may prove to be such a molecular diagnostic tool. Although current technology may not allow determination of c-di-GMP levels at the single-cell level in a complex sample, development of such technologies might be possible. To validate such an approach would require testing strains with carefully tuned c-di-GMP levels in different acute- and chronic-phase models. In addition, effects on relative in vitro biofilm formation would have to be measured. Such a tool would allow the researcher to ask some interesting questions. For example, does an entire population of *P. aeruginosa* in CF airways adopt the biofilm lifestyle? If not, may this allow us to correlate different ratios of biofilm versus planktonic organisms with a pathologic outcome?

Relationship between c-di-GMP and the Chronic- versus Acute-Phase Infection Paradigm

Many gram-negative bacterial opportunistic pathogens are capable of causing a wide range of infections. This is perhaps best illustrated by *P. aeruginosa,* which is capable of causing a number of nosocomial infections, including the chronic airway infections that afflict people suffering from CF (138). The pathogeneses of these infections vary from explosive septicemia that rapidly kills the host to CF airway infections that can persist for decades. Amazingly, these chronic lung infections can produce huge numbers of bacteria in the airways (10^9/ml of sputum) that do not disseminate and invade any tissue or organs outside the lungs. Why is *P. aeruginosa* able to cause such different types of infections? Part of the answer lies in the ability of the bacterium to inversely regulate the expression of factors that promote acute infection and those that promote persistence.

Several of the acute virulence factors produced by *P. aeruginosa,* such as secreted proteases, exotoxin A, other secreted toxins, motility functions, and type III secretion (T3S) have been documented for their role in acute-phase infection (138). In several different animal models of infection, mutating these factors produces an attenuated virulence phenotype. Additionally, intercellular signaling (or quorum sensing) controls several of these functions. The opposite is seen during the course of chronic infection. Adaptive mutations occur that many times reduce or eliminate the expression of acute virulence factors, while promoting expression of factors related to persistence. For example, in the airways, there is an accumulation of *lasR* mutations, which are known to reduce expression of several acute virulence functions (82). Often the motility capabilities are lost, while mutations leading to expression of polysaccharides are observed. For example, mutations in the anti-sigma factor *mucA* promote expression of the EPS alginate, producing the mucoid phenotype (145). In other words, the bacterium is acquiring mutations that dampen acute invasive virulence while accumulating mutations that protect it from antimicrobials and the host immune response.

It appears that c-di-GMP may be a focal point for this inverse regulation (33). High levels of c-di-GMP induce biofilm formation, while low levels of c-di-GMP promote motility (Fig. 1) and T3S. Highlighting this point are some of the adaptive mutations selected for in the CF airways that target c-di-GMP signaling. An example is the mutation of *wspF,* which is known to produce a rough small colony variant on solid medium (48). This mutation results in stimulation of the DGC WspR, which in turn produces high intracellular levels of c-di-GMP (23). This induces hyper biofilm formation and induction of biofilm matrix components such as the Pel and Psl polysaccharides. These variants also display enhanced resistance to aminoglycosides and oxidative stress. Thus, these variants have many traits that would contribute to persistence. Smith et al. recently verified that *wspF* mutations are selected for in the CF environment (124). It will be interesting to see if other gram-negative pathogens that produce chronic infection show adaptive mutations directed toward c-di-GMP signaling.

In summary, biofilm development is tighly intertwined with c-di-GMP signaling. A major role played by this mode of regulation is the positive control of EPS synthesis once attachment has been established. There is hope that with the development of new techniques we may be able to accurately quantify the biofilm lifestyle through changes in c-di-GMP levels. However, at the same time as promoting biofilm formation, c-di-GMP regulation is concomitantly down-regulating key planktonic phenotypes, specifically the ability of cells to undergo assisted movement. Finally, c-di-GMP signaling may also contribute to cells leaving a biofilm and entering the planktonic phase (Fig. 2).

BACTERIAL MOVEMENT AND c-di-GMP

Now that we have discussed the biofilm lifestyle, it is time to turn our attention to the planktonic lifestyle and, especially, the control of motility systems by c-di-GMP. The planktonic lifestyle is associated with the ability to mediate movement between two points. Bacteria are remarkable in their ability to utilize alternative mechanisms to achieve passive or assisted movement in different environments (57). Not every bacterial species will utilize an active mode of movement. A significant proportion of bacterial species utilize passive modes of movement between niches that exploit environmental conditions such as water flow and aerosal dispersal.

Assisted movement is mediated primarily by bacterial flagella and type IV pili (61, 86). A third mode of movement, requiring an ABC transporter, has also been described for members of the *Flavobacterium* group (57, 89). However, here we will focus on the well-characterized systems of flagella and type IV pili. Of these two systems, flagellum-mediated motility is used for covering large distances in liquid environments. Flagella, however, can also mediate surface-associated movement. In contrast, type IV pilus-mediated motility occurs only on a surface. In the following sections we will discuss how bacteria utilize these structures to move and how c-di-GMP regulates their function and the expression of the associated genes.

The Multiple Forms of Bacterial Motility

With respect to movement, bacteria are defined as being able to swim or swarm (Fig. 5). Importantly, the ability to utilize one or more of these modes of movement can dictate the efficiency of the transition from the planktonic to biofilm lifestyle. Swimming is defined as flagellum-mediated motility of individual cells in liquid environments and is driven by the rotation of a flexible whip-like filament (86). The flexibility of the filament results in a corkscrew rotation that generates the necessary thrust to propel the bacterial cell in any given direction (9, 11). There is significant diversity among flagellar systems with respect to the spatial organization of flagella on the cell surface and their numbers. The predominant types include (i) between 4 and 10 peritrichous flagella, randomly distributed over the cell surface, (ii) a single polar flagellum sometimes enclosed within a membraneous sheath, (iii) a tuft of flagella at a pole, and (iv) the periplasmic flagellum of the spirochetes (57). The rotation of the flagellum is generated by a proton or sodium gradient created by the static motor force generators that surround the base components of the flagellum (10, 71). The chemotaxis machinery that senses chemical gradients in the environment dictates the direction or speed of rotation. The perception of these gradients leads to a change in flagellar rotation, observed as the ability of the cell to either run or tumble. The switch between runs and tumbles is facilitated by the ability to rotate the flagellum in either a counterclockwise (run) or clockwise (tumble) manner (24). For multiflagellated bacteria, during a run, the helical nature of their flagellar filaments and the coordinated rotation of their flagella lead to a bundle being formed. The change in rotational direction leads to a disruption of the bundle, causing the cells to tumble. Not all flagellar systems generate runs and tumbles through a counterclockwise-clockwise rotational switch. Some systems use either (i) a stop-start mechanism or (ii) unidirectional rotation associated with either fast or slow torque generation (137). Irrespective of this, a random biased walk either up or down a perceived gradient results in a series of random runs and tumbles where the durations of runs are longer than tumbles.

Motility of individual cells in a liquid environment is relatively simple compared to surface-associated motility. Swarming motility is defined as the ability to move over surfaces in an organized community-based or social manner. The two most predominant types of swarming behavior are (i) flagellum-mediated swarming and (ii) type IV pilus-dependent gliding or twitching motility (43, 61). Swarming can occur over almost any moist surface. Both types of swarming behavior can be assisted by EPS and surfactants that promote the optimal conditions for movement. Microscopic evaluation of swarm edges from a number of species have identified the coordinated movement by groups of cells as well

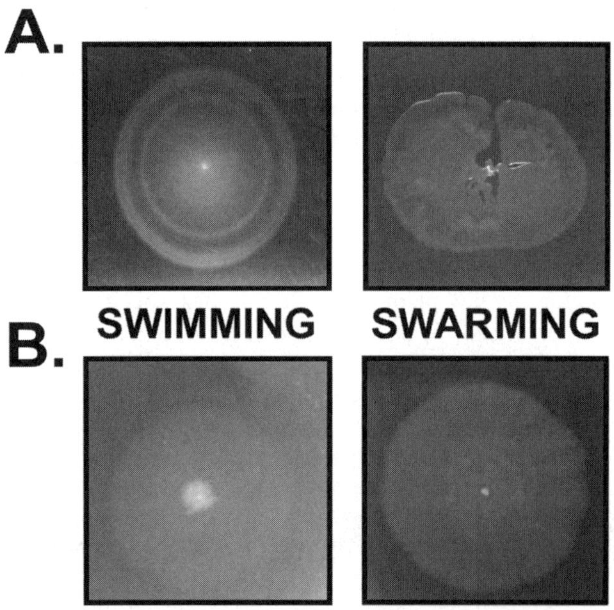

Figure 5. Contrasting patterns of flagellum-mediated swimming and swarming used to exemplify the difference between these two modes of movement. (A) *Salmonella enterica* serovar Typhimurium exhibiting flagellum-mediated swimming and swarming; (B) *Pseudomonas aeruginosa*. Motility assays for swimming are performed using 0.3% agar while swarming assays utilize between 0.5 and 0.7% agar. On 0.3% agar, the resulting swarm is of a uniform diameter containing characteristic rings embedded within the agar matrix. These rings are waves of bacteria utilizing chemotaxis to alternative nutrients in the medium. In contrast, swarms produced on 0.5 to 0.7% agar are much more spectacular in shape and diameter.

as individual cells escaping out of the leading edge. This behavior causes some swarming bacteria to exhibit nonuniform patterns on certain media, while others produce spectacular patterns (see reference 44 for examples) in contrast to the uniform pattern associated with swimming motility (Fig. 5).

It is generally accepted that swimming and swarming are utilized at different stages of the bacterial life cycle (Fig. 1). Logically, swimming motility promotes the movement of planktonic cells through aqueous environments in search of nutritional sources. Flagellar filaments play a critical role in the early stages of attachment to surfaces. Swarming is a crucial component of the early stages that lead to biofilm formation of many bacterial species. For example, Shrout et al. have demonstrated the importance of this behavior during biofilm development in *P. aeruginosa* (121). Albeit a social mode of motility, swarming does not always lead to the formation of a complex colony structure. In fact, this type of movement can promote dispersal over nutrient-rich surfaces where swimming is not possible (Fig. 5). Importantly, both modes of motility are subject to c-di-GMP control.

Flagellum-mediated swarming in enteric bacteria is characterized by an upregulation of flagellar gene expression, leading to hyperflagellation and an associated elongation of the swarmer cell (44). Studies of *Proteus mirabilis* have shown that these hyperflagellated cells form rafts of cells held together by their flagella and EPS (44). The role played by the EPS is to reduce surface tension, easing movement over the surface (4, 42, 45). Exactly how flagellar rotation mediates this coordinated movement is still not clearly understood. *P. mirabilis* swarms tend to have a wonderful rippled appearance. This results from these bacterial rafts moving in a coordinated manner for a period of time before a period of consolidation where they return to a single-cell vegetative state (30). The behavior of *P. mirabilis* is, however, just one example of flagellum-mediated swarming. Not all bacterial species capable of swarming do so like *P. mirabilis.* As a result, swarming motility patterns exhibit some variability in different species (Fig. 5).

Gliding and twitching motility was first described as a jerky motion that was flagellum independent, hence the term twitching (23). The best-studied systems for twitching motility are *P. aeruginosa, Neisseria gonorrhoeae,* and *Myxococcus xanthus* (60). *M. xanthus* utilizes two modes of motility over surfaces: (i) adventurous or A-motility that is still not well defined and (ii) social gliding motility (94). In principle, the mechanisms that drive twitching and social gliding motility are identical and will be discussed together. This mode of movement is controlled by a cycle of type IV pilus assembly, attachment to a surface, and then retraction of the pili (14, 109, 139). The current model is that multiple pili are assembled at the leading pole on the cell surface until they become attached to either the surface or another object such as a neighboring cell. Attachment signals then halt assembly and initiate pilus retraction. The net effect of this cycle is that the cell is physically pulled toward the site of attachment. Time-lapse microscopy of swarms has shown that cells can undergo localized individual twitching motility until they come into contact with other cells (105). Upon doing so, the cells become organized in a head-to-tail fashion. As a result and similar to flagellum-mediated swarming, swarming by type IV pilus-driven motility is observed to occur in groups of cells that can be 10 to 50 cells wide at the leading edge (43).

Type IV pilus-driven motility is controlled by a signal transduction cascade similar to the chemotaxis machinery of the flagellar system (94). During type IV pilus-driven motility, cells are observed to reverse their direction of movement, similar to runs and tumbles during swimming. Thus, just like when individual cells are swimming, bacteria engaged in surface motility are capable of perceiving environmental signals and alter the direction of movement. Illustrating this point, Kearns et al. demonstrated that *P. aeruginosa* can move up phosphotidyl ethanolamine gradients on solid medium using twitching motility (65). In *M. xanthus,* the control of direction results from the selective inhibition and activation of pilus assembly at the leading and lagging poles, respectively (94). The reversal rate is much slower than changes observed in swimming behavior. Under nonchemotactic conditions, *E. coli* will run for approximately 1 s with 0.1-s periods of tumbles (43). In contrast, directional reversion in *M. xanthus* has been calculated to occur every 8 s (60). However, these changes must also be placed into context with the speed of these types of movement. Swimming cells can reach speeds up to 50 μm/s. In contrast, swarming motility is much slower. Flagellum-mediated swarming has been calculated to occur at between 2 to 10 μm/s, while twitching motility only reaches speeds of between 0.06 and 0.3 μm/s. The lower speed of twitching motility is in part a result of its dependence upon the assembly-retraction cycle that occurs at approximately equal rates of 0.5 μm/s in *P. aeruginosa* (123).

The Concept of the Motile-to-Sessile Transition

Upon identifying a new niche, a bacterial cell must rapidly recognize this and switch to a sessile state to efficiently colonize the newfound environment. It is in this decision/transition that c-di-GMP plays its central role with respect to the control of motility. Conceptually, when passing through a motile-to-sessile transition (from now on defined as Mot^+ to Mot^-), a series of events must occur. All motile objects that utilize a motor can pass through one of four states: (i) moving in any given direction with the engine on (runs), (ii) stationary but with the engine on (tumbles or reversals), (iii) free movement with the engine off, and (iv) stationary with the engine off. When entering the equivalent of state 4 (Mot^- engine off), it is advantageous if no more motors are built and the existing ones are either functionally inhibited or destroyed. For pilus-mediated motility, this transition is a lot simpler than it is for flagella due to the nature of the assembly-retraction cycle. Thus, for pilus-mediated movement, all that is necessary is to inhibit either the assembly or retraction process, and almost immediately, twitching motility will come to a halt (Fig. 6). With respect to the flagellum, c-di-GMP has been shown to have several inputs into flagellar systems, including (i) the direct ability to stop flagellar rotation and (ii) altering flagellar gene expression through interactions with key flagellum-associated transcription factors (144). The result of the latter control (discussed below) is the gradual reduction in the rate of flagellar assembly until the rate of cell division is higher than the rate of assembly of a single flagellum, leading to the majority of daughter cells becoming aflagellate. To appreciate how c-di-GMP mediates this control, we must first understand how pili and flagella are assembled or removed.

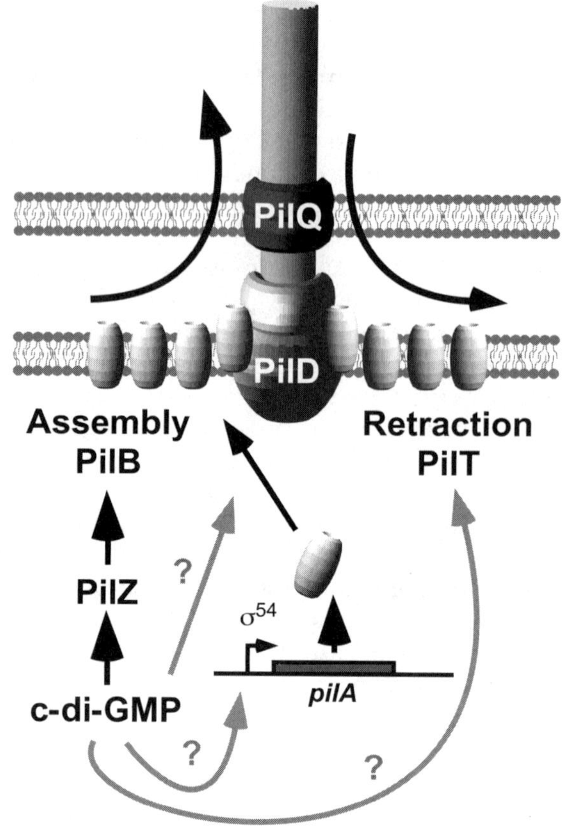

Figure 6. Schematic diagram of the type IV pilus and the assembly and retraction cycles mediated by the two ATPases PilB and PilT. The six major components are not all depicted in the structure for clarity, and the inner membrane anchor PilD and the proposed secretin PilQ are labeled. The models for assembly and retraction reflect the proposed models of Craig and Li (21) (assembly) and Kaiser (61) (retraction). An underlying principle of both cycles is that pilin subunits can be stored in the inner membrane until needed. The control of pilus assembly by c-di-GMP through PilZ is highlighted. The gray arrows indicate alternative routes for c-di-GMP regulation of the process, most of which require further investigation.

The Type IV Pilus Assembly-Retraction Pathway

Type IV pili are extremely strong, flexible filaments of a helical nature that are much thinner than the flagellar filament, having an average diameter of 6 to 9 nm and a 6- to 11-Å central channel (109). These filaments are able to withstand greater than 100 piconewtons of force. Pili are macromolecular structures requiring up to 40 gene products for efficient assembly and retraction (60). The best-characterized type IV pilus systems are those of *P. aeruginosa* and *N. gonorrhoeae*. Here, where necessary, we will use the *P. aeruginosa* gene nomenclature for ease (70). The core structure of a pilus requires six proteins: a major pilin, a prepillin peptidase, an assembly ATPase, a retraction ATPase, an inner membrane basal protein, and a secretin pore associated with the outer membrane (Fig. 6). Type IV pili are predominantly found within gram-negative bacteria and not only facilitate motility across surfaces but are also involved in DNA uptake, adhesion, protein secretion, and phage transduction (109).

The central channel of the pilus filament is far too small for unfolded proteins but has been proposed to aid flexibility (21). As a result of the tight packing of pilin subunits into the filament, assembly of the pilus is strikingly different from the flagellar filament (Fig. 6 and 7). Instead of incorporation at the growing tip, type IV pili are assembled by the addition of subsequent pilin subunits to the base of the structure at the periplasmic face of the inner membrane, pushing the entire structure outwards. Prepilin subunits are introduced into the inner membrane in a *sec*-dependent manner. These are then brought into the vicinity of the basal components of the pilus: PilC, the inner membrane anchor; PilD, the prepilin peptidase; and the assembly

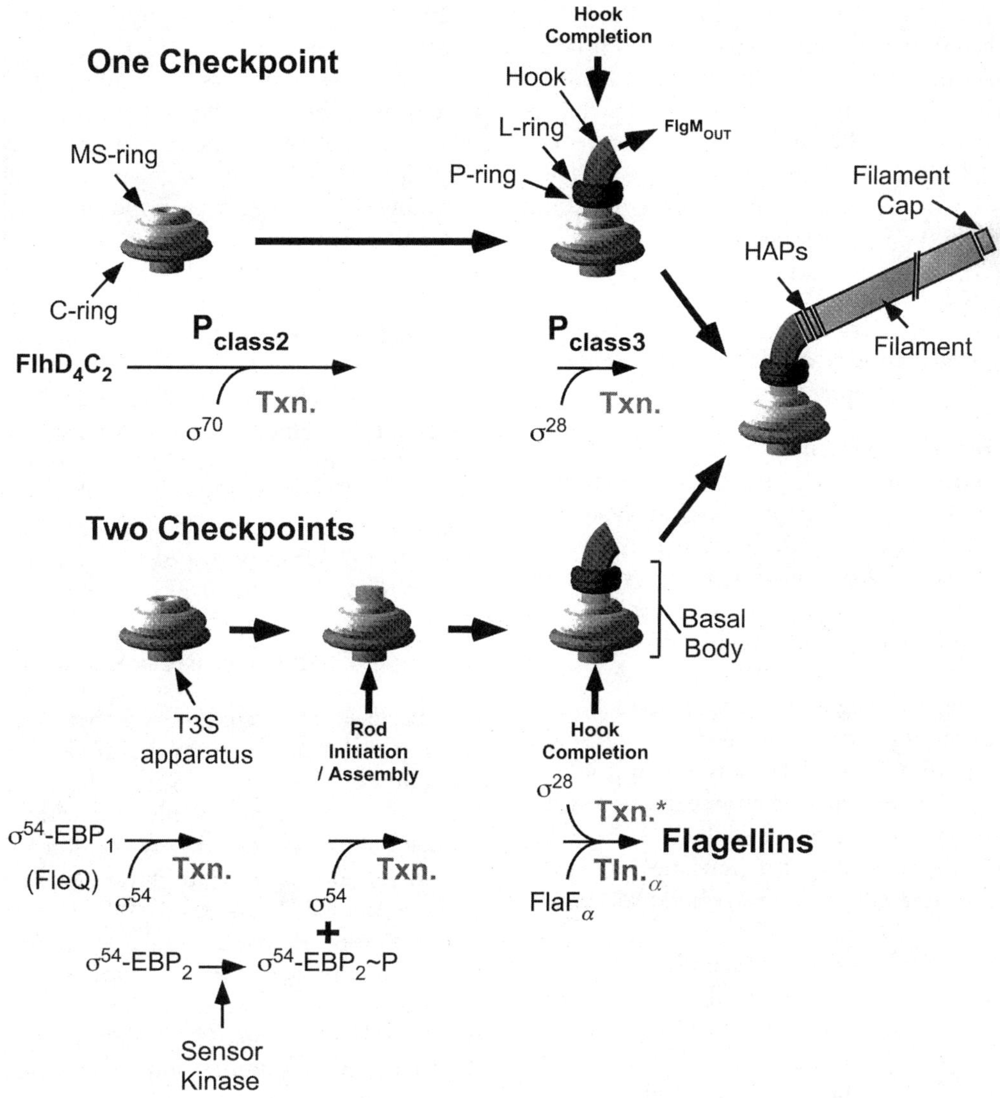

Figure 7. Schematic diagram comparing the coordination of flagellar gene expression and assembly when one or two assembly checkpoints are used. The important structural features of the flagellum are highlighted. One checkpoint: this pathway is utilized by the paradigm flagellar systems of *E. coli* and *S. enterica* serovar Typhimurium. The master regulator $FlhD_4C_2$ activates P_{class2} in conjunction with σ^{70}. This promoter class drives the expression of the HBB structural components and several regulatory proteins, including σ^{28} and FlgM. Upon HBB completion, FlgM is secreted, allowing σ^{28} to initiate transcription from P_{class3}. Two checkpoints: one example of this pathway is that used by *P. aeruginosa*. Here proximal basal body gene expression is activated by the atypical σ^{54} EBP, FleQ. A second σ^{54} EBP recognizes the initiation of T3S/rod assembly, activating transcription of genes that encode the distal HBB structural components and a number of other regulatory components. For *P. aeruginosa,* HBB completion is sensed in a similar manner to that of *E. coli* that results in σ^{28} activating flagellin gene expression. In *C. crescentus,* this checkpoint is coupled to flagellin translation, not transcription, as an added twist. Note, *C. crescentus* does not utilize a FleQ homologue to activate its transcription hierarchy but does utilize a σ^{54} EBP_2 sensory system to coordinate distal basal body gene expression and assembly. Txn., transcription; Tln., translation.

ATPase, PilB (Fig. 6). PilD cleaves the prepilin leader sequence and, in the model of Craig and Li, leaves the cleaved pilin associated with the inner membrane. These cleaved pilins are then incorporated into the growing pilus filament by utilizing the energy released from ATP hydrolysis by PilB. The type IV pilus requires a conduit to emerge from the outer membrane of gram-negative bacteria (21). It has been proposed that a protein, PilQ, similar to secretin, constitutes this ring.

The major difference between the type IV pilus and the flagellum is that the pilus facilitates movement through cycles of assembly and retraction. The retraction process is dependent upon energy released from ATP hydrolysis by the retraction ATPase PilT (93). The retraction mechanism has been compared to the function of a ratchet. In this mechanism, PilT acts to pull the pili back toward the cell surface. The energy released from ATP hydrolysis is used to remove the basal pilins from the pilus (Fig. 6). These pilins are not always recycled through proteolysis but are stored within the local proximity of the pilus base for a subsequent round of pilus assembly (21).

The Temporal Pathway of Flagellar Assembly

The bacterial flagellar system is a highly conserved macromolecular structure consisting of 29 structural subunits and, thus, much more complex than pili (85, 86) (Fig. 7). In total, flagellar assembly requires over 60 genes that include the structural subunits and regulators that control protein secretion and flagellar gene expression. Of the 600-plus genomes now annotated, approximately 45% possess all the necessary components to build a flagellum (62, 63). Structurally, the flagellum can be divided into three distinct components: the basal body, the hook, and the filament (Fig. 7) (85). The basal body anchors the structure in the bacterial membrane and also generates the necessary torque to rotate the filament. The hook acts as a flexible joint that translates the rotational energy generated by the basal body to the filament.

Flagellar assembly is a temporal pathway that begins with the insertion of the flagellum base in the inner membrane known as the MS-ring (Fig. 7) (85). The MS-ring acts as the foundation for the rest of the basal body that spans either the periplasmic space of gram-negative bacteria or the thick cell wall of gram-positive bacteria. On the cytoplasmic interface of the MS-ring, the C-ring and flagellum-specific T3S apparatus assemble (1, 2). The C-ring is involved in torque generation through interactions with the motor force generators and the chemotaxis protein CheY (137). The T3S apparatus is required to secrete all external structural subunits except for two rings known as the P- and L-rings. These two static rings form stabilizers that are associated with the peptidoglycan layer (P-ring) and the lipopolysaccharide outer membrane (L-ring) of gram-negative bacteria. After MS-ring/C-ring assembly, the rest of the structural subunits are secreted through a central channel and incorporated onto the growing tip of the structure. A rigid rod completes the basal body. From rod assembly onward, the growing structure is always capped with one of three capping proteins: the rod cap FlgJ, the hook cap FlgD, or the filament cap FliD (86). FlgJ possesses muramidase activity to allow the rod to pass through the peptidoglycan layer of gram-negative bacteria (97).

The hook is the first component of the flagellum that is assembled on the cell surface. Its completion represents an important regulatory checkpoint with respect to coordinating flagellar gene expression and assembly (Fig. 7) (2). A hook-filament junction is formed from three subunits, the hook-associated proteins, on the completed hook. The apical subunit of this junction is the filament cap protein FliD. FliD guides incorporation of the filament subunit, flagellin, into the growing filament (147). This incorporation event occurs below FliD, which remains at the growing tip. The filaments themselves are a helical assembly of flagellins with a width of approximately 20 nm. These filaments possess a hollow inner channel, approximately 3 nm wide, that extends all the way down to the MS-ring through which unfolded flagellin subunits travel up for assembly at the tip (118).

c-di-GMP Direct Regulation of Motility

Because of the nature of their assembly pathways, it is not surprising that c-di-GMP regulation of the pilus or flagellum is very different. Two direct ways to control type IV pilus function are to either reduce or increase the supply of pilin subunits (gene expression or stability) or modulate the activity of either PilB or PilT (the motor). One example of c-di-GMP control of type IV pili is through positive regulation of the assembly-retraction cycle by PilZ, described for *P. aeruginosa* and *Neisseria meningitidis* pilus systems (Fig. 6) (5). Deletion mutants of *pilZ* can make pilin subunits but cannot assemble them into the pilus. Importantly, this positive regulatory role of PilZ is consistent with the coordinated activation of twitching motility and increased c-di-GMP synthesis during biofilm development. In contrast, inhibition of flagellar function requires impairment of rotation. Two plausible mechanisms are by preventing the stator-basal body interaction or through inhibition of CheY-C-ring interaction.

The marine bacterium *Caulobacter crescentus* is a versatile model system for the study of regulation of the flagellar Mot^+-to-Mot^- transition due to the integration of this developmental switch into its cell cycle. In *C. crescentus,* the proteins DgrA and PleD regulate flagella rotation and function, while TipF regulates flagellar assembly (3, 17, 52, 108; see also chapter 9). The DgrA protein contains a PilZ domain, PleD is a DGC, and TipF is a PDE. Interestingly, the functional control of flagellar rotation by DgrA is potentially found across the bacterial kingdom. The DgrA homolog of *E. coli,* YcgR, was shown to have a similar effect on flagellum function (Fig. 8) (110,

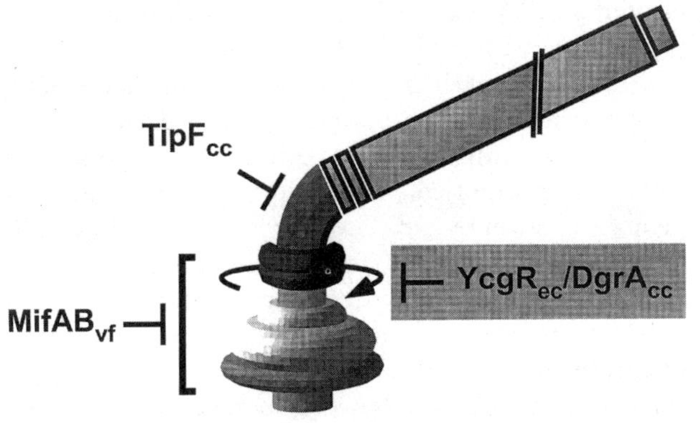

Figure 8. Summary of the multiple targets during flagellar assembly influenced by c-di-GMP regulation. (Gray box) Flagellar rotation is negatively controlled by the homologues YcgR and DgrA in *E. coli* (ec) and *C. crescentus* (cc), respectively. Flagellar assembly in *C. crescentus* and *V. fischeri* (vf) is controlled at a posttranscriptional level during HBB assembly, potentially to control the Mot$^-$-to-Mot$^+$ transition. A direct role for c-di-GMP has been shown for the *P. aeruginosa* flagellar master regulator FleQ (see Fig. 3).

116; see also chapter 4). This suggests that proteins such as DgrA and YcgR can interact with the flagellum to negatively regulate rotation and that this regulation is dependent upon the c-di-GMP-PilZ interaction.

Huitema et al. (52) have shown that, in *C. crescentus,* the PDE protein TipF regulates the assembly pathway. The exact mechanisms of this regulation are still unclear. Importantly, TipF was shown to arrest hook assembly (Fig. 8). Mutants of *tipF* exhibited reduced levels of hook and flagellin translation. As TipF is a PDE, this suggests that postbasal body assembly is inhibited by c-di-GMP and its breakdown to pGpG is required at a critical stage during assembly. For the reverse transition of sessile to motile, TipF regulation suggests that the cell can hold flagellar assembly in a state ready to promote a rapid exit from the sessile lifestyle by reducing inhibitory levels of c-di-GMP.

Vibrio fischeri is usually motile in high magnesium (Mg^{2+}) concentrations, indicative of its planktonic lifestyle in seawater. When entering the light organ (proposed to have a low Mg^{2+} concentration) of the squid *Euprymna scolopes, V. fischeri* enters a Mot$^+$-to-Mot$^-$ transition (see chapter 13). In a screen for motile mutants in low magnesium concentrations, two DGC, MifA and MifB, were isolated (102). Consistent with the role of TipF in *C. crescentus,* MifA and MifB regulate the reverse Mot$^-$-to-Mot$^+$ transition. MifA and MifB negatively regulate flagellar assembly at the posttranscriptional level (Fig. 6). The exact mechanism is still unclear.

Importantly, these studies show that c-di-GMP can regulate the inhibition of flagellar rotation during a Mot$^+$-to-Mot$^-$ transition. Alternatively, c-di-GMP can inhibit or stall flagellar assembly when bacteria are growing in a sessile lifestyle until developmental cues begin to initiate a transition back to the planktonic lifestyle.

Coordination of Flagellar Gene Expression and Assembly

An alternative mechanism to inhibiting flagellum or pilus assembly is to negatively regulate gene expression until these structures are needed. An advantage of this route is that energy will be conserved until entirely necessary. An elegant aspect of flagellar systems is the coordinated regulation of gene expression and assembly (Fig. 7) (2). The conserved feature of this coupling mechanism is a transcriptional hierarchy of flagellum-specific promoters controlled by a master regulator and two or three alternative sigma factors. The activity of the flagellar promoters, especially through a Mot$^-$-to-Mot$^+$ transition associated with a return to a planktonic state, is dictated by the stage of flagellar assembly.

The paradigms for regulation of flagellar gene expression are the *Salmonella enterica* serovar Typhimurium and *Escherichia coli* systems (16). In the enteric flagellar systems, all the genes can be grouped as early, middle, or late depending on when their products are needed during the assembly process. A transcriptional hierarchy comprising three promoter classes, P_{class1}, P_{class2}, and P_{class3}, coordinates the expression of these flagellar genes (Fig. 7) (16). P_{class1} controls the expression of the enteric flagellar master regulator *flhDC.* FlhD and FlhC are restricted to the enteric bacteria and the betaproteobacteria and form a heterocomplex of a FlhD tetramer and FlhC dimer (140). In conjunction with the housekeeping sigma factor, σ^{70}, $FlhD_4C_2$ activates P_{class2} that drives transcription of the middle genes. Two of these middle genes encode the regulators FlgM and the flagellum-specific sigma factor, σ^{28}. σ^{28} is required for the activation of P_{class3} (74, 100). P_{class3} is not activated during the Mot$^-$-to-Mot$^+$ transition until the first hook-basal body (HBB) is complete. The inhibition of σ^{28} activity prior to HBB completion is through a direct interaction with its anti-sigma factor, FlgM (36, 101). HBB completion represents an important assembly checkpoint. Completed HBB structures undergo a substrate specificity switch that results in FlgM secretion (Fig. 7) (51). All flagellar proteins secreted after this time point, including FlgM, are defined as late secretion substrates. Thus, FlgM secretion triggers P_{class3} activation and transcription of late flagellar genes, which include those that encode the flagellins, motor force generators, and chemotaxis machinery. This view of flagellar gene expression, however, considers only the activation of the hierarchy. For multiflagellated species (e.g., *Salmonella* spp. and *E. coli*), once the first few flagella have been assembled and the hierarchy is active, a further level of control to prevent overproduction is required.

In the majority of nonenteric gamma-proteobacteria and many other flagellated bacteria, the transcriptional hierarchy requires the activities of the housekeeping sigma factor (σ^{70}), the flagellum-specific sigma factor (σ^{28}), the alternate sigma factor (σ^{54}), or a combination thereof (2, 91). What is critical to appreciate in the context of this chapter is that although a number of alternative regulatory circuits are utilized, the underlying architecture of the network that coordinates gene expression and assembly in all flagellar systems is the same (Fig. 7) (2). Two alternative flagellar regulatory networks are those from *P. aeruginosa* and *C. crescentus*. An important difference from the enteric system, other than the recruitment of σ^{54} and the utilization of alternative master regulators, is the presence of two assembly checkpoints rather than one (Fig. 7) (26, 39). The second assembly checkpoint utilized by these two systems is the initiation of rod assembly that can also be seen as the time point at which T3S is activated. How these systems coordinate the sensing of this time point is still not understood, but it involves the activation of a group of response regulators via cognate sensor kinases that coordinate σ^{54} activity.

For a significant number of alphaproteobacteria (including *C. crescentus*) the σ^{28}-FlgM regulatory circuit is not used at all. Instead, this group of bacteria coordinates late gene translation with the assembly pathway rather than transcription (80, 88) (Fig. 7). In contrast to *C. crescentus, P. aeruginosa* still utilizes the σ^{28}-FlgM checkpoint to coordinate late gene expression and HBB completion. Importantly, the translational control observed in alpha-proteobacteria remains responsive to HBB completion.

c-di-GMP Regulation of Flagellar and Pilus Gene Expression

c-di-GMP has already been shown to impact expression and function of flagellar and pilus systems. There are, however, additional potential points where a regulatory input could influence these systems. The flagellar master regulators present a logical input point for negative regulation by c-di-GMP. By inhibiting the activity of this group of regulators, the knock-on effect is reduced flagellar gene expression. The activities of the alternative sigma factors recruited by flagellar and pilus systems are a second point of entry for c-di-GMP regulation.

At this point, it is worthwhile to consider exactly how the mechanisms of transcription activation by σ^{28} and σ^{54} differ. These differences become extremely important when considering the regulation of gene expression by c-di-GMP. σ^{28} is one of the smallest members of the σ^{70} superfamily (106). σ^{70} family members recognize unique but classical -35 and -10 promoter sequences. Upon recognizing promoters, this sigma family is able to initiate open complex formation with core RNA polymerase. A wide range of transcription factors can influence the activity of σ^{70} family members and RNA polymerase. For example, $FlhD_4C_2$ belongs to a class of transcriptional activators that interact with DNA directly upstream of the -35 box (53). $FlhD_4C_2$ directly interacts with the α-subunit of core RNA polymerase to promote transcription initiation from $FlhD_4C_2$-dependent promoters (79). In the context of c-di-GMP regulation of σ^{70}-dependent gene expression, therefore, it is not surprising that a major input route is through the modulation of the activity of the transcriptional activators and repressors rather than through a direct interaction with RNA polymerase, although such a level of control has not yet been ruled out.

In contrast to σ^{70} family members, σ^{54} belongs to its own sigma factor family. Unique among bacterial sigma factors, σ^{54} family members cannot form an open complex with core RNA polymerase (13, 130). Instead, to initiate transcription, σ^{54} requires the ATPase activity of an accessory protein called an enhancer binding protein (EBP). σ^{54} will interact with promoter sequences at the -24 and -12 sites upstream of the transcription start site. The requirement of the σ^{54} EBPs to initiate open complex formation results in σ^{54}-dependent promoters exhibiting much tighter control than σ^{70}-dependent promoters. If a σ^{54} EBP is rendered inactive, then the dependent promoters are effectively silent.

All σ^{54} EBPs belong to the NtrC family, which typically consist of three domains: an N-terminal activation domain, a central AAA ATPase domain, and a C-terminal DNA-binding domain. NtrC was the first σ^{54} EBP described. The activity of the central ATPase domain is controlled via activation of the N-terminal domain. For the majority of flagellum-associated σ^{54} EBPs, this N-terminal domain shares similarity to the phosphoryl receiver domain from a cognate two-component system (Fig. 7) (13). EBP binding sites can be up to 150 bp upstream of the promoter. σ^{54} EBPs can act as flagellar master regulators and accessory regulators of flagellar gene expression in nonenteric flagellar systems and are responsible for pilin expression in *P. aeruginosa*. For the pilus systems, it has been shown that only the major pilin subunit *pilA* is σ^{54} dependent, while the majority of other pilus components are either constitutively expressed or regulated in an alternate manner (70).

The most well studied example of a σ^{54} EBP flagellar master regulator is FleQ and its homologues (7, 25, 59; see chapter 11). FleQ regulates flagellar biosynthesis in *P. aeruginosa* and is known as FlrA and FlaK in *Vibrio* spp. (90). These σ^{54} EBPs belong to a subfamily of the NtrC family that do not possess the conserved aspartate that acts as the phosphate acceptor. Thus, their activity is potentially controlled via an alternate mechanism. Most significantly, FleQ has recently been shown to bind c-di-GMP (Fig. 4) (47). FleQ negatively regulates production of the Pel EPS in *P. aeruginosa.* This negative regulation is relieved when c-di-GMP binds to FleQ. Importantly, this regulatory mechanism does not influence FleQ regulation of flagellar gene expression significantly. Whether the regulation of FleQ activity by c-di-GMP sets a precedent for the regulation of other σ^{54} EBPs associated with lifestyle choices requires further research.

Interestingly, $FlhD_4C_2$ behaves in a somewhat similar manner to FleQ, although no direct input from c-di-GMP has yet been shown. $FlhD_4C_2$ negatively autoregulates its own expression and is a negative regulator of the expression of the *fim* operon (type I pili) in *S. enterica* serovar Typhimurium (18). Type I pili, like type IV pili, are involved in adhesion to surfaces. Conversely, the transcriptional activator of the *fim* operon, FimZ, has a reciprocal effect on *flhDC* expression when *S. enterica* serovar Typhimurium commits to *fim* expression. It is tempting to suggest that this dual control through FimZ and $FlhD_4C_2$ is responsive to c-di-GMP, although no direct evidence is as yet available.

A similar cross-regulation between planktonic and biofilm-associated lifestyles is seen in *E. coli* with respect to the differential expression of flagellar genes and the curli phenotype (110). Curli are pilus-like structures again implicated in surface-associated growth in enteric bacteria. In this case, the flagellar protein FliZ has been implicated in achieving a level of cross-regulation between the flagellar system and the expression of the genes responsible for curli. FliZ is a positive regulator of $FlhD_4C_2$ activity in enteric bacteria. The exact mechanism remains elusive. FliZ has been shown to bind DNA in the insect pathogen *Xenorhabdus nematophila* but also stabilizes at least FlhC in *S. entrica* serovar Typhimurium (76, 117). Furthermore, in *E. coli,* FliZ has been shown to alter the activity of the alternative sigma factor σ^S that in turn regulates curli via c-di-GMP levels through the DGC activity of YegE and YedQ and the PDE activity of YhjH (110; chapter 16). Interestingly, FliZ has been previously implicated in a degree of cross talk between the expression of flagellar genes and the virulent T3S system of *S. enterica* serovar Typhimurium SPI-1 (54, 81, 135).

CONCLUDING REMARKS

The examples we have discussed here highlight the role c-di-GMP plays to coordinate the shutdown of one lifestyle system while activating another. Researchers have begun to identify key components and delineate the mechanisms by which c-di-GMP exerts its regulatory input. In the chapters to follow, you will learn about c-di-GMP regulation in the context of a wide range of bacterial species. You will read about some important differences but also begin to see the common trends we have highlighted here.

The advantage of all motility systems to communicate with sessility-associated phenotypes to prevent unnecessary concomitant activity is obvious. What now remains to be determined is to what extent c-di-GMP plays a role in other systems in coordinating the transitions from motile to nonmotile lifestyles and vice versa. Furthermore, it is not only the expression of these systems that is being controlled by c-di-GMP. Another exciting area of exploration is identifying the environmental signals that control c-di-GMP signaling. This will provide important insight as to which lifestyle suits which environment. The fascinating flexibility of c-di-GMP has the net result of a versatile signaling molecule that can induce a gradual response, through changes in gene expression, as well as an immediate response by directly altering the functionality of macromolecular assembly pathways.

REFERENCES

1. **Aldridge, P., and K. T. Hughes.** 2001. How and when are substrates selected for type III secretion? *Trends Microbiol.* **9:** 209–214.
2. **Aldridge, P., and K. T. Hughes.** 2002. Regulation of flagellar assembly. *Curr. Opin. Microbiol.* **5:**160–165.
3. **Aldridge, P., R. Paul, P. Goymer, P. Rainey, and U. Jenal.** 2003. Role of the GGDEF regulator PleD in polar development of *Caulobacter crescentus. Mol. Microbiol.* **47:**1695–1708.
4. **Allison, C., and C. Hughes.** 1991. Bacterial swarming: an example of prokaryotic differentiation and multicellular behaviour. *Sci. Prog.* **75:**403–422.
5. **Alm, R. A., J. P. Hallinan, A. A. Watson, and J. S. Mattick.** 1996. Fimbrial biogenesis genes of *Pseudomonas aeruginosa: pilW* and *pilX* increase the similarity of type 4 fimbriae to the GSP protein-secretion systems and *pilY1* encodes a gonococcal PilC homologue. *Mol. Microbiol.* **22:**161–173.

6. **Altmann, D., P. Stief, R. Amann, D. de Beer, and A. Schramm.** 2003. *In situ* distribution and activity of nitrifying bacteria in freshwater sediment. *Environ. Microbiol.* **5**:798–803.
7. **Arora, S. K., B. W. Ritchings, E. C. Almira, S. Lory, and R. Ramphal.** 1997. A transcriptional activator, FleQ, regulates mucin adhesion and flagellar gene expression in *Pseudomonas aeruginosa* in a cascade manner. *J. Bacteriol.* **179**:5574–5581.
8. **Banin, E., M. L. Vasil, and E. P. Greenberg.** 2005. Iron and *Pseudomonas aeruginosa* biofilm formation. *Proc. Natl. Acad. Sci. USA* **102**:11076–11081.
9. **Berg, H. C.** 2008. Bacterial flagellar motor. *Curr. Biol.* **18**: R689–R691.
10. **Berg, H. C.** 2003. The rotary motor of bacterial flagella. *Annu. Rev. Biochem.* **72**:19–54.
11. **Berg, H. C., and R. A. Anderson.** 1973. Bacteria swim by rotating their flagellar filaments. *Nature* **245**:380–382.
12. **Bomchil, N., P. Watnick, and R. Kolter.** 2003. Identification and characterization of a *Vibrio cholerae* gene, *mbaA,* involved in maintenance of biofilm architecture. *J. Bacteriol.* **185**:1384–1390.
13. **Buck, M., M. T. Gallegos, D. J. Studholme, Y. Guo, and J. D. Gralla.** 2000. The bacterial enhancer-dependent σ^{54} (σ^{N}) transcription factor. *J. Bacteriol.* **182**:4129–4136.
14. **Burrows, L. L.** 2005. Weapons of mass retraction. *Mol. Microbiol.* **57**:878–888.
15. **Chang, W.-S., X. Li, and L. J. Halverson.** 12 February 2009. Influence of water limitation on endogenous oxidative stress and cell death within unsaturated *Pseudomonas putida* biofilms. *Environ. Microbiol.* [Epub ahead of print.]
16. **Chilcott, G., and K. T. Hughes.** 2000. Coupling of flagellar gene expression to flagellar assembly in *Salmonella enterica* serovar Typhimurium and *Escherichia coli. Microbiol. Mol. Biol. Rev.* **64**:694–708.
17. **Christen, M., B. Christen, M. G. Allan, M. Folcher, P. Jeno, S. Grzesiek, and U. Jenal.** 2007. DgrA is a member of a new family of cyclic diguanosine monophosphate receptors and controls flagellar motor function in *Caulobacter crescentus. Proc. Natl. Acad. Sci. USA* **104**:4112–4117.
18. **Clegg, S., and K. T. Hughes.** 2002. FimZ is a molecular link between sticking and swimming in *Salmonella enterica* serovar Typhimurium. *J. Bacteriol.* **184**:1209–1213.
19. **Costerton, J. W., Z. Lewandowski, D. E. Caldwell, D. R. Korber, and H. M. Lappin-Scott.** 1995. Microbial biofilms. *Annu. Rev. Microbiol.* **49**:711–745.
20. **Costerton, J. W., P. S. Stewart, and E. P. Greenberg.** 1999. Bacterial biofilms: a common cause of persistent infections. *Science* **284**:1318–1322.
21. **Craig, L., and J. Li.** 2008. Type IV pili: paradoxes in form and function. *Curr. Opin. Struct. Biol.* **18**:267–277.
22. **Danese, P. N., L. A. Pratt, and R. Kolter.** 2000. Exopolysaccharide production is required for development of *Escherichia coli* K-12 biofilm architecture. *J. Bacteriol.* **182**:3593–3596.
23. **D'Argenio, D. A., M. W. Calfee, P. B. Rainey, and E. C. Pesci.** 2002. Autolysis and autoaggregation in *Pseudomonas aeruginosa* colony morphology mutants. *J. Bacteriol.* **184**:6481–6489.
24. **Darnton, N. C., L. Turner, S. Rojevsky, and H. C. Berg.** 2007. On torque and tumbling in swimming *Escherichia coli. J. Bacteriol.* **189**:1756–1764.
25. **Dasgupta, N., E. P. Ferrell, K. J. Kanack, S. E. West, and R. Ramphal.** 2002. *fleQ,* the gene encoding the major flagellar regulator of *Pseudomonas aeruginosa,* is σ^{70} dependent and is downregulated by Vfr, a homolog of *Escherichia coli* cyclic AMP receptor protein. *J. Bacteriol.* **184**:5240–5250.
26. **Dasgupta, N., M. C. Wolfgang, A. L. Goodman, S. K. Arora, J. Jyot, S. Lory, and R. Ramphal.** 2003. A four-tiered transcriptional regulatory circuit controls flagellar biogenesis in *Pseudomonas aeruginosa. Mol. Microbiol.* **50**:809–824.
27. **Davey, M. E., N. C. Caiazza, and G. A. O'Toole.** 2003. Rhamnolipid surfactant production affects biofilm architecture in *Pseudomonas aeruginosa* PAO1. *J. Bacteriol.* **185**:1027–1036.
28. **de Kievit, T. R.** 2009. Quorum sensing in *Pseudomonas aeruginosa* biofilms. *Environ. Microbiol.* **11**:279–288.
29. **Fletcher, M., and G. Floodgate.** 1973. An electron-microscopic demonstration of an acidic polysaccharide involved in the adhesion of a marine bacterium to solid surfaces. *J. Gen. Microbiol.* **74**:325–334.
30. **Fraser, G. M., and C. Hughes.** 1999. Swarming motility. *Curr. Opin. Microbiol.* **2**:630–635.
31. **Friedman, L., and R. Kolter.** 2004. Genes involved in matrix formation in *Pseudomonas aeruginosa* PA14 biofilms. *Mol. Microbiol.* **51**:675–690.
32. **Friedman, L., and R. Kolter.** 2004. Two genetic loci produce distinct carbohydrate-rich structural components of the *Pseudomonas aerguinosa* biofilm matrix. *J. Bacteriol.* **186**:4457–4465.
33. **Furukawa, S., S. L. Kuchma, and G. A. O'Toole.** 2006. Keeping their options open: acute versus persistent infections. *J. Bacteriol.* **188**:1211–1217.
34. **Galperin, M. Y., A. N. Nikolskaya, and E. V. Koonin.** 2001. Novel domains of the prokaryotic two-component signal transduction systems. *FEMS Microbiol. Lett.* **203**:11–21.
35. **Gilbert, P., P. J. Collier, and M. R. Brown.** 1990. Influence of growth rate on susceptibility to antimicrobial agents: biofilms, cell cycle, dormancy, and stringent response. *Antimicrob. Agents Chemother.* **34**:1865–1868.
36. **Gillen, K. L., and K. T. Hughes.** 1991. Negative regulatory loci coupling flagellin synthesis to flagellar assembly in *Salmonella typhimurium. J. Bacteriol.* **173**:2301–2310.
37. **Gjermansen, M., P. Ragas, C. Sternberg, S. Molin, and T. Tolker-Nielsen.** 2005. Characterization of starvation-induced dispersion in *Pseudomonas putida* biofilms. *Environ. Microbiol.* **7**:894–906.
38. **Gjermansen, M., P. Ragas, amd T. Tolker-Nielsen.** 2006. Proteins with GGDEF and EAL domains regulate *Pseudomonas putida* biofilm formation and dispersal. *FEMS Microbiol. Lett.* **265**:215–224.
39. **Gober, J. W., J. C. England, Y. Brun, and L. J. Schimkets.** 2000. Regulation of flagellum biosynthesis and motility in *Caulobacter*, p. 319–399. *In* Y. Brun (ed.) *Prokaryotic Development.* ASM Press, Washington, DC.
40. **Gomez-Suarez, C., J. Pasma, A. J. van der Borden, J. Wingender, H.-C. Flemming, H. J. Busscher, and H. C. van der Mei.** 2002. Influence of extracellular polymeric substances on deposition and redeposition of *Pseudomonas aeruginosa* to surfaces. *Microbiology* **148**:1161–1169.
41. **Govan, J. R., and V. Deretic.** 1996. Microbial pathogenesis in cystic fibrosis: mucoid *Pseudomonas aeruginosa* and *Burkholderia cepacia. Microbiol. Rev.* **60**:539–574.
42. **Gygi, D., M. M. Rahman, H. C. Lai, R. Carlson, J. Guard-Petter, and C. Hughes.** 1995. A cell-surface polysaccharide that facilitates rapid population migration by differentiated swarm cells of *Proteus mirabilis. Mol. Microbiol.* **17**:1167–1175.
43. **Harshey, R. M.** 2003. Bacterial motility on a surface: many ways to a common goal. *Annu. Rev. Microbiol.* **57**:249–273.
44. **Harshey, R. M.** 1994. Bees aren't the only ones: swarming in gram-negative bacteria. *Mol. Microbiol.* **13**:389–394.
45. **Hay, N. A., D. J. Tipper, D. Gygi, and C. Hughes.** 1999. A novel membrane protein influencing cell shape and multicel-

lular swarming of *Proteus mirabilis*. *J. Bacteriol.* **181**:2008–2016.

46. **Hentzer, M., G. M. Teitzel, G. J. Balzer, A. Heydorn, S. Molin, M. Givskov, and M. R. Parsek.** 2001. Alginate overproduction affects *Pseudomonas aeruginosa* biofilm structure and function. *J. Bacteriol.* **183**:5395–5401.
47. **Hickman, J., and C. Harwood.** 2008. Identification of FleQ from *Pseudomonas aeruginosa* as a c-di-GMP-responsive transcription factor. *Mol. Microbiol.* **69**:376–389.
48. **Hickman, J. W., D. F. Tifrea, and C. S. Harwood.** 2005. A chemosensory system that regulates biofilm formation through modulation of cyclic diguanylate levels. *Proc. Natl. Acad. Sci. USA* **102**:14422–14427.
49. **Hinsa, S. M., M. Espinosa-Urgel, J. L. Ramos, and G. A. O'Toole.** 2003. Transition from reversible to irreversible attachment during biofilm formation by *Pseudomonas fluorescens* WCS365 requires an ABC transporter and a large secreted protein. *Mol. Microbiol.* **49**:905–918.
50. **Hoiby, N.** 2006. *P. aeruginosa* in cystic fibrosis patients resists host defenses, antibiotics. *Microbe* **1**:571–577.
51. **Hughes, K. T., K. L. Gillen, M. J. Semon, and J. E. Karlinsey.** 1993. Sensing structural intermediates in bacterial flagellar assembly by export of a negative regulator. *Science* **262**:1277–1280.
52. **Huitema, E., S. Pritchard, D. Matteson, S. K. Radhakrishnan, and P. H. Viollier.** 2006. Bacterial birth scar proteins mark future flagellum assembly site. *Cell* **124**:1025–1037.
53. **Ishihama, A.** 1993. Protein-protein communication within the transcription apparatus. *J. Bacteriol.* **175**:2483–2489.
54. **Iyoda, S., T. Kamidoi, K. Hirose, K. Kutsukake, and H. Watanabe.** 2001. A flagellar gene *fliZ* regulates the expression of invasion genes and virulence phenotype in *Salmonella enterica* serovar Typhimurium. *Microb. Pathog.* **30**:81–90.
55. **Jackson, D. W., J. W. Simecka, and T. Romeo.** 2002. Catabolite repression of *Escherichia coli* biofilm formation. *J. Bacteriol.* **184**:3406–3410.
56. **Jackson, K. D., M. Starkey, S. Kremer, M. R. Parsek, and D. J. Wozniak.** 2004. Identification of *psl*, a locus encoding a potential exopolysaccharide that is essential for *Pseudomonas aeruginosa* PAO1 biofilm formation. *J. Bacteriol.* **186**:4466–4475.
57. **Jarrell, K. F., and M. McBride.** 2008. The surprisingly diverse ways that prokaryotes move. *Nat. Rev. Microbiol.* **6**:466.
58. **Jenal, U., and J. Malone.** 2006. Mechanisms of cyclic-di-GMP signaling in bacteria. *Ann. Rev. Genet.* **40**:385–407.
59. **Jyot, J., N. Dasgupta, and R. Ramphal.** 2002. FleQ, the major flagellar gene regulator in *Pseudomonas aeruginosa*, binds to enhancer sites located either upstream or atypically downstream of the RpoN binding site. *J. Bacteriol.* **184**:5251–5260.
60. **Kaiser, D.** 2000. Bacterial motility: how do pili pull? *Curr. Biol.* **10**:R777–R780.
61. **Kaiser, D.** 2007. Bacterial swarming: a re-examination of cell-movement patterns. *Curr. Biol.* **17**:R561–R570.
62. **Kanehisa, M., M. Araki, S. Goto, M. Hattori, M. Hirakawa, M. Itoh, T. Katayama, S. Kawashima, S. Okuda, T. Tokimatsu, and Y. Yamanishi.** 2008. KEGG for linking genomes to life and the environment. *Nucleic Acids Res.* **36**:D480–D484.
63. **Kanehisa, M., S. Goto, S. Kawashima, Y. Okuno, and M. Hattori.** 2004. The KEGG resource for deciphering the genome. *Nucleic Acids Res.* **32**:D277–D280.
64. **Kaneko, Y., M. Thoendel, O. Olakanmi, B. E. Britigan, and P. K. Singh.** 2007. The transition metal gallium disrupts *Pseudomonas aeruginosa* iron metabolism and has antimicrobial and antibiofilm activity. *J. Clin. Investig.* **117**:877–888.
65. **Kearns, D. B., J. Robinson, and L. J. Shimkets.** 2001. *Pseudomonas aeruginosa* exhibits directed twitching motility up phosphatidylethanolamine gradients. *J. Bacteriol.* **183**:763–767.
66. **Kindaichi, T., T. Ito, and S. Okabe.** 2004. Ecophysiological interaction between nitrifying bacteria and heterotrophic bacteria in autotrophic nitrifying biofilms as determined by microautoradiography-fluorescence in situ hybridization. *Appl. Environ. Microbiol.* **70**:1641–1650.
67. **Kirisits, M. J., L. Prost, M. Starkey, and M. R. Parsek.** 2005. Characterization of colony morphology variants isolated from *Pseudomonas aeruginosa* biofilms. *Appl. Environ. Microbiol.* **71**:4809–4821.
68. **Klausen, M., A. Aaes-Jorgensen, S. Molin, and T. Tolker-Nielsen.** 2003. Involvement of bacterial migration in the development of complex multicellular structures in *Pseudomonas aeruginosa* biofilms. *Mol. Microbiol.* **50**:61–68.
69. **Klausen, M., A. Heydorn, P. Ragas, L. Lambertsen, A. Aaes-Jorgensen, S. Molin, and T. Tolker-Nielsen.** 2003. Biofilm formation by *Pseudomonas aeruginosa* wild type, flagella and type IV pili mutants. *Mol. Microbiol.* **48**:1511–1524.
70. **Koga, T., K. Ishimoto, and S. Lory.** 1993. Genetic and functional characterization of the gene cluster specifying expression of *Pseudomonas aeruginosa* pili. *Infect. Immun.* **61**:1371–1377.
71. **Kojima, S., and D. F. Blair.** 2004. The bacterial flagellar motor: structure and function of a complex molecular machine. *Int. Rev. Cytol.* **233**:93–134.
72. **Kolenbrander, P. E., R. N. Andersen, D. S. Blehert, P. G. Egland, J. S. Foster, and R. J. Palmer, Jr.** 2002. Communication among oral bacteria. *Microbiol. Mol. Biol. Rev.* **66**:486–505.
73. **Kulasakara, H., V. Lee, A. Brencic, N. Liberati, J. Urbach, S. Miyata, D. G. Lee, A. N. Neely, M. Hyodo, Y. Hayakawa, F. M. Ausubel, and S. Lory.** 2006. Analysis of *Pseudomonas aeruginosa* diguanylate cyclases and phosphodiesterases reveals a role for bis-(3′-5′)-cyclic-GMP in virulence. *Proc. Natl. Acad. Sci. USA* **103**:2839–2844.
74. **Kutsukake, K., Y. Ohya, and T. Iino.** 1990. Transcriptional analysis of the flagellar regulon of *Salmonella typhimurium*. *J. Bacteriol.* **172**:741–747.
75. **Landry, R. M., D. An, J. T. Hupp, P. K. Singh, and M. R. Parsek.** 2006. Mucin-*Pseudomonas aeruginosa* interactions promote biofilm formation and antibiotic resistance. *Mol. Microbiol.* **59**:142–151.
76. **Lanois, A., G. Jubelin, and A. Givaudan.** 2008. FliZ, a flagellar regulator, is at the crossroads between motility, haemolysin expression and virulence in the insect pathogenic bacterium *Xenorhabdus*. *Mol. Microbiol.* **68**:516–533.
77. **Lee, V. T., J. M. Matewish, J. L. Kessler, M. Hyodo, Y. Hayakawa, and S. Lory.** 2007. A cyclic-di-GMP receptor required for bacterial exopolysaccharide production. *Mol. Microbiol.* **65**:1474–1484.
78. **Lequette, Y., and E. P. Greenberg.** 2005. Timing and localization of rhamnolipid synthesis gene expression in *Pseudomonas aeruginosa* biofilms. *J. Bacteriol.* **187**:37–44.
79. **Liu, X., N. Fujita, A. Ishihama, and P. Matsumura.** 1995. The C-terminal region of the alpha subunit of *Escherichia coli* RNA polymerase is required for transcriptional activation of the flagellar level II operons by the FlhD/FlhC complex. *J. Bacteriol.* **177**:5186–5188.
80. **Llewellyn, M., R. J. Dutton, J. Easter, D. O'donnol, and J. W. Gober.** 2005. The conserved *flaF* gene has a critical role in coupling flagellin translation and assembly in *Caulobacter crescentus*. *Mol. Microbiol.* **57**:1127–1142.
81. **Lucas, R. L., C. P. Lostroh, C. C. DiRusso, M. P. Spector, B. L. Wanner, and C. A. Lee.** 2000. Multiple factors indepen-

dently regulate *hilA* and invasion gene expression in *Salmonella enterica* serovar Typhimurium. *J. Bacteriol.* **182**:1872–1882.

82. **Lujan, A. M., A. J. Moyano, I. Segura, C. E. Argarana, and A. M. Smania.** 2007. Quorum-sensing-deficient *(lasR)* mutants emerge at high frequency from a *Pseudomonas aeruginosa mutS* strain. *Microbiology* **153**:225–237.
83. **Ma, L., K. D. Jackson, R. M. Landry, M. R. Parsek, and D. J. Wozniak.** 2006. Analysis of *Pseudomonas aeruginosa* conditional Psl variants reveals roles for the Psl polysaccharide in adhesion and maintaining biofilm structure postattachment. *J. Bacteriol.* **188**:8213–8221.
84. **Mack, D., W. Fischer, A. Krokotsch, K. Leopold, R. Hartmann, H. Egge, and R. Laufs.** 1996. The intercellular adhesin involved in biofilm accumulation of *Staphylococcus epidermidis* is a linear β-1,6-linked glucosaminoglycan: purification and structural analysis. *J. Bacteriol.* **178**:175–183.
85. **Macnab, R. M.** 1992. Genetics and biogenesis of bacterial flagella. *Annu. Rev. Genet.* **26**:131–158.
86. **Macnab, R. M.** 1999. The bacterial flagellum: reversible rotary propellor and type III export apparatus. *J. Bacteriol.* **181**: 7149–7153.
87. **Mah, T. C., and G. A. O'Toole.** 2001. Mechanisms of biofilm resistance to antimicrobial agents. *Trends Microbiol.* **9**:34–39.
88. **Mangan, E. K., J. Malakooti, A. Caballero, P. Anderson, B. Ely, and J. W. Gober.** 1999. FlbT couples flagellum assembly to gene expression in *Caulobacter crescentus*. *J. Bacteriol.* **181**: 6160–6170.
89. **McBride, M. J., T. F. Braun, and J. L. Brust.** 2003. *Flavobacterium johnsoniae* GldH is a lipoprotein that is required for gliding motility and chitin utilization. *J. Bacteriol.* **185**: 6648–6657.
90. **McCarter, L. L.** 2001. Polar flagellar motility of the *Vibrionaceae*. *Microbiol. Mol. Biol. Rev.* **65**:445–462.
91. **McCarter, L. L.** 2006. Regulation of flagella. *Curr. Opin. Microbiol.* **9**:180–186.
92. **Merighi, M., V. T. Lee, M. Hyodo, Y. Hayakawa, and S. Lory.** 2007. The second messenger bis-(3′-5′)-cyclic-GMP and its PilZ domain-containing receptor Alg44 are required for alginate biosynthesis in *Pseudomonas aeruginosa*. *Mol. Microbiol.* **65**:876–895
93. **Merz, A. J., M. So, and M. P. Sheetz.** 2000. Pilus retraction powers bacterial twitching motility. *Nature* **407**:98–102.
94. **Mignot, T., and J. R. Kirby.** 2008. Genetic circuitry controlling motility behaviors of *Myxococcus xanthus*. *Bioessays* **30**: 733–743.
95. **Molin, S., and T. Tolker-Nielsen.** 2003. Gene transfer occurs with enhanced efficiency in biofilms and induces enhanced stabilisation of the biofilm structure. *Curr. Opin. Biotechnol.* **14**: 255–261.
96. **Morgan, R., S. Kohn, S.-H. Hwang, D. J. Hassett, and K. Sauer.** 2006. BdlA, a chemotaxis regulator essential for biofilm dispersion in *Pseudomonas aeruginosa*. *J. Bacteriol.* **188**:7335–7343.
97. **Nambu, T., T. Minamino, R. M. Macnab, and K. Kutsukake.** 1999. Peptidoglycan-hydrolyzing activity of the FlgJ protein, essential for flagellar rod formation in *Salmonella typhimurium*. *J. Bacteriol.* **181**:1555–1561.
98. **Nancharaiah, Y. V., P. Wattiau, S. Wuertz, S. Bathe, S. V. Mohan, P. A. Wilderer, and M. Hausner.** 2003. Dual labeling of *Pseudomonas putida* with fluorescent proteins for in situ monitoring of conjugal transfer of the TOL plasmid. *Appl. Environ. Microbiol.* **69**:4846–4852.
99. **Nivens, D. E., D. E. Ohman, J. Williams, and M. J. Franklin.** 2001. Role of alginate and its O acetylation in formation of *Pseudomonas aeruginosa* microcolonies and biofilms. *J. Bacteriol.* **183**:1047–1057.
100. **Ohnishi, K., K. Kutsukake, H. Suzuki, and T. Iino.** 1990. Gene *fliA* encodes an alternative sigma factor specific for flagellar operons in *Salmonella typhimurium*. *Mol. Gen. Genet.* **221**:139–147.
101. **Ohnishi, K., K. Kutsukake, H. Suzuki, and T. Lino.** 1992. A novel transcriptional regulation mechanism in the flagellar regulon of *Salmonella typhimurium:* an antisigma factor inhibits the activity of the flagellum-specific sigma factor, sigma F. *Mol. Microbiol.* **6**:3149–3157.
102. **O'Shea, T. M., A. H. Klein, K. Geszvain, A. J. Wolfe, and K. L. Visick.** 2006. Diguanylate cyclases control magnesium-dependent motility of *Vibrio fischeri*. *J. Bacteriol.* **188**:8196–8205.
103. **O'Toole, G., H. B. Kaplan, and R. Kolter.** 2000. Biofilm formation as microbial development. *Annu. Rev. Microbiol.* **54**:49–79.
104. **O'Toole, G. A.** 1998. Initiation of biofilm formation in *Pseudomonas fluorescens* WCS365 proceeds via multiple convergent signalling pathways: a genetic analysis. *Mol. Microbiol.* **28**:449–461.
105. **O'Toole, G. A., and R. Kolter.** 1998. Flagellar and twitching motility are necessary for *Pseudomonas aeruginosa* biofilm development. *Mol. Microbiol.* **30**:295–304.
106. **Paget, M. S., and J. D. Helmann.** 2003. The sigma70 family of sigma factors. *Genome Biol.* **4**:203.
107. **Parsek, M. R., and P. K. Singh.** 2003. Bacterial biofilms: an emerging link to disease pathogenesis. *Annu. Rev. Microbiol.* **57**:677–701.
108. **Paul, R., S. Weiser, N. C. Amiot, C. Chan, T. Schirmer, B. Giese, and U. Jenal.** 2004. Cell cycle-dependent dynamic localization of a bacterial response regulator with a novel diguanylate cyclase output domain. *Genes Dev.* **18**:715–727.
109. **Pelicic, V.** 2008. Type IV pili: e pluribus unum? *Mol. Microbiol.* **68**:827–837.
110. **Pesavento, C., G. Becker, N. Sommerfeldt, A. Possling, N. Tschowri, A. Mehlis, and R. Hengge.** 2008. Inverse regulatory coordination of motility and curli-mediated adhesion in *Escherichia coli*. *Genes Dev.* **22**:2434–2446.
111. **Pratt, L. A., and R. Kolter.** 1998. Genetic analysis of *Escherichia coli* biofilm formation: roles of flagella, motility, chemotaxis and type I pili. *Mol. Microbiol.* **30**:285–293.
112. **Prigent-Combaret, C., G. Prensier, T. T. Le Thi, O. Vidal, P. Lejeune, and C. Dorel.** 2000. Developmental pathway for biofilm formation in curli-producing *Escherichia coli* strains: role of flagella, curli and colanic acid. *Environ. Microbiol.* **2**: 450–464.
113. **Pruzzo, C., L. Vezzulli, and R. R. Colwell.** 2008. Global impact of *Vibrio cholerae* interactions with chitin. *Environ. Microbiol.* **10**:1400–1410.
114. **Purevdorj, B., J. W. Costerton, and P. Stoodley.** 2002. Influence of hydrodynamics and cell signaling on the structure and behavior of *Pseudomonas aeruginosa* biofilms. *Appl. Environ. Microbiol.* **68**:4457–4464.
115. **Ross, P., R. Mayer, H. Weinhouse, D. Amikam, Y. Huggirat, M. Benziman, E. de Vroom, A. Fidder, P. de Paus, and L. Sliedregt.** 1990. The cyclic diguanylic acid regulatory system of cellulose synthesis in *Acetobacter xylinum*. Chemical synthesis and biological activity of cyclic nucleotide dimer, trimer, and phosphothioate derivatives. *J. Biol. Chem.* **265**: 18933–18943.
116. **Ryjenkov, D. A., R. Simm, U. Romling, and M. Gomelsky.** 2006. The PilZ domain is a receptor for the second messenger c-di-GMP: the PilZ domain protein YcgR controls motility in enterobacteria. *J. Biol. Chem.* **281**:30310–30314.

117. **Saini, S., J. D. Brown, P. D. Aldridge, and C. V. Rao.** 2008. FliZ Is a posttranslational activator of $FlhD_4C_2$-dependent flagellar gene expression. *J. Bacteriol.* **190:**4979–4988.
118. **Samatey, F. A., K. Imada, S. Nagashima, F. Vonderviszt, T. Kumasaka, M. Yamamoto, and K. Namba.** 2001. Structure of the bacterial flagellar protofilament and implications for a switch for supercoiling. *Nature* **410:**331–337.
119. **Satoh, H., T. Yamakawa, T. Kindaichi, T. Ito, and S. Okabe.** 2006. Community structures and activities of nitrifying and denitrifying bacteria in industrial wastewater-treating biofilms. *Biotechnol. Bioeng.* **94:**762–772.
120. **Sauer, K., A. K. Camper, G. D. Ehrlich, J. W. Costerton, and D. G. Davies.** 2002. *Pseudomonas aeruginosa* displays multiple phenotypes during development as a biofilm. *J. Bacteriol.* **184:**1140–1154.
121. **Shrout, J. D., D. L. Chopp, C. L. Just, M. Hentzer, M. Givskov, and M. R. Parsek.** 2006. The impact of quorum sensing and swarming motility on *Pseudomonas aeruginosa* biofilm formation is nutritionally conditional. *Mol. Microbiol.* **62:** 1264–1277.
122. **Singh, P., M. R. Parsek, E. P. Greenberg, and M. J. Welsh.** 2002. A component of innate immunity prevents bacterial biofilm development. *Nature* **417:**552–555.
123. **Skerker, J. M., and H. C. Berg.** 2001. Direct observation of extension and retraction of type IV pili. *Proc. Natl. Acad. Sci. USA* **98:**6901–6904.
124. **Smith, E. E., D. G. Buckley, Z. Wu, C. Saenphimmachak, L. R. Hoffman, D. A. D'Argenio, S. I. Miller, B. W. Ramsey, D. P. Speert, S. M. Moskowitz, J. L. Burns, R. Kaul, and M. V. Olson.** 2006. Genetic adaptation by *Pseudomonas aeruginosa* to the airways of cystic fibrosis patients. *Proc. Natl. Acad. Sci. USA* **103:**8487–8492.
125. **Soutourina, O. A., and P. N. Bertin.** 2003. Regulation cascade of flagellar expression in gram-negative bacteria. *FEMS Microbiol. Rev.* **27:**505–523.
126. **Stewart, P. S.** 2003. Diffusion in biofilms. *J. Bacteriol.* **185:** 1485–1491.
127. **Stewart, P. S., and J. W. Costerton.** 2001. Antibiotic resistance of bacteria in biofilms. *Lancet* **358:**135–138.
128. **Stoodley, P., R. Cargo, C. J. Rupp, S. Wilson, and I. Klapper.** 2002. Biofilm material properties as related to shear-induced deformation and detachment phenomena. *J. Ind. Microbiol. Biotechnol.* **29:**361–367.
129. **Stoodley, P., K. Sauer, D. G. Davies, and J. W. Costerton.** 2002. Biofilms as complex differentiated communities. *Annu. Rev. Microbiol.* **56:**187–209.
130. **Studholme, D. J., and M. Buck.** 2000. The biology of enhancer-dependent transcriptional regulation in bacteria: insights from genome sequences. *FEMS Microbiol. Lett.* **186:** 1–9.
131. **Sudarsan, N., E. R. Lee, Z. Weinberg, R. H. Moy, J. N. Kim, K. H. Link, and R. R. Breaker.** 2008. Riboswitches in eubacteria sense the second messenger cyclic di-GMP. *Science* **321:**411–413.
132. **Teitzel, G. M., and M. R. Parsek.** 2003. Heavy metal resistance of biofilm and planktonic *Pseudomonas aeruginosa. Appl. Environ. Microbiol.* **69:**2313–2320.
133. **Thormann, K. M., S. Duttler, R. M. Saville, M. Hyodo, S. Shukla, Y. Hayakawa, and A. M. Spormann.** 2006. Control of formation and cellular detachment from *Shewanella oneidensis* MR-1 biofilms by cyclic di-GMP. *J. Bacteriol.* **188:** 2681–2691.
134. **Vallet, I., J. W. Olson, S. Lory, A. Lazdunski, and A. Filloux.** 2001. The chaperone/usher pathways of *Pseudomonas aeruginosa:* identification of fimbrial gene clusters *(cup)* and their involvement in biofilm formation. *Proc. Natl. Acad. Sci. USA* **98:**6911–6916.
135. **Van Immerseel, F., V. Eeckhaut, F. Boyen, F. Pasmans, F. Haesebrouck, and R. Ducatelle.** 2008. Mutations influencing expression of the *Salmonella enterica* serovar Enteritidis pathogenicity island I key regulator *hilA. Antonie Van Leeuwenhoek* **94:**455–461.
136. **Vasseur, P., I. Vallet-Gely, C. Soscia, S. Genin, and A. Filloux.** 2005. The pel genes of the *Pseudomonas aeruginosa* PAK strain are involved at early and late stages of biofilm formation. *Microbiology* **151:**985–997.
137. **Wadhams, G. H., and J. P. Armitage.** 2004. Making sense of it all: bacterial chemotaxis. *Nat. Rev. Mol. Cell. Biol.* **5:** 1024–1037.
138. **Wagner, V. E., and B. H. Iglewski.** 2008. *P. aeruginosa* biofilms in CF infection. *Clin. Rev. Allergy Immunol.* **35:**124–134.
139. **Wall, D., and D. Kaiser.** 1999. Type IV pili and cell motility. *Mol. Microbiol.* **32:**1–10.
140. **Wang, S., R. T. Fleming, E. M. Westbrook, P. Matsumura, and D. B. McKay.** 2006. Structure of the *Escherichia coli* FlhDC complex, a prokaryotic heteromeric regulator of transcription. *J. Mol. Biol.* **355:**798–808.
141. **Watnick, P. I., and R. Kolter.** 1999. Steps in the development of a *Vibrio cholerae* El Tor biofilm. *Mol. Microbiol.* **34:**586–595.
142. **Watnick, P. I., K. J. Fullner, and R. Kolter.** 1999. A role for the mannose-sensitive hemagglutinin in biofilm formation by *Vibrio cholerae* El Tor. *J. Bacteriol.* **181:**3606–3609.
143. **Whitchurch, C. B., R. A. Alm, and J. S. Mattick.** 1996. The alginate regulator AlgR and an associated sensor FimS are required for twitching motility in *Pseudomonas aeruginosa. Proc. Natl. Acad. Sci. USA* **93:**9839–9843.
144. **Wolfe, A. J., and K. L. Visick.** 2008. Get the message out: cyclic-di-GMP regulates multiple levels of flagellum-based motility. *J. Bacteriol.* **190:**463–475.
145. **Wu, W., H. Badrane, S. Arora, H. V. Baker, and S. Jin.** 2004. MucA-mediated coordination of type III secretion and alginate synthesis in *Pseudomonas aeruginosa. J. Bacteriol.* **186:**7575–7585.
146. **Yildiz, F. H., and G. K. Schoolnik.** 1999. *Vibrio cholerae* 01 El Tor: identification of a gene cluster required for the rugose colony type, exopolysaccharide production, chlorine resistance, and biofilm formation. *Proc. Natl. Acad. Sci. USA* **96:** 4028–4033.
147. **Yonekura, K., S. Maki, D. G. Morgan, D. J. DeRosier, F. Vonderviszt, K. Imada, and K. Namba.** 2000. The bacterial flagellar cap as the rotary promoter of flagellin self-assembly. *Science* **290:**2148–2152.

The Second Messenger Cyclic Di-GMP
Edited by A. J. Wolfe and K. L. Visick

Chapter 9

Role of Cyclic Di-GMP in *Caulobacter crescentus* Development and Cell Cycle Control

Sören Abel and Urs Jenal

The second messenger cyclic di-GMP (c-di-GMP) has emerged as a global regulator modulating multiple aspects of growth and behavior in many bacteria. The main components of this regulatory network are enzymes that synthesize (*di*guanylate *c*yclases, or DGCs) and hydrolyze (*p*hospho*die*sterases, or PDEs) c-di-GMP, as well as downstream effectors that bind the signaling molecule and, in response, generate a specific readout by interfering with a defined cellular target process. The molecular and structural properties of DGCs, PDEs, and c-di-GMP binding molecules are described in detail elsewhere in this volume (see chapters 4, 5, 7, and 15). In the past 10 years, genomics has brought to light the wide distribution and abundance of c-di-GMP signaling components in bacteria (23). The availability of whole-genome sequences has uncovered GGDEF, EAL, HD-GYP, and PilZ domain proteins as members of large and near ubiquitous families of bacterial signaling components. One of the hallmarks of these protein families is the astonishing number of components present in many individual bacterial species (23). This multiplicity can in principle be explained by the necessity to integrate a multitude of environmental and internal signals and by various diverging cellular functions being regulated by this network. But the presence of multiple homologs in the same species raises questions of signaling specificity and spatiotemporal organization of the network components involved. Genetic analyses of components of the c-di-GMP network have indicated that individual GGDEF, EAL, or PilZ domain proteins can have unique and distinctive biological functions within the same organism (for a review, see reference 32). This argues for the existence of discrete c-di-GMP regulatory modules with limited cross talk between each other. The complex cellular behavior regulated by c-di-GMP could be mediated through a common pool of c-di-GMP and a graded global response through effector molecules with different affinities for their ligand. Alternatively, signal transduction could occur in separated microdomains, where a defined set of c-di-GMP signaling components is functionally and maybe spatially linked to its target structure without communicating with input or output components outside the module (Fig. 1). Such a view is in line with the recent recognition that the interior of the bacterial cell is highly organized and that the physical arrangement and dynamic behavior of molecules represent an important aspect of bacterial cell physiology and regulation (27).

This chapter summarizes the current knowledge of c-di-GMP-mediated control in *Caulobacter crescentus*. *C. crescentus* is an aquatic alphaproteobacterium that divides asymmetrically to produce differentiated progeny, a motile swarmer (SW) and a sessile, surface-attached stalked (ST) cell, each with distinct morphological features and regulatory programs (Fig. 2). During the *Caulobacter* division cycle, the cell poles continuously change their identities through the successive assembly and removal of surface organelles that serve either the motile or sessile lifestyle. SW cells are equipped with a single flagellum (86) and polar type IV-like pili (72, 80), whereas ST cells possess an adhesive holdfast (13, 78, 79) located at the tip of a long extension of the cell body, the stalk (81). The newborn ST cell reinitiates the next round of DNA replication and cell division immediately. In contrast, the SW progeny is unable to

Sören Abel and Urs Jenal • Biozentrum, University of Basel, CH-4056 Basel, Switzerland.

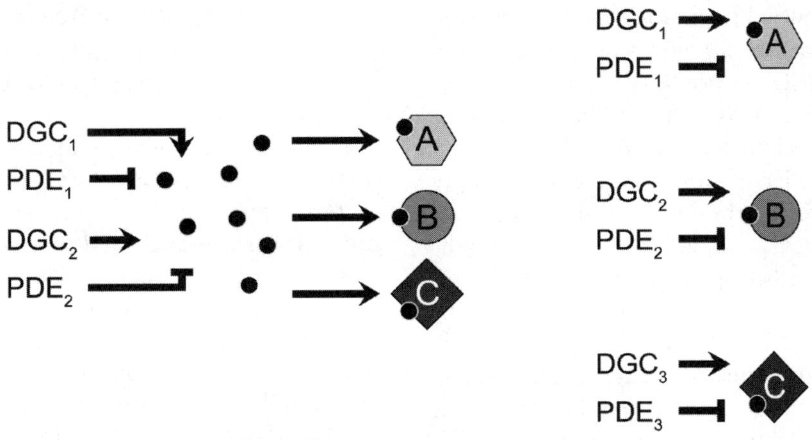

Figure 1. Global versus local c-di-GMP signaling modules. Two possible architectures of c-di-GMP signaling modules are schematically depicted. (Left) Several DGCs and PDEs can contribute to a common global pool of c-di-GMP (black dots). Different c-di-GMP effector proteins (A, B, and C) can bind c-di-GMP with different affinities and by that can stage a graded response as levels of the second messenger fluctuate. (Right) Temporally or spatially separated c-di-GMP signaling modules regulate specific cellular processes by exclusive interaction with one or several effector proteins.

initiate chromosome duplication and remains in a motile, but replication inert, state for a defined period of the cell cycle (G_1). Coincident with advancing into S phase and initiating DNA replication, the SW cell undergoes a striking morphogenetic transition by retracting the pili, shedding the flagellum and the chemotaxis apparatus, and replacing these organelles with the adhesive holdfast and the stalk. The holdfast is a strong polysaccharide-based adhesin, the precise composition of which is still unclear (78, 79). Al-

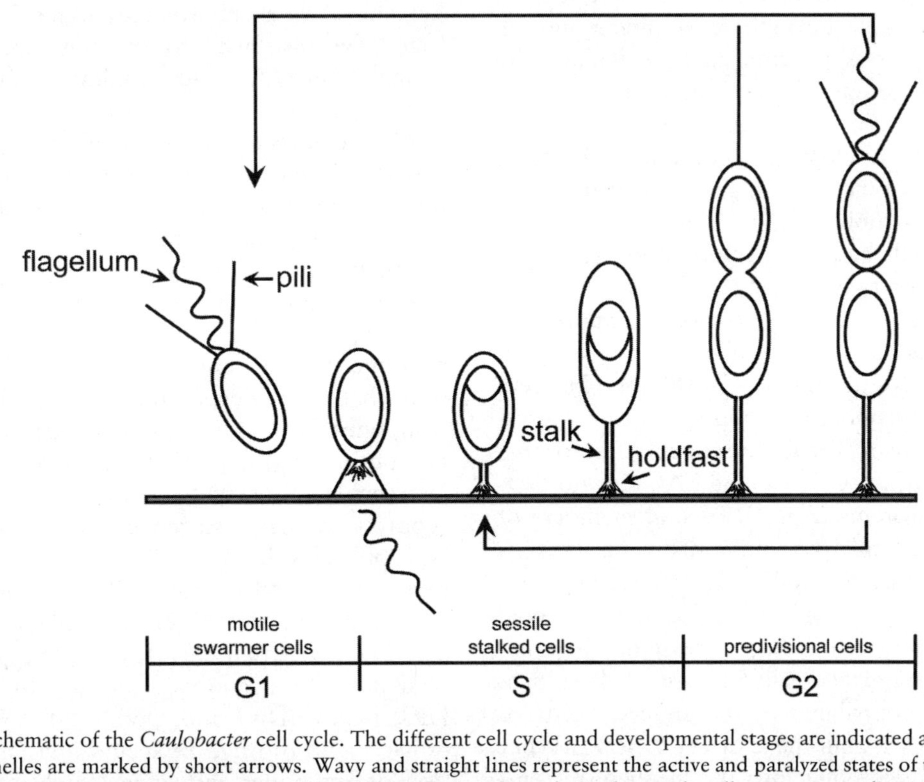

Figure 2. Schematic of the *Caulobacter* cell cycle. The different cell cycle and developmental stages are indicated at the bottom. Polar organelles are marked by short arrows. Wavy and straight lines represent the active and paralyzed states of the flagellum, respectively. Surface attachment of sessile ST and PD cells is indicated. Circles inside cells represent replication inert chromosomes, while replicating chromosomes are shown as θ symbols. Reentry of the newborn progeny into the cell cycle is indicated by long arrows.

though the genes required for holdfast polysaccharide export and anchoring are already transcribed in the predivisional (PD) cell, the holdfast biosynthesis machinery is posttranslationally activated but not before SW cells differentiate into ST cells (41, 49). The accurate timing and order of polar organelle assembly is of key importance for the behavioral response of the individual cell types, e.g., efficient surface attachment of motile SW cells requires an active flagellum, polar pili, and the newly synthesized holdfast (5, 20, 49). While the role of the flagellum is not fully understood yet, pili significantly increase surface adherence by mediating transient and reversible attachment (5, 20). Maximal attachment occurs during the SW-to-ST cell transition, when preformed pili and an active flagellum coincide at the same pole with the newly formed holdfast, which eventually anchors cells irreversibly to surfaces (5, 20, 49). Pole morphogenesis is tightly interlinked with the *Caulobacter* cell cycle, as different cell cycle events act as checkpoints for developmental processes. For example, DNA synthesis is necessary for the biosynthesis of the flagellum (70, 76) and completion of cytokinesis is required to initiate flagellar rotation (35) and pilus assembly (74).

To coordinate pole morphogenesis with the underlying cell cycle, positional information needs to be integrated with processes that determine the temporal progression of cells through development and the division cycle. This is accomplished by dynamically tagging proteins to specific subcellular sites, mainly the cell poles. Within the complex network regulating *Caulobacter* development, protein localization serves three main functions: (i) spatial control of polar organelle assembly; (ii) building of a gradient across the PD cell that, upon cell division, results in the production of two progeny with distinct developmental programs; and (iii) compartmentalization of enzymatic reactions and regulatory processes (47). Recent studies have identified c-di-GMP as a key regulator of pole morphogenesis and cell cycle progression in this organism. The *C. crescentus* chromosome encodes a total of 14 GGDEF and EAL (but no HD-GYP) domain proteins (Fig. 3). Several members of this family dynamically position to distinct polar sites during the *C. crescentus* cell cycle, where they contribute to the temporal and spatial regulation of pole morphogenesis and cell cycle progression. The analysis of the molecular and cellular basis of these processes not only contributes to the understanding of protein dynamics within the c-di-GMP network but also offers an entry point into the mechanisms employed to temporally and spatially separate functionally distinct c-di-GMP modules.

GENETIC DISSECTION OF *CAULOBACTER* POLE MORPHOGENESIS: A SHORT RETROSPECT

There is increasing evidence to suggest that c-di-GMP orchestrates and fine tunes the transition of bacterial cells from a motile, single-cell form to organized surface-anchored communities called biofilms (for a recent review, see reference 32; see also chapters 8, 11, 16, 17, and 19). This switch in cellular lifestyle is installed by simultaneously inhibiting cell motility and promoting the synthesis and exposure of adhesive organelles on the cell surface. Given that *Caulobacter* has integrated the very same motile-sessile transition into its asymmetric division cycle, it is, in hindsight, not surprising that the first c-di-GMP signaling component was genetically defined in this organism. Since the very early days of *Caulobacter* research, the field took a strong interest in the morphogenetic transitions experienced by the cells as they progress through the cell division cycle (58, 68, 69). To genetically uncover the regulatory basis of these changes in cell morphology, mutants with pleiotropic defects in polar organelle formation were of particular interest (21, 22, 51). A popular and very successful method was to first isolate bacteriophage-resistant mutants and subsequently screen for strains with additional pole development defects. In 1984, Ely and coworkers isolated mutants that were selected for resistance to the pilus-specific bacteriophage ϕCbK (72) and simultaneously had become nonmotile (19). One of the pleiotropic mutants, *pleC*, assembles a nonfunctional, paralyzed flagellum, has no pili, fails to synthesize a holdfast, and lacks a stalk except under conditions of phosphate limitation. Although *pleC* mutants divide normally, they fail to initiate the morphogenetic program that first leads to a motile SW and subsequently to a sessile ST cell (Fig. 4a). The observation that *pleC* encodes a histidine protein kinase (HPK) was the first indication that pole development is regulated by two-component signaling systems (82). Exploiting the nonmotile phenotype of *pleC* mutants, Sommer and Newton later isolated a series of motile extragenic suppressors that restored different aspects of wild-type pole development (73, 75). Three of the suppressors restored motility, pilus formation, and stalk biogenesis and were later mapped to the *divK*, *divJ*, and *divL* genes that encode a single domain response regulator (DivK) and two additional members of the HPK family (DivJ and DivL) (30, 59, 88). Conditional alleles of *divJ*, *divK*, and *divL* all show a division defect under restrictive conditions, and mutations in *divK* and *divL* are lethal (30, 88). Therefore, these genes are part of a signal transduction network

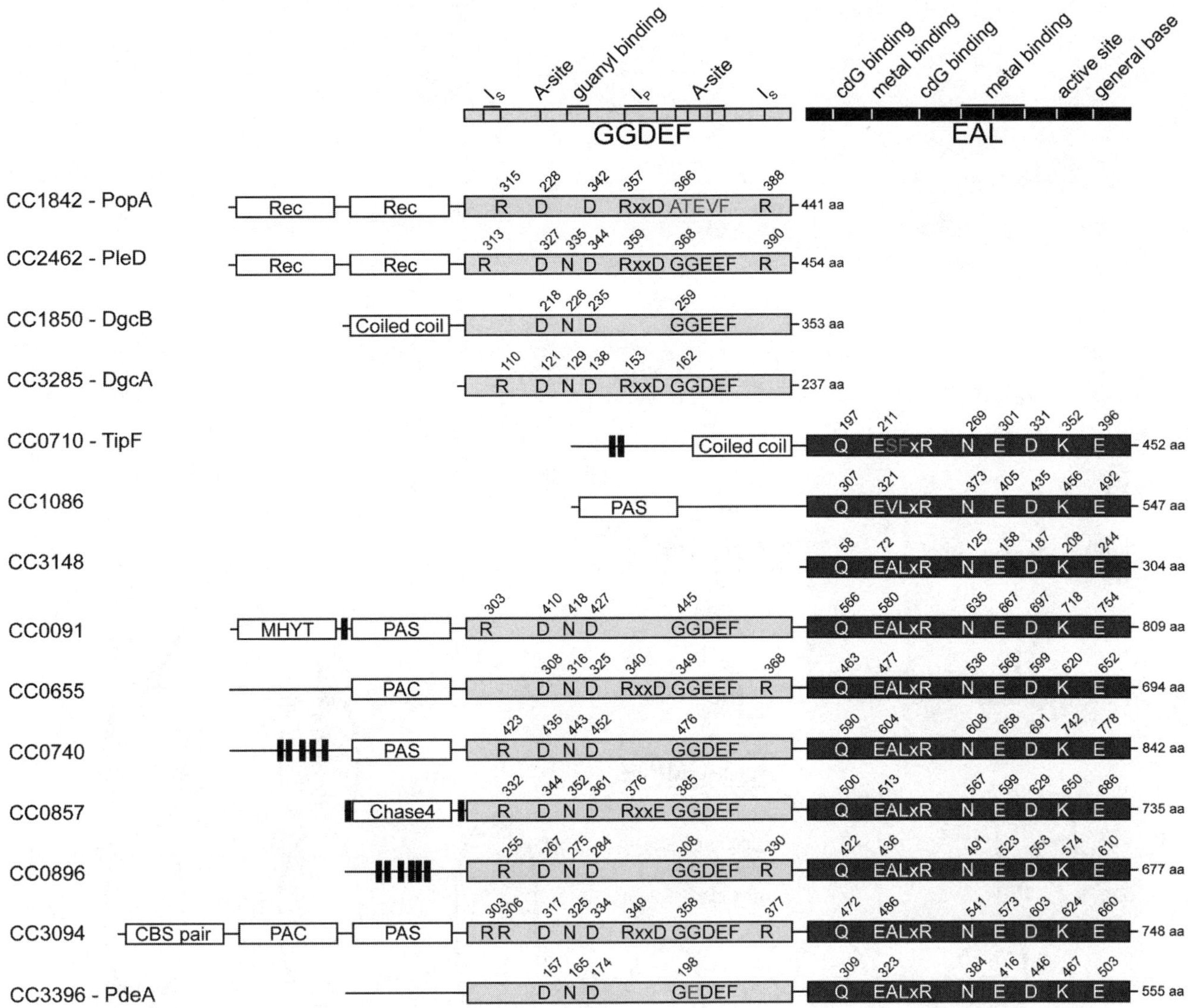

Figure 3. Domain organization of the *Caulobacter* GGDEF and EAL domain proteins. The CC gene numbers and the protein names are listed on the left. GGDEF and EAL domains are colored in grey and black, respectively. N-terminal domains are depicted in white and labeled. N-terminal regions larger than 100 amino acids (aa) without homology to known domains are shown as small black lines. Small black bars indicate predicted transmembrane helices. Domains are not drawn to scale, but the length of each protein is indicated on the right. Conserved amino acids involved in catalysis and/or c-di-GMP binding are indicated (7, 83) in the respective domain with their positions in the respective proteins. Nonconserved residues are shown in grey. The putative function of the conserved amino acids is indicated above the figure: I_P, primary I-site; I_S, secondary I-site; A-site, active site; cdG binding, c-di-GMP binding.

that coordinates polar differentiation with cell growth and division. A rationale for the genetic linkage of these genes was provided by in vitro experiments demonstrating that PleC~P and DivJ~P could both serve as efficient phosphodonors for DivK (30).

A fourth pleiotropic gene, *pleD,* was identified as a motile bypass suppressor of a mutant lacking PleC (75). In contrast to the other three suppressor classes, mutations in *pleD* do not restore the biogenesis of pili, holdfast, and stalk. Rather, *pleD* mutants are hypermotile, unable to shed the flagellum, and inefficient in forming the holdfast and stalk structures (49, 75) (Fig. 4a). Hypermotility of *pleD* mutants is a direct consequence of the failure to eject the flagellum and turn off motility during cell differentiation rather than the result of altered motor speed (1, 31). This was recently confirmed by quantitative analysis of different motility parameters. Whereas a considerably higher fraction of cells is motile in a Δ*pleD* mutant population than in the wild type, cells lacking PleD do not show an increased swimming speed (2) (A. Duerig and U. Jenal, unpublished data). The *pleD*

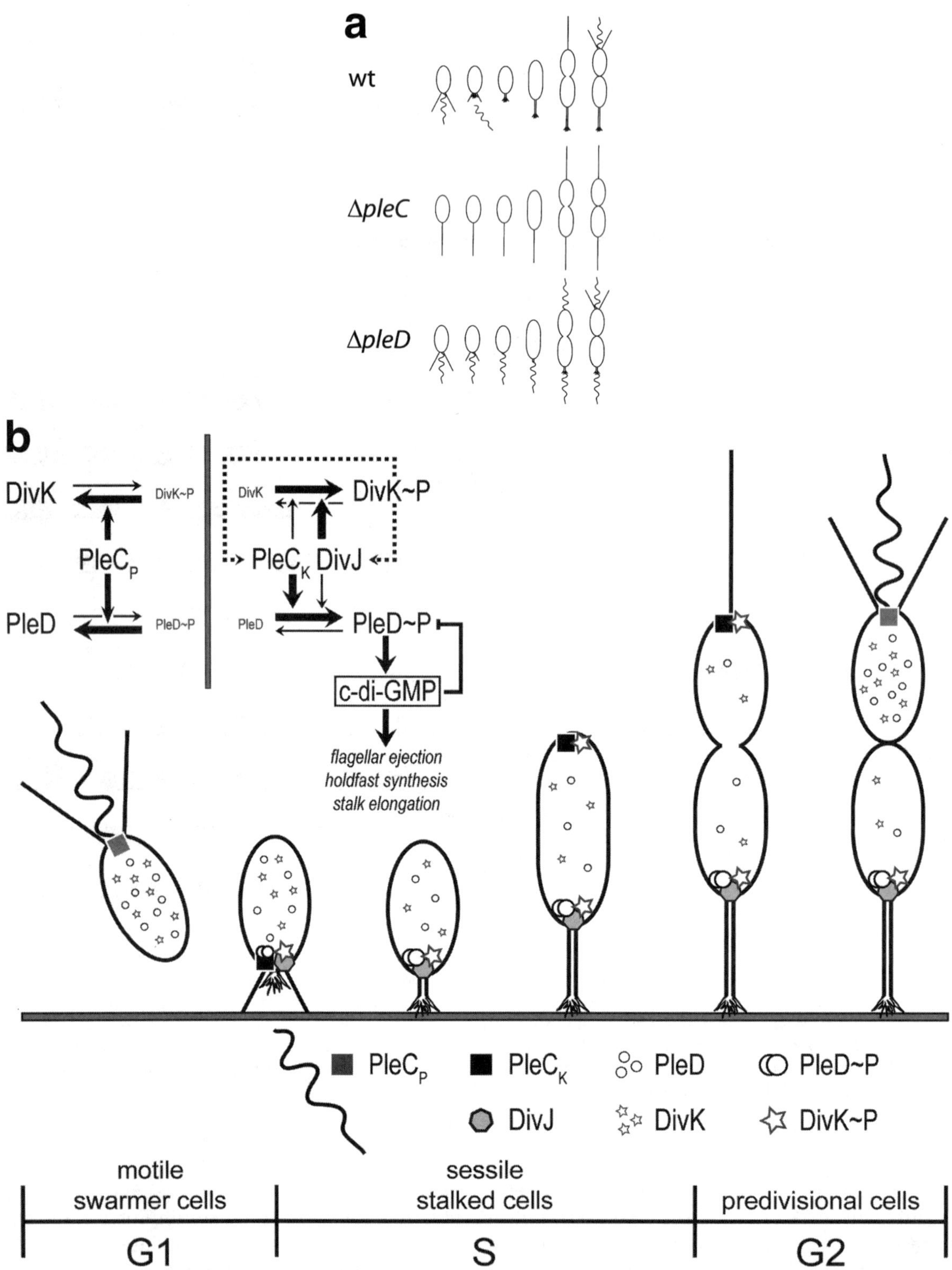

Figure 4. Signaling network controlling PleD activity during the *Caulobacter* cell cycle. (a) The sequence of developmental events in the *C. crescentus* wild type (wt) and in *pleC* and *pleD* mutants. Mutant phenotypes are explained in the text. The figure is adapted from references 1 and 31. (b) Diagram of the phosphorylation network controlling PleD-mediated c-di-GMP synthesis during the *C. crescentus* cell cycle. PleC kinase ($PleC_K$) and phosphatase states ($PleC_P$) are indicated. Stippled arrows highlight the allosteric activation of PleC and DivJ by the DivK response regulator. Product inhibition of PleD is indicated. The equilibrium of phosphorylation and dephosphorylation reactions of PleD and DivK are indicated by solid arrows. Cellular processes regulated by the pathway are shown in italics. The dynamic polar localization patterns of kinases and response regulators are marked with different symbols.

allele was mapped to a gene that is physically linked to *divK* and codes for an atypical response regulator with two receiver domains arranged in tandem (Rec-Rec′) and a C-terminal GGDEF output domain (31) (Fig. 3). Only the first of the two receiver domains (Rec) conforms to the consensus sequence of response regulator receiver domains, whereas the putative Asp phosphorylation site and a Lys residue involved in catalysis are not conserved in the Rec′ adaptor domain. The original finding that mutating the phosphoryl acceptor Asp53 of the Rec domain is able to suppress the motility defect of a *pleC* mutant strain indicated that PleD~P can block *C. crescentus* motility and that the paralyzed phenotype of a *pleC* mutant results from high levels of activated PleD~P, which results in unleashing of PleD activity (31). Hecht and Newton (31) were the first to coin the term GGDEF domain according to the most conserved stretch of amino acids in their alignment of the C-terminal domain of PleD with six other homologs that were present in the database in the mid-1990s. The authors correctly predicted that this newly discovered protein domain might have a catalytic function and made the visionary statement that "...although no function has been assigned to any of these sequences, the modular organization of response regulators into N-terminal regulatory and C-terminal effector domains leads us to speculate that the GGDEF sequence in PleD could function as an effector domain or, alternatively, a catalytic domain."

TEMPORAL AND SPATIAL REGULATION OF THE PleD DGC

The genetic and biochemical analyses carried out by Newton and colleagues suggested that PleC, DivJ, DivK, and PleD form a phosphorylation network controlling pole development in *C. crescentus.* The fact that the DivK response regulator lacks an output domain argued that this regulatory system at least in part operates through the GGDEF effector domain of PleD and that this domain plays a major role in determining *C. crescentus* cell fate. This was confirmed by demonstrating that *pleD* alleles with mutations altering the highly conserved GGDEF motif showed the same morphogenetic deficiencies as a Δ*pleD* null mutant (1). Moreover, phosphorylation-independent forms of WspR, a Rec-GGDEF protein from *Pseudomonas fluorescens,* are active in *Caulobacter* and are able to substitute for the role of PleD in pole development (2). To confirm that the GGDEF domain of WspR is indeed functional in *Caulobacter,* Aldridge and coworkers constructed hybrid proteins with the WspR GGDEF domain grafted onto the Rec-Rec′ receiver stem of PleD. The resulting hybrid proteins were active in *C. crescentus* and remained fully dependent on the phosphorylation of the first receiver domain (Rec) (2). These results not only suggested that the function of the GGDEF domain family is conserved in evolution but they also demonstrated that their members can easily be transferred between distantly related bacteria without losing their specific function. This was consistent with the proposed role of GGDEF domains in the turnover of the second messenger c-di-GMP. A few years earlier, Tal and coworkers had opened up the field with a breakthrough contribution that reported on the identification of several composite GGDEF-EAL domain proteins involved in the turnover of c-di-GMP in *Gluconacetobacter xylinus* (formerly *Acetobacter xylinum*) (77). This key observation, for the first time, established a connection between the GGDEF and EAL domains and the synthesis and hydrolysis of c-di-GMP (see chapter 3). The final confirmation that the GGDEF domain is a DGC came when purified PleD was shown to convert GTP into c-di-GMP and that an intact GGDEF domain was required for this activity (62). The finding that *C. crescentus* pole development is regulated by c-di-GMP raised several important questions. How is the activity of the PleD DGC regulated during the cell cycle to coordinate organelle assembly and removal? Are additional DGCs and PDEs involved in this process? And through which c-di-GMP-specific effector molecules and target processes do these factors control the morphogenetic changes occurring at the cell poles?

The activity of the PleD response regulator is controlled by phosphorylation of the first receiver domain, which leads to dimerization-mediated activation of PleD (60). Two crystal structures of PleD in its nonactivated and activated conformations suggested an activation mechanism that involves a rearrangement of the two receiver domains within one protomer and a consequential increase of dimerization surface of the Rec-Rec′ dimerization stem (7, 83). Ultimately, the formation of an antiparallel dimer brings two catalytic GGDEF domains into close proximity and positions the two enzyme-bound GTP molecules in the antiparallel arrangement necessary for the condensation reaction (for a detailed description of the PleD activation mechanism, see chapter 7). The developmental phenotypes of different *pleD* alleles suggested that the PleD DGC is inactive in SW cells but is activated upon differentiation of SW cells into ST cells (2, 75). In agreement with this, PleD dynamically localizes to the differentiating ST pole during the cell cycle as a function of its phosphorylation state (62) (Fig. 4b). PleD mutants that cannot be phosphorylated or are unable to dimerize fail to localize to the pole (60, 62). Based on this, it was proposed that PleD dimerization serves to sequester activated PleD to the differentiating *Caulobacter* cell pole. Such a mechanism not only implicates protein oligomerization in protein localization control but provides the molecular basis to couple PleD activation to subcellular localization (60).

The genetic interaction of *divJ*, *pleC*, and *pleD* had proposed PleC and DivJ as candidate kinases regulating PleD activity (2, 73, 75). This was first corroborated by in vitro phosphorylation experiments that confirmed that DivJ and PleC interact with PleD to modulate its phosphorylation state (62). Likewise, in vivo levels of PleD~P are reduced to about 10% in cells lacking DivJ and are undetectable in a mutant lacking both DivJ and PleC (2). The observation that relative levels of PleD~P were slightly increased in *pleC* single mutants indicates that PleC may also act as a phosphatase at some point during the cell cycle. In agreement with a direct role for DivJ and PleC in PleD phosphorylation, PleD localization to the differentiating pole is reduced in both *pleC* and *divJ* single mutants and was completely abolished in a mutant strain lacking both HPKs. In contrast, a phosphorylation-independent and constitutively active mutant, PleD*, exclusively localizes to the ST pole independent of PleC and DivJ (62). Together, these experiments led us to conclude that PleC and DivJ coactivate PleD.

Both DivJ and PleC undergo spatial regulation and are asymmetrically positioned to opposite poles of the *Caulobacter* PD cell. PleC localizes to the flagellated pole of PD and SW cells, but during the SW-to-ST cell transition, it is degraded and replaced by DivJ at the incipient ST pole (84). Importantly, during a short time window that coincides with the onset of pole morphogenesis and with the time of PleD activation, PleC and DivJ overlap at the same pole (61) (Fig. 4b). But how could this spatial distribution possibly explain the simultaneous activation of PleD by both kinases? A recent study identified DivK as an additional component of the signal transduction network controlling PleD activity during the cell cycle. Although a possible functional interaction between PleD and DivK had been postulated based on the coexpression of the *divK* and *pleD* genes and the high similarity of their phospho-accepting receiver domains, the molecular nature of this interaction came as a surprise. Like its counterpart response regulator PleD, DivK dynamically localizes to the *C. crescentus* cell poles in a phosphorylation-dependent manner (40). During the SW-to-ST cell transition, DivK~P first sequesters to the differentiating ST cell pole at a time that coincides with the arrival of PleD at the same subcellular site. Later in the cell cycle, DivK~P also localizes to the pole opposite the stalk but is released specifically from this site after completion of cytokinesis (Fig. 4b). While the DivJ kinase is the main phosphodonor for DivK and primarily responsible for its localization to the cell poles (44), PleC phosphatase activity displaces DivK from the flagellated pole by maintaining low DivK~P levels in the SW cell (44, 52). Compartmentalization of the DivJ kinase and the PleC phosphatase during cell division results in the sudden reduction of DivK~P levels in the SW cell, the release of DivK from the flagellated pole, and the consequential initiation of the SW cell-specific developmental program (52). Conversely, a rapid DivJ-mediated increase of DivK phosphorylation during the G_1-to-S transition is critical for SW-to-ST cell differentiation (37, 40, 87). PleC mutants that specifically lack kinase activity but have retained phosphatase activity are able to initiate the first step of development by turning on motility and directing pilus biogenesis in SW cells, but these mutants fail to facilitate the next, PleD-dependent, step of development, which includes loss of motility and efficient holdfast and stalk formation (52, 61). These experiments indicated that PleC is a bifunctional protein that alternates between phosphatase and kinase activity to catalyze consecutive steps of *C. crescentus* pole development. Whereas PleC phosphatase activity is critical to reduce DivK~P (and possibly PleD~P) levels in SW cells, PleC kinase activity is required for the activation of PleD during the SW-to-ST cell transition. But how is this switch in PleC activity controlled during the cell cycle?

Recent in vitro and in vivo studies provided convincing evidence that DivK acts as an allosteric regulator of PleC kinase activity (43, 61). DivK strongly stimulates PleC autophosphorylation but not phosphatase activity, arguing that it can switch PleC from the default phosphatase into the kinase state. Through allosteric activation of the DivJ autokinase, DivK also stimulates its own phosphorylation and polar localization (61). Together, these data suggested that DivK is the central effector of an integrated circuit that operates via spatially organized feedback loops to determine *C. crescentus* cell fate (Fig. 4b) (61). In SW cells, DivJ levels are low (84) and PleC is in its phosphatase mode. As a result, DivK and PleD are inactive and delocalized. As DivJ levels rise during the G_1-to-S transition, DivJ-mediated phosphorylation localizes DivK~P to the cell pole, where it forces PleC into the autokinase mode. At the same time, DivK-mediated activation of DivJ creates a positive feedback loop that may promote a sharp G_1-to-S transition and, through the activation of PleD, rapidly and robustly commit cells to the sessile and surface-adherent ST cell program. Conversely, the proposed feedback loops could explain how the system is rapidly reset in the motile SW progeny (Fig. 4b). During cytokinesis, DivK localized at the flagellated pole is separated from its main kinase DivJ. This results in a rapid reduction of DivK~P levels in this compartment and a decrease of the DivK concentration at the cell pole below a threshold level required for PleC kinase activation. As a consequence, PleC switches back into its default phosphatase mode, thereby installing the SW cell program (Fig. 4b). According to this model, the role of DivK is to facilitate long-range communication between the asymmetric DivJ and PleC antagonists and, through these kinase feedback loops, coordinate their activities. DivK reinforces the dominant role of the DivJ kinase over PleC and makes sure that PleC is rapidly set to the kinase mode whenever the two HPKs coincide within the same cellular compartment. Ultimately, this kinase feedback mechanism facilitates the periodic switch of the *C. crescentus* life cycle by regulating the activity and dynamic positioning of the PleD DGC (43).

An additional level of PleD control is provided by allosteric product inhibition. When Paul and coworkers originally identified PleD DGC activity in vitro, they observed a strong inhibition upon addition of chemically synthesized c-di-GMP (62). This phenomenon is due to PleD activity being feedback inhibited in a noncompetitive manner with an inhibition constant of about 1 μM, i.e., in the range of the cellular c-di-GMP concentration (10, 61, 71). The molecular mechanism of feedback control involves an allosteric binding site (I-site), which is functionally contained within the GGDEF domain and is located antipodal to the substrate-binding site (A-site) (7, 83). The core of this primary I-site (I_p) is defined by a conserved RXXD motif (Fig. 3) that makes specific contacts to a base-intercalated dimer of c-di-GMP. In the PleD crystal structures, the ligand bound to the I_p site cross-links the GGDEF domain either to a neighboring GGDEF or the Rec′ domain via distinct secondary binding sites. This inhibition by immobilization mechanism prevents the productive encounter of two GGDEF A-sites in the PleD dimer (for details, see chapter 7). I-site-mediated product inhibition was also found for other DGCs (10, 15), arguing that the general mechanism of allosteric control is conserved in this enzyme family. Consistent with this, more than 60% of all GGDEF domains in the database have a conserved RXXD I_p site and a conserved spacer length between the I_p and A-site, suggesting that the three-dimensional arrangement of the catalytic and allosteric pockets is similar in all of these enzymes (10). Although the physiological role of PleD product inhibition is unclear, there are several possibilities that could explain the need for this additional control (10). By setting an upper limit for the product concentration, feedback inhibition can prevent futile substrate consumption or establish precise threshold concentrations of the second messenger. This applies in particular to cases where regulatory circuits involving a DGC have been optimized for fast up-kinetics. In combination with counteracting PDEs, feedback-regulated DGC could produce short spikes or even generate oscillations of c-di-GMP. It is interesting to note that during the *C. crescentus* SW-to-ST cell differentiation, a characteristic PleD-dependent spike of c-di-GMP is observed. This is in contrast to the predicted situation, where levels of the second messenger would reach a constant plateau upon PleD activation in ST cells (61). Although the physiological significance of this transient increase of c-di-GMP remains unclear, such behavior could contribute to the coordination and directionality of PleD-mediated downstream processes.

ADDITIONAL c-di-GMP SIGNALING COMPONENTS INVOLVED IN *C. CRESCENTUS* POLE DEVELOPMENT

The primary function of the complex regulatory mechanism responsible for cell cycle-dependent PleD phosphorylation is to limit PleD DGC activity to the sessile ST cell and exclude it from the motile SW cell. The observation that motility can effectively be blocked by expressing a constitutively active mutant of PleD, PleD*, or by overexpressing the DGC DgcA (2, 10) (Fig. 3), emphasizes the importance of keeping c-di-GMP levels low in SW cells to sustain motility. Recent data demonstrate that in addition to the tight cell cycle-dependent regulation of PleD, a SW cell-specific PDE, PdeA, is required to establish the motile SW cell program. PdeA is a soluble PDE that harbors both a GGDEF and EAL domain and a PAS-like N-terminal extension (Fig. 3). The PDE activity is confined to the C-terminal EAL domain, which shows strict conservation of all residues required for c-di-GMP binding and hydrolysis (12, 54, 65). In contrast, the associated GGDEF domain contains an altered active site motif (GEDEF) and lacks detectable DGC activity (Fig. 3). Instead, this domain is able to bind GTP and, in response, activates the PDE activity in the neighboring EAL domain (12) (Fig. 5a). PDE activation is specific for GTP (K_d [dissociation constant], 4 μM) and operates by lowering the K_m for c-di-GMP of the EAL domain to a physiologically significant level (420 nM) (12). While the in vivo relevance of this control is unclear, this is a first example

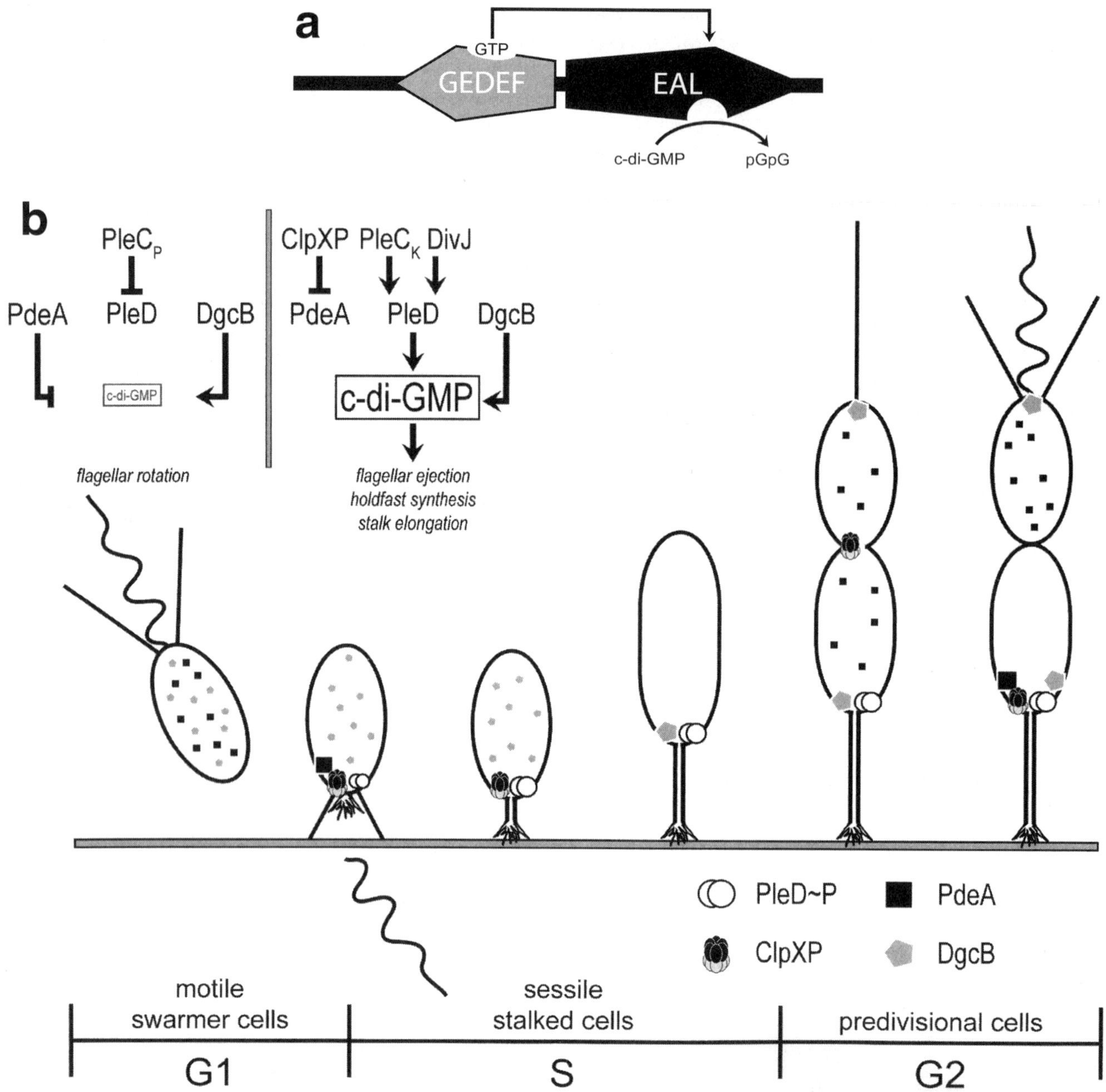

Figure 5. PdeA, DgcB, and PleD collectively control *Caulobacter* pole morphogenesis. (a) Schematic of the domain organization and molecular function of PdeA. The GGDEF domain (degenerate active site motif: GEDEF) functions to allosterically stimulate the neighboring PDE activity of the EAL domain. (b) Representation of the dynamic subcellular localization of PdeA, PleD, DgcB, and ClpXP and wiring of the components involved in c-di-GMP-mediated pole morphogenesis. PleC and DivJ are labeled as described in the legend to Fig. 4b. For reasons of simplicity, only the activated (phosphorylated) form of PleD is depicted. Fluctuating levels of c-di-GMP in SW and ST cells are indicated. Cellular processes regulated by the pathway are shown in italics.

for specific interaction of GGDEF and EAL domains in composite proteins to coordinate their activities. Furthermore, these experiments also demonstrated for the first time that some c-di-GMP signaling domains have evolved beyond their roles as enzymes and have adopted regulatory functions.

Mutants lacking PdeA have paralyzed flagella, show premature holdfast synthesis already in newborn SW cells, and display an increased propensity for surface colonization (S. Abel and U. Jenal, unpublished data). This phenotype is comparable to cells that have artificially increased c-di-GMP levels

(2, 10), arguing that the primary function of PdeA is to uphold the motile SW cell program against the antagonistic activity of one or several DGCs. Since PleD is inactive in SW cells, it cannot be considered for such a role. Consistent with this, mutations in *pleD* are unable to suppress the nonmotile phenotype of a Δ*pdeA* mutant. Instead, DgcB was identified as an antagonist of PdeA in a screen for motile suppressors of the Δ*pdeA* mutant. DgcB harbors a conserved GGDEF domain (Fig. 3) and shows DGC activity in vitro. Mutants lacking *dgcB* show increased motility, delayed holdfast synthesis, and lower propensity for surface colonization. Intriguingly, while both *pleD* and *dgcB* single mutants show a partial holdfast and surface attachment phenotype (Fig. 4a), mutants lacking both DGCs completely fail to assemble the primary cell adhesin and are unable to colonize surfaces (49; Abel and Jenal, unpublished). This led to the conclusion that DgcB and PleD together execute the sessile ST cell program and, in conjunction with PdeA, regulate the motile-sessile switch. While DgcB apparently operates in both cell types, the activities of PleD and PdeA are limited to the sessile and motile cell types, respectively. The following model integrates the activities of all three c-di-GMP metabolizing enzymes during the cell cycle (Fig. 5b). In the SW cell, DgcB is active but kept in check by the dominant PDE PdeA. As a result, c-di-GMP levels remain low, keeping the cell in its motile state. During the SW-to-ST cell transition, PdeA activity is removed, and as a result, DgcB takes over. At the same time, PleD is activated by phosphorylation, localizes to the incipient ST pole (see above), and together with DgcB, contributes to the increase in c-di-GMP, which ultimately triggers pole morphogenesis. Interestingly, like PleD, DgcB also changes its subcellular distribution during the course of the cell cycle (Fig. 5b) (Abel and Jenal, unpublished). The function of this dynamic localization is not clear yet.

If PdeA specifically functions in SW cells, how can its activity be confined to the motile cell type? As it turns out, the PdeA protein is present in SW cells but is rapidly degraded during the SW-to-ST cell transition. During differentiation, PdeA transiently localizes to the incipient ST pole, where it is degraded by the polar ClpXP protease complex (53, 66; Abel and Jenal, unpublished) (see below). Thus, both PdeA and PleD represent key regulatory activities for *C. crescentus* cell fate determination. Their cell type-specific roles are achieved by an elegant combination of proteolysis and phosphorylation control, which inversely regulates these two activities to shift the c-di-GMP balance during development and establish the characteristic cellular programs (Fig. 5b). It is interesting that the cell fate determinator DivK, which is critically involved in PleD activation during the SW-to-ST cell transition (see above), has also been implicated in ClpXP-dependent degradation control (4, 37). It is intriguing to speculate that DivK orchestrates the morphogenetic transitions in *C. crescentus* through a multilevel mechanism that involves the coordinated control of PdeA stability and PleD activity.

The role of PdeA in controlling flagellar motility in *C. crescentus* SW cells is reminiscent of the role described for the PDE YhjH in *Escherichia coli* and *Salmonella enterica* serovar Typhimurium (see chapters 10 and 16). The *yhjH* gene is expressed as part of the flagellar regulon, and its product keeps the c-di-GMP level low in actively swimming cells. Motility control is executed through YcgR, a c-di-GMP effector protein that, in its ligand-bound form, interferes with flagellar activity (25, 63, 67). YcgR is a member of the PilZ domain protein family, which represents one of a few known classes of c-di-GMP effectors (3) (see also chapter 4). The *C. crescentus* genome encodes two PilZ proteins, DgrA and DgrB, which both bind c-di-GMP specifically and with high affinity (K_d in the nanomolar range) (11). Both DgrA and DgrB are able to interfere with *C. crescentus* motility. In the presence of a plasmid-borne copy of *dgcA,* which encodes a highly active, soluble DGC (Fig. 3), *Caulobacter* SW cells assemble paralyzed flagella (10, 11). Whereas a mutation in *dgrB* partially suppresses this phenotype, motility is fully restored in the absence of DgrA. This indicates that both proteins, in their c-di-GMP-bound forms, can interfere with flagellar function. However, in the absence of a plasmid-borne copy of *dgcA,* motility was unaffected in Δ*dgrA* or Δ*dgrB* deletion mutants, indicating that DgrA and DgrB do not control motility under these conditions. Likewise, single and double mutations in *dgrA* and *dgrB* are unable to restore motility of mutant strains lacking the SW cell-specific PDE PdeA or of a strain expressing the constitutive form of PleD, PleD* (S. Abel, B. Christen, and U. Jenal, unpublished data). This indicates that the two PilZ domain proteins DgrA and DgrB do not play a major role in regulating motor function during the *C. crescentus* cell cycle in response to fluctuating levels of c-di-GMP.

The observation that PdeA, DgcB, and PleD modulate *Caulobacter* motility in a DgrA- and DgrB-independent manner, argues for the existence of distinct pathways and mechanisms to control flagellar function and removal in response to c-di-GMP. In addition, an EAL domain protein, TipF, controls *Caulobacter* motility as part of a signal transduction pathway that is required for localized flagellar assembly (36) (Fig. 3). About 50 genes are required for flagellar biogenesis in *C. crescentus*. Most of these genes are organized in a three-tier hierarchy, the expression of which is controlled by flagellar assembly checkpoints (18, 26, 55–57, 86) (Fig. 6a). Mutants lacking TipF are unable to efficiently express class III and class IV genes and fail to assemble a functional flagellum (36). Introducing a mutation in *flbT*, which encodes a translation regulator that links the expression of flagellar genes to the assembly status of the flagellum, restores class IV gene expression. However, secretion of hook and filament proteins is not restored, arguing that TipF primarily interferes with flagellar assembly (36). In addition to its flagellum-specific function, TipF is required for the efficient biosynthesis of pili and the localization of chemotaxis components to the flagellated pole (36) (Fig. 6a). This

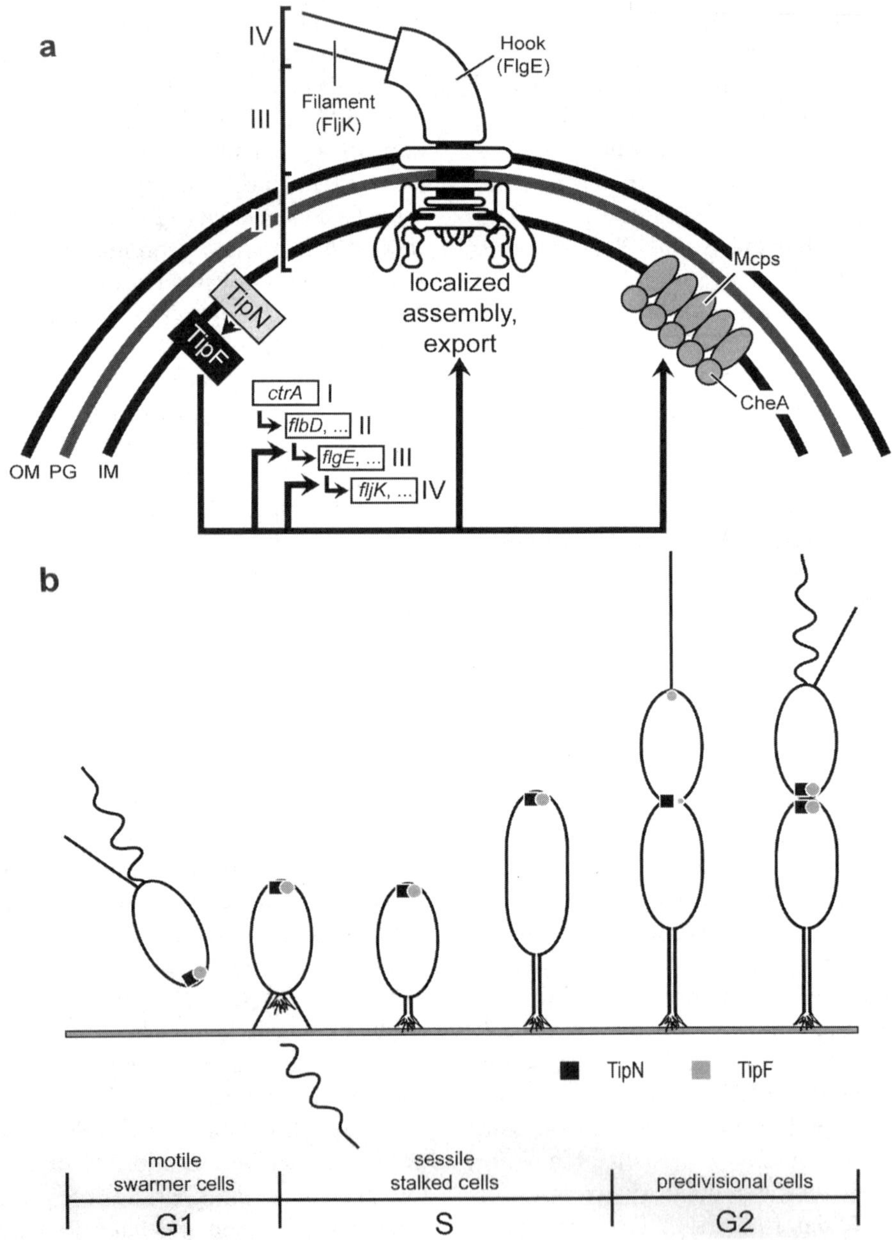

Figure 6. TipF-mediated spatial control of flagellar assembly. (a) Schematic diagram of the *C. crescentus* flagellum and chemotaxis apparatus at the SW cell pole. The flagellar gene regulatory hierarchy is indicated. TipF is recruited to the new cell pole by TipN. Arrows highlight the stimulatory effect of TipF on flagellar gene expression and assembly as well as positioning of the chemoreceptor complex. OM, outer membrane; IM, inner membrane; PG, peptidoglycan. (b) Illustration of the subcellular distribution of TipF and its localization factor TipN during the *Caulobacter* cell cycle.

suggests that TipF might have a general function in assisting the assembly and/or placement of different organelles at the flagellated pole.

TipF itself dynamically localizes to the newly formed cell poles during cytokinesis, the future site of flagellum biosynthesis (Fig. 6b). TipF placement to this subcellular site requires the birth scar protein TipN, which is deposited at the site of cytokinesis in a step that requires the cell division protein FtsZ (36, 45). In the absence of TipN, TipF and the flagellum are frequently misplaced, suggesting that, in addition to its role in organelle assembly, TipF is required for the correct positioning of the flagellum (36). Whether TipF functions as a SW pole-specific PDE is not clear yet. Residues known to be involved in substrate binding and catalysis are conserved in the TipF EAL domain (Fig. 3), and the highly conserved E211 of the EAL signature motif is required for TipF function (36). In summary, these data suggest that c-di-GMP can interfere with flagellum-based motility in *Caulobacter* on several levels, including motor activity and assembly. Multiple regulatory interactions might coordinate flagellar function with the asymmetric life cycle of *Caulobacter* and might help to adjust cell motility to a rapidly changing environment.

A ROLE FOR c-di-GMP SIGNALING IN *C. CRESCENTUS* CELL CYCLE REGULATION

The wide range of different cellular processes and molecular targets that are regulated by c-di-GMP reflects its remarkable versatility as a signaling device. Recently, cell cycle control and regulated proteolysis have been added to this growing list of cellular functions controlled by c-di-GMP. The *C. crescentus* cell cycle is controlled by a cascade of four master regulators that are activated sequentially and in a hierarchical manner (for recent reviews, see references 6 and 47). CtrA, GcrA, DnaA, and CcrM are successively synthesized to mediate a series of processes that execute chromosome replication, DNA segregation, and cytokinesis (50). CtrA, GcrA, and DnaA are transcription factors that together control about 200 genes within their respective cell cycle windows (14, 34, 46). In addition to their role as cell cycle timers of gene expression, CtrA and DnaA are directly involved in chromosome replication control. While DnaA is required for the initiation of DNA replication (28), CtrA has the opposite effect and inhibits replication initiation by directly binding to the chromosomal origin (64). Remarkably, all four master regulators are subject to controlled degradation during the cell cycle (16, 29, 34, 85). Degradation of the regulators contributes to the hierarchical behavior of the cascade, imposes directionality to the cell cycle, and directly influences key cell cycle steps, e.g., degradation of the replication inhibitor CtrA at the G_1/S boundary not only leads to the activation of *gcrA* transcription (the *gcrA* gene is negatively controlled by CtrA [34]) but also releases the replication block, thereby facilitating entry into S phase (64).

The CtrA response regulator is subject to a complex, multilevel control during the cell cycle, which includes transcription, activation, and degradation (47). Upon entry into S phase, the CtrA response regulator is inactivated via a dual mechanism involving dephosphorylation and degradation (16). Cell cycle-dependent degradation of CtrA entails a specific spatial arrangement of the proteolytic machinery. When differentiating SW cells proceed from G_1 into S and, likewise, when newborn ST cells reenter S phase, both CtrA and its cognate protease ClpXP (9, 42) sequester to the old cell pole where CtrA is rapidly degraded (53, 66). CtrA is stabilized in mutants that fail to localize either the protease or the substrate to this subcellular site, indicating that the timing of CtrA degradation is intimately linked to the dynamic spatial arrangement of these proteins during the cell cycle (17, 38, 53). Polar targeting of ClpXP is controlled by the single domain response regulator CpdR, which itself sequesters to the ST cell pole subject to its phosphorylation state (38, 39) (for a recent review, see reference 3) (Fig. 7). CpdR is kept in an inactive, phosphorylated form by the CckA-ChpT phosphorylation cascade that also activates CtrA. Through this elegant mechanism, CtrA activity and stability are directly interconnected (4).

Localization of the CtrA substrate to the ClpXP-occupied pole is regulated by PopA, a structural homolog of the PleD response regulator (Fig. 3). PopA functions as a spatial regulator and recruits CtrA to the cell pole via direct interaction with RcdA, a mediator protein that helps to localize CtrA (17, 53) (Fig. 7). Like PleD, PopA harbors two receiver-like domains that are fused to a GGDEF output domain. Despite this similar overall architecture and similar dynamic behavior (see below), the two proteins exhibit only 23% amino acid identity and their functions are fundamentally different. Although the first receiver domain of PopA has a conserved Asp phosphoryl acceptor site (Asp55), this residue is not required for PopA activity (17). Likewise, key residues important for phosphorylation-mediated intradomain signal transduction (24) are not conserved in this domain. The PopA output domain possesses a degenerate GGDEF motif (ATEVF) (Fig. 3) and lacks DGC activity. However, the PopA GGDEF domain has a conserved I-site, through which it is able to bind c-di-GMP (17). The affinity of PopA for c-di-GMP (K_d,

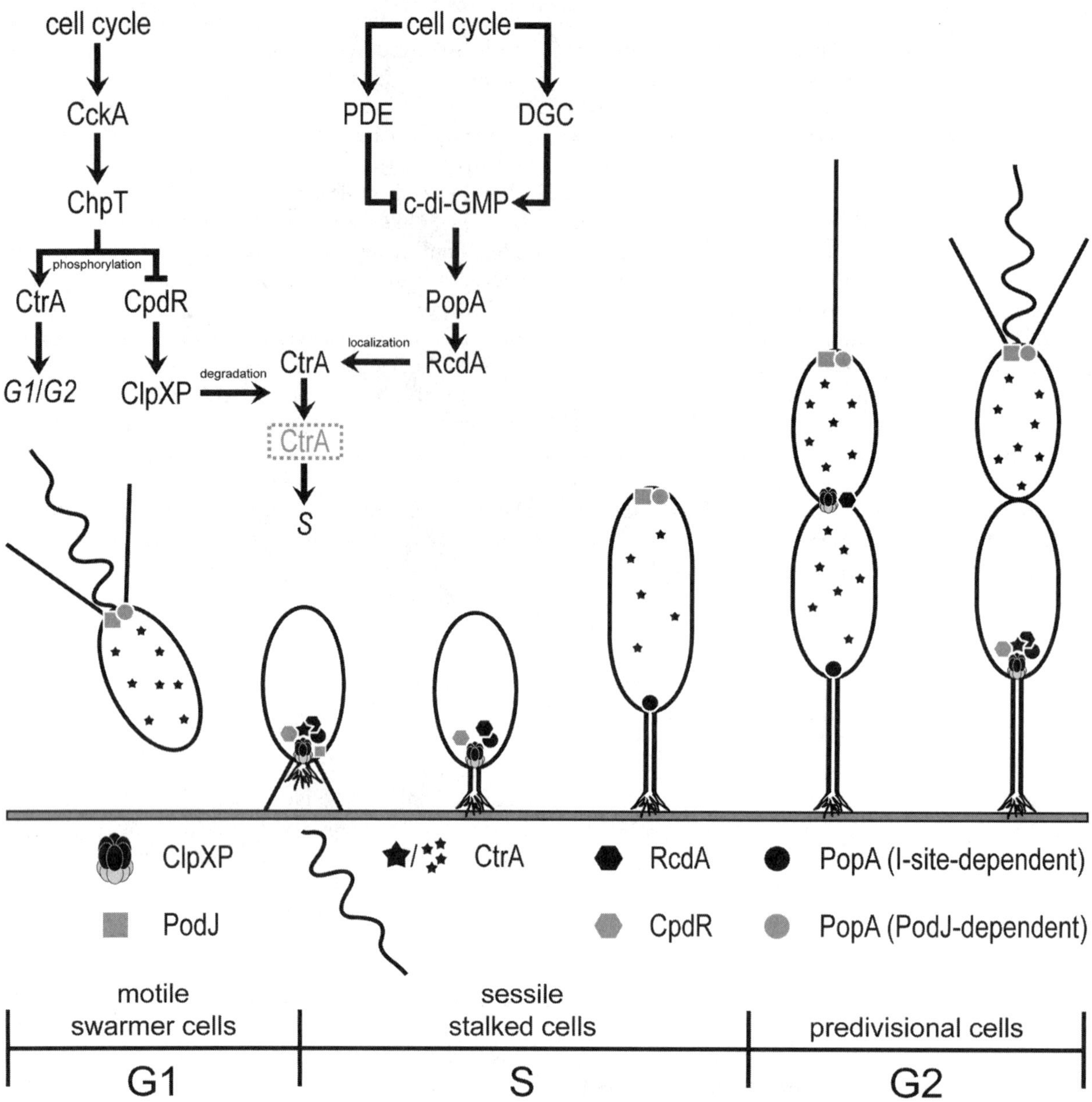

Figure 7. Spatiotemporal control of CtrA degradation during the *Caulobacter* cell cycle. The dynamic subcellular localization of the ClpXP protease, its substrate CtrA, and the localization factors PopA, PodJ, and CpdR are represented. The pathways regulating the phosphorylation, localization, and proteolysis of the cell cycle regulator CtrA are outlined schematically. The activation and stabilization of CtrA by the CckA-ChpT phosphorelay governs the G_1/G_2 program. PopA-dependent localization and degradation of CtrA facilitates S phase entry. PopA localization to the flagellated pole (grey circles) requires the spatial regulator PodJ. PopA localization to the ST pole (black circles) depends on c-di-GMP binding to the I-site.

2 μM) is similar to the allosteric I-sites of two catalytically active DGCs (7, 10, 83), suggesting that PopA binds c-di-GMP within a physiologically relevant concentration range (10, 71). Several experimental observations indicate that c-di-GMP binding to the I-site regulates PopA activity. PopA localizes to the new and the old cell poles at distinct times of the cell cycle (Fig. 7). Successive localization of PopA to these two subcellular sites relies on distinct mechanisms, which apparently serve different cellular functions. PopA localization to the incipient SW pole requires PodJ, a cell polarity determinant that also recruits the PleC histidine kinase/phosphatase and components of the pilus assembly machinery to this subcellular site (33, 48, 80). Mutants lacking PopA show reduced motility, indicating that PopA localization to the flagellated pole might contribute to motor function. Whereas PodJ is not required for CtrA

degradation and is cleared from the cell pole before cells enter S phase (8), PopA persists at the old pole to recruit RcdA and facilitate CtrA degradation. Because an intact I-site is required for PopA sequestration to the ST cell pole (both in the wild type and *podJ* mutants) but not for the interaction with RcdA, c-di-GMP binding appears to regulate the timing of PopA sequestration to the old cell pole, rather than its interaction with downstream components (17). Although PleD and PopA are activated by different mechanisms and have distinct cellular roles, a common polar localization mechanism could explain the conserved domain architecture of the two regulators.

Together, these data demonstrate that the PopA GGDEF protein has adopted a novel role as a c-di-GMP effector protein, which in response to ligand binding, localizes to the old cell pole where it functions as a spatial regulator to recruit RcdA and CtrA. This model predicts that a specific increase of c-di-GMP upon entry into S phase leads to PopA activation and polar sequestration. In line with this view, expression of a heterologous PDE rapidly and specifically interferes with PopA localization and CtrA degradation (17). However, CtrA degradation is unaltered in single mutants lacking any of the potential *C. crescentus* DGCs (Fig. 3). This leaves the possibility that several DGCs redundantly contribute to the c-di-GMP pool required for PopA activation and localization. Alternatively, a PDE that specifically reduces c-di-GMP levels in G_1 and G_2 might ultimately be responsible for shifting the balance of c-di-GMP during the cell cycle and localizing PopA to the ST cell pole. The observation that c-di-GMP levels peak during the G_1-to-S transition and that the PleD DGC is mainly responsible for this transient increase (61) suggests that PleD, together with other DGCs, could be involved in PopA activation. Recently, a *C. crescentus* mutant strain lacking all genes coding for potential DGCs was constructed (M. Nicollier, S. Abel, and U. Jenal, unpublished data). The analysis of PopA localization and CtrA degradation in this strain not only confirmed the critical role of c-di-GMP in the spatiotemporal control of the PopA/CtrA pathway but also suggested that PleD contributes to PopA activation. This offers the attractive possibility that PleD, by regulating both pole morphogenesis and cell cycle progression, contributes to the temporal coordination of the two processes.

PopA and CpdR define two distinct polar recruitment avenues for the CtrA substrate and the ClpXP protease, respectively (17), but how these pathways are interlinked is not clear yet. Mutants lacking CpdR, in addition to being unable to direct ClpXP to the cell pole, also fail to localize CtrA and show a partial defect in localizing RcdA to the cell pole (53). Conversely, CpdR and ClpX frequently mislocalize in *popA* mutants. Based on this and on the finding that CpdR weakly interacts with RcdA, it has been proposed that all of these proteins might be part of a macromolecular complex that transiently assembles at the cell pole to activate protein degradation in response to the appropriate signals during the cell cycle (Fig. 7) (17). Future studies will have to identify missing components involved in this process and elucidate the specific role of each of these proteins in delivering CtrA (and other substrates like PdeA) to the ClpXP degradation machinery.

CONCLUDING REMARKS

Several c-di-GMP signaling components dynamically localize to the cell poles at specific stages of the *Caulobacter* life cycle where they orchestrate different aspects of cell differentiation and cell cycle progression. But what is the significance of the subcellular localization of these proteins? Are they tagged to specific sites because they engage in a spatially confined c-di-GMP signaling compartment, or does their behavior simply reflect the inherent cellular asymmetry of *Caulobacter* and the associated need to compartmentalize key cell fate determinators? Are important steps of pole development and cell cycle progression regulated from a common cellular pool of c-di-GMP, which through carefully controlled fluctuations would serially activate targets with different affinities for the signaling molecule? Or are some of these processes modulated by insulated signaling modules that are transiently positioned to the cell poles? To clarify this, several issues need to be addressed, e.g., are the DGCs and PDEs that determine PopA localization to the ST cell pole present at the same subcellular site, and is such a spatial cooccurrence important for signal transduction? The finding that PleD might contribute to PopA activation opens up the possibility that the active forms of both proteins are directly linked in time and space. Also, are the target proteins that initiate flagellar ejection, holdfast biogenesis, and stalk formation in response to a PleD-mediated increase of c-di-GMP spatially connected to the DGC? Likewise, if TipF has PDE activity, is it acting on a c-di-GMP pool that is limited to the site of flagellum assembly? The only way to address these questions is by first identifying all the components of each of these signaling modules and then determining the mechanisms by which they interact. The main task with *Caulobacter* will be to find the effector molecules and target structures. Protein-protein interaction studies will help to understand how components are interconnected and how local signaling modules, if they exist at all, might operate. Finally, a major challenge will be to understand how the c-di-GMP-specific regulatory network that controls *Caulobacter* cell differentiation and propagation is able to respond to and integrate a multitude of signals from the environment. Several of the so far uncharacterized *Caulobacter* GGDEF and EAL domain proteins might be involved in this control (Fig. 3).

The *Caulobacter* life cycle with its characteristic sequence of spatially confined assembly and loss of adhesion and motility organelles, represents an ideal model system to address the issue of local versus global signaling within the c-di-GMP regulon and the significance of positional information and spatial control of its components. Although the developmental switch integrated in the *Caulobacter* life cycle is unique to this organism and its closest relatives, it is becoming increasingly clear that most, if not all, bacteria have evolved precisely tuned mechanisms to control spatial aspects of their physiology. The molecular and cellular mechanisms that underlie the characteristic behavior of *Caulobacter* cells and its regulation by c-di-GMP might thus be of general relevance for the understanding of processes involved in the motile-sessile transition in many other bacteria.

Acknowledgments. Work performed in the authors' laboratory and cited in this review has been supported by the Swiss National Science Foundation.

REFERENCES

1. **Aldridge, P., and U. Jenal.** 1999. Cell cycle-dependent degradation of a flagellar motor component requires a novel-type response regulator. *Mol. Microbiol.* **32:**379–391.
2. **Aldridge, P., R. Paul, P. Goymer, P. Rainey, and U. Jenal.** 2003. Role of the GGDEF regulator PleD in polar development of *Caulobacter crescentus. Mol. Microbiol.* **47:**1695–1708.
3. **Amikam, D., and M. Y. Galperin.** 2006. PilZ domain is part of the bacterial c-di-GMP binding protein. *Bioinformatics* **22:** 3–6.
4. **Biondi, E. G., S. J. Reisinger, J. M. Skerker, M. Arif, B. S. Perchuk, K. R. Ryan, and M. T. Laub.** 2006. Regulation of the bacterial cell cycle by an integrated genetic circuit. *Nature* **444:**899–904.
5. **Bodenmiller, D., E. Toh, and Y. V. Brun.** 2004. Development of surface adhesion in *Caulobacter crescentus. J. Bacteriol.* **186:**1438–1447.
6. **Bowers, L. M., E. B. Shapland, and K. R. Ryan.** 2008. Who's in charge here? Regulating cell cycle regulators. *Curr. Opin. Microbiol.* **11:**547–552.
7. **Chan, C., R. Paul, D. Samoray, N. C. Amiot, B. Giese, U. Jenal, and T. Schirmer.** 2004. Structural basis of activity and allosteric control of diguanylate cyclase. *Proc. Natl. Acad. Sci. USA* **101:**17084–17089.
8. **Chen, J. C., A. K. Hottes, H. H. McAdams, P. T. McGrath, P. H. Viollier, and L. Shapiro.** 2006. Cytokinesis signals truncation of the PodJ polarity factor by a cell cycle-regulated protease. *EMBO J.* **25:**377–386.
9. **Chien, P., B. S. Perchuk, M. T. Laub, R. T. Sauer, and T. A. Baker.** 2007. Direct and adaptor-mediated substrate recognition by an essential AAA+ protease. *Proc. Natl. Acad. Sci. USA* **104:**6590–6595.
10. **Christen, B., M. Christen, R. Paul, F. Schmid, M. Folcher, P. Jenoe, M. Meuwly, and U. Jenal.** 2006. Allosteric control of cyclic di-GMP signaling. *J. Biol. Chem.* **281:**32015–32024.
11. **Christen, M., B. Christen, M. G. Allan, M. Folcher, P. Jeno, S. Grzesiek, and U. Jenal.** 2007. DgrA is a member of a new family of cyclic diguanosine monophosphate receptors and controls flagellar motor function in *Caulobacter crescentus. Proc. Natl. Acad. Sci. USA* **104:**4112–4117.
12. **Christen, M., B. Christen, M. Folcher, A. Schauerte, and U. Jenal.** 2005. Identification and characterization of a cyclic di-GMP-specific phosphodiesterase and its allosteric control by GTP. *J. Biol. Chem.* **280:**30829–30837.
13. **Cole, J. L., G. G. Hardy, D. Bodenmiller, E. Toh, A. Hinz, and Y. V. Brun.** 2003. The HfaB and HfaD adhesion proteins of *Caulobacter crescentus* are localized in the stalk. *Mol. Microbiol.* **49:**1671–1683.
14. **Collier, J., S. R. Murray, and L. Shapiro.** 2006. DnaA couples DNA replication and the expression of two cell cycle master regulators. *EMBO J.* **25:**346–356.
15. **De, N., M. Pirruccello, P. V. Krasteva, N. Bae, R. V. Raghavan, and H. Sondermann.** 2008. Phosphorylation-independent regulation of the diguanylate cyclase WspR. *PLoS Biol.* **6:**e67.
16. **Domian, I. J., K. C. Quon, and L. Shapiro.** 1997. Cell type-specific phosphorylation and proteolysis of a transcriptional regulator controls the G1-to-S transition in a bacterial cell cycle. *Cell* **90:**415–424.
17. **Duerig, A., S. Abel, M. Folcher, M. Nicollier, T. Schwede, N. Amiot, B. Giese, and U. Jenal.** 2009. Second messenger-mediated spatiotemporal control of protein degradation regulates bacterial cell cycle progression. *Genes Dev.* **23:**93–104.
18. **Dutton, R. J., Z. Xu, and J. W. Gober.** 2005. Linking structural assembly to gene expression: a novel mechanism for regulating the activity of a sigma54 transcription factor. *Mol. Microbiol.* **58:**743–757.
19. **Ely, B., R. H. Croft, and C. J. Gerardot.** 1984. Genetic mapping of genes required for motility in *Caulobacter crescentus. Genetics* **108:**523–532.
20. **Entcheva-Dimitrov, P., and A. M. Spormann.** 2004. Dynamics and control of biofilms of the oligotrophic bacterium *Caulobacter crescentus. J. Bacteriol.* **186:**8254–8266.
21. **Fukuda, A., M. Asada, S. Koyasu, H. Yoshida, K. Yaginuma, and Y. Okada.** 1981. Regulation of polar morphogenesis in *Caulobacter crescentus. J. Bacteriol.* **145:**559–572.
22. **Fukuda, A., K. Miyakawa, H. Iida, and Y. Okada.** 1976. Regulation of polar surface structures in *Caulobacter crescentus:* pleiotropic mutations affect the coordinate morphogenesis of flagella, pili and phage receptors. *Mol. Gen. Genet.* **149:**167–173.
23. **Galperin, M. Y.** 2005. A census of membrane-bound and intracellular signal transduction proteins in bacteria: bacterial IQ, extroverts and introverts. *BMC Microbiol.* **5:**35.
24. **Gao, R., T. R. Mack, and A. M. Stock.** 2007. Bacterial response regulators: versatile regulatory strategies from common domains. *Trends Biochem. Sci* **32:**225–234.
25. **Girgis, H. S., Y. Liu, W. S. Ryu, and S. Tavazoie.** 2007. A comprehensive genetic characterization of bacterial motility. *PLoS Genet* **3:**1644–1660.
26. **Gober, J. W., and J. C. England.** 2000. Regulation of flagellum biosynthesis and motility in *Caulobacter*, p. 319–339. *In* Y. V.

Brun and L. J. Shimkets (ed.), *Prokaryotic Development*. ASM Press, Washington, DC.

27. **Goley, E. D., A. A. Iniesta, and L. Shapiro.** 2007. Cell cycle regulation in *Caulobacter:* location, location, location. *J. Cell. Sci.* **120:**3501–3507.
28. **Gorbatyuk, B., and G. T. Marczynski.** 2001. Physiological consequences of blocked *Caulobacter crescentus* dnaA expression, an essential DNA replication gene. *Mol. Microbiol.* **40:** 485–497.
29. **Gorbatyuk, B., and G. T. Marczynski.** 2005. Regulated degradation of chromosome replication proteins DnaA and CtrA in *Caulobacter crescentus. Mol. Microbiol.* **55:**1233–1245.
30. **Hecht, G. B., T. Lane, N. Ohta, J. M. Sommer, and A. Newton.** 1995. An essential single domain response regulator required for normal cell division and differentiation in *Caulobacter crescentus. EMBO J.* **14:**3915–3924.
31. **Hecht, G. B., and A. Newton.** 1995. Identification of a novel response regulator required for the swarmer- to-stalked-cell transition in *Caulobacter crescentus. J. Bacteriol.* **177:**6223–6229.
32. **Hengge, R.** 2009. Principles of c-di-GMP signalling in bacteria. *Nat. Rev. Microbiol.* **7:**263–273.
33. **Hinz, A. J., D. E. Larson, C. S. Smith, and Y. V. Brun.** 2003. The *Caulobacter crescentus* polar organelle development protein PodJ is differentially localized and is required for polar targeting of the PleC development regulator. *Mol. Microbiol.* **47:**929–941.
34. **Holtzendorff, J., D. Hung, P. Brende, A. Reisenauer, P. H. Viollier, H. H. McAdams, and L. Shapiro.** 2004. Oscillating global regulators control the genetic circuit driving a bacterial cell cycle. *Science* **304:**983–987.
35. **Huguenel, E. D., and A. Newton.** 1982. Localization of surface structures during procaryotic differentiation: role of cell division in *Caulobacter crescentus. Differentiation* **21:**71–78.
36. **Huitema, E., S. Pritchard, D. Matteson, S. K. Radhakrishnan, and P. H. Viollier.** 2006. Bacterial birth scar proteins mark future flagellum assembly site. *Cell* **124:**1025–1037.
37. **Hung, D. Y., and L. Shapiro.** 2002. A signal transduction protein cues proteolytic events critical to *Caulobacter* cell cycle progression. *Proc. Natl. Acad. Sci. USA* **99:**13160–13165.
38. **Iniesta, A. A., P. T. McGrath, A. Reisenauer, H. H. McAdams, and L. Shapiro.** 2006. A phospho-signaling pathway controls the localization and activity of a protease complex critical for bacterial cell cycle progression. *Proc. Natl. Acad. Sci. USA* **103:**10935–10940.
39. **Iniesta, A. A., and L. Shapiro.** 2008. A bacterial control circuit integrates polar localization and proteolysis of key regulatory proteins with a phospho-signaling cascade. *Proc. Natl. Acad. Sci. USA* **105:**16602–16607.
40. **Jacobs, C., D. Hung, and L. Shapiro.** 2001. Dynamic localization of a cytoplasmic signal transduction response regulator controls morphogenesis during the *Caulobacter* cell cycle. *Proc. Natl. Acad. Sci. USA* **98:**4095–4100.
41. **Janakiraman, R. S., and Y. V. Brun.** 1999. Cell cycle control of a holdfast attachment gene in *Caulobacter crescentus. J. Bacteriol.* **181:**1118–1125.
42. **Jenal, U., and T. Fuchs.** 1998. An essential protease involved in bacterial cell-cycle control. *EMBO J.* **17:**5658–5669.
43. **Jenal, U., and M. Y. Galperin.** 2009. Single domain response regulators: molecular switches with emerging roles in cell organization and dynamics. *Curr. Opin. Microbiol.* **12:**152–160.
44. **Lam, H., J. Y. Matroule, and C. Jacobs-Wagner.** 2003. The asymmetric spatial distribution of bacterial signal transduction proteins coordinates cell cycle events. *Dev. Cell* **5:**149–159.
45. **Lam, H., W. B. Schofield, and C. Jacobs-Wagner.** 2006. A landmark protein essential for establishing and perpetuating the polarity of a bacterial cell. *Cell* **124:**1011–1023.
46. **Laub, M. T., H. H. McAdams, T. Feldblyum, C. M. Fraser, and L. Shapiro.** 2000. Global analysis of the genetic network controlling a bacterial cell cycle. *Science* **290:**2144–2148.
47. **Laub, M. T., L. Shapiro, and H. H. McAdams.** 2007. Systems biology of *Caulobacter. Annu. Rev. Genet.* **41:**429–441.
48. **Lawler, M. L., D. E. Larson, A. J. Hinz, D. Klein, and Y. V. Brun.** 2006. Dissection of functional domains of the polar localization factor PodJ in *Caulobacter crescentus. Mol. Microbiol.* **59:**301–316.
49. **Levi, A., and U. Jenal.** 2006. Holdfast formation in motile swarmer cells optimizes surface attachment during *Caulobacter crescentus* development. *J. Bacteriol.* **188:**5315–5318.
50. **Li, S., P. Brazhnik, B. Sobral, and J. J. Tyson.** 2008. A quantitative study of the division cycle of *Caulobacter crescentus* stalked cells. *PLoS Comput. Biol.* **4:**e9.
51. **Marino, W., S. Ammer, and L. Shapiro.** 1976. Conditional surface structure mutants of *Caulobacter crescentus* temperature-sensitive flagella formation due to an altered flagellin monomer. *J. Mol. Biol.* **107:**115–130.
52. **Matroule, J. Y., H. Lam, D. T. Burnette, and C. Jacobs-Wagner.** 2004. Cytokinesis monitoring during development; rapid pole-to-pole shuttling of a signaling protein by localized kinase and phosphatase in *Caulobacter. Cell* **118:**579–590.
53. **McGrath, P. T., A. A. Iniesta, K. R. Ryan, L. Shapiro, and H. H. McAdams.** 2006. A dynamically localized protease complex and a polar specificity factor control a cell cycle master regulator. *Cell* **124:**535–547.
54. **Minasov, G., S. Padavattan, L. Shuvalova, J. S. Brunzelle, D. J. Miller, A. Basle, C. Massa, F. R. Collart, T. Schirmer, and W. F. Anderson.** 2009. Crystal structures of YkuI and its complex with second messenger c-di-GMP suggests catalytic mechanism of phosphodiester bond cleavage by EAL domains. *J. Biol. Chem.* **284:**13174–13184.
55. **Muir, R. E., and J. W. Gober.** 2002. Mutations in FlbD that relieve the dependency on flagellum assembly alter the temporal and spatial pattern of developmental transcription in *Caulobacter crescentus. Mol. Microbiol.* **43:**597–615.
56. **Muir, R. E., and J. W. Gober.** 2001. Regulation of late flagellar gene transcription and cell division by flagellum assembly in *Caulobacter crescentus. Mol. Microbiol.* **41:**117–130.
57. **Muir, R. E., T. M. O'Brien, and J. W. Gober.** 2001. The *Caulobacter crescentus* flagellar gene, fliX, encodes a novel trans-acting factor that couples flagellar assembly to transcription. *Mol. Microbiol.* **39:**1623–1637.
58. **Newton, A.** 1972. Role of transcription in the temporal control of development in *Caulobacter crescentus* (stalk-rifampin-RNA synthesis-DNA synthesis-motility). *Proc. Natl. Acad. Sci. USA* **69:**447–451.
59. **Ohta, N., T. Lane, E. G. Ninfa, J. M. Sommer, and A. Newton.** 1992. A histidine protein kinase homologue required for regulation of bacterial cell division and differentiation. *Proc. Natl. Acad. Sci. USA* **89:**10297–10301.
60. **Paul, R., S. Abel, P. Wassmann, A. Beck, H. Heerklotz, and U. Jenal.** 2007. Activation of the diguanylate cyclase pleD by phosphorylation-mediated dimerization. *J. Biol. Chem.* **282:** 29170–29177.
61. **Paul, R., T. Jaeger, S. Abel, I. Wiederkehr, M. Folcher, E. G. Biondi, M. T. Laub, and U. Jenal.** 2008. Allosteric regulation of histidine kinases by their cognate response regulator determines cell fate. *Cell* **133:**452–461.
62. **Paul, R., S. Weiser, N. C. Amiot, C. Chan, T. Schirmer, B. Giese, and U. Jenal.** 2004. Cell cycle-dependent dynamic localization of a bacterial response regulator with a novel diguanylate cyclase output domain. *Genes Dev.* **18:**715–727.
63. **Pesavento, C., G. Becker, N. Sommerfeldt, A. Possling, N. Tschowri, A. Mehlis, and R. Hengge.** 2008. Inverse regulatory

coordination of motility and curli-mediated adhesion in *Escherichia coli*. *Genes Dev.* **22**:2434–2446.

64. **Quon, K. C., B. Yang, I. J. Domian, L. Shapiro, and G. T. Marczynski.** 1998. Negative control of bacterial DNA replication by a cell cycle regulatory protein that binds at the chromosome origin. *Proc. Natl. Acad. Sci. USA* **95**:120–125.
65. **Rao, F., Y. Yang, Y. Qi, and Z. X. Liang.** 2008. Catalytic mechanism of cyclic di-GMP-specific phosphodiesterase: a study of the EAL domain-containing RocR from *Pseudomonas aeruginosa*. *J. Bacteriol.* **190**:3622–3631.
66. **Ryan, K. R., S. Huntwork, and L. Shapiro.** 2004. Recruitment of a cytoplasmic response regulator to the cell pole is linked to its cell cycle-regulated proteolysis. *Proc. Natl. Acad. Sci. USA* **101**:7415–7420.
67. **Ryjenkov, D. A., R. Simm, U. Romling, and M. Gomelsky.** 2006. The PilZ domain is a receptor for the second messenger c-di-GMP: the PilZ domain protein YcgR controls motility in enterobacteria. *J. Biol. Chem.* **281**:30310–30314.
68. **Shapiro, L., and N. Agabian-Keshishian.** 1970. Specific assay for differentiation in the stalked bacterium *Caulobacter crescentus*. *Proc. Natl. Acad. Sci. USA* **67**:200–203.
69. **Shapiro, L., N. Agabian-Keshishian, and I. Bendis.** 1971. Bacterial differentiation. *Science* **173**:884–892.
70. **Sheffery, M., and A. Newton.** 1981. Regulation of periodic protein synthesis in the cell cycle: control of initiation and termination of flagellar gene expression. *Cell* **24**:49–57.
71. **Simm, R., M. Morr, U. Remminghorst, M. Andersson, and U. Romling.** 2009. Quantitative determination of cyclic diguanosine monophosphate concentrations in nucleotide extracts of bacteria by matrix-assisted laser desorption/ionization-time-of-flight mass spectrometry. *Anal. Biochem.* **386**:53–58.
72. **Skerker, J. M., and L. Shapiro.** 2000. Identification and cell cycle control of a novel pilus system in *Caulobacter crescentus*. *EMBO J.* **19**:3223–3234.
73. **Sommer, J. M., and A. Newton.** 1991. Pseudoreversion analysis indicates a direct role of cell division genes in polar morphogenesis and differentiation in *Caulobacter crescentus*. *Genetics* **129**:623–630.
74. **Sommer, J. M., and A. Newton.** 1988. Sequential regulation of developmental events during polar morphogenesis in *Caulobacter crescentus:* assembly of pili on swarmer cells requires cell separation. *J. Bacteriol.* **170**:409–415.
75. **Sommer, J. M., and A. Newton.** 1989. Turning off flagellum rotation requires the pleiotropic gene *pleD: pleA, pleC,* and *pleD* define two morphogenic pathways in *Caulobacter crescentus*. *J. Bacteriol.* **171**:392–401.
76. **Stephens, C. M., and L. Shapiro.** 1993. An unusual promoter controls cell-cycle regulation and dependence on DNA replication of the *Caulobacter fliLM* early flagellar operon. *Mol. Microbiol.* **9**:1169–1179.
77. **Tal, R., H. C. Wong, R. Calhoon, D. Gelfand, A. L. Fear, G. Volman, R. Mayer, P. Ross, D. Amikam, H. Weinhouse, A. Cohen, S. Sapir, P. Ohana, and M. Benziman.** 1998. Three *cdg* operons control cellular turnover of cyclic di-GMP in *Acetobacter xylinum:* genetic organization and occurrence of conserved domains in isoenzymes. *J. Bacteriol.* **180**:4416–4425.
78. **Toh, E., H. D. Kurtz, Jr., and Y. V. Brun.** 2008. Characterization of the *Caulobacter crescentus* holdfast polysaccharide biosynthesis pathway reveals significant redundancy in the initiating glycosyltransferase and polymerase steps. *J. Bacteriol.* **190**:7219–7231.
79. **Tsang, P. H., G. Li, Y. V. Brun, L. B. Freund, and J. X. Tang.** 2006. Adhesion of single bacterial cells in the micronewton range. *Proc. Natl. Acad. Sci. USA* **103**:5764–5768.
80. **Viollier, P. H., N. Sternheim, and L. Shapiro.** 2002. A dynamically localized histidine kinase controls the asymmetric distribution of polar pili proteins. *EMBO J.* **21**:4420–4428.
81. **Wagner, J. K., S. Setayeshgar, L. A. Sharon, J. P. Reilly, and Y. V. Brun.** 2006. A nutrient uptake role for bacterial cell envelope extensions. *Proc. Natl. Acad. Sci. USA* **103**:11772–11777.
82. **Wang, S. P., P. L. Sharma, P. V. Schoenlein, and B. Ely.** 1993. A histidine protein kinase is involved in polar organelle development in *Caulobacter crescentus*. *Proc. Natl. Acad. Sci. USA* **90**:630–634.
83. **Wassmann, P., C. Chan, R. Paul, A. Beck, H. Heerklotz, U. Jenal, and T. Schirmer.** 2007. Structure of BeF(3)(−)-modified response regulator PleD: implications for diguanylate cyclase activation, catalysis, and feedback inhibition. *Structure* **15**:915–927.
84. **Wheeler, R. T., and L. Shapiro.** 1999. Differential localization of two histidine kinases controlling bacterial cell differentiation. *Mol. Cell.* **4**:683–694.
85. **Wright, R., C. Stephens, G. Zweiger, L. Shapiro, and M. R. Alley.** 1996. *Caulobacter* Lon protease has a critical role in cell-cycle control of DNA methylation. *Genes Dev.* **10**:1532–1542.
86. **Wu, J., and A. Newton.** 1997. Regulation of the *Caulobacter* flagellar gene hierarchy; not just for motility. *Mol. Microbiol.* **24**:233–239.
87. **Wu, J., N. Ohta, and A. Newton.** 1998. An essential, multicomponent signal transduction pathway required for cell cycle regulation in *Caulobacter*. *Proc. Natl. Acad. Sci. USA* **95**:1443–1448.
88. **Wu, J., N. Ohta, J. L. Zhao, and A. Newton.** 1999. A novel bacterial tyrosine kinase essential for cell division and differentiation. *Proc. Natl. Acad. Sci. USA* **96**:13068–13073.

The Second Messenger Cyclic Di-GMP
Edited by A. J. Wolfe and K. L. Visick

Chapter 10

Hierarchical Control of rdar Morphotype Development of *Salmonella enterica* by Cyclic Di-GMP

UTE RÖMLING, KRISTINA JONAS, ÖJAR MELEFORS, NINA GRANTCHAROVA, AND AGARISTI LAMPROKOSTOPOULOU

In *Salmonella enterica* serovar Typhimurium, the small molecule bis-(3′,5′)-cyclic di-GMP (c-di-GMP) was discovered as the output molecule of the enzymatic activity of the GGDEF domain protein AdrA (90). How could this finding be made? Ultimately, this discovery was the result of a number of coincidental findings which resulted from the desire to understand the molecular mechanisms leading to the expression of the rdar morphotype, a multicellular biofilm behavior of *S. enterica* serovar Typhimurium and other members of the *Enterobacteriaceae* that contributes to environmental survival and has a role in pathogen-host interaction (77, 83).

Initially, AdrA was identified in the search for components contributing to rdar morphotype expression (82). Further experiments demonstrated that AdrA posttranscriptionally activated the biosynthesis of cellulose, a newly identified extracellular matrix component of biofilms produced by *S. enterica* serovar Typhimurium (114). These findings could be related to the pioneering work of Moshe Benziman who some years earlier had discovered c-di-GMP as an allosteric activator of the cellulose synthase in the environmental bacterium *Gluconacetobacter xylinus* and had observed that GGDEF-EAL domains were important for the metabolism of this molecule (96). Subsequent studies unambiguously assigned diguanylate cyclase (DGC) activity to GGDEF domain proteins, such as AdrA, while the EAL domain was demonstrated to have c-di-GMP-specific phosphodiesterase (PDE) activity (90; see chapter 4 for a detailed description of the enzymatic activities of GGDEF and EAL domain proteins).

Simultaneously, it was recognized that c-di-GMP not only activated cellulose biosynthesis but also stimulated biofilm formation in general. Moreover, this observation did not only apply to pathogenic *S. enterica* serovar Typhimurium but also to the commensal bacterium *Escherichia coli* as well as the environmental bacterium and nosocomial pathogen *Pseudomonas aeruginosa* (90). Consistent with a role of c-di-GMP as a general activator of multicellular biofilm behavior, a number of previous studies had revealed functions of GGDEF and EAL domain proteins in the regulation of either attachment, autoaggregation, cellular development, or wrinkled colony formation in various bacteria (5, 21, 23, 34, 46). On the other hand, expression of AdrA and other DGCs, which synthesize c-di-GMP, inhibited motility in *S. enterica* serovar Typhimurium, *E. coli,* and *P. aeruginosa,* while expression of EAL domain proteins, which downregulate c-di-GMP levels, stimulated motility. In line with these findings, several other previously made accidental observations foreshadowed the role of c-di-GMP as a general repressor of motility (1, 36). Together with the discovery that GGDEF and EAL domain proteins are present in most sequenced bacterial genomes (26), these findings collectively supported the hypothesis that c-di-GMP functions as a global second messenger that regulates the sessility-motility lifestyle switch (90).

Compared to the previously known secondary messengers cAMP and ppGpp (see chapter 1), the mechanism of the novel second messenger c-di-GMP is much more complex in *S. enterica* serovar Typhimurium (75). *S. enterica* serovar Typhimurium pos-

Ute Römling, Kristina Jonas, Öjar Melefors, Nina Grantcharova, and Agaristi Lamprokostopoulou • Department of Microbiology, Tumor and Cell Biology, Karolinska Institutet, SE-171 77 Stockholm, Sweden.

sesses a c-di-GMP signaling network that includes 5 GGDEF, 7 EAL, and 7 GGDEF-EAL domains and at least two c-di-GMP binding proteins (PilZ domain proteins). In this chapter, we describe in detail the role of c-di-GMP signaling in the expression of the rdar morphotype, one of the major targets of c-di-GMP signaling in *S. enterica* serovar Typhimurium. We first describe the impact of this multicellular morphotype and the complex molecular mechanisms by which it is regulated. Further, we discuss the role of c-di-GMP in motility and characterize the upcoming molecular mechanisms that tightly control the switch between the sedentary state and motility at various levels in the respective regulatory cascades. In addition, the emerging role of c-di-GMP in virulence phenotypes of *S. enterica* serovar Typhimurium is addressed. The last paragraph describes the role of the global regulatory and RNA-binding protein CsrA in the regulation of the expression of c-di-GMP-metabolizing enzymes and EAL-like proteins. The regulation of distinct c-di-GMP-metabolizing proteins by global regulators provides a clue to how specificity is achieved in the complex modular network of c-di-GMP signaling.

PREVALENCE OF rdar MORPHOTYPE EXPRESSION IN *SALMONELLA ENTERICA* AND OTHER *ENTEROBACTERIACEAE*

S. enterica serovar Typhimurium, a pathogenic organism belonging to the family *Enterobacteriaceae,* causes self-limiting gastroenteritis in healthy humans. In the past, the arrangement of individual bacterial cells into multicellular communities, commonly called biofilm formation, was not considered to be a physiological trait in pathogenic organisms such as *S. enterica* serovar Typhimurium. However, when *S. enterica* serovar Typhimurium was grown on agar plates in the presence of the dyes Congo red and Coomassie brilliant blue, a characteristic rough morphotype was recognized (Fig. 1A and 2A). Since colonies grown on such indicator plates appeared not only rough but also red and dry, the morphotype was called rdar (*r*ed, *d*ry, *a*nd *r*ough). Not only was this morphotype distinct from the conventional smooth and shiny colony morphology but it also was distinct from colonies deficient in the O-antigen chain of the lipopolysaccharide, which are also termed rough (77, 83). In contrast to the smooth and rough colonies just described, rdar morphotype colonies are characterized by the expression of copious amounts of the adhesive extracellular matrix components curli fimbriae and cellulose (77, 113, 114). These extracellular matrix components and their regulatory components contribute to biofilm formation of *S. enterica* serovar Typhimurium in various forms as described in the next section.

The rdar morphotype is common among clinical isolates of *S. enterica* serovar Typhimurium and *Salmonella enterica* serovar Enteritidis; over 90% of isolates produce the morphotype (78). It can easily be lost under laboratory conditions, however (22). Other species of the family *Enterobacteriaceae,* such as *Escherichia coli, Citrobacter* spp., and *Enterobacter* spp., produce a rdar morphotype that is either indistinguishable from that of *S. enterica* serovar Typhimurium or very similar (12, 19, 113, 114). Since the rdar morphotype is conveniently visualized on agar plates, this bacterial behavior provides an easily screenable model system that can be characterized by genetic studies.

IMPACT OF rdar MORPHOTYPE EXPRESSION IN *SALMONELLA ENTERICA* AND OTHER *ENTEROBACTERIACEAE*

Initial characterization of the growth of *S. enterica* serovar Typhimurium in rdar morphotype colonies on agar plates demonstrated that the formation of this morphotype represents a specific form of multicellular behavior (biofilm) in bacteria (83). The bacterial cells within rdar colonies grow in an organized way to highly structured communities embedded in an extracellular matrix (Fig. 1B) (81), which tightly connects the cells such that the colonies can be lifted from the plates as whole communities (83).

Cells expressing the rdar morphotype also show other types of concerted growth, such as pellicle formation (bacterial lawns at the air-liquid interface), cell clumping (Fig. 1C and D), and adhesion to abiotic surfaces in steady-state culture (6, 44, 56, 81, 83). *S. enterica* serovar Typhimurium strains that express the rdar morphotype also form biofilms in a continuous-flow cell model (80; N. Grantcharova and U. Römling, unpublished data).

Under nonlaboratory conditions, expression of the rdar morphotype might have an impact on the persistence and survival of bacteria in the environment, as cells expressing the rdar morphotype are resistant to sanitizers and disinfectants (30, 47, 55, 95) and tolerate long-term desiccation (87, 92, 109). Although the contribution of the rdar morphotype to virulence in animal hosts is a matter of debate (82, 101, 108), there is consensus that the rdar morphotype aids not only transmission and survival outside the host but also host colonization. This consensus is supported by the observation that the rdar morpho-

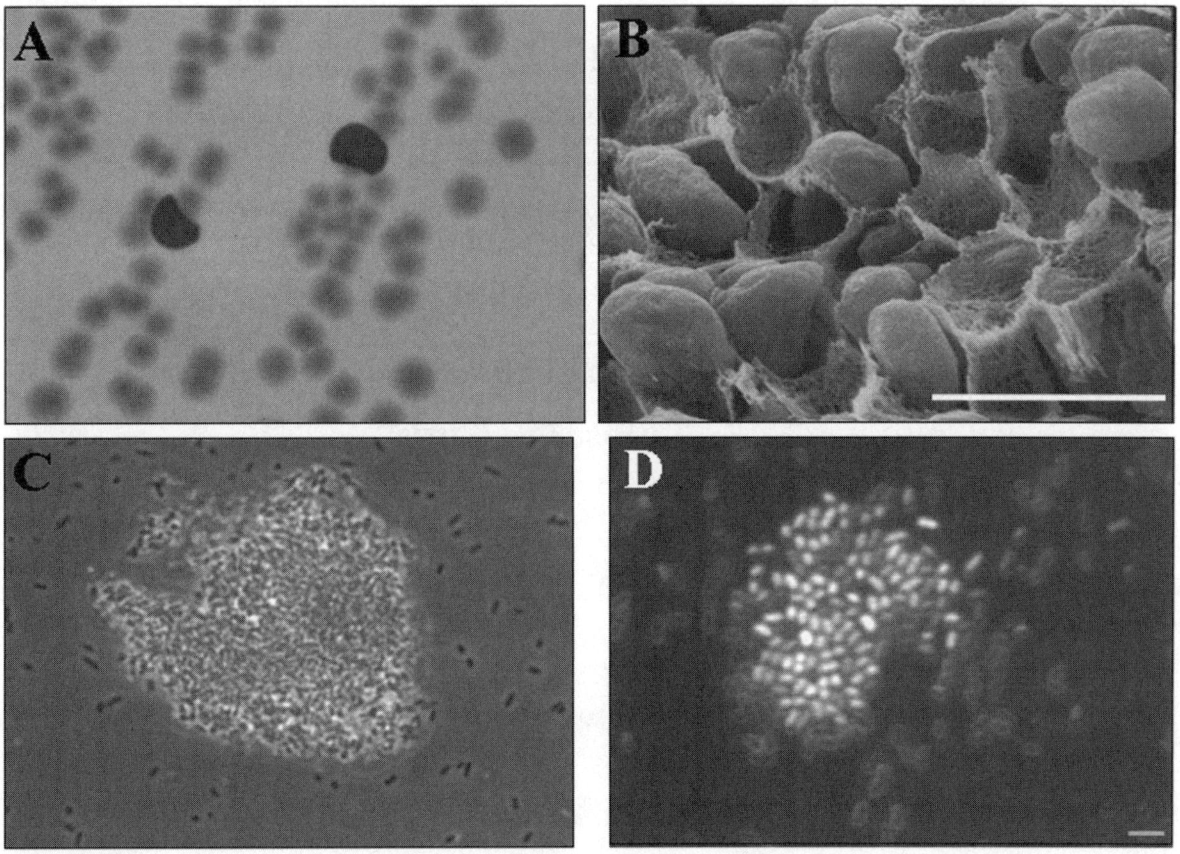

Figure 1. Appearance of the rdar morphotype of *S. enterica* serovar Typhimurium ATCC 14028. (A) rdar morphotype colonies among smooth colonies on Congo red agar plates. Streak-outs from the original *S. enterica* serovar Typhimurium ATCC 14028 culture were grown for 24 h at 37°C on Congo red agar plates using Luria broth (LB) without salt as growth medium. The extracellular matrix produced by rdar colonies bind the dye Congo red, which leads to a red-colored colony with extended dimensions. (B) Scanning electron microscopic analysis of the extracellular matrix structure and cell organization of rdar type *S. enterica* serovar Typhimurium ATCC 14028 (a Δ*fliC* Δ*fljB* mutant). The extracellular matrix surrounds the individual cell resembling a honeycomb-like structure. Magnification, ×15,000; bar, 2 μm. Picture taken by Manfred Rohde, Helmholtz Center for Infection Biology, Braunschweig, Germany. (C) rdar morphotype expression in liquid culture leads to cell clumping. Magnification, ×~473. (D) Differential expression of CsgD-GFP in rdar morphotype formation. GFP was fused to the *csgD* open reading frame on its native chromosomal location (Grantcharova and Römling, unpublished). Cell clumping is associated with the expression of the master regulator of rdar morphotype formation, CsgD, in the cytoplasm, while single cells usually do not express CsgD. Cells were grown for 24 h at 28°C on LB without salt agar plates. Cells were stained with FM4-64 membrane stain. Bar, 2 μm.

type enhances attachment and the colonization of plant tissue (8, 9), adherence to epithelial cells and intestinal tissue (58, 59, 103), invasion into epithelial cells (33, 53a, 56) and binding of host components (17, 37, 65, 66).

MOLECULAR CHARACTERIZATION OF THE rdar MORPHOTYPE

An intrinsic characteristic of biofilm formation, the formation of a self-produced extracellular matrix, is required to construct the complex three-dimensional architecture of the biofilm (20). The extracellular matrix of the rdar morphotype cells consists of at least four components: the two proteinaceous components curli fimbriae and BapA, the exopolysaccharide cellulose, and a capsule polysaccharide (Fig. 2A and B) (18, 32, 56, 77). These matrix components have various and sometimes even opposite functions in bacterial cell-cell interactions, resistance mechanisms, and interactions with the eukaryotic host. Curli fimbriae are a major component of the extracellular matrix of the rdar morphotype, the expression of which significantly changes the morphology and color of agar-grown colonies (Fig. 2A). Curli fimbriae are amyloid-like structures which mediate bacterial cell-cell interactions and which con-

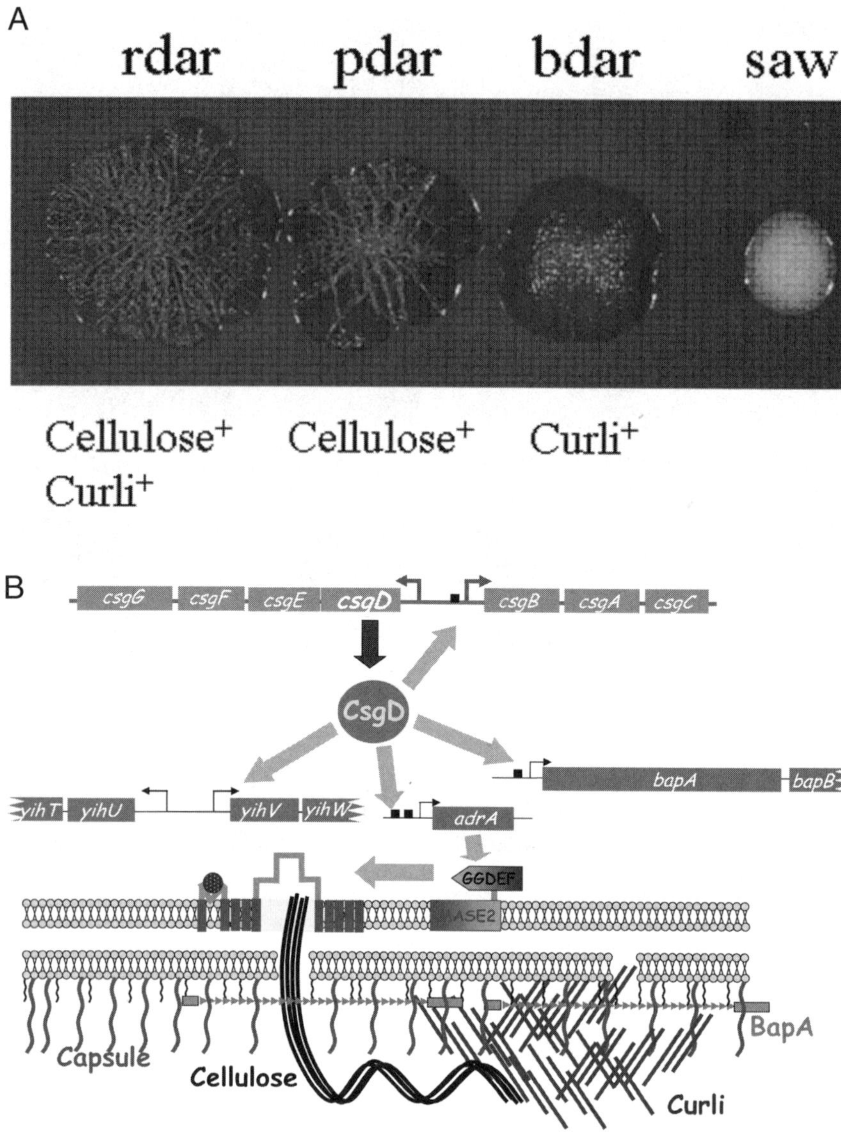

Figure 2. rdar morphotype expression in *S. enterica* serovar Typhimurium ATCC 14028. (A) Morphotypes of *S. enterica* serovar Typhimurium expressing the extracellular matrix components curli fimbriae and cellulose. For pronounced visualization, strain MAE52, an ATCC 14028 derivative with semiconstitutive expression of the rdar morphotype, and respective mutants were used. MAE52 has a single point mutation in the *csgD* promoter that confers threefold-enhanced transcription of *csgD,* which is RpoS and temperature independent (83). (B) The transcriptional regulator CsgD positively regulates the expression of at least four extracellular matrix components in *S. enterica* serovar Typhimurium, proteinaceous curli fimbriae, BapA, exopolysaccharide cellulose, and a capsule (18, 32, 56, 77, 82, 90). Thereby, CsgD activates the transcription of the target operons *csgBAC, adrA,* and *bapABCD.* The *csgBAC* operon encodes CsgA and CsgB, major and minor structural components of curli fimbriae, respectively. The *bapABCD* operon encodes the large surface protein BapA and its type I protein secretion apparatus. Expression of the DGC AdrA produces c-di-GMP required for posttranscriptional activation of cellulose biosynthesis. On the other hand, CsgD represses transcription of the *yihVW* operon, which leads to the activation of the divergently transcribed *yihU-OyshA* operon and expression of an O-antigen capsule.

tribute to adherence to abiotic surfaces in steady-state growth culture (6, 83). Curli fimbriae promote attachment to epithelial cells and are consequently also found in *Salmonella* biofilms that develop on epithelial surface layers (58).

The 386-kDa large surface protein BapA is loosely attached to the cell surface (56). Although the expression of BapA does not alter the visual appearance of the rdar morphotype, *bapA* mutants do not form pellicles at the air-liquid interface (44, 56). The deficiency in pellicle formation by *bapA* mutants can be complemented by overexpression of curli fimbriae. BapA, as curli fimbriae, contributes to the invasion of epithelial tissue (56).

The exopolysaccharide cellulose is a major component of the extracellular matrix of the rdar morphotype, and the absence of cellulose dramatically alters the morphology and color of agar-grown colonies (Fig. 2A) (114). Cellulose plays a structural role in bacterial cell-cell interactions, lending stability to the cell-cell interconnections, most probably by binding noncovalently to curli fimbriae. Notably, in bacterium-host interactions, cellulose can counteract the effect of curli fimbriae (as it inhibits curli-mediated adherence, invasion, and production of proinflammatory cytokines) (103), most likely through direct interaction with the fimbriae.

The capsule consists of repeated oligosaccharide units similar to the lipopolysaccharide O-antigen and is associated with the extracellular matrix consisting of curli fimbriae and cellulose (32, 107). The capsule does not contribute to multicellular aggregation, as it does not change the rdar morphotype morphology or biofilm formation but is a major factor mediating resistance to long-term desiccation.

CsgD AS THE MAJOR REGULATOR OF rdar MORPHOTYPE EXPRESSION

The expression of all four extracellular matrix components described above was reported to be positively regulated by CsgD, a transcriptional regulator of the LuxR family (32, 56, 77, 82, 114) (Fig. 2B). CsgD is actually part of the curli biosynthesis locus, which consists of the two divergently transcribed operons *csgDEFG* and *csgBAC* (77). Putative CsgD binding sequences have been identified upstream of both the *csgBAC* operon and the *adrA* gene, encoding a DGC, which is required for cellulose biosynthesis (14). It was suggested that CsgD also directly regulates the *bapABCD* operon as well as the *yihVW* operon, which is inversely expressed with the divergently transcribed *yihU-OyshA* genes encoding proteins required for biosynthesis of the capsule (32, 56). The fact that CsgD is involved in the regulation of curli, cellulose, BapA, and the capsule illustrates its central role in the regulation of multicellular aggregative behavior in *S. enterica* serovar Typhimurium. CsgD, however, not only activates the biosynthesis of extracellular matrix components but also adjusts the cells to the multicellular mode of growth. For example, in *E. coli,* CsgD inhibits the transcription of at least 24 genes that encode components that negatively influence biofilm formation (13, 14). In addition, CsgD upregulates the concentration of the stationary-phase sigma factor RpoS and consequently its own expression, as RpoS is required for *csgD* expression (35). CsgD upregulates RpoS protein levels indirectly by activating the expression of *iraP,* a stabilizer of RpoS. Thus, CsgD can act either directly or indirectly (via IraP and RpoS) to cause differential expression of CsgD-dependent genes.

REGULATION OF rdar MORPHOTYPE EXPRESSION

Expression of the rdar morphotype is usually highly controlled in response to environmental conditions. With few exceptions (78, 83), under laboratory conditions, the rdar morphotype is expressed on Luria broth agar plates without salt at temperatures below 30°C (77). In liquid culture, expression of the rdar morphotype occurs during stationary phase under microaerophilic conditions in rich medium without salt or under aerobic conditions in minimal medium (30). Besides temperature, the medium composition and oxygen tension, osmolarity, pH, iron availability, ethanol, starvation by phosphate, nitrogen, and other uncharacterized factors regulate the expression of CsgD (29, 83, 93).

This tight regulation of rdar morphotype expression occurs mainly at the level of *csgD* transcription. Consequently, in our model strain *S. enterica* serovar Typhimurium UMR1, a nalidixic acid-resistant ATCC 14028 derivative, expression of CsgD coincides with rdar morphotype expression and other biofilm behaviors, such as formation of clumps in liquid culture (Fig. 1A, C, and D). Once expressed, CsgD binds to the promoter region of the divergently transcribed *csgBAC* operon (K. Zakikhany and U. Römling, unpublished data). The *csgBAC* operon is separated from *csgD* by a 521-bp intergenic region, one of the largest intergenic regions in *S. enterica* serovar Typhimurium (31, 77). Binding of regulatory proteins that affect *csgD* expression occurs in a large part within this intergenic region (28, 29). In addition, the RNA transcript of *csgD* contains a large 174-bp 5′-untranslated leader sequence (28, 77), which contributes to a higher expression level of the *csgD* mRNA transcript throughout the growth phase.

The tight control of rdar morphotype expression by environmental conditions is reflected by the numerous regulatory proteins found to bind directly to the *csgD* promoter region and influence transcription (Fig. 3). At least four two-component systems are involved in the transcriptional regulation of *csgD:* the EnvZ-OmpR, the CpxAR, the RcsC/D-B/A, and the RstBA systems (29, 40, 48, 64). The response regulator OmpR is required for *csgD* transcription. At least seven OmpR binding sites upstream of the *csgD* promoter have been identified, each of which exerts different effects on the transcription of *csgD* (28, 29).

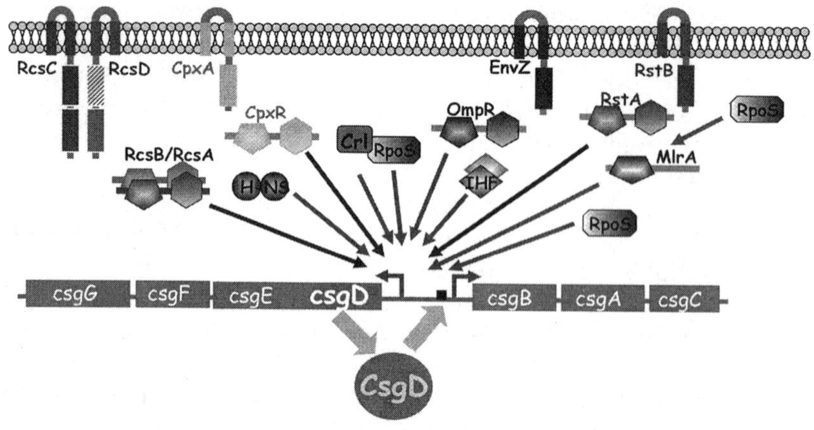

Figure 3. Regulatory network of rdar morphotype expression in *S. enterica* serovar Typhimurium with CsgD as central regulator. Regulation by c-di-GMP metabolic enzymes is shown in Fig. 5. Expression of CsgD is positively (gray arrows) and negatively (black arrows) regulated by two-component systems (EnvZ/OmpR, CpxA/CpxR, and RcsCD/RcsBA), sigma factors and associated proteins (RpoS and Crl), transcriptional regulators (MlrA), and nucleoid-binding proteins (H-NS and IHF).

For example, binding of OmpR to D1, the first binding site upstream of the promoter sequence, seems to be most important for the activation of *csgD* transcription. In contrast, binding to the sites D2 and D3 to D6 downregulates *csgD* expression. Phosphorylation of OmpR by the cognate sensor kinase EnvZ only partially contributes to the activation of *csgD* transcription (30), suggesting that alternative phosphodonors besides EnvZ phosphorylate OmpR in the stationary phase. In contrast to the overall effect of OmpR, the response regulators CpxR and RstA inhibit *csgD* transcription, likely because their DNA binding sites overlap the D1 binding site for OmpR (48, 64).

The DNA-binding and -bending protein integration host factor (IHF), a nucleoid protein that contributes to the folding and packaging of the genome, can also act locally, thereby influencing local DNA structural organization and thus transcriptional regulation in gram-negative bacteria (71). IHF binds to the *csgD* promoter region upstream of the transcriptional start site and has two additional binding sites in the *csgD* open reading frame (28, 29). Binding of IHF to the site located upstream of the promoter has been shown to be required for the upregulation of *csgD* transcription under microaerophilic conditions. Thereby, IHF competes with OmpR for binding at its upstream site, which overlaps with OmpR sites D3 to D6.

H-NS, another nucleoid protein, also exerts a net positive effect on *csgD* transcription (29). In vitro binding studies in combination with genetic analysis have demonstrated that the negative effect of H-NS on *csgD* is direct. The binding of H-NS to the *csgD* promoter reduces transcription. In contrast, all available evidence argues that the positive effect of H-NS is indirect.

Transcription of the *csgD* promoter requires the stationary phase and stress sigma factor RpoS, which directly recognizes the *csgD* promoter (72). RpoS-dependent transcription of *csgD* is enhanced by Crl, a small regulatory protein that favors association of RpoS with the core enzyme RNA polymerase and thereby increases RpoS activity (72). In addition to the factors that have been demonstrated to directly bind to the *csgD* promoter, several other components influence *csgD* expression. These include the RcsCB two-component system, the RpoS-dependent transcriptional regulator MlrA, and intracellular *N*-acetylglucosamine-6-phosphate levels (10, 15, 40).

DISCOVERY OF c-di-GMP IN *S. ENTERICA* SEROVAR TYPHIMURIUM

The discovery of the c-di-GMP signaling pathway in *S. enterica* serovar Typhimurium was a coincidence. Deletion of the gene(s) that encode the rdar matrix component curli fimbriae led to a bright pink colony morphology (pdar morphotype) on agar plates (83). Together with the detection of an unstructured matrix by scanning electron microscopy, this observation led to the hypothesis that the formation of the rdar morphotype involved another extracellular matrix component in addition to curli fimbriae (83). Known factors such as colanic acid or type 1 fimbriae, however, did not contribute to the extracellular matrix of rdar morphotype colonies (U. Römling, unpublished data). An unbiased genetic approach aimed at the identification of CsgD-regulated genes led to the discovery of the GGDEF domain protein AdrA, an integral membrane protein that was shown to act as a positive regulatory component of the unknown extracellular matrix (82). Shortly thereafter, the unknown extracellular matrix component was identified as the exopolysaccharide cellulose (114). Previously, proteins with GGDEF domains had been demonstrated to be required for the turnover of c-di-GMP, which was known to be an allosteric activator of the cellulose synthase in the fruit-degrading bacterium *G. xylinus* (96). Thus, the GGDEF domain protein AdrA was a candidate for a c-di-GMP-synthesizing enzyme, whose activity could influence the synthesis of cellulose (5, 82, 114).

Indeed, expression of AdrA led to the production of a small molecule that was structurally identical to chemically synthesized c-di-GMP as analyzed by mass spectrometry (90). Furthermore, the EAL domain protein STM3611 was shown to decrease c-di-GMP levels, and its EAL domain was identified as required for c-di-GMP-specific PDE activity in vivo (90). That the c-di-GMP-synthesizing and -degrading activity of GGDEF and EAL domains was complementary was confirmed by in vitro studies (67, 86, 88, 98) (described in detail in chapter 4).

THE c-di-GMP SIGNALING NETWORK IN *S. ENTERICA* SEROVAR TYPHIMURIUM

In *S. enterica* serovar Typhimurium, genes predicted to encode a total of 5 GGDEF, 7 GGDEF-EAL, and 7 EAL domain proteins were subsequently identified (Fig. 4) (75). Thus, the c-di-GMP signaling

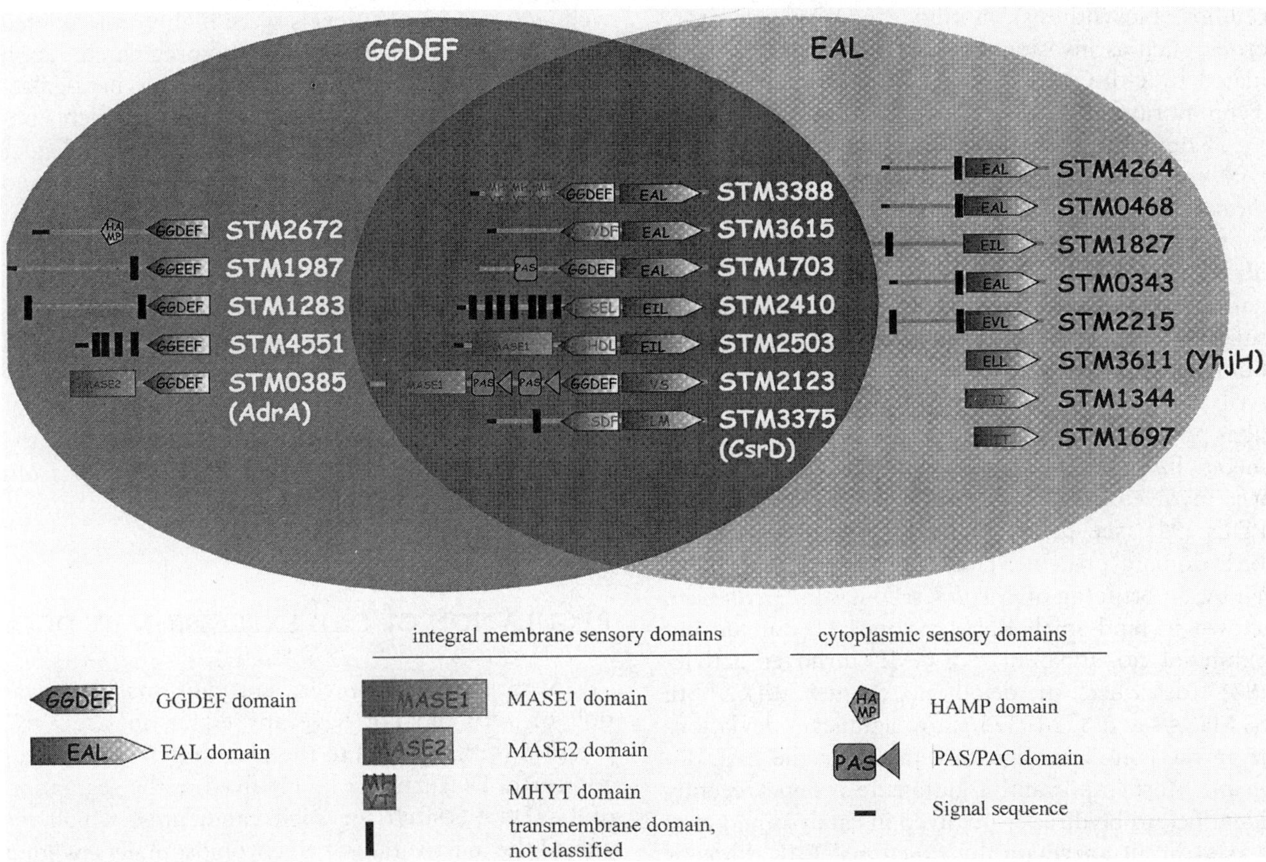

Figure 4. Panel of GGDEF-EAL domain proteins present in *S. enterica* serovar Typhimurium. The domain structures of the 5 GGDEF, 7 EAL, and 7 GGDEF-EAL domain proteins are shown. The GGDEF domain proteins STM2410, STM2503, STM3375, and STM3615 do not show a complete signature of conserved amino acids in the GGDEF motif (as indicated in the figure by pale letters) and other signature motifs indicative for enzymatic activity, suggesting that DGC activity is missing. The EAL domain proteins STM1344, STM1697, STM2123, and STM3375 do not show a complete signature of conserved amino acids in the EAL motif (as indicated in the figure by pale letters) and other signature motifs indicative for enzymatic activity, suggesting that c-di-GMP-specific PDE activity is missing. Although STM3611 shows several deviations from the PDE signature motif, considerable c-di-GMP-specific PDE activity was observed (69, 90, 91). Besides STM1344, STM1697, STM1703, and STM3611, the proteins are predicted to be integral membrane proteins, suggesting that the c-di-GMP signaling network responds mainly to external signals in *S. enterica* serovar Typhimurium.

network is considerably smaller than that of the closely related species *Escherichia coli,* whose chromosome encodes genes predicted to encode 12 GGDEF, 7 GGDEF-EAL, and 11 EAL domain proteins (26, 75, 104). Another gene, encoding an EAL domain protein, resides on the *E. coli* fertility plasmid (75). Most cytoplasmic GGDEF and EAL domain proteins in *E. coli* are not present in *S. enterica* serovar Typhimurium. Consequently, since the majority of GGDEF and EAL domain proteins are membrane bound, it is likely that c-di-GMP signaling and metabolism in *S. enterica* serovar Typhimurium are regulated primarily by extracellular stimuli. In vivo DGC activity (as proven by measurements of c-di-GMP concentrations in combination with mutational analysis and/or the assay of a relevant biological output) has been demonstrated for AdrA, STM1987, STM2123, and STM3388 (49, 90). Multicellular behavior (rdar morphotype expression and cellulose biosynthesis), motility, and virulence properties, such as invasion, have been shown to be regulated by c-di-GMP signaling in *S. enterica* serovar Typhimurium (43, 53a).

Comparison of the GGDEF and EAL domains with confirmed enzymatically functional domains indicated that several domains of either class do not show the complete signature of conserved amino acids in the GGDEF or EAL motifs and/or other signature motifs indicative for enzymatic activity. This suggested the possibility that some GGDEF and EAL domains have evolved novel functions (75, 88) (described in detail in chapter 4). Indeed, the EAL-like protein STM1344 does not have c-di-GMP-metabolizing activity nor does it bind c-di-GMP, but it is involved in the expression of c-di-GMP-specific PDEs (91) (see below). Furthermore, the GGDEF-EAL domain protein STM3375 (alternatively called YhdA, an ortholog of *E. coli* CsrD) has in *E. coli* been shown to bind small RNAs via its EAL domain but exhibited no apparent c-di-GMP turnover activity (94) (described in detail in chapter 14). Both STM1344 and STM3375 show significant deviations from the consensus signature motifs in the EAL domain. Most important, a glutamate residue recently identified to be directly involved in catalysis and conserved in all enzymatically functional EAL domains (63, 70) is missing in STM1344 and STM3375.

REGULATION OF CsgD EXPRESSION BY THE c-di-GMP SIGNALING NETWORK

CsgD expression is regulated by conventional global regulatory proteins as described above. In addition, CsgD expression is also regulated by the second messenger c-di-GMP (49, 89). Indeed, current studies indicate that CsgD expression is a major target for the c-di-GMP signaling pathway (Fig. 5). At least two DGCs and four PDEs affect CsgD expression as described in detail below (Table 1). Thereby, the enzymes act nonredundantly, as each protein shows a distinct temporal pattern of up- or downregulation of CsgD expression.

Experiments in which *adrA*, encoding a highly potent DGC, and STM3611, encoding a highly potent PDE, were overexpressed first indicated c-di-GMP-dependent regulation of CsgD expression (49). Overexpression of *adrA* from a plasmid under the control of an arabinose-inducible promoter upregulated not only cellulose biosynthesis but also the rdar morphotype on Congo red agar plates, while overexpression of STM3611 had the opposite effect. Subsequent analysis showed that *csgD* mRNA levels as well as CsgD protein levels were highly upregulated upon *adrA* expression and downregulated upon STM3611 expression. The fact that a beta-galactosidase transcriptional fusion to *csgD,* which contains the entire intergenic region and the leader sequence of the *csgD* mRNA (29), does not respond to changes in c-di-GMP concentrations (unpublished data), suggests that c-di-GMP does not directly act on the transcriptional activation of *csgD* but rather influences RNA stability. This finding is surprising, considering the complex regulatory network required for *csgD* transcription as described above. In addition, CsgD might also be regulated beyond RNA stability as c-di-GMP concentrations influenced the CsgD protein level when CsgD was expressed from a plasmid (48).

REGULATION OF CsgD EXPRESSION BY DGCs

Surprisingly, however, chromosomal inactivation of *adrA* did not have any effect on CsgD expression (49). This led to the idea that one or several additional DGCs might be involved in the regulation of CsgD. A search for such candidates, which involved the inactivation of all chromosomally encoded GGDEF domain proteins, led to the identification of STM2123 and STM3388, two complex membrane-bound GGDEF-EAL domain proteins that were required for positive regulation of CsgD expression (Fig. 5). The finding that STM2123 and STM3388 acted in an additive manner on CsgD expression indicated that the proteins either produce distinct c-di-GMP pools or act additively to build up one common c-di-GMP pool.

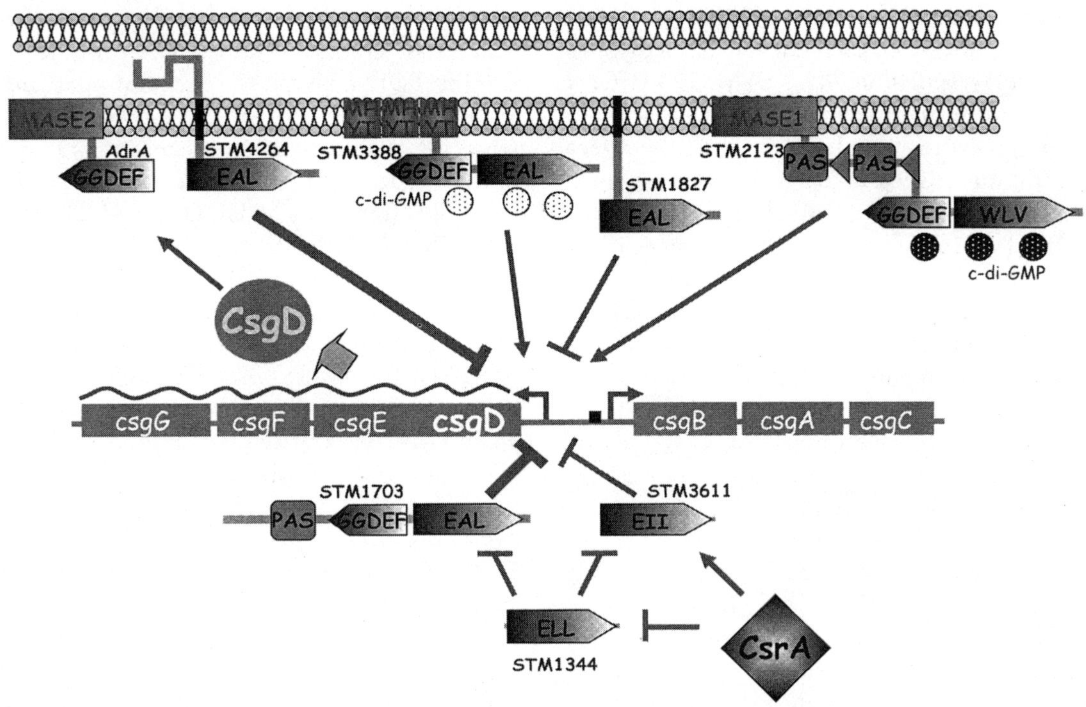

Figure 5. c-di-GMP metabolic network affecting expression of CsgD, the master regulator of rdar morphotype formation. At least eight GGDEF and/or EAL domain proteins are involved in the regulation of the rdar morphotype in *Salmonella* serovar Typhimurium growing on agar plates (49, 82, 89). STM2123, STM3388, and other unidentified DGCs produce c-di-GMP (indicated by dots). STM1703, STM4264, STM1827, and STM3611 lead to c-di-GMP degradation. STM4264 and STM1703 have the most drastic effect on CsgD expression (indicated by thick lines). STM4264 acts upstream of STM1703, since STM1703 can complement an STM4264 defect by downregulation of CsgD but not vice versa (89). STM1827 and STM3611 have a less dramatic effect on CsgD expression. *adrA* regulated on the transcriptional level by CsgD is necessary for cellulose biosynthesis and is partially involved in the production of curli fimbriae (49). The EAL-like protein STM1344 does not have PDE activity but represses expression of the PDEs STM1703 and STM3611 (91). STM1344 and STM3611 are direct targets of the carbon storage regulatory protein CsrA (42a). Thereby, CsrA represses expression of STM1344 and activates expression of STM3611.

Table 1. Characteristics of GGDEF-EAL domain proteins affecting CsgD expression

Protein	Domain structure	Function	% CsgD expression[a]	% Intracellular c-di-GMP[b]
STM2123	MASE1-PAS-PAS-GGDEF-EAL	DGC	63	Not statistically significant
STM3388	MHYT-GGDEF-EAL	DGC	63	Not statistically significant
STM2123 and STM3388			46	Not statistically significant
STM1703	PAS-GGDEF-EAL	c-di-GMP PDE	950	184
STM4264	Unknown-TM-EAL	c-di-GMP PDE	450	550
STM1703 and STM4264			1,050	NM
STM3611	EAL	c-di-GMP PDE	240	Not statistically significant
STM1827	TM-unknown-EAL	c-di-GMP PDE	140	184
STM1344	EAL	Unknown, represses expression of STM3611 and STM1703	48	48

[a] Percentage of wild-type level found in mutant. The CsgD concentrations for the STM2123 and STM3388 mutants were determined after 16 h of growth on Luria broth without salt agar plates at 28°C (49), while the CsgD concentrations for the PDE mutants were determined after 20 h of growth (89).
[b] Percentage of wild-type level found in mutant. The c-di-GMP concentrations for the STM2123 and STM3388 mutants were determined after 16 h of growth on Luria broth without salt agar plates at 28°C (49), while the c-di-GMP concentrations for the PDE mutants were determined after 20 h of growth (89). NM, not measured.

Although STM2123 and STM3388 act additively on CsgD expression, each enzyme has a distinct pattern of CsgD regulation (49). STM2123 affects CsgD expression primarily in the early growth phase on agar plate growth (10 and 16 h), while the effect is significantly diminished at 24 h. In contrast, in a deletion mutant of STM3388, CsgD expression was downregulated at 16 and 24 h (49). Notably, at 10 h of growth, the mutation in STM3388 caused CsgD expression to increase. STM3388 has both a GGDEF and an EAL domain. Inspecting the EAL domain for consensus residues required for PDE activity indicates that the EAL domain of STM3388 might be functional. Therefore, the oppositional effects of STM3388 in the regulation of CsgD expression might be explained by the possibility that STM3388 can act both as a DGC and as a PDE. The finding that deletion of either STM2123 or STM3388 did not significantly alter c-di-GMP levels in the cell suggests that the DGC activities of the two proteins exert a much localized effect. Furthermore, overexpression of STM2123 and STM3388 only led to a three- to fivefold increase in c-di-GMP levels compared to wild-type levels (49). In comparison, when the DGC gene *adrA* was overexpressed under the same conditions from the same vector, c-di-GMP levels changed 1,000-fold (90).

In the double mutant STM2123 STM3388, CsgD expression was moderately decreased to 40 to 60% of that of the wild-type parent. As depletion of the c-di-GMP pool by overexpression of the PDE STM3611 leads to an almost complete downregulation of CsgD expression (49), it is hypothesized that not only does c-di-GMP modulate CsgD expression but some basal level of c-di-GMP is required for CsgD to be expressed in the cell. The concerted action of many GGDEF domain proteins might contribute to CsgD expression, as deletion of each of the remaining GGDEF protein-encoding genes individually had no influence upon CsgD expression. Alternatively, but not contradictorily, the effect of other DGCs might only be observed when c-di-GMP degradation is inhibited through deletion of the genes that encode specific PDEs.

REGULATION OF CsgD EXPRESSION BY c-di-GMP-SPECIFIC PDEs

The net biological effect of a second messenger is determined by the delicate balance between synthesis and degradation of the compound (39; described in detail in chapters 1 and 4). *S. enterica* serovar Typhimurium is predicted to express a total of 12 EAL or GGDEF-EAL domain proteins that could exhibit PDE activity (79). Mutational analysis revealed that four EAL or GGDEF-EAL domain proteins significantly and consistently caused the rdar morphotype and CsgD expression to be downregulated (89) (Fig. 5). The GGDEF-EAL domain protein STM1703 and the EAL domain protein STM4264 exerted the strongest effect on *csgD* expression throughout growth with *csgD* expression upregulated four- to sevenfold in the deletion mutants. The effect of STM1703 and STM4264 on *csgD* expression was, in part, additive. The implication is that STM1703 and STM4264 control *csgD* expression by degrading overlapping c-di-GMP pools. STM4264, however, seems to act upstream of STM1703; whereas overexpression of STM1703 could complement the STM4264 mutant, overexpression of STM4264 could not complement the STM1703 mutant.

Besides STM1703 and STM4264, mutational analysis indicated that two EAL proteins, STM3611 and STM1827, caused the rdar morphotype and *csgD* expression to be downregulated. However, there was a distinct temporal difference in the effect of the two EAL domain proteins on rdar morphotype expression. Deletion of STM3611 led to upregulation of the rdar morphotype and CsgD expression up to 20 h, while STM1827 upregulated the rdar morphotype and CsgD expression after 20 h.

Determination of the c-di-GMP concentrations in the deletion mutants demonstrated nonlinearity between PDE activity and CsgD expression. Deletion of STM4264 led to a strong (5.5-fold) increase in total intracellular c-di-GMP concentration, which triggered a 4.5-fold-enhanced CsgD expression. Thus, CsgD expression responded in a linear manner to the c-di-GMP pool present in the STM4264 mutant. On the other hand, deletion of STM1703 led to a modest increase in the c-di-GMP pool (184% of the wild-type concentration) but a strong 9.5-fold enhancement in CsgD expression. This finding indicated that the c-di-GMP pool present in the STM1703 mutant is dedicated to CsgD expression to a larger extent than the c-di-GMP pool present in the STM4264 mutant. Indeed, a recent microarray analysis of *E. coli* has shown that deletion of *yciR*, the STM1703 ortholog, affected almost exclusively the expression of genes of the *csg* operon (104). The nonlinear relationship between PDE activity and the output response (CsgD expression) shows that the response to c-di-GMP is complex and, in contrast to the DGCs of the same system, cannot be explained strictly in terms of absolute activities (see above and reference 49).

REGULATION OF CsgD EXPRESSION BY THE EAL-LIKE PROTEIN STM1344

The GGDEF and EAL domains constitute two superfamilies of bacterial domains with DGC and c-di-GMP-specific PDE activity, respectively. Several members of these superfamilies have not retained their enzymatic activity but have instead evolved alternative functions, including the sensing of GTP or the guidance of small RNAs for degradation (16, 94). In such unorthodox proteins, a significant fraction of residues that are required for the enzymatic activity of the GGDEF or EAL domains are altered. In some cases, these alterations even comprise novel consensus motifs conserved in orthologous proteins (94; see also chapters 4 and 14).

S. enterica serovar Typhimurium contains two closely related EAL domain proteins, STM1697 and STM1344, that show high homology to the PDE STM3611 (75). All three of these proteins are standalone EAL domains that do not contain any other known domains. When the effect of STM1344 on rdar morphotype expression was investigated, atypical phenotypes were observed (89, 91). Unlike deletions of other EAL domain proteins (which result in the upregulation of the rdar morphotype), deletion of STM1697 had no effect on rdar morphotype expression and deletion of STM1344 resulted in a distinct downregulation of rdar morphotype expression. In agreement with the STM3144 mutant phenotype, *csgD* expression and the synthesis of both curli fimbriae and cellulose were also downregulated. This phenotype of the STM1344 mutant is exactly the opposite of that expected from a PDE and rather resembles the phenotype of a DGC mutant.

In agreement with the effect on rdar morphotype expression, subsequent biochemical analysis showed that STM1344 does not act as a c-di-GMP-dependent PDE (91). Furthermore, STM1344 possessed neither DGC activity nor c-di-GMP binding capacity. The exact function of STM1344 remains to be resolved.

Genetic studies showed that the effect of STM1344 on *csgD* expression required the action of STM1703 and STM3611 but not that of STM4264 or STM1827 (91). Subsequent expression studies showed that the PDEs STM1703 and STM3611 are negatively regulated by STM1344, which is consistent with the observed positive effect of STM1344 on *csgD* expression (Fig. 5). It remains to be investigated whether STM1344 has other functions besides the regulation of STM1703 and STM3611 PDEs.

REGULATION OF CELLULOSE BIOSYNTHESIS IN *S. ENTERICA* SEROVAR TYPHIMURIUM BY c-di-GMP

The activation of cellulose biosynthesis in *S. enterica* serovar Typhimurium is not restricted to the DGC AdrA. Other DGCs may also be involved. For example, the GGDEF domain protein STM1987 was required for the activation of cellulose biosynthesis when cells were suspended at 37°C in a starvation medium devoid of trace elements but rich in glucose (92). Surprisingly, under these conditions, AdrA was unable to complement the STM1987 mutant (27), even though AdrA was demonstrated to produce c-di-GMP under a variety of other environmental conditions (R. Simm and U. Römling, unpublished results). In contrast, when overexpressed, a panel of overlapping, but not identical, GGDEF domain-containing proteins activated cellulose biosynthesis in this starvation medium and under growth conditions that are optimal for rdar morphotype expression (e.g., STM1987 was able to complement an *adrA* knockout) (27). Overexpression of GGDEF and EAL domain proteins must be taken with care, as abundant production of the potentially freely diffusible c-di-GMP molecule might saturate all c-di-GMP-affected cellular processes (111). However, several studies have shown that certain GGDEF and EAL domain proteins, when overexpressed, did not specifically influence bacterial behavior (53). Therefore, the above-mentioned findings indicate that possibly some other GGDEF domain proteins have the potential to activate cellulose biosynthesis under certain environmental conditions when expressed from the chromosome.

In contrast to the biofilm regulator CsgD, c-di-GMP is not required for expression of the *bcsABZC* operon encoding proteins required for cellulose biosynthesis (49). Instead, c-di-GMP activates cellulose biosynthesis on a posttranslational level. The cellulose synthase BcsA harbors a PilZ domain at the C terminus (Fig. 6) (4). PilZ domains, among them the PilZ domain of the cellulose synthase from *G. xylinus,* have been demonstrated to bind c-di-GMP (85). Consequently, c-di-GMP binding to the PilZ domain of *S. enterica* serovar Typhimurium cellulose synthase BcsA presumably activates cellulose biosynthesis.

REGULATION OF MOTILITY IN *S. ENTERICA* SEROVAR TYPHIMURIUM BY c-di-GMP

Elevated c-di-GMP concentrations inhibit motility (11, 90, 111). In *S. enterica* serovar Typhimur-

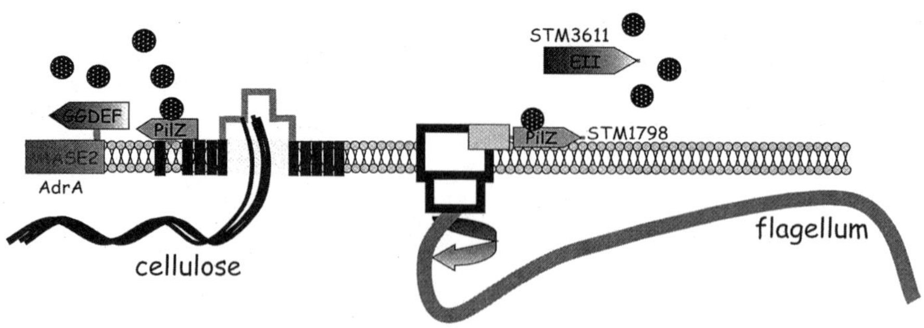

Figure 6. c-di-GMP binding proteins identified in *S. enterica* serovar Typhimurium. The PilZ domain is a c-di-GMP binding domain. The PilZ domain containing proteins STM1798 (YcgR) and BcsA have been identified in *S. enterica* serovar Typhimurium, which regulate motility and perform cellulose biosynthesis, respectively. The PilZ domain containing protein YcgR regulates motility. Upon high c-di-GMP levels, YcgR binds c-di-GMP (85) and presumably inhibits the flagellar motor function (52). c-di-GMP is required for cellulose biosynthesis in *S. enterica* serovar Typhimurium (90, 114). BcsA, the catalytic subunit of the cellulose synthase, contains a PilZ domain at the C terminus. As the PilZ domain of the cellulose synthase of *G. xylinus* has been demonstrated to bind c-di-GMP (85), it is hypothesized that PilZ binds c-di-GMP, which allosterically activates cellulose biosynthesis in *S. enterica* serovar Typhimurium.

ium, inhibition of flagellum-mediated swimming and swarming motility by c-di-GMP requires the c-di-GMP-binding protein STM1798 (YcgR) (Fig. 6) (85). Thereby, c-di-GMP affects flagellar motor function but not flagellar biogenesis (52). The effect of YcgR is observed in strains with elevated cytoplasmic c-di-GMP levels but not in the wild-type strain *S. enterica* serovar Typhimurium UMR1 (85). YcgR has been shown to bind c-di-GMP via its PilZ domain, thereby adopting a more compact ligand-bound structure. This conformational change also creates a novel interface which productively can interact with components of the flagellar apparatus to inhibit motor function. Of the *S. enterica* serovar Typhimurium c-di-GMP metabolizing proteins, only the PDE STM3611 affected swimming and swarming motility in the wild-type background (89). The DGC(s) involved in swimming and swarming motility have not been identified yet.

Some non-DGC proteins, however, also appear to influence motility. For example, the GGDEF-EAL protein STM3375 and the EAL-like protein STM1344 each had an effect on swimming and swarming motility (89, 91). The STM3375 mutant displayed an apparent PDE phenotype in swimming and swarming motility in contrast to its effect on biofilm formation in liquid culture (89). As described above, deletion of the EAL-like protein STM1344, which does not exhibit c-di-GMP-metabolizing activity, enhanced swimming and swarming motility. Such a phenotype is normally associated with deletion of a DGC. However, as described above, the STM1344 mutant phenotype is consistent with the observation that STM1344 inhibits the expression of PDEs such as STM3611, which is required for swimming and swarming motility (91).

REGULATION OF THE TRANSITION BETWEEN SESSILITY AND MOTILITY IN *S. ENTERICA* SEROVAR TYPHIMURIUM

The transition between motility and sessility is a behavioral pattern regulated by c-di-GMP signaling in many bacteria. High c-di-GMP levels stimulate the production of adhesive extracellular matrix components that enhance biofilm development, while low c-di-GMP levels stimulate motility, which is associated with the dispersion of biofilms (11, 90, 100); described in detail in chapter 8). Most likely, c-di-GMP signaling acts downstream of other regulatory mechanisms and/or can override existing regulatory circuits. This hypothesis is supported by early experiments that used overexpressed DGCs and PDEs to globally manipulate the c-di-GMP levels (11). Since motile and sessile cells coexist in close proximity in biofilms (51), it is likely that the switch between sessility and motility occurs at the single-cell level. In line with that idea, we have observed a bistable pattern of CsgD expression in different kinds of biofilms, such as in rdar morphotype colonies and continuous-flow cell biofilms. In these cases, CsgD expression is associated with cell aggregation and expression of the exopolysaccharide cellulose (Fig. 1D) (Grantcharova and Römling, unpublished). The switch between sessility and motility is tightly controlled at many different levels. In *S. enterica* serovar Typhimurium, a major player seems to be the carbon storage regulator CsrA, which not only directly affects the expression of GGDEF-EAL domain proteins (see below) (42a) but is also involved in the regulation of key regulatory genes such as the master regulator of motility FlhDC (105). STM1344, which downregulates the PDEs STM1703 and STM3611 (91) and genes in the flagellar regulon (112), is actually a direct target of CsrA (42a). Furthermore, STM1344 connects sessility with motility as STM3611, when expressed from the chromosome, affects both CsgD expression and motility (89).

REGULATION OF VIRULENCE PROPERTIES BY THE c-di-GMP SIGNALING NETWORK IN *S. ENTERICA* SEROVAR TYPHIMURIUM

In animal and plant pathogens, virulence is a phenotype frequently found to be affected by c-di-GMP signaling (53, 84; reviewed in reference 97). The contribution of c-di-GMP signaling to virulence properties seems to be complex, involving processes stimulated and inhibited by c-di-GMP signaling (76, 97). In *S. enterica* serovar Typhimurium, using a mouse model of systemic infection, the EAL-like protein STM1344 has been shown to be required for virulence (38). In particular, STM1344 was required for resistance against oxidative stress and to reduce macrophage killing. As the EAL-like protein STM1344 does not show c-di-GMP-dependent PDE activity but has, in contrast, been demonstrated to inhibit the expression of c-di-GMP-specific PDEs (91), it is possibile that c-di-GMP is required for the expression of these virulence phenotypes.

In agreement with the current paradigm that c-di-GMP inhibits the acute virulence phenotypes (as described in detail in chapter 20), we recently found that c-di-GMP has an inhibitory effect on virulence properties relevant in the gastrointestinal tract. c-di-GMP signaling inhibits invasion of a gastrointestinal epithelial cell line and the gut epithelium of mice (53a). Induction of the proinflammatory cytokine interleukin-8 in the gastrointestinal epithelial cell line is also abolished by high c-di-GMP levels. A more detailed analysis showed that c-di-GMP signaling inhibits invasion through expression of rdar morphotype components CsgD and cellulose, while inhibition of interleukin-8 production at high c-di-GMP levels is performed through CsgD.

GLOBAL REGULATION OF PROTEINS WITHIN THE c-di-GMP SIGNALING NETWORK

The fact that numerous GGDEF and EAL domain proteins regulate the expression of the rdar morphotype and CsgD expression raises the question of the relevance of this redundancy in the regulatory system. A likely explanation is that not all of the proteins that act on a specific target are present and active at the same time but rather individual or certain subsets of GGDEF-EAL domain proteins are expressed and activated under distinct environmental conditions. What are the factors and signals that orchestrate the expression and activity of GGDEF-EAL domain proteins in response to certain environmental conditions?

In *S. enterica* serovar Typhimurium, as well as in *E. coli,* the carbon storage regulator protein CsrA seems to play an important role in the tight regulation of several genes encoding proteins with GGDEF-EAL domains (42, 42a) (Fig. 5 and 7). Members of the CsrA family are RNA-binding proteins, which are widely distributed among eubacteria and which control various phenotypes, including biofilm formation, motility, and carbon flux, as well as numerous functions in the interactions with animal and plant hosts (55, 62, 73). By binding to target mRNAs, CsrA can either down- or upregulate the expression of genes at the posttranscriptional level. The activity of CsrA is regulated at the posttranslational level by small noncoding RNAs (sRNAs) (7, 55; described in detail in chapter 14). In *E. coli,* the two sRNAs, CsrB and CsrC, have been demonstrated to antagonize CsrA activity by binding to the protein and thereby sequestering it (7, 61, 73, 106). *S. enterica* serovar Typhimurium possesses homologs of CsrB and CsrC, which are predicted to function as in *E. coli* (2, 25). Transcription of *csrB* and *csrC* is controlled by the BarA-SirA two-component system, through which unidentified external signals are integrated into the Csr regulatory network (3, 99). The unconventional GGDEF-EAL domain protein CsrD (YhdA, STM3375 in *S. enterica* serovar Typhimurium) was recently identified as another Csr component in *E. coli* (45, 94) and was found to target CsrB and CsrC sRNAs for degradation by RNase E (94).

In *E. coli,* CsrA has been recently demonstrated to directly regulate the expression of genes encoding proteins with GGDEF domains, providing a link between CsrA and c-di-GMP signaling (42). The effect on *ycdT* and *ydeH,* both of which encode DGCs that inhibit motility, was the strongest. YdeH is also involved in the positive regulation of biofilm forma-

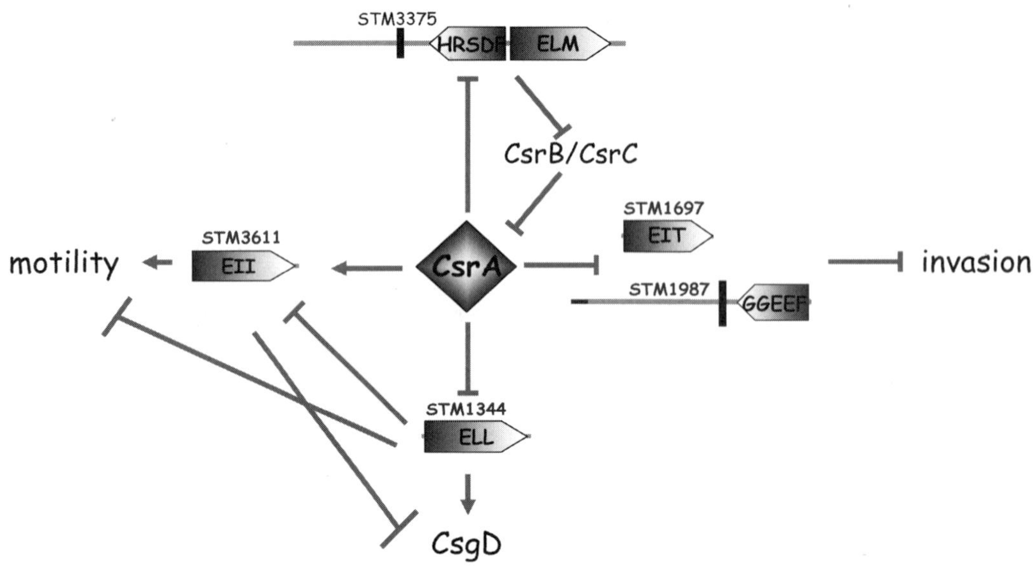

Figure 7. Direct control of regulation of GGDEF-EAL domain proteins by the carbon storage regulator CsrA in *S. enterica* serovar Typhimurium ATCC 14028 controls biofilm formation, motility, and invasion. CsrA directly represses the expression of the GGDEF-EAL protein STM3375 and the EAL-like domain proteins STM1344 and STM1697, whereas the PDE STM3611 is activated through CsrA. CsrA is predicted to inversely control rdar morphotype expression and motility through STM1344 and STM3611. CsrA is suggested to control invasion by regulating STM1697 and STM1987 by an unknown mechanism. CsrA regulates the expression of invasion genes (2). CsrA controls its own activity by regulating STM3375, which in turn destabilizes the CsrB and CsrC sRNAs, the antagonists of CsrA activity (94).

tion. Besides regulating motility and biofilm behavior in a c-di-GMP-dependent manner, CsrA has been previously demonstrated to control these two phenotypes directly. By stabilizing the transcript of the flagellar master regulator FlhDC (105) and by destabilizing the *pgaA* transcript (102), encoding the synthesis apparatus of the PGA biofilm polysaccharide, CsrA enhances motility and, reciprocally, inhibits biofilm formation in *E. coli.*

S. enterica serovar Typhimurium contains a CsrA homolog that is identical to the *E. coli* one. However, *S. enterica* serovar Typhimurium does not contain homologs of *ycdT, ydeH,* or most of the other GGDEF-EAL genes that are affected by CsrA in *E. coli.* To investigate the effect of CsrA on c-di-GMP metabolism in *S. enterica* serovar Typhimurium, a screen was set up in which the effect of a *csrA* mutation on the expression of all 20 GGDEF and/or EAL domain proteins present in *S. enterica* serovar Typhimurium was systematically analyzed (42a). A subset of these genes was indeed up- or downregulated by a *csrA* mutation; among the most strongly affected genes were STM1697, STM1344, STM1987, STM3611, and STM3375 (the CsrD ortholog). CsrA bound specifically and with high affinity to the transcripts of the five genes as determined by in vitro gel mobility shift assays using purified CsrA protein. For STM1697, STM1344, STM1987, and STM3375, CsrA binding was detected as multiple shifted bands, and the mRNA leader sequence of each transcript contained at least two putative binding sites as predicted by sequence analysis. For STM3611, only one shifted complex was observed. Expression of STM3611 was more than 10-fold downregulated by the *csrA* mutation, implicating CsrA as a positive regulator of the PDE. CsrA mediated its effect on STM3611 not only by directly interacting with the transcript but also by indirectly involving the FlhDC/FliA flagellar cascade and STM1344, which downregulates STM3611 (see above). Presumably, this multilayer control of STM3611 expression allows for a precise regulation of the levels and thereby the activity of the PDE in response to multiple intra- and extracellular input signals. Expression of STM3375, STM1987, STM1344, and STM1697 showed an upregulation in the *csrA* mutant (42a). The fact that CsrA upregulates the PDE STM3611 and, at the same time, downregulates the DGC STM1987 and STM1344, which represses STM3611, confirms that CsrA accomplishes its central role in bacterial physiology by inversely regulating oppositional phenotypes at multiple levels.

As in *E. coli,* the unorthodox GGDEF-EAL protein STM3375 might function in *Salmonella* by targeting the CsrB and CsrC sRNAs for degradation.

The fact that CsrA controls the expression of STM3375 by a direct mechanism implies that CsrA regulates its own activity by an autoregulatory loop involving STM3375. This illustrates that, despite its apparent inability to directly synthesize or degrade c-di-GMP, STM3375 is involved in c-di-GMP metabolism by controlling a global regulator of GGDEF and EAL domain-containing proteins. The tight regulation of CsrA activity ensures that the downstream targets controlled by CsrA are precisely coordinated in response to multiple input signals. STM1697 and STM1987 do not show a distinct phenotype in motility or biofilm formation (89). The ability to invade epithelial cells, however, was increased in mutants in STM1697 and STM1987 (A. Lamprokostopoulou, E. Streck, and U. Römling, unpublished results). It had been reported that invasion is strongly controlled by CsrA, as the expression of most invasion genes was drastically downregulated in a *csrA* mutant (2, 57). Furthermore, STM1344 had also been associated with virulence in *Salmonella* (38) (described above). Together, these data suggest that CsrA-mediated regulation of GGDEF-EAL domain proteins leads not only to the coordinated and inverse control of motility and biofilm phenotypes but also to the tight regulation of CsrA autoregulation and virulence. Besides its global functions in the control of physiological phenotypes, CsrA has a role in the regulation of carbon metabolism, which is well documented in *E. coli* (73), but less understood in *Salmonella*. Originally, CsrA was identified in *E. coli* by its ability to repress glycogen biosynthesis (74). Later, CsrA was found to also repress gluconeogenesis and, on the other hand, to activate glycolysis and acetate metabolism (73). Thus, CsrA couples carbon and energy metabolism with the regulation of bacterial lifestyles. High levels of CsrA enhance glycolysis and, thus, the supply of energy that is needed for flagellum-mediated motility. In contrast, when CsrA activity is low, gluconeogenesis and the formation of polysaccharides, some of which build up the biofilm matrix, is favored. The chemical structures of the signals that control the Csr regulatory system, e.g., through the BarA-UvrY (BarA-SirA in *S. enterica* serovar Typhimurium) two-component system have not been identified. However, findings of several studies suggest that the metabolic status of the cells might play a role in this regulation (68). The control of physiological processes in response to the metabolic status of the cells might be important for enteric bacteria to switch between an environmental state and the survival and colonization of the intestine in host organisms (110).

Connecting carbon metabolism with the control of bacterial lifestyles might be a strategy that is exploited by a number of other global regulators. For example, the cAMP repressor protein (CRP) has been found to control biofilm formation in various bacteria (24, 41, 50, 60). CRP is allosterically activated by the small molecule cAMP, which accumulates in the absence of sugars (such as glucose) usually imported by the phosphotransferase sugar uptake system (as described in detail in chapter 1). In analogy to CsrA, in *Vibrio cholerae,* cAMP-CRP was found to control biofilm formation by directly controlling the synthesis of a biofilm polysaccharide and, in combination, by regulating the expression of a set of GGDEF-EAL domain proteins (24). These results demonstrated, for the first time, the interconnection between the networks of two major bacterial second messengers.

CONCLUSIONS AND PERSPECTIVES

Although *S. enterica* serovar Typhimurium was one of the first organisms in which second messenger signaling by c-di-GMP had been described, there are still more obstacles than answers. Although a large panel of c-di-GMP-metabolizing enzymes exists in *S. enterica* serovar Typhimurium, there are yet surprisingly few phenotypes that are known to be affected by c-di-GMP signaling. In addition, even for most of the known phenotypes, the direct targets of c-di-GMP signaling (i.e., the c-di-GMP binding proteins), still must be identified. In the case of rdar morphotype expression through CsgD, a major challenge is to identify the mechanisms by which c-di-GMP controls CsgD expression. As there are several GGDEF and EAL domain proteins that affect CsgD expression in distinct ways, there may be more than one c-di-GMP-associated mechanism that contributes to the control of CsgD expression. Indeed, our studies have shown that c-di-GMP might affect CsgD expression at several levels (49). In addition, the interconnections between the c-di-GMP signaling network and other components (such as the two-component systems) that influence CsgD expression merit investigation. Certainly, another factor contributing to the variability, flexibility, and even specificity of c-di-GMP signaling is the concerted expression of subsets of c-di-GMP-metabolizing enzymes in response to major global regulators (43). These global regulators control the transition between different lifestyles such as the switch from motility and sessility or from sessility to acute virulence. It is a major challenge to define these different regulatory networks and the interactions that tightly control the transition between different regulatory networks.

Acknowledgments. Work in the laboratory of U.R. as described in this review was supported by the Karolinska Institutet (elitforskartjänst to U.R.), the European Commission (MEST-CT-2004-

008475), the Swedish Research Council Natural Science (621-2004-3979 and 621-2007-6509), the Swedish Research Council Medicine (521-2001-6463), the Carl Trygger Foundation (CTS-07:306), the Swedisch International Development Cooperation Agency (Sida; 5999-2005), and Längmanska Kulturfonden.

REFERENCES

1. **Aldridge, P., R. Paul, P. Goymer, P. Rainey, and U. Jenal.** 2003. Role of the GGDEF regulator PleD in polar development of *Caulobacter crescentus*. *Mol. Microbiol.* **47:**1695–1708.
2. **Altier, C., M. Suyemoto, and S. D. Lawhon.** 2000. Regulation of *Salmonella enterica* serovar Typhimurium invasion genes by *csrA*. *Infect. Immun.* **68:**6790–6797.
3. **Altier, C., M. Suyemoto, A. I. Ruiz, K. D. Burnham, and R. Maurer.** 2000. Characterization of two novel regulatory genes affecting *Salmonella* invasion gene expression. *Mol. Microbiol.* **35:**635–646.
4. **Amikam, D., and M. Y. Galperin.** 2006. PilZ domain is part of the bacterial c-di-GMP binding protein. *Bioinformatics* **22:**3–6.
5. **Ausmees, N., R. Mayer, H. Weinhouse, G. Volman, D. Amikam, M. Benziman, and M. Lindberg.** 2001. Genetic data indicate that proteins containing the GGDEF domain possess diguanylate cyclase activity. *FEMS Microbiol. Lett.* **204:**163–167.
6. **Austin, J. W., G. Sanders, W. W. Kay, and S. K. Collinson.** 1998. Thin aggregative fimbriae enhance *Salmonella enteritidis* biofilm formation. *FEMS Microbiol. Lett.* **162:**295–301.
7. **Babitzke, P., and T. Romeo.** 2007. CsrB sRNA family: sequestration of RNA-binding regulatory proteins. *Curr. Opin. Microbiol.* **10:**156–163.
8. **Barak, J. D., L. Gorski, P. Naraghi-Arani, and A. O. Charkowski.** 2005. *Salmonella enterica* virulence genes are required for bacterial attachment to plant tissue. *Appl. Environ. Microbiol.* **71:**5685–5691.
9. **Barak, J. D., C. E. Jahn, D. L. Gibson, and A. O. Charkowski.** 2007. The role of cellulose and O-antigen capsule in the colonization of plants by *Salmonella enterica*. *Mol. Plant-Microbe Interact.* **20:**1083–1091.
10. **Barnhart, M. M., J. Lynem, and M. R. Chapman.** 2006. GlcNAc-6P levels modulate the expression of curli fibers by *Escherichia coli*. *J. Bacteriol.* **188:**5212–5219.
11. **Beyhan, S., A. D. Tischler, A. Camilli, and F. H. Yildiz.** 2006. Transcriptome and phenotypic responses of *Vibrio cholerae* to increased cyclic di-GMP level. *J. Bacteriol.* **188:**3600–3613.
12. **Bokranz, W., X. Wang, H. Tschäpe, and U. Römling.** 2005. Expression of cellulose and curli fimbriae by *Escherichia coli* isolated from the gastrointestinal tract. *J. Med. Microbiol.* **54:**1171–1182.
13. **Brombacher, E., A. Baratto, C. Dorel, and P. Landini.** 2006. Gene expression regulation by the curli activator CsgD protein: modulation of cellulose biosynthesis and control of negative determinants for microbial adhesion. *J. Bacteriol.* **188:**2027–2037.
14. **Brombacher, E., C. Dorel, A. J. Zehnder, and P. Landini.** 2003. The curli biosynthesis regulator CsgD co-ordinates the expression of both positive and negative determinants for biofilm formation in *Escherichia coli*. *Microbiology* **149:**2847–2857.
15. **Brown, P. K., C. M. Dozois, C. A. Nickerson, A. Zuppardo, J. Terlonge, and R. Curtiss III.** 2001. MlrA, a novel regulator of curli (AgF) and extracellular matrix synthesis by *Escherichia coli* and *Salmonella enterica* serovar Typhimurium. *Mol. Microbiol.* **41:**349–363.
16. **Christen, M., B. Christen, M. Folcher, A. Schauerte, and U. Jenal.** 2005. Identification and characterization of a cyclic di-GMP-specific phosphodiesterase and its allosteric control by GTP. *J. Biol. Chem.* **280:**30829–30837.
17. **Collinson, S. K., P. C. Doig, J. L. Doran, S. Clouthier, T. J. Trust, and W. W. Kay.** 1993. Thin, aggregative fimbriae mediate binding of *Salmonella enteritidis* to fibronectin. *J. Bacteriol.* **175:**12–18.
18. **Collinson, S. K., L. Emody, K. H. Muller, T. J. Trust, and W. W. Kay.** 1991. Purification and characterization of thin, aggregative fimbriae from *Salmonella enteritidis*. *J. Bacteriol.* **173:**4773–4781.
19. **Collinson, S. K., L. Emody, T. J. Trust, and W. W. Kay.** 1992. Thin aggregative fimbriae from diarrheagenic *Escherichia coli*. *J. Bacteriol.* **174:**4490–4495.
20. **Costerton, J. W., Z. Lewandowski, D. E. Caldwell, D. R. Korber, and H. M. Lappin-Scott.** 1995. Microbial biofilms. *Annu. Rev. Microbiol.* **49:**711–745.
21. **D'Argenio, D. A., M. W. Calfee, P. B. Rainey, and E. C. Pesci.** 2002. Autolysis and autoaggregation in *Pseudomonas aeruginosa* colony morphology mutants. *J. Bacteriol.* **184:**6481–6489.
22. **Davidson, C. J., A. P. White, and M. G. Surette.** 2008. Evolutionary loss of the rdar morphotype in *Salmonella* as a result of high mutation rates during laboratory passage. *ISME J.* **2:**293–307.
23. **Drenkard, E., and F. M. Ausubel.** 2002. *Pseudomonas* biofilm formation and antibiotic resistance are linked to phenotypic variation. *Nature* **416:**740–743.
24. **Fong, J. C., and F. H. Yildiz.** 2008. Interplay between cyclic AMP-cyclic AMP receptor protein and cyclic di-GMP signaling in *Vibrio cholerae* biofilm formation. *J. Bacteriol.* **190:**6646–6659.
25. **Fortune, D. R., M. Suyemoto, and C. Altier.** 2006. Identification of CsrC and characterization of its role in epithelial cell invasion in *Salmonella enterica* serovar Typhimurium. *Infect. Immun.* **74:**331–339.
26. **Galperin, M. Y., A. N. Nikolskaya, and E. V. Koonin.** 2001. Novel domains of the prokaryotic two-component signal transduction systems. *FEMS Microbiol. Lett.* **203:**11–21.
27. **Garcia, B., C. Latasa, C. Solano, F. G. Portillo, C. Gamazo, and I. Lasa.** 2004. Role of the GGDEF protein family in *Salmonella* cellulose biosynthesis and biofilm formation. *Mol. Microbiol.* **54:**264–277.
28. **Gerstel, U., A. Kolb, and U. Römling.** 2006. Regulatory components at the *csgD* promoter—additional roles for OmpR and integration host factor and role of the 5′ untranslated region. *FEMS Microbiol. Lett.* **261:**109–117.
29. **Gerstel, U., C. Park, and U. Römling.** 2003. Complex regulation of *csgD* promoter activity by global regulatory proteins. *Mol. Microbiol.* **49:**639–654.
30. **Gerstel, U., and U. Römling.** 2001. Oxygen tension and nutrient starvation are major signals that regulate *agfD* promoter activity and expression of the multicellular morphotype in *Salmonella typhimurium*. *Environ. Microbiol.* **3:**638–648.
31. **Gerstel, U., and U. Römling.** 2003. The *csgD* promoter, a control unit for biofilm formation in *Salmonella typhimurium*. *Res. Microbiol.* **154:**659–667.
32. **Gibson, D. L., A. P. White, S. D. Snyder, S. Martin, C. Heiss, P. Azadi, M. Surette, and W. W. Kay.** 2006. *Salmonella* produces an O-antigen capsule regulated by AgfD and important for environmental persistence. *J. Bacteriol.* **188:**7722–7730.
33. **Gophna, U., M. Barlev, R. Seijffers, T. A. Oelschlager, J. Hacker, and E. Z. Ron.** 2001. Curli fibers mediate internali-

zation of *Escherichia coli* by eukaryotic cells. *Infect. Immun.* 69:2659–2665.

34. **Gronewold, T. M., and D. Kaiser.** 2001. The *act* operon controls the level and time of C-signal production for *Myxococcus xanthus* development. *Mol. Microbiol.* **40:**744–756.
35. **Gualdi, L., L. Tagliabue, and P. Landini.** 2007. Biofilm formation-gene expression relay system in *Escherichia coli:* modulation of σ^S-dependent gene expression by the CsgD regulatory protein via σ^S protein stabilization. *J. Bacteriol.* **189:** 8034–8043.
36. **Hecht, G. B., and A. Newton.** 1995. Identification of a novel response regulator required for the swarmer-to-stalked-cell transition in *Caulobacter crescentus. J. Bacteriol.* **177:**6223–6229.
37. **Herwald, H., M. Morgelin, A. Olsen, M. Rhen, B. Dahlback, W. Muller-Esterl, and L. Bjorck.** 1998. Activation of the contact-phase system on bacterial surfaces—a clue to serious complications in infectious diseases. *Nat. Med.* **4:**298–302.
38. **Hisert, K. B., M. MacCoss, M. U. Shiloh, K. H. Darwin, S. Singh, R. A. Jones, S. Ehrt, Z. Zhang, B. L. Gaffney, S. Gandotra, D. W. Holden, D. Murray, and C. Nathan.** 2005. A glutamate-alanine-leucine (EAL) domain protein of *Salmonella* controls bacterial survival in mice, antioxidant defence and killing of macrophages: role of cyclic diGMP. *Mol. Microbiol.* **56:**1234–1245.
39. **Houslay, M. D., and G. Milligan.** 1997. Tailoring cAMP-signalling responses through isoform multiplicity. *Trends Biochem. Sci.* **22:**217–224.
40. **Huang, Y. H., L. Ferrieres, and D. J. Clarke.** 2006. The role of the Rcs phosphorelay in Enterobacteriaceae. *Res. Microbiol.* **157:**206–212.
41. **Jackson, D. W., J. W. Simecka, and T. Romeo.** 2002. Catabolite repression of *Escherichia coli* biofilm formation. *J. Bacteriol.* **184:**3406–3410.
42. **Jonas, K., A. N. Edwards, R. Simm, T. Romeo, U. Römling, and Ö. Melefors.** 2008. The RNA binding protein CsrA controls cyclic di-GMP metabolism by directly regulating the expression of GGDEF proteins. *Mol. Microbiol.* **70:**236–257.

42a. **Jonas, K., A. N. Edwards, I. Ahmad, T. Romeo, U. Römling, and Ö. Melefors.** 2009. Complex regulatory network encompassing the Csr, c-di-GMP and motility systems of *Salmonella* Typhimurium. *Environ. Microbiol.*, in press.

43. **Jonas, K., Ö. Melefors, and U. Römling.** 2009. Regulation of c-di-GMP metabolism in biofilms. *Future Microbiol.* **4:**341–358.
44. **Jonas, K., H. Tomenius, A. Kader, S. Normark, U. Römling, L. M. Belova, and Ö. Melefors.** 2007. Roles of curli, cellulose and BapA in Salmonella biofilm morphology studied by atomic force microscopy. *BMC Microbiol.* **7:**70.
45. **Jonas, K., H. Tomenius, U. Römling, D. Georgellis, and Ö. Melefors.** 2006. Identification of YhdA as a regulator of the *Escherichia coli* carbon storage regulation system. *FEMS Microbiol. Lett.* **264:**232–237.
46. **Jones, H. A., J. W. Lillard, Jr., and R. D. Perry.** 1999. HmsT, a protein essential for expression of the haemin storage (Hms+) phenotype of *Yersinia pestis. Microbiology* **145:**2117–2128.
47. **Joseph, B., S. K. Otta, and I. Karunasagar.** 2001. Biofilm formation by *Salmonella* spp. on food contact surfaces and their sensitivity to sanitizers. *Int. J. Food Microbiol.* **64:**367–372.
48. **Jubelin, G., A. Vianney, C. Beloin, J. M. Ghigo, J. C. Lazzaroni, P. Lejeune, and C. Dorel.** 2005. CpxR/OmpR interplay regulates curli gene expression in response to osmolarity in *Escherichia coli. J. Bacteriol.* **187:**2038–2049.
49. **Kader, A., R. Simm, U. Gerstel, M. Morr, and U. Römling.** 2006. Hierarchical involvement of various GGDEF domain proteins in rdar morphotype development of *Salmonella enterica* serovar Typhimurium. *Mol. Microbiol.* **60:**602–616.
50. **Kalivoda, E. J., N. A. Stella, D. M. O'Dee, G. J. Nau, and R. M. Shanks.** 2008. The cyclic AMP-dependent catabolite repression system of *Serratia marcescens* mediates biofilm formation through regulation of type 1 fimbriae. *Appl. Environ. Microbiol.* **74:**3461–3470.
51. **Klausen, M., A. Aaes-Jorgensen, S. Molin, and T. Tolker-Nielsen.** 2003. Involvement of bacterial migration in the development of complex multicellular structures in *Pseudomonas aeruginosa* biofilms. *Mol. Microbiol.* **50:**61–68.
52. **Ko, M., and C. Park.** 2000. Two novel flagellar components and H-NS are involved in the motor function of *Escherichia coli. J. Mol. Biol.* **303:**371–382.
53. **Kulesekara, H., V. Lee, A. Brencic, N. Liberati, J. Urbach, S. Miyata, D. G. Lee, A. N. Neely, M. Hyodo, Y. Hayakawa, F. M. Ausubel, and S. Lory.** 2006. Analysis of *Pseudomonas aeruginosa* diguanylate cyclases and phosphodiesterases reveals a role for bis-(3′-5′)-cyclic-GMP in virulence. *Proc. Natl. Acad. Sci. USA* **103:**2839–2844.

53a. **Lamprokostopoulou, A., C. Monteiro, M. Rhen, and U. Römling.** 2009. Cyclic di-GMP signalling controls virulence properties of *Salmonella enterica* serovar Typhimurium at the mucosal lining. *Environ. Microbiol.*, in press.

54. **Lapidot, A., U. Römling, and S. Yaron.** 2006. Biofilm formation and the survival of *Salmonella* Typhimurium on parsley. *Int. J. Food Microbiol.* **109:**229–233.
55. **Lapouge, K., M. Schubert, F. H. Allain, and D. Haas.** 2008. Gac/Rsm signal transduction pathway of gamma-proteobacteria: from RNA recognition to regulation of social behaviour. *Mol. Microbiol.* **67:**241–253.
56. **Latasa, C., A. Roux, A. Toledo-Arana, J. M. Ghigo, C. Gamazo, J. R. Penades, and I. Lasa.** 2005. BapA, a large secreted protein required for biofilm formation and host colonization of *Salmonella enterica* serovar Enteritidis. *Mol. Microbiol.* **58:** 1322–1339.
57. **Lawhon, S. D., J. G. Frye, M. Suyemoto, S. Porwollik, M. McClelland, and C. Altier.** 2003. Global regulation by CsrA in *Salmonella typhimurium. Mol. Microbiol.* **48:**1633–1645.
58. **Ledeboer, N. A., J. G. Frye, M. McClelland, and B. D. Jones.** 2006. *Salmonella enterica* serovar Typhimurium requires the Lpf, Pef, and Tafi fimbriae for biofilm formation on HEp-2 tissue culture cells and chicken intestinal epithelium. *Infect. Immun.* **74:**3156–3169.
59. **Ledeboer, N. A., and B. D. Jones.** 2005. Exopolysaccharide sugars contribute to biofilm formation by *Salmonella enterica* serovar Typhimurium on HEp-2 cells and chicken intestinal epithelium. *J. Bacteriol.* **187:**3214–3226.
60. **Liang, W., A. J. Silva, and J. A. Benitez.** 2007. The cyclic AMP receptor protein modulates colonial morphology in *Vibrio cholerae. Appl. Environ. Microbiol.* **73:**7482–7487.
61. **Liu, M. Y., G. Gui, B. Wei, J. F. Preston III, L. Oakford, U. Yuksel, D. P. Giedroc, and T. Romeo.** 1997. The RNA molecule CsrB binds to the global regulatory protein CsrA and antagonizes its activity in *Escherichia coli. J. Biol. Chem.* **272:** 17502–17510.
62. **Lucchetti-Miganeh, C., E. Burrowes, C. Baysse, and G. Ermel.** 2008. The post-transcriptional regulator CsrA plays a central role in the adaptation of bacterial pathogens to different stages of infection in animal hosts. *Microbiology* **154:**16–29.
63. **Minasov, G., S. Padavattan, L. Shuvalova, J. S. Brunzelle, D. J. Miller, A. Basle, C. Massa, F. R. Collart, T. Schirmer, and W. F. Anderson.** 2009. Crystal structures of *YkuI* and its complex with second messenger cyclic di-GMP suggest catalytic mechanism of phosphodiester bond cleavage by EAL domains. *J. Biol. Chem.* **284:**13174–13184.

64. **Ogasawara, H., A. Hasegawa, E. Kanda, T. Miki, K. Yamamoto, and A. Ishihama.** 2007. Genomic SELEX search for target promoters under the control of the PhoQP-RstBA signal relay cascade. *J. Bacteriol.* **189:**4791–4799.
65. **Olsen, A., A. Jonsson, and S. Normark.** 1989. Fibronectin binding mediated by a novel class of surface organelles on *Escherichia coli. Nature* **338:**652–655.
66. **Olsen, A., M. J. Wick, M. Morgelin, and L. Bjorck.** 1998. Curli, fibrous surface proteins of *Escherichia coli,* interact with major histocompatibility complex class I molecules. *Infect. Immun.* **66:**944–949.
67. **Paul, R., S. Weiser, N. C. Amiot, C. Chan, T. Schirmer, B. Giese, and U. Jenal.** 2004. Cell cycle-dependent dynamic localization of a bacterial response regulator with a novel diguanylate cyclase output domain. *Genes Dev.* **18:**715–727.
68. **Pernestig, A. K., D. Georgellis, T. Romeo, K. Suzuki, H. Tomenius, S. Normark, and O. Melefors.** 2003. The *Escherichia coli* BarA-UvrY two-component system is needed for efficient switching between glycolytic and gluconeogenic carbon sources. *J. Bacteriol.* **185:**843–853.
69. **Pesavento, C., G. Becker, N. Sommerfeldt, A. Possling, N. Tschowri, A. Mehlis, and R. Hengge.** 2008. Inverse regulatory coordination of motility and curli-mediated adhesion in *Escherichia coli. Genes Dev.* **22:**2434–2446.
70. **Rao, F., Y. Yang, Y. Qi, and Z. X. Liang.** 2008. Catalytic mechanism of cyclic di-GMP-specific phosphodiesterase: a study of the EAL domain-containing RocR from *Pseudomonas aeruginosa. J. Bacteriol.* **190:**3622–3631.
71. **Rice, P. A.** 1997. Making DNA do a U-turn: IHF and related proteins. *Curr. Opin. Struct. Biol.* **7:**86–93.
72. **Robbe-Saule, V., V. Jaumouille, M. C. Prevost, S. Guadagnini, C. Talhouarne, H. Mathout, A. Kolb, and F. Norel.** 2006. Crl activates transcription initiation of RpoS-regulated genes involved in the multicellular behavior of *Salmonella enterica* serovar Typhimurium. *J. Bacteriol.* **188:**3983–3994.
73. **Romeo, T.** 1998. Global regulation by the small RNA-binding protein CsrA and the non-coding RNA molecule CsrB. *Mol. Microbiol.* **29:**1321–1330.
74. **Romeo, T., M. Gong, M. Y. Liu, and A. M. Brun-Zinkernagel.** 1993. Identification and molecular characterization of *csrA,* a pleiotropic gene from *Escherichia coli* that affects glycogen biosynthesis, gluconeogenesis, cell size, and surface properties. *J. Bacteriol.* **175:**4744–4755.
75. **Römling, U.** 2005. Characterization of the rdar morphotype, a multicellular behaviour in Enterobacteriaceae. *Cell. Mol. Life Sci.* **62:**1234–1246.
76. **Römling, U.** 2009. Cyclic di-GMP (c-di-GMP) goes into host cells—c-di-GMP signaling in the obligate intracellular pathogen *Anaplasma phagocytophilum. J. Bacteriol.* **191:**683–686.
77. **Römling, U., Z. Bian, M. Hammar, W. D. Sierralta, and S. Normark.** 1998. Curli fibers are highly conserved between *Salmonella typhimurium* and *Escherichia coli* with respect to operon structure and regulation. *J. Bacteriol.* **180:**722–731.
78. **Römling, U., W. Bokranz, W. Rabsch, X. Zogaj, M. Nimtz, and H. Tschäpe.** 2003. Occurrence and regulation of the multicellular morphotype in *Salmonella* serovars important in human disease. *Int. J. Med. Microbiol.* **293:**273–285.
79. **Römling, U., M. Gomelsky, and M. Y. Galperin.** 2005. c-di-GMP: the dawning of a novel bacterial signalling system. *Mol. Microbiol.* **57:**629–639.
80. **Römling, U., D. Pesen, and S. Yaron.** 2007. Biofilms of *Salmonella enterica,* p. 127–145. *In* M. Rhen, D. Maskell, P. Mastroeni, and J. Threlfall. (ed.), *Salmonella Molecular Biology and Pathogenesis.* Horizon Press, Norfolk, United Kingdom.
81. **Römling, U., and M. Rohde.** 1999. Flagella modulate the multicellular behavior of *Salmonella typhimurium* on the community level. *FEMS Microbiol. Lett.* **180:**91–102.
82. **Römling, U., M. Rohde, A. Olsen, S. Normark, and J. Reinköster.** 2000. AgfD, the checkpoint of multicellular and aggregative behaviour in *Salmonella typhimurium* regulates at least two independent pathways. *Mol. Microbiol.* **36:**10–23.
83. **Römling, U., W. D. Sierralta, K. Eriksson, and S. Normark.** 1998. Multicellular and aggregative behaviour of *Salmonella typhimurium* strains is controlled by mutations in the *agfD* promoter. *Mol. Microbiol.* **28:**249–264.
84. **Ryan, R. P., Y. Fouhy, J. F. Lucey, B. L. Jiang, Y. Q. He, J. X. Feng, J. L. Tang, and J. M. Dow.** 2007. Cyclic di-GMP signalling in the virulence and environmental adaptation of *Xanthomonas campestris. Mol. Microbiol.* **63:**429–442.
85. **Ryjenkov, D. A., R. Simm, U. Römling, and M. Gomelsky.** 2006. The PilZ domain is a receptor for the second messenger c-di-GMP. The PilZ domain protein YcgR controls motility in enterobacteria. *J. Biol. Chem.* **281:**30310–30314.
86. **Ryjenkov, D. A., M. Tarutina, O. V. Moskvin, and M. Gomelsky.** 2005. Cyclic diguanylate is a ubiquitous signaling molecule in bacteria: insights into biochemistry of the GGDEF protein domain. *J. Bacteriol.* **187:**1792–1798.
87. **Scher, K., U. Römling, and S. Yaron.** 2005. Effect of heat, acidification, and chlorination on *Salmonella enterica* serovar Typhimurium cells in a biofilm formed at the air-liquid interface. *Appl. Environ. Microbiol.* **71:**1163–1168.
88. **Schmidt, A. J., D. A. Ryjenkov, and M. Gomelsky.** 2005. The ubiquitous protein domain EAL is a cyclic diguanylate-specific phosphodiesterase: enzymatically active and inactive EAL domains. *J. Bacteriol.* **187:**4774–4781.
89. **Simm, R., A. Lusch, A. Kader, M. Andersson, and U. Römling.** 2007. Role of EAL-containing proteins in multicellular behavior of *Salmonella enterica* serovar Typhimurium. *J. Bacteriol.* **189:**3613–3623.
90. **Simm, R., M. Morr, A. Kader, M. Nimtz, and U. Römling.** 2004. GGDEF and EAL domains inversely regulate cyclic di-GMP levels and transition from sessility to motility. *Mol. Microbiol.* **53:**1123–1134.
91. **Simm, R., U. Remminghorst, I. Ahmad, K. Zakikhany, and U. Römling.** 2009. A role for the EAL-like protein STM1344 in regulation of CsgD expression and motility in *Salmonella enterica* serovar Typhimurium. *J. Bacteriol.* **191:**3928–3937.
92. **Solano, C., B. Garcia, J. Valle, C. Berasain, J. M. Ghigo, C. Gamazo, and I. Lasa.** 2002. Genetic analysis of *Salmonella enteritidis* biofilm formation: critical role of cellulose. *Mol. Microbiol.* **43:**793–808.
93. **Sukupolvi, S., A. Edelstein, M. Rhen, S. J. Normark, and J. D. Pfeifer.** 1997. Development of a murine model of chronic *Salmonella* infection. *Infect. Immun.* **65:**838–842.
94. **Suzuki, K., P. Babitzke, S. R. Kushner, and T. Romeo.** 2006. Identification of a novel regulatory protein (CsrD) that targets the global regulatory RNAs CsrB and CsrC for degradation by RNase E. *Genes Dev.* **20:**2605–2617.
95. **Tabak, M., K. Scher, E. Hartog, U. Römling, K. R. Matthews, M. L. Chikindas, and S. Yaron.** 2006. The effect of triclosan on *Salmonella* Typhimurium at different growth stages. *FEMS Microbiol. Lett.* **267:**200–206.
96. **Tal, R., H. C. Wong, R. Calhoon, D. Gelfand, A. L. Fear, G. Volman, R. Mayer, P. Ross, D. Amikam, H. Weinhouse, A. Cohen, S. Sapir, P. Ohana, and M. Benziman.** 1998. Three *cdg* operons control cellular turnover of cyclic di-GMP in *Acetobacter xylinum:* genetic organization and occurrence of conserved domains in isoenzymes. *J. Bacteriol.* **180:**4416–4425.
97. **Tamayo, R., J. T. Pratt, and A. Camilli.** 2007. Roles of cyclic diguanylate in the regulation of bacterial pathogenesis. *Annu. Rev. Microbiol.* **61:**131–148.

98. **Tamayo, R., A. D. Tischler, and A. Camilli.** 2005. The EAL domain protein VieA is a cyclic diguanylate phosphodiesterase. *J. Biol. Chem.* **280:**33324–33330.
99. **Teplitski, M., R. I. Goodier, and B. M. Ahmer.** 2003. Pathways leading from BarA/SirA to motility and virulence gene expression in *Salmonella. J. Bacteriol.* **185:**7257–7265.
100. **Thormann, K. M., S. Duttler, R. M. Saville, M. Hyodo, S. Shukla, Y. Hayakawa, and A. M. Spormann.** 2006. Control of formation and cellular detachment from *Shewanella oneidensis* MR-1 biofilms by cyclic di-GMP. *J. Bacteriol.* **188:** 2681–2691.
101. **Tukel, C., M. Raffatellu, A. D. Humphries, R. P. Wilson, H. L. Andrews-Polymenis, T. Gull, J. F. Figueiredo, M. H. Wong, K. S. Michelsen, M. Akcelik, L. G. Adams, and A. J. Baumler.** 2005. CsgA is a pathogen-associated molecular pattern of *Salmonella enterica* serotype Typhimurium that is recognized by Toll-like receptor 2. *Mol. Microbiol.* **58:**289–304.
102. **Wang, X., A. K. Dubey, K. Suzuki, C. S. Baker, P. Babitzke, and T. Romeo.** 2005. CsrA post-transcriptionally represses *pgaABCD,* responsible for synthesis of a biofilm polysaccharide adhesin of *Escherichia coli. Mol. Microbiol.* **56:**1648–1663.
103. **Wang, X., M. Rochon, A. Lamprokostopoulou, H. Lünsdorf, M. Nimtz, and U. Römling.** 2006. Impact of biofilm matrix components on interaction of commensal *Escherichia coli* with the gastrointestinal cell line HT-29. *Cell. Mol. Life Sci.* **63:**2352–2363.
104. **Weber, H., C. Pesavento, A. Possling, G. Tischendorf, and R. Hengge.** 2006. Cyclic-di-GMP-mediated signalling within the sigma network of *Escherichia coli. Mol. Microbiol.* **62:** 1014–1034.
105. **Wei, B. L., A. M. Brun-Zinkernagel, J. W. Simecka, B. M. Pruss, P. Babitzke, and T. Romeo.** 2001. Positive regulation of motility and *flhDC* expression by the RNA-binding protein CsrA of *Escherichia coli. Mol. Microbiol.* **40:**245–256.
106. **Weilbacher, T., K. Suzuki, A. K. Dubey, X. Wang, S. Gudapaty, I. Morozov, C. S. Baker, D. Georgellis, P. Babitzke, and T. Romeo.** 2003. A novel sRNA component of the carbon storage regulatory system of *Escherichia coli. Mol. Microbiol.* **48:**657–670.
107. **White, A. P., D. L. Gibson, S. K. Collinson, P. A. Banser, and W. W. Kay.** 2003. Extracellular polysaccharides associated with thin aggregative fimbriae of *Salmonella enterica* serovar Enteritidis. *J. Bacteriol.* **185:**5398–5407.
108. **White, A. P., D. L. Gibson, G. A. Grassl, W. W. Kay, B. B. Finlay, B. A. Vallance, and M. G. Surette.** 2008. Aggregation via the red, dry, and rough morphotype is not a virulence adaptation in *Salmonella enterica* serovar Typhimurium. *Infect. Immun.* **76:**1048–1058.
109. **White, A. P., D. L. Gibson, W. Kim, W. W. Kay, and M. G. Surette.** 2006. Thin aggregative fimbriae and cellulose enhance long-term survival and persistence of *Salmonella. J. Bacteriol.* **188:**3219–3227.
110. **Winfield, M. D., and E. A. Groisman.** 2003. Role of nonhost environments in the lifestyles of *Salmonella* and *Escherichia coli. Appl. Environ. Microbiol.* **69:**3687–3694.
111. **Wolfe, A. J., and K. L. Visick.** 2008. Get the message out: cyclic-di-GMP regulates multiple levels of flagellum-based motility. *J. Bacteriol.* **190:**463–475.
112. **Wozniak, C. E., C. Lee, and K. T. Hughes.** 2008. T-POP array identifies EcnR and PefI-SrgD as novel regulators of flagellar gene expression. *J. Bacteriol.* **191:**1498–1508.
113. **Zogaj, X., W. Bokranz, M. Nimtz, and U. Römling.** 2003. Production of cellulose and curli fimbriae by members of the family *Enterobacteriaceae* isolated from the human gastrointestinal tract. *Infect. Immun.* **71:**4151–4158.
114. **Zogaj, X., M. Nimtz, M. Rohde, W. Bokranz, and U. Römling.** 2001. The multicellular morphotypes of *Salmonella typhimurium* and *Escherichia coli* produce cellulose as the second component of the extracellular matrix. *Mol. Microbiol.* **39:**1452–1463.

The Second Messenger Cyclic Di-GMP
Edited by A. J. Wolfe and K. L. Visick

Chapter 11

Role of Cyclic Di-GMP in *Pseudomonas aeruginosa* Biofilm Development

CAROLINE S. HARWOOD

High intracellular cyclic di-GMP [bis (3′-5′)-cyclic di-GMP] (c-di-GMP) promotes biofilm formation and reduces motility in many bacterial species. Conversely, low concentrations of intracellular c-di-GMP are correlated with increased motility (27, 35, 68, 79, 86). Such is the case for the opportunistic pathogen and model biofilm organism *Pseudomonas aeruginosa*. Like other gram-negative bacteria, *P. aeruginosa* has many genes encoding proteins with predicted GGDEF domains responsible for the synthesis of c-di-GMP. It also has genes with predicted EAL and HD-GYP domains for c-di-GMP hydrolysis (42, 70). GGDEF domains have diguanylate cyclase (DGC) activities that catalyze the synthesis of c-di-GMP from GTP, whereas EAL and HD-GYP domains have phosphodiesterase (PDE) activities that catalyze the degradation of c-di-GMP to pGpG, which spontaneously converts to GTP (8, 71, 73) (see chapters 4 and 5 for a discussion of the biochemistry).

The general parameters of c-di-GMP activity have been established, but environmental inputs that stimulate c-di-GMP synthesis, intracellular trafficking of c-di-GMP, and mechanisms of c-di-GMP action are just beginning to be explored. As with other bacteria, only a small subset of *P. aeruginosa* EAL-GGDEF gene mutants has discernible mutant phenotypes. However, such mutants have provided valuable clues about how cells produce and traffic c-di-GMP intracellularly. Also, several different novel c-di-GMP receptor proteins have been identified in *P. aeruginosa,* and this has expanded our understanding of the range of functions that c-di-GMP can influence. In this chapter, I will review c-di-GMP metabolism and activity in *P. aeruginosa* with a focus on how c-di-GMP modulates the expression and activities of genes and proteins required for biofilm formation.

PSEUDOMONAS AERUGINOSA CAN ADOPT EITHER A BIOFILM OR A PLANKTONIC EXISTENCE

P. aeruginosa is an opportunistic human pathogen that persistently colonizes the lungs of people with cystic fibrosis (CF) and eventually kills them (21, 61). It also frequently causes septicemia in burn patients and urinary tract infections in catheterized patients (46). It is widely distributed in nature and can be isolated from most soil and water environments. This versatile organism is notable for its ability to adopt either a planktonic or a biofilm lifestyle, and it is a principal model organism used for studies of bacterial biofilm formation. When in its planktonic form, *P. aeruginosa* is highly motile in liquid by means of a single polar flagellum. Planktonic cells have strong chemotactic responses to amino acids and other organic compounds, which allow them to sense and to swim to sources of soluble nutrients (2, 57, 78). *P. aerguinosa* also moves on solid surfaces by means of flagellum-mediated swarming (39) and pilus-mediated twitching (85). In some situations, which are not well defined, surface-associated cells stop moving, adhere to surfaces, and adopt a sessile existence. Sessile cells tend to proliferate on surfaces, eventually differentiating to form structured communities called biofilms that are encased in a self-produced matrix comprised of several different macromolecules including exopolysaccharides (EPS) (58). Biofilm cells have certain advantages over planktonic

Caroline S. Harwood • Department of Microbiology, The University of Washington, Seattle, WA 98195.

cells. One is an ability to use insoluble polymers as food sources through the action of exoenzymes produced by a concentrated population of cells (5). Another characteristic of biofilms that is advantageous to the microbe, but disadvantageous to humans, is resistance to antibiotics and biocides (48). *P. aeruginosa* infections are notoriously recalcitrant to antibiotic treatment (32). Part of the reason for this is that *P. aeruginosa* cells often exist as biofilms in infected patients. They form biofilms on indwelling medical devices, such as urinary catheters, and there is good evidence that *P. aeruginosa* cells exist as biofilms in the lungs of CF patients (33, 74, 82).

The first evidence for an involvement of c-di-GMP in *P. aeruginosa* biofilms came from the realization that several genes that had previously been shown to influence cell adhesion and colony morphology, two manifestations of biofilm, had EAL or GGDEF domains (10, 15). It was against this backdrop that studies were initiated to systematically examine the influence of intracellular c-di-GMP on the lifestyle of *P. aeruginosa*.

DGCs AND PDEs IN *P. AERUGINOSA*

The genome sequence of *P. aeruginosa* suggests that this species has a network of sophisticated c-di-GMP signaling systems that could potentially impact biofilm formation. *P. aeruginosa* strain PAO1 has genes for a total of 40 proteins that likely influence c-di-GMP metabolism (42, 70). Strain PAO1 encodes 17 proteins with a GGDEF domain predicted to have DGC activity. It also encodes five proteins with an EAL domain and two proteins with an HD-GYP domain, each predicted to have PDE activity. In addition, there are 16 predicted composite proteins that have both a GGDEF and an EAL domain. The majority of these genes are present in 19 other clinical and environmental isolates of *P. aeruginosa* as analyzed by microarray (42). Strain PA14 and a number of other strains examined are missing one GGDEF and one EAL gene, each of which is located near regions that are hypervariable in *P. aeruginosa*. PA14 encodes an additional EAL domain protein, named PvrR (15), which is not present in strain PAO1. The domain architectures of many of the proteins discussed in this chapter are shown in Fig. 1.

In *P. aeruginosa,* as in other bacteria, a high percentage of GGDEF and EAL domains are present as modules in proteins that have N-terminal domains implicated in ligand binding. These include PAS domains, which may bind flavins or hemes; GAF domains, which are known to bind phytochromes; and Chase domains, which in bacteria are thought to serve as receptors for as yet unidentified compounds (19, 88). Several of the proteins have CheY-like receiver (REC) domains. Of the 17 GGDEF proteins, 7 have predicted transmembrane domains. A higher proportion, 11 of 16, of the predicted hybrid GGDEF-EAL domain proteins are predicted to be membrane associated. Two of the five PAO1 EAL proteins have predicted transmembrane regions. Just one of the five has an extra domain, in the form of a REC domain. This high proportion of predicted membrane proteins supports that idea that GGDEF and EAL domain-containing proteins are present at the cell surface and produce or degrade c-di-GMP in response to diverse extracellular stimuli. The DGC and PDE gene inventory of *P. aeruginosa* paints a picture of an organism that likely senses and responds to a great variety of environmental stimuli to modulate intracellular c-di-GMP. One caveat to the bioinformatic identification of EAL, GGDEF, and composite proteins is that some of them lack conserved catalytic residues for c-di-GMP synthesis or degradation and may have an alternative role as a c-di-GMP- or GTP-binding effector protein. FimX from *P. aeruginosa* is an example of this (37, 54a).

ELEVATED INTRACELLULAR c-di-GMP CAUSED BY OVEREXPRESSION OF DGCs IS ASSOCIATED WITH INCREASED BIOFILM FORMATION, WHEREAS OVEREXPRESSION OF PDEs MINIMIZES BIOFILM FORMATION

Phenotypes controlled by c-di-GMP in *Pseudomonas* include biofilm formation, twitching motility, swimming reversals, and swarming (30, 40, 53, 86). Biofilm formation is complex and is assayed by several different means in the *Pseudomonas* literature (see also chapter 8). Here we describe three different assays used to visualize different but overlapping aspects of biofilms and then how they have been used to assess the effects of high and low intracellular c-di-GMP. The first and most commonly used assay is often referred to as the microtiter dish attachment assay (56). It measures attachment as well as biofilm formation and involves adding cell cultures to test tubes or to the wells of microtiter dishes and incubating them statically for various times, typically from 1 h to overnight. The planktonic and loosely adherent cells are then washed off, and remaining cells that have stayed attached to the walls of the tubes or microtiter dish wells are stained by the addition of 0.1% crystal violet. The stain is then solubilized in ethanol, and the color is measured spectrophotometrically (56). A second assay, often referred to as a continuous flow assay, involves growing bio-

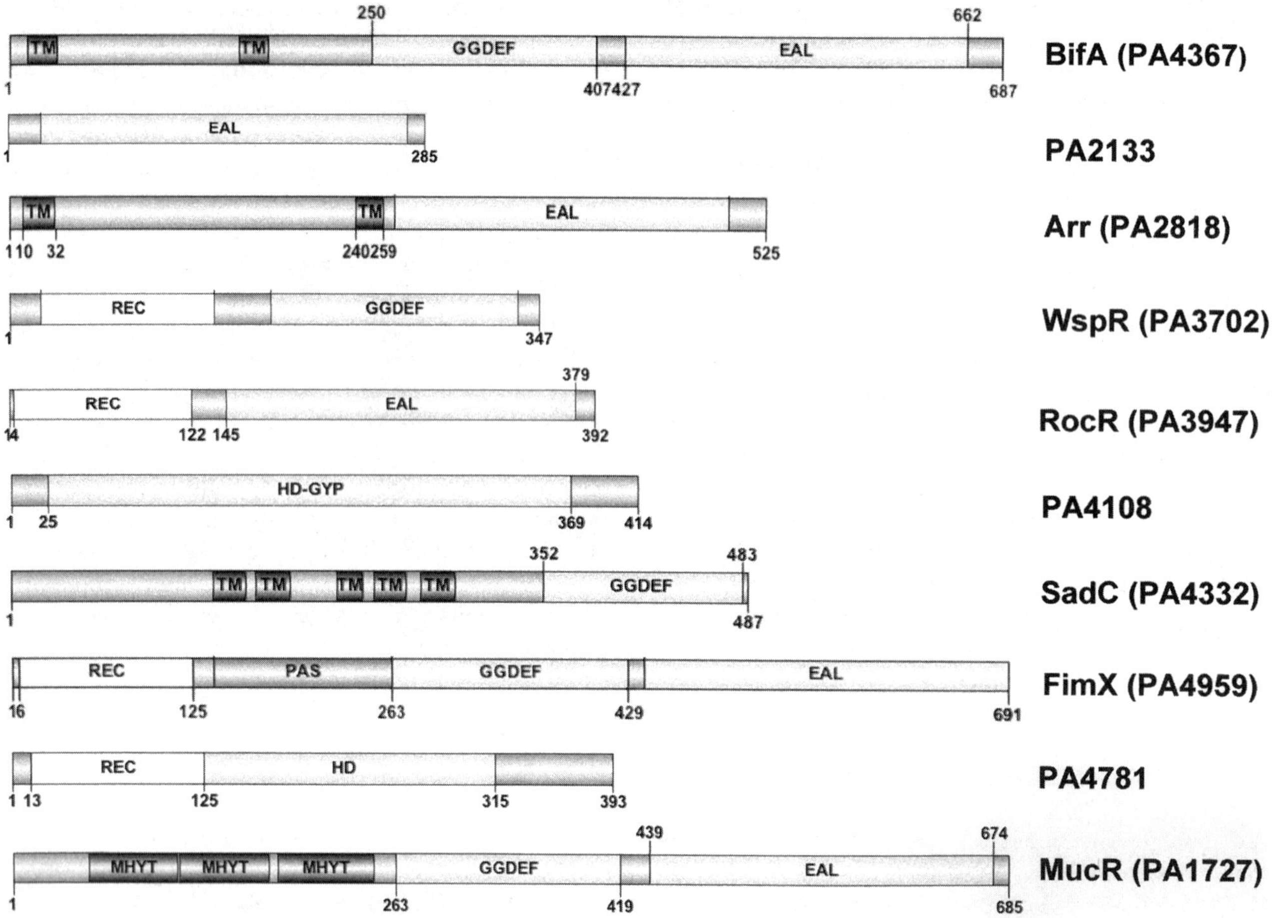

Figure 1. Domain architectures of selected DGC and PDE proteins from *P. aeruginosa*.

films on a surface of a small chamber through which growth medium is continuously flowed (28). Typically, cells inoculated into a chamber are allowed to attach to a glass coverslip that constitutes the top of the chamber for an hour or two before the flow of growth medium is started. Cells continue to attach to and grow on the coverslip and develop into biofilms over a period of days. Strains usually carry a constitutively expressed green fluorescent protein gene, and biofilm images are obtained using a confocal laser-scanning microscope. Continuous flow assays yield detailed information about biofilm thickness, biomass, and topology. This assay is generally not as sensitive, however, as the attachment assay in revealing slight differences in biofilm formation. A third measure of biofilm formation is colony morphology and dye binding. Normally, *P. aeruginosa* forms smooth colonies. However rough small colony variants (RSCVs), also known as wrinkly colonies, occur and reflect cells that form thick biofilms (38, 77). Increased absorption of the dye Congo red by colonies is also correlated with increased production of EPS (17).

Overexpression of EAL and GGDEF genes in *P. aeruginosa* has major effects on biofilm formation. Overexpression of several genes with only a GGDEF domain caused dramatic increases in biofilm formation, whereas overexpression of several genes with only an EAL domain caused a dramatic decrease in the ability of cells to form biofilms (15, 30, 42, 45, 53). Many of these proteins have DGC or PDE activities that correlate with increased or decreased biofilm (30, 42). There are mutations that stimulate the DGC activity of cognate GGDEF proteins. The best example of this is a Δ*wspF* mutation in *Pseudomonas* species. This mutation stimulates the activity of a REC-GGDEF domain protein called WspR, which results in elevated intracellular c-di-GMP (4, 10, 30). The mechanism for this will be explained below. Δ*wspF* mutants of *P. aeruginosa* strains PAO1 and PA14 form wrinkly colonies, show increased attachment to the walls of microtiter plate wells, and form thicker biofilms in flow cell chambers than the wild type (9, 30). All of these effects were reversed by overexpression of the EAL domain gene, PA2133. A Δ*wspF* strain carrying a PA2133 gene, overexpressed from a plasmid, had undetectable intracellular c-di-GMP and was unable to initiate biofilm formation in flow cells (30).

INCREASED EPS SYNTHESIS IN RESPONSE TO ELEVATED c-di-GMP IS RESPONSIBLE FOR INCREASED BIOFILM FORMATION

In many bacteria, elevated c-di-GMP levels are associated with increased EPS synthesis. Such is the case for three known EPS loci encoded by various *P. aeruginosa* strains: alginate, Pel, and Psl (72).

Alginate is an EPS that is important for the pathology of *P. aeruginosa* in the CF lung (22). *P. aeruginosa* strains isolated from the lungs of CF patients have often converted to a mucoid colony morphology due to overproduction of alginate. The most common mutations leading to mucoid conversion occur in *mucA*, which encodes an anti-sigma factor that inhibits the activity of AlgU, an alternative RNA polymerase sigma factor that activates the expression of alginate biosynthesis genes (6, 49, 72). Such mutations seem to be selected for in the CF lung environment. Evidence that c-di-GMP levels modulate alginate production was first presented by Merighi et al. who showed that overexpression of the DGC PA1107 was associated with an increase in alginate production by a mucoid *mucA* mutant of strain PA14, whereas overexpression of the PDE PA2133 resulted in decreased alginate (52).

Two *P. aeruginosa* strains that are commonly used for laboratory biofilm studies, PAO1 and PA14, will become mucoid and produce alginate if they have *mucA* mutations (52, 69). However, wild-type PAO1 and PA14 strains do not express or contain detectable alginate when grown in the laboratory (87). Instead, *P. aeruginosa* PAO1 relies on *pel* (PA3058-3064) and *psl* (PA2231-2245) operons to direct the synthesis of glucose- and mannose-rich EPS, respectively (16, 17, 34, 50, 81). Wild-type strain PA14 carries a deletion mutation in its *psl* locus and thus synthesizes only Pel EPS (17). In contrast to a Δ*wspF* strain of PAO1, which forms wrinkly colonies and has elevated levels of c-di-GMP, a Δ*psl* Δ*pel* Δ*wspF* mutant forms smooth colonies and does not attach in a microtiter dish assay, even though its intracellular c-di-GMP levels are high (Fig. 2). Either Pel or Psl alone is sufficient to support biofilm formation by strain PAO1; both operons must be deleted to abolish biofilm formation. In *P. aeruginosa*

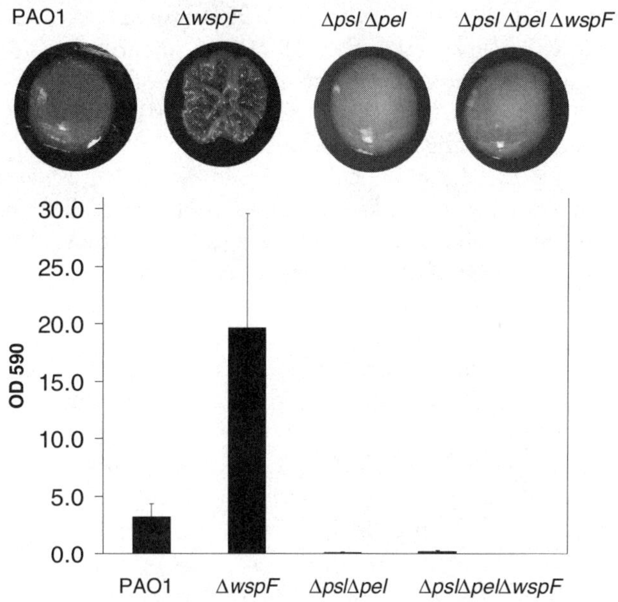

Figure 2. Loss of EPS production reverses the colony and attachment phenotypes of a *wspF* mutant. EPS is encoded by *psl* and *pel* genes in *P. aeruginosa*. (Top) Colony morphologies. (Bottom) Attachment assays were carried out in microtiter dish wells (J. W. Hickman and C. S. Harwood, unpublished data). OD, optical density.

strain PA14, which forms only Pel EPS, a mutation in the PDE BifA results in increased biofilm formation and wrinkly colonies, but Δ*bifA pelA* colonies are no longer wrinkly (40). The hyperadherent and wrinkly phenotype of a Δ*wspF* mutant of PA14 disappears when *pel* gene mutations are introduced (9).

These results showing that Psl, Pel, and alginate are responsible for the major biofilm-related effects of c-di-GMP in *P. aeruginosa* are important because they tell us that these EPS synthesis complexes and the genes and mRNA transcripts that encode them are likely to be major targets of c-di-GMP action in cells.

DGC AND PDE MUTANTS TEND TO BE DEFECTIVE IN SOME, BUT NOT ALL, c-di-GMP-ASSOCIATED TRAITS

The relatively large number of proteins that are potentially involved in c-di-GMP metabolism suggests that there is redundancy of function and that many proteins may contribute to a total intracellular c-di-GMP pool. Consistent with this, many *P. aeruginosa* DGC and PDE mutants do not have obvious mutant phenotypes (42). However, a few mutants do have phenotypes, and interestingly, all such mutants are affected in some, but not all, c-di-GMP-associated traits. For example, *sadC* and *bifA* mutants, described below, have several c-di-GMP-related phenotypes, including alterations in swarming and biofilm formation (40, 53). However, the c-di-GMP-associated trait of twitching motility was unaffected in these mutants. A *sadC* mutant has increased swimming reversal frequency, but a *bifA* mutant is not altered in this trait. Similarly, *wspR* mutants are defective in biofilm formation (40, 42) but have normal twitching and swimming motility (E. Lim, V. Huangyutitham, and C. S. Harwood, unpublished data). The WspR (PA3702) protein forms subcellular clusters that are distributed throughout the cytoplasm but are not generally found at cell poles (23). These data suggest that at least some DGCs and PDEs modulate subcellular pools of c-di-GMP that could have localized effects at specific sites in cells.

P. AERUGINOSA GGDEF AND EAL DOMAIN PROTEINS THAT INFLUENCE BIOFILM FORMATION

A systematic analysis of a *P. aeruginosa* transposon mutant library by the Lory laboratory and other more directed studies carried out by other investigators have identified just 8 of the 40 DGC and PDE genes present in strain PAO1 as influencing biofilm formation. In this section, we describe each of these genes and studies that have addressed their functions.

Arr (PA2818)

In one study, subinhibitory concentrations of tobramycin and several other aminoglycoside antibiotics stimulated biofilm formation in laboratory and clinical strains of *P. aeruginosa* in attachment assays, whereas other antibiotics had little effect (31). A subsequent screen of transposon mutants revealed that PA2818, which encodes a predicted membrane-associated PDE, was responsible for tobramycin-induced biofilm formation. This gene was named *arr* for aminoglycoside regulator. Arr is found in strain PAO1 but not in 16 of 19 other *P. aeruginosa* strains that were examined, including strain PA14 (42). Membranes of *arr* mutant cells had about half the c-di-GMP PDE activity of wild-type cell membranes. The authors suggest that tobramycin enhances Arr activity, and this results in decreased c-di-GMP and enhanced biofilm formation (31). This is inconsistent with many other reports in which increased PDE activity is associated with decreased intracellular c-di-GMP and decreased biofilm formation and suggests that this interesting gene requires further study.

SadC (PA4332)

SadC (surface attachment defective) is a membrane-bound DGC that has a C-terminal GGDEF domain and five transmembrane regions in the N-terminal region of the protein (53). *sadC* mutants were initially identified in a screen for strain PA14 mutants defective in biofilm formation (56). Subsequently, Merritt et al. (53) found that a *sadC* mutant has a hyperswarming phenotype and that mutant cells reverse swimming direction more frequently than wild-type cells. These authors suggested that these characteristics may explain why a *sadC* mutant is defective in the transition from reversible to irreversible attachment to surfaces, a characteristic of early biofilm formation. A *sadC* mutant absorbed less Congo red dye than wild-type cells, suggesting that it is also defective in the synthesis of Pel EPS. It turns out that SadC-mediated stimulation of early biofilm formation depends on SadB (PA5346), suggesting that the SadB protein is part of a signal transduction pathway from SadC to SadB to the Pel biosynthetic machinery. Reverse transcription assays indicated that the effects of *sadC* on Pel EPS production are not at the level of gene expression.

BifA (PA4367)

BifA (biofilm formation) is a hybrid protein with a GGDEF domain (the core motif is altered and is GGDQF) and an active EAL domain. The *bifA* mutation was identified in a screen of a *P. aeruginosa* strain PA14 transposon mutant library for strains with altered biofilm formation. *bifA* mutants are hyperbiofilm formers that have increased Congo red binding and decreased swarming motility, phenotypes that are the opposite of those of a *sadC* mutant. BifA has PDE activity in vitro as well as in vivo but no detectable DGC activity (40). However, the full-length gene is required to complement the *bifA* mutant phenotype, indicating that both domains are required for the proper function of the protein. BifA has two predicted transmembrane regions and appears to localize to the inner membrane, but membrane localization does not seem to be essential for function. The hyperbiofilm/Congo red phenotype was abolished in a *bifA pel* mutant, indicating that BifA PDE activity specifically affects Pel EPS synthesis. Recall that strain PA14, the background in which the *bifA* mutation was identified, does not synthesize Psl EPS. Further experiments indicated that EPS stimulation in a *bifA* mutant was not at the level of Pel transcription. A *bifA* mutant is defective in swarming, even in a *pel* background, indicating that Pel EPS has no effect on this phenotype. Mutational analysis suggests that SadB also contributes to increased Pel EPS synthesis in the *bifA* mutant. Kuchma and coworkers and Wolfe and Visick suggest that SadB modulates Pel EPS production in response to a possibly localized pool of c-di-GMP that is generated by the opposing activities of BifA and SadC in response to unknown environmental signals (40, 86).

WspR (PA3702)

Wsp (wrinkly spreader phenotype) mutants were initially discovered in *Pseudomonas fluorescens* and subsequently in a transposon screen for small wrinkly colonies of *P. aeruginosa* (10, 60). WspR is a hybrid REC-GGDEF domain protein that is encoded in an operon with genes that encode a chemotaxis-like signal transduction complex (Fig. 3). *wspR* mutants are defective in biofilm formation (40, 42). c-di-GMP produced by WspR appears to have specific effects on Pel EPS synthesis because a *wspR* Δ*psl* mutant of PAO1 is defective in attachment to surfaces relative to a Δ*psl* mutant but a *wspR pelA* mutant has an attachment phenotype similar to that of a *pelA* mutant (E. Lim, J. W. Hickman, and C. S. Harwood, unpublished data).

In comparing the Wsp system to the well-studied *Escherichia coli* chemotaxis signaling complex, there is a membrane-bound methyl-accepting chemotaxis protein-like receptor (WspA), CheW-like adaptor proteins (WspB and WspD), a CheA/Y hybrid histidine sensor kinase with a receiver domain (WspE), a methyltransferase CheR homolog (WspC), a methylesterase CheB homolog (WspF), and WspR, which catalyzes the synthesis of c-di-GMP (Fig. 3). By analogy with chemotaxis, one would predict that WspA senses an environmental stimulus and then rapidly signals WspE to autophosphorylate and subsequently alter the phosphorylation status of WspR by phosphotransfer. WpsE likely also controls the phosphorylation status of a second response regulator receiver protein, WspF, with slower kinetics. WspF-P is predicted to be active as a methylesterase, which would modify the methylation state of glutamate residues in WspA in conjunction with a constitutively active WspC methyltransferase. Changes in the methylation state of WspA are predicted to reset its signaling activity, thereby allowing the Wsp system to adapt to the original input stimulus. By analogy with chemotaxis, a *wspF* mutation likely has the effect of locking the Wsp chemosensory system into a configuration where it is constantly signaling, so that the output protein, WspR, is constantly phosphorylated and constantly producing c-di-GMP. Consistent with this model, a *P. aeruginosa wspF* mutant, as mentioned above, produces elevated levels of intracellular c-di-GMP, and disruption of the Wsp system by mutation of *wspA* or *wspR* causes the *wspF* wrinkly colony phenotype to convert to smooth (23). However, virtually all other aspects of this model remain to be tested.

A recent study analyzed the subcellular localization of Wsp proteins as an avenue to expand our knowledge of the behavior of a GGDEF protein and accompanying c-di-GMP production in individual cells (23). A yellow fluorescent protein (YFP)-tagged version of WspA, a predicted membrane-bound protein, forms multiple lateral patches at the peripheries of cells. This pattern of localization differs from that of methyl-accepting chemotaxis proteins involved in chemotaxis, which localize mainly at cell poles in a large patch (24, 47, 76).

A YFP-tagged version of WspR is dispersed in the cytoplasm of liquid-grown wild-type cells rather than present at a particular subcellular location. By contrast, WspR-YFP localizes in clusters visualized as bright spots in liquid-grown Δ*wspF* mutant cells, a condition where WspR is probably continuously phosphorylated (Fig. 4). A mutant *wspR* allele carrying an Asn change at Asp70, the predicted site of phosphorylation, was dispersed in Δ*wspF* mutant cells. Also, WspR-YFP clusters did not form in Δ*wspF* Δ*wspA* mutant cells. This indicates that WspA is essential for WspR-YFP clustering, probably through WspA-dependent phosphorylation. Interestingly, a mutant form of WspR (E253A) that does not produce c-di-GMP does form clusters. That WspR forms visible clusters in cells when it is phosphorylated suggests that the subcellular localization characteristics of WspR-YFP can be used to identify conditions that stimulate Wsp signaling. In fact, it was found that although WspR-YFP was dispersed in wild-type cells growing in LB broth, it formed fluorescent spots in cells grown to confluence on the surface of LB agar for 20 hours (23) (Fig. 5). Thus, it appears that growth on an agar surface stimulates Wsp signal transduction. This supports the notion that a condition associated with cell-to-cell contact or cell-surface interactions stimulates *P. aeruginosa* to form biofilms.

The Sondermann group solved crystal structures of *P. aeruginosa* WspR with c-di-GMP bound at an inhibitory site (the I-site) (13a, 14) (Fig. 6). Their studies suggest that the active form of WspR is a tetramer. In each monomer, a CheY homology domain is connected to a GGDEF domain by a coil-coil-like linker region. The GGDEF motifs (GGEEF in WspR) in a dimer are oriented so that the two active sites are close to each other. A straightforward model for DGC activity based on studies of PleD from *Caulobacter crescentus* (59, 83) and WspR (14) is that di-

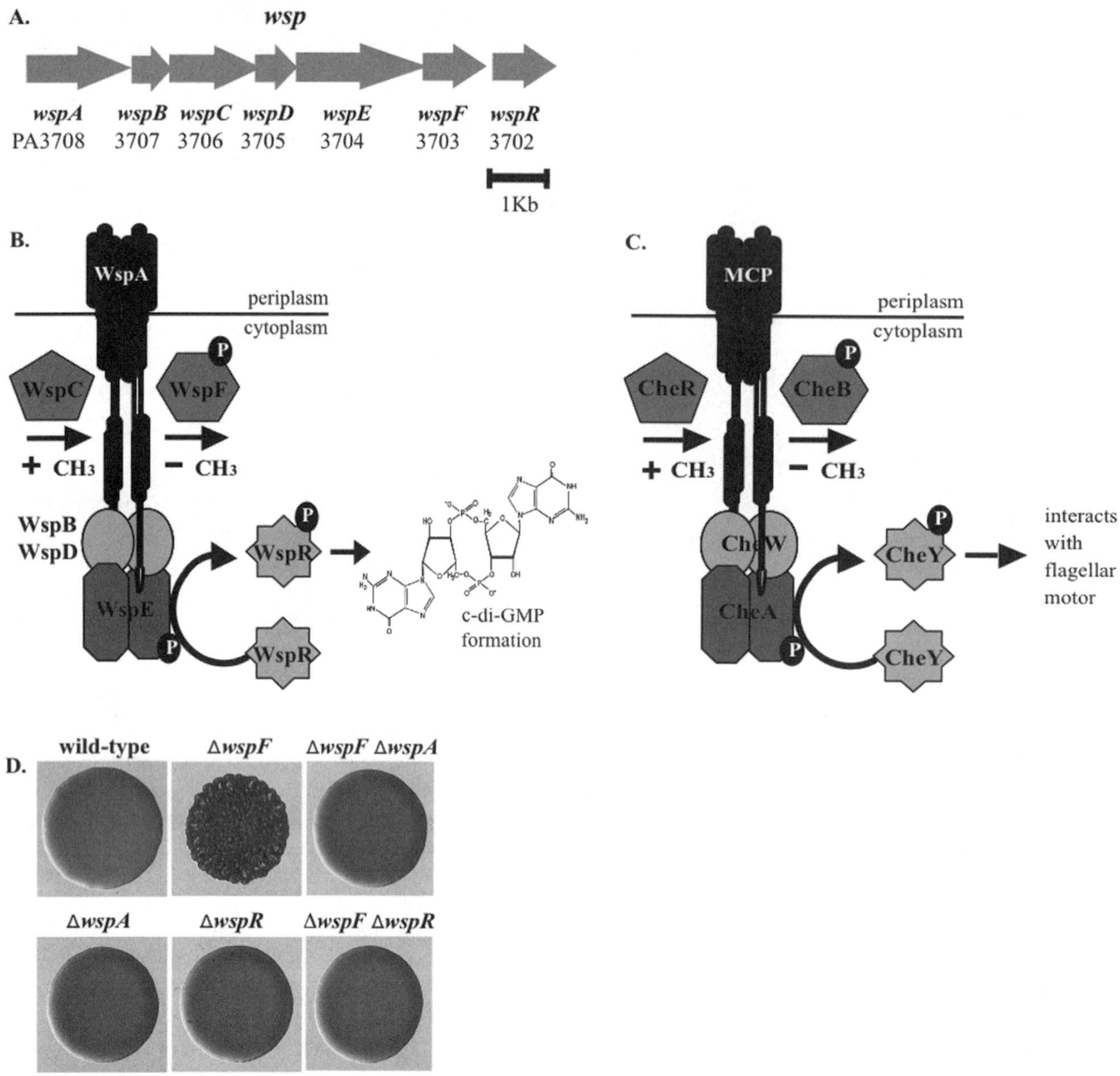

Figure 3. Characteristics of the Wsp signal transduction system. (A) Organization of the genes encoding the Wsp system. (B) The Wsp proteins are predicted to be a membrane-bound methyl-accepting chemotaxis protein (WspA), a CheR-like methyltransferase (WspC), a CheB-like methylesterase (WspF), and two CheW homologues (WspB and WspD) that are predicted to serve as adapters between WspA and a hybrid histidine kinase response regulator (WspE). The response regulator protein, WspR, has a GGDEF domain and catalyzes the synthesis of c-di-GMP when phosphorylated. As described in the text, a *wspF* mutation is predicted to lock the Wsp system into a configuration where WspR is constantly phosphorylated and thus constantly producing c-di-GMP. (C) Physical organization of homologous Che proteins. (D) Colony morphologies of *P. aeruginosa* PAO1 wild-type and *wsp* deletion strains. *wspF* mutants have high levels of intracellular c-di-GMP relative to wild-type cells. Reprinted from Güvener and Harwood (23) with permission.

merization brings two GGDEF domains, each loaded with a GTP, together in an antiparallel arrangement such that the GTP 3′ hydroxyl groups deprotonate and undergo intermolecular nucleophilic attack to generate c-di-GMP (see chapter 7 for an in-depth description of crystal structures).

De et al. proposed a novel mechanism for modulation of WspR activity by product inhibition by c-di-GMP, where the stalks of adjacent molecules in nucleotide-bound tetramers block the active sites (13a, 14). They also present evidence that WspR can be released from its product inhibited state by PDEs,

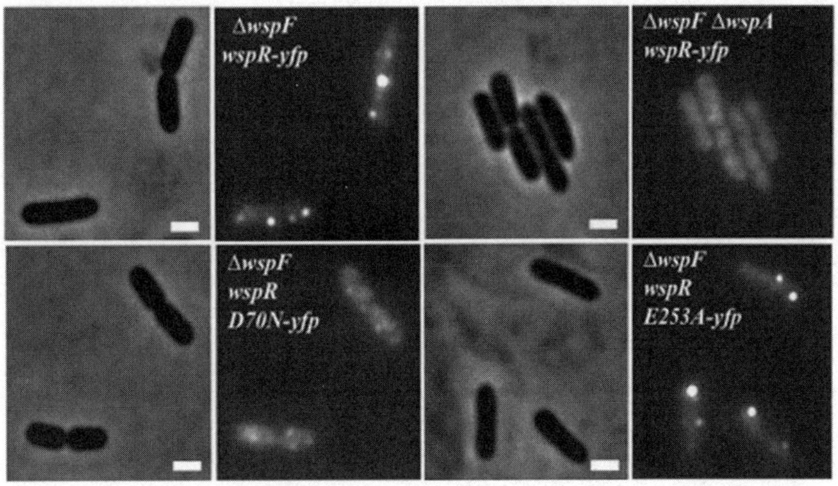

Figure 4. Phosphorylation is required for WspR cluster formation. The localization of WspR-YFP in Δ*wspF wspR-yfp* or Δ*wspF* Δ*wspA wspR-yfp* mutant cells are shown in the top panels, and the localization of two WspR-YFP mutant proteins in Δ*wspF* cells are shown in the bottom panels. Phase-contrast images of cells are shown on the left. Bar, 1 μm. Reprinted from Güvener and Harwood (23) with permission.

suggesting that this could be an important mechanism by which WspR activity is regulated in vivo.

In recent work the Sondermann group not only demonstrated that a nucleotide-free WspR tetramer is the active form of the enzyme, but also elucidated a role for phosphorylation in facilitating tetramer formation (13a).

MucR (PA1727)

The conversion of nonmucoid to mucoid variants that overproduce the EPS alginate, an unbranched polymer of mannuronic and guluronic acids (67), seems to be important for persistence of *P. aeruginosa* in the CF lung (22, 63). Typically, in the CF lung, mutations in the negative regulatory gene, *mucA*, led to overproduction of alginate (72) (Fig. 7). Although most studies of effects of c-di-GMP on EPS production have centered on Pel and Psl EPS, c-di-GMP is also important for alginate production (26, 52). MucR is a hybrid GGDEF-EAL protein that has MHYT domains that consist of six predicted transmembrane segments, connected by short arginine-rich cytoplasmic and periplasmic loops rich in charged residues. The MHYT domain has been proposed to serve as a sensor domain in bacterial signaling proteins (18). MucR has DGC activity in vitro, and a *mucR* deletion mutant of the *P. aeruginosa*

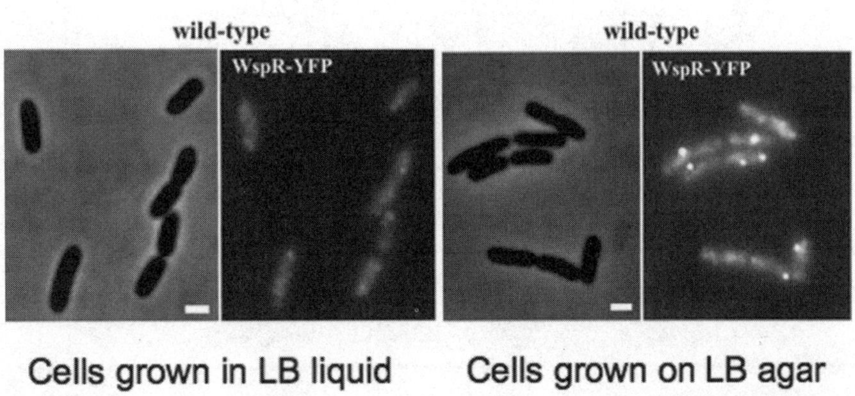

Figure 5. Surface growth stimulates Wsp signal transduction as assayed by cluster formation of WspR-YFP. WspR-YFP localization in wild-type cells grown in liquid or on agar is shown. The phase-contrast (left) and fluorescence (right) images of cells expressing *wspR-yfp* are shown. Bar, 1 μm. Reprinted from Güvener and Harwood (23) with permission.

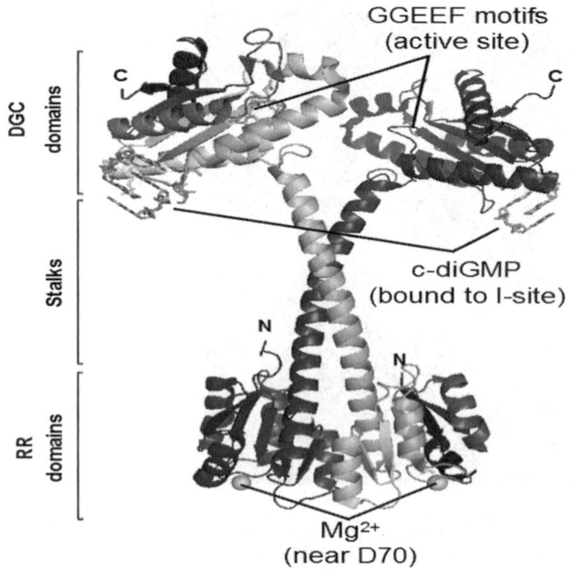

Figure 6. Structure of WspR in its active conformation. The site of c-di-GMP binding that leads to an inhibited conformation is also shown. From De et al. (14).

alginate-overproducing strain PDO300 is nonmucoid (26). This strain is also defective in alginate production, as assayed by uronic acid content of purified extracellular material. A *mucR* mutant has a wild-type attachment phenotype in *P. aeruginosa* PA14, a strain that does not produce alginate (42). These data are consistent with a model whereby MucR controls a localized pool of c-di-GMP that binds to Alg44, a c-di-GMP binding protein (52, 66) (see below), to activate alginate biosynthesis.

RocR (PA3947)

P. aeruginosa has *cupA, cupB,* and *cupC* gene clusters that encode fimbrae that are involved in cell adhesion (80). Overexpression of any one of the three clusters results in increased attachment of a strain with deletions of chromosomal copies of all *cup* genes. Overexpression strains also form thicker pellicles. The *cup* genes are not very well expressed

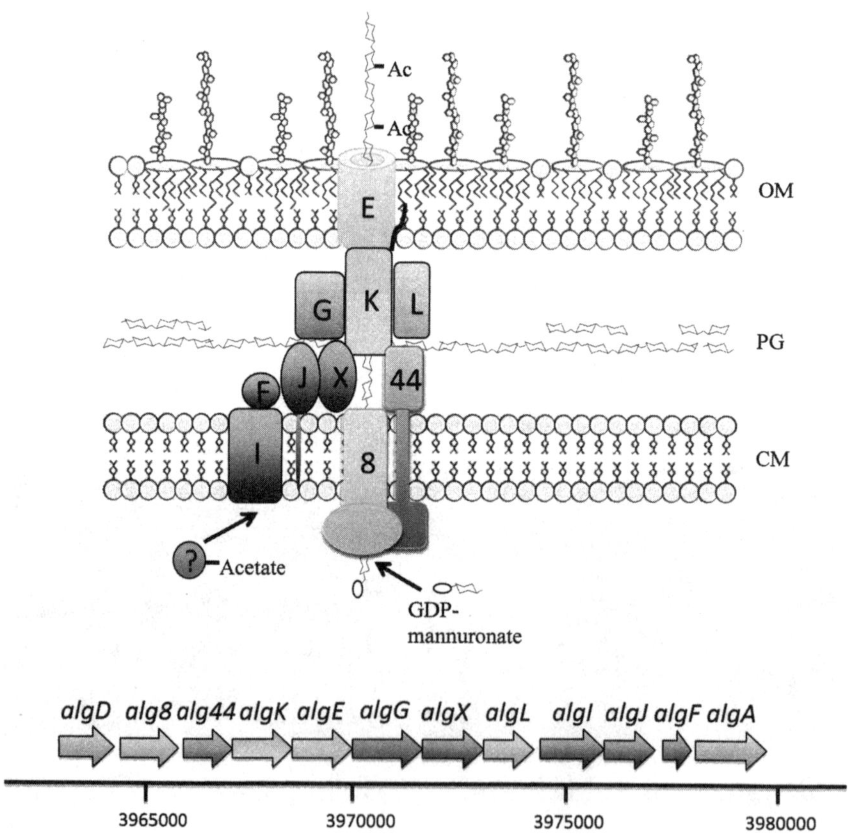

Figure 7. Predicted organization of the alginate biosynthesis complex. OM, outer membrane; PG, peptidoglycan; CM, cell membrane. Adapted from Ramsey and Wozniak (63) by J. Weadge, J. Whitney, C.-K. Keiski, A. N. Neculai, and P. L. Howell.

under standard laboratory conditions. A screen for transposon mutants with increased expression of the *cup* loci led to the identification of the *rocSAR* locus encoded by PA3946-3948 (43). RocS1 is a sensor histidine kinase with multiple domains; RocA1 has REC and helix-turn-helix DNA binding domains and is involved in activating *cupC* expression in response to activated RocS1. RocR is a REC-EAL domain protein that was proposed to antagonize the effects of RocA1 and repress *cupC* expression by an unknown mechanism that might involve competition for phosphotransfer from RocS1 (43). The authors of this study speculated that RocR, which has an EVL sequence rather than an EAL sequence, might act in a c-di-GMP-independent manner. At about the same time, Kuchma et al. described the same set of genes in *P. aeruginosa* strain PA14 and named them *sadARS* (41). *sadARS* mutants have wild-type motility and attachment phenotypes but exhibit slight alterations in mature biofilm architecture. Microarray studies suggested that the *sadARS* system regulates the expression of type III secretion genes. The authors have not yet investigated how the SadR EAL domain may affect biofilm formation or type III secretion.

Subsequent studies showed that RocR is indeed an active c-di-GMP PDE. Most recently, RocR has been used as a model to investigate the catalytic mechanism of a PDE (64, 65). The protein functions as a tetramer and hydrolyzes c-di-GMP to produce 5′-pGpG. It has no activity with cAMP or cGMP. Its K_m of 3.2 ± 03 μM is close to the estimated intracellular concentration of c-di-GMP in *P. aeruginosa* as well as in other bacteria. A bound Mg^{2+} ion assists a general base-catalyzed hydrolysis of c-di-GMP. Studies of RocR in conjunction with structural studies of two EAL protein domains led to the identification of amino acid residues and a subdomain that are essential for the dimerization of EAL proteins as well as for binding c-di-GMP and Mg^{2+} ion (54, 64, 65) (see chapter 7 for a discussion of crystal structures).

PA4108 and PA4781

PA4108 and PA4781 are HD-GYP domain proteins that caused a reduction in intracellular c-di-GMP when expressed in *Xanthomonas campestris,* suggesting that they have c-di-GMP PDE activity (70). Consistent with this, PA4108 and PA4781 mutants have elevated intracellular c-di-GMP and reduced swarming and twitching motility. These mutants were less virulent to the larvae of the caterpillar *Galleria mellonella* than the wild-type parent. Mutant cells formed biofilms that had altered architecture and were generally flatter than the wild type. This observation differs from other observations showing a correlation between elevated c-di-GMP and the formation of thicker biofilms. PA4781 has a REC domain at its N terminus but is not located next to a sensor kinase in the genome.

c-di-GMP RECEPTOR PROTEINS: REGULATORY MECHANISMS THAT STIMULATE BIOFILM FORMATION IN *P. AERUGINOSA*

Three mechanisms for c-di-GMP-mediated regulation of biofilm initiation and development have been put forth: (i) posttranslational (allosteric) control, (ii) posttranscriptional control, and (iii) transcriptional control. There is evidence for each of these mechanisms in bacteria (27). So far, only the first and the third mechanisms have been described for *P. aeruginosa*.

POSTTRANSLATIONAL CONTROL OF EPS SYNTHESIS BY c-di-GMP

The first evidence for c-di-GMP-mediated effects on EPS production came from studies of cellulose synthesis in *Gluconacetobacter xylinus*. These studies indicated that c-di-GMP binds to the BcsA1 component of cellulose synthase to allosterically activate cellulose synthase (84). BcsA contains a conserved sequence subsequently named PilZ, which was proposed to be a c-di-GMP binding domain (3). *P. aeruginosa* has eight PilZ domain proteins, all of which bind c-di-GMP (45, 52, 62). Most are conserved hypothetical proteins of unknown function, but one, PilZ itself (PA2960), is required for normal pilus assembly (1). Another PilZ domain protein with a known function is the membrane-associated protein Alg44 (PA3542) (52, 66) (Fig. 7). *P. aeruginosa* alginate-producing strains with mutations in the PilZ domain of Alg44 fail to bind c-di-GMP and also fail to produce alginate (52). Alg44 binds c-di-GMP specifically with a K_d (dissociation constant) in the 10-micromolar range. GTP and GMP are able to outcompete c-di-GMP binding at Alg44 but only at millimolar concentrations. Nucleotide binding and alginate production depend on several conserved residues in the PilZ domain. It was suggested that Alg44 may respond to c-di-GMP produced by the DGC MucR, as *mucR* mutants were defective in alginate production but not in other c-di-GMP-associated phenotypes (26). It is likely that upon binding c-di-GMP, Alg44 has a regulatory function in controlling the polymerization or transport of alginate (55) (Fig. 7).

Early on, it was recognized that c-di-GMP receptor proteins in addition to PilZ domain-containing proteins must exist because not all bacterial species that encode DGCs and PDEs have PilZ genes. In fact, a systematic analysis of the c-di-GMP binding properties of recombinant *P. aeruginosa* Pel protein components revealed that the PelD protein binds c-di-GMP (45) through a domain that resembles the inhibitory I-site in the GGDEF domain of DGCs. Several PelD residues were identified as essential for c-di-GMP binding and they were also essential in vivo for Pel EPS synthesis. The Pel biosynthesis machinery consists of seven proteins that are predicted to be located in the cytosol (PelF, a predicted glycosyltransferase), the inner membrane (PelD, PelG, and PelE), the periplasm (PelA), and the outer membrane (PelB and PelC). The predicted locations of these proteins suggest that they are involved in the transport of nascent glycan chains to the cell surface (P. L. Howell, personal communication) (81) (Fig. 8). The function of PelD and if and how it physically interacts with other Pel proteins are not known. Although it is a predicted inner membrane protein, it has a large cytoplasmic domain that includes the region that binds c-di-GMP (81). At this writing, there is no evidence that any of the components of the Psl biosynthetic machinery bind c-di-GMP.

The identification of c-di-GMP receptor proteins is an important step toward uncovering the mechanism of c-di-GMP action in controlling the activities of large protein complexes like cellulose syntheses, alginate syntheses, and Pel synthase. However, it is not known how c-di-GMP binding proteins participate in EPS synthesis in any bacterium or whether there is a common mechanism by which c-di-GMP controls EPS synthases.

c-di-GMP-MEDIATED REGULATION OF GENE EXPRESSION

When the GGDEF protein WspR is constitutively active, it causes a dramatic increase in intracellular c-di-GMP, which is associated with increased expression of genes involved in biofilm formation and decreased flagellar gene expression (30, 77). In another study, high-level expression of three different DGCs led to increases in EPS gene expression, pre-

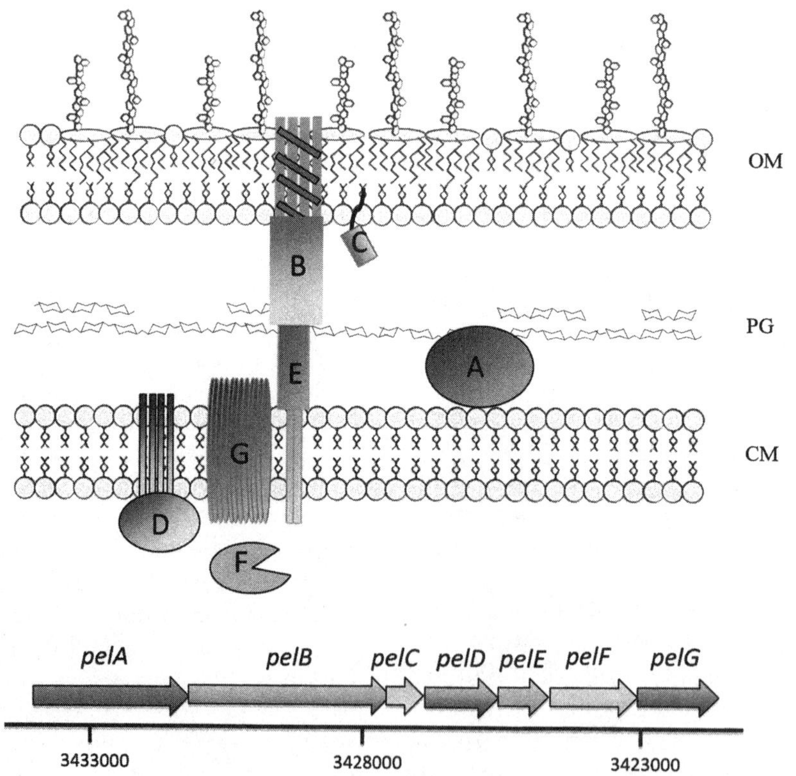

Figure 8. Predicted organization of the Pel biosynthesis complex. OM, outer membrane; PG, peptidoglycan; CM, cell membrane. Courtesy of J. Weadge, J. Whitney, C.-K. Keiski, A. N. Neculai, and P. L. Howell.

sumably as a consequence of elevated intracellular c-di-GMP (45). Gene expression might be controlled at the level of transcription by a transcriptional regulator or at the posttranscriptional level.

c-di-GMP Does Not Appear To Control Gene Expression Posttranscriptionally

A mechanism for posttranscriptional control of gene expression in *P. aeruginosa* is that the RNA binding protein RsmA binds to and represses the translation of certain mRNAs that have an appropriate secondary structure and conserved GGA trinucleotide RsmA binding site in the 5′ leader region. The repressive activity of RsmA on translation can be counteracted by two small untranslated RNAs, RsmZ and RsmY, which can bind to RsmA and sequester it from its target mRNAs (7, 44). Microarray data indicate that *psl* and *pel* gene expression is higher in an *rsmA* mutant, as is the expression of the GGDEF gene, PA0338 (7). In *E. coli,* the RNA binding protein CsrA, a homologue of RsmA, influences c-di-GMP metabolism by regulating the expression of several GGDEF-EAL genes (36). However, there is currently no evidence that c-di-GMP is directly involved in posttranscriptional regulation by affecting the activities or levels of RsmA or untranslated RNAs in *P. aeruginosa* (see chapter 14 for a more complete description of the relationship between RNA binding proteins and c-di-GMP metabolism).

Transcriptional Control of Gene Expression by c-di-GMP

A recent report showed that c-di-GMP derepresses gene expression by binding to the transcriptional regulator FleQ (29). FleQ is a well-documented master activator of flagellar biosynthesis and chemotaxis gene expression in *P. aeruginosa* (12, 13). It is an enhancer-binding protein that has an N-terminal FleQ domain, an AAA-σ^{54} interaction domain, and a helix-turn-helix DNA binding domain (Fig. 9). In addition to its role in regulating flagellar gene expression, FleQ represses expression of the *psl* and *pel* operons for EPS synthesis as well as the transcription of several other genes that are expressed in response to high levels of c-di-GMP (13, 29). Purified FleQ binds poorly to *pelA* promoter DNA in electrophoretic mobility shift assays, but the further addition of a second protein, FleN, and 10 μM ATP or ADP caused a supershift of *pelA* promoter DNA. Importantly, this supershift was abrogated by the addition of c-di-GMP (29). FleN is an antiactivator of FleQ-mediated flagellar gene expression, and apparently, it

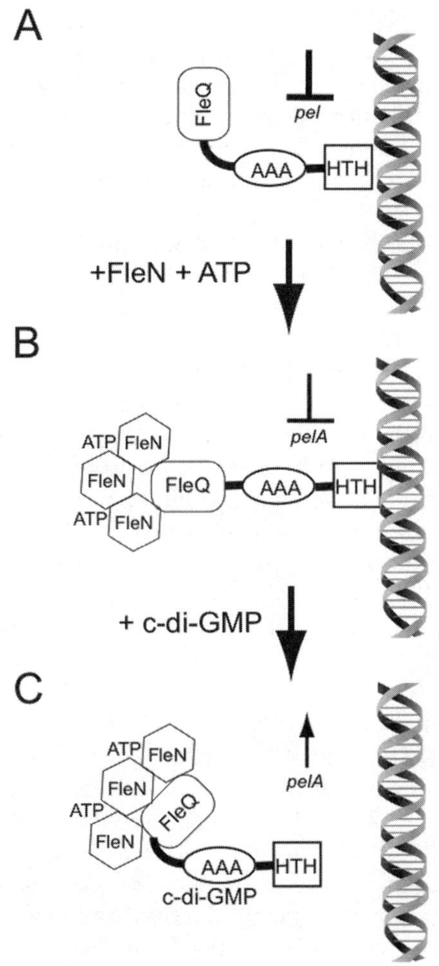

Figure 9. Model for the regulation of gene expression by FleQ, FleN, and c-di-GMP. (A) FleQ in the absence of FleN or c-di-GMP maximally represses *pel* transcription. (B) Situation in wild-type cells. FleQ binding at the *pelA* promoter is reduced by FleN and ATP/ADP, resulting in less *pel* repression than the situation in panel A. (C) c-di-GMP binds to FleQ to cause it to dissociate from DNA, thereby causing derepression of transcription from the *pel* promoter. Reprinted from Hickman and Harwood (29) with permission.

also interacts with FleQ to modulate EPS gene expression (11, 29). These results and genetic data show that FleQ, FleN, and c-di-GMP function together to relieve transcriptional repression of EPS genes necessary for biofilm formation (Fig. 9). Interestingly, in contrast to the situation with the full-length protein, a truncated version of FleQ that lacks its N-terminal domain binds quite well to *pelA* promoter DNA on its own without FleN, and its ability to bind to the *pel* promoter is inhibited by c-di-GMP (29). The data suggest that in the absence of FleN or ATP, FleQ maximally represses *pelA* expression. When FleN and ATP/ADP are present, they partially inhibit the ability of FleQ to repress transcription of *pelA*. c-di-GMP binding to FleQ/FleN apparently lowers the affinity of the complex for DNA sufficiently to allow for derepression of transcription. c-di-GMP did not abrogate binding of FleQ/FleN to the *fleSR* promoter, indicating that c-di-GMP might not regulate flagellar gene expression through FleQ (29). It may be, as suggested by Wolfe and Visick (86), that the small effects that c-di-GMP has on flagellar gene expression are indirect. Also, whereas the RNA polymerase sigma factor σ^{54} is required for activation of FleQ-mediated expression of flagellar genes, repression of EPS gene expression by FleQ does not depend on this sigma factor (29). Although FleQ binds radiolabeled c-di-GMP, the precise c-di-GMP binding site has not been defined. However, it is likely to be a new type of binding domain because FleQ lacks a predicted secondary structure that resembles known c-di-GMP binding domains found in DGCs, PilZ proteins, or other c-di-GMP receptor proteins (M. Galperin, personal communication).

FleQ homologues regulate flagellar gene expression in other polarly flagellated gammaproteobacteria including other *Pseudomonas, Vibrio,* and *Legionella* species (51). Some of these proteins may also regulate gene expression in response to c-di-GMP and share the same c-di-GMP binding properties as *P. aeruginosa* FleQ. In support of this idea, *P. fluorescens* FleQ negatively regulates expression of the *wss* operon necessary for cellulose production, and expression of these genes is also activated by high levels of c-di-GMP (20).

GLOBAL VERSUS LOCALIZED EFFECTS OF c-di-GMP

That individual EAL or GGDEF gene mutants tend to be defective in only a subset of c-di-GMP-related traits supports the concept that individual proteins control compartmentalized pools of c-di-GMP that have localized subcellular effects in *P. aeruginosa*. On the other hand, the observation that *P. aeruginosa* responds to elevated intracellular c-di-GMP to modulate gene expression would seem to support the idea that proteins with GGDEF and EAL domains integrate a variety of environmental input signals to regulate the total intracellular concentration of c-di-GMP, which in turn has global effects. It is possible that c-di-GMP has both global and localized subcellular effects on cell physiology. However, we do not know if in normal physiological situations a transcriptional regulator like FleQ responds to the total intracellular pool of c-di-GMP or whether it responds to a localized high concentration of c-di-GMP that may be generated by a particular DGC that specifically interacts with FleQ.

CLINICAL RELEVANCE OF c-di-GMP IN CHRONIC INFECTIONS

During the course of the chronic infections in the CF lung that can persist for decades, *P. aeruginosa* undergoes phenotypic and genetic diversification. One manifestation of this is the appearance of mucoid colonies. In addition, *P. aeruginosa* tends to generate RSCVs whose appearance is often correlated with a progression of lung disease (25). One might expect RSCVs to cause chronic infections that are more damaging than those caused by wild-type cells because such cells are very adherent and form thicker biofilms than cells with smooth colony morphologies. RSCVs also arise in laboratory-grown biofilms (38). That the RSCV phenotype is due to elevated c-di-GMP is supported by direct measurements of this dinucleotide in cells and by the finding that overexpression of the EAL protein, PA2133, in such strains reversed the wrinkly colony phenotype to smooth in every case tested (38, 77). One route to the appearance of RSCVs is *wspF* mutations. In one study, 6 of 91 longitudinally collected *P. aeruginosa* isolates from 29 different CF patients had an RSCV morphology and a *wspF* mutation (75). Clinical RSCVs have also been described that do not have *wspF* mutations and that are not complemented by a *wspF*-bearing plasmid (77). To investigate the degree to which *wspF* mutations are responsible for the RSCV phenotype, 50 laboratory biofilm-derived RSCVs were provided with a *wspF* gene in *trans*. This gene complemented 34 (68%) of the isolates as assayed by the formation of smooth colonies. The rest of the RSCVs apparently had mutations in other genes that lead to elevated intracellular c-di-GMP and increased EPS synthesis (77). The nature of other mutations that can give rise to RSCVs is not known.

That high levels of c-di-GMP stimulate biofilm formation, a trait associated with chronic virulence in *P. aeruginosa,* suggests that the development of therapeutic agents that lower c-di-GMP levels in cells might be a promising avenue for preventing or treating chronic biofilm infections. In this context, the observation that the degradation of c-di-GMP in wild-type *P. aeruginosa* (through expression of the EAL protein PA2133) to negligible intracellular levels prevents even the initiation of biofilm development is encouraging (30).

CONCLUSIONS

Although the basic actions of c-di-GMP signaling in *P. aeruginosa* are coming into focus, many questions remain. For example, what are the functions of those GGDEF and EAL genes that do not have apparent phenotypes? These genes may have distinctive but as yet unknown functions, and it is also likely that some of them are expressed under specific growth conditions that have not yet been tested. This has yet to be explored. How does c-di-GMP binding alter the functions of the EPS biosynthetic protein PelD and the transcriptional regulator FleQ? We still do not know how c-di-GMP controls Psl biosynthesis. The two- to threefold transcriptional effects of FleQ and c-di-GMP on *psl* operon transcription are insufficient to account for the dramatic effect that high intracellular c-di-GMP levels have on Psl EPS synthesis. Although there is good evidence that DGCs like WspR localize in subcellular compartments and synthesize c-di-GMP at specific sites, it is not known what factors direct the movements of these proteins to specific places in cells, nor is it known if DGCs and PDEs physically interact with specific target proteins. What is the interplay of DGCs and PDEs in modulating localized c-di-GMP pools? What is the range of environmental signals that are sensed by DGCs and PDEs? And how are apparent physical signals like cell surface touching communicated from the outside to the inside of cells to alter DGC/PDE activities? Clearly there is much work to be done and many new discoveries to be made in the fascinating area of c-di-GMP signal transduction in *P. aeruginosa*.

Acknowledgments. C.S.H. is supported by Public Health Service grant GM56665 from the National Institute of General Medical Sciences.

I thank P. L. Howell and her colleagues from the Hospital for Sick Children, Toronto, Canada, for supplying Fig. 7 and 8.

REFERENCES

1. **Alm, R. A., A. J. Bodero, P. D. Free, and J. S. Mattick.** 1996. Identification of a novel gene, *pilZ*, essential for type 4 fimbrial biogenesis in *Pseudomonas aeruginosa*. *J. Bacteriol.* **178**:46–53.
2. **Alvarez-Ortega, C., and C. S. Harwood.** 2007. Identification of a malate chemoreceptor in *Pseudomonas aeruginosa* by screening for chemotaxis defects in an energy taxis-deficient mutant. *Appl. Environ. Microbiol.* **73**:7793–7795.
3. **Amikam, D., and M. Y. Galperin.** 2006. PilZ domain is part of the bacterial c-di-GMP binding protein. *Bioinformatics* **22**: 3–6.
4. **Bantinaki, E., R. Kassen, C. G. Knight, Z. Robinson, A. J. Spiers, and P. B. Rainey.** 2007. Adaptive divergence in experimental populations of *Pseudomonas fluorescens*. III. Mutational origins of wrinkly spreader diversity. *Genetics* **176**:441–453.
5. **Baty, A. M., Z. Diwu, G. Dunham, C. C. Eastburn, G. G. Geesey, A. E. Goodman, P. A. Suci, and S. Techkarnjanaruk.** 2001. Characterization of extracellular chitinolytic activity in biofilms. *Methods Enzymol.* **336**:279–301.
6. **Boucher, J. C., H. Yu, M. H. Mudd, and V. Deretic.** 1997. Mucoid *Pseudomonas aeruginosa* in cystic fibrosis: characterization of *muc* mutations in clinical isolates and analysis of clearance in a mouse model of respiratory infection. *Infect. Immun.* **65**:3838–3846.
7. **Brencic, A., and S. Lory.** 2009. Determination of the regulon and identification of novel mRNA targets of *Pseudomonas aeruginosa* RsmA. *Mol. Microbiol.* **72**:612–632.
8. **Christen, M., B. Christen, M. Folcher, A. Schauerte, and U. Jenal.** 2005. Identification and characterization of a cyclic di-GMP-specific phosphodiesterase and its allosteric control by GTP. *J. Biol. Chem.* **280**:30829–30837.
9. **Chung, I. Y., K. B. Choi, Y. J. Heo, and Y. H. Cho.** 2008. Effect of PEL exopolysaccharide on the *wspF* mutant phenotypes in *Pseudomonas aeruginosa* PA14. *J. Microbiol. Biotechnol.* **18**:1227–1234.
10. **D'Argenio, D. A., M. W. Calfee, P. B. Rainey, and E. C. Pesci.** 2002. Autolysis and autoaggregation in *Pseudomonas aeruginosa* colony morphology mutants. *J. Bacteriol.* **184**:6481–6489.
11. **Dasgupta, N., S. K. Arora, and R. Ramphal.** 2000. *fleN,* a gene that regulates flagellar number in *Pseudomonas aeruginosa*. *J. Bacteriol.* **182**:357–364.
12. **Dasgupta, N., E. P. Ferrell, K. J. Kanack, S. E. West, and R. Ramphal.** 2002. *fleQ,* the gene encoding the major flagellar regulator of *Pseudomonas aeruginosa,* is sigma70 dependent and is downregulated by Vfr, a homolog of *Escherichia coli* cyclic AMP receptor protein. *J. Bacteriol.* **184**:5240–5250.
13. **Dasgupta, N., M. C. Wolfgang, A. L. Goodman, S. K. Arora, J. Jyot, S. Lory, and R. Ramphal.** 2003. A four-tiered transcriptional regulatory circuit controls flagellar biogenesis in *Pseudomonas aeruginosa*. *Mol. Microbiol.* **50**:809–824.

13a. **De, N., M. V. A. S. Navarro, R. V. Raghavan, and H. Sondermann.** 2009. Determinants for the activation and autoinhibition of the diguanylate cyclase response regulator WspR. *J. Mol. Biol.* doi:10.1016/j.jmb.2009.08.030.

14. **De, N., M. Pirruccello, P. V. Krasteva, N. Bae, R. V. Raghavan, and H. Sondermann.** 2008. Phosphorylation-independent regulation of the diguanylate cyclase WspR. *PLoS Biol.* **6**:e67.
15. **Drenkard, E., and F. M. Ausubel.** 2002. *Pseudomonas* biofilm formation and antibiotic resistance are linked to phenotypic variation. *Nature* **416**:740–743.
16. **Friedman, L., and R. Kolter.** 2004. Genes involved in matrix formation in *Pseudomonas aeruginosa* PA14 biofilms. *Mol. Microbiol.* **51**:675–690.
17. **Friedman, L., and R. Kolter.** 2004. Two genetic loci produce distinct carbohydrate-rich structural components of the *Pseudomonas aeruginosa* biofilm matrix. *J. Bacteriol.* **186**:4457–4465.
18. **Galperin, M. Y., T. A. Gaidenko, A. Y. Mulkidjanian, M. Nakano, and C. W. Price.** 2001. MHYT, a new integral membrane sensor domain. *FEMS Microbiol. Lett.* **205**:17–23.
19. **Galperin, M. Y., A. N. Nikolskaya, and E. V. Koonin.** 2001. Novel domains of the prokaryotic two-component signal transduction systems. *FEMS Microbiol. Lett.* **203**:11–21.
20. **Giddens, S. R., R. W. Jackson, C. D. Moon, M. A. Jacobs, X. X. Zhang, S. M. Gehrig, and P. B. Rainey.** 2007. Muta-

tional activation of niche-specific genes provides insight into regulatory networks and bacterial function in a complex environment. *Proc. Natl. Acad. Sci. USA* **104:**18247–18252.

21. **Gomez, M. I., and A. Prince.** 2007. Opportunistic infections in lung disease: *Pseudomonas* infections in cystic fibrosis. *Curr. Opin. Pharmacol.* 7:244–251.
22. **Govan, J. R., and V. Deretic.** 1996. Microbial pathogenesis in cystic fibrosis: mucoid *Pseudomonas aeruginosa* and *Burkholderia cepacia. Microbiol. Rev.* **60:**539–574.
23. **Güvener, Z. T., and C. S. Harwood.** 2007. Subcellular location characteristics of the *Pseudomonas aeruginosa* GGDEF protein, WspR, indicate that it produces cyclic-di-GMP in response to growth on surfaces. *Mol. Microbiol.* **66:**1459–1473.
24. **Güvener, Z. T., D. F. Tifrea, and C. S. Harwood.** 2006. Two different *Pseudomonas aeruginosa* chemosensory signal transduction complexes localize to cell poles and form and remould in stationary phase. *Mol. Microbiol.* **61:**106–118.
25. **Haussler, S., B. Tummler, H. Weissbrodt, M. Rohde, and I. Steinmetz.** 1999. Small-colony variants of *Pseudomonas aeruginosa* in cystic fibrosis. *Clin. Infect. Dis.* **29:**621–625.
26. **Hay, I. D., U. Remminghorst, and B. H. Rehm.** 2009. MucR, a novel membrane-associated regulator of alginate biosynthesis in *Pseudomonas aeruginosa. Appl. Envrion. Microbiol.* **75:** 1110–1120.
27. **Hengge, R.** 2009. Principles of c-di-GMP signaling in bacteria. *Nat. Rev. Microbiol.* **7:**263–273.
28. **Heydorn, A., B. K. Ersboll, M. Hentzer, M. R. Parsek, M. Givskov, and S. Molin.** 2000. Experimental reproducibility in flow-chamber biofilms. *Microbiology* **146:**2409–2415.
29. **Hickman, J. W., and C. S. Harwood.** 2008. Identification of FleQ from *Pseudomonas aeruginosa* as a c-di-GMP-responsive transcription factor. *Mol. Microbiol.* **69:**376–389.
30. **Hickman, J. W., D. F. Tifrea, and C. S. Harwood.** 2005. A chemosensory system that regulates biofilm formation through modulation of cyclic diguanylate levels. *Proc. Natl. Acad. Sci. USA* **102:**14422–14427.
31. **Hoffman, L. R., D. A. D'Argenio, M. J. MacCoss, Z. Zhang, R. A. Jones, and S. I. Miller.** 2005. Aminoglycoside antibiotics induce bacterial biofilm formation. *Nature* **436:**1171–1175.
32. **Hoiby, N.** 2002. New antimicrobials in the management of cystic fibrosis. *J. Antimicrob. Chemother.* **49:**235–238.
33. **Hoiby, N.** 2002. Understanding bacterial biofilms in patients with cystic fibrosis: current and innovative approaches to potential therapies. *J. Cyst. Fibros.* **1:**249–254.
34. **Jackson, K. D., M. Starkey, S. Kremer, M. R. Parsek, and D. J. Wozniak.** 2004. Identification of *psl,* a locus encoding a potential exopolysaccharide that is essential for *Pseudomonas aeruginosa* PAO1 biofilm formation. *J. Bacteriol.* **186:**4466–4475.
35. **Jenal, U., and J. Malone.** 2006. Mechanisms of cyclic-di-GMP signaling in bacteria. *Annu. Rev. Genet.* **40:**385–407.
36. **Jonas, K., A. N. Edwards, R. Simm, T. Romeo, U. Romling, and O. Melefors.** 2008. The RNA binding protein CsrA controls cyclic di-GMP metabolism by directly regulating the expression of GGDEF proteins. *Mol. Microbiol.* **70:**236–257.
37. **Kazmierczak, B. I., M. B. Lebron, and T. S. Murray.** 2006. Analysis of FimX, a phosphodiesterase that governs twitching motility in *Pseudomonas aeruginosa. Mol. Microbiol.* **60:**1026–1043.
38. **Kirisits, M. J., L. Prost, M. Starkey, and M. R. Parsek.** 2005. Characterization of colony morphology variants isolated from *Pseudomonas aeruginosa* biofilms. *Appl. Environ. Microbiol.* **71:**4809–4821.
39. **Kohler, T., L. K. Curty, F. Barja, C. van Delden, and J. C. Pechere.** 2000. Swarming of *Pseudomonas aeruginosa* is dependent on cell-to-cell signaling and requires flagella and pili. *J. Bacteriol.* **182:**5990–5996.
40. **Kuchma, S. L., K. M. Brothers, J. H. Merritt, N. T. Liberati, F. M. Ausubel, and G. A. O'Toole.** 2007. BifA, a cyclic-di-GMP phosphodiesterase, inversely regulates biofilm formation and swarming motility by *Pseudomonas aeruginosa* PA14. *J. Bacteriol.* **189:**8165–8178.
41. **Kuchma, S. L., J. P. Connolly, and G. A. O'Toole.** 2005. A three-component regulatory system regulates biofilm maturation and type III secretion in *Pseudomonas aeruginosa. J. Bacteriol.* **187:**1441–1454.
42. **Kulasekara, H., V. Lee, A. Brencic, N. Liberati, J. Urbach, S. Miyata, D. G. Lee, A. N. Neely, M. Hyodo, Y. Hayakawa, F. M. Ausubel, and S. Lory.** 2006. Analysis of *Pseudomonas aeruginosa* diguanylate cyclases and phosphodiesterases reveals a role for bis-(3′-5′)-cyclic-GMP in virulence. *Proc. Natl. Acad. Sci. USA* **103:**2839–2844.
43. **Kulasekara, H. D., I. Ventre, B. R. Kulasekara, A. Lazdunski, A. Filloux, and S. Lory.** 2005. A novel two-component system controls the expression of *Pseudomonas aeruginosa* fimbrial *cup* genes. *Mol. Microbiol.* **55:**368–380.
44. **Lapouge, K., M. Schubert, F. H. Allain, and D. Haas.** 2008. Gac/Rsm signal transduction pathway of gamma-proteobacteria: from RNA recognition to regulation of social behaviour. *Mol. Microbiol.* **67:**241–253.
45. **Lee, V. T., J. M. Matewish, J. L. Kessler, M. Hyodo, Y. Hayakawa, and S. Lory.** 2007. A cyclic-di-GMP receptor required for bacterial exopolysaccharide production. *Mol. Microbiol.* **65:**1474–1484.
46. **Lyczak, J. B., C. L. Cannon, and G. B. Pier.** 2000. Establishment of *Pseudomonas aeruginosa* infection: lessons from a versatile opportunist. *Microbes Infect.* **2:**1051–1060.
47. **Maddock, J. R., and L. Shapiro.** 1993. Polar location of the chemoreceptor complex in the *Escherichia coli* cell. *Science* **259:**1717–1723.
48. **Mah, T. F., and G. A. O'Toole.** 2001. Mechanisms of biofilm resistance to antimicrobial agents. *Trends Microbiol.* **9:**34–39.
49. **Martin, D. W., M. J. Schurr, M. H. Mudd, J. R. Govan, B. W. Holloway, and V. Deretic.** 1993. Mechanism of conversion to mucoidy in *Pseudomonas aeruginosa* infecting cystic fibrosis patients. *Proc. Natl. Acad. Sci. USA* **90:**8377–8381.
50. **Matsukawa, M., and E. P. Greenberg.** 2004. Putative exopolysaccharide synthesis genes influence *Pseudomonas aeruginosa* biofilm development. *J. Bacteriol.* **186:**4449–4456.
51. **McCarter, L. L.** 2006. Regulation of flagella. *Curr. Opin. Microbiol.* **9:**180–186.
52. **Merighi, M., V. T. Lee, M. Hyodo, Y. Hayakawa, and S. Lory.** 2007. The second messenger bis-(3′-5′)-cyclic-GMP and its PilZ domain-containing receptor Alg44 are required for alginate biosynthesis in *Pseudomonas aeruginosa. Mol. Microbiol.* **65:**876–895.
53. **Merritt, J., K. Brothers, S. L. Kuchma, and G. O'Toole.** 2007. SadC reciprocally influences biofilm formation and swarming motility via modulation of exopolysaccharide production and flagellar function. *J. Bacteriol.* **189:**8154–8164.
54. **Minasov, G., S. Padavattan, L. Shuvalova, J. S. Brunzelle, D. J. Miller, A. Basle, C. Massa, F. R. Collart, T. Schirmer, and W. F. Anderson.** 2009. Crystal structures of YkuI and its complex with second messenger c-di-GMP suggests catalytic mechanism of phosphodiester bond cleavage by EAL domains. *J. Biol. Chem.* **284:**13174–13184.

54a. **Navarro, M. V. A. S., N. De, N. Bae, Q. Wang, and H. Sondermann.** 2009. Structural analysis of the GGDEF-EAL domain-containing c-di-GMP receptor FimX. *Structure* **17:** 1104–1116.

55. **Oglesby, L. L., S. Jain, and D. E. Ohman.** 2008. Membrane topology and roles of *Pseudomonas aeruginosa* Alg8 and Alg44 in alginate polymerization. *Microbiology* **154:**1605–1615.

56. **O'Toole, G. A., and R. Kolter.** 1998. Flagellar and twitching motility are necessary for *Pseudomonas aeruginosa* biofilm development. *Mol. Microbiol.* **30:**295–304.

57. **Parales, R. E., A. Ferrandez, and C. S. Harwood.** 2004. Chemotaxis in pseudomonads, p. 793–815. *In* J.-L. Ramos (ed.), *Pseudomonas,* vol. 1. *Genomics, Life Style and Molecular Architecture.* Kluwer Academic/Plenum Publishers, New York, NY.

58. **Parsek, M. R., and P. K. Singh.** 2003. Bacterial biofilms: an emerging link to disease pathogenesis. *Annu. Rev. Microbiol.* **58:**677–701.

59. **Paul, R., S. Abel, P. Wassmann, A. Beck, H. Heerklotz, and U. Jenal.** 2007. Activation of the diguanylate cyclase PleD by phosphorylation-mediated dimerization. *J. Biol. Chem.* **282:** 29170–219177.

60. **Rainey, P. B., and M. Travisano.** 1998. Adaptive radiation in a heterogeneous environment. *Nature* **394:**69–72.

61. **Rajan, S., and L. Saiman.** 2002. Pulmonary infections in patients with cystic fibrosis. *Semin. Respir. Infect.* **17:**47–56.

62. **Ramelot, T. A., A. Yee, J. R. Cort, A. Semesi, C. H. Arrowsmith, and M. A. Kennedy.** 2007. NMR structure and binding studies confirm that PA4608 from *Pseudomonas aeruginosa* is a PilZ domain and a c-di-GMP binding protein. *Proteins* **66:** 266–271.

63. **Ramsey, D. M., and D. J. Wozniak.** 2005. Understanding the control of *Pseudomonas aeruginosa* alginate synthesis and the prospects for management of chronic infections in cystic fibrosis. *Mol. Microbiol.* **56:**309–322.

64. **Rao, F., Y. Yang, Y. Qi, and Z. X. Liang.** 2008. Catalytic mechanism of cyclic di-GMP-specific phosphodiesterase: a study of the EAL domain-containing RocR from *Pseudomonas aeruginosa. J. Bacteriol.* **190:**3622–3631.

65. **Rao, F., Y. Qi, H. S. Chong, M. Kotaka, B. Li, J. Li, J. Lescar, K. Tang, and Z. X. Liang.** 17 April 2009. The functional role of a conserved loop in EAL domain-based cyclic di-GMP-specific phosphodiesterase. *J. Bacteriol.* **191:**4722–4731.

66. **Remminghorst, U., and B. H. Rehm.** 2006. Alg44, a unique protein required for alginate biosynthesis in *Pseudomonas aeruginosa. FEBS Lett.* **580:**3883–3888.

67. **Remminghorst, U., and B. H. Rhem.** 2006. Bacterial alginates: from biosynthesis to applications. *Biotechnol. Lett.* **28:**1701–1712.

68. **Romling, U., M. Gomelsky, and M. Y. Galperin.** 2005. C-di-GMP: the dawning of a novel bacterial signalling system. *Mol. Microbiol.* **57:**629–639.

69. **Rowen, D. W., and V. Deretic.** 2000. Membrane-to-cytosol redistribution of ECF sigma factor AlgU and conversion to mucoidy in *Pseudomonas aeruginosa* isolates from cystic fibrosis patients. *Mol. Microbiol.* **36:**314–327.

70. **Ryan, R. P., J. Lucey, K. O'Donovan, Y. McCarthy, L. Yang, T. Tolker-Nielsen, and J. M. Dow.** 2009. HD-GYP domain proteins regulate biofilm formation and virulence in *Pseudomonas aeruginosa. Environ. Microbiol.* **11:**1126–1136.

71. **Ryan, R. P., Y. Fouhy, J. F. Lucey, L. C. Crossman, S. Spiro, Y. W. He, L. H. Zhang, S. Heeb, M. Camara, P. Williams, and J. M. Dow.** 2006. Cell-cell signaling in *Xanthomonas campestris* involves an HD-GYP domain protein that functions in cyclic di-GMP turnover. *Proc. Natl. Acad. Sci. USA* **103:**6712–6717.

72. **Ryder, C., M. Byrd, and D. Wozniak.** 2007. Role of polysaccharides in *Pseudomonas aeruginosa* biofilm development. *Curr. Opin. Microbiol.* **10:**664–648.

73. **Schmidt, A. J., D. A. Ryjenkov, and M. Gomelsky.** 2005. The ubiquitous protein domain EAL is a cyclic diguanylate-specific phosphodiesterase: enzymatically active and inactive EAL domains. *J. Bacteriol.* **187:**4774–4781.

74. **Singh, P. K., A. L. Schaefer, M. R. Parsek, T. O. Moninger, M. J. Welsh, and E. P. Greenberg.** 2000. Quorum-sensing signals indicate that cystic fibrosis lungs are infected with bacterial biofilms. *Nature* **407:**762–764.

75. **Smith, E. E., D. G. Buckley, Z. Wu, C. Saenphimmachak, L. R. Hoffman, D. A. D'Argenio, S. I. Miller, B. W. Ramsey, D. P. Speert, S. M. Moskowitz, J. L. Burns, R. Kaul, and M. V. Olson.** 2006. Genetic adaptation by *Pseudomonas aeruginosa* to the airways of cystic fibrosis patients. *Proc. Natl. Acad. Sci. USA* **103:**8487–8492.

76. **Sourjik, V., and H. C. Berg.** 2000. Localization of components of the chemotaxis machinery of *Escherichia coli* using fluorescent protein fusions. *Mol. Microbiol.* **37:**740–751.

77. **Starkey, M., J. H. Hickman, L. Ma, N. Zhang, S. De Long, A. Hinz, S. Palacios, C. Manoil, M. J. Kirisits, T. D. Starner, D. J. Wozniak, C. S. Harwood, and M. R. Parsek.** 27 March 2009. *Pseudomonas aeruginosa* rugose small-colony variants have adaptations that likely promote persistence in the cystic fibrosis lung. *J. Bacteriol.* **191:**3492–3503.

78. **Taguchi, K., H. Fukutomi, A. Kuroda, J. Kato, and H. Ohtake.** 1997. Genetic identification of chemotactic transducers for amino acids in *Pseudomonas aeruginosa. Microbiology* **143:** 3223–3229.

79. **Tamayo, R., J. T. Pratt, and A. Camilli.** 2007. Roles of cyclic diguanylate in the regulation of bacterial pathogenesis. *Annu. Rev. Microbiol.* **61:**131–148.

80. **Vallet, I., J. W. Olson, S. Lory, A. Lazdunski, and A. Filloux.** 2001. The chaperone/usher pathways of *Pseudomonas aeruginosa:* identification of fimbrial gene clusters *(cup)* and their involvement in biofilm formation. *Proc. Natl. Acad. Sci. USA* **98:**6911–6916.

81. **Vasseur, P., I. Vallet-Gely, C. Soscia, S. Genin, and A. Filloux.** 2005. The *pel* genes of the *Pseudomonas aeruginosa* PAK strains are involved in early and late stages of biofilm formation. *Microbiology* **151:**985–997.

82. **Wagner, V. E., and B. H. Iglewski.** 2008. *P. aeruginosa* biofilms in CF infection. *Clin. Rev. Allergy Immunol.* **35:**124–134.

83. **Wassmann, P., C. Chan, R. Paul, A. Beck, H. Heerklotz, U. Jenal, and T. Schirmer.** 2007. Structure of BeF3-modified response regulator PleD: implications for diguanylate cyclase activation, catalysis, and feedback inhibition. *Structure* **15:**915–927.

84. **Weinhouse, H., S. Sapir, D. Amikam, Y. Shilo, G. Volman, P. Ohana, and M. Benziman.** 1997. C-di-GMP-binding protein, a new factor regulating cellulose synthesis in *Acetobacter xylinum. FEBS Lett.* **416:**207–211.

85. **Whitchurch, C. B., A. J. Leech, M. D. Young, D. Kennedy, J. L. Sargent, J. J. Bertrand, A. B. Semmler, A. S. Mellick, P. R. Martin, R. A. Alm, M. Hobbs, S. A. Beatson, B. Huang, L. Nguyen, J. C. Commolli, J. N. Engel, A. Darzins, and J. S. Mattick.** 2004. Characterization of a complex chemosensory signal transduction system which controls twitching motility in *Pseudomonas aeruginosa. Mol. Microbiol.* **52:**873–893.

86. **Wolfe, A. J., and K. L. Visick.** 2008. Get the message out: cyclic-di-GMP regulates multiple levels of flagellum-based motility. *J. Bacteriol.* **190:**463–475.
87. **Wozniak, D. J., T. J. Wyckoff, M. Starkey, R. Keyser, P. Azadi, G. A. O'Toole, and M. R. Parsek.** 2003. Alginate is not a significant component of the extracellular polysaccharide matrix of PA14 and PAO1 *Pseudomonas aeruginosa* biofilms. *Proc. Natl. Acad. Sci. USA* **100:**7907–7912.
88. **Zhulin, I. B., A. N. Nikolskaya, and M. Y. Galperin.** 2003. Common extracellular sensory domains in transmembrane receptors for diverse signal transduction pathways in *Bacteria* and *Archaea*. *J. Bacteriol.* **185:**285–294.

The Second Messenger Cyclic Di-GMP
Edited by A. J. Wolfe and K. L. Visick

Chapter 12

The Scr Circuit in *Vibrio parahaemolyticus* Modulates Swarming and Sticking

Linda L. McCarter

The marine bacterium and human pathogen *Vibrio parahaemolyticus* readily adapts to life on surfaces. When grown on a surface or in a viscous layer, the organism can induce a large gene system and differentiate to the swarmer cell, which is an elongated, hyperflagellated cell type; this differentiation allows highly efficient movement over surfaces. However, *V. parahaemolyticus* also proficiently adheres to surfaces and forms robust biofilms. Thus, there are times when growing on a surface that it must be advantageous to rapidly translocate over that surface and move into a new environment, while under different circumstances it may be a better survival strategy to be able to stick, i.e., to be a more adhesive and sessile cell type (Fig. 1). Scr (*s*warming and *c*apsular polysaccharide gene *r*egulation) genes influence the lifestyle adaptation between swarming and sticking by modulating the cellular nucleotide pool. Some *scr* genes encode protein products containing the highly conserved GGDEF and EAL domains. These two domains, named after signature amino acid motifs, are found in diguanylate cyclases and phosphodiesterases and are responsible for the formation and degradation, respectively, of the nucleotide bis-3′, 5′ cyclic di-GMP (c-di-GMP) (reviewed in references 18 and 24) (see also chapter 4). As this book clearly provides evidence, the function of c-di-GMP as a signaling molecule is being described in an expanding list of organisms. c-di-GMP seems most often implicated in regulating the production of a variety of cell surface structures (such as flagella and pili) and extracellular matrixes (such as cellulose and polysaccharide) (4, 6, 7, 12, 21, 22, 28, 40, 42, 47). Thus, the role of c-di-GMP in determining the switch between motile and sessile lifestyles appears to be an emerging theme in many bacteria (29, 35, 38, 43, 47, 55) (see also chapter 8); in some bacteria, one important consequence is an impact on virulence (reviewed in (references 11, 14, and 50).

scrABC SIGNALING LOCUS

The *scrABC* operon encodes three proteins: ScrA, a predicted class V aminotransferase containing a pyridoxal phosphate attachment site (468 amino acids [aa]; E value = 3.1e−12); ScrB, a predicted extracellular solute-binding protein (321 aa; E value = 4.5e−10); and ScrC, a sensory protein (774 aa). Fractionation experiments localize ScrA to the cytoplasm, ScrB in the periplasm, and ScrC with the cell membrane. The ScrC N terminus contains two predicted transmembrane domains flanking a potential periplasmic region (~300 aa in length). An ScrC-LacZ fusion protein created at the 309th aa residue of ScrC provides evidence that the C terminus of ScrC is cytoplasmically located. It contains both GGDEF (E value = 2.5e−60) and EAL (E-value = 1.4e−104) conserved protein domains.

Mutants with defects in the *scrABC* locus are swarm defective and sticky. Disruption of any of the three genes in the *scrABC* operon reduces swarming and produces a crinkly colony morphology; furthermore, the phenotype of a mutant with a deletion of the entire operon is similar to a mutant with a deletion of *scrC* or simply a deletion of the portion of the gene encoding the EAL domain. The swarming of all these mutants is greatly decreased, and their colony morphology on Congo red medium is very wrinkled due to the overproduction of capsular polysaccharide (CPS) (Fig. 2A and B). Double *scr cps* mutants are smooth (due to an inability to produce

Linda L. McCarter • Department of Microbiology, The University of Iowa, Iowa City, IA 52242.

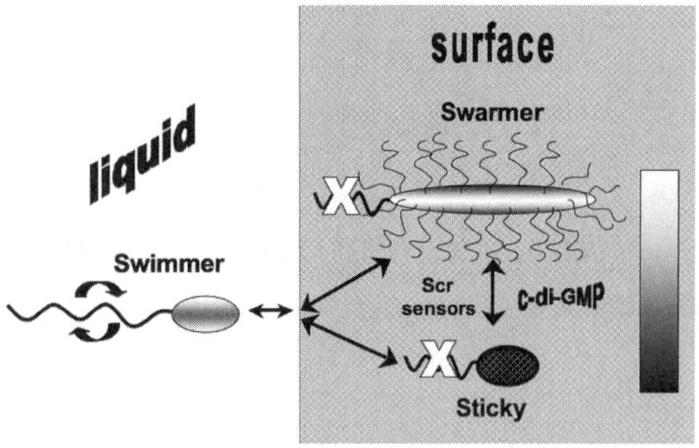

Figure 1. Lifestyle decision for *Vibrio parahaemolyticus:* to swim, swarm, or stick? When growing in liquid, the highly motile swimmer cell is propelled by a single polar flagellum. Upon contact with surfaces, this polar flagellum is no longer an effective propulsive organelle. Unable to rotate its polar flagellum, the organism must decide whether to induce a second flagellar system and differentiate to the elongated, hyperflagellated swarmer cell or to orchestrate the production of sticky cell surface molecules promoting adhesion and biofilm development. This decision is modulated by Scr sensors that are thought to integrate environmental information by modulating the level of c-di-GMP. High levels of c-di-GMP promote the sessile lifestyle, whereas low levels of c-di-GMP enhance surface motility.

CPS), but they retain an inability to move on surfaces (Fig. 2C). Hence, the failure to swarm is not simply the consequence of being too sticky to move. Mutants with *scrABC* lesions produce decreased levels of lateral flagellin, which is the structural subunit for the flagellar propeller. These phenotypes are the direct consequence of reduced lateral flagellar *(laf)* gene expression and increased *cps* gene expression (Fig. 2D and E, respectively).

SWARMING AND STICKING ARE INFLUENCED BY c-di-GMP

The GGDEF and EAL domains of ScrC are both highly conserved, and each possesses the signature amino acids that have been shown to be critical for enzymatic activity in other proteins (5, 28, 47, 52). The EAL domain, known to be associated with the phosphodiesterase activity, is critical for the activity of ScrABC; deletion of the portion of the gene coding for the EAL domain produces a mutant phenotype similar to deletion of the entire gene (Fig. 2). Overexpression of *scrABC* is sufficient to induce expres-

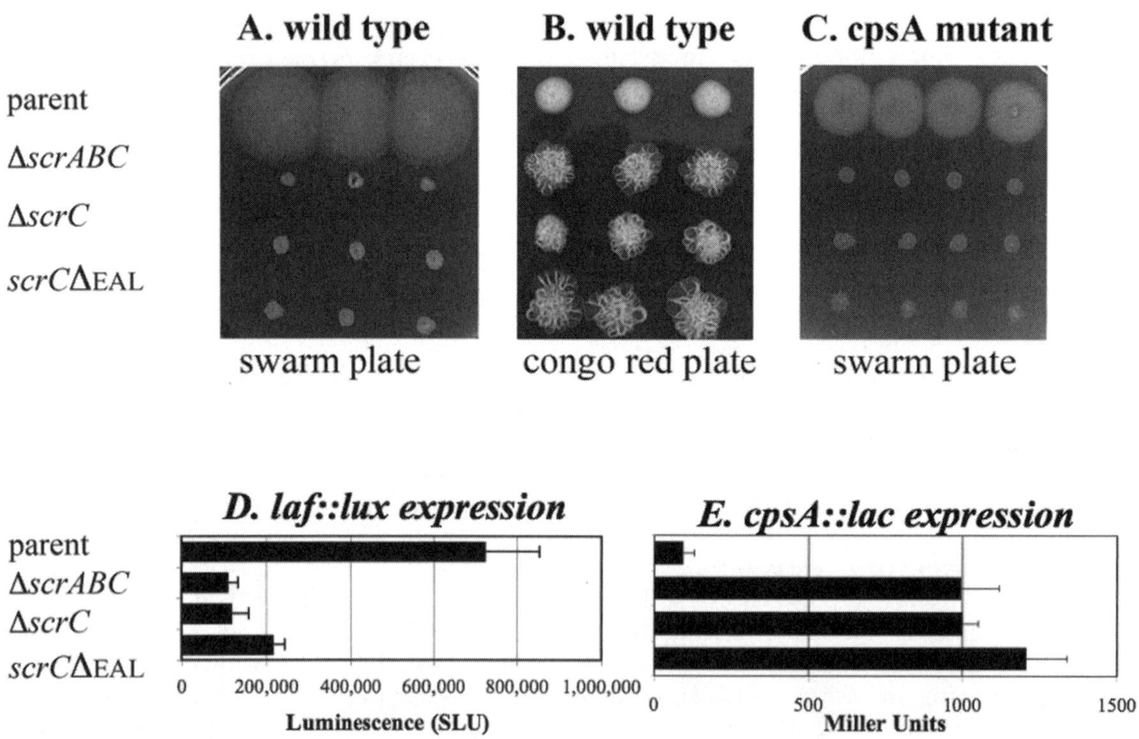

Figure 2. Swarming and sticking are reciprocally influenced at the level of transcription by the *scrABC* operon. Mutants with defects in the *scrABC* operon fail to swarm on solid surfaces (A) and produce crinkly colonies on Congo red medium (B). The inability of these mutants to swarm is not the simple consequence of being too sticky to move, as they also fail to swarm upon introduction of a lesion in the CPS biosynthetic locus (C). The altered phenotypes are a consequence of altered lateral flagellar *(laf)* and *cps* gene expression. Transcription was measured in *laf::lux* (D) and *cps::lacZ* (E) reporter strains.

sion of swarming genes in liquid (expression which normally occurs only when the bacteria are growing on a surface). Removal of the EAL domain or mutation of a key amino acid (EAL→AAL) known to be important for activity in other phosphodiesterases (5, 28, 52) completely inactivates the ability of ScrABC to induce lateral flagellar gene expression. Examination of ^{32}P-labeled nucleotide pools by using two-dimensional thin-layer chromatography reveals that this overproduction of ScrABC decreases the c-di-GMP spot, whereas overproduction of $ScrABC_{\Delta EAL}$ does not (Fig. 3A). Furthermore, expression of the *cps* biosynthetic locus is depressed upon ectopic expression of *scrABC*. Thus, it seems that a low concentration of this nucleotide promotes swarming and prevents sticking. Conversely, a high concentration of c-di-GMP prevents swarming on surfaces and induces the production of CPS. Removal of the EAL domain does not simply inactivate ScrABC: overexpression of $scrABC_{\Delta EAL}$ increases the amount of c-di-GMP in the cell (Fig. 3A), decreases swarming, and increases CPS production (Fig. 3B). As a consequence, biofilm development of *scrABC* mutants is premature and more robust than that of the wild type.

ScrC IS A BIFUNCTIONAL ENZYME, AND ITS ACTIVITY IS MODULATED BY ScrA AND ScrB

The complete reversal of the activity of ScrC upon removal of the EAL domain, from being able to induce swarming to becoming a strong repressor of swarming (as well as from being able to suppress stickiness to becoming an inducer of sticking), sug-

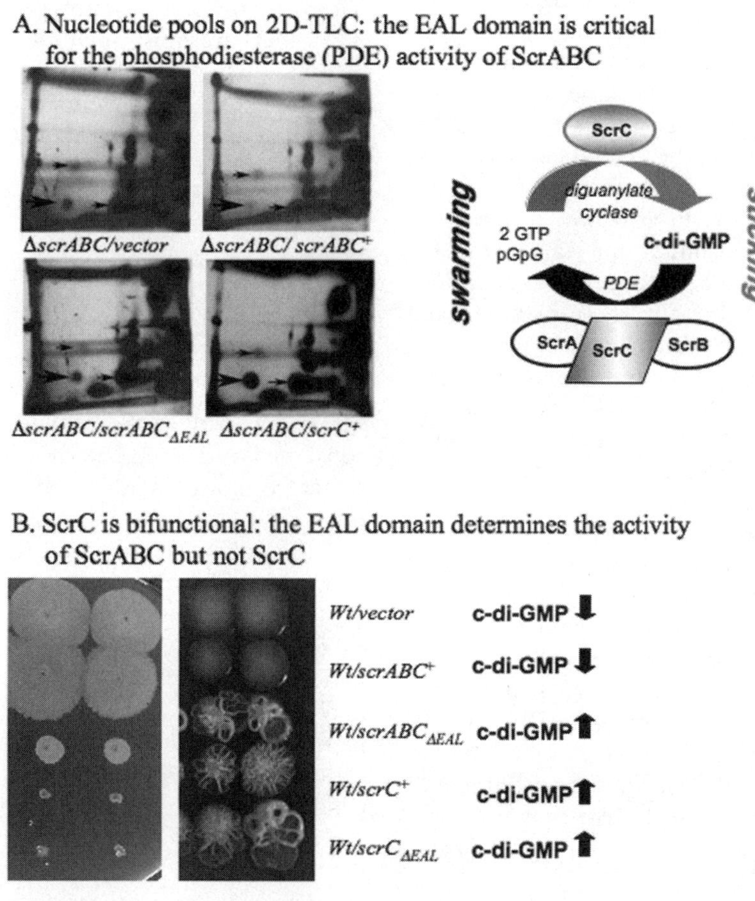

Figure 3. ScrC is a bifunctional enzyme whose activity is influenced by ScrA and ScrB. (A) Examination of ^{32}P-labeled nucleotides extracted from a Δ*scrABC* strain carrying expression clones by using two-dimensional thin-layer chromatography (2D-TLC). The small arrows indicate control spots to provide orientation, and the large arrow indicates c-di-GMP. ScrABC decreases the c-di-GMP spot, the EAL domain is critical for the phosphodiesterase (PDE) activity of ScrABC, and ScrC produced by itself (without the context of functional ScrA and ScrB) increases the level of this nucleotide. (B) The EAL domain determines the PDE activity of ScrABC but not the diguanylate cyclase activity of ScrC: phenotypes of the wild-type (Wt) strain on swarm plates and Congo red medium upon ectopic expression of *scrC* alleles.

gests that ScrC might have two enzymatic activities. Curiously, overproduction of ScrC without coproduction of ScrA and ScrB does not simply result in loss of increased *laf* expression but rather recapitulates the ScrABC$_{\Delta EAL}$ phenotype (Fig. 3B) by preventing *laf* expression while enhancing *cps* transcription. For many proteins with dual GGDEF and EAL domains, only one domain seems catalytically active and the second domain is inactive or regulatory (37, 46). In contrast, both the GGDEF and the EAL domains of ScrC possess catalytic activity: by itself (in the absence of ScrA and ScrB), ScrC can synthesize c-di-GMP (in *V. parahaemolyticus* as well as in *Escherichia coli*), whereas in the context of functional ScrA and ScrB, ScrC is capable of degrading this secondary messenger.

In having the capacity to both degrade and form c-di-GMP, ScrC resembles BphG1, a bifunctional GGDEF-EAL protein of *Rhodobacter sphaeroides* (51). The EAL domain of BphG1 seems to lock down the activity of the GGDEF domain: only upon physical cleavage of the phosphodiesterase domain does the diguanylate cyclase activity become liberated. In comparison, the activity of ScrC seems to be modulated by ScrA and ScrB. In their absence, ScrC has diguanylate cylase activity; in their presence, ScrC acts as a phosphodiesterase. In vivo (i.e., in the context of ScrA and ScrB and under our standard laboratory swarm conditions), ScrC appears to function as a phosphodiesterase, although there may be particular conditions, yet to be discovered, in which the diguanylate cyclase activity of ScrC predominates. How ScrA and ScrB work to modulate activity is not known. Coproduction of ScrB, the periplasmic solute-binding protein, with ScrC is sufficient to neutralize ScrC's diguanylate cyclase activity, but ScrA is needed to complete the switch to the phosphodiesterase mode (L. McCarter, unpublished data). MbaA (VCA0703) is a *Vibrio cholerae* protein that, although not an ortholog of ScrC, has similar architecture with respect to possessing a large periplasmic domain and cytoplasmically located GGDEF and EAL domains. MbaA plays a role in biofilm formation (7), and its activity is modulated by the NspS periplasmic binding protein for norspermidine (25). Thus, environmental input signals might similarly activate a signaling cascade that would influence the activity of the GGDEF-EAL domains of ScrC.

ScrG IS A SECOND GGDEF-EAL PROTEIN PARTICIPATING IN RECIPROCAL SWARMING AND STICKING GENE CONTROL

The *scrABC* operon is not the sole GGDEF-EAL regulatory element controlling swarming and sticking: another GGDEF-EAL encoding gene, *scrG*, has been studied (27). The *scrG* gene was discovered similarly to the *scrABC* locus, i.e., it was found in a screen for low-copy-number cosmids from a *V. parahaemolyticus* library that promoted lateral flagellar gene expression during growth in liquid. It encodes a potential 568-aa cytoplasmic protein containing PAS (cd00130; E value = 6e−06), GGDEF (cd01949; E value = 4e−12), and EAL (cd01048; E value = 2e−59) domains. Overexpression of *scrG* is sufficient to induce lateral flagellar gene expression in liquid, prevent biofilm formation, decrease *cps* gene expression, and suppress the *scrABC* phenotype. Overexpression of *scrG* decreases the intensity of the ^{32}P-labeled nucleotide spot comigrating with the c-di-GMP standard. Mutants with defects in *scrG* show altered swarming, lateral flagellin production, and colony morphology (but not swimming motility), although the defects are not as severe as those displayed by the *scrC* phenotype. For example, swarming defects are most apparent on low-salt, swarming medium, which is a more restrictive condition for swarming and thus better articulates more subtle swarming defects (Fig. 4A). *scrG* mutants can be complemented by ectopic expression of *scrABC* (Fig. 4A). Thus, the in vivo activity of ScrG, like ScrC, appears to be that of a phosphodiesterase.

Similar to the experiments with ScrC, the removal of the EAL domain (or introduction of a point mutation in the signature motif, ESL→AAL) reverses ScrG activity, converting ScrG to an inhibitor of swarming and activator of *cps* expression. This is a surprising result as the GGDEF domain is not greatly conserved and in fact shows little identity with the signature motif (although there is conservation with other sites known to be critical in the active site but not the inhibitory site) of the diguanylate cyclase PleD, which has been cocrystallized with c-di-GMP (9) (see also chapter 7). The potential synthetic activity for ScrG$_{\Delta EAL}$ could be corroborated upon examination of the cellular nucleotide pools. Thus, ScrG may be like ScrC and have dual and opposing enzymatic activities; however, final proof requires purification and in vitro assays. An alternate but intriguing possibility for the phenotype is that ScrG$_{\Delta EAL}$ may negatively interfere with other c-di-GMP signaling proteins in the cell. Precedent for such interactions exists in *Xanthomonas axonopodis* pv. citri (1); it has been shown that an HD-GYP phosphodiesterase physically interacts with other GGDEF proteins in the cell. The prospect that certain GGDEF, EAL, and HD-GYP proteins might form a physical array of interacting sensory signal transducers could have important and exciting implications for signal reception, integration, and amplification of c-di-GMP circuits.

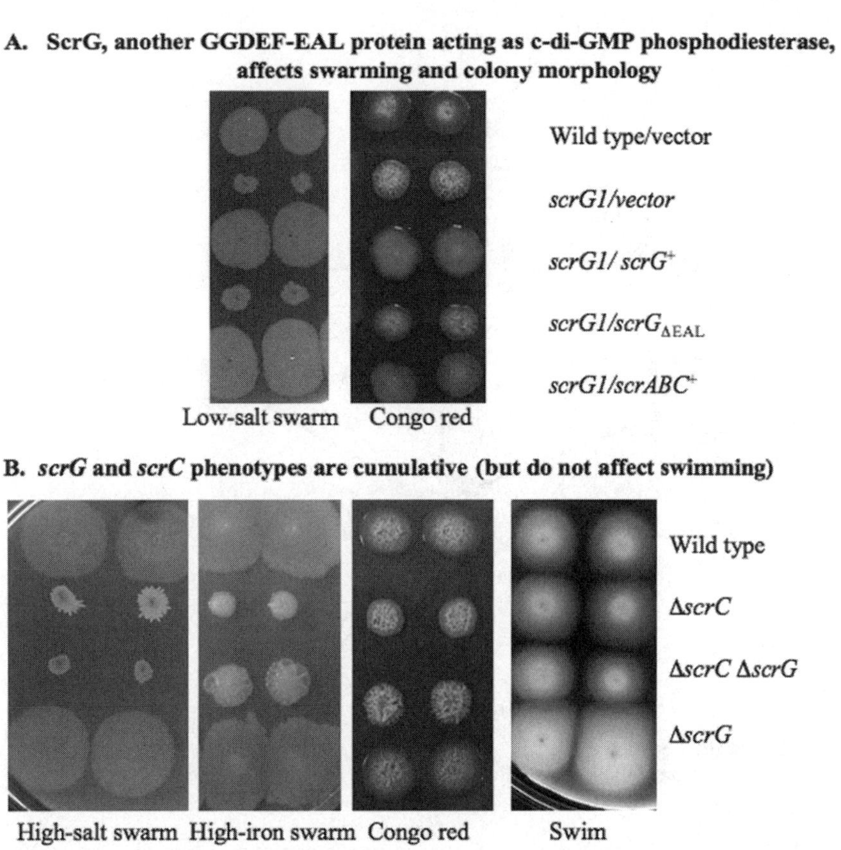

Figure 4. ScrG is a second GGDEF-EAL protein participating in the control of swarming and sticking. (A) Although the *scrG* phenotype is less profound than the *scrC* phenotype on standard swarming agar, defects become apparent on low-salt swarming medium, which is less permissive for swarming than the standard high-salt swarming medium. The *scrG* phenotype can be complemented by *scrG*$^+$ or *scrABC*$^+$ but not by $scrG_{\Delta EAL}$. (B) The effects of *scrC* and *scrG* are cumulative on swarming and colony morphology but not on swimming. Wild-type and mutant strains were inoculated in high-salt swarm plates (1.5% agar), high-salt, iron-supplemented (enhances colony rugosity) swarm plates (1.5% agar), Congo red medium (2% agar), and semisolid swim medium (0.3% agar). ScrG mutants (row 4) show slight impairment of swarming on high-salt agar; however, in combination with an *scrC* deletion, the ability of the double mutant (row 3) to swarm is more defective than that of the *scrC* single mutant strain (row 2). In addition, the colony morphology of the double mutant strain is more crinkly than either of the single mutant strains (shown on high-iron and Congo red media). The *scr* system has a negligible impact on swimming motility: all *scr* mutants show little difference in their ability to swim compared to the wild type.

Scr c-di-GMP CIRCUIT: TRANSCRIPTIONAL CONTROL OF SWARMING AND STICKING

Lateral flagellar and *cps* gene expression is altered in *scr* mutants: (i) *scrABC* mutants are dramatically affected with respect to swarming and sticking; (ii) *scrG* mutants have mildly defective swarming phenotypes; (iii) the *scrG scrC* double mutant has a more profoundly defective swarming phenotype than the single mutant strains measured with respect to swarming speed (Fig. 4B), lateral flagellin protein production, and *laf* gene expression; (iv) *scrG scrC* double mutants also display super crinkly colony morphologies compared to the single-mutant parents (Fig. 4B); and (v) ScrG and ScrC work through a common *cpsA* regulator (CpsR, described below) to affect crinkly colony morphology (and *cpsA::lacZ* transcription). Taken together, such information suggests there may be a circuit of Scr proteins capable of integrating diverse input signals, e.g., through the periplasmic domains of ScrB and ScrC and the PAS domain of ScrG and influencing the cellular nucleotide pool: a low concentration of c-di-GMP favors surface mobility, whereas high levels of this nucleotide promote a more adhesive cell type (Fig. 5).

The original identification of the roles for *scrABC* and *scrG* resulted from screening a cosmid library for clones stringently regulating—either up or down—the expression of a *laf::lux* fusion. The cosmid vector for the library had a low copy number (~5) and contained very large inserts of genomic DNA (~35 kb); thus, most genes on each clone were probably expressed from their native promoters. Approximately 15 clones were obtained that showed

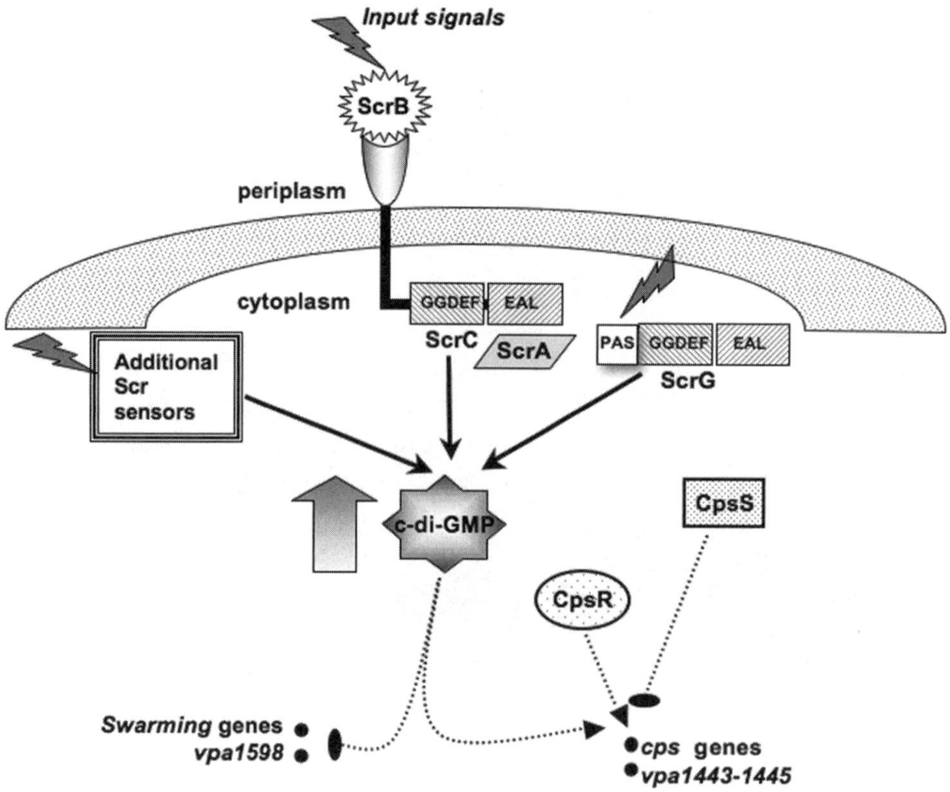

Figure 5. Model of the Scr c-di-GMP circuit in *V. parahaemolyticus:* transcriptional control of swarming and sticking. Two GGDEF-EAL proteins, ScrC and ScrG, regulate swarming and sticking gene expression by acting as phosphodiesterases to modulate the level of c-di-GMP. The phosphodiesterase activity of ScrC is controlled by interaction with ScrA, a predicted pyridoxal phosphate-dependent aminotransferase, and ScrB, a predicted periplasmic solute-binding protein. Potentially, input signals transmitted via the periplasmic sensing domain of ScrC and the PAS domain of ScrG may moderate the activity of the output GGDEF-EAL domains. High levels of c-di-GMP promote production of CPS production and biofilm formation, and low levels of this signaling molecule promote swarming motility. Additional output targets include a predicted cell surface adhesin binding chitin or *N*-acetylglucosamine (VPA1598) and the *mfp* operon, which encodes a type I membrane fusion transport system that is known to play a role in biofilm development (VPA1443-5). The nature of the input signals and how alterations in the cellular c-di-GMP pool affect transcription of lateral flagellar and *cps* genes are not known, although Cps-specific transcriptional regulators (CpsR and CpsS) participating in this circuit have been identified. This model also predicts the existence of additional GGDEF-, EAL-, or HD-GYP-type Scr sensors, and current work is focused on identifying new members of the Scr sensory array.

strong alteration of *laf* gene expression (producing either light in liquid or being dark on plates). Of these, approximately half encoded candidate regulators with GGDEF, EAL, or HD domains, and it seems likely that more candidate c-di-GMP-pertinent regulatory clones would have been identified had the criteria for affecting regulation been less stringent. In addition, some of these same genes as well as new candidates have been identified in mutant suppressor screens for the *scrABC* phenotype (McCarter, unpublished). Thus, we suspect the existence of a network of potential *scr* members capable of modulating the levels of c-di-GMP and influencing swarming and sticking.

Curiously and different from other systems in which this has been examined, even in other *Vibrio* such as *V. cholerae* and *Vibrio fischeri* (55), alterations of the c-di-GMP pool in *V. parahaemolyticus* have little effect on swimming motility (Fig. 4B) (27). Perhaps this is because *V. parahaemolyticus* possesses two distinct flagellar systems: the polar, which is dedicated to swimming in liquid, and the lateral, which is designed for optimal movement over surfaces and through viscous environments. This capacity resulting from dual modes of locomotion is unlike the abovementioned *Vibrio* species, which have only polar flagella, as well as many of the other organisms, which also have only one flagellar system, in which the relationship between c-di-GMP and motility has been investigated. The polar motility organelle of *V. parahaemolyticus* is also a sensory organ: by measuring rotation speed, it serves to inform the bacterium of contact with surfaces (26, 33). Perhaps because of this additional, special role for the polar flagellum, its transcription and production are relatively constant and insulated from regulatory effects of c-di-GMP.

How alterations in the cellular c-di-GMP pool affect transcription of lateral flagellar and *cps* genes is not yet known, although one Cps-specific transcriptional regulator (CpsR) necessary for this circuit has been characterized. CpsR was identified in a screen for suppressor mutations that would revert the Scr phenotype (19). The double mutant *scrA cpsR* strain produces a smooth colony morphology but is still unable to swarm. CpsR is a 444-aa predicted transcription factor; it contains an AAA-type ATPase domain (cd00009; E value = 2e−12) and a DNA-binding regulatory domain in the FIS family (cl01091; E value = 3e−04) domains. CpsR is not a σ^{54}-dependent activator (although it is annotated as such) because introduction of an *rpoN* mutation into *scr* strains has no consequence on the crinkly colony phenotype or *cpsA* transcription. CpsR is homologous to *V. cholerae* VpsR (78% identities/90% positives in a BLASTP alignment). There are striking similarities and differences between these two regulators. Both CpsR and VpsR modulate extracellular polysaccharide gene expression, and both appear to be necessary for c-di-GMP-mediated events (19, 31). *V. cholerae* VpsR is essential for *V. cholerae* extracellular polysaccharide production and, hence, biofilm formation, as it seems is the *V. fischeri* homolog, also named VpsR (13, 56). However, CpsR is not essential for *V. parahaemolyticus* extracellular polysaccharide production or biofilm formation. Introduction of a *cpsR* lesion into the wild-type strain has little effect on CPS production (assessed by colony morphotype and adhesion to glass) or *cps* expression (measured by using a *lacZ* transcriptional fusion). A clear role for CpsR becomes apparent only when c-di-GMP levels are elevated: transcription of a *cpsA::lacZ* reporter is increased about fivefold in an *scrA* mutant strain and is similar to the wild-type level in the *cpsR scrA* double mutant (19). Why this is so is not known, for c-di-GMP has no effect on transcription of *cpsR*. Perhaps c-di-GMP affects the translation or activity of CpsR.

Alternately, CpsR may only be an incidental or upstream player in the complex web of *cps* regulation. Another known regulator of *cps* in *V. parahaemolyticus* is CpsS (19). CpsS also has a *V. cholerae* homolog, VpsT, that has been implicated in the c-di-GMP circuit (3, 53). CpsS and VpsT provide another interesting wrinkle of difference among members of the *Vibrionaceae* in the cascade of *cps* regulation. These two proteins are homologous CsgD-like transcription factors (with the conserved domain PRK10100) that regulate extracellular polysaccharide gene expression in entirely opposite ways. CpsS is a negative regulator (19), and VpsT is a positive regulator (8, 31). Mutants with lesions in *cpsS* have very crinkly colony phenotypes, which are more extreme than the *scrABC* phenotype, the consequence of *cps* expression being elevated ~30-fold over wild-type levels. Such a striking phenotype reveals that *V. parahaemolyticus* has a wide capacity to regulate its levels of CPS production. *cpsR* is epistatic to *cpsS* as the level of *cpsA::lacZ* expression in the double *cpsR cpsS* is similar to that of the *cpsR* mutant; however, *cpsS* does not affect transcription of *cpsR* (19). How and where c-di-GMP enters into this nest of *cps* regulators is currently being unraveled. Similarly, little is known about c-di-GMP control of lateral flagellar gene expression, except that it must occur at a level prior to the regulation of the master flagellar transcriptional regulator LafK because transcription of a *lafK::lacZ* fusion is influenced by c-di-GMP fluctuations (McCarter, unpublished).

POTENTIAL SCOPE OF Scr SIGNALING

How Many Potential Scr Sensors Exist?

The enzymes controlling the level of c-di-GMP include proteins with GGDEF and EAL domains as well as the phosphohydrolase-associated HD-GYP domain (17, 44). The *scrABC* locus appears to be one major participant in setting the c-di-GMP balance in the swarmer cell, and mutants with defects in this locus have a strong phenotype. The *scrABC* operon is specifically induced by growth on a surface (16). However, *V. parahaemolyticus* possesses about 60 proteins that can potentially form or degrade c-di-GMP (Table 1). The genes encoding these enzymes occur about equally on both chromosomes. A little more than half are predicted to encode membrane proteins; and the architecture of the majority of these proteins also includes potential interaction or signal reception domains. Why are there so many? Do many of these proteins have a dedicated function or can they be integrated and capable of controlling a common cellular nucleotide pool? Certainly, c-di-GMP has been shown to influence a number of activities, and it does so in many different ways. It was first discovered to allosterically control activity of a cellulose biosynthetic enzyme (32). c-di-GMP has since been shown to bind to the PilZ domain of numerous proteins to control their activities (2, 10, 34, 39, 45, 54), to other non-PilZ domain-containing proteins such as PelD, which has a fold common to the inhibitory site of diguanylate cyclases (30) and to the flagellar transcription factor FleQ (20), or to mRNA

Table 1. *V. parahaemolyticus* proteins with GGDEF, EAL, or HD domains[a]

Protein	Size (aa)	Location[b]	Position of indicated conserved domain (E value)			
			GGDEF[c]	EAL[c]	HD-GYP[d]	Other
VP tag (chromosome 1)						
71	1,046	M (2)	614–775 (1.4e−40)	787–1033 (7.1e−59)		PAS_4, 502–609 (0.0012); ABC_sub_bind, 27–324 (0.11)
92	811	C	390–548 (8.8e−55)	562–803 (1.3e−113)		PAS_4, 152–263 (1e−07) and 275–385 (5.1e−05)
117	848	M (2)	423–578 (3.1e−25)	592–834 (1.1e−95)		HAMP, 348–418 (1.2e−15); CACHE_1, 184–267 (0.0075)
354	626	M (1)	461–615 (2e−45)			PAS_3, 362–454 (4.8e−13)
376	250	C	10–232 (1.5e−07)			
486	581	M (1)	408–573 (1.6e−65)			PAS_4, 164–255 (0.089) and 285–396 (5.7e−11)
699	521	C	358–517 (1.1e−35)			priB_priC, 168–332 (0.48)
953	667	M (1)	507–657 (7.8e−25)			SBP_bac_3, 250–473 (7.5e−06)
1255	445	M (1)	274–430 (1.4e−41)			CIA30, 29–213 (1.3e−05)
1289	381	C	202–363 (6.6e−52)			Hemerythrin, 13–70 (0.00049) and 72–131 (0.00064)
1377 (ScrG)	599	C	173–331 (1.4e−06)	345–586 (5.9e−79)		PAS_3, 75–166 (9.1e−19)
1378	403	C			35–360 (2e−56)	
1423	421	C			200–421 (2.0e−44)	
1427	374	C			170–360 (6e−16)	
1441	513	M (2)	353–508 (9e−40)			CHASE, 78–299 (2.7e−09)
1483	320	C	152–313 (3.9e−56)			PAS_4, 18–140 (0.2)
1637	658	M (1)	479–640 (1.9e−34)			TPR_1, 256–289 (0.0015), 296–329 (0.1), 336–369 (0.78), and 376–409 (0.27)
1754	719	M (2)	272–438 (1.8e−26)	452–694 (4.9e−109)		HAMP, 199–267 (2.6e−12)
1768	502	M (2)		256–489 (1e−66)		
1881	571	M (3)		343–568 (8.1e−21)		trans_reg_C, 28–112 (0.26)
1979	818	C	256–402 (0.52)	417–658 (5.5e−68)		PAS, 128–238 (7.4e−06); CBS_pair, 700–814 (5e−07)
2076	486	M (2)	328–483 (3.5e−43)			
2366	306	C	137–297 (1.3e−52)			PAS_4, 12–132 (0.028)
2446	679	C	249–406 (3.2e−55)	420–661 (1.1e−109)		PAS_4, 14–121 (0.00085)
2631	417	C			15–360 (1e−55)	
2708	669	M (2)	224–382 (2.8e−10)	397–639 (1.1e−41)		PriCT_2, 328–408 (0.09)
2888	396	C	210–371 (1.4e−52)			PAS_4, 81–205(0.07)
2979	828	C	409–568 (0.017)	590–824 (2.8e−47)		DUF1745, 60–270 (3.6e−61)

VPA tag (chromosome 2)						
59	358	M (5)	187–358 (4e−52)			7TMR-DISM_7TM, 10–175 (0.4)
68	363	M (2)	11–363 (3.7e−33)			
184	418	M (7)	238–396 (3.4e−44)			7TMR-DISM_7TM, 25–217 (0.99)
198	451	M (2)	290–451 (2.3e−47)			CHASE, 42–248 (3.9e−16)
202	337	C	74–333 (4.4e−49)			
360	338	M (6)	290–451 (2.3e−47)			
476	355	C	179–340 (4.1e−53)			GAF, 36–170 (2.5e−19)
518	606	C	173–327 (2e−21)	345–588 (1.3e−81)		GAF, 18–153 (0.62)
556	480	M (2)	306–461 (1.8e−40)			SBP_bac_3, 41–262 (6.6e−07)
594	488	M (2)		239–476 (5e−78)		
609	533	M (3)	101–256 (1.3e−37)	270–510 (2.4e−80)		DUF1145, 7–85 (0.89)
713	248	C		13–226 (3.1e−07)		
737	388	C		137–374 (9.6e−75)		Response_reg, 5–122 (4.7e−19)
739	456	M (2)	302–455 (4.5e−34)			
818	272	C		13–255 (1.1e−25)		
846	262	C		18–250 (2.7e−12)		
869	636	C	199–358 (2.8e−05)	372–612 (2.5e−96)		GAF, 14–151 (7.3e−13)
878	652	M (2)	480–640 (5e−62)			HAMP, 160–230 (1.3e−10); PAS_4, 253–353 (0.053) and 365–476 (0.18)
925	414	M (6)	234–408 (1.7e−40)			YibE_F, 43–238 (0.41)
927	323	C	151–316 (1.4e−47)			Response_reg, 12–125 (7.8e−28)
1069	637	M (2)	468–630 (5.6e−49)			TPPR_1, 195–228 (0.01)
1088	787	M (2)	336–502 (5.2e−40)	516–754 (1.8e−68)		HAMP, 262–331 (0.00011)
1104	907	M (2)	475–629 (2.5e−07)	646–891 (9.1e−100)		CHASE4, 58–223 (1.2e−06); PAS, 358–465 (5.2e−06)
1115	682	M (2)	463–675 (1.2e−43)			TPR_1, 152–185 (0.76) and 232–265 (0.33)
1176	763	M (2)	332–488 (1.8e−47)	502–742 (5.4e−107)		
1324	230	C		3–226 (1.9e−32)		
1429	681	M (9)		437–670 (5.2e−57)		PTS_EIIC, 34–330 (3.1e−05)
1457	710	M (1)	528–694 (2.3e−42)			GAF, 195–333 (0.45) and 357–494 (0.37)
1478	527	C	365–522 (6.6e−34)			TPR_1, 122–155 (0.66)
1511 (ScrC)	774	M (2)	343–504 (2.5e−60)	518–758 (1.4e−104)		
1735	637	M (2)	233–388 (3e−09)	403–637 (7e−19)		HAMP, 152–221 (0.00049); DUF1655, 45–95 (0.19)

[a] Predictions were made using the *Vibrio parahaemolyticus* RIMD 2210633 genome.
[b] The predicted cellular location is indicated by C for cytoplasm and M for membrane. The number of potential transmembrane domains is given in parentheses. Predictions of the number of transmembrane regions were performed at http://www.ch.embnet.org/software/TMPRED_form.html.
[c] Conserved domain searches of the pfam database were performed at http://motif.genome.jp/ (GGDEF, PF00990; EAL, PF00563).
[d] Conserved domain searches for the HD-GYP domain (COG2206) were performed at http://imgweb.jgi-psf.org/cgi-bin/w/main.cgi?section=FindFunctions&page=findFunctions.

and acting to flip a riboswitch (49). There most probably is segregation of function for many of the GGDEF, EAL, and HD-GYP type proteins in the cell, via timing of gene expression, stimulation of activity by highly specific signaling events, spatial localization and sequestration, and product/substrate allosteric control of activity. So there may be different available pools of c-di-GMP and highly specific c-di-GMP-mediated events, and yet, there is also the potential for interaction via signal integration—the activities of these many enzymes may directly or indirectly influence a common nucleotide pool, which in turn may modulate certain activities.

The latter scenario could be most pertinent to the reciprocal regulation of swarming and sticking by c-di-GMP in *V. parahaemolyticus* and perhaps other bacteria. c-di-GMP signaling is also implicated in the switch between swarming and biofilm formation in *Pseudomonas aeruginosa* (29, 35). Differentiation to the swarmer cell state is a significant commitment for the cell, and using c-di-GMP as a true second messenger would be an excellent strategy to monitor and integrate diverse environmental inputs to make a decision as to whether it is advantageous to commit to the swarming differentiation pathway. Swarming for many organisms, not just for *V. parahaemolyticus,* is clearly affected by numerous cellular processes, e.g., in *E. coli* almost 10% of the nonessential genome can contribute to swarming. This large number was found upon screening a complete transposon library for mutants with defects in swarming (23). Similarly, many genes have been shown by mutational analyses to affect swarming in *Pseudomonas aeruginosa* and *Proteus mirabilis* (36, 41). Clearly some of the many genes within the *V. parahaemolyticus* genome encoding enzymes capable of modulating the c-di-GMP pool may have more importance than others with respect to swarming. Some genes may have quite dedicated and distinct functions, yet the potential for any enzyme that forms or degrades c-di-GMP to affect the cellular pool exists.

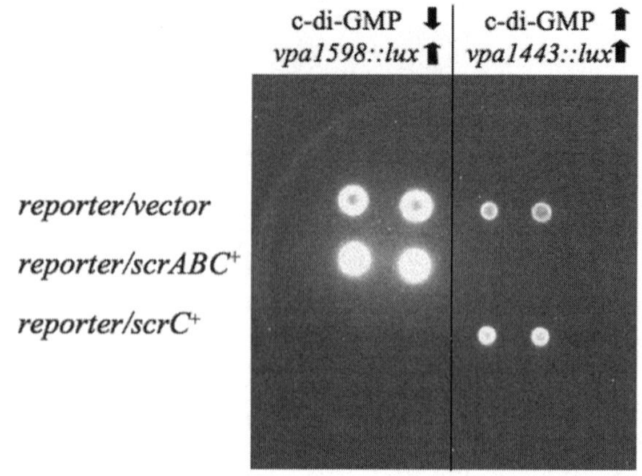

Figure 6. Additional output targets of the Scr circuit. c-di-GMP affects transcription of two known additional targets. Colonies containing bioluminescence *(lux)* fusions were grown overnight on plates and photographed in the dark. All colonies grew equally well on the plate, but only the luminous colonies appear in the photograph. The luminescence of *vpa1598::lux* is enhanced by low levels of c-di-GMP (the consequence of overexpression of *scrABC*) and repressed by high levels of c-di-GMP (upon overexpression of *scrC*) compared to strains carrying a vector control. In contrast, the *lux* fusion in *vpa1443* is regulated oppositely, i.e., light production is enhanced by high levels of c-di-GMP and repressed by low levels of the nucleotide signal. Both genes encode predicted cell surface moieties. VPA1598 encodes a chitin- or *N*-acetylglucosamine-binding protein. VPA1443 is in the *mfp* operon, which encodes an RTX transporter system known to participate in biofilm development and affect colony morphology. Thus the Scr circuit, originally named for *s*warming and *c*ps *r*egulation, influences *s*warming and *c*ell surface *r*egulation.

How Many *scr* Targets Exist?

Although the full scope of output for the *scr* system is not yet known, some additional target genes have been discovered, and they encode molecules pertinent to the cell surface. These genes show altered regulation in an *scrABC* mutant compared to the wild type, and their transcription is influenced by overexpression of *scrABC* and *scrC* (Fig. 6) (McCarter, unpublished). These new targets include VPA1598, which although annotated as a chitinase is more likely a chitin-binding or *N*-acetylglucosamine-binding protein as it lacks the enzymatic domain of a true chitinase, and the *mfp* operon (VPA1443-45), which encodes a type I secretion ABC transporter, a type I secretion *m*embrane *f*usion *p*rotein, and a potential calcium-binding secreted substrate. VPA1598 was first identified in a search for nonflagellar genes that were induced by growth on a surface (48); the *mfp* operon was identified as a genetic determinant participating in biofilm formation and contributing to colony morphology on Congo red (15). Transcription of VPA1598, encoding the chitin-binding protein, is enhanced by low levels of c-di-GMP and repressed by high levels of c-di-GMP; whereas expression of the *mfp* operon is regulated oppositely. Thus, the meaning of the *scr* gene designation, which originally denoted *s*warming and *c*ps *r*egulation, can be expanded to encompass *s*warming and *c*ell surface *r*egulation.

What Signals Are Being Integrated?

Presently, little is known about the nature of the input signals. We know that the Scr circuit is not integral to surface sensing per se, as *scr* mutants retain their ability to discriminate between life in liquid and life on a surface (6). A more likely role for Scr sensing is modulatory, to influence the propensity to be motile or adhesive once the bacterium is on a surface as a function of diverse environmental cues. This is intriguing to think about in the context of the cueing that must occur in the cycles of swarmer cell differentiation and dedifferentiation producing terraced, bull's-eye swarming patterns on plates. With respect to the specific signals influencing this decision between swarming and sticking, we suspect that ScrB is the receptor for a small molecule, perhaps an amino acid, as it looks like a periplasmic binding protein. ScrG, with its PAS domain, could be responsive to redox potential, cellular energy, oxygen, or other ligands. Although the phenotype of *scrG* is more severe under low-salt than high-salt conditions, this does not necessarily indicate an input cue. Environmental conditions produce potential signature input signals but can also influence the expression of different sets of sensor enzymes, and both will ultimately affect some shared c-di-GMP pool. Thus, signal integration is rather sophisticated because the network of Scr sensors is not static. For example, the expression of the *scrABC* operon is induced 8- to 10-fold by growth on a surface (16), and expression of some other genes encoding GGDEF- and EAL-type proteins also changes between growth in liquid and on a surface (McCarter, unpublished). This adds another dimension of regulation whereby input signals can have differing weights in a signal integration scheme as a direct consequence of their presence, absence, or concentration or as a function of input recognition and signal transduction, which would depend upon the particular panoply of sensors being produced at a given time. A variable array of Scr sensors would produce altered output responses in response to similar input cues; this translates into complex and more flexible decision making.

SUMMARY AND PERSPECTIVES

In possessing a multiplicity of cell types appropriate for life under different circumstances, the swimmer, swarmer, and sticky cell types, the ubiquitous marine bacterium and human pathogen *Vibrio parahaemolyticus* must make certain critical lifestyle determinations. c-di-GMP plays a key role by influencing this decision-making process. The *scr* genes represent a GGDEF-EAL circuit integrating multiple regulators influencing global programs of gene control that are pertinent to growth on surfaces, in particular, swarmer cell differentiation and biofilm development. The *scr* genes modulate expression of gene sets both positively and negatively; these gene sets are pertinent to surface motility and colonization. Although only two sensors have been characterized in detail, genomic and genetic analyses suggest a wide capacity to detect and integrate diverse stimuli by modulating a pool of c-di-GMP; thus, we suspect a network of Scr sensors. The advantage may be to allow a rapid and coordinated response to diverse, contradictory, or oscillating inputs. Such a sensory array of c-di-GMP modulators would enable a hierarchical decision-making or a conflict-resolving strategy to allow the bacterium to make the most appropriate choice between moving and sticking.

Acknowledgments. I thank all the members of my laboratory who have contributed to this project—by working directly on the c-di-GMP project or through cross-fertilization of ideas.

This work has been funded by the National Science Foundation (MCB0315617 and MCB0817593).

REFERENCES

1. **Andrade, M. O., M. C. Alegria, C. R. Guzzo, C. Docena, M. C. Rosa, C. H. Ramos, and C. S. Farah.** 2006. The HD-GYP domain of RpfG mediates a direct linkage between the Rpf quorum-sensing pathway and a subset of diguanylate cyclase proteins in the phytopathogen *Xanthomonas axonopodis* pv citri. *Mol. Microbiol.* **62:**537–551.
2. **Benach, J., S. S. Swaminathan, R. Tamayo, S. K. Handelman, E. Folta-Stogniew, J. E. Ramos, F. Forouhar, H. Neely, J. Seetharaman, A. Camilli, and J. F. Hunt.** 2007. The structural basis of cyclic diguanylate signal transduction by PilZ domains. *EMBO J.* **26:**5153–5166.
3. **Beyhan, S., K. Bilecen, S. R. Salama, C. Casper-Lindley, and F. H. Yildiz.** 2007. Regulation of rugosity and biofilm formation in *Vibrio cholerae:* comparison of VpsT and VpsR regulons and epistasis analysis of *vpsT, vpsR,* and *hapR. J. Bacteriol.* **189:**388–402.
4. **Beyhan, S., A. D. Tischler, A. Camilli, and F. H. Yildiz.** 2006. Transcriptome and phenotypic responses of *Vibrio cholerae* to increased cyclic di-GMP level. *J. Bacteriol.* **188:**3600–3613.
5. **Bobrov, A. G., O. Kirillina, and R. D. Perry.** 2005. The phosphodiesterase activity of the HmsP EAL domain is required for negative regulation of biofilm formation in *Yersinia pestis. FEMS Microbiol. Lett.* **247:**123–130.
6. **Boles, B. R., and L. L. McCarter.** 2002. *Vibrio parahaemolyticus scrABC,* a novel operon affecting swarming and capsular polysaccharide regulation. *J. Bacteriol.* **184:**5946–5954.
7. **Bomchil, N., P. Watnick, and R. Kolter.** 2003. Identification and characterization of a *Vibrio cholerae* gene, *mbaA,* involved in maintenance of biofilm architecture. *J. Bacteriol.* **185:**1384–1390.
8. **Casper-Lindley, C., and F. H. Yildiz.** 2004. VpsT is a transcriptional regulator required for expression of *vps* biosynthesis genes and the development of rugose colonial morphology in *Vibrio cholerae* O1 El Tor. *J. Bacteriol.* **186:**1574–1578.
9. **Chan, C., R. Paul, D. Samoray, N. C. Amiot, B. Giese, U. Jenal, and T. Schirmer.** 2004. Structural basis of activity and allosteric control of diguanylate cyclase. *Proc. Natl. Acad. Sci. USA* **101:**17084–17089.
10. **Christen, M., B. Christen, M. G. Allan, M. Folcher, P. Jeno, S. Grzesiek, and U. Jenal.** 2007. DgrA is a member of a new

family of cyclic diguanosine monophosphate receptors and controls flagellar motor function in *Caulobacter crescentus*. *Proc. Natl. Acad. Sci. USA* **104:**4112–4117.
11. **Cotter, P. A., and S. Stibitz.** 2007. c-di-GMP-mediated regulation of virulence and biofilm formation. *Curr. Opin. Microbiol.* **10:**17–23.
12. **D'Argenio, D. A., M. W. Calfee, P. B. Rainey, and E. C. Pesci.** 2002. Autolysis and autoaggregation in *Pseudomonas aeruginosa* colony morphology mutants. *J. Bacteriol.* **184:**6481–6489.
13. **Darnell, C. L., E. A. Hussa, and K. L. Visick.** 2008. The putative hybrid sensor kinase SypF coordinates biofilm formation in *Vibrio fischeri* by acting upstream of two response regulators, SypG and VpsR. *J. Bacteriol.* **190:**4941–4950.
14. **Dow, J. M., Y. Fouhy, J. F. Lucey, and R. P. Ryan.** 2006. The HD-GYP domain, cyclic di-GMP signaling, and bacterial virulence to plants. *Mol. Plant-Microbe Interact.* **19:**1378–1384.
15. **Enos-Berlage, J. L., Z. T. Guvener, C. E. Keenan, and L. L. McCarter.** 2005. Genetic determinants of biofilm development of opaque and translucent *Vibrio parahaemolyticus*. *Mol. Microbiol.* **55:**1160–1182.
16. **Ferreira, R. B., L. C. Antunes, E. P. Greenberg, and L. L. McCarter.** 2008. *Vibrio parahaemolyticus* ScrC modulates cyclic dimeric GMP regulation of gene expression relevant to growth on surfaces. *J. Bacteriol.* **190:**851–860.
17. **Galperin, M. Y.** 2004. Bacterial signal transduction network in a genomic perspective. *Environ. Microbiol.* **6:**552–567.
18. **Galperin, M. Y., and A. N. Nikolskaya.** 2007. Identification of sensory and signal-transducing domains in two-component signaling systems. *Methods Enzymol.* **422:**47–74.
19. **Guvener, Z. T., and L. L. McCarter.** 2003. Multiple regulators control capsular polysaccharide production in *Vibrio parahaemolyticus*. *J. Bacteriol.* **185:**5431–5441.
20. **Hickman, J. W., and C. S. Harwood.** 2008. Identification of FleQ from *Pseudomonas aeruginosa* as a c-di-GMP-responsive transcription factor. *Mol. Microbiol.* **69:**376–389.
21. **Hickman, J. W., D. F. Tifrea, and C. S. Harwood.** 2005. A chemosensory system that regulates biofilm formation through modulation of cyclic diguanylate levels. *Proc. Natl. Acad. Sci. USA* **102:**14422–14427.
22. **Huang, B., C. B. Whitchurch, and J. S. Mattick.** 2003. FimX, a multidomain protein connecting environmental signals to twitching motility in *Pseudomonas aeruginosa*. *J. Bacteriol.* **185:**7068–7076.
23. **Inoue, T., R. Shingaki, S. Hirose, K. Waki, H. Mori, and K. Fukui.** 2007. Genome-wide screening of genes required for swarming motility in *Escherichia coli* K-12. *J. Bacteriol.* **189:**950–957.
24. **Jenal, U., and J. Malone.** 2006. Mechanisms of cyclic-di-GMP signaling in bacteria. *Annu. Rev. Genet.* **40:**385–407.
25. **Karatan, E., T. R. Duncan, and P. I. Watnick.** 2005. NspS, a predicted polyamine sensor, mediates activation of *Vibrio cholerae* biofilm formation by norspermidine. *J. Bacteriol.* **187:**7434–7443.
26. **Kawagishi, I., M. Imagawa, Y. Imae, L. McCarter, and M. Homma.** 1996. The sodium-driven polar flagellar motor of marine *Vibrio* as the mechanosensor that regulates lateral flagellar expression. *Mol. Microbiol.* **20:**693–699.
27. **Kim, Y. K., and L. L. McCarter.** 2007. ScrG, a GGDEF-EAL protein, participates in regulating swarming and sticking in *Vibrio parahaemolyticus*. *J. Bacteriol.* **189:**4094–4107.
28. **Kirillina, O., J. D. Fetherston, A. G. Bobrov, J. Abney, and R. D. Perry.** 2004. HmsP, a putative phosphodiesterase, and HmsT, a putative diguanylate cyclase, control Hms-dependent biofilm formation in *Yersinia pestis*. *Mol. Microbiol.* **54:**75–88.
29. **Kuchma, S. L., K. M. Brothers, J. H. Merritt, N. T. Liberati, F. M. Ausubel, and G. A. O'Toole.** 2007. BifA, a cyclic-di-GMP phosphodiesterase, inversely regulates biofilm formation and swarming motility by *Pseudomonas aeruginosa* PA14. *J. Bacteriol.* **189:**8165–8178.
30. **Lee, V. T., J. M. Matewish, J. L. Kessler, M. Hyodo, Y. Hayakawa, and S. Lory.** 2007. A cyclic-di-GMP receptor required for bacterial exopolysaccharide production. *Mol. Microbiol.* **65:**1474–1484.
31. **Lim, B., S. Beyhan, J. Meir, and F. H. Yildiz.** 2006. Cyclic-diGMP signal transduction systems in *Vibrio cholerae:* modulation of rugosity and biofilm formation. *Mol. Microbiol.* **60:**331–348.
32. **Mayer, R., P. Ross, H. Weinhouse, D. Amikam, G. Volman, P. Ohana, R. D. Calhoon, H. C. Wong, A. W. Emerick, and M. Benziman.** 1991. Polypeptide composition of bacterial cyclic diguanylic acid-dependent cellulose synthase and the occurrence of immunologically crossreacting proteins in higher plants. *Proc. Natl. Acad. Sci. USA* **88:**5472–5476.
33. **McCarter, L., M. Hilmen, and M. Silverman.** 1988. Flagellar dynamometer controls swarmer cell differentiation of *V. parahaemolyticus*. *Cell* **54:**345–351.
34. **Merighi, M., V. T. Lee, M. Hyodo, Y. Hayakawa, and S. Lory.** 2007. The second messenger bis-(3′-5′)-cyclic-GMP and its PilZ domain-containing receptor Alg44 are required for alginate biosynthesis in *Pseudomonas aeruginosa*. *Mol. Microbiol.* **65:**876–895.
35. **Merritt, J. H., K. M. Brothers, S. L. Kuchma, and G. A. O'Toole.** 2007. SadC reciprocally influences biofilm formation and swarming motility via modulation of exopolysaccharide production and flagellar function. *J. Bacteriol.* **189:**8154–8164.
36. **Overhage, J., S. Lewenza, A. K. Marr, and R. E. Hancock.** 2007. Identification of genes involved in swarming motility using a *Pseudomonas aeruginosa* PAO1 mini-Tn*5*-*lux* mutant library. *J. Bacteriol.* **189:**2164–2169.
37. **Paul, R., S. Weiser, N. C. Amiot, C. Chan, T. Schirmer, B. Giese, and U. Jenal.** 2004. Cell cycle-dependent dynamic localization of a bacterial response regulator with a novel diguanylate cyclase output domain. *Genes Dev.* **18:**715–727.
38. **Pesavento, C., G. Becker, N. Sommerfeldt, A. Possling, N. Tschowri, A. Mehlis, and R. Hengge.** 2008. Inverse regulatory coordination of motility and curli-mediated adhesion in *Escherichia coli*. *Genes Dev.* **22:**2434–2446.
39. **Pratt, J. T., R. Tamayo, A. D. Tischler, and A. Camilli.** 2007. PilZ domain proteins bind cyclic diguanylate and regulate diverse processes in *Vibrio cholerae*. *J. Biol. Chem.* **282:**12860–12870.
40. **Rashid, M. H., C. Rajanna, A. Ali, and D. K. Karaolis.** 2003. Identification of genes involved in the switch between the smooth and rugose phenotypes of *Vibrio cholerae*. *FEMS Microbiol. Lett.* **227:**113–119.
41. **Rather, P. N.** 2005. Swarmer cell differentiation in *Proteus mirabilis*. *Environ. Microbiol.* **7:**1065–1073.
42. **Romling, U.** 2005. Characterization of the rdar morphotype, a multicellular behaviour in *Enterobacteriaceae*. *Cell. Mol. Life Sci.* **62:**1234–1246.
43. **Romling, U., and D. Amikam.** 2006. Cyclic di-GMP as a second messenger. *Curr. Opin. Microbiol.* **9:**218–228.
44. **Ryan, R. P., Y. Fouhy, J. F. Lucey, and J. M. Dow.** 2006. Cyclic di-GMP signaling in bacteria: recent advances and new puzzles. *J. Bacteriol.* **188:**8327–8334.
45. **Ryjenkov, D. A., R. Simm, U. Romling, and M. Gomelsky.** 2006. The PilZ domain is a receptor for the second messenger c-di-GMP: the PilZ domain protein YcgR controls motility in enterobacteria. *J. Biol. Chem.* **281:**30310–30314.

46. **Schmidt, A. J., D. A. Ryjenkov, and M. Gomelsky.** 2005. The ubiquitous protein domain EAL is a cyclic diguanylate-specific phosphodiesterase: enzymatically active and inactive EAL domains. *J. Bacteriol.* **187:**4774–4781.

47. **Simm, R., M. Morr, A. Kader, M. Nimtz, and U. Romling.** 2004. GGDEF and EAL domains inversely regulate cyclic di-GMP levels and transition from sessility to motility. *Mol. Microbiol.* **53:**1123–1134.

48. **Stewart, B. J.** 2002. *Investigating Surface-Induced Gene Expression: the Lateral Flagellar Gene System of* Vibrio parahaemolyticus. Ph.D. dissertation. The University of Iowa, Iowa City.

49. **Sudarsan, N., E. R. Lee, Z. Weinberg, R. H. Moy, J. N. Kim, K. H. Link, and R. R. Breaker.** 2008. Riboswitches in eubacteria sense the second messenger cyclic di-GMP. *Science* **321:**411–413.

50. **Tamayo, R., J. T. Pratt, and A. Camilli.** 2007. Roles of cyclic diguanylate in the regulation of bacterial pathogenesis. *Annu. Rev. Microbiol.* **61:**131–148.

51. **Tarutina, M., D. A. Ryjenkov, and M. Gomelsky.** 2006. An unorthodox bacteriophytochrome from *Rhodobacter sphaeroides* involved in turnover of the second messenger c-di-GMP. *J. Biol. Chem.* **281:**34751–34758.

52. **Tischler, A. D., and A. Camilli.** 2004. Cyclic diguanylate (c-di-GMP) regulates *Vibrio cholerae* biofilm formation. *Mol. Microbiol.* **53:**857–869.

53. **Waters, C. M., W. Lu, J. D. Rabinowitz, and B. L. Bassler.** 2008. Quorum sensing controls biofilm formation in *Vibrio cholerae* through modulation of cyclic di-GMP levels and repression of *vpsT*. *J. Bacteriol.* **190:**2527–2536.

54. **Weinhouse, H., S. Sapir, D. Amikam, Y. Shilo, G. Volman, P. Ohana, and M. Benziman.** 1997. c-di-GMP-binding protein, a new factor regulating cellulose synthesis in *Acetobacter xylinum*. *FEBS Lett.* **416:**207–211.

55. **Wolfe, A. J., and K. L. Visick.** 2008. Get the message out: cyclic-di-GMP regulates multiple levels of flagellum-based motility. *J. Bacteriol.* **190:**463–475.

56. **Yildiz, F. H., N. A. Dolganov, and G. K. Schoolnik.** 2001. VpsR, a member of the response regulators of the two-component regulatory systems, is required for expression of *vps* biosynthesis genes and EPS^{ETr}-associated phenotypes in *Vibrio cholerae* O1 El Tor. *J. Bacteriol.* **183:**1716–1726.

The Second Messenger Cyclic Di-GMP
Edited by A. J. Wolfe and K. L. Visick

Chapter 13

Roles of Diguanylate Cyclases and Phosphodiesterases in Motility and Biofilm Formation in *Vibrio fischeri*

ALAN J. WOLFE AND KAREN L. VISICK

Cyclic di-GMP (c-di-GMP) plays important roles in lifestyle choices, notably in the decision to be motile and/or to form a biofilm (reviewed in chapter 8). This is true for *Vibrio fischeri,* a marine microorganism that encodes over 50 proteins predicted to synthesize, degrade, or bind c-di-GMP. A subset of these c-di-GMP metabolism genes are known to be involved in influencing flagellar biogenesis, while others have been shown to influence biofilm formation. Here, we will first describe the biology of *V. fischeri* and the environmental influences that may dictate its need for large numbers of proteins dedicated to c-di-GMP control and function. We then will present a bioinformatic analysis of these c-di-GMP-associated proteins. Next, we will provide a brief description of flagellar biogenesis and the roles played in controlling that process by both the magnesium cation (Mg^{2+}) and specific c-di-GMP-associated enzymes. We then will outline the current knowledge concerning the role of c-di-GMP in cellulose biosynthesis and biofilm formation in *V. fischeri*. Finally, we will present models for the roles of c-di-GMP in biofilm formation and flagellar biogenesis.

INTRODUCTION TO THE BIOLOGY OF *V. FISCHERI*

V. fischeri is a marine bioluminescent microorganism that exists both as a free-living organism in seawater and in associations with animals, including fishes and squids. For a long time, *V. fischeri* was best known for its ability to bioluminesce using a cell density detection system now termed quorum sensing. This fame was well deserved, as studies of *V. fischeri* bioluminescence resulted in the discovery of the canonical LuxI/LuxR quorum-sensing regulators now known to exist in most gram-negative bacteria studied to date (10).

In the past 20 years, *V. fischeri* has gained additional fame as an important model symbiont: this microbe forms an exclusive relationship with the small Hawaiian squid *Euprymna scolopes,* an association that is experimentally tractable and used as a natural model of bacterium-animal interactions (33, 46, 50, 53). Though not a pathogen itself, *V. fischeri* is closely related to important human pathogens, including *Vibrio cholerae* (which causes cholera), *Vibrio parahaemolyticus* (which causes gastroenteritis), and *Vibrio vulnificus* (which produces septicemia and is responsible for the majority of *Vibrio*-associated deaths in the United States) (29). Thus, studies of the *V. fischeri*-squid association have significant potential for illuminating both symbiotic and pathogenic associations.

Adult *E. scolopes* contain a symbiotic organ, termed a "light organ," where *V. fischeri* cells exist at a cell density greater than 10^{10} cells per cm^3 (32). The bioluminescence from *V. fischeri* is used by the squid to avoid detection by predators using a mechanism known as counterillumination (18, 55). In this phenomenon, the nocturnal animals direct the bacterial bioluminescence downward to avoid casting a shadow, due to the down-welling moonlight, that can be detected by predators. In return, *E. scolopes* provides *V. fischeri* with nutrients, in the form of peptides, and a protected niche (13).

In contrast to the adult, newly hatched juveniles are not colonized. Within hours of hatching, however, the nascent light organs of these animals become colonized by *V. fischeri* (Fig. 1). Only *V. fischeri* cells colonize, despite the presence of numerous

Alan J. Wolfe and Karen L. Visick • Department of Microbiology and Immunology, Loyola University Chicago, Maywood, IL 60153.

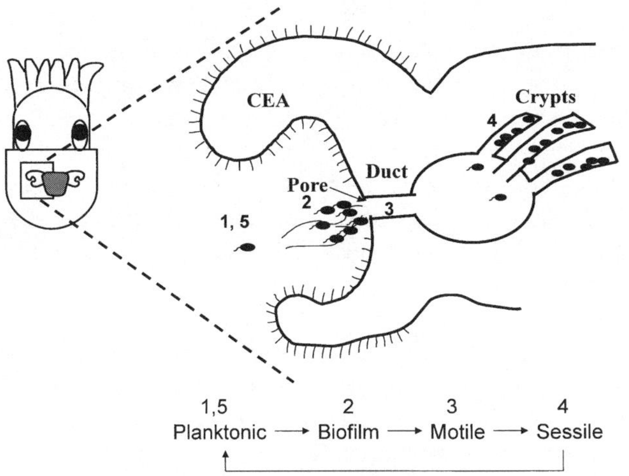

Figure 1. Lifestyle transitions during symbiosis. *V. fischeri* undergoes several switches between motile/planktonic and biofilm/sessile forms during symbiotic colonization of its host, *E. scolopes*. A cartoon of the juvenile squid with its light organ is depicted on the left. A portion of the light organ is enlarged on the right, representing the path to colonization of one of six crypts found in the juvenile light organ. The numbers represent different stages as follows: 1, motile, planktonic *V. fischeri* cells are present in seawater; 2, *V. fischeri* cells aggregate in squid-secreted mucus near the pores of the light organ; 3, motile bacteria enter the light organ; 4, the bacteria establish colonization in the crypts, where they lose their flagella; 5, an expulsion event releases *V. fischeri* cells back into the seawater, where they can become motile again. CEA, ciliated epithelial appendages.

other bacteria in seawater. The factors accounting for this specificity remain under investigation. However, a number of genes and traits that are necessary for symbiotic colonization have been identified, and the process by which colonization occurs has been examined microscopically (33, 54). These studies have revealed that both flagellar motility and biofilm formation—traits known to be impacted by c-di-GMP—are critical to colonization by *V. fischeri*.

Indeed, the process of colonization (and release) appears to occur through a series of transitions between the motile and sessile forms of *V. fischeri*. Newly hatched juveniles ventilate seawater containing *V. fischeri* and other bacteria at a rapid rate: it has been estimated that a single *V. fischeri* cell enters and exits a juvenile squid every 0.3 second (53). Thus, there is little opportunity for *V. fischeri* to simply find its way into one of the six pores that lead into the light organ. Rather, it appears that the initial stage of colonization proceeds via a biofilm-like attachment to the surface of the light organ. Studies with green fluorescent protein-labeled cells revealed that *V. fischeri* cells first aggregate in a biofilm-like structure on the surface of the light organ by embedding in mucus secreted by the squid (34, 59). Other bacteria, such as the closely related *V. parahaemolyticus,* also can adhere to the light organ. In this respect, however, *V. fischeri* is superior: an equal mixture of *V. parahaemolyticus* and *V. fischeri* resulted in an aggregate of greater than 80% *V. fischeri* cells (31). This ability to aggregate depends upon a large polysaccharide locus, termed *syp:* loss of this locus disrupts aggregation and colonization, while overexpression of this locus leads to increased biofilm formation in culture and increased symbiotic aggregation (59, 60). The similar genetic requirements in vitro and in vivo support the conclusion that aggregation represents a form of biofilm formation.

Following aggregation, the bacteria migrate into the light organ through one of the six pores (Fig. 1). This stage of colonization requires motility. Nonmotile strains of *V. fischeri* successfully aggregate on the light organ surface but never migrate to the pores and, thus, fail to colonize (12, 26, 34). A number of hypermotile *V. fischeri* strains also colonize poorly; in this case, the strains fail to aggregate (25). Thus, control of motility by *V. fischeri* is an essential feature of symbiotic initiation.

The pores lead into ducts (Fig. 1), which contain mucus and cilia that beat outward (24, 53). Thus, bacterial motility also is likely to be required for combating these outwardly directed forces. Beyond the ducts lie the crypts (Fig. 1), the sites of bacterial colonization. Here, multiplication of *V. fischeri* occurs rapidly (47), coincident with the induction of bioluminescence. Surprisingly, flagella, which facilitate bacterial entry, appear to be superfluous inside the crypts. A majority of crypt-residing *V. fischeri* cells are nonflagellated within about 48 h (47); whether loss of flagella is required for efficient colonization by *V. fischeri* remains unknown. These now-sessile bacteria are embedded in a matrix of host and/or bacterial origin (32), making it likely that they exist in the form of a biofilm inside the light organ.

Finally, for the duration of the symbiosis, the bacteria face the daily challenge of a changing environment. Each dawn, the squid expels 90 to 95% of its symbionts from the light organ (20, 32) through a muscle-controlled contraction. At least some of the expelled bacteria can become flagellated again within 45 min of their release into seawater (47). These flagellated bacteria may be primed for entry into another host.

Less is known about the life cycle of *V. fischeri* outside its squid host. It can be found in the gut tracts of fishes as well as in a specific symbiosis with the fish *Monocentris japonica* (30, 48). In seawater, it is more abundant in areas near or adjacent to squid habitats than in distal regions, a finding that is consistent with the known expulsion behavior of the squid host (20). Outside the squid, the bacteria tend to be found in the sediment: at the same sampling location, *V. fischeri* cells were much more abundant in sediment samples than in seawater (20). Given that small particles like those found in sediment are ideal surfaces for biofilm formation, these observations suggest that *V. fischeri* may form a host-independent, potentially abiotic, biofilm as part of its life cycle. Thus, *V. fischeri* represents an excellent model for understanding transitions between the motile/planktonic and sessile/biofilm forms of bacterial life, transitions known to correlate with c-di-GMP metabolism (see chapter 8).

BIOINFORMATICS-BASED INVESTIGATION OF c-di-GMP-SYNTHESIZING AND -DEGRADING PROTEINS

In 2001, M. Y. Galperin and colleagues inventoried complete prokaryotic genomes for the numbers and types of signaling domains (11) (see also chapter 3). In this study, they identified GGDEF, EAL, and HD-GYP domains, associated with the synthesis and degradation of c-di-GMP. Although *V. fischeri* was not part of their data set, the related microbe *V. cholerae* was. At that time, *V. cholerae* contained more predicted proteins with a GGDEF domain, 41, than any of the other 35 bacteria and archaea in the study. Although the numbers of the others varied from none to 33, most contained 11 or fewer. Similarly, *V. cholerae* contained the largest number of EAL domain proteins, with 22. Because some of the proteins contained both domains, the numbers for *V. cholerae* domains are as follows: 31 GGDEF, 12 EAL, and 10 GGDEF/EAL. We now known that *V. fischeri* contains a similarly large number of such proteins with 28 GGDEF, 9 EAL, and 11 GGDEF/EAL domains (Table 1; Fig. 2) (58). *V. fischeri* also contains two predicted HD-GYP domain proteins.

Table 1. Genes predicted to control c-di-GMP levels in *V. fischeri*

Domain(s)	No. of genes in: VF (chromosome 1)	No. of genes in: VFA (chromosome 2)	Total no. of genes
GGDEF	10	18	28
EAL	2	7	9
GGDEF/EAL	6	5	11
HD-GYP	1	1	2
Total	19	31	50

The sequenced vibrios each contain a larger chromosome that is relatively conserved among the different species and a smaller chromosome that is quite divergent. Thus, it is thought that the smaller chromosome is responsible for the differences in lifestyle and niche selection for the vibrio species (9). For *V. fischeri,* the sizes of the two chromosomes are 2.9 and 1.3 Mb, respectively (49). Interestingly, a recent examination of two-component response regulators encoded by *V. fischeri* revealed an uneven distribution of this regulatory class between the two chromosomes. Despite being twice as large, chromosome 1 contained 22, or slightly more than half, of the 40 response regulators, while the much smaller chromosome 2 contained almost as many (i.e., 18) response regulators (17). Similarly, there is an uneven distribution of genes with GGDEF and EAL domains (Table 1): the smaller chromosome accounts for 30 of the 48 genes. The two HD-GYP genes are split between the two chromosomes. The role(s) of all of these genes in niche selection for *V. fischeri* remains to be determined.

For each of the 28 proteins predicted to contain only a GGDEF domain, that domain resides at the C terminus, taking up about one-third of the protein (Fig. 2A). This organization is consistent with what is known for other GGDEF proteins. For five of the nine EAL domain proteins, the EAL domain could be found in the N terminus or the C terminus (Fig. 2B). In the other four proteins, the EAL domain encompassed the majority of the protein; these proteins averaged less than 300 amino acids in length, with the EAL domain including about 200 of them. Finally, in 9 of 11 proteins that contained both domains, the EAL domain was located at the C terminus, with the GGDEF domain immediately adjacent (Fig. 2C). In the other two proteins, the EAL domain was at the N terminus while the GGDEF domain was at the C terminus. It will be interesting to determine whether the locations of the domains in the dual-domain proteins will be an accurate predictor of diguanylate cyclase (DGC) or phosphodiesterase (PDE) activity. In support of this possibility, the GGDEF domains in the dual-domain proteins tend to be less well conserved. If most dual-domain proteins function as PDEs, this would result in a more even balance between GGDEF proteins and EAL proteins.

In many cases, GGDEF and/or EAL domains are combined with sensory domains such as HAMP (5), PAS (52), Cache (4), and CHASE (61) (Fig. 2) (see also chapter 3). Nine of the 28 GGDEF-only proteins contain sequences that are associated with other known or predicted domains, including CHASE (and

CHASE 3), Cache_1, CIA30, HAMP, and PAS, another 5 contain a portion of a conserved motif, and the remainder contain sequences dissimilar to any known (searchable) motif. Many of the 28 proteins appear to contain putative transmembrane domains, suggesting that the novel domains detect specific environmental signals that influence DGC activity. In one case, the GGDEF domain is combined with a response regulator receiver domain (REC), indicating that this protein responds to an upstream signaling event, presumably generated through a sensor kinase.

For the EAL-domain-only proteins that contain sequences other than the EAL domain (four of nine), only a single protein contains a fully conserved motif, a REC domain. Two contain truncated HDc superfamily motifs, and another contains a truncated PtsG superfamily motif. Finally, one contains a region with no similarity to known motifs.

For the GGDEF/EAL proteins, seven are combined with full motifs, including DUF1745, HAMP, Cache_1, PAS_4, Cache_2, GAF, and PRK11059, while the other four (including the two with the EAL domain located at the N terminus) contain no additional conserved sequences. As with the GGDEF proteins, many of these dual-domain proteins contain predicted transmembrane segments, indicating that the proteins may localize to the membrane where they would be positioned to receive an environmental signal that modulates their activity.

Of the two HD-GYP proteins, one contains two regions with similarity to periplasmic binding protein domains, while the other has a region with no similarity to known motifs (Fig. 2D). Each has putative transmembrane regions as well. Therefore, like the GGDEF and EAL proteins, the putative HD-GYP proteins in *V. fischeri* appear positioned to integrate signals.

In summary, the bioinformatics data show a wide variety of sensory domains associated with GGDEF and EAL domains, indicating that these proteins may recognize a variety of signal inputs that presumably modulate their activities. The truncation of such sensory domains or, in many cases, the absence of known sensory domains, however, suggests that the study of these proteins will reveal additional and novel sensory domains. This hypothesis is consistent with the finding that a high percentage of GGDEF and EAL domain genes reside on the less well conserved chromosome 2 of *V. fischeri,* the chromosome thought to be involved in niche selection. Finally, the potential association of many of these proteins with the inner membrane indicate that their positioning may play important roles in activity and/or signal exchange.

BIOINFORMATICS-BASED INVESTIGATION OF c-di-GMP RECEPTOR PROTEINS

Less is known about proteins predicted to be involved in c-di-GMP binding in *V. fischeri*. In 2006, Amikam and Galperin used bioinformatics, with sequenced bacterial genomes, to identify genes with a PilZ domain, one of several domains now known to bind c-di-GMP (3). In contrast to the abundant GGDEF and EAL domain genes, the *V. fischeri* genome was predicted to contain only four PilZ domain genes (a number similar to that seen for *V. cholerae,* which is five). Two other c-di-GMP-binding domains are known: a degenerate GGDEF domain that retains the c-di-GMP inhibition site (see chapter 4), such as in PelD (21), and FleQ (16) (see also chapter 11). *V. fischeri* possesses at least one degenerate GGDEF that retains the inhibition site (VF0355) and one true homolog of FleQ, termed FlrA (VF1856). Both FleQ and FlrA are known to function as the master activator of the flagellar regulon (6, 26). FleQ is now known to directly bind to and repress transcription from the *pel* promoter, which is required for polysaccharide biosynthesis; c-di-GMP inhibits binding and relieves repression. Since a FleQ variant lacking its N-terminal domain still inhibits *pel* transcription, the c-di-GMP-binding site must be located in the C-terminal two-thirds of the protein (16). Much of this part of the FleQ is conserved in FlrA; thus, it is distinctly possible that FlrA also binds c-di-GMP. In addition, there are 11 other proteins with significant sequence similarity to FleQ, largely to the σ^{54} interaction domain, which is included in the truncated variant. Thus, other proteins could be sensitive to c-di-GMP.

Three of the four PilZ domain genes are located on the large chromosome. A bioinformatic examination of the genes and genomic context of the four PilZ domain genes (Fig. 3) immediately suggests a role for one, VFA0884, in cellulose biosynthesis and biofilm formation. The protein encoded by VFA0884 is similar to the cellulose synthase catalytic subunit protein, BcsA, that was originally identified as a c-di-GMP-binding protein (42, 45). Our recent evidence suggests that the protein encoded by VFA0884 does indeed play a role in cellulose biosynthesis (7) (see "Role for c-di-GMP in controlling cellulose biosynthesis" below). The genomic context of the other PilZ genes may also provide clues to function. For example, VF1838 is embedded in the part of the flagellar gene cluster dedicated to protein secretion,

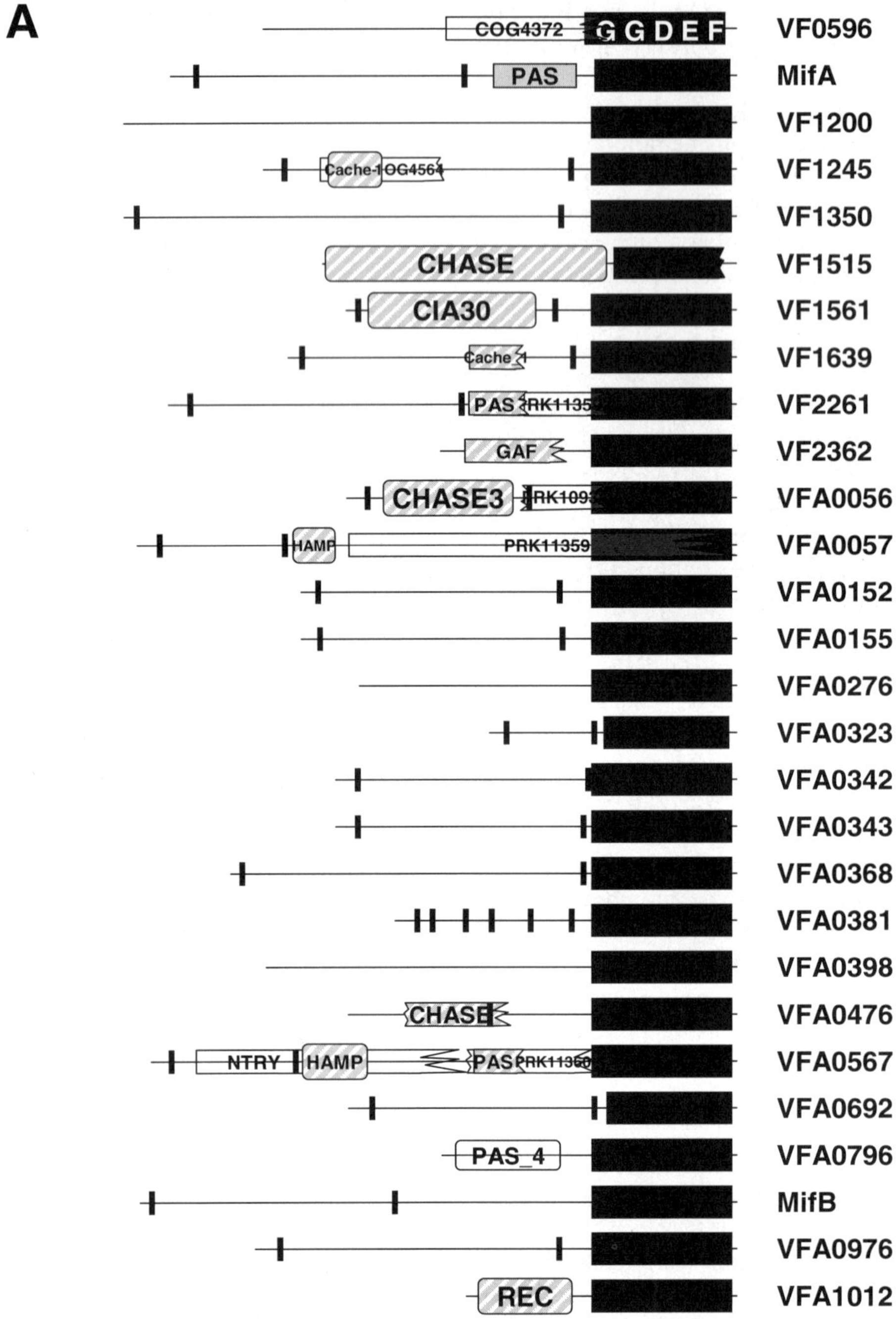

Figure 2. Bioinformatic analysis of the GGDEF, EAL, and HD-GYP domain proteins of *V. fischeri*. Domain structures of the 28 putative GGDEF proteins (A), 9 putative EAL proteins (B), 11 putative GGDEF/EAL proteins (C), and 2 putative HD-GYP proteins (D) encoded by *V. fischeri*. Each GGDEF domain is indicated by a black box, while each EAL domain is indicated by a gray box; poorly conserved GGDEF and EAL domains contain white hatch marks within the black and gray boxes. Other domains are as indicated. Broken boxes indicate a truncated domain. Possible transmembrane segments are indicated by small black rectangles. The MifA, MifB, and MifD proteins are encoded by VF0989, VFA0959, and VF0087, respectively.

suggesting a role in flagellar biogenesis. VF0556 is flanked by genes that encode a putative transglycosylase-associated protein and a murein transglycosylase, suggesting a role in peptidoglycan remodeling. VF0527 may represent the first gene in an operon that also encodes a sensor kinase-response regulator pair, suggesting a role in signal transduction. Further work will be necessary to determine if these proteins actually bind c-di-GMP and, if so, to elucidate their precise functions.

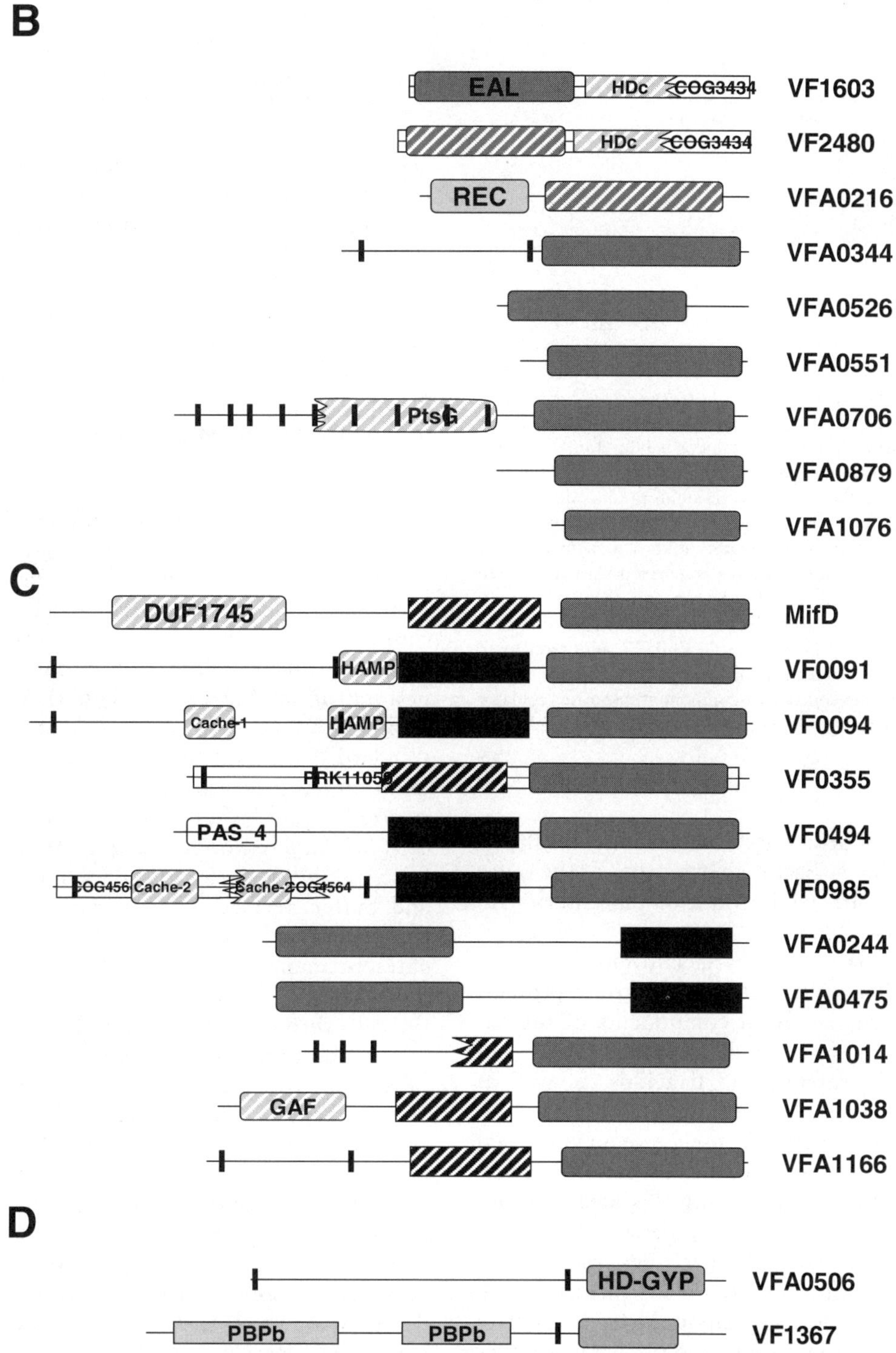

Figure 2. *Continued*

NATURE OF THE FLAGELLAR AND CHEMOTAXIS APPARATUS OF *V. FISCHERI*

To appreciate the current understanding of the role played by c-di-GMP in controlling motility of *V. fischeri,* it is first necessary to review what we know concerning flagellar biosynthesis and regulatory control in this microbe. Motile *V. fischeri* cells contain a tuft of flagella at one pole (2, 30, 47). This arrangement contrasts with that of enterics (e.g., *Escherichia coli* and *Salmonella enterica*), which assemble flagella randomly across their surface, a distribution termed

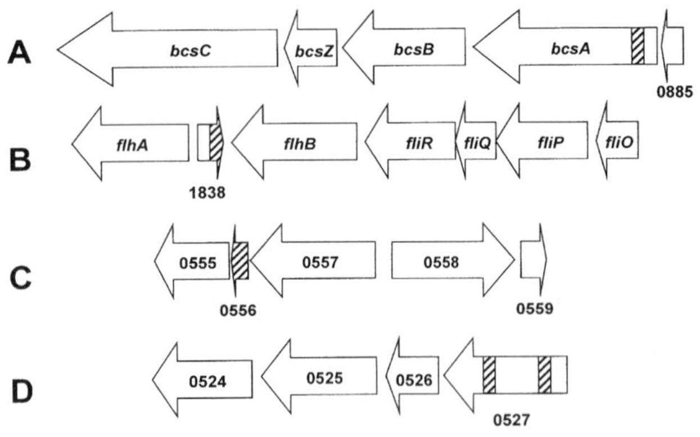

Figure 3. PilZ domain proteins of *Vibrio fischeri*. (A) VFA0884 *(bcsA)*, a member of the cellulose biosynthesis operon *(bcsABZC)*, encodes a glycosyltransferase that includes a PilZ domain (cross-hatching). (B) VF1838, embedded in the part of the large flagellar cluster devoted to flagellar protein secretion, is predicted to encode a PilZ domain protein whose function remains unknown. (C) VF0556, whose function also remains unknown, is flanked by genes predicted to encode a putative transglycosylase-associated protein (VF0555), a predicted ABC transporter (VF0557), a putative soluble lytic murein transglycosylase (VF0558), and a predicted transcription factor (VF0559). (D) VF0527, annotated as an ATP-dependent serine protease with two PilZ domains, is part of the locus predicted to encode a two-component response regulator (VF0526) and two two-component sensor kinases (VF0525 and VF0524).

peritrichous (1). Furthermore, each *V. fischeri* flagellum is enclosed by a sheath, an extension of the outer membrane (27, 30), while those of the enterics remain unsheathed.

V. fischeri carries on its large chromosome a single, large flagellar gene cluster that encodes proteins similar to well-characterized components of the flagellum, including basal body proteins, the hook, hook-associated proteins, and flagellins, as well as motility regulators such as FlrA and FlrCB (49). It has been hypothesized that *V. fischeri* regulates its flagellar genes in a hierarchy similar to that of *V. cholerae* (26, 38, 39). Consistent with this idea, the motility of *V. fischeri* absolutely depends on σ^{54}, encoded by *rpoN*, and FlrA, a σ^{54}-dependent regulator predicted to sit at the top of the hierarchy (26, 56). Motility also depends upon FlrC, a putative σ^{54}-dependent response regulator predicted to control a subset of flagellar genes (17).

Up to six different flagellins (FlaA to F) are assembled into the flagellar filament, although it is not known at this time whether a single filament can contain multiple flagellins (27). The flagellin FlaA appears to be an important component of the flagellum. In contrast to the loss of the FlaC flagellin, which did not noticeably impair motility or flagellation, the loss of FlaA greatly reduced motility. Fewer cells contained flagella, and those that were flagellated elaborated reduced numbers of flagella (26). These data suggest that while FlaA plays a key role, in its absence, other flagellins can be assembled to produce at least partially functional flagella.

The large flagellar gene cluster also includes genes that encode the chemotaxis machinery, including CheY. As is true for other organisms such as *E. coli* and *Salmonella,* loss of *cheY* results in smooth-swimming cells (17). Like its relative *V. cholerae, V. fischeri* also encodes a large number (about 40) of putative methyl-accepting chemoreceptors distributed to both chromosomes (8, 15, 49). This wealth of methyl-accepting chemoreceptors suggests that *V. fischeri* is capable of sensing and responding to many different environmental stimuli.

MOTILITY AND DIVALENT CATIONS

Experiments designed to examine the motility and chemotaxis of *V. fischeri* yielded a surprising result: the motility of *V. fischeri* depends upon the presence of divalent cations and, in particular, the magnesium cation (Mg^{2+}) (Fig. 4) (35). *V. fischeri* grows optimally in a growth medium that is similar to the standard *E. coli* LB medium. Termed LBS, this medium contains tryptone, yeast extract, sodium chloride (2%), and buffer. In this medium, however, the cells are largely nonmotile (Fig. 4). This is also true for a simplified version of LBS that lacks yeast extract and buffer, termed TBS. In contrast, growth of *V. fischeri* in the seawater-based, tryptone- and yeast extract-containing medium SWT permits substantial motility of *V. fischeri,* as monitored by migration of the cells through soft agar motility plates.

In addition to NaCl, SWT contains the following salts: $MgSO_4$, $CaCl_2$, and KCl. Introduction of each of these salts into TBS revealed that $MgSO_4$ is sufficient to induce migration, that $CaCl_2$ exerts a small positive effect on motility, and that KCl enhances growth of *V. fischeri* (35). Migration through soft motility agar occurred efficiently with the addition of as little as 0.2 mM $MgSO_4$. It was maximal in the presence of about 35 mM $MgSO_4$, a level that is similar to that which is present in seawater (about 50 mM). The effect was not limited to $MgSO_4$; $MgCl_2$ similarly exerted a strong positive effect on migration.

The fact that $CaCl_2$ could enhance migration suggested that the response is not specific to the Mg^{2+} cation but could encompass other cations. Indeed, 2 mM concentrations of salts containing either the Mg^{2+} or Ca^{2+} cations could promote similar rates of migration through soft agar. At concentrations greater than 10 mM, however, Ca^{2+} adversely affected the growth and, therefore, motility of *V. fis-*

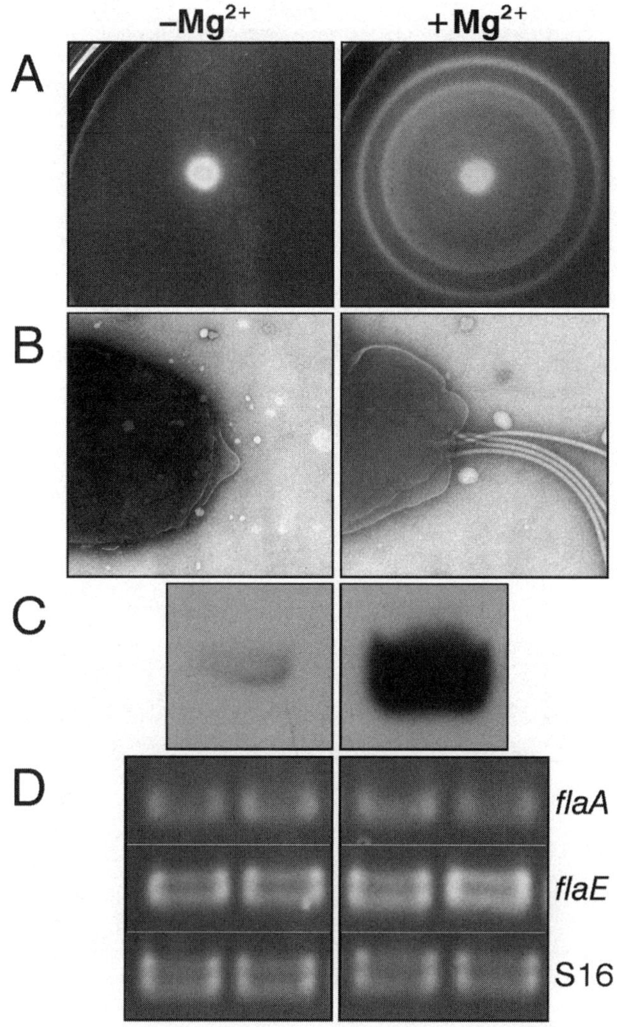

Figure 4. Mg^{2+}-dependent flagellation is controlled posttranscriptionally. In the presence (+) of Mg^{2+} (right), cells tend to migrate well in semisolid agar, forming dense bands of cells characteristic of chemotaxis (A), display a tuft of polar flagella (B), express large amounts of flagellins (C), and transcribe flagellin genes (D). In contrast, in the absence (−) of Mg^{2+} (left), cells tend to migrate poorly in semisolid agar (A), display no flagella (B), express small amounts of flagellins (C), but yet transcribe flagellin genes about as well as cells exposed to Mg^{2+} (D).

cheri, making it difficult to compare its relative efficacy to that of Mg^{2+}. The level of Ca^{2+} in seawater is about 10 mM; thus, higher levels of Ca^{2+} are unlikely to be physiologically relevant. Similarly, other divalent cations were toxic at the high concentrations that are optimal for Mg^{2+}-mediated induction of migration. Given the relevance of high Mg^{2+} concentrations to the biology of *V. fischeri* (in seawater), it seems likely that this cation is the primary cation sensed by the cell to control motility.

Swapping experiments, in which *V. fischeri* cells were grown in the presence or absence of Mg^{2+} and then inoculated onto the surface of soft agar motility plates in the presence or absence of Mg^{2+}, revealed that the impact of Mg^{2+} could be lasting. Cells grown with Mg^{2+} migrated sooner and farther on Mg^{2+}-free plates than those grown without Mg^{2+}. However, cells inoculated onto Mg^{2+}-containing plates exhibited the greatest rates of migration. Thus, exposure to Mg^{2+} at an early time provides an advantage, but continued Mg^{2+} exposure provides an even greater advantage.

Together, these data are consistent with the hypothesis that Mg^{2+} could impact the function and/or production of flagella. Indeed, subsequent analysis with transmission electron microscopy (Fig. 4B) revealed that while Mg^{2+}-grown *V. fischeri* cells display flagella, those grown without Mg^{2+} lack flagella. Mg^{2+}-grown cells carry on average one to three flagella and as many as eight flagella, while the majority of those grown without Mg^{2+} possess no flagella (35). Similarly, Western analysis detected large amounts of flagellin protein from Mg^{2+}-grown cells but little from those grown in the absence of Mg^{2+} (Fig. 4C).

What is the significance of the role of Mg^{2+} in flagellar motility? Mg^{2+} can play both structural and signaling roles. Because of its unique size, hydration, and valence properties (reviewed in references 19, 22, 28, and 37), Mg^{2+} tends to play structural roles. Most commonly, it binds ATP or some other nucleotide triphosphate (e.g., c-di-GMP) in the catalytic pocket of an enzyme. Indeed, Mg^{2+} plays a significant structural role in c-di-GMP-regulated processes: it is required for both the DGC and PDE activities (42–44). However, Mg^{2+} also can function as an environmental signal. For example, it controls the activity of the PhoQP two-component signaling pathway (14). Whereas high Mg^{2+} conditions trigger PhoQP to upregulate one subset of genes and downregulate another, low Mg^{2+} conditions exert the opposite effect. It is hypothesized that this two-component system is used by *S. enterica* serovar Typhimurium to determine whether it is in an extracellular environment (where high millimolar concentrations of Mg^{2+} exist) or an intracellular one (where Mg^{2+} is limiting) and to control virulence gene expression accordingly.

We hypothesize that, in *V. fischeri* motility, Mg^{2+} plays a signaling role in addition to its required structural one. We propose that a scenario similar to that of *S. enterica* serovar Typhimurium *pho* control could occur for the marine microbe: the ability to sense Mg^{2+} could distinguish seawater with its high [50 mM] Mg^{2+} concentration from the interior of its host *E. scolopes*. If so, then this ability could help dictate the transition from the motile, free-living state to the sessile, host-associated state and vice versa. Little is known about Mg^{2+} levels in cephalopods, let alone in *E. scolopes*. However, it is clear that these invertebrates can control the osmolarity and levels of inorganic ions in various body locations (40). Furthermore, recent evidence indicates that *E. scolopes* controls the osmolarity of its light organ environment (51). If symbiotic Mg^{2+} concentrations are reduced below 0.2 mM, then the mechanism controlling motility in response to Mg^{2+} may act to inhibit motility during colonization (Fig. 1).

ROLE FOR DGCs IN CONTROLLING MOTILITY

We hypothesized that Mg^{2+} could enhance the activity of a positive regulator of motility or inhibit the activity of a negative regulator. We further predicted that loss of the putative negative regulator would permit motility of *V. fischeri* even in the absence of Mg^{2+}. Using a random transposon mutagenesis, we isolated a number of mutants that migrated on soft agar plates despite the absence of Mg^{2+} (36). This screen repeatedly uncovered mutants with an insertion in VF0989, a gene predicted to encode a protein with a GGDEF domain and, thus, with DGC activity. It also revealed a mutant with an insertion in VFA0959, a second gene predicted to encode a GGDEF domain protein. These genes were subsequently termed *mifA* and *mifB*.

Loss of either *mifA* or *mifB* promotes migration on soft agar motility plates that lack Mg^{2+} (Fig. 5A). On this medium, the wild-type parent forms only fuzzy, indistinct ring patterns, indicating poor motility. In contrast, both *mif* mutants form concentric ring patterns consistent with chemotaxis and, thus, considerably better motility. However, the addition of Mg^{2+} further enhances the motility of the *mif* mutants. One possible explanation is that the two putative DGCs carry out redundant functions. To test this hypothesis, a double *mifA mifB* mutant was constructed (36). In the absence of Mg^{2+}, this double mutant migrates at a slightly faster rate than do the two parental single mutants but, again, does not reach the rate that occurs in the presence of Mg^{2+}. Thus, at least in part, Mg^{2+} must operate through a c-di-GMP-independent pathway, or another Mg^{2+}-sensitive factor(s) remains to be identified.

Beyond identifying MifA and MifB as putative DGCs, bioinformatic analyses revealed few clues to function. They did, however, predict MifA to be a 631-amino-acid protein with three putative domains: a periplasmic domain, a cytoplasmic PAS domain, and a cytoplasmic GGDEF domain (Fig. 2). Hydrophobicity analysis predicted two transmembrane segments (amino acids 9 to 27 and 314 to 333). The

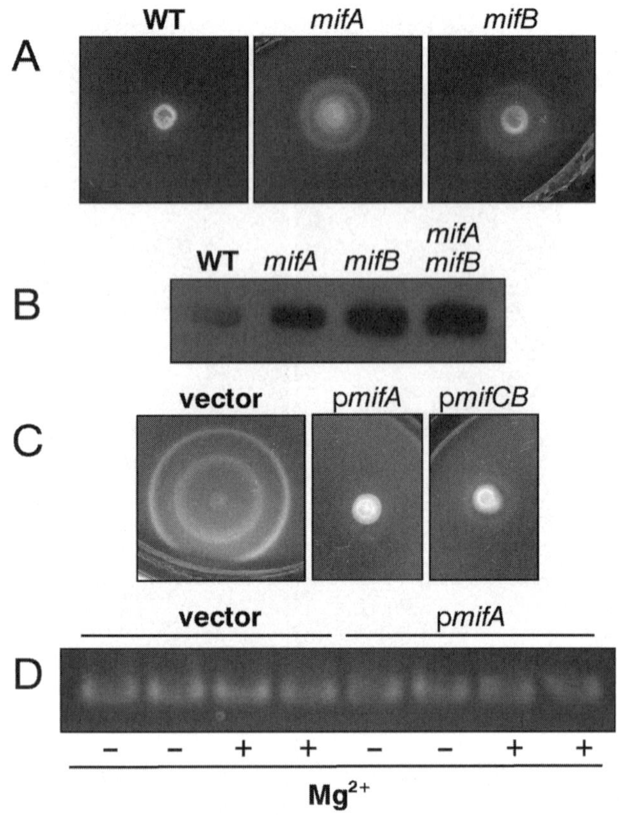

Figure 5. c-di-GMP-dependent flagellation is controlled postranscriptionally. (A) Relative to the wild type (WT), in the absence of Mg^{2+}, loss of *mifA* or *mifB* increases motility. (B) Loss of *mifA*, *mifB*, or both increases flagellin protein levels relative to WT, as observed by Western immunoblot analysis. (C) Overexpression of *mifA* or *mifCB* inhibits motility, even in the presence of Mg^{2+}. (D) The levels of transcript of the major flagellin gene, *flaA*, are unaltered by either Mg^{2+} addition or *mifA* overexpression. +, present; –, absent.

intervening domain, therefore, would be predicted to project into the periplasm and thus be positioned to receive an environmental signal. The putative PAS domain is predicted to immediately follow the second transmembrane segment with a well-conserved GGDEF domain (with the signature amino acids being GGEEF) residing at the C terminus. These analyses yielded a similar arrangement for MifB, a putative 656-amino-acid protein predicted by hydrophobicity analysis to possess two transmembrane regions (amino acids 2 to 31 and 276 to 296); thus, the N-terminal portion may exist as a periplasmic domain that receives environmental signals. The approximately 200 amino acids that lie between these two putative domains do not exhibit similarity to known motifs in the databases.

Like MifA, MifB is predicted to possess a well-conserved GGDEF domain at its C terminus (Fig. 2). Unlike MifA, which is encoded by a single gene operon, the MifB protein is encoded by the second gene of a two-gene operon. Because the two genes overlap by 1 bp, their protein products may have a common function. The first gene, VFA0960, encodes a protein with a putative Pbp domain (23), suggesting that it may function as a periplasmic binding protein. Neither the localization of this protein nor that of MifA and MifB has yet been determined.

Consistent with the role of MifA as a DGC, increased levels of c-di-GMP could be detected using two-dimensional thin-layer chromatography when the *mifA* gene was overexpressed in *E. coli,* and overexpression of *mifA* decreased motility in both *E. coli* and *V. fischeri* (36). Furthermore, alterations of specific amino acids within the GGEEF motif (for example, to AGEEF) disrupted the negative impact on motility generated by *mifA* overexpression (A. J. Wolfe and K. L. Visick, unpublished data). Finally, overexpression of *mifA* led to apparently nonspecific effects on biofilm formation (see cellulose biosynthesis section below), results that match well with those obtained with the overexpression of other DGCs (and PDEs, with opposite results) that have been studied (reviewed in reference 41).

Overexpression of the *mifCB* operon did not result in a detectable increase in c-di-GMP levels (36). Given the conservation of the GGDEF domain and the phenotype of the *mifB* mutant, it seems likely, however, that MifB indeed functions as a DGC. Consistent with this hypothesis, overexpression of *mifCB* in wild-type cells caused a substantial decrease in motility, regardless of exogenous Mg^{2+} concentration (Fig. 5C). Surprisingly, when *mifC* was deleted from the overexpression construct, the impact of *mifB* on motility became dependent on the presence of Mg^{2+} in the medium (36). When Mg^{2+} was present, overexpression of *mifB* exerted a weak influence over motility. In contrast, in the absence of Mg^{2+}, *mifB* overexpression reduced motility to levels similar to those obtained by overexpression of *mifCB*. These data suggest that MifB activity can be modulated by Mg^{2+}, at least in the absence of MifC. A deeper understanding of the interactions between MifC and MifB, and their roles in motility, awaits further investigation.

ROLE FOR A PDE IN CONTROLLING MOTILITY

Because DGCs (demonstrated and putative) controlled motility in the absence of Mg^{2+}, we hypothesized the existence of both c-di-GMP-associated PDEs and c-di-GMP-binding proteins. To identify the latter, we considered the four predicted PilZ genes to be an obvious place to start (Fig. 3). Indeed, as it is embedded in the flagellar gene locus, VF1838 seemed an excellent candidate to be involved in motility. However, under our conditions, we found no role for VF1838 in motility (Wolfe and Visick, unpublished). Similarly, our preliminary data also do not support major roles in motility for the other PilZ genes, VF0527, VF0556, and VFA0884 (Wolfe and Visick, unpublished). While it is possible that these proteins exhibit functional redundancy, it seems more likely that this pathway requires a protein(s) with a non-PilZ domain c-di-GMP-binding motif.

In contrast, we have succeeded in identifying a Mif-specific PDE. To search for this protein, we introduced a library of plasmid-borne chromosomal fragments into wild-type *V. fischeri* and evaluated motility (A. J. Wolfe, B. Zemaitaitis, S. Shibata, and K. L. Visick, unpublished data). Of five plasmids that appeared to increase motility, three encoded GGDEF and/or EAL proteins. The plasmid that exerted the greatest impact on motility encoded two GGDEF/EAL proteins, VF0087 and VF0091. Deletion of a portion of VF0091 did not exert a large impact on motility, while disruption of VF0087 largely abolished the increased motility caused by its parent plasmid. Thus, we focused our attention on VF0087.

Because it is well-known that overexpression of proteins that contain GGDEF and/or EAL domains causes nonphysiological effects (57), we sought verification of a role for VF0087 in *V. fischeri* motility by disrupting the gene in the chromosome, which resulted in decreased migration both in the absence of Mg^{2+} and in its presence. Because these motility phenotypes were consistent with our expectations for a *mif*-specific PDE, we designated this gene *mifD* (Fig. 2B) (Wolfe et al., unpublished). That this motility effect is relatively specific is suggested by the lack of motility defects exhibited by mutants defective for several other putative c-di-GMP PDE mutants, including VFA0216 (a REC/EAL domain protein) (17) and VFA1038 (a putative GGDEF/EAL protein) (C. M. Bassis and K. L. Visick, submitted for publication) (Fig. 2B and C).

We next evaluated the role of MifD in the Mif pathway by evaluating the motility of a double *mifD mifB* mutant. Because the two mutations exert opposite effects on motility-associated phenotypes (loss of *mifB* increases flagellation and thus migration in the absence of Mg^{2+}, while loss of *mifD* decreases both), we investigated the motility phenotypes of the double mutant. We found that flagellation and thus migration of the *mifD mifB* double mutant approached that of wild-type cells (Wolfe et al., unpublished). These data suggest that the loss of *mifD* compensates for the loss of *mifB* and support our hypothesis that these two proteins function in the same pathway.

Our previous results, showing that the combined loss of *mifA* and *mifB* does not permit the same high rates of motility in the absence of magnesium as those observed with wild-type cells in the presence of magnesium, suggested the existence of additional regulators. We predicted that if a third DGC that modulates motility exists, then the overexpression of a PDE in the *mifA mifB* double mutant might further increase motility in the absence of Mg^{2+}. Thus, we overexpressed *mifD* in the *mifA mifB* double mutant. However, none of these strains exhibited increased motility relative to the vector control (Wolfe et al., unpublished). Thus, either the overexpressed genes do not produce sufficient or correctly localized PDE activity or c-di-GMP cannot fully account for the impact of magnesium on motility of *V. fischeri*.

POSTTRANSCRIPTIONAL CONTROL OF MOTILITY

How do Mg^{2+} and c-di-GMP impact the motility of *V. fischeri?* Do they function in the same pathway or in parallel pathways? With the current data, it appears that Mg^{2+} exerts the greater impact on motility. To assemble its polar tuft of flagella, this marine symbiont requires Mg^{2+} at concentrations similar to those found in seawater. Under Mg^{2+}-limited conditions, the vast majority of cells do not assemble flagella, apparently due to a paucity of flagellar proteins: Mg^{2+}-limited cells possess very little of the highly abundant flagellin subunits that normally comprise the filament (Fig. 4C) (35). However, we have found that Mg^{2+} does not substantially impact transcription of a variety of flagellar genes, including those for flagellin, as assayed by reverse transcription-PCR (Fig. 4D) (36). Because Mg^{2+} does not seem to affect transcription but clearly influences steady-state protein levels (35, 36), we propose that Mg^{2+} exerts its influence somewhere after transcript synthesis and stability. Because transmission electron microscopy found no obvious basal body-like structures at the poles of the nonflagellated Mg^{2+}-limited cells (35), the block likely occurs just before or during an early stage of assembly, e.g., translation, protein stability, and/or export of basal body components.

The impact of c-di-GMP on motility is not yet understood for *V. fischeri*. Like the addition of Mg^{2+}, loss of *mifA* or *mifB* appeared to increase the steady-state levels of flagellin protein (Fig. 5B) but not to substantially alter flagellin gene transcription (36). Of greater significance, overexpression of *mifA* substantially decreased motility but did not decrease transcription (evaluated by reverse transcription-PCR) (compare Fig. 5A and D) (36). Thus, like Mg^{2+}, the Mif pathway appeared to function somewhere after transcript synthesis and stability. Although some evidence supported the hypothesis that Mg^{2+} interacts with the Mif pathway, i.e., the observation that Mg^{2+} can influence MifB activity but only when MifB is overexpressed and thus present in excess over MifC (36), other evidence supports the hypothesis that Mg^{2+} works independently of the Mif pathway. When exposed to Mg^{2+}, each *mif* mutant dramatically increased the mean number of flagella per cell. Thus, the response to Mg^{2+} did not require each of the known Mif pathway components. Yet, even in the presence of Mg^{2+}, the impact of each *mif* mutation was still observed, e.g., *mifA* and *mifB* mutants displayed more flagella per cell than did their wild-type parent, which assembled more flagella per cell than did the *mifD* mutant (Wolfe et al., unpublished). On the basis of these data, we propose that Mif and Mg^{2+} work independently, at least in part, to exert posttranscriptional effects on motility. Thus, we propose to change the name of Mif from the original designation *m*agnesium-*i*nduced *f*lagellation to *m*agnesium-*i*ndependent *f*lagellation.

ROLE FOR c-di-GMP IN CONTROLLING CELLULOSE BIOSYNTHESIS

Overexpression of DGCs and PDEs tends to exert global impacts on cell physiology in various bacteria (41). This is true for *V. fischeri* as well. Overexpression of the DGC MifA not only inhibits motility but also results in an increase in biofilm formation (36). When MifA is overexpressed, the resulting colonies exhibit a wrinkled colony morphology, a trait that correlates with increased biofilm formation. Furthermore, those colonies exhibit increased dye-binding properties. Notably, when grown on plates that contain Congo red, a dye that is often used as an indicator of cellulose biosynthesis, the *mifA* overexpression strain shows an increase in red color relative to the vector control (36). In addition, when grown on plates that contain calcofluor, a fluorescent indicator of β-1,4 linkages like those in cellulose, and exposed to ultraviolet light, the *mifA* overexpression strain becomes distinctly fluorescent relative to the wild type, which remains dark. The increased Congo red dye binding property of the *mifA* overexpression cell depends upon the cellulose locus, as its disruption eliminates that phenotype (C. L. Darnell and K. L. Visick, unpublished data).

The cellulose locus contains a PilZ domain gene, VFA0884 (Fig. 3 and 6). The product of VFA0884 is predicted to bind c-di-GMP, permitting its activity in cellulose biosynthesis. The production of cellulose in *V. fischeri* appears to be under complex regulation. Overexpression of either of two regulators, the sensor kinase SypF (encoded by the *syp* polysaccharide locus) or the putative response regulator VpsR, causes an increase in cellulose biosynthesis (as measured by Congo red binding) and biofilm formation (as evaluated by a crystal violet-based assay of attachment to glass) (Fig. 6) (7). Loss of cellulose biosynthesis eliminates those phenotypes. In addition, a GGDEF/EAL protein, VFA1038, impacts cellulose biosynthesis: loss of VFA1038 increases Congo red binding and glass attachment measured by the crystal violet assay (Bassis and Visick, submitted). Like overexpression of MifA, SypF, and VpsR, disruption of the cellulose gene locus eliminates the increase in biofilm formation observed with the VFA1038 mutant. The roles for these genes remain uncertain, particularly in the case of overexpressed regulators. However, our data (particularly those for disruption of VFA1038 but also in the case of MifA overexpression) are consistent with a role for c-di-GMP in promoting cellulose biosynthesis, perhaps by binding to VFA0884, the PilZ protein. Because MifA is unlikely to be the cellulose-specific regulator, additional work is necessary to decipher the regulatory network

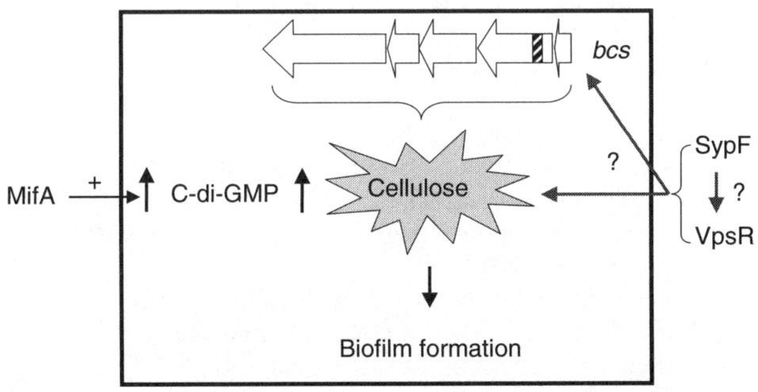

Figure 6. Model for c-di-GMP and cellulose. The *bcs* genes in *V. fischeri* (VFA0885 to 0881) encode the enzymes necessary for cellulose biosynthesis. The production of cellulose is enhanced by overexpression of either of the two-component regulators SypF and VpsR. It is also increased by overexpression of the DGC MifA. For MifA, it is likely that a rise in c-di-GMP levels increases the cellulose synthetic activity of the PilZ domain protein, BcsA. For SypF and VpsR, the level at which these proteins impact cellulose production is unknown, as indicated by the question marks. Increased production of cellulose results in enhanced biofilm formation.

controlling c-di-GMP production and cellulose biosynthesis (Fig. 6).

MODEL FOR MOTILITY

We have developed a working model that takes into account all of our motility results to date (Fig. 7). Under conditions of low Mg^{2+}, flagellar biogenesis is impaired. MifA and MifB (perhaps in concert with MifC) catalyze the production of c-di-GMP, while MifD catalyzes its degradation. One or more (non-PilZ) c-di-GMP-binding proteins (X) bind the Mif-produced c-di-GMP. This causes a conformational change that interferes with flagellar biogenesis by impairing translation or protein stability, export, and/or assembly.

Under conditions of high Mg^{2+}, flagellar biogenesis proceeds. Several alternatives can be envisioned: Mg^{2+} could inhibit the DGC activity of MifA and MifB or enhance the PDE activity of MifD, by altering their expression, localization, and/or active conformation. Alternatively, Mg^{2+} could inhibit the c-di-GMP-binding protein X and/or override the c-di-GMP-mediated inhibition of translation or protein stability, export, and/or assembly. Finally, it is possible that Mg^{2+} acts through a Mif-independent pathway to promote flagellar biogenesis.

CONCLUDING REMARKS

There is now no doubt that c-di-GMP plays a key role in the transition of many bacterial species from the motile, planktonic form to the sessile, biofilm state—the so called "stick-or-swim" decision. Although much effort has been invested in understanding how c-di-GMP is synthesized and degraded and how c-di-GMP influences production of the extracellular polysaccharides that comprise the bulk of biofilm matrices, there exist only a limited number of studies that have explored the role of c-di-GMP in regulating flagellar biogenesis. Nonetheless, clear evidence suggests that the mechanisms used by c-di-GMP to impact motility are as diverse as the levels of control that are known to exist (57). The details of those mechanisms, however, remain unclear. The work described in this chapter is one of the few studies, along with those investigating YcgR (chapter 4) and PleD (chapter 9), where a detailed investigation of c-di-GMP in motility is being pursued.

Acknowledgments. Work in our laboratories is funded by NIH R01 grants GM066130 to A.J.W. and GM59690 to K.L.V. We also thank the LUMC Research Funding Council for funding our work on the role of c-di-GMP in the motility of *V. fischeri*.

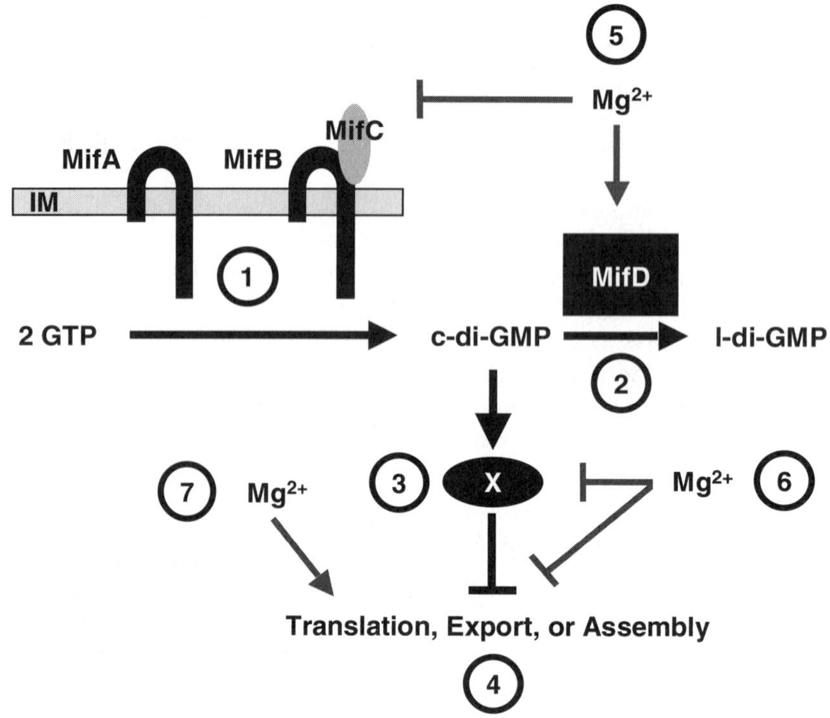

Figure 7. Model for c-di-GMP and motility. The DGCs MifA and MifB (in concert with MifC) (1) and the PDE MifD (2) set the steady-state levels of c-di-GMP, which binds to an unknown c-di-GMP-binding protein, X (3). This complex interferes with the translation, export, and/or assembly of very early flagellar components (4). Several possibilities exist for the role of Mg^{2+} in promoting motility. It could block this process upstream of c-di-GMP, either by inhibiting the DGCs MifA and MifB or by activating the PDE MifD (5). It could act downstream of c-di-GMP either by inhibiting the binding of c-di-GMP to its binding protein(s) X or by inhibiting the action of the c-di-GMP/X complex (6). Finally, Mg^{2+} could override the inhibitory action of c-di-GMP via an independent pathway (7). Combinations of these alternatives could operate. IM, inner membrane; l-di-GMP, linear di-GMP.

REFERENCES

1. **Aizawa, S.-I., and T. Kubori.** 1998. Bacterial flagellation and cell division. *Genes Cells* **3:**1–10.
2. **Allen, R. D., and P. Baumann.** 1971. Structure and arrangement of flagella in species of the genus *Beneckea* and *Photobacterium fischeri*. *J. Bacteriol.* **107:**295–302.
3. **Amikam, D., and M. Y. Galperin.** 2006. PilZ domain is part of the bacterial c-di-GMP binding protein. *Bioinformatics* **22:** 3–6.
4. **Anantharaman, V., and L. Aravind.** 2000. Cache—a signaling domain common to animal Ca(2+)-channel subunits and a class of prokaryotic chemotaxis receptors. *Trends Biochem. Sci.* **25:**535–537.
5. **Aravind, L., and C. P. Ponting.** 1999. The cytoplasmic helical linker domain of receptor histidine kinase and methyl-accepting proteins is common to many prokaryotic signalling proteins. *FEMS Microbiol. Lett.* **176:**111–116.
6. **Arora, S. K., B. W. Ritchings, E. C. Almira, S. Lory, and R. Ramphal.** 1997. A transcriptional activator, FleQ, regulates mucin adhesion and flagellar gene expression in *Pseudomonas aeruginosa* in a cascade manner. *J. Bacteriol.* **179:**5574–5581.
7. **Darnell, C. L., E. A. Hussa, and K. L. Visick.** 2008. The putative hybrid sensor kinase SypF coordinates biofilm formation in *Vibrio fischeri* by acting upstream of two response regulators, SypG and VpsR. *J. Bacteriol.* **190:**4941–4950.
8. **DeLoney-Marino, C. R., A. J. Wolfe, and K. L. Visick.** 2003. Chemoattraction of *Vibrio fischeri* to serine, nucleosides and *N*-acetylneuraminic acid, a component of squid light-organ mucus. *Appl. Environ. Microbiol.* **69:**7527–7530.
9. **Dryselius, R., K. Kurokawa, and T. Iida.** 2007. *Vibrionaceae,* a versatile bacterial family with evolutionarily conserved variability. *Res. Microbiol.* **158:**479–486.
10. **Fuqua, C., S. C. Winans, and E. P. Greenberg.** 1996. Census and consensus in bacterial ecosystems: the LuxR-LuxI family of quorum-sensing transcriptional regulators. *Annu. Rev. Microbiol.* **50:**727–751.
11. **Galperin, M. Y., A. N. Nikolskaya, and E. V. Koonin.** 2001. Novel domains of the prokaryotic two-component signal transduction systems. *FEMS Microbiol. Lett.* **203:**11–21.
12. **Graf, J., P. V. Dunlap, and E. G. Ruby.** 1994. Effect of transposon-induced motility mutations on colonization of the host light organ by *Vibrio fischeri*. *J. Bacteriol.* **176:**6986–6991.
13. **Graf, J., and E. G. Ruby.** 1998. Host-derived amino acids support the proliferation of symbiotic bacteria. *Proc. Natl. Acad. Sci. USA* **95:**1818–1822.
14. **Groisman, E. A.** 2001. The pleiotropic two-component regulatory system PhoP-PhoQ. *J. Bacteriol.* **183:**1835–1842.
15. **Heidelberg, J. F., J. A. Eisen, W. C. Nelson, R. A. Clayton, M. L. Gwinn, R. J. Dodson, D. H. Haft, E. K. Hickey, J. D. Peterson, L. A. Umayam, S. R. Gill, K. E. Nelson, T. D. Read,**

H. Tettelin, D. Richardson, M. D. Ermolaeva, J. Vamathevan, S. Bass, H. Qin, I. Dragoi, P. Sellers, L. McDonald, T. Utterback, R. D. Fleishmann, W. C. Nierman, O. White, S. L. Salzberg, H. O. Smith, R. R. Colwell, J. J. Mekalanos, J. C. Venter, and C. M. Fraser. 2000. DNA sequence of both chromosomes of the cholera pathogen *Vibrio cholerae*. *Nature* **406:** 477–483.

16. **Hickman, J. W., and C. S. Harwood.** 2008. Identification of FleQ from *Pseudomonas aeruginosa* as a c-di-GMP-responsive transcription factor. *Mol. Microbiol.* **69:**376–389.
17. **Hussa, E. A., T. M. O'Shea, C. L. Darnell, E. G. Ruby, and K. L. Visick.** 2007. Two-component response regulators of *Vibrio fischeri:* their identification, mutagenesis and characterization. *J. Bacteriol.* **189:**5825–5838.
18. **Jones, B. W., and M. K. Nishiguchi.** 2004. Counterillumination in the Hawaiian bobtail squid, *Euprymna scolopes* Berry (Mollusca: Cephalopoda). *Mar. Biol.* **144:**1151–1155.
19. **Kehres, D. G., and M. E. Maguire.** 2002. Structure, properties and regulation of magnesium transport proteins. *Biometals* **15:** 261–270.
20. **Lee, K.-H., and E. G. Ruby.** 1994. Effect of the squid host on the abundance and distribution of symbiotic *Vibrio fischeri* in nature. *Appl. Environ. Microbiol.* **60:**1565–1571.
21. **Lee, V. T., J. M. Matewish, J. L. Kessler, M. Hyodo, Y. Hayakawa, and S. Lory.** 2007. A cyclic-di-GMP receptor required for bacterial exopolysaccharide production. *Mol. Microbiol.* **65:**1474–1484.
22. **Maguire, M. E., and J. A. Cowan.** 2002. Magnesium chemistry and biochemistry. *Biometals* **15:**203–210.
23. **Marchler-Bauer, A., J. B. Anderson, M. K. Derbyshire, C. DeWeese-Scott, N. R. Gonzales, M. Gwadz, L. Hao, S. He, D. I. Hurwitz, J. D. Jackson, Z. Ke, D. Krylov, C. J. Lanczycki, C. A. Liebert, C. Liu, F. Lu, S. Lu, G. H. Marchler, M. Mullokandov, J. S. Song, N. Thanki, R. A. Yamashita, J. J. Yin, D. Zhang, and S. H. Bryant.** 2007. CDD: a conserved domain database for interactive domain family analysis. *Nucleic Acids Res.* **35:**D237–D240.
24. **McFall-Ngai, M. J., and E. G. Ruby.** 1998. Sepiolids and vibrios: when first they meet. Reciprocal interactions between host and symbiont lead to the creation of a complex light-emitting organ. *BioScience* **48:**257–265.
25. **Millikan, D. S., and E. G. Ruby.** 2002. Alterations in *Vibrio fischeri* motility correlate with a delay in symbiosis initiation and are associated with additional symbiotic colonization defects. *Appl. Environ. Microbiol.* **68:**2519–2528.
26. **Millikan, D. S., and E. G. Ruby.** 2003. FlrA, a σ^{54}-dependent transcriptional activator in *Vibrio fischeri,* is required for motility and symbiotic light-organ colonization. *J. Bacteriol.* **185:** 3547–3557.
27. **Millikan, D. S., and E. G. Ruby.** 2004. *Vibrio fischeri* flagellin A is essential for normal motility and for symbiotic competence during initial squid light organ colonization. *J. Bacteriol.* **186:**4315–4325.
28. **Moncrief, M. B. C., and M. E. Maguire.** 1999. Magnesium transport in prokaryotes. *J. Biol. Inorg. Chem.* **4:**523–527.
29. **Morris, J. G., Jr.** 2003. Cholera and other types of vibriosis: a story of human pandemics and oysters on the half shell. *Clin. Infect. Dis.* **37:**272–280.
30. **Nealson, K. H.** 1978. Isolation, identification, and manipulation of luminous bacteria. *Methods Enzymol.* **57:**153–166.
31. **Nyholm, S. V., and M. J. McFall-Ngai.** 2003. Dominance of *Vibrio fischeri* in secreted mucus outside the light organ of *Euprymna scolopes:* the first site of symbiont specificity. *Appl. Environ. Microbiol.* **69:**3932–3937.
32. **Nyholm, S. V., and M. J. McFall-Ngai.** 1998. Sampling the light-organ microenvironment of *Euprymna scolopes:* description of a population of host cells in association with the bacterial symbiont *Vibrio fischeri*. *Biol. Bull.* **195:**89–97.
33. **Nyholm, S. V., and M. J. McFall-Ngai.** 2004. The winnowing: establishing the squid-*Vibrio* symbiosis. *Nat. Rev. Microbiol.* **2:** 632–642.
34. **Nyholm, S. V., E. V. Stabb, E. G. Ruby, and M. J. McFall-Ngai.** 2000. Establishment of an animal-bacterial association: recruiting symbiotic vibrios from the environment. *Proc. Natl. Acad. Sci. USA* **97:**10231–10235.
35. **O'Shea, T. M., C. R. DeLoney-Marino, S. Shibata, S.-I. Aizawa, A. J. Wolfe, and K. L. Visick.** 2005. Magnesium promotes flagellation of *Vibrio fischeri*. *J. Bacteriol.* **187:**2058–2065.
36. **O'Shea, T. M., A. H. Klein, K. Geszvain, A. J. Wolfe, and K. L. Visick.** 2006. Diguanylate cyclases control magnesium-dependent motility of *Vibrio fischeri*. *J. Bacteriol.* **188:**8196–8205.
37. **Papp-Wallace, K. M., and M. E. Maguire.** 25 September 2008, posting date. Chapter 5.4.4.2, Magnesium transport and magnesium homeostasis. *In* A. Bock, R. Curtiss III, J. B. Kaper, F. C. Neidhardt, T. Nystrom, K. E. Rudd, and C. L. Squires (ed.), *EcoSal—*Escherichia coli *and* Salmonella: *Cellular and Molecular Biology.* ASM Press, Washington, DC. http://www.ecosal.org.
38. **Prouty, M. G., N. E. Correa, and K. E. Klose.** 2001. The novel σ^{54}- and σ^{28}-dependent flagellar gene transcription hierarchy of *Vibrio cholerae*. *Mol. Microbiol.* **39:**1595–1609.
39. **Prüβ, B. M., D.-J. Kim, S. Forst, R. T. Fleming, K. L. Visick, and A. J. Wolfe.** 2005. Genomics of flagella, p. 1–12. *In* B. M. Prüβ (ed.), *Global Regulatory Networks in Enteric Bacteria.* Research Signpost, Kerala, India.
40. **Robertson, J. D.** 1965. Studies on the chemical composition of muscle tissue. III. The mantle muscle of cephalopod molluscs. *J. Exp. Biol.* **42:**153–175.
41. **Romling, U., M. Gomelsky, and M. Y. Galperin.** 2005. C-di-GMP: the dawning of a novel bacterial signaling system. *Mol. Microbiol.* **57:**629–639.
42. **Ross, P., Y. Aloni, H. Weinhouse, D. Michaeli, P. Ohana, R. Mayer, and M. Benziman.** 1986. Control of cellulose synthesis in *A. xylinum*. A unique guanyl oligonucleotide is the immediate activator of cellulose synthase. *Carbohydr. Res.* **149:**101–117.
43. **Ross, P., Y. Aloni, H. Weinhouse, D. Michaeli, P. Ohana, R. Mayer, and M. Benziman.** 1985. The unusual guanyl oligonucleotide regulates cellulose synthesis in *Acetobacter xylinum*. *FEBS Lett.* **186:**191–196.
44. **Ross, P., R. Mayer, and M. Benziman.** 1991. Cellulose biosynthesis and function in bacteria. *Microbiol. Rev.* **55:**35–58.
45. **Ross, P., H. Weinhouse, Y. Aloni, D. Michaeli, P. Weinberger-Ohana, R. Mayer, S. Braun, E. de Vroom, G. A. van der Marel, J. H. van Boom, and M. Benziman.** 1987. Regulation of cellulose synthesis in *Acetobacter xylinum* by cyclic diguanylic acid. *Nature* **325:**279–281.
46. **Ruby, E. G.** 2008. Symbiotic conversations are revealed under genetic interrogation. *Nat. Rev. Microbiol.* **6:**752–762.
47. **Ruby, E. G., and L. M. Asato.** 1993. Growth and flagellation of *Vibrio fischeri* during initiation of the sepiolid squid light organ symbiosis. *Arch. Microbiol.* **159:**160–167.
48. **Ruby, E. G., and K. H. Nealson.** 1976. Symbiotic association of *Photobacterium fischeri* with the marine luminous fish *Monocentris japonica:* a model of symbiosis based on bacterial studies. *Biol. Bull.* **151:**574–586.
49. **Ruby, E. G., M. Urbanowski, J. Campbell, A. Dunn, M. Faini, R. Gunsalus, P. Lostroh, C. Lupp, J. McCann, D. Millikan,**

A. Schaefer, E. Stabb, A. Stevens, K. Visick, C. Whistler, and E. P. Greenberg. 2005. Complete genome sequence of *Vibrio fischeri:* a symbiotic bacterium with pathogenic congeners. *Proc. Natl. Acad. Sci. USA* **102:**3004–3009.

50. **Stabb, E. V.** 2006. The *Vibrio fischeri-Euprymna scolopes* light organ symbiosis, p. 204–218. *In* F. L. Thompson et al. (ed.), *The Biology of Vibrios.* ASM Press, Washington, DC.

51. **Stabb, E. V., M. S. Butler, and D. M. Adin.** 2004. Correlation between osmolarity and luminescence of symbiotic *Vibrio fischeri* strain ES114. *J. Bacteriol.* **186:**2906–2908.

52. **Taylor, B. L., and I. B. Zhulin.** 1999. PAS domains: internal sensors of oxygen, redox potential, and light. *Microbiol. Mol. Biol. Rev.* **63:**479–506.

53. **Visick, K. L., and M. J. McFall-Ngai.** 2000. An exclusive contract: specificity in the *Vibrio fischeri-Euprymna scolopes* partnership. *J. Bacteriol.* **182:**1779–1787.

54. **Visick, K. L., and E. G. Ruby.** 2006. *Vibrio fischeri* and its host: it takes two to tango. *Curr. Opin. Microbiol.* **9:**632–638.

55. **Wei, S. L., and R. E. Young.** 1989. Development of symbiotic bacterial bioluminescence in a nearshore cephalopod, *Euprymna scolopes. Mar. Biol.* **103:**541–546.

56. **Wolfe, A. J., D. S. Millikan, J. M. Campbell, and K. L. Visick.** 2004. *Vibrio fischeri* σ^{54} controls motility, biofilm formation, luminescence and colonization. *Appl. Environ. Microbiol.* **70:** 2520–2524.

57. **Wolfe, A. J., and K. L. Visick.** 2008. Get the message out: cyclic-di-GMP regulates multiple levels of flagellum-based motility. *J. Bacteriol.* **190:**463–475.

58. **Yildiz, F. H., and K. L. Visick.** 2009. *Vibrio* biofilms: so much the same yet so different. *Trends Microbiol.* **17:**109–118.

59. **Yip, E. S., K. Geszvain, C. R. DeLoney-Marino, and K. L. Visick.** 2006. The symbiosis regulator RscS controls the *syp* gene locus, biofilm formation and symbiotic aggregation by *Vibrio fischeri. Mol. Microbiol.* **62:**1586–1600.

60. **Yip, E. S., B. T. Grublesky, E. A. Hussa, and K. L. Visick.** 2005. A novel, conserved cluster of genes promotes symbiotic colonization and σ^{54}-dependent biofilm formation by *Vibrio fischeri. Mol. Microbiol.* **57:**1485–1498.

61. **Zhulin, I. B., A. N. Nikolskaya, and M. Y. Galperin.** 2003. Common extracellular sensory domains in transmembrane receptors for diverse signal transduction pathways in *Bacteria* and *Archaea. J. Bacteriol* **185:**285–294.

The Second Messenger Cyclic Di-GMP
Edited by A. J. Wolfe and K. L. Visick

Chapter 14

Csr (Rsm) System and Its Overlap and Interplay with Cyclic Di-GMP Regulatory Systems

Tony Romeo and Paul Babitzke

Free-living bacteria must respond rapidly to changing environmental and physiological conditions to survive stresses, maintain homeostasis and growth, and compete with other species. Coordination of gene expression and metabolism during such times involves pathways and cascades of factors, as well as signaling networks that span multiple regulatory circuits and determine global phenotypic properties (lifestyles). Here, we describe recent evidence of shared regulatory targets and interconnections between two global regulatory systems that antagonistically influence bacterial lifestyle choices by governing surface properties, biofilm development, motility, and in many species, virulence factors. Csr (Rsm) is a post-transcriptional regulatory system comprised of an RNA binding protein, CsrA, which regulates mRNA translation and stability, and noncoding regulatory RNAs, CsrB and CsrC, which bind to and antagonize CsrA activity. The secondary messenger, cyclic di-GMP (c-di-GMP), and proteins that are responsible for its synthesis and turnover, GGDEF and EAL or HD-GYP domain proteins, respectively, also regulate motility and biofilm development. Although information about its mechanism of action is limited, c-di-GMP serves as a regulatory ligand of glycosyltransferases, transcription factors, and riboswitches. The CsrD protein, which triggers the turnover of CsrB and CsrC RNAs by RNaseE, contains degenerate GGDEF and EAL domains but does not synthesize or degrade c-di-GMP. Instead, CsrD apparently has evolved to function as an RNA binding protein. Finally, there is recent evidence that CsrA directly regulates expression of GGDEF domain proteins and c-di-GMP levels and that the Csr and c-di-GMP systems interact via the stationary-phase sigma factor, σ^s, the master regulator of flagellum synthesis, FlhDC, and other factors.

GLOBAL REGULATORY SYSTEMS AND LIFESTYLE CHOICES

Bacterial global regulatory networks coordinate extensive phenotypic changes in response to physiological and environmental conditions and thereby permit the bacterium to choose between alternative lifestyles. For example, a species can opt for rapid growth and metabolism or slow growth and high resistance to stresses, including antibiotic exposure (for examples, see references 36 and 76). The choice between a motile or planktonic form of growth versus growth as a sessile biofilm is often available and is influenced in an inverse fashion by Csr and c-di-GMP levels in a variety of species, including *Escherichia coli* (47, 74, 84, 100, 108). Similarly, global regulatory circuitry involving the Csr (Rsm) system has been proposed to determine, at least in part, whether certain pathogens cause an acute, fulminant infection involving the production and secretion of destructive virulence factors versus less destructive, though not entirely benign, chronic association with the host (25, 30).

Although global regulons were originally defined in the context of transcriptional control, the discovery of alternative global regulatory mechanisms, such as translation control and regulated mRNA decay, requires a broader definition. Here, we consider a global regulon to be a large and functionally diverse set of genes and/or proteins whose expression levels or activities are subject to a common regulator. This

Tony Romeo • Department of Microbiology and Cell Science, University of Florida, Gainesville, FL 32611. **Paul Babitzke** • Department of Biochemistry and Molecular Biology, The Pennsylvania State University, University Park, PA 16802.

broader definition is applicable to a regulator such as c-di-GMP, which acts as a ligand in disparate molecular mechanisms involving riboswitches (94), transcription factors (39), and polysaccharide polymerases (115).

This chapter will illustrate the interactions of Csr and its interplay with c-di-GMP global regulatory systems. Various features of the Csr (Rsm) system have been the subjects of previous reviews (4, 8, 55, 61, 62, 103). Here, we will present a historical overview and describe the components and circuitry of Csr, along with the details of its role in biofilm formation in *E. coli.* This is needed to understand the role of CsrD, an enzymatically inactive member of the GGDEF-EAL domain superfamily, and will facilitate a discussion of the interactions of Csr with c-di-GMP-based regulatory systems. The reader should be aware that our present understanding of these interactions probably represents only a glimpse of the full picture.

CsrA: REGULATORY ROLES AND PHYLOGENETIC DISTRIBUTION

The key component of the Csr system, CsrA (*c*arbon *s*torage *r*egulator) or RsmA (*r*epressor of *s*tationary-phase *m*etabolites), was discovered through a genetic screen designed to identify regulators of glycogen biosynthesis in *E. coli* (81). The structural genes for this metabolic pathway are induced during the transition from the exponential to the stationary phase of growth under cAMP and ppGpp control (80, 83). The effects of a *csrA* mutation were pleiotropic, and CsrA was proposed to be a global regulatory protein, whose disruption dramatically affects glycogen accumulation and *glg* gene expression, cell surface properties (biofilm formation), expression of *pckA,* a gene involved in gluconeogenesis, and cell size. Subsequent studies revealed additional roles for CsrA in glycogen catabolism as well as biosynthesis, glycolysis, gluconeogenesis, peptide uptake, acetate metabolism, exopolysaccharide synthesis, motility, biofilm formation and dispersal, and regulation of the RNA chaperone Hfq in *E. coli* K-12 (Fig. 1 and 2) (4–6, 47, 79, 89, 108, 111, 112, 119). Recently, CsrA was shown to repress c-di-GMP levels and the expression of genes involved in its synthesis (49). CsrA often regulates opposing processes in an inverse fashion, which is also characteristic of c-di-GMP (74, 85). This RNA binding protein represses stationary-phase processes and secondary metabolism while it activates processes and metabolism needed for rapid growth: it activates glycolysis and inhibits gluconeogenesis, it activates motility and inhibits sessility, and it activates biofilm dispersion and represses biofilm formation.

CsrA regulates virulence factor expression, host-microbe interactions, quorum sensing, and other important functions in a variety of species (for examples, see references 7, 8, 15, 38, 55, 57, 69, 77, 100, and 103). Preliminary studies indicate that these functions likely represent only a small part of the Csr regulon, which may include over 700 genes in *E. coli,* whose mRNAs were copurified with a hexahistidine-tagged CsrA protein and were detected by cDNA synthesis and 454 pyrosequencing (21).

Homologues of *csrA* are not present in eukaryotic or archeal genomes but are broadly distributed in the eubacterial superkingdom. Genomic sequencing data have revealed *csrA* in proteobacteria (gamma-, epsilon-, and deltaproteobacteria), firmicutes *(Clostridiales* and *Bacillales),* spirochetes, planctomycetes, *Thermotogales,* actinobacteria, and the *Cytophaga-Flavobacteria-Bacteroides* group *("Bacteroidales").* Furthermore, *csrA*-like genes have been identified in bacteriophage and plasmid DNA (for examples, see references 64 and 116; NCBI website). Some species encode more than one CsrA homolog (e.g., *Pseudomonas fluorescens* and *Legionella pneumophila*), and certain degenerate pathogens possess *csrA* as one of their few evolutionarily preserved regulatory genes (e.g., *Borrelia burgorferi*). The role of the Csr system, or at least the *csrA* gene, has been experimentally examined in a limited number of gram-negative gamma- and epsilonproteobacteria and in *Bacillus subtilis,* a gram-positive representative of the firmicutes (for examples, see references 7, 24, 55, 65, and 118). The RNA binding function of CsrA is conserved in all species that have been examined to date.

CsrA STRUCTURE AND FUNCTION

Cross-linking and analytical centrifugation studies demonstrated that CsrA functions as a dimer (20). Solution nuclear magnetic resonance and X-ray crystallography studies (34, 35, 78) revealed an unusual structure for this dimer, in which the five β-strands of the two CsrA polypeptides are interdigitated (Fig. 1A). Comprehensive alanine scanning mutagenesis showed that the dimer contains two identical surfaces that are required for RNA binding and regulation, located on opposite sides of the symmetrical protein. These two surfaces are composed primarily of amino acid residues on the $\beta 1$ and $\beta 5$ strands of opposite polypeptides, which lie adjacent and parallel to each other in the three-dimensional structure of the protein, as well as some residues of the C-terminal α-helix (Fig. 1) (64). These conclusions were confirmed

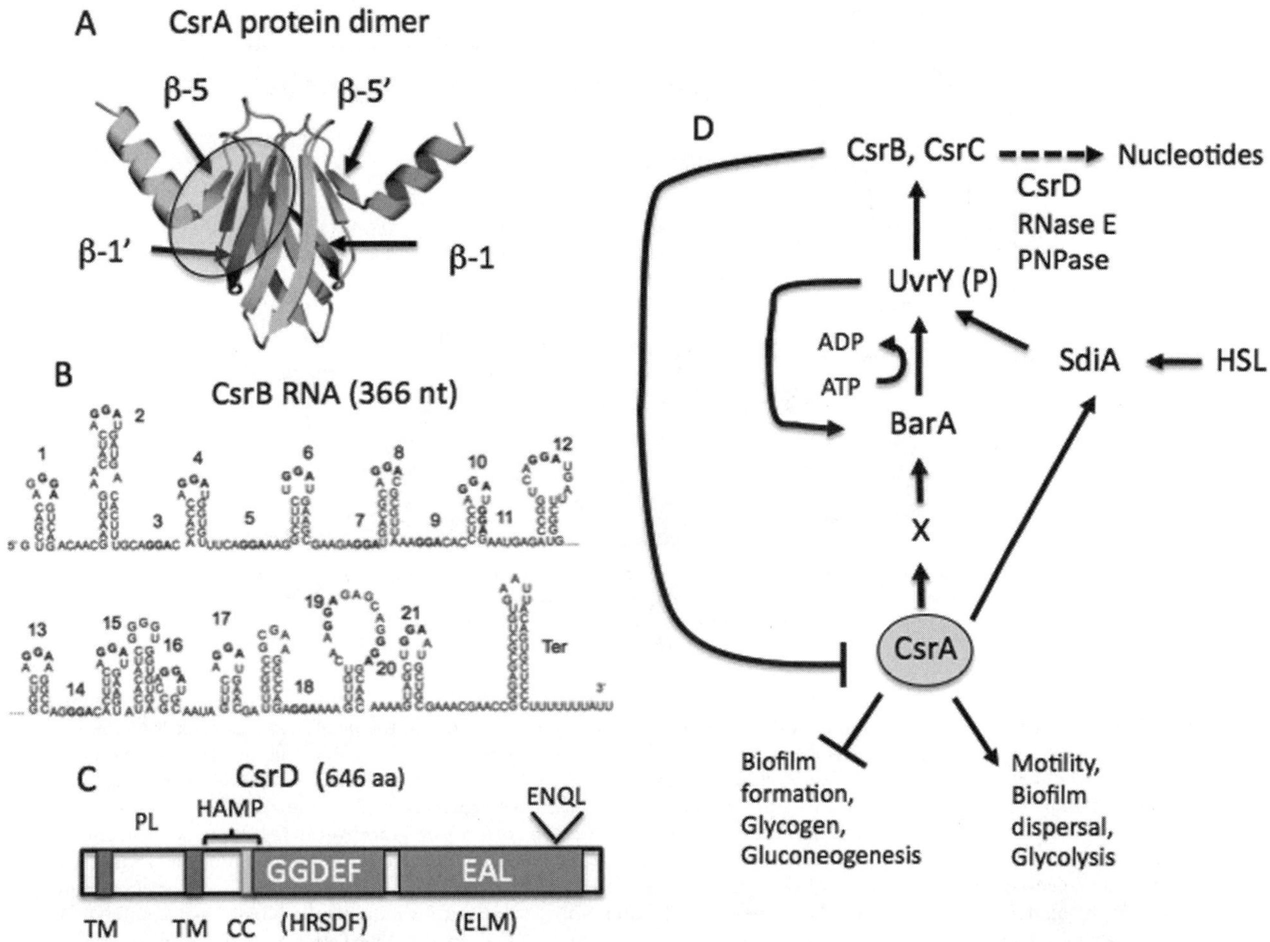

Figure 1. Csr components and circuitry based on the *E. coli* system. (A) Ribbon diagram of the CsrA dimer. The two polypeptides are interdigitated within the dimer. The first and fifth beta strands of opposite polypeptides (β-1 and β-5′; β-1′ and β-5) are parallel in the dimer. These parallel strands form two surfaces for RNA binding (one of which is circled), located on opposite sides of the protein (64, 92). (B) Predicted secondary structure and CsrA binding sites of CsrB RNA. The boldface and numbered GGA-containing sequences (most frequent, CAGGAUG) in the predicted loops or unstructured RNA, respectively, may serve as binding sites for CsrA (4, 58). The terminal stem-loop is a putative factor-independent terminator (Ter). nt, nucleotides. (C) Domain predictions for CsrD protein. TM, transmembrane; CC, coiled coil; PL, periplasmic loop; HAMP, HAMP-like domain; aa, amino acids. GGDEF and EAL domains are shown along with the noncanonical sequences in CsrD. The sequence ENQL at residues 581 to 584 is 100% conserved in apparent CsrD orthologs but is not present in other EAL domain proteins and is important for CsrD function (95). (D) Wire diagram of Csr circuitry. X is an unknown regulator of BarA activity, which is regulated by CsrA (73, 94). BarA and UvrY are the two component signal transduction sensor kinase and response regulator, respectively, which activate *csrB* and *csrC* transcription (96, 113). SdiA is the *E. coli* homolog of LuxR, which responds to HSL and activates *uvrY* transcription (96, 107). RNase E and polynucleotide phosphorylase (PNPase), are ribonucleases involved in CsrD-dependent decay of CsrB and CsrC (95). Activation and repression are depicted using arrowheads and perpendicular lines, respectively. Two apparent autoregulatory loops are shown for CsrA (via SdiA and BarA) and one for UvrY (via BarA). Such regulation is evidence of tight and finely tuned control of CsrA activity within this circuitry.

by structural studies of an RNA complex formed by a CsrA homolog of *Pseudomonas aeruginosa,* RsmE, which additionally revealed RNA contacts with the peptide backbone (91). Such a structure should increase the avidity of CsrA interactions with RNA substrates containing two or more binding sites and may affect the binding cooperativity that CsrA exhibits with certain RNAs. The CsrA-RNA structure also accounts for the stoichiometry of the globular CsrA-CsrB ribonucleoprotein complex, which consists of ~18 to 20 CsrA subunits (9 or 10 dimers containing 2 binding sites per dimer) in complex with an RNA containing 18 to 22 predicted binding sequences for CsrA (Fig. 1B) (4, 58).

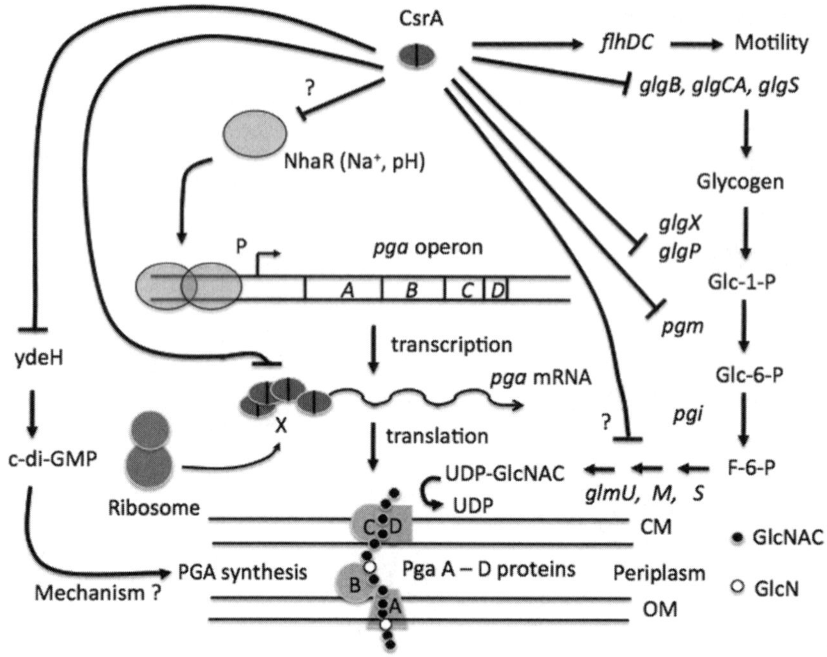

Figure 2. Multitier repression of biofilm formation by CsrA. The regulatory circuitry by which Csr regulates or is posited to regulate biofilm formation at several levels in *E. coli* is shown. The binding of CsrA to the *pgaABCD* transcript leader at six sites represses translation of *pgaA* and destabilizes this transcript. This operon is needed for production, covalent modification, and secretion of the adhesive polysaccharide PGA (43, 44, 108, 109). CsrA binds specifically to transcripts for GGDEF proteins, including YdeH and YcdT, which synthesize c-di-GMP. Thus, CsrA represses levels of c-di-GMP in *E. coli* (49). c-di-GMP somehow activates PGA production (29). CsrA represses glycogen synthesis and turnover (81, 119), which also affects biofilm formation (47). A possible biochemical pathway for this effect is shown, in which carbon flow into the synthesis of UDP-*N*-acetylglucosamine, the precursor of the adhesin PGA, is limited by CsrA repression of *glg* genes, by *pgm* (89), and perhaps, by regulation of the *glmU* and *glmS* genes (21). Transcription of *pga* mRNA requires the NhaR LysR family DNA binding protein, which may represent another point of Csr regulation, based on the observation that the *nhaR* transcript copurifies with CsrA (21). Conversely, CsrA activates motility by stabilizing *flhDC* mRNA (111). OM, outer membrane; CM, cytoplasmic membrane.

RNA TARGETS OF CsrA BINDING

The first evidence that CsrA regulates gene expression posttranscriptionally was that *csrA* was found to repress *glgCAP* expression (in part) by causing decay of this transcript (60). This regulation required the translation initiation region of *glgC*, as determined by deletion analysis of a *glgC′-′lacZ* translational fusion. Shortly thereafter, purified CsrA protein was found to bind specifically to *glg* mRNA as well as to the small noncoding RNA (sRNA) CsrB (57, 58). CsrA binds to *glgC* mRNA and other CsrA-repressed messages at binding sites that partially resemble the Shine-Dalgarno sequence (discussed below), competes with 30S ribosomal subunit binding to the translation initiation region of these RNAs, and thus inhibits translation (6, 20, 108). Inhibition of translation may account for mRNA destabilization that was observed for *glgCAP* and other CsrA-repressed genes (60, 108). In contrast, translational repression by CsrA is not invariably accompanied by message destabilization, as demonstrated recently for *hfq*. Nevertheless, *hfq* mRNA levels in the *csrA* mutant are elevated, apparently because of indirect effects of CsrA on *hfq* transcription (5). Such findings warrant caution in making assumptions about the mode of CsrA action from incomplete data.

CsrB AND CsrC: NONCODING RNAs THAT SEQUESTER CsrA

CsrB was identified during the isolation of CsrA as the major RNA component of a CsrA-RNA complex. Furthermore, CsrA binding to CsrB was determined to sequester this protein and antagonize its activity (58). This finding offered the first example of a bacterial noncoding RNA that binds to and regulates the activity of a regulatory protein. CsrC is a second sRNA that also binds to and antagonizes CsrA in *E. coli*. This RNA was discovered by a molecular genetic approach designed to identify genes that regulate glycogen biosynthesis (82, 113). More recently, species such as *P. fluorescens* and *Vibrio cholerae* have been shown to produce three Csr (Rsm) regulatory RNAs (51, 57).

CsrA is an abundant protein and is present in the *E. coli* cytoplasm at ~6,000 to 17,000 dimers per cell, depending on the growth phase (33). This suggests an intracellular concentration of 6 to 17 μM or 2 to 3 orders of magnitude above the apparent K_d (dissociation constant) of CsrA for most of its mRNA targets and 3 to 4 orders of magnitude above its K_d for CsrC and CsrB, respectively (113). These values explain how the noncoding RNAs can serve as effective antagonists of CsrA. There must be little free CsrA protein in the cell at any given time, and CsrA binds with high affinity to the Csr RNAs. Thus far, there is no known role for Csr RNAs other than binding and sequestration of CsrA.

CsrA interaction with CsrB and CsrC sRNAs has at most a minor effect on their stability in *E. coli* (31, 95). In contrast, in *P. fluorescens,* binding of the CsrA homlogs (RsmA and RsmE) to their sRNA antagonists stabilizes these RNAs (for examples, see reference 106). Furthermore, the RNAs of the Csr and Rsm systems have distinctly different decay kinetics. CsrB and CsrC sRNAs of *E. coli* have half-lives of 1.5 to 4 min, compared to >20 min for the Rsm RNAs of *P. fluorescens*. Because RNA half-life limits the rate at which transcript levels can be changed in response to physiological demands (13, 54), these values imply that levels of the *E. coli* RNAs, and thus CsrA activity in this species, are able to respond much more rapidly to changing physiological or environmental conditions.

BINDING SITE RECOGNITION BY CsrA

The binding sites for CsrA in CsrB RNA were proposed to be related to the sequence CAGGA(U/A/C)G, with the first C and last G closing a hairpin loop (58). This was based on the predominance of this sequence in CsrB RNA and its approximate one-to-one correlation with the number of CsrA polypeptides estimated to be present in the CsrA-CsrB complex. Subsequently, several CsrA binding sites were experimentally demonstrated in CsrA-repressed mRNAs (5, 6, 20, 108), and SELEX analysis and nucleotide substitution studies defined the characteristics of a high-affinity ligand for CsrA binding (19). These studies revealed a GGA as being almost invariant within target sites. A consensus sequence of RUACARGGAUGU (underlined residues were 100% conserved) was derived from the SELEX analysis. The presence of secondary structure was beneficial but not essential for CsrA binding (4, 19). Regulated mRNAs typically contain multiple CsrA binding sites, although a single site suffices for modest regulation of *hfq* (5). While translational repression is better understood, CsrA can also activate gene expression (89, 111). In the case of *flhDC,* the master operon that initiates the cascade of gene expression for motility and chemotaxis, CsrA binds at more than one location within the *flhDC* mRNA leader and stabilizes this transcript (111). The precise mechanism for this process has not been determined. In summary, the regulatory Csr RNAs perform their antagonistic roles by containing multiple repeated sequences that mimic the binding sites of target mRNAs. In the case of CsrB, this RNA sequesters CsrA within a large globular ribonucleoprotein complex (58). Based on our present understanding of CsrA structure and function, we propose that the interaction of the dual RNA binding surfaces of 9 to 10 CsrA dimers with an equal number of binding sites in CsrB allows CsrA to tether this RNA into the observed globular complex (58).

Csr CIRCUITRY

CsrA activity is detectable by genetic analyses under all growth conditions that have been examined. Thus, it is apparent that the Csr circuitry does not function strictly as an on-off switch but instead functions as a governor to modulate gene expression and is superimposed upon the effects of other regulatory systems (for examples, see references 2 and 81). CsrA levels vary modestly with respect to the growth cycle, increasing at the transition to stationary-phase growth (33). No doubt, regulation of *csrA* expression plays an important role in determining CsrA activity. However, no factors have been reported to directly regulate transcription or translation of *E. coli csrA*. RpoS (σ^s) negatively affects *csrA* transcript levels (threefold) in minimal media in the stationary phase of growth, although the basis of this effect and its impact on CsrA protein levels remain to be determined (16). In rich growth media, *rpoS* did not affect CsrA protein levels (33). In the *E. coli* relative *Salmonella enterica,* Crp exhibited conditional modest effects on a *csrA* reporter fusion (101).

CsrB and CsrC RNAs mediate environmental control of the Csr system. CsrA is present in excess throughout the growth curve (in rich media) with respect to CsrB, the major Csr sRNA, and is thus available to interact with other RNAs (33). While CsrA has only weak effects on CsrB/C stability, it is necessary for normal accumulation of these RNAs (95, 113). This role of CsrA requires a two-component signal transduction system (TCS), BarA-UvrY, which activates transcription of these genes (Fig. 1D) (73, 95, 113). An interesting consequence of this circuitry is that the loss of either CsrB or CsrC leads to an increase in CsrA activity in the cell and thus a compensatory increase in the remaining RNA (113). This type of response is evidence of the robust functioning of this regulatory circuitry.

BarA-UvrY homologs, which are present in many gram-negative bacteria, are variously known as Gac, Var, Exp, and Let TCS, and also work in conjunction with Csr systems (for examples, see references 15, 57, and 65). Importantly, the BarA-UvrY system also activates *rpoS* transcription in *E. coli,* possibly through DNA binding by the response regulator UvrY (36, 68). This is one illustration of the ways in which σ^s and CsrA appear to be poised as antagonistic regulators, with UvrY activating both the synthesis of CsrA antagonists (CsrB/C) and the transcription of *rpoS*. Neither the signal for the sensor-kinase activity of BarA nor the way in which CsrA regulates this pathway is known, which unfortunately limits our understanding of the significance of this circuitry. However, BarA signaling is pH dependent in *E. coli* (66), and signaling of this TCS appears to respond to (unknown) quorum-sensing signals in *Pseudomonas* and *Vibrio* species (51, 57).

A critical feedback loop in which CsrA activates the transcription of its RNA antagonists (Fig. 1D) suggests that Csr may function as a homeostatic mechanism for the control of CsrA activity (95). Another interesting feature of the Csr circuitry is that *uvrY* expression in *E. coli* responds to quorum-sensing molecules, *N*-hexanoyl-L-homoserine lactones (HSL), (Fig. 1D) via the DNA binding protein SdiA (95, 107, 111). While *E. coli* and *Salmonella* do not produce HSL, they are able to respond to such compounds produced by other species of bacteria (2). These findings predict that CsrA activity should be decreased, and consequently, secondary metabolism and its influence on resistance properties should be increased, in the presence of HSL-producing bacteria. Additional autoregulatory circuits are evident within the Csr circuitry (Fig. 1D), e.g., between BarA and UvrY and between CsrA, UvrY, and SdiA (22, 95).

In view of its important global regulatory role, the tight control of CsrA activity, which is evident in the design of the Csr circuitry, should not be surprising. Some of the key questions and issues to be resolved concerning the functioning of this circuitry include (i) the nature of the signal for modulation of BarA kinase and the pathway for its production, (ii) the means by which CsrA influences this pathway, (iii) the connectivity, both afferent and efferent, of this circuitry to other regulatory circuits, including those affecting c-di-GMP regulation, (iv) the factors that control CsrA expression levels, and (v) the need for additional studies to evaluate the hypothesis that the Csr circuitry functions as a homeostatic device.

c-di-GMP: INTRACELLULAR SIGNAL FOR GLOBAL PHENOTYPIC SHIFTS

The nucleotide c-di-GMP serves as an activator for the production of adhesins and exopolysaccharides and an inhibitor of motility in numerous gram-negative bacteria. In this respect, c-di-GMP and CsrA function as the antitheses of each other. c-di-GMP, the systems that produce and degrade it, and the processes that it governs have been extensively reviewed (for examples, see references 48, 85, 86, 92, 97, and 117). Here we present salient features of these systems to provide context for understanding their interactions with the Csr system in *E. coli*.

Moshe Benziman and his colleagues discovered c-di-GMP for its role as an allosteric activator of cellulose synthase of *Gluconacetobacter xylinus* (87, 115). Based largely on genetic studies, c-di-GMP is known or presumed to stimulate the synthesis of secreted polysaccharides of many gram-negative bacteria, which serve as biofilm adhesins (for examples, see references 10, 23, 52, 56, and 70). Dispersal of preformed biofilms appears to be triggered by depletion of c-di-GMP in a number of species (for examples, see references 27, 67, and 102), although the underlying processes remain to be defined, and it is premature to conclude that this is a widespread theme.

Synthesis of c-di-GMP from two GTPs is catalyzed by diguanylate cyclases (DGC) that contain the GGDEF (DUF-1) domain (for an example, see reference 86). Turnover of c-di-GMP by cleavage into an inactive linear product (linear di-GMP) is catalyzed by specific phosphodiesterases (PDE) containing either the EAL (DUF-2) or HD-GYP domain (for examples, see references 18, 90, and 98). The GGDEF and EAL superfamilies of proteins are found in all major branches of eubacteria but, similar to Csr systems, are absent from the *Archaea* (48). Gram-negative species possess variable numbers of such proteins, which can rival even the TCS in abundance. These domains are found singly or together and are often located distally with respect to sensory domains, which can govern their activities (26). *E. coli* possesses 29 proteins that contain GGDEF (19), EAL (17), or both (7) domains.

Proteins containing both GGDEF and EAL domains are common. In most examples, these domains have evolved such that only one of them is catalytically active in c-di-GMP metabolism (for examples, see references 12 and 90). In contrast, *Mycobacterium smegmatis* produces a single GGDEF-EAL domain protein that possesses both DGC and PDE activities (53). Unlike its function in gram-negative species, c-di-GMP production by *Mycobacterium smegmatis* did not affect biofilm formation but enhanced starvation survival. Thus, these systems have evolved to function in ways that are not always predictable.

Proteins with degenerate or evolved GGDEF and/or EAL domains, which do not participate in c-di-GMP metabolism, also have been described. The GGDEF-EAL domain protein, CsrD, which functions as a specificity factor in CsrB/C RNA decay in *E. coli*, was the first such example (95) and is discussed below. The BLUF-EAL domain protein YcgF of *E. coli* uses blue light to elicit its activity as an antirepressor. The EAL domain of this protein does not bind to or metabolize c-di-GMP but mediates protein-protein interactions with a MerR family DNA binding protein (104). The GGDEF protein GdpS of staphylococci affects biofilm formation by an undetermined regulatory mechanism involving *icaADBC* expression, which is necessary for poly-β-1,6-*N*-acetylglucosamine production (31, 42). Because GdpS does not metabolize c-di-GMP and is the only well-conserved GGDEF domain protein of staphylococci, these species might not produce c-di-GMP. The inner membrane protein, LapD, of *Pseudomonas fluorescens* is a GGDEF-EAL domain protein that does not metabolize c-di-GMP. The EAL domain of LapD binds to c-di-GMP and thereby undergoes an apparent conformational change, causing the protein adhesin LapA to be retained in a functional state on the cell surface instead of being secreted (71). No doubt, additional roles of evolved GGDEF and EAL domains remain to be discovered.

c-di-GMP TARGETS

In contrast to TCS, in which signal detection is coupled to response regulation via specific protein-protein interactions, the outputs of typical GGDEF and EAL domain proteins are mediated by a soluble molecule, which has the potential to diffuse throughout the cell. Nevertheless, specificity in these systems is maintained, at least in part, by temporal control and spatial localization of these proteins near c-di-GMP targets (discussed in reference 48). Studies employing overexpression of genes affecting c-di-GMP metabolism must be interpreted with caution because of the potential to globally influence c-di-GMP levels and thereby affect noncognate targets.

The first known target of c-di-GMP action was the cellulose synthase of *G. xylinus,* which is allosterically activated by c-di-GMP (87). This apparently involves binding to the α-subunit of cellulose synthase (BcsA), a multidomain protein containing a PilZ domain, a c-di-GMP binding domain present in a variety of proteins (3, 97). The three-dimensional structure of a PilZ domain protein from *V. cholerae* was recently determined in the presence and absence of c-di-GMP (9). Binding of c-di-GMP caused this protein to undergo a structural shift, suggestive of an allosteric regulatory transition, although the downstream signaling mechanism of this protein remains to be determined. The PilZ protein of *E. coli,* YcgR, somehow regulates flagellar motor function in response to c-di-GMP binding (74, 88).

c-di-GMP binding motifs/proteins in addition to PilZ and evolved EAL domains have been uncovered. The inhibitory site (I-site) of allosteric GGDEF domain proteins is a surface within the GGDEF domain, which binds to c-di-GMP and mediates noncompetitive product inhibition (12). The I-site contains a signature GXXD core sequence motif. The PelD protein of *P. aeruginosa* contains another kind of c-di-GMP binding element, which somehow activates the synthesis of the PEL exopolysaccharide, a biofilm adhesin that apparently is repressed by CsrA (RsmA) in this bacterium (30, 56). While many studies have observed apparent effects of c-di-GMP on transcription, the FleQ protein of *P. aeruginosa* was the first c-di-GMP binding transcription factor to be reported (39). In response to c-di-GMP binding, FleQ reciprocally regulates motility and expression of *pel* structural genes for the PEL polysaccharide adhesin. c-di-GMP binds to a region of FleQ that appears to be distinct from all of the other c-di-GMP binding domains, PilZ, PelD, and the I-sites of DGC, and thereby relieves *pel* gene repression. There is no reason to presume that the list of c-di-GMP binding domains is anywhere near completion.

A recently discovered c-di-GMP binding element is not a protein but a riboswitch (92). This c-di-GMP binding RNA domain or aptamer was originally identified as a conserved GEMM sequence motif (*g*enes for the *e*nvironment, *m*embranes, and *m*otility) in the 5′-untranslated segment of numerous mRNAs involved in c-di-GMP metabolism, virulence, motility, and pilus formation (114). Example GEMM-containing RNAs were shown to respond positively or negatively to c-di-GMP (93). While the c-di-GMP riboswitch is found in both gram-positive *(Firmicutes)* and gram-negative species, including the *Gammaproteobacteria,* it is apparently absent from the *Enterobacteriaceae* and many other families. Whether convergent evolution has produced other c-di-GMP binding riboswitches besides the GEMM motif is an open question.

CsrD: A NOVEL FUNCTION FOR A GGDEF-EAL DOMAIN PROTEIN AND A NOVEL POINT OF REGULATION IN THE Csr SYSTEM

The intracellular concentration of an RNA molecule is dictated by its synthesis and decay (13, 54). Interestingly, decay of the regulatory RNAs CsrB and CsrC was found to be dependent upon CsrD, a protein whose expression is repressed by CsrA itself (Fig. 1C) (95). Thus, CsrD participates in a Csr feedback loop (not shown in Fig. 1), located between CsrA and CsrD, in which CsrA represses *csrD,* while CsrD triggers turnover of CsrB and CsrC RNAs (95). This feedback loop illustrates yet another example of inverse regulation of opposing processes by CsrA. Transcriptional activation of CsrB/C by BarA-UvrY is stimulated by CsrA, while expression of CsrD, which targets CsrB/C turnover, is repressed. CsrB/C decay also requires the essential endonuclease RNase E and the exonuclease PNPase (polynucleotide phosphorylase), key components of the degradosome (95). CsrD does not appear to be a nuclease and did not affect decay of other RNAs (one sRNA and two mRNAs were examined) that are degraded by RNase E-dependent mechanisms, suggesting that CsrD functions chiefly as a specificity factor for CsrB/C decay. Epistasis analysis showed that CsrD works through its role in CsrB/C turnover, as CsrD regulatory effects on several genes and systems were eliminated in strains lacking *csrB* and *csrC* genes or in a *csrA::kanR* mutant. However, it is formally possible that CsrD might regulate decay of transcripts other than those that have been tested.

The structure of CsrD was predicted to contain two membrane-spanning domains, a HAMP-like domain, and GGDEF and EAL domains, the latter three being essential for its function (Fig. 1C) (95). The HAMP-like domain, named for the diverse proteins in which it is found (*h*istidine kinases, *a*denylyl cyclases, *m*ethyl-accepting chemotaxis proteins, and *p*hosphatases), may mediate protein-protein interactions that are required for CsrD to function. Biochemical and genetic studies revealed that the GGDEF and EAL domains of CsrD are not involved in c-di-GMP metabolism. In addition, regulation by CsrD was not affected by genetic manipulations designed to alter intracellular c-di-GMP levels, indicating that CsrD activity is not regulated by this nucleotide. Furthermore, the GGDEF and EAL domains of CsrD have diverged from the consensus domain sequences at crucial residues, and amino acid residues that are predicted to be essential for their catalytic activities were not involved in CsrD function (95). CsrD orthologs from several species contain conserved sequences that are not characteristic of c-di-GMP-metabolizing proteins (95). Interestingly, CsrD binds to CsrB RNA at high affinity, although its precise mechanism of action in CsrB/C decay remains to be determined. The GGDEF domain itself apparently evolved from the Palm domain family of nucleotidyltransferases and RNA binding proteins (discussed in reference 95). These findings raise the possibility that this domain of CsrD may have reacquired a primordial RNA binding function. Because c-di-GMP itself is essentially a small circular RNA, such putative evolutionary transitions from RNA binding to c-di-GMP cyclase and back again should not be difficult.

MULTITIER REGULATORY INTERACTIONS OF CsrA AND c-di-GMP IN BIOFILM DEVELOPMENT

The role of Csr in governing alternative phenotypic choices is perhaps best understood in the context of motility versus biofilm formation. Furthermore, the role of Csr in biofilm development can be used to illustrate the complexity of this regulatory network and its connections to c-di-GMP signaling systems (Fig. 2 and 3). The initial studies of *csrA* suggested that it represses biofilm formation under static growth conditions (81). However, a clear understanding of this role awaited the elucidation of its regulatory mechanism, the development of quantitative assays for its effects on biofilm development, and laser scanning confocal microscopic imaging of biofilm structure (47). *csrA* was found to regulate biofilm development under all growth conditions and in all *E. coli* strains and enteric species that were tested (46, 47). Its effect on biofilm formation did not depend on any adhesins or surface factors that were known at that time.

A transposon mutagenesis screen for gene(s) responsible for the adhesive phenotype of a *csrA* mutant revealed an operon, *pgaABCD* (formerly *ycdSRQP*), which is necessary for *E. coli* biofilm formation (109). A Pga-dependent polysaccharide adhesin was isolated and determined to be poly-β-1,6-*N*-acetyl-D-glucosamine (PGA). At that time, this polymer was known only from staphylococci (31). However, subsequent studies have shown that PGA is produced by diverse species, where it functions as a biofilm adhesin and is important in host-microbe interaction (for examples, see references 41, 43–45, and 72). In *E. coli,* PGA mediates cellular attachment

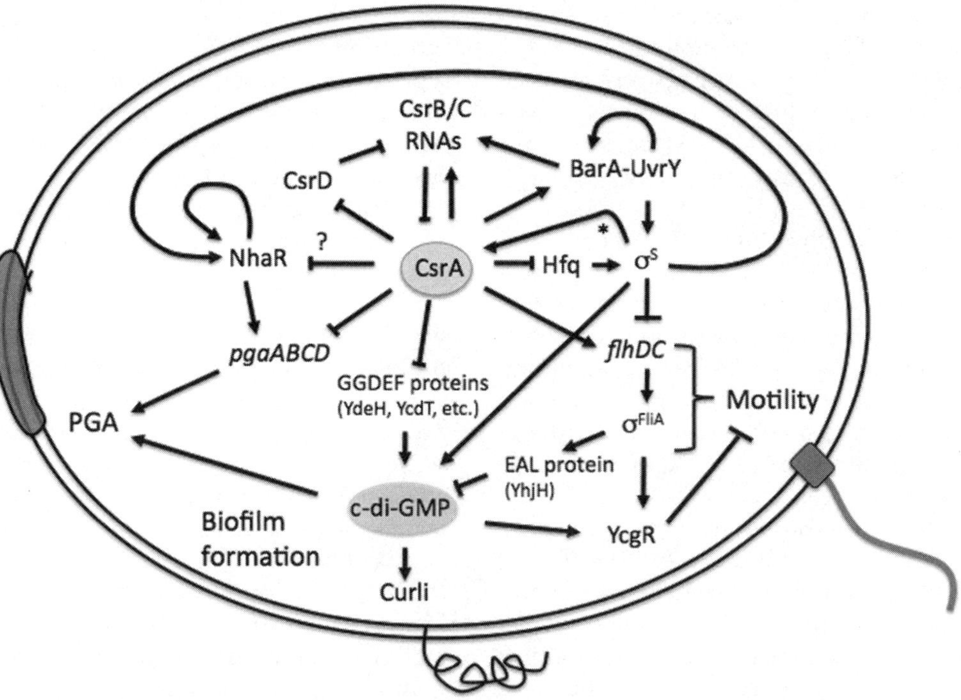

Figure 3. Convergence of Csr and c-di-GMP regulation in *E. coli*. CsrA and c-di-GMP have inverse effects on motility versus sessility. CsrA activates motility and represses biofilm formation, while c-di-GMP represses motility and activates biofilm formation. CsrA posttranscriptionally activates *flhDC*, the master operon of the motility cascade, and is required for flagellum biosynthesis (111), while it represses *pgaABCD*, the structural genes for production, covalent modification, and secretion of the biofilm adhesin PGA (108). CsrA represses genes for c-di-GMP biosynthesis (48), including *ydeH*, which is important for biofilm formation and PGA synthesis (29). Because CsrA activates *flhDC*, it should indirectly activate σ^{FliA}, the motility sigma factor that is under FlhDC control, and in turn should activate *yhjH* (74), although this indirect effect was not apparent in array studies (49). The latter gene encodes a c-di-GMP-specific PDE, which represses biofilm formation (29, 95). Because CsrA represses c-di-GMP production (49), it might activate motility indirectly through effects on the PilZ domain protein, YcgR, although this remains to be seen. c-di-GMP activates expression of curli fimbriae (74) and PGA production (29) by undetermined mechanisms. Several feedback loops are apparent in this system: CsrA–BarA-UvrY–CsrB/C (94), BarA-UvrY (94), CsrA-CsrD-CsrB/C (95), and NhaR, which activates its own transcription (17) and that between CsrA and σ^s. In principle, σ^s may be regulated by CsrA through its effects on BarA-UvrY signaling (95), which affects *rpoS* transcription (36, 68), or its effect on *hfq* translation (5), as Hfq functions as an RNA chaperone that mediates the positive effects of two antisense RNA regulators on *rpoS* translation (62, 105). In turn, σ^s has multiple regulatory effects on c-di-GMP and expression of genes for curli fimbriae (74, 93). The asterisk between σ^s and CsrA indicates conditional modest effects of *rpoS* on *csrA* transcripts (16). Not shown are the inverse effects of CsrA and σ^s on glycogen biosynthesis (36, 81, 119). The question mark indicates that CsrA is predicted to repress *nhaR* translation (21).

to abiotic surfaces and cell-cell adhesion. PGA is needed for the conversion of an initial temporary polar cell-surface attachment to permanent attachment during the initiation of biofilm development and for the formation of the nonrandom cellular spatial patterns of biofilms, which may affect diffusion processes (1).

CsrA binds to the untranslated leader of the *pgaABCD* operon and the *pgaA* initiation codon, blocks *pgaA* translation, and destabilizes the *pgaABCD* transcript (118). Binding occurs at six sites, the most of any mRNA target of CsrA that has been examined to date. Functionality of the two downstream CsrA binding sites, which overlap the Shine-Dalgarno sequence and translation initiation codon, was demonstrated by site-directed mutagenesis of these sites (118). Additional studies will be required to elucidate the details of CsrA (RsmA) effects on biofilm formation and polysaccharide production in species that lack the *pgaABCD* operon (for examples, see references 24, 30, 50, and 57).

In addition to its direct role in regulating translation and stability of *pga* mRNA, there is evidence that CsrA regulates carbon flux pathways that are important for biofilm formation. Glycogen biosynthesis and turnover and central carbon flux are broadly regulated by CsrA (81, 89, 119). In turn, glycogen biosynthesis is needed for optimal biofilm formation, as structural gene mutations that prevent glycogen synthesis decreased biofilm formation in *E. coli* (47), while inhibitors and activators of glycogen synthesis reduced and increased biofilm formation by *Salmonella enteritidis,* respectively (11). Historically, the glycogen biosynthesis pathway was the first example of CsrA and σ^s antagonism (37, 119).

As glycogen is an intracellular polymer, it was unlikely to be involved directly in cell-cell or cell-surface interactions and was proposed instead to serve as a possible carbon and energy source for biosynthesis of a surface adhesin (47). The identification of PGA as the key adhesin of *E. coli* K-12, which is derepressed in a *csrA* mutant, is consistent with this idea (109). In principle, PGA can be derived directly from products of glycogen metabolism (Fig. 2). Recently, transcripts of two additional genes, *glmU* and *glmS,* in the pathway from glycogen to PGA (Fig. 2) were isolated in complex with a His-tagged CsrA protein (21). While the potential of CsrA to control biochemical pathways that affect biofilm development is established, the biological relevance of the latter observation requires further investigation.

Transcription of *pgaABCD* is driven by a single promoter, which is activated by the LysR-family DNA binding protein NhaR (28). *nhaAR* was identified by transposon screening (109) as an operon that is important for biofilm formation. This operon encodes NhaR itself and the NhaA Na^+/H^+ antiporter and responds to intracellular sodium levels and high pH (17). NhaR triggers PGA synthesis and biofilm formation by *E. coli* in response to conditions of high pH (28), such as might be encountered in infected urine. In fact, NhaR was recently found to be a virulence factor of the urinary tract pathogen *Proteus mirabilis,* known for producing urease, which increases the pH of urine during infection (40). The *nhaAR* operon is transcribed from two promoters, one of which is responsible for expression in the stationary phase of growth and is σ^s dependent (17) (Fig. 3). Interestingly, *nhaR* mRNA was recently identified as a CsrA-bound transcript (21). The translation initiation region of *nhaR* mRNA contains a putative CsrA binding site, identified by a position-specific weight matrix search program (5). Further studies will be required to determine whether CsrA represses *nhaR* translation and represents another example of CsrA and σ^s antagonism, as predicted by these findings.

CsrA directly affects c-di-GMP production in the cell. It binds to and regulates mRNAs for GGDEF domain-containing DGC, the most dramatic examples being *ydeH* and *ycdT* (49). In turn, evidence indicates that c-di-GMP activates both biofilm formation and PGA production, as genes that affect the synthesis and turnover of this nucleotide have strong effects on both processes (29, 95). The gene *ydeH* is a particularly strong activator of PGA production and biofilm formation (Fig. 2) (28, 49). The mechanism of action of c-di-GMP in PGA production has not been determined.

CsrA regulates at least two pathways controlling *rpoS* expression, which might indirectly influence c-di-GMP levels (Fig. 2 and 3) (110). First, CsrA activity is critical for BarA-UvrY signaling (96, 113), which apparently activates *rpoS* transcription (36, 68). Second, CsrA has modest effects on *hfq* expression (5), which in turn is required for translational activation of *rpoS* by *dsrA* and *rprA* antisense RNAs (62). The biological significance of these indirect effects of CsrA remains to be clearly established. In fact, we observed only minor effects of *rpoS* and its apparent biofilm target, curli fimbriae, on biofilm formation by *E. coli* K-12 strains such as MG1655 and its *csrA* mutant (47, 109). Nevertheless, biofilm formation depends on curli in certain *E. coli* strain backgrounds, such as those containing conjugal plasmids, OmpR*, or overexpressing the curli activator *csgD* (for examples, see references 32 and 63). While σ^s has been clearly tied to c-di-GMP signaling in *E. coli* K-12 (for examples, see references 74 and 110), the role of σ^s in biofilm formation itself was not examined in those studies, and efforts to define the role of σ^s in biofilm formation have not reached consistent conclusions (for examples, see references 14 and 75) and warrant further study.

CsrA may indirectly repress c-di-GMP levels through its activation of *flhDC* expression and the motility cascade (Fig. 3) (111), which induces *yhjH,* encoding an EAL domain protein (74). Unfortunately, *yhjH* mRNA was too low to be evaluated in the previous analyses of CsrA effects on GGDEF and EAL genes (49), so this hypothesis remains to be tested. Additional potential regulatory interactions of CsrA and c-di-GMP with the motility cascade are presented in Fig. 3 and its legend.

SUMMARY

There are now several examples of genes and processes that are governed antagonistically by CsrA (RsmA) and c-di-GMP, and this list is probably nowhere near completion. In *E. coli,* CsrA represses biofilm formation at multiple levels, by posttranscriptionally repressing genes for production of PGA synthesis, glycogen biosynthesis and catabolism, and c-di-GMP synthesis, and it may regulate many other processes that affect biofilm, including the *glmS* and *glmU* genes for the biosynthesis of UDP-*N*-

acetylglucosamine, and *nhaR*, the activator of *pga* transcription. Thus far, c-di-GMP appears to have more focused (less global) effects on biofilm factors. In *E. coli*, it activates biofilm formation by stimulating the production of the polysaccharide adhesin PGA and curli fimbria gene expression, although the mechanisms for these effects remain to be defined. CsrA and c-di-GMP also antagonistically regulate (activate and repress, respectively) motility in *E. coli*. The regulatory interactions of these two systems are incompletely established but include the direct effects of CsrA on *ydeH* and *ycdT* expression and thus c-di-GMP levels, as well as apparent reciprocal influences of CsrA and σ^s. The relative dearth of information on the environmental signals and cues for *csr* gene expression and c-di-GMP production and turnover clearly limits our present appreciation of these important biological systems and is sure to be a subject of much future investigation.

Acknowledgments. Research from our laboratories that is related to the subject matter of this chapter was funded in part by the National Institutes of Health (GM066794 and GM59969).

We are deeply indebted to the students, postdoctoral fellows, and colleagues who contributed to these studies.

REFERENCES

1. **Agladze, K., X. Wang, and T. Romeo.** 2005. Spatial periodicity of *Escherichia coli* K-12 biofilm microstructure initiates during a reversible, polar attachment phase of development and requires the polysaccharide adhesin PGA. *J. Bacteriol.* **187:**8237–8246.
2. **Ahmer, B. M.** 2004. Cell-to-cell signalling in *Escherichia coli* and *Salmonella enterica*. *Mol. Microbiol.* **52:**933–945.
3. **Amikam, D., and M. Y. Galperin.** 2006. PilZ domain is part of the bacterial c-di-GMP binding protein. *Bioinformatics* **22:** 3–6.
4. **Babitzke, P., and T. Romeo.** 2007. CsrB sRNA family: sequestration of RNA-binding regulatory proteins. *Curr. Opin. Microbiol.* **10:**156–163.
5. **Baker, C. S., L. A. Eory, H. Yakhnin, J. Mercante, T. Romeo, and P. Babitzke.** 2007. CsrA inhibits translation initiation of *Escherichia coli hfq* by binding to a single site overlapping the Shine-Dalgarno sequence. *J. Bacteriol.* **189:**5472–5481.
6. **Baker, C. S., I. Morozov, K. Suzuki, T. Romeo, and P. Babitzke.** 2002. CsrA regulates glycogen biosynthesis by preventing translation of *glgC* in *Escherichia coli*. *Mol. Microbiol.* **44:**1599–1610.
7. **Barnard, F. M., M. F. Loughlin, H. P. Fainberg, M. P. Messenger, D. W. Ussery, P. Williams, and P. J. Jenks.** 2004. Global regulation of virulence and the stress response by CsrA in the highly adapted human gastric pathogen *Helicobacter pylori*. *Mol. Microbiol.* **51:**15–32.
8. **Bejerano-Sagie, M., and K. B. Xavier.** 2007. The role of small RNAs in quorum sensing. *Curr. Opin. Microbiol.* **10:**189–198.
9. **Benach, J., S. S. Swaminatha, R. Tamayo, S. K. Handelman, E. Folta-Stogniew, J. E. Ramos, F. Forouhar, H. Neely, J. Seetharaman, A. Camilli, and J. F. Hunt.** 2007. The structural basis of cyclic diguanylate signal transduction by PilZ domains. *EMBO J.* **26:**5153–5166.
10. **Bobrov, A. G., O. Kirillina, S. Forman, D. Mack, and R. D. Perry.** 2008. Insights into *Yersinia pestis* biofilm development: topology and co-interaction of Hms inner membrane proteins involved in exopolysaccharide production. *Environ. Microbiol.* **10:**1419–1432.
11. **Bonafonte, M. A., C. Solano, B. Sesma, M. Alvarez, L. Montuenga, D. García-Ros, and C. Gamazo.** 2000. The relationship between glycogen synthesis, biofilm formation and virulence in *Salmonella enteritidis*. *Microbiol. Lett.* **191:**31–36.
12. **Christen, B., M. Christen, R. Paul, F. Schmid, M. Folcher, P. Jenoe, M. Meuwly, and U. Jenal.** 2006. Allosteric control of cyclic di-GMP signaling. *J. Biol. Chem.* **281:**32015–32024.
13. **Condon, C.** 2007. Maturation and degradation of RNA in bacteria. *Curr. Opin. Microbiol.* **10:**271–278.
14. **Corona-Izquierdo, F. P., and Membrillo-Hernández J.** 2002. A mutation in *rpoS* enhances biofilm formation in *Escherichia coli* during exponential phase of growth. *FEMS Microbiol. Lett.* **211:**105–110.
15. **Cui, Y., A. Chatterjee, and A. K. Chatterjee.** 2001. Effects of the two-component system comprising GacA and GacS of *Erwinia carotovora* subsp. carotovora on the production of global regulatory *rsmB* RNA, extracellular enzymes, and harpinEcc. *Mol. Plant-Microbe Interact.* **14:**516–526.
16. **Dong, T., and H. E. Schellhorn.** 2008. Control of RpoS in global gene expression of *Escherichia coli* in minimal media. *Mol. Genet. Genomics* **281:**19–33.
17. **Dover, N., and E. Padan.** 2001. Transcription of *nhaA*, the main Na^+/H^+ antiporter of *Escherichia coli*, is regulated by Na^+ and growth phase. *J. Bacteriol.* **183:**644–653.
18. **Dow, J. M., Y. Fouhy, J. F. Lucey, and R. P. Ryan.** 2006. The HD-GYP domain, cyclic di-GMP signaling, and bacterial virulence to plants. *Mol. Plant-Microbe Interact.* **19:**1378–1384.
19. **Dubey, A. K., C. S. Baker, T. Romeo, and P. Babitzke.** 2005. RNA sequence and secondary structure participate in high-affinity CsrA-RNA interaction. *RNA* **11:**1579–1587.
20. **Dubey, A. K., C. S. Baker, K. Suzuki, A. D. Jones, P. Pandit, T. Romeo, and P. Babitzke.** 2003. CsrA regulates translation of the *Escherichia coli* carbon starvation gene, *cstA*, by blocking ribosome access to the *cstA* transcript. *J. Bacteriol.* **185:** 4450–4560.
21. **Edwards, A. N., J. W. Mercante, P. Babitzke, and T. Romeo.** 2008. Determination of the global regulatory role of CsrA in *Escherichia coli*, abstr. H-067, p. 295. *Abstr. 108th Gen. Meet. Am. Soc. Microbiol.*, Boston, MA.
22. **Evangelista, M., C. S. Baker, T. Romeo, and P. Babitzke.** 2008. CsrA activates expression of *sdiA*, the gene encoding the *Escherichia coli* homoserine lactone receptor, abstr. H-141, p. 310. *Abstr. 108th Gen. Meet. Am. Soc. Microbiol.*, Boston, MA.
23. **Ferreira, R. B., L. C. Antunes, E. P. Greenberg, and L. L. McCarter.** 2008. *Vibrio parahaemolyticus* ScrC modulates cyclic dimeric GMP regulation of gene expression relevant to growth on surfaces. *J. Bacteriol.* **190:**851–860.
24. **Fields, J. A., and S. A. Thompson.** 2008. *Campylobacter jejuni* CsrA mediates oxidative stress responses, biofilm formation, and host cell invasion. *J. Bacteriol.* **190:**3411–3416.
25. **Furukawa, S., S. L. Kuchma, and G.A. O'Toole.** 2006. Keeping options open: acute versus persistent infections. *J. Bacteriol.* **188:**1211–1217.
26. **Galperin, M. Y., A. N. Nikolskaya, and E. V. Koonin.** 2001. Novel domains of the prokaryotic two-component signal transduction systems. *FEMS Microbiol. Lett.* **203:**11–21.
27. **Gjermansen, M., P. Ragas, and T. Tolker-Nielsen.** 2006. Proteins with GGDEF and EAL domains regulate *Pseudomonas putida* biofilm formation and dispersal. *FEMS Microbiol. Lett.* **265:**215–224.

28. **Goller, C., X. Wang, Y. Itoh, and T. Romeo.** 2006. The cation-responsive protein NhaR of *Escherichia coli* activates *pgaABCD* transcription, required for production of the biofilm adhesin poly-β-1,6-*N*-acetyl-D-glucosamine. *J. Bacteriol.* **188:** 8022–8032.
29. **Goller, C. C., A. Pannuri, T. Romeo, Y. Itoh, and K. Suzuki.** 2008. PGA accumulation and biofilm formation in *Escherichia coli:* modulation by c-di-GMP, abstr. H-057, p. 293. *Abstr. 108th Gen. Meet. Am. Soc. Microbiol.,* Boston, MA.
30. **Goodman, A. L., B. Kulasekara, A. Rietsch, D. Boyd, R. S. Smith, and S. A. Lory.** 2004. A signaling network reciprocally regulates genes associated with acute infection and chronic persistence in *Pseudomonas aeruginosa. Dev. Cell* 7:745–754.
31. **Götz, F.** 2002. *Staphylococcus* and biofilms. *Mol. Microbiol.* **43:**1367–1378.
32. **Gualdi, L., L. Tagliabue, S. Bertagnoli, T. Ieranò, C. De Castro, and P. Landini.** 2008. Cellulose modulates biofilm formation by counteracting curli-mediated colonization of solid surfaces in *Escherichia coli. Microbiology* **154:**2017–2024.
33. **Gudapaty, S., K. Suzuki, X. Wang, P. Babitzke, and T. Romeo.** 2001. Regulatory interactions of Csr components: the RNA binding protein CsrA activates *csrB* transcription in *Escherichia coli. J. Bacteriol.* **183:**6017–6027.
34. **Gutiérrez, P., Y. Li, M. J. Osborne, E. Pomerantseva, Q. Liu, and K. Gehring.** 2005. Solution structure of the carbon storage regulator protein CsrA from *Escherichia coli. J. Bacteriol.* **187:** 3496–3501.
35. **Heeb, S., S. A. Kuehne, M. Bycroft, S. Crivii, M. D. Allen, D. Haas, M. Camara, and P. Williams.** 2006. Functional analysis of the post-transcriptional regulator RsmA reveals a novel RNA-binding site. *J. Mol. Biol.* **355:**1026–1036.
36. **Hengge, R.** 2008. The two-component network and the general stress sigma factor RpoS (σ^s) in *Escherichia coli. Adv. Exp. Med. Biol.* **631:**40–53.
37. **Hengge-Aronis, R., and D. Fischer.** 1992. Identification and molecular analysis of *glgS,* a novel growth-phase-regulated and *rpoS*-dependent gene involved in glycogen synthesis in *Escherichia coli. Mol. Microbiol.* **6:**1877–1886.
38. **Heroven, A. K., K. Böhme, M. Rohde, and P. Dersch.** 2008. A Csr-type regulatory system, including small non-coding RNAs, regulates the global virulence regulator RovA of *Yersinia pseudotuberculosis* through RovM. *Mol. Microbiol.* **68:** 1179–1195.
39. **Hickman, J. W., and C. S. Harwood.** 2008, Identification of FleQ from *Pseudomonas aeruginosa* as a c-di-GMP-responsive transcription factor. *Mol. Microbiol.* **69:**376–389.
40. **Himpsl, S. D., C. V. Lockatell, J. R. Hebel, D. E. Johnson, and H. L. Mobley.** 2008. Identification of virulence determinants in uropathogenic *Proteus mirabilis* using signature-tagged mutagenesis. *J. Med. Microbiol.* **57:**1068–1078.
41. **Hinnebusch, B. J., and D. L. Erickson.** 2008. *Yersinia pestis* biofilm in the flea vector and its role in the transmission of plague. *Curr. Top. Microbiol. Immunol.* **322:**229–248.
42. **Holland, L. M., S. T. O'Donnell, D. A. Ryjenkov, L. Gomelsky, S. R. Slater, P. D. Fey, M. Gomelsky, and J. P. O'Gara.** 2008. A staphylococcal GGDEF domain protein regulates biofilm formation independently of cyclic dimeric GMP. *J. Bacteriol.* **190:**5178–5189.
43. **Itoh, Y., J. D. Rice, C. Goller, A. Pannuri, J. Taylor, J. Meisner, T. J. Beveridge, J. F. Preston III, and T. Romeo.** 2008. Roles of *pgaABCD* genes in synthesis, modification, and export of the *Escherichia coli* biofilm adhesin poly-β-1,6-*N*-acetyl-D-glucosamine. *J. Bacteriol.* **190:**3670–3680.
44. **Itoh, Y., X. Wang, B. J. Hinnebusch, J. F. Preston III, and T. Romeo.** 2005. Depolymerization of β-1,6-*N*-acetyl-D-glucosamine disrupts the integrity of diverse bacterial biofilms. *J. Bacteriol.* **187:**382–387.
45. **Izano, E. A., I. Sadovskaya, E. Vinogradov, M. H. Mulks, K. Velliyagounder, C. Ragunath, W. B. Kher, N. Ramasubbu, S. Jabbouri, M. B. Perry, and J. B. Kaplan.** 2007. Poly-N-acetylglucosamine mediates biofilm formation and antibiotic resistance in *Actinobacillus pleuropneumoniae. Microb. Pathog.* **43:**1–9.
46. **Jackson, D. W., J. W. Simecka, and T. Romeo.** 2002. Catabolite repression of *Escherichia coli* biofilm formation. *J. Bacteriol.* **184:**3406–3410.
47. **Jackson, D. W., K. Suzuki, L. Oakford, J. W. Simecka, M. E. Hart, and T. Romeo.** 2002. Biofilm formation and dispersal under the influence of the global regulator CsrA of *Escherichia coli. J. Bacteriol.* **184:**290–301.
48. **Jenal, U., and J. Malone.** 2006. Mechanisms of cyclic-di-GMP signaling in bacteria. *Annu. Rev. Genet.* **40:**385–407.
49. **Jonas, K., A. N. Edwards, R. Simm, T. Romeo, U. Römling, and O. Melefors.** 2008. The RNA binding protein CsrA controls cyclic di-GMP metabolism by directly regulating the expression of GGDEF proteins. *Mol. Microbiol.* **70:**236–257.
50. **Jones, M. K., E. B. Warner, and J. D. Oliver.** 2008. *csrA* inhibits the formation of biofilms by *Vibrio vulnificus. Appl. Environ. Microbiol.* **74:**7064–7066.
51. **Kay, E., C. Dubuis, and D. Haas.** 2005. Three small RNAs jointly ensure secondary metabolism and biocontrol in *Pseudomonas fluorescens* CHA0. *Proc. Natl. Acad. Sci. USA* **102:** 17136–17141.
52. **Kuchma, S. L., K. M. Brothers, J. H. Merritt, N. T. Liberati, F. M. Ausubel, and G. A. O'Toole.** 2007. BifA, a cyclic-di-GMP phosphodiesterase, inversely regulates biofilm formation and swarming motility by *Pseudomonas aeruginosa* PA14. *J. Bacteriol.* **189:**8165–8178.
53. **Kumar, M., and D. Chatterji.** 2008. Cyclic di-GMP: a second messenger required for long-term survival, but not for biofilm formation, in *Mycobacterium smegmatis. Microbiology* **154:** 2942–2955.
54. **Kushner, S. R.** 2002. mRNA decay in *Escherichia coli* comes of age. *J. Bacteriol.* **184:**4658–4665.
55. **Lapouge, K., M. Schubert, F. H. Allain, and D. Haas.** 2008. Gac/Rsm signal transduction pathway of gamma-Proteobacteria: from RNA recognition to regulation of social behaviour. *Mol. Microbiol.* **67:**241–253.
56. **Lee, V. T., J. M. Matewish, J. L. Kessler, M. Hyodo, Y. Hayakawa, and S. Lory.** 2007. A cyclic-di-GMP receptor required for bacterial exopolysaccharide production. *Mol. Microbiol.* **65:**1474–1484.
57. **Lenz, D. H., M. B. Miller, J. Zhu, R. V. Kulkarni, and B. L. Bassler.** 2005. CsrA and three redundant small RNAs regulate quorum sensing in *Vibrio cholerae. Mol. Microbiol.* **58:**1186–1202.
58. **Liu, M. Y., G. Gui, B. Wei, J. F. Preston III, L. Oakford, U. Yuksel, D. P. Giedroc, and T. Romeo.** 1997. The RNA molecule CsrB binds to the global regulatory protein CsrA and antagonizes its activity in *Escherichia coli. J. Biol. Chem.* **272:** 17502–17510.
59. **Liu, M. Y., and T. Romeo.** 1997. The global regulator CsrA of *Escherichia coli* is a specific mRNA-binding protein. *J. Bacteriol.* **179:**4639–4642.
60. **Liu, M. Y., H. Yang, and T. Romeo.** 1995. The product of the pleiotropic *Escherichia coli* gene *csrA* modulates glycogen biosynthesis via effects on mRNA stability. *J. Bacteriol.* **177:** 2663–2672.
61. **Lucchetti-Miganeh, C., E. Burrowes, C. Baysse, and G. Ermel.** 2008. The post-transcriptional regulator CsrA plays a central

role in the adaptation of bacterial pathogens to different stages of infection in animal hosts. *Microbiology* **154:**16–29.

62. **Majdalani, N., C. K. Vanderpool, and S. Gottesman.** 2005. Bacterial small RNA regulators. *Crit. Rev. Biochem. Mol. Biol.* **40:**93–113.
63. **May, T., and S. Okabe.** 2008. *Escherichia coli* harboring a natural IncF conjugative F plasmid develops complex mature biofilms by stimulating synthesis of colanic acid and curli. *J. Bacteriol.* **190:**7479–7490.
64. **Mercante, J., K. Suzuki, X. Cheng, P. Babitzke, and T. Romeo.** 2006. Comprehensive alanine-scanning mutagenesis of *Escherichia coli* CsrA defines two subdomains of critical functional importance. *J. Biol. Chem.* **281:**31832–31842.
65. **Molofsky, A. B., and S. Swanson.** 2004. Differentiate to thrive: lessons from the *Legionella pneumophila* life cycle. *Mol. Microbiol.* **53:**29–40.
66. **Mondragon, V., B. Franco, K. Jonas, K. Suzuki, T. Romeo, Ö. Melefors, and D. Georgellis.** 2006. pH dependent activation of the BarA-UvrY two-component system in *Escherichia coli*. *J. Bacteriol.* **188:**8303–8306.
67. **Morgan, R., S. Kohn, S. H. Hwang, D. J. Hassett, and K. Sauer.** 2006. BdlA, a chemotaxis regulator essential for biofilm dispersion in *Pseudomonas aeruginosa*. *J. Bacteriol.* **188:**7335–7343.
68. **Mukhopadhyay, S., J. P. Audia, R. N. Roy, and H. E. Schellhorn.** 2000. Transcriptional induction of the conserved alternative sigma factor RpoS in *Escherichia coli* is dependent on BarA, a probable two-component regulator. *Mol. Microbiol.* **37:**371–381.
69. **Mulcahy, H., J. O'Callaghan, E. P. O'Grady, M. D. Maciá, N. Borrell, C. Gómez, P. G. Casey, C. Hill, C. Adams, C. G. Gahan, A. Oliver, and F. O'Gara.** 2008. *Pseudomonas aeruginosa* RsmA plays an important role during murine infection by influencing colonization, virulence, persistence, and pulmonary inflammation. *Infect. Immun.* **76:**632–638.
70. **Nakhamchik, A., C. Wilde, and D. A. Rowe-Magnus.** 2008. Cyclic-di-GMP regulates extracellular polysaccharide production, biofilm formation, and rugose colony development by *Vibrio vulnificus*. *Appl. Environ. Microbiol.* **74:**4199–4209.
71. **Newell, P. D., R. D. Monds, and G. A. O'Toole.** 2009. LapD is a bis-(3′,5′)-cyclic dimeric GMP-binding protein that regulates surface attachment by *Pseudomonas fluorescens* Pf0-1. *Proc. Natl. Acad. Sci. USA* **106:**3461–3466.
72. **Parise, G., M. Mishra, Y. Itoh, T. Romeo, and R. Deora.** 2007. Role of a putative polysaccharide locus in *Bordetella* biofilm development. *J. Bacteriol.* **189:**750–760.
73. **Pernestig, A. K., D. Georgellis, T. Romeo, K. Suzuki, H. Tomenius, S. Normark, and O. Melefors.** 2003. The *Escherichia coli* BarA-UvrY two-component system is needed for efficient switching between glycolytic and gluconeogenic carbon sources. *J. Bacteriol.* **185:**843–853.
74. **Pesavento, C., G. Becker, N. Sommerfeldt, A. Possling, N. Tschowri, A. Mehlis, and R. Hengge.** 2008. Inverse regulatory coordination of motility and curli-mediated adhesion in *Escherichia coli*. *Genes Dev.* **22:**2434–2446.
75. **Prigent-Combaret, C., E. Brombacher, O. Vidal, A. Ambert, P. Lejeune, P. Landini, and C. Dorel.** 2001. Complex regulatory network controls initial adhesion and biofilm formation in *Escherichia coli* via regulation of the *csgD* gene. *J. Bacteriol.* **183:**7213–7223.
76. **Rand, J. D., S. G. Danby, D. L. Greenway, and R. R. England.** 2002. Increased expression of the multidrug efflux genes *acrAB* occurs during slow growth of *Escherichia coli*. *FEMS Microbiol. Lett.* **207:**91–95.
77. **Reimmann, C., C. Valverde, E. Kay, and D. Haas.** 2005. Posttranscriptional repression of GacS/GacA-controlled genes by the RNA-binding protein RsmE acting together with RsmA in the biocontrol strain *Pseudomonas fluorescens* CHA0. *J. Bacteriol.* **187:**276–285.
78. **Rife, C., R. Schwarzenbacher, D. McMullan, P. Abdubek, E. Ambing, H. Axelrod, T. Biorac, J. M. Canaves, H. J. Chiu, A. M. Deacon, M. DiDonato, M. A. Elsliger, A. Godzik, C. Grittini, S. K. Grzechnik, J. Hale, E. Hampton, G. W. Han, J. Haugen, M. Hornsby, L. Jaroszewski, H. E. Klock, E. Koesema, A. Kreusch, P. Kuhn, S. A. Lesley, M. D. Miller, K. Moy, E. Nigoghossian, J. Paulsen, K. Quijano, R. Reyes, E. Sims, G. Spraggon, R. C. Stevens, H. van den Bedem, J. Velasquez, J. Vincent, A. White, G. Wolf, Q. Xu, K. L. O. Hodgson, J. Wooley, and I. A. Wilson.** 2005. Crystal structure of the global regulatory protein CsrA from *Pseudomonas putida* at 2.05 A resolution reveals a new fold. *Proteins* **61:**449–453.
79. **Romeo, T.** 1998. Global regulation by the small RNA-binding protein CsrA and the non-coding RNA molecule CsrB. *Mol. Microbiol.* **29:**1321–1330.
80. **Romeo, T., J. Black, and J. Preiss.** 1990. Genetic regulation of glycogen biosynthesis in Escherichia coli: *in vivo* effects of the catabolite repression and stringent response systems. *Curr. Microbiol.* **21:**131–137.
81. **Romeo, T., M. Gong, M. Y. Liu, and A. M. Brun-Zinkernagel.** 1993. Identification and molecular characterization of *csrA,* a pleiotropic gene from *Escherichia coli* that affects glycogen biosynthesis, gluconeogenesis, cell size, and surface properties. *J. Bacteriol.* **175:**4744–4755.
82. **Romeo, T., J. Moore, and J. Smith.** 1991. A simple method for cloning genes involved in glucan biosynthesis: application to the isolation of structural and regulatory genes for glycogen synthesis in *Escherichia coli*. *Gene* **108:**23–29.
83. **Romeo, T., and J. Preiss.** 1989. Genetic regulation of glycogen biosynthesis in *Escherichia coli:* in vitro effects of cyclic AMP and guanosine 5′-diphosphate 3′-diphosphate and analysis of in vivo transcripts. *J. Bacteriol.* **171:**2773–2782.
84. **Römling, U.** 2005. Characterization of the rdar morphotype, a multicellular behaviour in Enterobacteriaceae. *Cell. Mol. Life Sci.* **62:**1234–1246.
85. **Römling, U., and D. Amikam.** 2006. Cyclic di-GMP as a second messenger. *Curr. Opin. Microbiol.* **9:**218–228.
86. **Römling, U., M. Gomelsky, and M. Y. Galperin.** 2005. C-di-GMP: the dawning of a novel bacterial signalling system. *Mol. Microbiol.* **57:**629–639.
87. **Ross, P., H. Weinhouse, Y. Aloni, D. Michaeli, P. Weinberger-Ohana, R. Mayer, S. Braun, E. de Vroom, G. A. van der Marel, J. H. van Boom, and M. Benziman.** 1987. Regulation of cellulose synthesis in *Acetobacter xylinum* by cyclic diguanylic acid. *Nature* **325:**279–281.
88. **Ryjenkov, D. A., R. Simm, U. Römling, and M. Gomelsky.** 2006. The PilZ domain is a receptor for the second messenger c-di-GMP: the PilZ domain protein YcgR controls motility in enterobacteria. *J. Biol. Chem.* **281:**30310–30314.
89. **Sabnis, N. A., H. Yang, and T. Romeo.** 1995. Pleiotropic regulation of central carbohydrate metabolism in *Escherichia coli* via the gene *csrA*. *J. Biol. Chem.* **270:**29096–29104.
90. **Schmidt, A. J., D. A. Ryjenkov, and M. Gomelsky.** 2005. The ubiquitous protein domain EAL is a cyclic diguanylate-specific phosphodiesterase: enzymatically active and inactive EAL domains. *J. Bacteriol.* **187:**4774–4781.
91. **Schubert, M., K. Lapouge, O. Duss, F. C. Oberstrass, I. Jelesarov, D. Haas, and F. H. Allain.** 2007. Molecular basis of messenger RNA recognition by the specific bacterial repressing clamp RsmA/CsrA. *Nat. Struct. Mol. Biol.* **14:**807–813.
92. **Simm, R., M. Morr, A. Kader, M. Nimtz, and U. Römling.** 2004. GGDEF and EAL domains inversely regulate cyclic di-

GMP levels and transition from sessility to motility. *Mol. Microbiol.* **53**:1123–1134.
93. **Sommerfeldt, N., A. Possling, G. Becker, C. Pesavento, N. Tschowri, and R. Hengge.** 2009. Gene expression patterns and differential input into curli fimbriae regulation of all GGDEF/EAL domain proteins in *Escherichia coli. Microbiology* **155**:1318–1331. [Epub ahead of print.]
94. **Sudarsan, N., E. R. Lee, Z. Weinberg, R. H. Moy, J. N. Kim, K. H. Link, and R. R. Breaker.** 2008. Riboswitches in eubacteria sense the second messenger cyclic di-GMP. *Science* **321**:411–413.
95. **Suzuki, K., P. Babitzke, S. R. Kushner, and T. Romeo.** 2006. Identification of a novel regulatory protein (CsrD) that targets the global regulatory RNAs CsrB and CsrC for degradation by RNase E. *Genes Dev.* **20**:2605–2617.
96. **Suzuki, K., X. Wang, T. Weilbacher, A. K. Pernestig, Ö. Melefors, D. Georgellis, P. Babitzke, and T. Romeo.** 2002. Regulatory circuitry of the CsrA/CsrB and BarA/UvrY systems of *Escherichia coli. J. Bacteriol.* **184**:5130–5140.
97. **Tamayo, R., J. T. Pratt, and A. Camilli.** 2007. Roles of cyclic diguanylate in the regulation of bacterial pathogenesis. *Annu. Rev. Microbiol.* **61**:131–148.
98. **Tamayo, R., A. D. Tischler, and A. Camilli.** 2005. The EAL domain protein VieA is a cyclic diguanylate phosphodiesterase. *J. Biol. Chem.* **280**:33324–33330.
99. **Taylor, B. L.** 2007. Aer on the inside looking out: paradigm for a PAS-HAMP role in sensing oxygen, redox and energy. *Mol. Microbiol.* **65**:1415–1424.
100. **Teplitski, M., A. Al-Agely, and B. M. Ahmer.** 2006. Contribution of the SirA regulon to biofilm formation in *Salmonella enterica* serovar Typhimurium. *Microbiology* **152**: 3411–3424.
101. **Teplitski, M., R. I. Goodier, and B. M. Ahmer.** 2006. Catabolite repression of the SirA regulatory cascade in *Salmonella enterica. Intl. J. Med. Microbiol.* **296**:449–466.
102. **Thormann, K. M., S. Duttler, R. M. Saville, M. Hyodo, S. Shukla, Y. Hayakawa, and A. M. Spormann.** 2006. Control of formation and cellular detachment from *Shewanella oneidensis* MR-1 biofilms by cyclic di-GMP. *J. Bacteriol.* **188**: 2681–2691.
103. **Toledo-Arana, A., F. Repoila, and P. Cossart.** 2007. Small noncoding RNAs controlling pathogenesis. *Curr. Opin. Microbiol.* **10**:182–188.
104. **Tschowri, N., S. Bussy, and R. Hengge.** 2009. The BLUF-EAL protein YcgF acts as a direct anti-repressor in a blue-light response of *Escherichia coli. Genes Dev.* **23**:522–534.
105. **Updegrove, T., N. Wilf, X. Sun, and R. M. Wartell.** 2008. Effect of Hfq on RprA-rpoS mRNA pairing: Hfq-RNA binding and the influence of the 5′ rpoS mRNA leader region. *Biochemistry* **47**:11184–11195.
106. **Valverde, C., M. Lindell, E. G. H. Wagner, and D. Haas.** 2004. A repeated GGA motif is critical for the activity and stability of the riboregulator RsmY of *Pseudomonas fluorescens. J. Biol. Chem.* **279**:25066–25074.
107. **Van Houdt, R., A. Aertsen, P. Moons, K. Vanoirbeek, and C. W. Michiels.** 2006. N-acyl-l-homoserine lactone signal interception by *Escherichia coli. FEMS Microbiol. Lett.* **256**: 83–89.
108. **Wang, X., A. K. Dubey, K. Suzuki, C. S. Baker, P. Babitzke, and T. Romeo.** 2005. CsrA post-transcriptionally represses *pgaABCD,* responsible for synthesis of a biofilm polysaccharide adhesin of *Escherichia coli. Mol. Microbiol.* **56**:1648–1663.
109. **Wang, X., J. F. Preston III, and T. Romeo.** 2004. The *pgaABCD* locus of *Escherichia coli* promotes the synthesis of a polysaccharide adhesin required for biofilm formation. *J. Bacteriol.* **186**:2724–2734.
110. **Weber, H., C. Pesavento, A. Possling, G. Tischendorf, and R. Hengge.** 2006. Cyclic-di-GMP-mediated signalling within the sigma network of *Escherichia coli. Mol. Microbiol.* **62**: 1014–1034.
111. **Wei, B. L., A. M. Brun-Zinkernagel, J. W. Simecka, B. M. Prüβ, P. Babitzke, and T. Romeo.** 2001. Positive regulation of motility and *flhDC* expression by the RNA-binding protein CsrA of *Escherichia coli. Mol. Microbiol.* **40**:245–256.
112. **Wei, B. L., S. Shin, D. LaPorte, A. J. Wolfe, and T. Romeo.** 2000. Global regulatory mutations in *csrA* and *rpoS* cause severe central carbon stress in *Escherichia coli* in the presence of acetate. *J. Bacteriol.* **182**:1632–1640.
113. **Weilbacher, T., K. Suzuki, A. K. Dubey, X. Wang, S. Gudapaty, I. Morozov, C. S. Baker, D. Georgellis, P. Babitzke, and T. Romeo.** 2003. A novel sRNA component of the carbon storage regulatory system of *Escherichia coli. Mol. Microbiol.* **48**:657–670.
114. **Weinberg, Z., J. E. Barrick, Z. Yao, A. Roth, J. N. Kim, J. Gore, J. X. Wang, E. R. Lee, K. F. Block, N. Sudarsan, S. Neph, M. Tompa, W. L. Ruzzo, and R. R. Breaker.** 2007. Identification of 22 candidate structured RNAs in bacteria using the CMfinder comparative genomics pipeline. *Nucleic Acids Res.* **35**:4809–4819.
115. **Weinhouse, H., S. Sapir, D. Amikam, Y. Shilo, G. Volman, P. Ohana, and M. Benziman.** 1997. c-di-GMP-binding protein, a new factor regulating cellulose synthesis in *Acetobacter xylinum. FEBS Lett.* **416**:207–211.
116. **White, D., M. E. Hart, and T. Romeo.** 1996. Phylogenetic distribution of the regulatory gene *csrA* among Eubacteria. *Gene* **182**: 221–223.
117. **Wolfe, A. J., and K. L. Visick.** 2008. Get the message out: cyclic-di-GMP regulates multiple levels of flagellum-based motility. *J. Bacteriol.* **190**:463–475.
118. **Yakhnin, H., P. Pandit, T. J. Petty, C. S. Baker, T. Romeo, and P. Babitzke.** 2007. CsrA of *Bacillus subtilis* regulates translation initiation of the gene encoding the flagellin protein *(hag)* by blocking ribosome binding. *Mol. Microbiol.* **64**: 1605–1620.
119. **Yang, H., M. Y. Liu, and T. Romeo.** 1996. Coordinate genetic regulation of glycogen catabolism and biosynthesis in *Escherichia coli* via the CsrA gene product. *J. Bacteriol.* **178**: 1012–1017.

The Second Messenger Cyclic Di-GMP
Edited by A. J. Wolfe and K. L. Visick

Chapter 15

Riboswitches That Sense Cyclic Di-GMP

ELAINE R. LEE, NARASIMHAN SUDARSAN, AND RONALD R. BREAKER

Changes in concentrations of cyclic di-GMP (c-di-GMP) are known to signal a variety of bacterial lifestyle transformations including biofilm formation, cell differentiation, cellular motility, and the expression of virulence genes (23, 48, 61). The second messenger is generated by the action of specific diguanylate cyclase (DGC) enzymes that couple two GTP molecules to form each cyclic RNA dimer (23, 48, 61). Concentrations are lowered by the action of specific phosphodiesterase (PDE) enzymes that hydrolyze c-di-GMP to form a linear pGpG, thereby reducing intracellular concentrations of the circular form of the dimer (8, 50). Although c-di-GMP is thought to regulate many genes at the transcriptional, posttranscriptional, and translational levels, the identities of most of the participants in the genetic signaling network remain undiscovered (23, 61).

It is known that c-di-GMP can bind several protein enzymes and regulate their functions. In *Gluconacetobacter xylinus,* c-di-GMP has been shown to activate cellulose biosynthesis proteins allosterically (46, 47). This protein and a number of others that are regulated by c-di-GMP carry PilZ domains (23, 49), which serve as the binding site for the compound and can be used as a predictor of proteins that are regulated by this second messenger. Detailed studies of c-di-GMP binding to a DGC in *Caulobacter crescentus,* PleD, have demonstrated that the second messenger can allosterically repress its own biosynthesis by a noncompetitive inhibition mechanism (10, 68). It is likely that there are many other proteins which are regulated by c-di-GMP but lack these distinguishing binding sites. For example, PelD from *Pseudomonas aeruginosa* does not have a PilZ or an inhibitory site (I-site) (23) domain but c-di-GMP modulates the activity of the protein and thereby regulates the synthesis of PEL polysaccharides (29).

Although the discoveries of c-di-GMP regulation of protein enzymes have begun to reveal the diverse function of the second messenger, these observations cannot explain how c-di-GMP controls so many different processes in such a great variety of bacterial species. Some of these c-di-GMP binding domains may reside in gene regulation factors whose actions can influence a greater collection of genes. For example, the FleQ protein from *P. aeruginosa* is modulated by c-di-GMP and can work as a transcription factor that activates flagellum biosynthesis genes or represses exopolysaccharide biosynthesis genes (22). However, even if the modulation of the activities of protein enzymes and genetic factors could explain the larger physiological changes brought about by c-di-GMP signaling, not all bacteria that utilize c-di-GMP encode proteins carrying known c-di-GMP binding domains such as PilZ (61). Such organisms may carry undiscovered protein domains that bind to the second messenger (61), but other mechanisms for physiological regulation are also likely to be involved.

It has been suggested that the repertoire of regulatory mechanisms for c-di-GMP may involve its direct interaction with other nucleic acids (23, 61). Interestingly, small-molecule receptors made of RNA, called riboswitches, are commonly used in bacteria to modulate gene expression in response to changing metabolite concentrations (2, 9, 64, 74). Riboswitches are almost always located in the 5′ untranslated regions (UTRs) of messenger RNAs, where they typically control either transcription elongation or translation initiation of genes whose protein products synthesize or transport the regulatory ligand (2,

Elaine R. Lee • Department of Molecular, Cellular and Developmental Biology, Yale University, New Haven, CT 06520. **Narasimhan Sudarsan** • Howard Hughes Medical Institute, Yale University, New Haven, CT 06520. **Ronald R. Breaker** • Department of Molecular, Cellular and Developmental Biology, Department of Molecular Biophysics and Biochemistry, and Howard Hughes Medical Institute, Yale University, New Haven, CT 06520.

9, 64, 74). Indeed, we have recently discovered a widespread riboswitch class that selectively responds to the second messenger c-di-GMP (57). Numerous representatives of the riboswitch class are associated with genes that are expected to respond to c-di-GMP, and therefore, the existence of these riboswitches can explain the larger effects of this second messenger in some bacteria. Below we discuss various aspects of the structures, mechanisms, and genetic distributions of this new-found riboswitch class, which reveals much about how RNA is involved in sensing and responding to c-di-GMP in bacteria.

RIBOSWITCHES AS METABOLITE SENSORS AND GENE CONTROL ELEMENTS

Metabolite-sensing riboswitches are present in organisms from all bacterial lineages, but most of the experimentally validated classes exist in gram-positive organisms. For example, it is estimated that ~2% of the genes in *Bacillus subtilis* are regulated by riboswitches (33). Though most of the known riboswitch classes have been found solely in bacteria, riboswitches that bind to thiamine pyrophosphate (TPP) are also found in plants and fungi (9, 11, 58, 59, 65), and bioinformatics searches also have revealed candidate TPP riboswitches in some species of archaea (58).

Riboswitches are typically composed of two functional domains, an aptamer and an expression platform. Each aptamer forms a complex folded receptor that binds to ligands with high affinity and specificity (2, 9, 74). The most highly conserved portion of riboswitches is the aptamer domain, although some compounds are sensed by different structural classes of aptamers. For example, three major classes of riboswitch aptamers have been identified that bind to *S*-adenosylmethionine (SAM), a coenzyme used for methylation reactions (66). A fourth distinct aptamer class has been identified that binds to *S*-adenosylhomocysteine (SAH), a compound that differs from SAM by the absence of a single methyl group and associated positive charge (67).

Although the SAM and SAH riboswitch classes carry structurally diverse aptamers that sense the same or similar compounds, other riboswitches carry near-identical aptamers that have accrued more subtle structural differences to recognize very different compounds. For example, a riboswitch class that recognizes guanine adopts a three-stem junction that carries a conserved cytosine residue in its ligand-binding pocket. This cytosine residue forms a Watson/Crick base pair with the guanine ligand (4, 26). By changing this single nucleotide from a cytidine to a uridine in the aptamer, the specificity of the riboswitch was changed so that it no longer binds guanine with high affinity but instead selectively binds adenine (26). The diversity of aptamer structures and molecular recognition specificities already established suggests that RNA molecules have the sophistication necessary to form binding pockets for numerous biological molecules. Given these observations, it seems reasonable to speculate that second messenger molecules such as c-di-GMP could be candidate ligands for undiscovered riboswitch classes (61).

Aptamers are commonly coupled to expression platforms that modulate gene expression at the level of transcription or translation (2, 9, 74). One common mechanism for transcriptional gene control involves the formation of a rho-independent transcription terminator stem (2, 9, 74). Riboswitches control gene expression by folding into one of two different structures that favor either continued transcription or transcription termination. An "off" switch is defined as a riboswitch that forms the terminator stem when ligand binds to the aptamer, thus terminating transcription before the open reading frame (ORF) of the gene is transcribed. In contrast, an "on" switch reverses this genetic logic by forming the antiterminator structure when ligand binds, which allows RNA polymerase to complete transcription of the full-length mRNA.

Riboswitches can also have expression platforms that regulate translation and involve occlusion of the ribosome binding site or Shine-Dalgarno (SD) site (2, 9, 26, 74). Ligand binding controls the formation of either an anti-SD stem to repress translation or an anti-anti-SD stem to promote translation. As with expression platforms that control transcription termination, both off and on riboswitch function is possible in response to ligand binding. In the great majority of instances, riboswitches will bind ligands and repress expression of proteins that make or import more of the ligand or its precursors. However, some riboswitches activate gene expression, as is observed for an adenine riboswitch (34) from *B. subtilis,* for all known SAH riboswitches (67), and for most glycine riboswitches (35).

Although genetic off switches dominate, this certainly is not due to the inability of riboswitches to adopt arrangements that activate gene expression upon ligand binding. Rather, this bias in favor of ligand-triggered gene repression is likely due to the more common need of cells to repress genes that make or transport essential metabolites. However, riboswitches that sense second messengers such as c-di-GMP should reflect a more equal distribution of on and off switch function, as cells are more likely to use signaling ligands to activate some processes and inactivate others.

DISCOVERING RIBOSWITCHES

Over 20 different classes of riboswitches have been identified to date that bind to a variety of ligands including coenzymes, purines and purine derivatives, amino acids, and a sugar molecule (2, 9, 33, 74). The first examples of riboswitches were identified by closer examination of the long 5′ UTRs of mRNAs whose expression was known to be controlled by a particular metabolite, but no protein effectors could be found. For example, the flavin mononucleotide (FMN) riboswitch is often located in the 5′ UTRs of genes that encode biosynthesis proteins for riboflavin, a precursor of FMN. Because RNA aptamers that bound to FMN had been isolated by in vitro selection (16), it was hypothesized that FMN aptamers could exist naturally (18, 55). It also was known that mutations in the 5′ UTR of an operon that encodes riboflavin biosynthesis genes in *Bacillus subtilis* disrupted downstream gene regulation (24). Upon closer examination of the UTR, a highly conserved structured RNA element was identified (18) and shown to bind to FMN (40, 75).

At least seven riboswitch classes have been validated by resolving long-standing genetic mysteries like that for FMN metabolism described above. However, the discovery and validation of additional riboswitch classes pose greater challenges. Specifically, how can new riboswitch candidates be identified among the vastness of bacterial genomes and, once a candidate is identified, how can researchers determine what ligand is recognized by the riboswitch?

The identification of RNA structures that are candidate riboswitches is aided by bioinformatics tools, which can be used to search for conserved sequences and structures that may represent riboswitch aptamers (2, 3, 12, 17, 69). Since bacterial riboswitches frequently are located in the 5′ UTRs of biosynthesis genes for the ligand to which they bind, bioinformatics analyses can be focused on these regions of numerous bacterial genomes. Recent bioinformatics searches have used a covariance model to search for sequences that might encode structured RNAs in all UTRs of genomic DNA (70, 78, 79). A covariance model is ideal for finding noncoding RNAs because it can identify similarities between RNAs when the base-pairing is conserved, even if the identities of the nucleotides are not. This type of analysis is superior to BLAST analysis because it allows for the discovery of conserved structures as well as conserved sequences (15). These bioinformatics searches have identified a wide variety of noncoding RNAs, some of which are candidate riboswitches (for examples, see reference 69).

When riboswitch candidates are found upstream of biosynthesis genes, in vitro structural probing assays can be performed to determine whether the RNA element is capable of binding to any metabolite in the biosynthetic pathway. In these cases, there may be a large panel of compounds to test, but the identity of the associated genes provides a great deal of assistance in choosing ligands to analyze. Sometimes, however, the genes associated with a candidate riboswitch code for proteins whose functions are obscure or the collections of functions do not immediately point to a likely ligand for the riboswitch. For this reason, there are many potential riboswitches with unknown ligand specificities (3, 12, 69). Unfortunately, signaling compounds such as c-di-GMP are likely to control the expression of a great diversity of genes, and therefore, candidate riboswitches for second messengers may not immediately be recognized.

IDENTIFICATION OF A RIBOSWITCH CLASS THAT BINDS c-di-GMP

We conducted a bioinformatics search (69) using the CMFinder comparative genomics pipeline (78, 79) and identified 22 different bacterial RNA motifs. Four of these motifs were quickly recognized and validated as metabolite-binding riboswitches for SAM (70), SAH (6), pre-queuosine 1 ($preQ_1$) (39), and molybdenum cofactor (44). Some of the remaining motifs exhibit characteristics typical of riboswitches but remained unvalidated. One of these elements, initially called GEMM (Fig. 1), was found in the 5′ UTR of a number of seemingly unrelated genes (69) but has since been confirmed as a riboswitch aptamer for c-di-GMP (57).

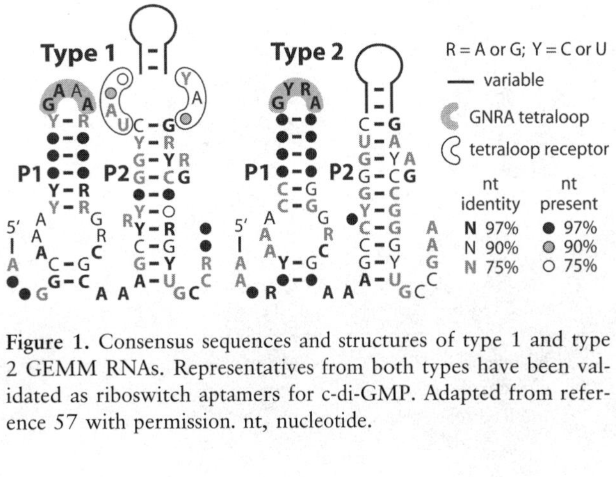

Figure 1. Consensus sequences and structures of type 1 and type 2 GEMM RNAs. Representatives from both types have been validated as riboswitch aptamers for c-di-GMP. Adapted from reference 57 with permission. nt, nucleotide.

GEMM RNAs were very promising candidate riboswitches because, in part, they contain regions of high sequence conservation interspersed with secondary structure elements supported by covariation (Fig. 1). Two types of GEMM elements were identified among 503 recognized representatives, each with two adjacent hairpins called pairing elements P1 and P2 that are almost always linked by two adenosines. Examples were identified in both gram-positive and gram-negative bacteria and were predicted to be present in the 5′ UTRs of numerous genes (69). This latter arrangement suggested these RNAs are *cis*-acting domains with regulatory function. Consistent with this hypothesis was the observation that many GEMM RNAs were associated with sequences downstream of the elements that could serve as expression platforms. For example, a representative GEMM RNA from *Geobacter sulfurreducens* carries a strong hairpin followed by a run of U residues that conforms to the typical arrangement for an intrinsic transcription terminator (Fig. 2). The formation of the conserved P2 hairpin is antagonistic with terminator stem formation, which is an arrangement suggestive of a ligand-responsive "on" switch.

The two types of GEMM elements are distinguished by differences in the tetraloop sequences that cap the P1 stem and by differences near the top of the P2 stem (57, 69). Most of the tetraloop sequences capping P1 are of the GNRA class. Type 1 RNAs (303 examples) carry a GAAA sequence whereas type 2 RNAs (171 examples) carry a GYRA sequence (where N, R, and Y represent any nucleotide, purine, and pyrimidine, respectively) (Fig. 1 and 2). Of the 29 motifs that did not fit the consensus for type 1 or type 2 elements, 24 had a sequence that matched the GNRA tetraloop consensus, but they did not have nucleotides involved in base-pairing interactions immediately adjacent to the sequence on both sides as is typically seen in GNRA tetraloops (57).

GNRA tetraloops are known to bestow remarkable stability to RNA hairpins (13, 21) and are found in many structured RNAs including ribosomal RNAs, ribozymes, and signal recognition particle RNAs (13, 21). Recently, GNRA tetraloops have also been found in several riboswitch classes (2), including a class that likely binds to molybdenum cofactor (44). GNRA tetraloops have several key characteristics that give rise to their high stability. The nucleotides in the element stack on top of each other, and two hydrogen bonds are formed between the conserved guanosine and adenosine nucleotides at the first and last positions of the sequence (21). The first guanosine also forms a hydrogen bond with a phosphate group on the 5′ side of the terminal adenosine, an interaction that is thought to help stabilize the noncanonical G-A base-pairing. Additionally, a hydrogen bond is formed between a 2′-hydroxyl group of the sugar on the first guanosine and the base of the third nucleotide in the sequence (21).

In addition to their stable structures, GNRA tetraloops can form tertiary interactions with tetraloop receptor domains. In type 1 GEMM RNAs, a tetraloop receptor that conforms to the typical 11-nucleotide motif that is known to dock with GAAA tetraloops (13) is observed near the top of the P2 hairpin (Fig. 1). In contrast, some type 2 RNAs carry

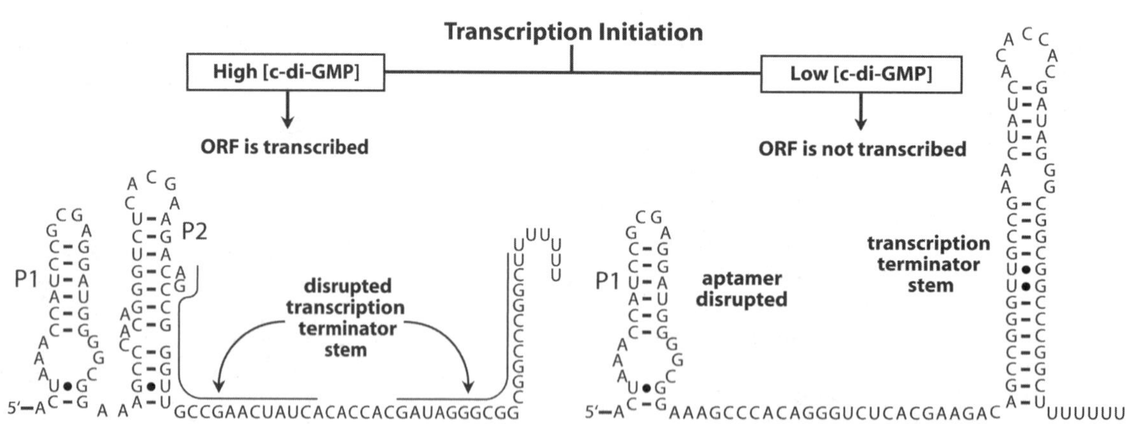

Figure 2. Predicted mechanism of a c-di-GMP riboswitch from *Geobacter sulfurreducens* located in the 5′ UTR of a putative cytochrome *c* gene. When c-di-GMP concentrations are high (left), ligand binding is predicted to stabilize the tertiary structure involving both stems P1 and P2, which should preclude formation of an intrinsic transcription terminator stem. When c-di-GMP concentrations are low (right), the aptamer will not be bound to ligand and the terminator stem will form rather than the unstable P2 stem. This arrangement of aptamer and expression platform is predicted to function as a genetic on switch, wherein ligand binding prevents premature transcription termination and thus permits expression of the adjoining ORF. Adapted from reference 69.

sequences in P2 that are known to dock with GYRA tetraloops (13). These arrangements strongly suggest that type 1 and type 2 RNAs both exploit distinct GNRA tetraloop and tetraloop receptors in the P1 and P2 stems to adopt similar tertiary structures. The presence of these commonly occurring structural motifs in many GEMM representatives added further support to the hypothesis that the function of GEMM requires the formation of a more complex tertiary structure.

On closer inspection of the genes associated with GEMM RNAs, an intriguing pattern of genetic affiliation was recognized (Fig. 3). Many of the genes downstream of GEMM RNAs were DGCs or PDEs, proteins that synthesize or degrade c-di-GMP. Additionally, many genes encoding proteins involved in pathways that are predicted to be controlled by c-di-GMP were found in close proximity to the element. Because the riboswitch candidate was frequently located upstream of genes that are associated with c-di-GMP metabolism or signaling pathways, we hypothesized that GEMM RNAs may sense and respond to changing concentrations of this second messenger.

VALIDATION OF c-di-GMP BINDING TO A CANDIDATE RIBOSWITCH

The second of two GEMM RNAs from *Vibrio cholerae,* termed Vc2, was examined in greater detail to determine whether it was capable of binding to c-di-GMP (Fig. 4) (57). The first technique employed to test the candidate riboswitch was an assay called in-line probing, which can be used to assess structural changes in riboswitches in response to ligand binding (43, 54). In-line probing takes advantage of the slow spontaneous cleavage of RNA via an internal phosphoester transfer reaction involving the attack by a 2′ oxygen of ribose on the adjacent phosphorus center (Fig. 4A) (53). More rapidly cleaving RNA phosphodiester linkages tend to reside in structurally flexible parts of an RNA molecule, which permits each linkage to adopt an in-line attack conformation more frequently than linkages between RNA nucleotides that participate in secondary or tertiary structures (53).

Vc2 RNA radiolabeled with ^{32}P at its 5′ terminus was incubated in the absence or presence of a candidate ligand under in-line probing conditions (50 mM Tris-HCl [pH 8.3 at 23°C], 100 mM KCl, 20 mM $MgCl_2$), typically for 24 to 48 hours at 23°C (43). The products of spontaneous RNA cleavage were then separated by polyacrylamide gel electrophoresis (Fig. 4B). The intensities of the spontaneous cleavage product bands were quantitatively assessed, and the locations of Vc2 RNA chain cleavage are mapped onto a secondary structure model of the RNA (Fig. 4C). The results reveal that regions of the RNA predicted to participate in base-pairing interactions or tertiary interactions exhibit far less spontaneous cleavage. In the absence of c-di-GMP, more than 30 internucleotide linkages exhibit a high level of spontaneous cleavage. Importantly, nearly half of

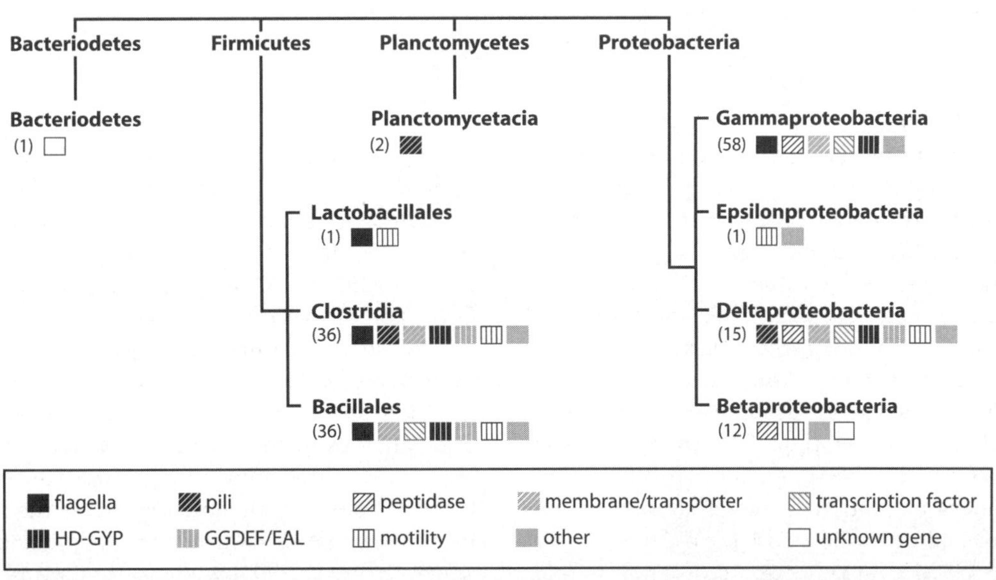

Figure 3. Riboswitches for c-di-GMP are widespread in eubacteria. The phylogenetic tree depicts the number of c-di-GMP aptamers found in various bacterial lineages. Types of genes predicted to be controlled by c-di-GMP riboswitches are identified.

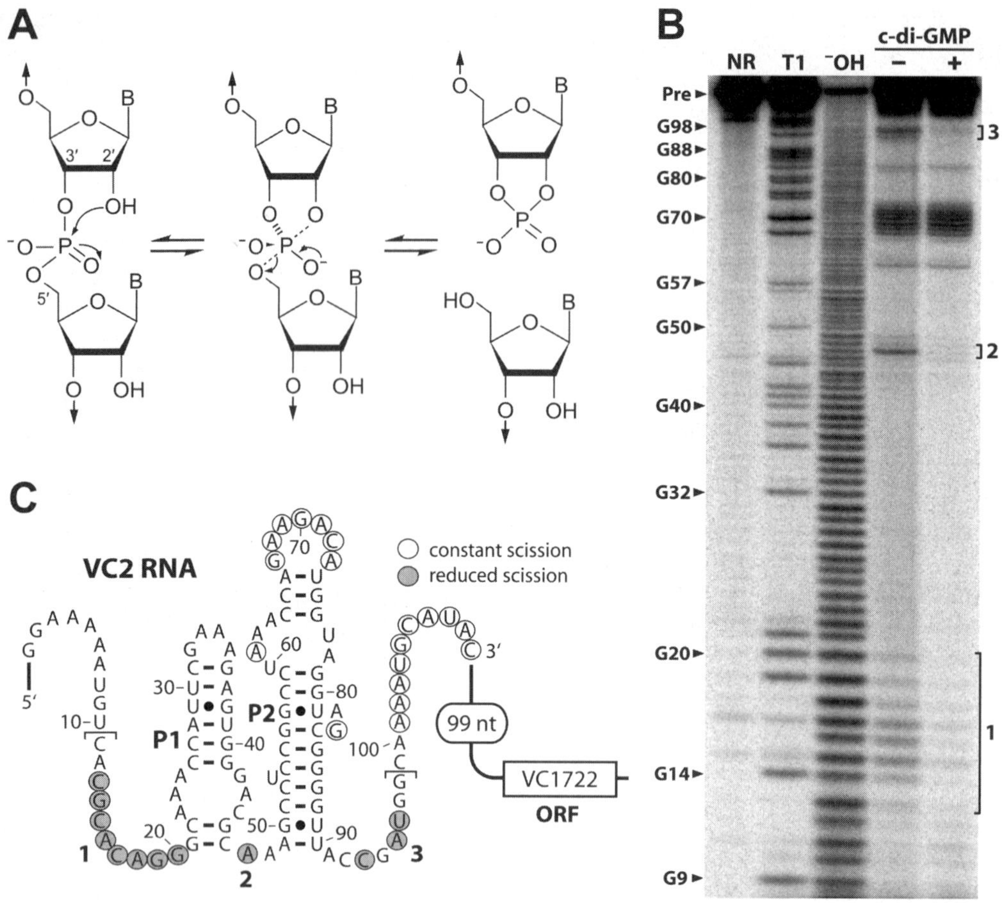

Figure 4. A c-di-GMP riboswitch representative from *V. cholerae* undergoes ligand-induced structural modulation. (A) Mechanism of spontaneous RNA phosphoester transfer and chain cleavage exploited by in-line probing assays. (B) Polyacrylamide gel electrophoresis separation of RNA products generated by in-line probing of 5′ ^{32}P-labeled 110 Vc2 RNA: NR, no reaction; T1, partial digest with RNase T1; $^{-}$OH, partial digest with alkali. RNA was incubated in the absence (−) or presence (+) of 100 μM c-di-GMP. (C) Sequence and secondary structure model of the Vc2 RNA from *V. cholerae* chromosome 2 and its proximity to the ORF of VC1722. Bold numbers identify regions of ligand-mediated structure modulation as observed in panel B. Brackets identify the minimal 5′ or 3′ terminus (when the opposing terminus for 110 Vc2 RNA is retained) that exhibits structural modulation when tested with 1 μM c-di-GMP. Adapted from reference 57 with permission. nt, nucleotide.

these linkages become stabilized when 100 μM c-di-GMP is present during the in-line probing reaction.

By conducting a series of in-line probing assays at various c-di-GMP concentrations, the apparent dissociation constant (K_D) can be established for the second messenger (54). A K_D value of ~1 nM for the complex formed between Vc2 RNA and c-di-GMP was established by plotting the relative extents of cleavage product band modulation at different ligand concentrations, which serves as an indicator of ligand saturation of the aptamer (Fig. 5A). In comparison, other characterized representatives of other riboswitch aptamers exhibit K_D values that range from about 100 pM (28a) to 200 μM (76). Thus, the affinity of the GEMM aptamer for c-di-GMP is relatively high compared to many other riboswitches. Furthermore, ligand binding by the aptamer is tighter than binding by a PilZ domain protein from *Escherichia coli* which binds c-di-GMP with a K_D of approximately 840 nM (49). Similar c-di-GMP-mediated modulation of RNA structures was observed for other GEMM representatives, suggesting that GEMM RNAs are aptamer components of c-di-GMP-responsive riboswitches.

Additional in-line probing analyses using various derivatives of c-di-GMP and guanosine compounds revealed that Vc2 strongly discriminates against even close analogs of the second messenger (Fig. 5B). Perhaps most important is the observation that c-di-GMP is bound by the aptamer more than 2 orders of magnitude tighter than the linear dinucleotide pGpG, which is generated by the action of PDE enzymes with EAL domains (23, 48, 61). Likewise, c-di-GMP binds more than 5 orders of magnitude tighter than

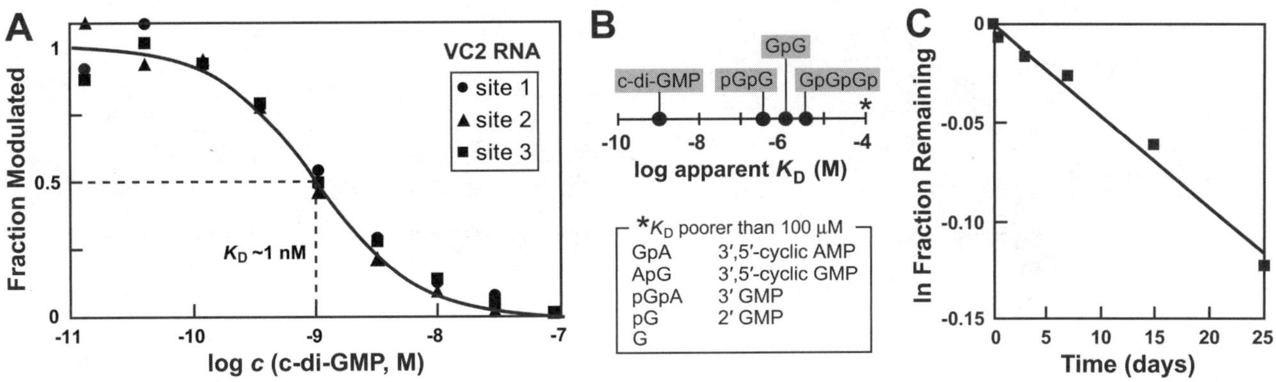

Figure 5. Vc2 RNA binds c-di-GMP with high affinity and specificity. (A) Plot of the normalized fraction of 110 Vc2 aptamer cleaved versus c-di-GMP concentration. Sites of structural modulation are as depicted in Fig. 4B. (B) Comparison of K_D values exhibited by the 110 Vc2 aptamer for c-di-GMP and various analogs. G, guanosine; pG, pGpG, pGpA, 5′ phosphorylated mono- and dinucleotides; GpGpG, trinucleotide. (C) Plot of the natural logarithm of the fraction of c-di-GMP remaining intact versus incubation time under in-line probing conditions. The negative slope of the line reflects the uncatalyzed rate constant (k_{uncat}) for cleavage of the second messenger. Adapted from reference 57 with permission.

the monomer pG, which is generated by the action of PDE enzymes with HD-GYP domains (23, 48, 61). These results indicate that c-di-GMP, and not a breakdown product of the second messenger, is likely to serve as the biologically relevant ligand for Vc2 RNA.

Several other analogs of c-di-GMP were also tested for binding to Vc2 RNA. Whereas the dinucleotide GpG binds with a K_D value of ~1 μM, no binding is observed for the analog dinucleotides ApG or GpA at concentrations as high as 100 μM (Fig. 5B). This finding indicates that both guanine moieties of c-di-GMP are recognized by the aptamer. There also appears to be a steric constraint for the size of RNA recognized by the aptamer, as the trinucleotide GpGpGp binds less tightly than GpG and pGpG. This apparent restriction of ligand size may allow the aptamer to discriminate against large RNA transcripts that inevitably will carry many GpG dinucleotides embedded within their polynucleotide chains.

To further verify that the aptamer does not bind a breakdown product of c-di-GMP, the compound was incubated for nearly one month under in-line probing assay conditions and the fraction of intact second messenger remaining was monitored by high-pressure liquid chromatography (Fig. 5C). The rate constant for the spontaneous cleavage (k_{uncat}) of c-di-GMP degradation was determined to be 3.2×10^{-6} min^{-1}. This rate constant is similar to that predicted for internucleotide linkages in noncircularized RNAs incubated under these conditions (30). This also corresponds to a half-life for c-di-GMP of ~150 days under in-line probing conditions, and a half-life of approximately four years under conditions that are typical for bacterial cells. Very little c-di-GMP is degraded during a typical in-line probing assay, and therefore, it is unlikely that trace amounts of c-di-GMP breakdown products are causing Vc2 RNA structure modulation. Moreover, the second messenger is not transiently stable and will require the action of PDE enzymes to degrade c-di-GMP on a time scale that is relevant to individual bacterial cells.

APTAMERS FOR c-di-GMP FUNCTION AS GENE CONTROL ELEMENTS

Riboswitches typically have expression platforms that control either transcription or translation of downstream genes (2, 9, 33, 74). As noted above, some c-di-GMP aptamers are associated with RNA structures that are predicted to serve as expression platforms (Fig. 2). However, no intrinsic terminator stem could be recognized in the nucleotide sequence downstream of Vc2 RNA. We hypothesized that a putative expression platform may regulate translation of the associated gene. To test this, sequences encoding wild-type (WT) and mutant Vc2 aptamers (Fig. 6A) as well as the associated promoter and expression platforms were fused to an *E. coli lacZ* reporter gene in a translational reporter vector. The reporter constructs were transformed into *E. coli,* which, like *V. cholerae,* is a gram-negative bacterium, and β-galactosidase assays were used to measure the reporter gene expression (57).

The reporter gene fused to the WT Vc2 aptamer was expressed at a high level (Fig. 6B). In contrast, the reporter gene fused to the M1 aptamer variant that carries mutations disrupting P1 exhibited less

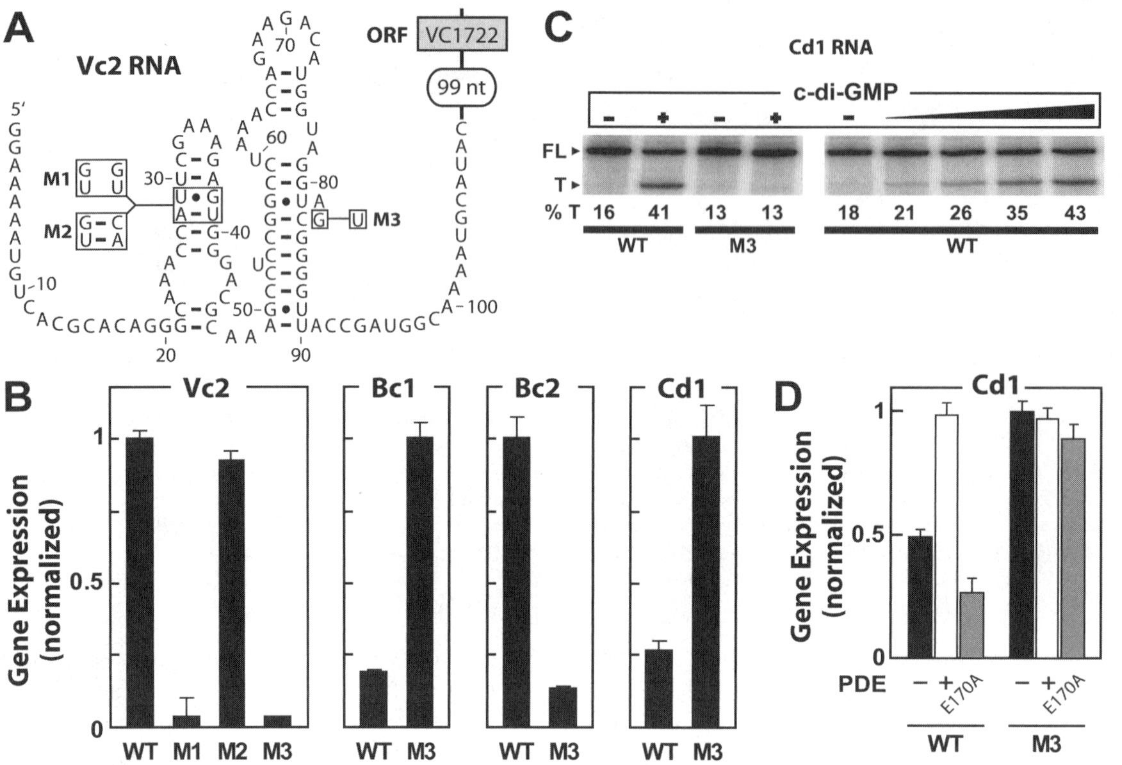

Figure 6. Representative c-di-GMP aptamers are components of gene control elements. (A) Reporter fusion constructs carry WT or mutant (M1 through M3) riboswitches from *V. cholerae* (Vc2) or carry the equivalent WT and M3 riboswitches from *B. cereus* (Bc1 and Bc2) or *C. difficile* (Cd1). (B) β-Galactosidase reporter gene assays for the constructs depicted in panel A when transformed into *E. coli*. Maximum Miller units measured for the four representatives were 436, 47, 5, and 51, respectively. (C) Products of single-round transcription of DNA templates coding for the WT and disruption mutant M3 Cd1 riboswitches. The M3 mutant carries a single G-to-U mutation in the c-di-GMP aptamer as noted in panel A. FL and T identify full-length (runoff) transcripts and terminated transcripts, respectively, and % T represents the percentage of transcripts that terminate via riboswitch action. Reactions were conducted in the absence (−) or presence (+) of 100 μM c-di-GMP or in the presence of increasing concentrations of c-di-GMP (last four lanes from left to right: 0.01, 0.1, 1, or 10 μM). (D) β-Galactosidase reporter assays of *B. subtilis* cells carrying the reporter gene fused to WT or M3 riboswitches as indicated and transformed with plasmids as described for panel A. The normalized gene expression value of 1 for Cd1 represents 102 Miller units. Adapted from reference 57 with permission. nt, nucleotide.

than 10% of the expression observed with WT. These results were supported by in-line probing data of the M1 RNA, which revealed that mutant aptamer is partially misfolded and does not bind c-di-GMP. The M2 RNA carries mutations that reestablish the P1 stem structure despite four nucleotide differences compared to the WT. These mutations restore gene expression to levels nearly as high as WT levels (Fig. 6B), indicating that base-pairing of P1 is essential for gene control activity, whereas the identities of the nucleotides forming the P1 stem are not critical. In-line probing of the M2 construct also revealed that the aptamer is not misfolded and binds to c-di-GMP with a K_D similar to that observed for the WT aptamer. A third mutant, M3, carries a mutation at an otherwise strictly conserved G residue, and this mutation causes a substantial loss of reporter gene expression. No appreciable binding was observed when 100 μM c-di-GMP was added to an in-line probing assay containing M3 (57). These data indicate that the Vc2 aptamer is the ligand-binding domain of a riboswitch that activates gene expression when c-di-GMP binds.

Similarly, 5′ UTRs containing WT and mutant c-di-GMP aptamers from the gram-positive bacteria *Bacillus cereus* and *Clostridium difficile* that were predicted to transcriptionally regulate gene expression were fused to an *E. coli lacZ* gene and inserted into the *amyE* locus of the gram-positive bacterium *B. subtilis*. One of the WT *B. cereus* riboswitches, Bc1, yields a low level of reporter gene expression, whereas a variant aptamer carrying the M3 mutation exhibits nearly a fivefold increase in expression (Fig. 6B). A similar result is observed for the WT Bc2 construct based on a c-di-GMP riboswitch from *C. difficile*. In contrast, a second *B. cereus* riboswitch, Bc2, exhibits a pattern of gene regulation for WT and M3 constructs that is similar to that observed for Vc2. These results indicate that Vc2 and Bc2 RNAs are genetic on switches, while Bc1 and Cd1 RNAs are genetic off switches.

In vitro transcription termination assays were used to assess whether c-di-GMP induces transcription termination in the Cd1 riboswitch (57). Transcription termination was assessed by conducting single-round in vitro transcription termination assays (Fig. 6C) A construct coding for the WT Cd1 riboswitch when transcribed in the presence of 100 μM c-di-GMP produced a short transcript that is not seen in identical reactions either lacking c-di-GMP or containing the M3 RNA variant. These results suggested that c-di-GMP caused termination of transcription upon binding to Cd1 riboswitches, which is consistent with its genetic action as an off switch (Fig. 6B).

REGULATION OF RIBOSWITCH-ASSOCIATED GENES BY c-di-GMP

Both hosts, *E. coli* and *B. subtilis,* contain many DGCs and PDEs that control the intracellular levels of c-di-GMP by synthesizing and degrading the compound. In the reporter expression experiments described above, these enzymes are not altered in the host organisms, and therefore, c-di-GMP should be present at its natural levels. To determine whether a representative c-di-GMP riboswitch indeed responds to changes in intracellular c-di-GMP levels, a plasmid containing *vieA* from *V. cholerae* was transformed into *B. subtilis* cells containing the WT Cd1 riboswitch-reporter fusion. VieA is a PDE containing an EAL domain that degrades c-di-GMP to produce pGpG (62). By expressing the VieA protein in the host bacteria, it is expected that the levels of c-di-GMP will be reduced.

When WT VieA is expressed in the reporter strain, c-di-GMP levels are expected to drop and cause the Cd1 off switch to yield higher reporter gene expression. Indeed, reporter gene expression is increased when WT VieA is expressed in cells carrying a WT Cd1 riboswitch-reporter fusion construct (Fig. 6D). In contrast, expression of a functionally defective VieA mutant does not cause an increase in expression. Furthermore, the M3 riboswitch variant does not respond to the expression of VieA. These results indicate that the Cd1 riboswitch binds c-di-GMP to deactivate gene expression and that the intracellular concentration of the second messenger affects the expression levels of riboswitch-associated genes (57).

THE REGULONS OF c-di-GMP RIBOSWITCHES

The recognition that GEMM RNAs function as c-di-GMP riboswitches facilitates the identification of entire regulons of genes that are likely controlled by changing concentrations of this second messenger. Additional genes that are controlled by c-di-GMP can be identified by using comparative RNA sequence and structure analysis, wherein portions of a bacterial genome that conform to the consensus c-di-GMP aptamer architecture are considered likely to function as c-di-GMP riboswitches. Of course, some riboswitches of this class may conform only poorly to the consensus architecture and, thus, could be overlooked. Also, some variant RNAs that are similar to c-di-GMP riboswitches in sequence and structure may have accrued mutations that switch ligand specificity or biological function of the element. Regardless of these complications, the association of genes with sequences that appear homologous to validated c-di-GMP riboswitches may help researchers identify and examine processes that are likely to be regulated by this second messenger.

V. cholerae carries two elements that conform to the c-di-GMP riboswitch consensus (Fig. 7). The Vc2 representative, which has been biochemically and genetically validated as a c-di-GMP riboswitch, resides upstream of VC1722. This gene is homologous to *tfoX* genes found in *V. cholerae* and other bacteria, which are known to express protein products that regulate bacterial competence (37). *V. cholerae* is naturally found on chitinous exoskeletons of aquatic organisms where it forms biofilms or surface-attached colonies (32, 37). In the presence of chitin, TfoX proteins are thought to induce the production of pseudopilus structures that increase bacterial compentence. Interestingly, *V. cholerae* colonies that exhibit a rugose phenotype, which is characterized by wrinkled colony morphology, have been shown to have elevated c-di-GMP concentrations in addition to having a higher copy number of VC1722 mRNAs (5, 31). This is consistent with the hypothesis that the Vc2 riboswitch is an on switch wherein higher c-di-GMP concentrations should yield more VC1722 mRNA (57).

When *V. cholerae* is found in aquatic environments, its ability to attach to surfaces is thought to be essential for survival (27). When the pathogen is ingested by mammals, intracellular levels of c-di-GMP have been shown to decrease (62) and, consequently, changes in protein expression allow the bacteria to attach to the small intestines of the host and produce the cholera toxin (27). The Vc1 representative thus appears to control the expression of a protein called GbpA that is thought to mediate adhesion

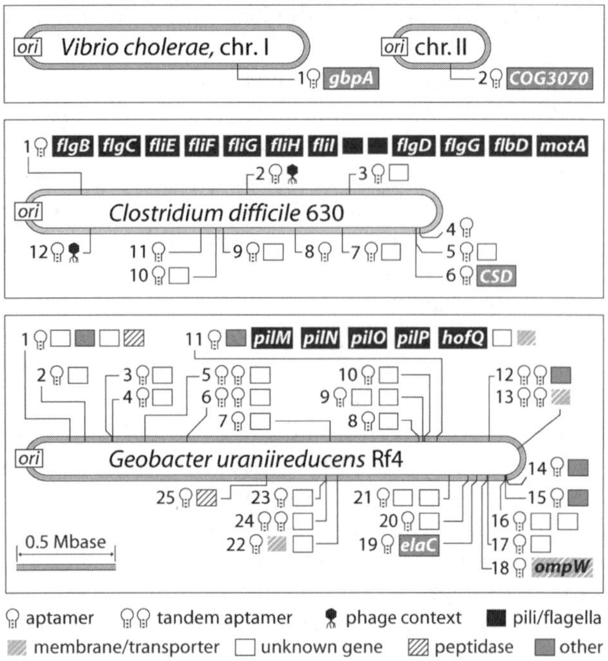

Figure 7. Genomic locations of c-di-GMP riboswitches or aptamers for representative organisms. Genes residing immediately downstream of representatives are depicted where multiple genes indicate predicted operons. An absence of genes indicates that the c-di-GMP aptamer is not located 5′ proximal to any ORF, possibly indicating an association with a noncoding RNA. COG3070 is joined to a pfam04994 domain, which is a protein arrangement similar to TfoX (12, 17). Chromosomes are represented by shaded lines. ori, origin of replication; chr., chromosome. Adapted from reference 57 with permission.

to both chitin in aquatic environments and epithelial cells in host intestines by binding sugars (27). Because binding to the intestinal tract is necessary to cause disease, and changes in *gbpA* expression could be involved in the decision to colonize intestines, the protein may be an important virulence factor (27). If a riboswitch associated with GbpA is used to detect the decrease in c-di-GMP concentrations and modulate gene expression accordingly, the riboswitch could be an important factor in the genetic changes in bacteria needed to exhibit pathogenic characteristics in mammals.

C. difficile carries at least 12 elements that conform to the c-di-GMP aptamer consensus, and most are associated with genes of unknown function (Fig. 7). However, the Cd1 representative is associated with the large operon encoding proteins known to be important for the production of flagella (57). This arrangement suggests that the Cd1 riboswitch, a genetic off switch, participates in the change between motile and sessile forms of the bacterium by controlling the expression of flagellum-related genes in response to decreased c-di-GMP levels.

Interestingly, 2 of the 12 elements are flanked by bacteriophage genes that are embedded in the genome of *C. difficile*. For example, Cd2 is located within the lysis module of the PhiCD119 bacteriophage genome and DNA encoding the element has been shown to be packaged into infectious bacteriophage particles (20). To date, over 20 different classes of riboswitches have been found, and there are thousands of examples of riboswitches identified in bacteria (2, 9, 74). However, c-di-GMP riboswitches are the first to be found in association with phage genes, which suggests that this phenomenon may be more than simple coincidence. Perhaps bacteriophages may not gain an evolutionary advantage by monitoring the levels of common metabolites within their bacterial hosts. In contrast, c-di-GMP riboswitches may allow a bacteriophage to monitor physiological changes being made by a bacterium, such as switching between motile and sessile lifestyles, and recognition of these major changes may help a bacteriophage optimize its propagation in the environment.

Some deltaproteobacteria including *Geobacter sulfurreducens* and *Geobacter uraniireducens* can oxidize organic compounds to form ATP using metal ions (38). This ability allows some *Geobacter* species to use metal ions like uranium as electron acceptors, and this process causes the metal ions to precipitate. There is interest in the possibility of using these bacteria in bioremediation efforts to precipitate heavy metal pollutants from groundwater (38). Interestingly, *G. uraniireducens* carries 30 elements that conform to the c-di-GMP aptamer consensus. Although most of the genes associated with these elements code for proteins of unknown function, some elements are located in the 5′ UTRs of genes that encode proteins for pili and cytochrome *c* enzymes that are thought to be important in the reduction of metal ions and electron transport (25, 36, 42). Since the riboswitch regulons of these organisms include many genes of unknown function, there is far more to learn about the processes regulated by this second messenger.

COMPLEX RIBOSWITCH ARRANGEMENTS AND FUNCTIONS OF c-di-GMP RIBOSWITCHES

c-di-GMP levels are predicted to constantly be in flux in response to signals from the environment and physiological changes, and it is important for bacteria to distinguish these signals and respond appropriately (23, 48, 61). Riboswitches are highly selective and sensitive receptors capable of responding to small changes in ligand concentrations, and therefore, it seems reasonable to find that bacteria entrust riboswitches to accurately sense and respond to changes in c-di-GMP concentrations. The fact that bacteria often have multiple genes controlled by riboswitches (57, 69) (Fig. 7) is not surprising given that c-di-GMP is involved in many signaling pathways that result in significant morphological changes. However, sensing and responding to a second messenger like c-di-GMP may pose some challenges that are not commonly encountered by riboswitch classes that sense other metabolites.

Most bacteria have multiple copies of DGCs and PDEs, suggesting that production and destruction of the second messenger are tightly controlled in response to numerous signals. Furthermore, it has been proposed that DGCs and PDEs may be spatially localized in cells to control local concentrations of c-di-GMP, while other areas may be experiencing different concentrations (23, 48, 61). Perhaps the locality of a c-di-GMP riboswitch on a genome that also is spatially organized in a cell may allow the riboswitch to respond to local rather than global changes in c-di-GMP. For example, in *Caulobacter crescentus,* the PleD protein, a DGC enzyme, is localized to the pole that develops the stalk (23). Perhaps only riboswitches positioned in the genome near the pole would perceive changes in c-di-GMP levels caused by PleD activity. Alternatively, each c-di-GMP riboswitch may be tuned to respond to a different c-di-GMP concentration than other representatives in the same organism. This could be achieved either by exhibiting a distinctive K_D value for each aptamer, or by exhibiting unique kinetics of RNA transcription or translation that could alter the concentration of ligand needed to trigger riboswitch function (19, 45, 72, 73).

Sometimes riboswitches employ more sophisticated architectures to create riboswitch arrangements that exhibit complex gene control characteristics (9). Riboswitches that sense glycine usually carry two aptamers in tandem, followed by a single expression platform (35). This arrangement allows cooperative binding between the two aptamers and produces a more digital gene control response wherein smaller changes in ligand concentration lead to greater changes in gene expression compared to riboswitches that carry a single aptamer (35, 71). Glycine riboswitches are located upstream of the *gcvT* operon which encodes proteins responsible for the degradation of glycine and routing its carbon through the citric acid cycle. The necessity for a cooperative riboswitch is thought to be due to the selective advantage that is likely gained by cells that can most efficiently detect small increases in glycine concentration and can generate greater gene expression of the *gcvT* operon as a result. Furthermore, cells using the cooperative glycine riboswitch arrangement would also respond most efficiently to deficiencies in glycine concentration, more effectively suppress *gcvT* operon expression, and prevent degradation of glycine to levels that would preclude protein synthesis (35). Only one example, in the 5′ UTR of the fimbrial operon in *Idiomarina baltica,* has been found that carries a single expression platform associated with two complete c-di-GMP aptamers.

Geobacter uraniireducens has a large number of c-di-GMP riboswitches associated with genes that are widely dispersed in its genome (Fig. 7). Twenty-five different gene clusters are located downstream of c-di-GMP aptamers, but five of these transcriptional units carry tandem aptamer arrangements. However, it is not clear whether any of these tandem arrangements permit c-di-GMP to bind cooperatively. Even if these examples do not function cooperatively, tandem riboswitch aptamers can still generate a more digital gene control response to ligand changes. In *Bacillus anthracis,* two complete TPP riboswitches composed of two aptamers and two expression platforms using transcription terminator stems have been observed to function independently of each other (71). Many examples of candidate c-di-GMP riboswitches occur in the 5′ UTRs of mRNAs in the genomes of firmicutes and deltaproteobacteria that have an expression platform associated with each aptamer. Such arrangements inherently increase the dynamic range of gene control, but the more likely biological role is to yield a more digital gene control response than could be achieved by a single riboswitch.

Although it is not yet known whether the tandem arrangements of c-di-GMP aptamers increase the responsiveness to ligand, the fact that other riboswitches have been shown to respond more digitally when arranged in a tandem fashion suggests that the arrangements could be biologically relevant. Since concentrations of c-di-GMP are so tightly regulated within bacteria, it is possible that tandem arrangements of c-di-GMP aptamers have evolved to diversify the genetic responses of some genes to c-di-GMP relative to others in the same organism that carry single-aptamer architectures.

Other tandem arrangements of riboswitches also have been found where the two aptamers bind to different compounds (60). In *Bacillus clausii,* a tandem riboswitch is assembled from a representative that binds SAM and another that binds coenzyme B_{12}. The binding of either compound represses the levels of the downstream gene, *metE.* This example illustrates how tandem arrays of riboswitch aptamers can add sophistication to the way they can respond to changing ligand concentrations. However, there are currently no validated examples of c-di-GMP riboswitches residing in tandem with other riboswitch classes.

Though it is thought that differential expression of c-di-GMP is important for regulating cellular changes, it is possible that the activation of specific promoters could also play an important role. For example, in *B. subtilis* and the closely related bacterium *Bacillus licheniformis,* biosurfactant production is coupled to the development of natural bacterial competence (1). Surfactin from *B. subtilis* and lichenysin from *B. licheniformis* are nonribosomal lipopeptide antibiotics that are synthesized by multisubunit enzyme complexes. The *srfA* and *lch* operons encode genes required for surfactin and lichenysin biosynthesis, respectively (1), and ComA protein is a positive activator of these operons (77). The levels of ComA are modulated by a multitude of signals including quorum sensing and the DegS-DegU two-component system (1). The 5′ UTR of the *lch* operon carries a c-di-GMP sensing riboswitch that presumably turns off the expression of the operon under elevated c-di-GMP levels. If true, then *lch* operon expression would be possible only when adequate ComA protein is present and when c-di-GMP concentrations are at a low level. Such a condition is possible when the cell density has the required quorum and the environmental cues are such that they activate PDE enzymes to degrade c-di-GMP.

In addition to those found in *B. licheniformis,* operons specific for nonribosomal peptide synthesis in *Bacillus pumilus, B. cereus,* and *Bacillus thuringiensis* are expected to be regulated by c-di-GMP riboswitches (57). Interestingly, the *B. subtilis* genome does not have a recognizable c-di-GMP riboswitch in the *srfA* operon, even though it is highly conserved in its regulatory architecture with the *lch* operon of *B. licheniformis*. This may be due to the domestication of the *B. subtilis* 168 genome, as it is considerably attenuated in its capability to form biofilm and fruiting bodies in comparison to *B. subtilis* strains isolated from the wild (14).

Bacteria use a variety of small RNAs to recognize and respond to various environmental stresses such as osmotic shock and oxidative damage (56). Although experimental evidence currently is lacking, it seems plausible that a small noncoding RNA could interact with a riboswitch to preclude triggering by a second messenger. In this way, multiple-input control systems could be created that integrate the genetic circuit controlled by a small noncoding RNA with that for c-di-GMP.

POTENTIAL APPLICATIONS FOR c-di-GMP RIBOSWITCHES

The identification of c-di-GMP riboswitches will allow researchers to more easily identify processes that are regulated in response to changing concentrations of this second messenger. Furthermore, some of these riboswitches already have proven to function in new hosts and, therefore, may function as useful biosensors that can report the concentrations of c-di-GMP in cells in different genetic backgrounds or under different environmental conditions. Given the widespread distribution of c-di-GMP riboswitches in some pathogens, perhaps compounds that mimic the chemical structure of c-di-GMP could be developed to disregulate the expression of genes involved in critical processes such as biofilm formation or virulence.

It has already been suggested that riboswitches may be potential drug targets for antimicrobial compounds (6). An analog of TPP called pyrithiamine pyrophosphate, which has been known for decades to inhibit bacterial and fungal growth, is bound by TPP riboswitches (52, 59, 63). Similarly, toxic analogs of lysine have been found that inhibit expression of riboswitch-controlled genes upon binding to the aptamer (7, 51). Another compound called roseoflavin is a natural analog of FMN that has been shown to bind to FMN riboswitches and downregulate the expression of reporter genes, suggesting that the antimicrobial effects of roseoflavin may be related to binding of FMN riboswitches in bacteria (28a). Roseoflavin is produced by *Streptomyces davawensis* (41), and this first example of a naturally occurring antibiotic that binds to a riboswitch may in the future be joined by other riboswitch-targeting compounds isolated from natural sources.

Given these findings and the fact that c-di-GMP riboswitches are associated with virulence gene expression and with the expression of genes for major physiological changes in pathogenic bacteria, this riboswitch class could also potentially serve as an antimicrobial drug target. A toxic analog of c-di-GMP has been identified already (28), but it is not known whether this compound targets c-di-GMP-sensing riboswitches.

CONCLUDING REMARKS

Riboswitches that bind to c-di-GMP have been identified that appear to control a great diversity of genes associated with second messenger signaling (57, 69). Although the examples of c-di-GMP riboswitches discovered to date are not enough to explain c-di-GMP regulation in all bacteria, the recently identified c-di-GMP riboswitch helps explain part of a signaling network that has long remained mysterious. Undoubtedly, more proteins will be identified that respond to c-di-GMP, and such discoveries will help elucidate the regulatory circuitry of signaling by this second messenger.

It seems very likely that additional riboswitch classes will be discovered that sense c-di-GMP. This second messenger is ubiquitous, yet the recently discovered class of c-di-GMP riboswitches is present in a limited number of gram-positive bacteria. There are some metabolites such as SAM and prequeuosine I that are bound by multiple aptamer classes (4, 5). Given the importance of this second messenger to so many bacteria and given that many nucleotide-like ligands have been found to trigger riboswitch activity, we speculate that evolution has given rise to multiple different classes of riboswitches that sense and respond to c-di-GMP.

REFERENCES

1. **Auchtung, J. M., C. A. Lee, and A. D. Grossman.** 2006. Modulation of the ComA-dependent quorum response in *Bacillus subtilis* by multiple rap proteins and Phr peptides. *J. Bacteriol.* **188:**5273–5285.
2. **Barrick, J. E., and R. R. Breaker.** 2007. The distributions, mechanisms, and structures of metabolite-binding riboswitches. *Genome Biol.* **8:**R239.
3. **Barrick, J. E., K. A. Corbino, W. C. Winkler, A. Nahvi, M. Mandal, J. Collins, M. Lee, A. Roth, N. Sudarsan, I. Jona, J. K. Wickiser, and R. R. Breaker.** 2004. New RNA motifs suggest an expanded scope for riboswitches in bacterial genetic control. *Proc. Natl. Acad. Sci. USA* **101:**6421–6426.
4. **Batey, R. T., S. D. Gilbert, and R. K. Montange.** 2004. Structure of a natural guanine-responsive riboswitch complexed with the metabolite hypoxanthine. *Nature* **432:**411–415.
5. **Beyhan, S., and F. H. Yildiz.** 2007. Smooth to rugose phase variation in *Vibrio cholerae* can be mediated by a single nucleotide change that targets c-di-GMP signaling pathway. *Mol. Microbiol.* **63:**995–1007.
6. **Blount, K. F., and R. R. Breaker.** 2006. Riboswitches as antibacterial drug targets. *Nat. Biotechnol.* **24:**1558–1564.
7. **Blount, K. F., J. X. Wang, J. Lim, N. Sudarsan, and R. R. Breaker.** 2007. Antibacterial lysine analogs that target lysine riboswitches. *Nat. Chem. Biol.* **3:**44–49.
8. **Bobrov, A. G., O. Kirillina, and R. D. Perry.** 2005. The phosphodiesterase activity of the HmsP EAL domain is required for negative regulation of biofilm formation in *Yersinia pestis. FEMS Microbiol. Lett.* **247:**123–130.
9. **Breaker, R. R.** 2008. Complex riboswitches. *Science* **319:** 1795–1797.
10. **Chan, C., R. Paul, D. Samoray, N. C. Amiot, B. Giese, U. Jenal, and T. Schirmer.** 2004. Structural basis of activity and allosteric control of diguanylate cyclase. *Proc. Natl. Acad. Sci. USA* **101:**17084–17089.
11. **Cheah, M. T., A. Wachter, N. Sudarsan, and R. R. Breaker.** 2007. Control of alternative RNA splicing and gene expression by eukaryotic riboswitches. *Nature* **447:**497–500.
12. **Corbino, K. A., J. E. Barrick, J. Lim, R. Welz, B. J. Tucker, I. Puskarz, M. Mandal, N. D. Rudnick, and R. R. Breaker.** 2005. Evidence for a second class of S-adenosylmethionine riboswitches and other regulatory RNA motifs in alpha-proteobacteria. *Genome Biol.* **6:**R70.
13. **Costa, M., and F. Michel.** 1995. Frequent use of the same tertiary motif by self-folding RNAs. *EMBO J.* **14:**1276–1285.
14. **Earl, A. M., R. Losick, and R. Kolter.** 2008. Ecology and genomics of *Bacillus subtilis. Trends Microbiol.* **16:**269–275.
15. **Eddy, S. R., and R. Durbin.** 1994. RNA sequence analysis using covariance models. *Nucleic Acids Res.* **22:**2079–2088.
16. **Fan, P., A. K. Suri, R. Fiala, D. Live, and D. J. Patel.** 1996. Molecular recognition in the FMN-RNA aptamer complex. *J. Mol. Biol.* **258:**480–500.
17. **Fuchs, R. T., F. J. Grundy, and T. M. Henkin.** 2006. The S_{MK} box is a new SAM-binding RNA for translational regulation of SAM synthetase. *Nat. Struct. Mol. Biol.* **13:**226–233.
18. **Gelfand, M. S., A. A. Mironov, J. Jomantas, Y. I. Kozlov, and D. A. Perumov.** 1999. A conserved RNA structure element involved in the regulation of bacterial riboflavin synthesis genes. *Trends Genet.* **15:**439–442.
19. **Gilbert, S. D., C. D. Stoddard, S. J. Wise, and R. T. Batey.** 2006. Thermodynamic and kinetic characterization of ligand binding to the purine riboswitch aptamer domain. *J. Mol. Biol.* **359:**754–768.
20. **Govind, R., J. A. Fralick, and R. D. Rolfe.** 2006. Genomic organization and molecular characterization of *Clostridium difficile* bacteriophage φCD119. *J. Bacteriol.* **188:**2568–2577.
21. **Heus, H. A., and A. Pardi.** 1991. Structural features that give rise to the unusual stability of RNA hairpins containing GNRA loops. *Science* **253:**191–194.
22. **Hickman, J. W., and C. S. Harwood.** 2008. Identification of FleQ from Pseudomonas aeruginosa as a c-di-GMP-responsive transcription factor. *Mol. Microbiol.* **69:**376–389.
23. **Jenal, U., and J. Malone.** 2006. Mechanisms of cyclic-di-GMP signaling in bacteria. *Annu. Rev. Genet.* **40:**385–407.
24. **Kil, Y. V., V. N. Mironov, I. Y. Gorishin, R. A. Kreneva, and D. A. Perumov.** 1992. Riboflavin operon of *Bacillus subtilis:* unusual symmetric arrangement of the regulatory region. *Mol. Gen. Genet.* **233:**483–486.
25. **Kim, B., X. Qian, C. Leang, M. V. Coppi, and D. R. Lovley.** 2006. Two putative c-type multiheme cytochromes required for the expression of OmcB, an outer membrane protein essential for optimal Fe(III) reduction in *Geobacter sulfurreducens. J. Bacteriol.* **188:**3138–3142.
26. **Kim, J. N., and R. R. Breaker.** 2008. Purine sensing by riboswitches. *Biol. Cell* **100:**1–11.
27. **Kirn, T. J., B. A. Jude, and R. K. Taylor.** 2005. A colonization factor links *Vibrio cholerae* environmental survival and human infection. *Nature* **438:**863–866.
28. **Kline, T., S. R. Jackson, W. Deng, C. L. M. J. Verlinde, and S. I. Miller.** 2008. Design and synthesis of bis-carbamate analogs of cyclic bis-(3′-5′)-diguanylic acid (c-di-GMP) and the acyclic dimer PGPG. *Nucleosides Nucleotides Nucleic Acids* **27:**1282–1300.

28a. **Lee, E. R., K. F. Blount, and R. R. Breaker.** 2009. Roseoflavin is a natural antibacterial compound that binds to FMN riboswitches and regulates gene expression. *RNA Biol.* **6:**187–194.

29. **Lee, V. T., J. M. Matewish, J. L. Kessler, M. Hyodo, Y. Hayakawa, and S. Lory.** 2007. A cyclic-di-GMP receptor required for bacterial exopolysaccharide production. *Mol. Microbiol.* **65:**1474–1484.
30. **Li, Y., and R. R. Breaker.** 1999. Kinetics of RNA degradation by specific base catalysis of transesterification involving the 2′-hydroxyl group. *J. Am. Chem. Soc.* **121:**5364–5372.

31. **Lim, B., S. Beyhan, J. Meir, and F. H. Yildiz.** 2006. Cyclic di-GMP signal transduction systems in *Vibrio cholerae:* modulation of rugosity and biofilm formation. *Mol. Microbiol.* **60:** 331–348.
32. **Lipp, E. K., A. Huq, and R. R. Colwell.** 2002. Effects of global climate on infectious disease: the cholera model. *Clin. Microbiol. Rev.* **15:**757–770.
33. **Mandal, M., B. Boese, J. E. Barrick, W. C. Winkler, and R. R. Breaker.** 2003. Riboswitches control fundamental biochemical pathways in *B. subtilis* and other bacteria. *Cell* **113:**577–586.
34. **Mandal, M., and R. R. Breaker.** 2004. Adenine riboswitches and gene activation by disruption of a transcription terminator. *Nat. Struct. Mol. Biol.* **11:**29–35.
35. **Mandal, M., M. Lee, J. E. Barrick, Z. Weinberg, G. M. Emilsson, W. L. Ruzzo, and R. R. Breaker.** 2004. A glycine-dependent riboswitch that uses cooperative binding to control gene expression. *Science* **306:**275–279.
36. **Mehta, T., M. V. Coppi, S. E. Childers, and D. R. Lovley.** 2005. Outer membrane c-type cytochromes required for Fe(III) and Mn(IV) oxide reduction in *Geobacter sulfurreducens. Appl. Environ. Microbiol.* **71:**8634–8641.
37. **Meibom, K. L., M. Blokesch, N. A. Dolganov, C. Wu, and G. K. Schoolnik.** 2005. Chitin induces natural competence in *Vibrio cholerae. Science* **310:**1824–1827.
38. **Methé, B. A., K. E. Nelson, J. A. Eisen, I. T. Paulsen, W. Nelson, J. F. Heidelberg, D. Wu, M. Wu, N. Ward, M. J. Beanan, R. J. Dodson, R. Madupu, L. M. Brinkac, S. C. Daugherty, R. T. DeBoy, A. S. Durkin, M. Gwinn, J. F. Kolonay, S. A. Sullivan, D. H. Haft, J. Selengut, T. M. Davidsen, N. Zafar, O. White, B. Tran, C. Romero, H. A. Forberger, J. Weidman, H. Khouri, T. V. Feldblyum, T. R. Utterback, S. E. Van Aken, D. R. Lovley, and C. M. Fraser.** 2003. Genome of *Geobacter sulfurreducens:* metal reduction in subsurface environments. *Science* **302:**1967–1969.
39. **Meyer, M. M., A. Roth, S. M. Chervin, G. A. Garcia, and R. R. Breaker.** 2008. Confirmation of a second natural preQ1 aptamer class in *Streptococcaceae* bacteria. *RNA* **14:**685–695.
40. **Mironov, A. S., I. Gusarov, R. Rafikov, L. E. Lopez, K. Shatalin, R. A. Kreneva, D. A. Perumov, and E. Nudler.** 2002. Sensing small molecules by nascent RNA: a mechanism to control transcription in bacteria. *Cell* **111:**747–756.
41. **Otani, S., M. Takatsu, M. Nakano, S. Kasai, and R. Miura.** 1974. Roseoflavin, a new antimicrobial pigment from *Streptomyces. J. Antibiot.* (Tokyo) **27:**86–87. (Letter.)
42. **Reguera, G., K. D. McCarthy, T. Mehta, J. S. Nicoll, M. T. Tuominen, and D. R. Lovley.** 2005. Extracellular electron transfer via microbial nanowires. *Nature* **435:**1098–1101.
43. **Regulski, E. E., and R. R. Breaker.** 2008. In-line probing analysis of riboswitches. *Methods Mol. Biol.* **419:**53–67.
44. **Regulski, E. E., R. H. Moy, Z. Weinberg, J. E. Barrick, Z. Yao, W. L. Ruzzo, and R. R. Breaker.** 2008. A widespread riboswitch candidate that controls bacterial genes involved in molybdenum cofactor and tungsten cofactor metabolism. *Mol. Microbiol.* **68:**918–932.
45. **Rieder, R., K. Lang, D. Graber, and R. Micura.** 2007. Ligand-induced folding of the adenosine deaminase A-riboswitch and implications on riboswitch translational control. *Chembiochem* **8:**896–902.
46. **Ross, P., H. Weinhouse, Y. Aloni, D. Michaeli, P. Weinberger-Ohana, R. Mayer, S. Braun, E. de Vroom, G. A. van der Marel, J. H. van Boom, and M. Benziman.** 1987. Regulation of cellulose synthesis in *Acetobacter xylinum* by cyclic diguanylic acid. *Nature* **325:**279–281.
47. **Ross, P., Y. Aloni, C. Weinhouse, D. Michaeli, P. Weinberger-Ohana, R. Meyer, and M. Benziman.** 1985. An unusual guanyl oligonucleotide regulates cellulose synthesis in *Acetobacter xylinum. FEBS Lett.* **186:**191–196.
48. **Ryan, R. P., Y. Fouhy, J. F. Lucey, and J. M. Dow.** 2006. Cyclic di-GMP signaling in bacteria: recent advances and new puzzles. *J. Bacteriol.* **188:**8327–8334.
49. **Ryjenkov, D. A., R. Simm, U. Romling, and M. Gomelsky.** 2006. The PilZ domain is a receptor for the second messenger c-di-GMP: the PilZ domain protein YcgR controls motility in enterobacteria. *J. Biol. Chem.* **281:**30310–30314.
50. **Schmidt, A. J., D. A. Ryjenkov, and M. Gomelsky.** 2005. The ubiquitous protein domain EAL is a cyclic diguanylate-specific phosphodiesterase: enzymatically active and inactive EAL domains. *J. Bacteriol.* **187:**4774–4781.
51. **Serganov, A., L. Huang, and D. J. Patel.** 2008. Structural insights into amino acid binding and gene control by a lysine riboswitch. *Nature* **455:**1263–1267.
52. **Serganov, A., A. Polonskaia, A. T. Phan, R. R. Breaker, and D. J. Patel.** 2006. Structural basis for gene regulation by a thiamine pyrophosphate-sensing riboswitch. *Nature* **441:** 1167–1171.
53. **Soukup, G. A., and R. R. Breaker.** 1999. Relationship between internucleotide linkage geometry and the stability of RNA. *RNA* **5:**1308–1325.
54. **Soukup, G. A., E. C. DeRose, M. Koizumi, and R. R. Breaker.** 2001. Generating new ligand-binding RNAs by affinity maturation and disintegration of allosteric ribozymes. *RNA* **7:**524–536.
55. **Stormo, G. D., and Y. Ji.** 2001. Do mRNAs act as direct sensors of small molecules to control their expression? *Proc. Natl. Acad. Sci. USA* **98:**9465–9467.
56. **Storz, G., S. Altuvia, and K. M. Wassarman.** 2005. An abundance of RNA regulators. *Annu. Rev. Biochem.* **74:**199–217.
57. **Sudarsan, N., E. R. Lee, Z. Weinberg, R. H. Moy, J. N. Kim, K. H. Link, and R. R. Breaker.** 2008. Riboswitches in eubacteria sense the second messenger cyclic di-GMP. *Science* **321:** 411–413.
58. **Sudarsan, N., J. E. Barrick, and R. R. Breaker.** 2003. Metabolite-binding RNA domains are present in the genes of eukaryotes. *RNA* **9:**644–647.
59. **Sudarsan, N., S. Cohen-Chalamish, S. Nakamura, G. M. Emilsson, and R. R. Breaker.** 2005. Thiamine pyrophosphate riboswitches are targets for the antimicrobial compound pyrithiamine. *Chem. Biol.* **12:**1325–1335.
60. **Sudarsan, N., M. C. Hammond, K. F. Block, R. Welz, J. E. Barrick, A. Roth, and R. R. Breaker.** 2006. Tandem riboswitch architectures exhibit complex gene control functions. *Science* **314:**300–304.
61. **Tamayo, R., J. T. Pratt, and A. Camilli.** 2007. Roles of cyclic diguanylate in the regulation of bacterial pathogenesis. *Annu. Rev. Microbiol.* **61:**131–148.
62. **Tamayo, R., A. D. Tischler, and A. Camilli.** 2005. The EAL domain protein VieA is a cyclic diguanylate phosphodiesterase. *J. Biol. Chem.* **280:**33324–33330.
63. **Thore, S., M. Leibundgut, and N. Ban.** 2006. Structure of the eukaryotic thiamine pyrophosphate riboswitch with its regulatory ligand. *Science* **312:**1208–1211.
64. **Vitreschak, A. G., D. A. Rodionov, A. A. Mironov, and M. S. Gelfand.** 2004. Riboswitches: the oldest mechanism for the regulation of gene expression? *Trends Genet.* **20:**44–50.
65. **Wachter, A., M. Tunc-Ozdemir, B. C. Grove, P. J. Green, D. K. Shintani, and R. R. Breaker.** 2007. Riboswitch control of gene expression in plants by splicing and alternative 3′ end processing of mRNAs. *Plant Cell* **19:**3437–3450.
66. **Wang, J. X., and R. R. Breaker.** 2008. Riboswitches that sense *S*-adenosylmethionine and *S*-adenosylhomocysteine. *Biochem. Cell Biol.* **86:**157–168.

67. **Wang, J. X., E. R. Lee, D. R. Morales, J. Lim, and R. R. Breaker.** 2008. Riboswitches that sense *S*-adenosylhomocysteine and activate genes involved in coenzyme recycling. *Mol. Cell* **29:**691–702.
68. **Wassmann, P., C. Chan, R. Paul, A. Beck, H. Heerklotz, U. Jenal, and T. Schirmer.** 2007. Structure of BeF3$^-$-modified response regulator PleD: implications for diguanylate cyclase activation, catalysis, and feedback inhibition. *Structure* **15:** 915–927.
69. **Weinberg, Z., J. E. Barrick, Z. Yao, A. Roth, J. N. Kim, J. Gore, J. X. Wang, E. R. Lee, K. F. Block, N. Sudarsan, S. Neph, M. Tompa, W. L. Ruzzo, and R. R. Breaker.** 2007. Identification of 22 candidate structured RNAs in bacteria using the CMfinder comparative genomics pipeline. *Nucleic Acids Res.* **35:**4809–4819.
70. **Weinberg, Z., E. E. Regulski, M. C. Hammond, J. E. Barrick, Z. Yao, W. L. Ruzzo, and R. R. Breaker.** 2008. The aptamer core of SAM-IV riboswitches mimics the ligand-binding site of SAM-I riboswitches. *RNA* **14:**822–828.
71. **Welz, R., and R. R. Breaker.** 2007. Ligand binding and gene control characteristics of tandem riboswitches in *Bacillus anthracis. RNA* **13:**573–582.
72. **Wickiser, J. K., M. T. Cheah, R. R. Breaker, and D. M. Crothers.** 2005. The kinetics of ligand binding by an adenine-sensing riboswitch. *Biochemistry* **44:**13404–13414.
73. **Wickiser, J. K., W. C. Winkler, R. R. Breaker, and D. M. Crothers.** 2005. The speed of RNA transcription and metabolite binding kinetics operate an FMN riboswitch. *Mol. Cell* **18:**49–60.
74. **Winkler, W. C., and R. R. Breaker.** 2005. Regulation of bacterial gene expression by riboswitches. *Annu. Rev. Microbiol.* **59:**487–517.
75. **Winkler, W. C., S. Cohen-Chalamish, and R. R. Breaker.** 2002. An mRNA structure that controls gene expression by binding FMN. *Proc. Natl. Acad. Sci. USA* **99:**15908–15913.
76. **Winkler, W. C., A. Nahvi, A. Roth, J. A. Collins, and R. R. Breaker.** 2004. Control of gene expression by a natural metabolite-responsive ribozyme. *Nature* **428:**281–286.
77. **Yakimov, M. M., A. Kröger, T. N. Slepak, L. Giuliano, K. N. Timmis, and P. N. Golyshin.** 1998. A putative *lichenysin A* synthetase operon in *Bacillus licheniformis:* initial characterization. *Biochim. Biophys. Acta* **1399:**141–153.
78. **Yao, Z., J. Barrick, Z. Weinberg, S. Neph, R. Breaker, M. Tompa, and W. L. Ruzzo.** 2007. A computational pipeline for high-throughput discovery of *cis*-regulatory noncoding RNA in prokaryotes. *PLoS Comput. Biol.* **3:**1212–1223.
79. **Yao, Z., Z. Weinberg, and W. L. Ruzzo.** 2006. CMfinder—a covariance model based RNA motif finding algorithm. *Bioinformatics* **22:**445–452.

The Second Messenger Cyclic Di-GMP
Edited by A. J. Wolfe and K. L. Visick

Chapter 16

Role of Cyclic Di-GMP in the Regulatory Networks of *Escherichia coli*

REGINE HENGGE

Recent research with *Escherichia coli* has substantially contributed to our understanding of the fundamental principles of cyclic di-GMP (c-di-GMP) signaling in bacteria. Every bacterial species, including *E. coli*, uses its own unique and often rather large set of diguanylate cyclases (DGCs) and phosphodiesterases (PDEs) to "make and break" this second messenger molecule. Equally diverse are c-di-GMP-binding effector components and their specific targets. In *E. coli* and related bacteria, c-di-GMP signaling is tightly integrated with the regulation of the growth cycle and the general stress response. This involves switching from the motile-planktonic state of postexponentially growing cells to the adhesive lifestyle during entry into the stationary phase, which corresponds to a first step in the formation of complex biofilms. At the molecular level, several distinct c-di-GMP control modules contribute to the precise inverse coordination of motility and curli fimbriae-mediated adhesion. *E. coli* also has several proteins with GGDEF and EAL domains, in which key amino acids required for DGC and PDE activities, respectively, are not conserved. Two of these, YcgF and YhdA (CsrD), are the prototypical degenerate proteins, for which alternative molecular functions based on direct macromolecular interactions have recently been demonstrated. The goal of this review is to provide an integrative picture that reflects general principles of c-di-GMP signaling as well as specific molecular functions of the 29 GGDEF/EAL domain proteins of *E. coli* and that can serve as a framework to elucidate the multiple functions of c-di-GMP signaling in a well-characterized model organism.

INTRODUCTION TO *ESCHERICHIA COLI* AND GENERAL CONCEPTS OF c-di-GMP SIGNALING

Adaptive Versatility of *E. coli*, Stress Responses, and the Role of c-di-GMP

E. coli is an environmentally versatile gram-negative bacterium that belongs to the gamma-proteobacteria and comes in a wide variety of strains that include commensals as well as important pathogens. As an enteric bacterium it is an inhabitant of the mammalian intestine, but certain pathogenic strains are specialized to living in other locations in the host. Moreover, *E. coli* also thrives in freshwater or marine aquatic environments or in the soil. *E. coli* can adapt to these very different and variable environmental conditions due to a complex general stress response, which is triggered by many different stresses, as well as numerous single stress-specific responses (123). *E. coli* is also used to a feast-or-famine existence, as it can efficiently transit between rapid growth phases, slow growth, and stationary phase depending on the availability of resources.

Furthermore, *E. coli* can switch between a planktonic-motile single-cell state and an adhesive multicellular lifestyle, i.e., living in a biofilm (7) (see also chapter 8). As in many other bacteria, this lifestyle control is dependent on the signaling molecule c-di-GMP and involves a whole series of GGDEF/EAL domain proteins, i.e., enzymes that make and break c-di-GMP (see chapter 4 and recent reviews [53, 105, 109]). In addition, these lifestyle decisions

Regine Hengge • Institut für Biologie–Mikrobiologie, Freie Universität Berlin, 14195 Berlin, Germany.

(swim or stick) are linked to different growth phases (Fig. 1). Thus, the postexponential growth phase, where resources are no longer optimal but not yet completely exhausted, is a foraging phase, in which *E. coli* cells become highly motile (1, 4, 91, 146). However, if this strategy is not successful, i.e., when resources dwindle further and cells finally enter into the stationary phase, motility is reduced again, adhesins (e.g., the autoaggregative curli fimbriae) are expressed, and exopolysaccharides can be produced and secreted. In liquid medium, cellular aggregation can be observed; on agar plates, this leads to the wrinkled or rdar colony morphotype, which represents a biofilm at a wet surface-air interface (this morphotype is similar and has been particularly well studied in the close relative of *E. coli, Salmonella enterica* serovar Typhimurium, as described in chapter 10; note that this organism is denoted hereafter as *Salmonella*).

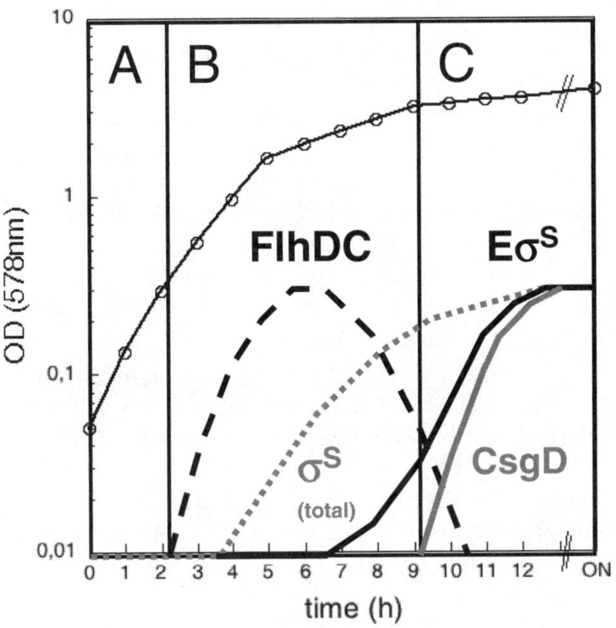

Figure 1. Growth phases and corresponding master regulators in *E. coli* K-12. In a culture of the *E. coli* K-12 strain W3110 growing in complex Luria-Bertani medium (LB), three growth phases can be distinguished: exponential or log phase (A), postexponential phase (B), and stationary phase (C). In the first part of the postexponential phase (starting at an optical density [OD] of approximately 0.3), the flagellar master regulator FlhDC and, therefore, flagella are expressed, but later on, further expression of FlhDC is shut down and existing FlhDC is degraded. As a consequence, further synthesis of flagella (and other proteins under FlhDC control) also comes to an end, but assembled flagella are active and cells are highly motile. In parallel, the master regulator of the stationary phase, σ^S, begins to accumulate but initially is only slowly assembling into active RNAP holoenzyme ($E\sigma^S$). Upon entry into the stationary phase (at an OD of approximately 3), the regulator CsgD is induced, which is essential for the expression of adhesive curli fimbriae and other biofilm-related functions. Note that only relative amounts of the various regulatory proteins or complexes are shown, which cannot be compared directly. OD (578 nm), optical density of the culture measured at 578 nm; ON, overnight, i.e., at approximately 24 h. For further details and references, see the text.

Consequently, c-di-GMP signaling in *E. coli* is intimately connected to the complex regulatory circuitry that coordinates the transition from postexponential to stationary phase (91, 121, 137). The main components of this circuitry are the flagellar control cascade directed by the master regulator FlhDC (21) and the stationary phase and general stress response coordinated by the σ^S (RpoS) subunit of RNA polymerase (RNAP) (48) (Fig. 1). These master regulators receive major input from two other second messenger molecules, i.e., cAMP (9) and guanosine penta/tetraphosphate [(p)ppGpp] (94) (see also chapter 1). *E. coli* is the organism for which this integration of c-di-GMP signaling into the global signaling and regulatory network of the cell is already most apparent (92), but also in other respects, work with *E. coli* has significantly contributed to current concepts of c-di-GMP signaling (47).

GGDEF/EAL Domain Proteins and the Concept of the c-di-GMP Control Module

Regulation by c-di-GMP requires (i) controlled production and degradation of this second messenger, (ii) effector components that sense c-di-GMP by directly binding it, and (iii) targets, whose output activity is controlled by direct interaction with the effectors. As described in detail elsewhere in this volume (see chapters 1, 4, and 5), c-di-GMP is synthesized by DGCs characterized by GGDEF domains (90, 113) and degraded by specific PDEs, which can feature either EAL or HD-GYP domains (19, 24, 110, 114, 127) (*E. coli* has only EAL type PDEs). c-di-GMP-binding effectors can be either proteins that, for instance, belong to the PilZ family (3) or RNAs, which operate as riboswitches (124) (see also chapters 4 and 15). Targets can be (i) promoter regions if the effector is a transcription factor (49); (ii) an RNA, with the ability to be further elongated or translated, in cases where the effector is a riboswitch (124); (iii) enzymes with the effector being a regulatory subunit or protein domain (3); or (iv) complex molecular structures such as the flagellar basal body (38, 91, 112), an exopolysaccharide synthesis and excretion apparatus (81, 86), or proteolytic targeting machinery (31) (see also chapter 9).

In a given species, these components can constitute one or more c-di-GMP control modules that are defined by the group of DGCs and PDEs (one or more of each) that, via a common pool of c-di-GMP, cooperate with a specific effector and specific target to control a common output (47). This obviously requires all components of a module to be coexpressed, which sometimes, but not necessarily, is reflected by coregulation. Knockout mutations that eliminate the DGCs, PDEs, and effector or target components that belong to the same module should affect the same output phenotype(s). If a group of coexpressed DGCs and PDEs together control the cellular level of freely diffusible c-di-GMP, this c-di-GMP can be bound by more than one effector and thereby affect several targets. Alterations in the expression patterns of DGCs and PDEs may easily reassign these enzymes to another functional module, which probably contributes to the evolutionary plasticity of c-di-GMP signaling.

Multiplicity of GGDEF/EAL Domain Proteins and the Concept of Sequestration of c-di-GMP Control Modules

The genomics era has revealed a striking abundance of genes encoding GGDEF and EAL domains in the genomes of many bacterial species (genes encoding HD-GYP domains are less frequent but can also be present several times in certain genomes). In particular, the gamma-proteobacteria often have dozens of genes encoding GGDEF/EAL domains (35). This multiplicity of putative DGCs and PDEs raises questions of signaling specificity but may also provide a new flexibility and variations of signaling not present with a simple second messenger system operating with a minimal set of components (for instance, the cAMP system in enteric bacteria) (for a detailed discussion, see reference 47). If it is hypothesized that all DGCs and PDEs made by a certain species contribute to the cellular pool of freely diffusible c-di-GMP, how can single GGDEF/EAL gene knockout mutations have the strong phenotypes observed (e.g., in the control of adhesive curli fimbriae, as described below)? How is stochastic noise in the expression and activity of all these DGCs and PDEs buffered? Also, would it make sense to have the activities of so many DGCs and PDEs converging into a single parameter, i.e., the cellular c-di-GMP pool, and consequently into a common output? Or could multiplicity be used to generate nonconverging pathways operating in parallel, which would link different outputs to different DGCs and PDEs and thereby greatly enhance the flexibility of c-di-GMP signaling?

The solution to the signaling specificity problem, which would also overcome the limitations of a convergent common output system, is the sequestration of GGDEF/EAL domain proteins and entire c-di-GMP control modules. In other words, not all these proteins should be present and active at the same time and place (for a detailed discussion, see reference 47). Sequestration can be temporal, i.e., by differentially regulating cellular levels and activities of DGCs and PDEs in response to intra- or extracellular signals that change over time. In addition, sequestration may be functional, i.e., entire c-di-GMP control modules (with different regulatory outputs) could operate at the same time but be physically separated from each other. In practical terms, this means that two DGCs that can be shown to be present and active under the same conditions differentially affect distinct target processes. Functional sequestration implies local action and therefore most likely direct interaction of the DGC, PDE, effector, and target component involved. This may result in temporarily higher local concentrations of c-di-GMP or even the generation of a local c-di-GMP pool due to some structural confinement (forming a microcompartment or microdomain). Colocalization and direct interactions of components of c-di-GMP signaling pathways have indeed been observed in many systems (summarized in reference 47). This would also allow spatial sequestration of distinct c-di-GMP control modules that are targeted to specific cellular locations, for instance, the cell poles, where they exert a specific function (see chapter 9).

It should be noted that c-di-GMP control modules that regulate the cellular pool of freely diffusible c-di-GMP and functionally highly specialized local c-di-GMP control modules may easily coexist (evidence is described below; see also Fig. 4). Moreover, in other species, other second messenger systems seem to show a similar multiplicity of components. For instance, certain mycobacterial species have few genes encoding GGDEF/EAL domains but high numbers of adenylate cyclases, and certain alphaproteobacterial species have multiple DGCs and adenylate cyclases (5, 35, 118, 119). Moreover, eukaryotic cells also express multiple adenylate cyclases that have been shown to operate in cAMP microdomains (140).

Evolutionary Loss of c-di-GMP Signaling and the Emergence of New Functions for GGDEF/EAL Domain Proteins

Functional sequestration and local operation of certain c-di-GMP control modules based on direct interactions in protein complexes can open new pathways for evolution (47). Initially, such protein interactions may simply play a scaffolding role, i.e., just serve to keep the components together and perhaps provide some barrier for c-di-GMP diffusion. However, it is conceivable that interactions acquire a regulatory role, i.e., the interacting proteins may affect each other's activity. In some cases, the enzymatic activity of one partner protein may get lost, but the former substrate or product molecule may still be able to bind and now act as a ligand for allosteric control. A nice example is the PdeA protein in *Caulobacter crescentus,* where PDE activity is provided by the C-terminal EAL domain, which is activated by GTP binding to a slightly degenerate and enzymatically inactive N-terminal GGDEF domain (24). If a GGDEF or EAL domain is still able to bind c-di-GMP, it may also function as an effector in a c-di-GMP control module, as recently observed for the EAL domain of LapD in *Pseudomonas fluorescens* (84). Other GGDEF or EAL proteins that are part of a locally acting complex may lose the ability to bind GTP or c-di-GMP completely. Thus, the resulting signaling systems would have given up using c-di-GMP and now completely rely on macromolecular interactions (47). This scenario provides an explanation for the existence of many GGDEF/EAL domain proteins in which specific amino acid residues required for DGC or PDE activity are not conserved but that nevertheless control distinct molecular and cellular functions in many bacteria. The best example studied at the molecular level so far is the EAL protein YcgF of *E. coli,* which does not have PDE activity but acts as a direct antirepressor (130) (see below).

GGDEF/EAL DOMAIN PROTEINS IN *E. COLI*

The number of proteins with GGDEF/EAL domains encoded in the genomes of various enteric bacteria is quite variable. While nonpathogenic *E. coli* K-12 strains usually have 29 such proteins, the closely related *Salmonella enterica* serovar Typhimurium utilizes only 19 and 3 of these proteins are absent in *E. coli* (Table 1). In particular, *E. coli* has a higher number of proteins with intact GGDEF motifs, i.e., known or putative DGCs. This could indicate that *E. coli* is better adapted to forming biofilms in the environment than *Salmonella.* Also, acute virulence genes of pathogens tend to be negatively regulated by c-di-GMP, which fits the observation that the pathogenic *E. coli* O157:H7 features a homolog of a *Salmonella* EAL protein (encoded by STM0343) not present in *E. coli* K-12 strains (Table 1). These variations in the numbers of GGDEF/EAL domain-encoding genes in different strains suggest that systematic comparisons in pathogenic and nonpathogenic strains may reveal interesting trends.

The 29 GGDEF/EAL Domain Proteins of *E. coli* K-12

Twelve of the 29 GGDEF/EAL domain proteins in *E. coli* carry GGDEF domains only, 7 are composite proteins with both domains in a single polypeptide, and 10 are EAL-only proteins (Tables 1 and 2). It should be noted that certain commonly used *E. coli* K-12 laboratory strains (e.g., MC4100) do not have the EAL domain-encoding *yahA* gene (93). Currently, nearly all these genes are denoted with "y" designations, and it is suggested here to maintain this nomenclature to avoid confusion. Once direct, unequivocal evidence for the enzymatic or other activities of all the corresponding proteins is available, one might then consider switching to a unifying nomenclature (e.g., *dgc* and *pde* genes).

Up to now, only a few of these *E. coli* GGDEF/EAL domain proteins have been purified and tested for enzymatic activity in vitro. The GGDEF-only proteins YdaM (137) and YeaP (113) exhibit DGC activity. PDE activity was shown for the EAL-only proteins YahA (114) and YhjH (91) as well as for the GGDEF plus EAL proteins YciR (137) and YddU(Dos) (114). However, based on crystal structures and detailed knowledge about the molecular functions of distinct conserved amino acids for DGC and PDE activity (18, 28, 82, 102, 136), enzymatic activities can be predicted (see also chapter 7). Thirteen proteins in *E. coli* carry an intact GGDEF/GGEEF motif (Table 2), which is known to provide the DGC active site (A-site), i.e., residues involved in GTP binding and catalysis of c-di-GMP formation (18). However, one of these proteins, YciR, also carries an EAL domain with demonstrated PDE activity and does not exhibit any DGC activity in vitro (137) (H. Weber and R. Hengge, unpublished data). The remaining 12 intact GGDEF motif proteins would be predicted to be active DGCs (YaiC, YcdT, YdaM, YddV, YdeH, YeaJ, YeaP, YedQ, YfiN, YliF, YneF, and YegE; experimentally confirmed already for YdaM and YeaP, see above). Among these, seven also display the I-site, i.e., an RXXD motif separated from the GGDEF motif by five amino acids (Table 2), which represents the secondary c-di-GMP binding site responsible for allosteric product feedback inhibition (22; see chapters 7 and 9). There is also a single GGDEF protein, YeaI, which features an intact I-site but a degenerate A-site. As shown for the *Caulobacter crescentus* I^+A^- protein PopA (31), YeaI may have switched sides during evolution, i.e., evolved from a former DGC into a c-di-GMP-binding effector protein (of still unknown function).

Sequence-based prediction of PDE activity is somewhat less straightforward as the amino acid residues involved in binding of c-di-GMP and Mg^{2+} and in catalysis of the cleavage reaction are scattered along the primary sequence (102). The essential residues seem present in 13 EAL domain proteins, which

Table 1. GGDEF/EAL genes in *Escherichia coli* and *Salmonella enterica* serovar Typhimurium[a]

Domain(s)	*Escherichia coli* K-12				*Salmonella enterica* serovar Typhimurium	
	Gene	b no.	N-terminal domain(s)	Alternative designation	Gene	Alternative designation
GGDEF only	*yaiC*	b0385*	?		STM0385*	*adrA*
	ycdT	b1025*	?		—	
	ydaM	b1341	?-PAS		—	
	yddV	b1490	Rec-Rec		—	
	ydeH	b1535	?		—	
	yeaI	b1785*	?		—	
	yeaJ	b1786*	?		STM1283*	
	yeaP	b1794	GAF		—	
	yedQ	b1956*	?		STM1987*	*gcpA*
	yfiN	b2604*	?		STM2672*	
	yliF	b0834*	?		—	
	yneF	b1522*	?		—	
	—				STM4551*	
GGDEF and EAL	***yciR***	b1285	PAS	*gmr*	**STM1703**	*gcpE*
	yddU	b1489	PAS-PAS	*dos*	—	
	yegE	b2067*	MASE1-PAS_3		**STM2123***	
	yfeA	b2395*	?		STM2410*	
	yfgF	b2503*	MASE1		STM2503*	
	yhdA	b3252*	HAMP-like	*csrD*	STM3375*	
	yhjK	b3529*	?		STM3615*	
	—				**STM3388***	
EAL only	*rtn*	b2176*	?		STM2215*	
	yahA	b0315	LuxR		—	
	ycgF	b1163	BLUF		—	
	ycgG	b1168*	?		—	
	ydiV	b1707	—		STM1344	
	yhjH	b3525	—		STM3611	
	yjcC	b4061*	?		**STM4264***	
	ylaB	b0457*	?		STM0468*	
	yliE	b0833*	?		—	
	yoaD	b1815*	?		STM1827*	
	—				STM0343*	

[a]An asterisk indicates membrane localization of the gene product; genes listed in boldface are involved in the regulation of curli fimbria expression. For references, see main text. ?, N-terminal domain of unknown structure and/or function; —, no N-terminal domain present or no homolog present in either *E. coli* or *Salmonella*.

are therefore highly likely to represent active PDEs (the GGDEF plus EAL proteins YciR, YddU, YfeA, YfgF, and YhjK as well as the EAL-only proteins Rtn, YahA, YcgG, YhjH, YjcC, YlaB, YliE, and YoaD) (experimentally confirmed for YddU, YciR, YahA, and YhjH; see above). Interestingly, YhjH does not feature all the amino acids that have been implicated in enzymatic function but obviously the essential ones (Table 2), as it is a highly active PDE in vitro (91).

In addition, this analysis revealed GGDEF/EAL domain proteins in which multiple functionally important residues are not conserved (hereafter referred to as degenerate GGDEF/EAL domain proteins). These are highly unlikely to have enzymatic activities. Degenerate domains include the GGDEF domains of the composite proteins YddU, YfeA, YfgF, and YhjK, all of which display the highly conserved residues in their EAL domains and therefore probably act as PDEs. The degenerate GGDEF domains in these composite proteins could play regulatory roles. In the GGDEF plus EAL protein YhdA (CsrD), both domains are degenerate, and among the EAL-only proteins, YcgF and YdiV are degenerate. These three proteins are therefore expected to be enzymatically inactive (for alternative functions of these proteins, see below).

In conclusion, direct biochemical and sequence analyses indicate that *E. coli* has (i) 12 active DGCs, (ii) 13 active PDEs, (iii) one putative c-di-GMP effector protein of the GGDEF(I^+A^-) type, and (iv) three enzymatically inactive degenerate GGDEF/EAL domain proteins in *E. coli* K-12. Which of the DGCs and PDEs belong to common functionally defined c-di-GMP control modules has become apparent only for a few so far and will be described in further detail below.

Regulation of Expression of the GGDEF/EAL Genes

Consistent with having functions in different physiological contexts, the *E. coli* genes encoding GGDEF/EAL domains are differentially expressed and regulated. In a systematic study (121), it was observed that only 7 of 28 GGDEF/EAL domain-encoding genes (*yahA* was absent in the strain used) are not expressed in standard LB medium. All the other genes are expressed at levels that vary by more than 2 orders of magnitude (data summarized in Table 2). Parameters that substantially affect expression include growth phase, temperature, and growth on a surface.

During exponential and postexponential growth, most GGDEF/EAL domain-encoding genes are expressed at low levels or not at all. Two genes, however, show higher expression and therefore probably dominate c-di-GMP signaling in this phase of the growth cycle (121). These encode the DGC YeaJ and the PDE YhjH, which negatively and positively control motility, respectively (38, 91). During entry into the stationary phase, however, this pattern changes drastically: while expression of *yeaJ* and *yhjH* is shut off, a whole series of other GGDEF/EAL domain-encoding genes is now activated. These include genes for DGCs and PDEs at approximately equal numbers, with the most strongly expressed genes being *ydaM, yddV,* and *yddU* (Table 2). With one exception *(yhjK),* stationary-phase activation of all these genes depends on the σ^S subunit of RNAP, i.e., their induction is part of the general stress response (121, 137). Thus, fine-tuned c-di-GMP-mediated signal transduction, which integrates multiple signals via the sensory domains of the many GGDEF/EAL domain proteins deployed, seems particularly important in slowly growing or stationary-phase cells.

Stationary-phase induction of these genes is not strictly in parallel, but there are early and late genes. Early induction can be observed, e.g., for *ydaM,* the *yddV-yddU* operon, *yedQ, yciR, yegE, yjcC,* and *ylaB,* whereas *yaiC* and *yoaD* are induced several hours later (121, 137). Such temporal succession can reflect a regulatory cascade: the DGC/PDE pair YdaM/YciR controls the expression of the regulatory protein CsgD, which in turn is required for activating the *yaiC* gene (137).

Interestingly, σ^S not only activates nearly all of these stationary-phase-induced genes (directly or indirectly) but also negatively regulates five GGDEF/EAL domain-encoding genes, i.e., they exhibit higher expression in an *rpoS* mutant. This is probably due to higher levels of the vegetative and flagellar RNAP holoenzymes in the absence of σ^S. Negative regulation by σ^S is most prominent for the DGC-encoding *ydeH* gene, which is even induced during entry into the stationary phase in the absence of σ^S only, suggesting that YdeH is involved in some stress response that reduces growth but does not induce σ^S (121). *ydeH* is also directly activated by copper and a shift to microanaerobiosis, with both responses being mediated by the CpxAR two-component system (88, 145). In addition, *ydeH* is negatively regulated by the RNA-binding protein CsrA (58).

Another general pattern in the regulation of genes encoding GGDEF/EAL domain proteins is temperature regulation. Among the 21 genes expressed in LB medium, 7 show increased or even exclusive expression at reduced temperature (28°C) (Table 2). This is consistent with a major target of c-di-GMP control, i.e., the expression of adhesive curli fimbriae, exhibiting similar temperature regulation (87). Only two genes, *yeaJ* and *yahA,* are somewhat less expressed at 28°C than at 37°C (121) (G. Becker and R. Hengge, unpublished data). At 28°C, the role of YeaJ in motility control is taken over by two other DGCs (YegE and YedQ) (91).

A small set of GGDEF/EAL domain-encoding genes *(ydaM, yciR, yaiC,* and *yoaD)* is more strongly expressed in cells grown on a surface (i.e., on an agar plate) than in liquid-grown cells (121). These genes encode the DGCs and PDEs that in *E. coli* are most important for the expression of adhesive curli fimbriae and cellulose (Fig. 2) and, therefore, for the complex rdar morphotype of plate-grown colonies (see chapter 10). On the other hand, *yhjH,* which encodes a PDE that promotes motility and downregulates curli formation (91), is the only GGDEF/EAL domain-encoding gene that is significantly downregulated in plate-grown cells (121).

A number of studies have found particular GGDEF/EAL domain-encoding genes to be members of specific regulons (Fig. 2 and Table 2). The PDE-encoding *yhjH* gene is a class 3 gene in the flagellar gene hierarchy, as it requires the flagellar master regulator FlhDC and the flagellar sigma factor FliA for expression, which is in line with its function in keeping cellular c-di-GMP levels low to allow motility (33, 38, 67, 91, 111). Activation of the DGC-encoding *yaiC* gene by the curli/cellulose regulator CsgD has already been mentioned above (12, 13). The DGC-encoding *ydaM* gene is one of the rare genes under positive control of the histone-like protein H-NS (137). *ydiV,* which encodes a degenerate EAL protein of unknown function, is under direct

Table 2. Summary of sequence features and expression data of GGDEF/EAL genes in *E. coli* K-12[a]

Domain(s)	Gene	b no.[b]	N-terminal domain(s)	GGDEF domain[c] (RXXD GGDEF)	EAL domain[d] EALXR	EAL domain[d] c-di-GMP (Q/R/D/D)	EAL domain[d] Catalysis (T/E)	Expression in LB[e]	Highest expression at: Temp (°C)	Highest expression at: Growth phase[f]	Regulated by: σ^s	Regulated by: Other regulator(s)[g]
GGDEF only	*yaiC*	b0385*	MASE2	RGSDVIGRFGGDEF				++	28	SP	pos	CsgD(pos)
	ycdT	b1025*	?	RPDDLLARVGGDEF				+	28/37	LP	—	CsrA(neg)
	ydaM	b1341	?-PAS	RKGDLVFRWGGDEF				+++	28/37	SP	pos	H-NS(pos)
	yddV	b1490	Sensor globin	RSSDYVFRYGGDEF				+++	28/37	SP	pos	
	ydeH	b1535	?	RDYETVYRYGGEEF				++	28/37	LP/SP	neg	CpxR(pos), CsrA(neg)
	yeaI	b1785*	?	RPDDILARLEGEVF				–	—	—		
	yeaJ	b1786*	?	RKSDYAIRLGGDEF				+++	28/37	Late LP/SP	—	
	yeaP	b1794	GAF	QNGEVIGRLGGDEF				++	28/37	LP	—	
	yedQ	b1956*	?	RAQDVAGRVGGDEF				++	28/37	LP/SP	pos	
	yfiN	b2604*	?	GLRHKAYRLGGDEF				+	28/37	LP	—	
	yliF	b0834*	?	VDKGKVYRFGGDEF				–	—	—		
	yneF	b1522*	?	GDKGLVARMGGDEF				–	—	—		
GGDEF and EAL	*yciR*	b1285	PAS	EHDQVLARPGGDEF	EALVR	Q/R/D/D	T/E	++	28/37	LP/SP	pos	
	yddU	b1489	PAS-PAS	KPDQYLCRIEGTQF	EALAR	Q/R/D/D	T/E	+++	28/37	SP	pos	
	yegE	b2067*	MASE1-PAS$_3$	RSSDVLARLGGDEF	EARNL	–/–/–/–	–/–	++	28/37	LP/SP	pos	
	yfeA	b2395*	MASE1	QENEKLYQLPGSEL	EILAR	Q/R/D/D	T/E	++	28/37	LP	neg	
	yfgF	b2503*	MASE1	EPGEDVYQLSGNDL	EILLR	Q/R/D/D	T/E	–	—	—		
	yhdA	b3252*	?	YPGALLARYHRSDF	ELMCR	Q/R/–/–	–/–	+	28/37	LP	—	
	yhjK	b3529*	?	SPRMILAQISGYDF	EVLLR	Q/R/D/D	T/E	++	28	SP	—	

EAL only	*rtn*	b2176*	?			EVLLR	Q / R / D / D	T / E	++	28/37	LP	neg	
	yahA	b0315	LuxR			EVLVR	Q / R / D / D	T / E	++[h]	28/37	LP	—	
	ycgF	b1163	BLUF			EAIVQ	– / – / – / –	T / –	+++	28	LP/SP	neg	
	ycgG	b1168*	?			EVLAR	Q / R / D / D	T / E	+	28/37	LP	—	
	ydiV	b1707	—			–––––	– / – / – / –	– / –	–	—	—		SdiA(pos)
	yhjH	b3525	—			ELLTV	Q / – / D / –	– / E	+++	28/37	Late LP	neg	FlhDC/ FliA(pos)
	yjcC	b4061*	?			EALLR	Q / R / D / D	T / E	++	28/37	SP	pos	
	ylaB	b0457*	?			EALAR	Q / R / D / D	T / E	++	28/37	LP/SP	pos	
	yliE	b0833*	?			EALCR	Q / R / D / D	T / E	–	—	—		
	yoaD	b1815*	?			EILLR	Q / R / D / D	T / E	++	28/37	SP	pos	

[a] The *E. coli* K-12 strain used was W3110Δ*lacU169* (data are based on reference 121). ?, unknown N-terminal domain; —, not present.
[b] An asterisk indicates membrane location of the gene product.
[c] I-site, inhibitory binding site for c-di-GMP, is on the left (RXXD); A-site, GTP-binding and enzyme active site is on the right (consensus is GGDEF/GGEEF).
[d] According to reference 102, the EAL motif is involved in coordinating Mg^{2+}. Four amino acids (Q, R, D, and D) dispersed along the primary sequence are required for c-di-GMP binding (the R is the one that follows the EAL motif with one nonconserved amino acid in between), and two amino acids (T and E) are directly involved in catalysis.
[e] Relative expression was determined with translational single copy *lacZ* reporter fusions inserted early into the coding sequence. For full expression patterns along the growth curve, see reference 121. Maximal specific β-galactosidase activity measured was >0.1 (+++), 0.01 (++), or 0.002 (+) μmol $min^{-1}mg^{-1}$ total cellular protein. –, not expressed.
[f] LP, logarithmic growth phase; SP, stationary phase. Underscores indicate conditions of predominant or exclusive expression.
[g] For references, see the text. pos, positive; neg, negative.
[h] Determined by G. Becker and R. Hengge (unpublished data). The strain used in reference 121 did not have *yahA*, which was eliminated by the Δ*lacU169* deletion.

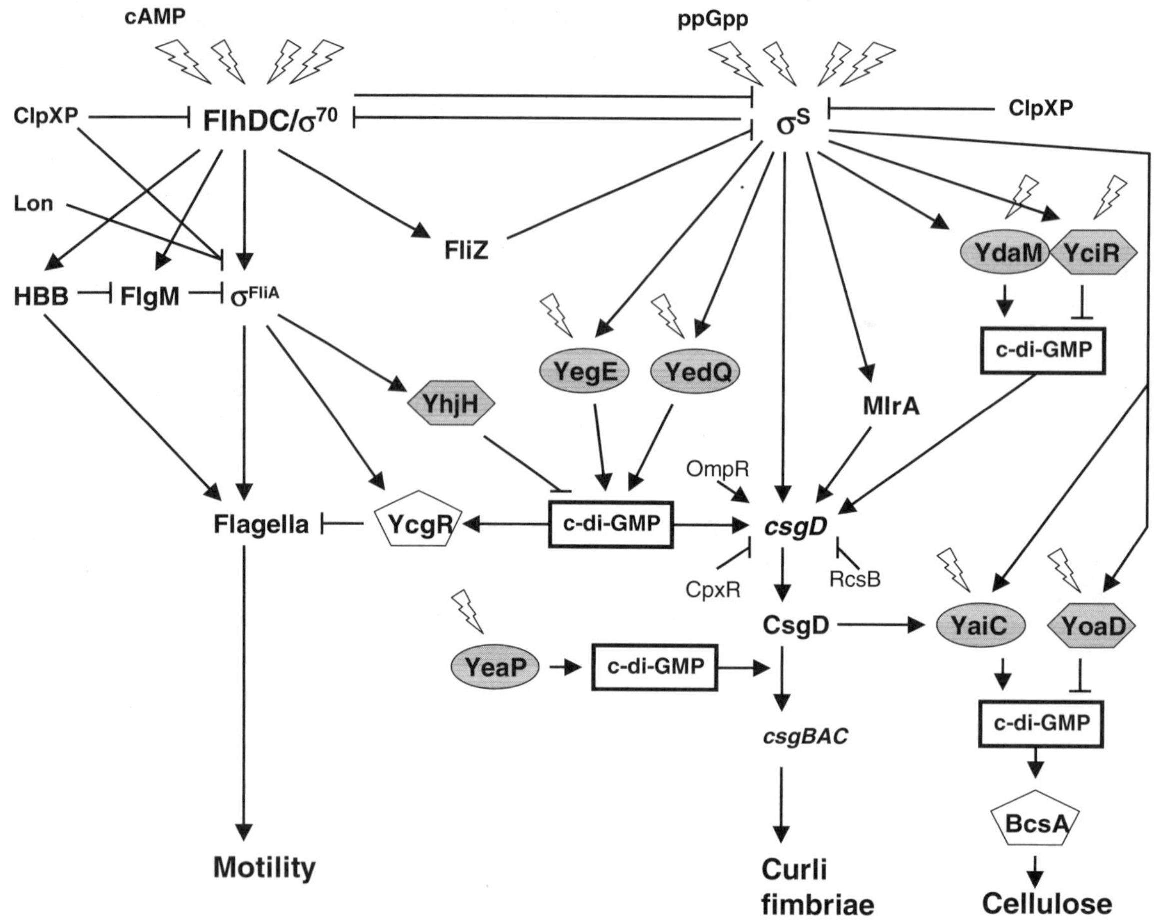

Figure 2. Inverse coordination of motility and curli-mediated adhesion in *E. coli* K-12. The flagellar control cascade (FlhDC/FliA) interferes with the activity of the σ^S/CsgD/curli control cascade at two levels: (i) FliZ, which is expressed from a class 2 gene in the flagellar hierarchy, is an inhibitor of σ^S activity at many σ^S-dependent promoters, including those of *ydaM, mlrA,* and *csgD;* and (ii) the PDE YhjH, which is expressed from a class 3 gene (under σ^{FliA} control), degrades c-di-GMP and thereby keeps motility going while not allowing the activation of transcription of *csgD* and, therefore, curli expression. When the flagellar control cascade (including *yhjH* expression) shuts down in mid-postexponential phase, the DGCs YegE and YedQ, which are increasingly expressed due to the now accumulating σ^S, outbalance the PDE activity of YhjH, and c-di-GMP can accumulate. Via YcgR, this c-di-GMP interferes with flagellar activity and, via an unknown effector, stimulates *csgD* transcription. In essence, this c-di-GMP control module acts as a checkpoint that allows curli expression only after flagellar gene expression has ceased. In parallel, a second DGC/PDE system, YdaM/YciR, is expressed under σ^S control. The latter system exclusively acts on *csgD* transcription in a way which is not additive with the effect of the YegE/YhjH system (but both systems are essential for activation). Additional c-di-GMP control modules operate downstream of CsgD expression and affect the expression of the curli structural operon *csgBAC* and cellulose biosynthesis. In principle, the activity of all DGCs and PDEs (probably with the exception of YhjH, which basically consists of an EAL domain only) is likely to be modulated by additional unknown signals (lightning bolts) perceived by their N-terminal sensor domains. Note, only relevant genes or proteins under FlhDC and σ^S control are shown here; overall, FlhDC and σ^S activate more than 60 and approximately 500 genes, respectively. For further details and references, see the text. HBB, hook basal body.

positive control of the quorum-sensing activator SdiA, which responds to autoinducers produced by other species (*E. coli* does not produce a corresponding autoinducer itself) (147). Finally, the RNA-binding protein CsrA interferes with expression of two GGDEF domain-encoding genes, *ycdT* and *ydeH* (58).

The overall picture emerging from these regulatory data is that c-di-GMP signaling becomes more sophisticated (i.e., more DGCs and PDEs with different sensory input domains are expressed), when *E. coli* cells go into the stationary phase and are exposed to temperatures below 30°C. This suggests that c-di-GMP signaling is especially important for growth-restricted cells in an extrahost environment. In addition, some GGDEF/EAL domain-encoding genes are expressed under very specific conditions (as for instance *ydeH*), which can provide hints to the yet unknown physiological functions of the corresponding proteins.

Regulation of Activity of the GGDEF/EAL Domain Proteins

The actual total DGC and PDE activities inside a cell, and therefore the cellular pool(s) of c-di-GMP, depend on the numbers of enzyme molecules present as well as on their specific activities. The latter can be controlled by specific signals perceived by the N-terminal sensory domains of most DGCs and PDEs (see chapter 3). Of the 29 GGDEF/EAL domain proteins, 27 have significant N-terminal extensions (only the PDE YhjH and the degenerate EAL protein YdiV basically consist of the EAL domain only) (Table 2).

Nineteen of the 29 GGDEF/EAL domain proteins contain one or more transmembrane domains in their N-terminal parts, which in some cases (YaiC, YegE, YfeA, and YfgF) combine into specific membrane-associated sensory domains (MASE1, MASE2) (Table 2) (85). MASE1 and MASE2 domains have also been found in histidine sensor kinases and adenylate cyclases, respectively (34), but the signals transduced by them are still elusive.

The 10 soluble cytoplasmic GGDEF/EAL domain proteins feature PAS, GAF, sensor globin, BLUF, and LuxR-like domains (Table 2). PAS domains, which are also common in histidine sensor kinases, bind various ligands such as heme, flavin, or adenine and can sense light, oxygen, and other parameters (36, 128). PAS domains are present both in YdaM and YciR, i.e., the DGC/PDE pair that controls expression of the curli/cellulose regulator CsgD. In the PAS domain of the PDE YddU (Dos), the presence of oxygen results in a shift in the position of the heme molecule, thereby generating a conformational change of the PAS domain (29). Its likely antagonist, the putative DGC YddV, which is encoded in an operon together with YddU (80), features an N-terminal sensor globin domain, which can be involved in sensing oxygen, NO, or CO (134), but the actual parameter sensed by YddV is unclear.

GAF domain can bind to various ligands, including cAMP, cGMP, or different chromophores (51, 149), can sense nitric oxide via a nonheme iron center (76), and can even have enzymatic activity (76). What is sensed by the GAF domain of the *E. coli* DGC YeaP, is unclear, but it seems to provide an additional signal input into the expression of the curli genes (121).

The BLUF domain is a blue light sensor that uses flavin adenine dinucleotide as a cofactor (*b*lue *l*ight-sensing *u*sing *f*lavin adenine dinucleotide) (41). Blue light-induced conformational changes not only in the BLUF domain but also in the entire protein have been observed for YcgF, and the protein is believed to dimerize upon photo-excitation (46, 83, 101, 115, 144). *E. coli* YcgF is a degenerate EAL domain protein without PDE activity that by direct interaction releases a repressor from its operator site upon blue light irradiation (130) (and see below).

Finally, the PDE YahA features an N-terminal domain with similarity to the DNA-binding C-terminal domain of LuxR and the output domains in several response regulators. This suggests that this domain in YahA serves for DNA binding rather than as a sensory domain (unless the availability of putative binding sites on the DNA serves as an input signal). The role of potential binding of this PDE to DNA is currently obscure.

In summary, as in other bacterial species, nearly all GGDEF/EAL domain proteins in *E. coli* seem to contain N-terminal sensory domains and their activities can therefore be expected to be modulated by a variety of input signals. Only in two cases, however, have the actual input signals been identified, i.e., oxygen sensed by the PAS domain of YddU (Dos) and blue light sensed by the BLUF domain of YcgF.

c-di-GMP-BINDING EFFECTORS AND THEIR DIRECT MOLECULAR TARGETS IN *E. COLI*

The currently best-studied family of c-di-GMP-binding effector proteins is the PilZ domain proteins that were named after a type IV pilus control protein in *Pseudomonas aeruginosa* (3) (see also chapters 4 and 7). PilZ domains can occur alone or in combination with GGDEF/EAL/HD-GYP domains or some output domains (3, 81, 86). *E. coli* has only two PilZ domain proteins, YcgR and BcsA, which are involved in the control of motility and cellulose biosynthesis, respectively (38, 91, 112).

The *ycgR* gene was first identified as a knockout mutation that suppressed the nonmotility phenotype of a mutant deficient in the histone-like protein H-NS (67). It is still not clear why an *hns* mutant is nonmotile, but given the now-known function of YcgR as a c-di-GMP binding effector, one has to assume that the absence of the global repressor/silencer H-NS shifts the overall cellular DGC/PDE balance in favor of higher c-di-GMP accumulation. Accordingly, *ycgR* null mutations also suppress impaired motility of a mutant deficient in the major PDE YhjH in both *E. coli* and *Salmonella* (38, 91, 112). *ycgR* and *yhjH* are class 3 genes in the flagellar gene hierarchy (33, 38, 91).

YcgR, which has a small N-terminal domain besides the PilZ domain, is a monomeric protein that binds c-di-GMP in vitro with a K_D (equilibrium dissociation constant) of 840 nM and interferes with motility (112). While this output activity generates clear effects in mutant backgrounds with artificially increased c-di-GMP accumulation as demonstrated in the suppressor studies mentioned above, knocking out *ycgR* alone produces subtle phenotypes only. These are an altered flagellar rotation switching pattern (38) and a delay in the reduction in cellular swimming speed that can be observed during entry into the stationary phase (91). In *E. coli,* mutations in *ycgR* (or *yhjH*) do not significantly affect the expression of flagellar genes (91). This suggests that YcgR directly controls flagellar activity by interacting with some component at the cytoplasmic face of the flagellar basal body. Similar conclusions have been drawn for the *C. crescentus* PilZ domain protein DgrA, which downregulates flagellar activity during entry into the flagellar-to-stalked cell differentiation program (23). In other bacterial species, however, c-di-GMP can also control the expression of the flagellar genes (141). Thus, while the PDE YhjH is a factor that promotes motility, YcgR represents a built-in break in the motility system, which is activated as soon as certain DGCs are expressed or activated sufficiently to overcome YhjH activity. The DGCs involved are YeaJ (at 37°C), YegE (at 28°C), and to a minor extent and independent of the growth temperature, YedQ (38, 91).

The PilZ domain protein BcsA has not been investigated experimentally in *E. coli,* but orthologous proteins occur widely in gram-negative bacteria and a study of its function in the cellulose biosynthesis of *Gluconoacetobacter xylinum* in fact started research on c-di-GMP more than 20 years ago (108, 139) (see chapter 2). BcsA is a large protein with cellulose synthase activity controlled by its C-terminal c-di-GMP-binding regulatory PilZ domain (3). Cellulose is the extracellular matrix polysaccharide responsible for the rdar, or wrinkled, morphotype of bacterial colonies on agar surfaces (148) (see chapter 10).

The two PilZ domain proteins YcgR and BcsA are clearly not sufficient to generate all c-di-GMP-controlled effects in *E. coli.* A major target of c-di-GMP control in *E. coli* is the transcription of the *csgD* gene (91, 137), which encodes a key regulator for the expression of adhesive curli fimbriae, cellulose, and other functions of importance during biofilm formation (12, 13, 97). Two DGC/PDE systems affect *csgD* transcription: YegE/YhjH, which also controls motility, and YdaM/YciR, which strongly and apparently exclusively regulates *csgD* transcription (91, 137). The c-di-GMP-binding effectors mediating this input into CsgD expression are currently elusive, but it is clear that YcgR is not involved (91).

Curli fimbriae expression is also positively modulated by the DGC YeaP. However, YeaP does not significantly affect the cellular level of the curli regulator CsgD. Rather, YeaP seems to play a role in the expression of the curli structural operon *csgCAB* (121). The effector served by YeaP-produced c-di-GMP and the direct target controlling the expression of curli proteins is currently unknown.

YeaI is the single *E. coli* GGDEF protein with a defective A-site but an intact I-site and, therefore, is another candidate for a c-di-GMP-binding effector protein. While *yeaI* is located right upstream of the DGC-encoding *yeaJ* gene, the two genes do not constitute an operon. In contrast to *yeaJ,* which is one of the most strongly expressed GGDEF/EAL domain-encoding genes, *yeaI* is not expressed in LB medium and is also not present in the genome of *Salmonella* (Table 1). At present, no phenotype is known for an *E. coli yeaI* knockout mutant (121).

In conclusion, our current knowledge about c-di-GMP-binding effector components and their cognate targets in *E. coli* seems largely incomplete and certainly represents a rewarding field for future research.

PHYSIOLOGICAL FUNCTIONS OF c-di-GMP SIGNALING IN *E. COLI*

Physiological Context of c-di-GMP Signaling

As pointed out in the introduction to this chapter, c-di-GMP signaling is tightly interconnected with the *E. coli* growth cycle and its regulatory circuits (Fig. 1). Key players in these circuits are at least three sigma subunits of RNAP (σ^{70}, σ^{FliA}, and σ^{S}) (43, 48), the flagellar master regulator FlhDC (21), and the second messengers cAMP and (p)ppGpp besides c-di-GMP (92). The interplay of all these factors is crucial for metabolic adaptation to available resources and the control of motility, stress resistances, and biofilm functions during the exponential, postexponential, and stationary phases of the growth cycle (Fig. 2).

When growing in complex medium (LB), exponentially growing cells devote most of their resources to the synthesis of ribosomes, i.e., the machinery necessary for driving rapid growth and proliferation. This is reflected by rather condensed nucleoids and the concentration of RNAP in a few cellular foci at the ribosomal gene clusters (54). When carbon/energy sources become less optimal, cells enter the post-exponential growth phase (in an LB-grown *E. coli* K-12 culture this happens around an optical density of approximately 0.3) (Fig. 1) (116). Now cAMP begins to accumulate and, via cAMP receptor protein (CRP) as its effector protein, activates numerous genes, many of which are involved in alternative carbon source scavenging (50). Overall, gene expression now strongly diversifies, the nucleoid becomes less compact, and RNAP becomes more dispersed (54). One of the many operons activated by cAMP-CRP encodes the master regulator for flagella expression, FlhDC (122). Flagellar structural proteins and secondary regulators (including the flagellar sigma factor σ^{FliA} and its antisigma factor FlgM) are expressed in a fine-tuned temporal order (6, 21, 62) (see chapter 8). This process culminates in the assembly of several flagella per cell and entry into the motile foraging phase of the growth cycle (Fig. 2).

In parallel to reduced growth rate and as a consequence of nutrient limitation, RelA/SpoT-synthesized (p)ppGpp begins to accumulate (94), which contributes to increased expression of the general stress and stationary-phase sigma factor σ^S (37, 70). However, σ^S does not yet seem to gain significant access to the RNAP core (as the large majority of σ^S-dependent genes are still not expressed). Rather, the sigma subunits that determine overall gene expression during the postexponential phase are the vegetative σ^{70} and the flagellar sigma factor σ^{FliA} (also known as σ^{28}), which specializes in expressing flagellar class 3 genes (Fig. 2). Moreover, the regulatory factor FliZ that is expressed under the control of FlhDC and thus belongs to the flagellar gene hierarchy, interferes with the expression of many σ^S-dependent genes in a not yet understood manner. FliZ thereby gives priority to FlhDC-driven motility over the σ^S-dependent general stress response (91).

This picture changes at a distinct point of the growth cycle that corresponds to an optical density of approximately 2 in an LB-grown *E. coli* culture (Fig. 1). At this point, σ^S is approaching the level typical for stationary-phase cells (71) and accessory factors such as Crl (10, 96, 131), the anti-σ^{70} factor Rsd (55, 56), and (p)ppGpp (57) influence sigma factor competition for the RNAP core in favor of σ^S-containing RNAP holoenzyme formation. This results in the activation of σ^S-dependent genes in a fine-tuned temporal order that probably reflects the affinity of their promoters for σ^S-containing RNAP (91). Due to RNAP core limitation (42, 43, 77), the cellular levels of vegetative and flagellar RNAP holoenzymes must be reduced in parallel, and this process may actually be further enhanced by the (p)ppGpp-driven induction of the stress sigma subunit σ^E (26).

Concomitantly and possibly as a consequence of the now low levels of vegetative RNAP, the expression of FlhDC comes to an end (6, 91). The existing FlhDC protein complex is degraded by ClpXP (91, 129). Together with FlgM-dependent inactivation (21) and Lon-mediated degradation of flagellar sigma factor σ^{FliA} (6), this results in a cessation of flagellar gene expression (91). Moreover, cells begin to reduce their swimming speed, i.e., modify the activity of existing flagella (see below), and finally become nonmotile later in the stationary phase, which probably also reflects energy starvation (1, 4, 91).

In parallel, σ^S-driven gene expression (68, 89, 138) results in (i) morphological changes (cells become shorter and ovoid rather than rod-shaped) (72), (ii) drastic metabolic alterations (toward a more anaerobic/fermentative energy metabolism) (138), (iii) the development of multiple and strong stress resistances (73, 79), and (iv) cellular adhesion via curli fimbriae (if cells are grown below 30°C) (44, 87, 98).

Role of c-di-GMP in the Inverse Coordination of Motility and Curli-Mediated Adhesion

Where in this complex growth cycle scenario does c-di-GMP come into play? Downregulation of motility and induction of adhesive factors by c-di-GMP is a very general theme in c-di-GMP-mediated regulation (see various chapters in part III of this book). This already suggested that c-di-GMP plays a key role during transition from the postexponential to the stationary phase, which is fully consistent with increased expression of many GGDEF/EAL domain-encoding genes during this period of the growth cycle (121, 137; see above). Moreover, it suggested that downregulation of motility and induction of adhesion may not just coincide but may be directly coupled by c-di-GMP-mediated regulation. This mutual exclusivity of motility and adhesion would make obvious sense. For cells that have to be motile, being sticky would clearly be counterproductive, whereas becoming sessile on a surface may require switching off the propellers at some point (91). Also, in developing biofilms, motile and adhesive subpopulations of cells coexist and drive the formation of the complex biofilm architecture (66), and a regulatory mechanism that ensures that motility and adhesion are mutually exclusive may be crucial to establish and organize these subpopulations.

At the molecular level, this means that the regulatory cascades that control flagellar expression and function on the one hand and the expression of adhesive factors on the other should be inversely coordinated by c-di-GMP (Fig. 2). In *E. coli,* the flagellar master regulator FlhDC (expressed from the class 1 *flhDC* operon) activates the expression of class 2 genes, which encode the inner part of the flagellum (i.e., the hook basal body), the flagellar sigma factor σ^{FliA}, its antisigma factor FlgM that binds to and inactivates σ^{FliA}, the FliZ protein (which interferes with

σ^S-dependent gene expression; see above), and other proteins of unknown function. When the hook basal body is fully assembled, FlgM is exported and σ^{FliA} is released to now drive the expression of class 3 genes, which encode the outer parts of the flagella as well as accessory components required to render flagella functional (e.g., the Mot and Che proteins and chemoreceptors, etc.). In addition, FlgM expression is boosted by a class 3 promoter, such that σ^{FliA} becomes inactivated again (for reviews, see references 20, 21, and 141 and chapter 8). These processes happen with a precise temporal order during the first half of the postexponential growth phase (6). c-di-GMP control is built into this regulatory cascade by two genes with flagellar class 3, i.e., σ^{FliA}-recognized promoters (Fig. 2): *yhjH,* which encodes an EAL protein with high PDE activity, and *ycgR,* which encodes the c-di-GMP-binding effector that can interfere with flagellar activity (see above). Due to high expression during the first half of the postexponential phase, YhjH can keep c-di-GMP low and YcgR therefore remains inactive, i.e., YhjH promotes motility (Fig. 2). Accordingly, a *yhjH* mutant of *E. coli* shows strongly reduced motility, and this phenotype can be fully suppressed by a secondary mutation in *ycgR* (38, 67, 91, 112).

However, this situation changes when flagellar gene expression is shut off during the second half of the postexponential phase, i.e., when FlhDC expression ceases and FlhDC and σ^{FliA} (present in excess over FlgM) are degraded (6, 91, 129). This also means that de novo expression of YhjH and YcgR no longer takes place and cellular concentrations of these proteins are reduced by further cellular mass increase (about twofold until the stationary phase) (91). This allows certain DGCs to outbalance YhjH. c-di-GMP can now accumulate and, via YcgR, alter the flagellar rotation pattern and swimming speed (38, 91, 112). At 37°C, the main DGC involved is YeaJ, which is already expressed during the postexponential phase. At 28°C (which allows the expression of adhesive curli fimbriae slightly later), the main DGC is YegE, and YedQ plays a minor role independent of growth temperature. The *yegE* and *yedQ* genes exhibit basal expression in growing cells but are further activated by σ^S during the transition from the postexponential to the stationary phase, which contributes to outbalancing the PDE activity provided by YhjH (91) (Fig. 2).

In parallel, the expression of the curli/cellulose activator CsgD is activated. A *yegE* mutant is strongly reduced in the transcription of *csgD* and, consequently, the curli structural operon *csgBAC,* whereas a *yhjH* mutation produces the opposite phenotype. Thus, YegE/YhjH-controlled c-di-GMP not only downregulates motility but also stimulates *csgD* transcription (Fig. 2). This regulation is relatively direct, as *yegE* or *yhjH* mutations do not affect σ^S activity in general nor the expression of any factors known to be essential for *csgD* transcription. Interestingly, the c-di-GMP-binding effector that controls motility, YcgR, does not affect *csgD* transcription, indicating that the c-di-GMP involved serves more than one effector. Therefore, and consistent with several DGCs contributing to this pool, this c-di-GMP control module most likely operates via the cytoplasmic pool of freely diffusible c-di-GMP (91).

In conclusion, switching from motility to curli-mediated adhesion during the transition from the postexponential to the stationary phase of *E. coli* crucially involves a shift in the balance between c-di-GMP production and degradation. This shift depends on the major PDE (YhjH) being under control of the FlhDC/σ^{FliA} flagellar regulation cascade, which is switched off, and on the DGCs (mainly YegE) that are induced under the control of the general stress sigma factor σ^S (Fig. 2). As YegE/YhjH-controlled c-di-GMP inhibits flagellar activity and activates curli expression, this mechanism provides a checkpoint that allows curli expression only after successful shutoff of the flagellar control cascade and thereby results in an obligatory inverse coordination of motility and curli-mediated adhesion.

Integration of Several c-di-GMP Control Modules in the Expression of Adhesive Curli Fimbriae

Besides the YegE/YhjH system, a second c-di-GMP control module regulates CsgD and curli expression (Fig. 2). This system consists of the DGC YdaM and the PDE YciR. During entry into the stationary phase, the *ydaM* gene is activated under exclusive control of σ^S, whereas *yciR* exhibits earlier basal expression already, which, however, is also further enhanced by σ^S (121, 137). Interestingly, the YegE/YhjH and YdaM/YciR systems do not act just additively. Whereas the former affects both motility and *csgD* transcription, the latter seems to exclusively control *csgD* transcription (unlike the *yegE* mutation, a *ydaM* mutation does not suppress the reduced motility of a *yhjH* mutant; also, the *yciR* mutant does not have a motility phenotype as does the *yhjH* mutant). Moreover, a *yegE* mutation reduces *csgD* expression by about 60 to 80%, whereas the *ydaM* knockout virtually eliminates it (even though YegE is present in this strain and is active, as it is involved in downregulation of motility also in a *ydaM yhjH* double mutant). Also, single mutations in either *yhjH* or *yciR* maximally activate *csgD* transcription, with no further stimulation in the double mutant. Taken together, the DGC YdaM is essential and highly specific for CsgD expression, and it was therefore suggested that the YdaM/YciR system directly controls *csgD* transcription in a local manner (in contrast to the YegE/YhjH system which seems to control freely diffusible c-di-GMP; see above) (47, 91). This is consistent with a direct interaction of YdaM and YciR, as recently observed by two-hybrid analysis (S. Lindenberg and R. Hengge, unpublished data). The c-di-GMP-binding effector in this module has not been identified, but a candidate is MlrA. This MerR-like transcription factor (14) is equally essential for and exclusively acting on *csgD* transcription (15, 138). Currently, it is unknown how the two c-di-GMP control modules are integrated in the control of CsgD and therefore curli expression, but the data summarized above suggest that YegE/YhjH (and therefore freely diffusible c-di-GMP) somehow acts upstream of YdaM/YciR (91).

c-di-GMP-mediated regulation is also observed downstream of *csgD* transcription (Fig. 2). A mutation in *yeaP,* which encodes an active DGC (113), reduces the expression of *csgB::lacZ,* i.e., the curli structural gene operon, by about 50% but has no significant effect on CsgD levels or the kinetics of CsgD accumulation during entry into the stationary phase (121). At the moment, it is unclear how YeaP, and therefore probably YeaP-synthesized c-di-GMP, can stimulate the expression of the *csgBAC* operon without affecting *csgD* transcription, as usually *csgBAC* expression immediately follows CsgD expression.

Another DGC, YaiC, is itself expressed under the control of CsgD and σ^S-containing RNAP. The *yaiC* gene is activated approximately 2 hours later than *ydaM* when cells enter into the stationary phase (121, 137). As its homolog in *Salmonella,* AdrA (106, 107), YaiC probably provides the c-di-GMP required to activate cellulose biosynthesis.

In summary, a general picture is currently emerging (Fig. 2) in which the signaling activities of several c-di-GMP control modules are integrated above and toward the expression of CsgD (91, 121). As a master regulator for entry into the adhesive state and, consequently, biofilm formation (12, 13, 59, 97), CsgD appears like an informational bottleneck. Further regulatory input that converges onto CsgD expression comes from OmpR and the Cpx and Rcs two-component systems (60, 97, 132, 133). Below CsgD, additional c-di-GMP control modules then seem to control various specific functions required for building up and shaping a complex biofilm.

Finally, it should be noted that the general picture of the regulation of CsgD and curli fimbriae expression is similar in *Salmonella* (104). However, there are interesting differences in the details of c-di-GMP signaling as can already be expected from the different numbers of GGDEF/EAL domain proteins present in the two organisms (Table 1) (see also chapter 10). *Salmonella* does not possess the DGC YdaM, which seems functionally replaced by a GGDEF+EAL protein (STM3388) that is not present in *E. coli* (61). Also, a *Salmonella* mutant deficient in YhjH is only marginally reduced in motility, and another PDE, YjcC (STM4264), but not YhjH, is involved in the control of CsgD expression (120). In *E. coli,* YjcC does not play such a role, perhaps because it is expressed under exclusive σ^S control only later during entry into the stationary phase (121). The DGC YeaP, which seems to modulate curli expression downstream of CsgD (121), does not exist in *Salmonella,* but there is evidence for CsgD-independent fine-tuning of curli expression by the DGC AdrA (61).

Role of c-di-GMP in Exopolysaccharide Biosynthesis

A variety of exopolysaccharides contribute to the bacterial biofilm matrix (7, 11). In *E. coli,* these include cellulose, β-1,6-*N*-acetyl-D-glucosamine (poly-GlcNAc or PGA) (2, 52), and colanic acid, a complex polymer consisting of glucose, galactose, fucose, and glucuronic acid (27, 45, 99). While cellulose and PGA production are regulated by c-di-GMP, currently there seems to be no such evidence for colanic acid, which is produced under the control of the Rcs two-component system during the maturation of biofilms (78, 98).

Cellulose biosynthesis was the very first c-di-GMP-controlled function discovered (108, 126, 139), with the BcsA protein providing a paradigm for a PilZ-like c-di-GMP-binding effector that directly affects an enzymatic activity, i.e., cellulose synthesis (3). In *E. coli,* the GGDEF domain protein and putative DGC YaiC (AdrA in *Salmonella*) provides the c-di-GMP required for cellulose biosynthesis, with the EAL domain protein and putative PDE YoaD playing an antagonistic role (12, 13). The expression of *yaiC* occurs during entry into the stationary phase and only below 30°C, is further stimulated by surface growth, and is under the control of the σ^S/YdaM/MlrA/CsgD control cascade (Fig. 2) (12, 13, 121, 137). The similarly stationary phase-induced *yoaD* gene equally requires σ^S for expression, but while CsgD overproduction apparently downregulates *yoaD* (12), a *csgD* knockout does not seem to affect its expression (N. Sommerfeldt and R. Hengge, unpublished data).

The *pgaABCD* operon encodes the proteins required for synthesis, modification, and secretion of the adhesive exopolysaccharide PGA (2, 39, 52) (see chapter 14). PGA formation is strongly inhibited by CsrA, which binds to *pgaABCD* mRNA where it interferes with translation and promotes degradation (135). The production of PGA seems to be inversely coordinated with that of cellulose. CsgD and *yaiC*, which control cellulose synthesis, are downregulated by high osmolarity via the OmpR and CpxR response regulators (60), whereas the *pgaABCD* operon is activated under high-salt conditions by NhaR (40). Similarly, PGA-dependent biofilm formation is induced in a ppGpp-free strain or by sublethal doses of translation-inhibitory antibiotics, which lower the cellular ppGpp level (8). By contrast, σ^S and, therefore, the CsgD/curli/cellulose pathway are under positive ppGpp control (70). Also, the low-temperature activation of CsgD and *yaiC* suggests a role for cellulose mainly in biofilms formed in an extrahost environment, whereas PGA is produced at 37°C (2) and contributes to virulence within a mammalian host (17).

In addition, PGA production is positively controlled by c-di-GMP. Besides its direct effect on *pgaABCD* mRNA (see above), CsrA also downregulates the expression of two GGDEF domain proteins, YcdT and YdeH (39, 58). YdeH is essential for increased PGA formation in response to translation-inhibitory antibiotics (8). The gene encoding YcdT is located next to the *pgaABCD* locus on the *E. coli* chromosome with a divergent regulatory region in between. Interestingly, *Salmonella* (as well as some other gammaproteobacteria) does not have the *pga* genes nor YcdT but a distant homolog of YcdT with a different sensory input region (STM4551) (Table 1).

In conclusion, a whole series of global regulatory factors (including CsrA, σ^S, CsgD, and the Cpx system) antagonistally regulate PGA and curli/cellulose and, therefore, the different types of *E. coli* biofilms formed in a mammalian host or in an outside environment. c-di-GMP, however, always seems to play a positive role, but different DGCs are involved, the expression of which is again antagonistically controlled.

Role of c-di-GMP in Virulence-Associated Functions

In some pathogens, a role for c-di-GMP in the control of virulence genes is well established (e.g., see chapter 20). In pathogenic *E. coli*, c-di-GMP signaling has hardly been studied so far. A recent exception is a Crohn disease-associated adherent-invasive *E. coli* strain, in which type 1 pili were found to be coregulated with flagella. Their expression depends on the PDE YhjH (25), which is part of the flagellar gene hierarchy. Thus, type 1 pili seem to require low c-di-GMP levels for expression. Flagella and type 1 fimbriae also contribute to virulence of uropathogenic *E. coli* (63), but in these pathogens, manipulation of flagellar gene control does not alter the expression of type 1 fimbriae (whereas overexpression of type 1 fimbriae strongly interferes with motility and flagellum expression) (69). It should be noted that flagella and type 1 fimbriae are also involved in the initiation of biofilm formation on certain abiotic surfaces (95, 100), which is a reminder that not all of the complex processes that contribute to biofilm formation are necessarily positively c-di-GMP controlled. In any case, as motility, adhesion, and biofilm formation contribute in various ways to the virulence of pathogenic *E. coli,* both in acute and chronic infections (7), the role of c-di-GMP in pathogenicity should be studied in more detail.

BIOLOGICAL ACTIVITIES OF ENZYMATICALLY INACTIVE GGDEF/EAL DOMAIN PROTEINS

E. coli has four GGDEF/EAL proteins in which the key amino acids required for DGC or PDE activities are not conserved (Table 2) and which, therefore, should have alternative molecular functions. The GGDEF domain protein YeaI, which features the secondary binding site for c-di-GMP (I-site) but not the DGC active site (A-site), may act as a c-di-GMP-binding effector controlling an unknown function. By contrast, molecular functions for the EAL-only protein YcgF and the GGDEF+EAL protein YhdA have recently been identified. For the most strongly degenerate EAL domain protein, YdiV, at least a physiological context of function is currently becoming apparent.

The Blue Light Sensor and EAL Protein YcgF Acts as a Direct Antirepressor

The N-terminal BLUF domain of the EAL protein YcgF is a well-studied blue light sensor (46, 83, 101, 115, 144), but the direct regulatory output and physiological function of YcgF remained unknown until recently. All four amino acids involved in c-di-GMP binding of EAL domains, an essential catalytic glutamic acid as well as several other amino acids involved in PDE activity (102), are not conserved in *E. coli* YcgF (130). Consistently, purified YcgF does not degrade or bind to c-di-GMP but instead can directly bind to the MerR-like repressor YcgE, which is encoded downstream of *ycgF* in the *E. coli* chromosome. YcgE represents a paralog of the regulator MlrA, which in CsgD/curli control cooperates with the DGC/PDE system YdaM/YciR (see above). Upon blue light irradiation, YcgF induces a release of YcgE from its operator DNA in vitro (130). The blue light-induced conformational change of YcgF (46, 83, 144) probably increases the affinity of YcgF for YcgE. In addition, derepression of YcgE-controlled genes in vivo is also observed when the YcgF:YcgE stoichiometry increases, i.e., upon YcgF overproduction or shift to low temperature (16°C), where above all YcgF is strongly induced (YcgE is induced to a minor extent). Blue light irradiation and shift to low temperature act additively in this system (130). Thus, the EAL protein YcgF is a blue light-modulated direct antagonist of the repressor YcgE (Fig. 3).

Interestingly, in the bacterial species that are known to possess YcgF, several evolutionary intermediates between an active PDE and the antirepressor protein are present (*E. coli* YcgF protein is the most degenerate with respect to residues essential for PDE activity). In *Bordetella avium* and *Alteromonas macleodii,* where YcgE and the YcgE-controlled target genes (see below) do not exist, the YcgF sequence suggests it to be a blue light-controlled PDE. In *Klebsiella pneumoniae* and *Enterobacter sakazakii,* which do have YcgE, YcgF is degenerate, but less so than in *E. coli,* and possibly could still have residual PDE activity (see supplement to reference 130).

In *E. coli,* the YcgF/YcgE system controls a distinct set of eight genes that encode small proteins (71 to 126 amino acids) (Fig. 3). YcgE directly binds to the promoter region of the *ycgZ-ymgABC* operon (which is also σ^S controlled). For *yliL* and *ynaK,* the mode of regulation is not yet clarified, and the equally σ^S-dependent genes *bdm* and *ybgS* are regulated indirectly via YmgB (Fig. 3) (130). Mutants defective in *ymgA, ymgB,* and *ymgC* exhibit increased biofilm formation, and the *ymgB* mutant also shows reduced survival at pH 2.5 (75). YmgB and, to a mi-

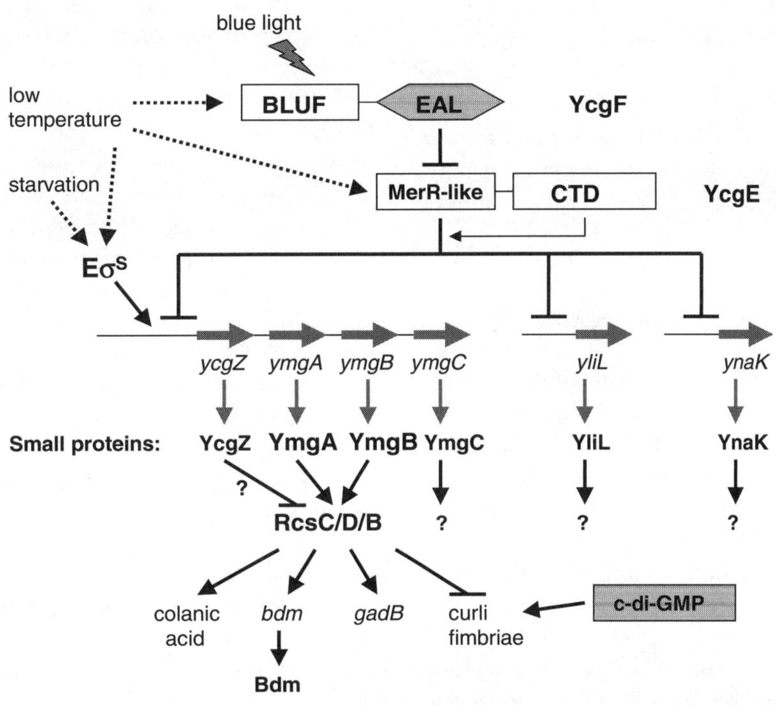

Figure 3. The degenerate EAL protein YcgF is a blue light-modulated antirepressor and controls a small-protein and two-component network that modulates biofilm formation. By direct protein-protein interaction, blue light-irradiated YcgF (or excess nonirradiated YcgF) releases the MerR-like repressor YcgE from its operator site upstream of the *ycgZ-ymgABC* operon (also *yliL* and *ynaK* are derepressed, but there the mechanism has not yet been studied). All gene products of these target genes are small proteins (78 to 90 amino acids). YmgB (and YmgA and YcgZ, to minor extents) modulates the biofilm-associated functions indicated via the two-component phosphorelay system RcsC/RcsD/RcsB. The *bdm* gene also encodes a small protein of unknown molecular function (71 amino acids). In addition, *ybgS* (encoding a small protein of 121 amino acids) is also under control of the YcgF/YcgE/YmgB cascade (*ybgS* is not included in the figure, as the mechanism of this regulation has not yet been clarified). At the levels of protein activity as well as gene expression, this system integrates blue light, low temperature, starvation, and probably other stress signals and, therefore, seems important for *E. coli* in an extrahost and probably aquatic environment. For further details and references, see the text. Adapted from reference 130 with permission.

nor degree, YmgA activate the production of the biofilm matrix substance colanic acid as well as the expression of acid resistance genes and downregulate adhesive curli fimbriae. All of these effects require the RcsC/RcsD/RcsB two-component phosphorelay system, suggesting that the small proteins YmgB and YmgA modulate the activity of this two-component signaling pathway (130). Small regulatory proteins with such activity have been termed connectors, as they provide additional signal input into two-component systems (32, 64, 65), and it is conceivable that other small proteins under YcgF/YcgE control have similar activities. As observed by microarray analysis, the *ycgZ-ymgABC* genes also belong to a group of genes with intermediate timing of expression during biofilm development (30). This would be consistent with YmgB and YmgA inducing cessation of curli expression and stimulating colanic acid production during biofilm maturation. Also, the *ycgZ-ymgABC* operon is among the genes downregulated by the indole/SdiA pathway, which interferes with the formation of thick biofilms (74), for which colanic acid is crucial (98).

In conclusion, the YcgF/YcgE/small protein signaling pathway modulates biofilm formation via the two-component signaling network of *E. coli.* As blue light, cold, and starvation signals are integrated in this pathway, one may predict that this system plays a role in biofilm development in an extrahost aquatic environment (130).

Role of YhdA (CsrD) in Small RNA Turnover

YhdA (CsrD) is the only *E. coli* GGDEF+EAL domain protein which lacks key amino acids for DGC/PDE activity in both domains (Table 2). In vivo, YhdA(CsrD) stimulates rapid degradation of the small RNAs CsrB and CsrC, which can sequester the RNA-binding protein CsrA. Via this pathway, YhdA interferes with PGA production and biofilm formation and activates expression of CsrB and CsrC in a feedback circuit. As the turnover of CsrB and CsrC RNAs requires RNaseE and YhdA does not have RNase activity but can bind CsrB and CsrC RNAs with high affinity, it is possible that YhdA targets these small RNAs to RNaseE (125) (see also chapter 14). In addition, a knockout mutation in *yhdA* reduces the expression of curli fimbriae, as it affects the expression of the DGC YdaM as well as the regulator CsgD (121).

There is tentative evidence suggesting that some EAL domain proteins that do have PDE activity may also be involved in RNA turnover. A mutation in YciR (termed *gmr*) was found to increase RNase II levels by eliminating proteolytic turnover of this RNase, which was proposed to have consequences for cellular regulation (16). Given our current knowledge that YciR is a PDE and negatively controls expression of the regulator CsgD (137), elevated levels of either c-di-GMP or CsgD could be responsible for slowing down RNase II proteolysis in the *yciR* mutant. Finally, it should be mentioned that the EAL domain protein and putative PDE YcgG copurified with the degradosome under a variety of conditions (103). The molecular and physiological role of YcgG has not been studied, but after all, c-di-GMP is a cyclic dinucleotide RNA molecule and the possibility that members of a PDE family that cleaves c-di-GMP may have evolved to adopt additional roles in RNA turnover would not be too surprising.

Is YdiV a Modulator of Biofilm Architecture?

Knockout mutations eliminating the degenerate EAL-only protein YdiV have a minor stimulatory effect on the motility of planktonic cells of *E. coli* or *Salmonella* (91, 120, 121). However, ectopic (over)expression of *ydiV* can downregulate flagellar gene expression both in *Salmonella* (143) and in *E. coli* (N. Tschowri and R. Hengge, unpublished data). YdiV seems to act at a step after class 1 (i.e., *flhDC*) transcription by an unknown mechanism (143). The natural expression of *ydiV* is under positive control of the quorum-sensing regulator SdiA (147). In response to indole, which is produced by *E. coli* during the stationary phase and activates *sdiA* expression, and to quorum-sensing signals produced by other species, SdiA can downregulate motility. During biofilm formation, this leads to flat colonies rather than a thick biofilm characterized by scattered towers of bacteria (74, 142). In future studies, it should be tested whether YdiV plays a role in this SdiA-mediated modulation of biofilm architecture.

CONCLUSIONS AND PERSPECTIVES

c-di-GMP signaling in bacterial lifestyle control has been a focus of research for only a few years but has already provided fundamental new insights into the molecular biology, physiology, and even ecology of second messenger signaling in bacteria (recently summarized in reference 92). The striking multiplicity of the enzymes that make and break c-di-GMP has led to the concepts of (i) the c-di-GMP module, (ii) local signaling based on protein-protein interactions within modules that can act in parallel, and (iii) the evolutionary emergence of degenerate GGDEF/EAL domain proteins that no longer operate via c-di-

GMP but rely on macromolecular interactions only (47). Thus, a simple, diffusible second messenger (like cAMP in enteric bacteria) seems to represent an ancient signaling system, and the analysis of c-di-GMP signaling is revealing how complex and sophisticated systems can evolve from this simple origin (Fig. 4). The multiplicity of adenylate cyclases in mycobacteria or certain alphaproteobacteria (35, 117, 119) suggests that in these bacteria, cAMP signaling may have evolved to similar complexity.

Bacteria precisely tailor their use of c-di-GMP and therefore swim-or-stick lifestyle decisions to their ecological niches. It can be integrated into the control of the growth cycle (as, e.g., in *E. coli*), of a cell cycle that is obligatorily coupled to lifestyle switching (as, e.g., in *C. crescentus*), or of complex adaptation to journeys between a mammalian host and the environment (as, e.g., in *V. cholerae*) (see chapters 8, 9, 17, and 20). Interestingly, even degenerate GGDEF/EAL domain proteins, which do not have DGC/PDE activities but exhibit alternative biochemical activities, seem to remain within the same physiological context, i.e., still modulate motility and/or biofilm formation.

With a wealth of genome sequences now available, it has become apparent that c-di-GMP signaling indeed undergoes rapid evolution. Closely related species and even strains of the same species exhibit different sets of GGDEF/EAL domain proteins and use orthologs of these proteins in different regulatory circuits, suggesting that horizontal gene transfer and changes in regulation of expression are crucial for the rapid evolution of c-di-GMP signaling. Also, species that can flexibly adapt to many environments tend to have higher numbers of GGDEF/EAL domain proteins than related pathogenic species that are more tightly associated with their hosts.

A generalist such as *E. coli,* which equally thrives within a mammalian host and in various outside environments and, with 29, has a large, but still easily tractable number of GGDEF/EAL domain proteins, seems perfect for elucidating the complete systems biology of c-di-GMP signaling in a well-characterized model organism.

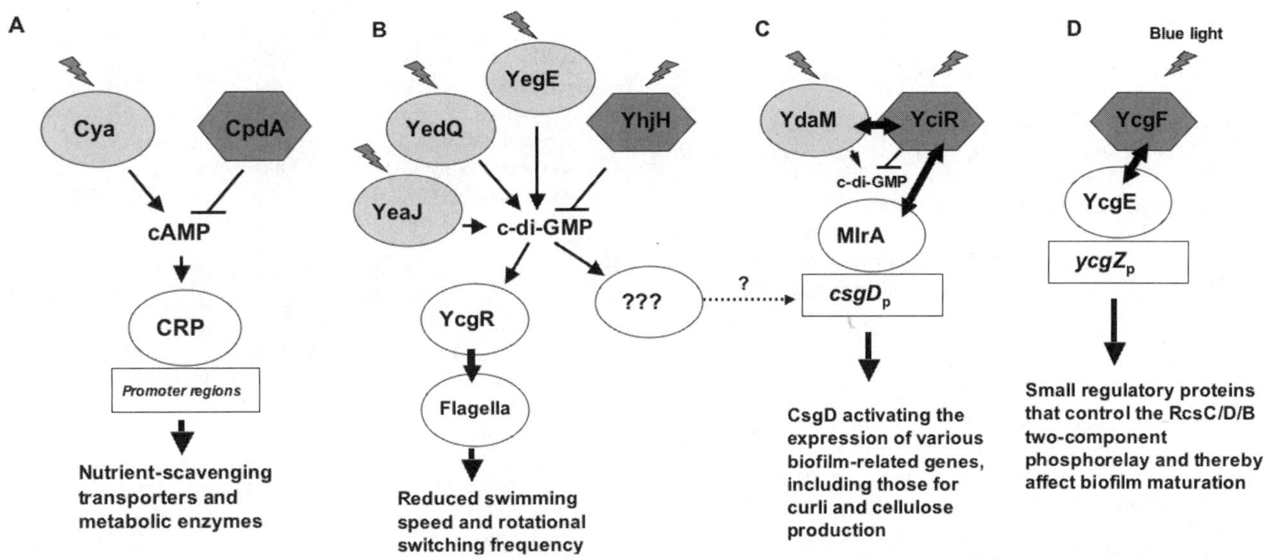

Figure 4. Model for the evolution of second messenger signaling as illustrated by systems operating in *E. coli*. The simplest and probably most ancient form of second messenger signaling is represented by the *E. coli* cAMP signaling system (A) (9, 92). Here, only one second messenger-producing enzyme (the adenylate cyclase Cya), one second messenger-degrading enzyme (the cAMP PDE CpdA), and one effector component (the transcription factor CRP) control a variety of promoter regions that all feature similar cAMP-CRP binding sites. More complex systems that integrate many more signals use multiples of all the components involved, as illustrated by c-di-GMP signaling in the control of flagellar activity and *csgD* transcription (B). In this system, several DGCs generate c-di-GMP, which in turn serves at least two effector components, indicating that this c-di-GMP is freely diffusible (91). The multiplicity of DGC, PDEs, effectors, and targets also allows the evolution of locally acting systems, in which complexes consisting of a DGC, a PDE, and an effector highly specifically interact with distinct targets (C). An example seems to be the YdaM/YciR system which specifically regulates *csgD* transcription (91, 137) (the tentative protein interactions indicated in the figure are based on unpublished in vivo and in vitro data by S. Lindenberg, H. Weber, and R. Hengge). Finally, the YcgF/YcgE system (which is a paralog of YciR/MlrA) is an example of a locally acting system in which c-di-GMP signaling has been lost and which relies entirely on protein-protein interactions (130) (D). Adenylate cyclases or DGCs are shown as light gray ovals; PDEs (or proteins derived thereof, such as YcgF) are shown as dark gray hexagons.

REFERENCES

1. **Adler, J., and B. Templeton.** 1967. The effect of environmental conditions on the motility of *Escherichia coli. J. Gen. Microbiol.* **46:**175–184.
2. **Agladze, K., X. Wang, and T. Romeo.** 2005. Spatial periodicity of *Escherichia coli* K-12 biofilm microstructure initiates during a reversible, polar attachment phase of development and requires the polysaccharide adhesin PGA. *J. Bacteriol.* **187:**8237–8246.
3. **Amikam, D., and M. Y. Galperin.** 2006. PilZ domain is part of the bacterial c-di-GMP binding protein. *Bioinformatics* **22:** 3–6.
4. **Amsler, C. D., M. Cho, and P. Matsumura.** 1993. Multiple factors underlying the maximum motility of *Escherichia coli* as cultures enter post-exponential growth. *J. Bacteriol.* **175:** 6238–6244.
5. **Baker, D. A., and J. M. Kelly.** 2004. Structure, function and evolution of microbial adenylyl and guanylyl cyclases. *Mol. Microbiol.* **52:**1229–1242.
6. **Barembruch, C., and R. Hengge.** 2007. Cellular levels and activity of the flagellar sigma factor FliA of *Escherichia coli* are controlled by FlgM-modulated proteolysis. *Mol. Microbiol.* **65:**76–89.
7. **Beloin, C., A. Roux, and J.-M. Ghigo.** 2008. *Escherichia coli* biofilms. *Curr. Top. Microbiol. Immunol.* **322:**249–289.
8. **Boehm, A., S. Steiner, F. Zaehringer, A. Casanova, F. Hamburger, D. Ritz, W. Keck, M. Ackerman, T. Schirmer, and U. Jenal.** 2009. Second messenger signaling governs *Escherichia coli* biofilm induction upon ribosomal stress. *Mol. Microbiol.* **72:**1500–1516.
9. **Botsford, J. L., and J. G. Harman.** 1992. Cyclic AMP in prokaryotes. *Microbiol. Rev.* **56:**100–122.
10. **Bougdour, A., C. Lelong, and J. Geiselmann.** 2004. Crl, a low temperature-induced protein in *Escherichia coli* that binds directly to the stationary phase sigma subunit of RNA polymerase. *J. Biol. Chem.* **279:**19540–19550.
11. **Branda, S. S., A. Vik, L. Friedman, and R. Kolter.** 2005. Biofilms: the matrix revisited. *Trends Microbiol.* **13:**20–26.
12. **Brombacher, E., A. Baratto, C. Dorel, and P. Landini.** 2006. Gene expression regulation by the curli activtor CsgD protein: modulation of cellulose biosynthesis and control of negative determinants for microbial adhesion. *J. Bacteriol.* **188:**2027–2037.
13. **Brombacher, E., C. Dorel, A. J. B. Zehnder, and P. Landini.** 2003. The curli biosynthesis regulator CsgD co-ordinates the expression of both positive and negative determinants for biofilm formation in *Escherichia coli. Microbiology* **149:**2847–2857.
14. **Brown, N. L., J. V. Stoyanov, S. P. Kidd, and J. L. Hobman.** 2003. The MerR family of transcriptional regulators. *FEMS Microbiol. Rev.* **27:**145–163.
15. **Brown, P. K., C. M. Dozois, C. A. Nickerson, A. Zuppardo, J. Terlonge, and R. Curtiss III.** 2001. MlrA, a novel regulator of curli (Agf) and extracellular matrix synthesis by *Escherichia coli* and *Salmonella enterica* serovar Typhimurium. *Mol. Microbiol.* **41:**349–363.
16. **Cairrao, F., A. Chora, R. Zilhao, A. J. Carpousis, and C. M. Arraiano.** 2001. RNase II levels change according to the growth conditions: characterization of *gmr,* a new *Escherichia coli* gene involved in the modulation of RNase II. *Mol. Microbiol.* **39:**1550–1561.
17. **Cerca, N., T. Maira-Litrán, K. K. Jefferson, M. Grout, D. A. Goldmann, and G. B. Pier.** 2007. Protection against *Escherichia coli* infection by antibody to the *Staphylococcus aureus* poly-N-acetylglucosamine surface polysaccharide. *Proc. Natl. Acad. Sci. USA* **104:**7528–7533.
18. **Chan, C., R. Paul, D. Samoray, N. Amiot, B. Giese, U. Jenal, and T. Schirmer.** 2004. Structural basis of activity and allosteric control of diguanylate cyclase. *Proc. Natl. Acad. Sci. USA* **101:**17084–17089.
19. **Chang, A. L., J. R. Tuckerman, G. Gonzalez, R. Mayer, H. Weinhouse, G. Volman, D. Amikam, M. Benziman, and M.-A. Gilles-Gonzalez.** 2001. Phosphodiesterase A1, a regulator of cellulose synthesis in *Acetobacter xylinum,* is a heme-based sensor. *Biochemistry* **40:**3420–3426.
20. **Chevance, F. F. V., and K. T. Hughes.** 2008. Coordinating assembly of a bacterial macromolecular machine. *Nat. Rev. Microbiol.* **6:**455–465.
21. **Chilcott, G. S., and K. T. Hughes.** 2000. Coupling of flagellar gene expression to flagellar assembly in *Salmonella enterica* serovar Typhimurium and *Escherichia coli. Microbiol. Mol. Biol. Rev.* **64:**694–708.
22. **Christen, B., M. Christen, R. Paul, F. Schmid, M. Folcher, P. Jenoe, M. Meuwly, and U. Jenal.** 2006. Allosteric control of cyclic di-GMP signaling. *J. Biol. Chem.* **281:**32015–32024.
23. **Christen, M., B. Christen, M. G. Allan, M. Folcher, P. Jenö, S. Grzesiek, and U. Jenal.** 2007. DgrA is a member of a new family of cyclic diguanosine monophosphate receptors and controls flagellar motor function in *Caulobacter crescentus. Proc. Natl. Acad. Sci. USA* **104:**4112–4117.
24. **Christen, M., B. Christen, M. Folcher, A. Schauerte, and U. Jenal.** 2005. Identification and characterization of a cyclic di-GMP-specific phosphodiesterase and its allosteric control by GTP. *J. Biol. Chem.* **280:**30829–30837.
25. **Claret, L., S. Miquel, N. Vieille, D. A. Ryjenkov, L. Gomelsky, and A. Darfeuille-Michaud.** 2007. The flagellar sigma factor FliA regulates adhesion and invasion of Crohn disease-associated *Escherichia coli* via a cyclic dimeric GMP-dependent pathway. *J. Biol. Chem.* **282:**33275–33283.
26. **Costanzo, A., and S. Ades.** 2006. Growth phase-dependent regulation of the extracytoplasmic stress factor, sigmaE by guanosine 3′,5′-bispyrophosphate (ppGpp). *J. Bacteriol.* **188:** 4627–4634.
27. **Danese, P., L. A. Pratt, and R. Kolter.** 2000. Exopolysaccharide production is required for development of *Escherichia coli* K-12 biofilm architecture. *J. Bacteriol.* **182:**3593–3596.
28. **De, N., M. Pirruccello, P. V. Krasteva, N. Bae, R. V. Raghavan, and H. Sondermann.** 2008. Phosphorylation-independent regulation of the diguanylate cyclase WspR. *PLoS Biol.* **6:**e67.
29. **Delgado-Nixon, V. M., G. Gonzalez, and M.-A. Gilles-Gonzalez.** 2000. Dos, a heme-binding PAS protein from *Escherichia coli,* is a direct oxygen sensor. *Biochemistry* **39:** 2685–2691.
30. **Domka, J., J. Lee, T. Bansal, and T. K. Wood.** 2007. Temporal gene expression in *Escherichia coli* K-12 biofilms. *Environ. Microbiol.* **9:**332–346.
31. **Duerig, A., M. Folcher, S. Abel, M. Nicollier, T. Schwede, N. Amiot, B. Giese, and U. Jenal.** 2009. Second messenger-mediated spatiotemporal control of protein degradation regulates bacterial cell cycle progression. *Genes Dev.* **23:**93–104.
32. **Eguchi, Y., J. Itou, M. Yamane, R. Demizu, F. Yamato, A. Okada, H. Mori, A. Kato, and R. Utsumi.** 2007. B1500, a small membrane protein, connects the two-component systems EvgS/EvgA and PhoQ/PhoP in *Escherichia coli. Proc. Natl. Acad. Sci. USA* **104:**18712–18717.
33. **Frye, J., J. E. Karlinsey, H. R. Felise, B. Marzolf, N. Dowidar, M. McClelland, and K. T. Hughes.** 2006. Identification of new flagellar genes of *Salmonella enterica* serovar Typhimurium. *J. Bacteriol.* **188:**2233–2243.

34. **Galperin, M. Y.** 2004. Bacterial signal transduction network in a genomic perspective. *Environ. Microbiol.* **6:**552–567.
35. **Galperin, M. Y.** 2005. A census of membrane-bound and intracellular signal transduction proteins in bacteria: bacterial IQ, extroverts and introverts. *BMC Microbiol.* **5:**35.
36. **Galperin, M. Y., A. N. Nikolskaya, and E. V. Koonin.** 2001. Novel domains of the prokaryotic two-component signal transduction systems. *FEMS Microbiol. Lett.* **203:**11–21.
37. **Gentry, D. R., V. J. Hernandez, L. H. Nguyen, D. B. Jensen, and M. Cashel.** 1993. Synthesis of the stationary-phase sigma factor σ^S is positively regulated by ppGpp. *J. Bacteriol.* **175:** 7982–7989.
38. **Girgis, H. S., Y. Liu, W. S. Ryu, and S. Tavazoie.** 2007. A comprehensive genetic characterization of bacterial motility. *PLoS Genetics* **3:**e154.
39. **Goller, C., and T. Romeo.** 2008. Environmental influences on biofilm development. *Curr. Top. Microbiol. Immunol.* **322:**37–66.
40. **Goller, C., X. Wang, Y. Itoh, and T. Romeo.** 2006. The cation-responsive protein NhaR of *Escherichia coli* activates pgaABCD transcription, required for production of the biofilm adhesin poly-beta-1,6-*N*-acetyl-D-glucosamine. *J. Bacteriol.* **188:**8022–8032.
41. **Gomelsky, M., and G. Klug.** 2002. BLUF: a novel FAD-binding domain involved in sensory transduction in microorganisms. *Trends Biochem. Sci.* **27:**497–500.
42. **Grigorova, I. R., N. J. Phleger, V. K. Mutalik, and C. A. Gross.** 2006. Insights into transcriptional regulation and sigma competition from an equilibrium model of RNA polymerase binding to DNA. *Proc. Natl. Acad. Sci. USA* **103:**5332–5337.
43. **Gruber, T., and C. A. Gross.** 2003. Multiple sigma subunits and the partitioning of bacterial transcription space. *Annu. Rev. Microbiol.* **57:**441–466.
44. **Hammar, M., A. Arnquist, Z. Bian, A. Olsén, and S. Normark.** 1995. Expression of two *csg* operons is required for production of fibronectin- and Congo red-binding curli polymers in *Escherichia coli* K-12. *Mol. Microbiol.* **18:**661–670.
45. **Hanna, A., M. Berg, V. Stout, and A. Razatos.** 2003. Role of capsular colanic acid in adhesion of uropathogenic *Escherichia coli*. *Appl. Environ. Microbiol.* **69:**4474–4481.
46. **Hasegawa, K., S. Masuda, and T. A. Ono.** 2006. Light induced structural changes of a full-length protein and its BLUF domain in YcgF (Blrp), a blue-light sensing protein that uses FAD (BLUF). *Biochemistry* **45:**3785–3793.
47. **Hengge, R.** 2009. Principles of cyclic-di-GMP signaling. *Nat. Rev. Microbiol.* **7:**263–273.
48. **Hengge-Aronis, R.** 2000. The general stress response in *Escherichia coli,* p. 161–178. *In* G. Storz and R. Hengge-Aronis (ed.), *Bacterial Stress Responses*. ASM Press, Washington, DC.
49. **Hickman, J. W., and C. S. Harwood.** 2008. Identification of FleQ from *Pseudomonas aeruginosa* as a c-di-GMP-responsive transcription factor. *Mol. Microbiol.* **69:**376–389.
50. **Holland, K., S. J. Busby, and G. S. Lloyd.** 2007. New targets for the cyclic AMP receptor protein in the *Escherichia coli* K-12 genome. *FEMS Microbiol. Lett.* **274:**89–94.
51. **Hurley, J. H.** 2003. GAF domains: cyclic nucleotides come full circle. *Sci. STKE* **164:**pe1.
52. **Itoh, Y., J. D. Rice, C. Goller, A. Pannuri, J. Taylor, J. Meisner, T. J. Beveridge, J. F. Preston III, and T. Romeo.** 2008. Roles of *pgaABCD* genes in synthesis, modification, and export of the *Escherichia coli* biofilm adhesin poly-β-1,6-*N*-acetyl-D-glucosamine. *J. Bacteriol.* **190:**3670–3680.
53. **Jenal, U., and J. Malone.** 2006. Mechanisms of cyclic-di-GMP signaling in bacteria. *Annu. Rev. Genet.* **40:**385–407.
54. **Jin, D. J., and J. E. Cabrera.** 2006. Coupling the distribution of RNA polymerase to global gene regulation and the dynamic structure of the bacterial nucleoid in *Escherichia coli*. *J. Struct. Biol.* **156:**284–291.
55. **Jishage, M., and A. Ishihama.** 1998. A stationary phase protein in *Escherichia coli* with binding activity to the major sigma subunit of RNA polymerase. *Proc. Natl. Acad. Sci. USA* **95:** 4953–4958.
56. **Jishage, M., and A. Ishihama.** 1999. Transcriptional organization and in vivo role of the *Escherichia coli rsd* gene, encoding the regulator of RNA polymerase sigma D. *J. Bacteriol.* **181:**3768–3776.
57. **Jishage, M., K. Kvint, V. Shingler, and T. Nyström.** 2002. Regulation of sigma factor competition by the alarmone ppGpp. *Genes Dev.* **16:**1260–1270.
58. **Jonas, K., A. N. Edwards, R. Simm, T. Romeo, U. Römling, and O. Melefors.** 2008. The RNA binding protein CsrA controls c-di-GMP metabolism by directly regulating the expression of GGDEF proteins. *Mol. Microbiol.* **70:**236–257.
59. **Jonas, K., H. Tomenius, A. Kader, S. Normark, U. Römling, L. B. Belova, and O. Melefors.** 2007. Roles of curli, cellulose and BapA in *Salmonella* biofilm morphology studied by atomic force microscopy. *BMC Microbiol.* **7:**70.
60. **Jubelin, G., A. Vianney, C. Beloin, J. M. Ghigo, J. C. Lazzaroni, P. Lejeune, and C. Dorel.** 2005. CpxR/OmpR interplay regulates curli gene expression in response to osmolarity in *Escherichia coli*. *J. Bacteriol.* **187:**2038–2049.
61. **Kader, A., R. Simm, U. Gerstel, M. Morr, and U. Römling.** 2006. Hierarchical involvement of various GGDEF domain proteins in rdar morphotype development of *Salmonella enterica* serovar *typhimurium*. *Mol. Microbiol.* **60:**602–616.
62. **Kalir, S., and U. Alon.** 2004. Using a quantitative blueprint to reprogram the dynamics of the flagella gene network. *Cell* **117:**713–720.
63. **Kaper, J. B., J. P. Nataro, and H. L. T. Mobley.** 2004. Pathogenic *Escherichia coli*. *Nat. Rev. Microbiol.* **2:**123–140.
64. **Kato, A., and E. A. Groisman.** 2004. Connecting two-component regulatory systems by a protein that protects a response regulator from dephosphorylation by its cognate sensor. *Genes Dev.* **18:**2302–2313.
65. **Kato, A., A. Y. Mitrophanov, and E. A. Groisman.** 2007. A connector of two-component regulatory systems promotes signal amplification and persistence of expression. *Proc. Natl. Acad. Sci. USA* **104:**12063–12068.
66. **Klausen, M., M. Gjermansen, J. U. Kreft, and T. Tolker-Nielsen.** 2006. Dynamics of development and dispersal in sessile microbial communities: examples from *Pseudomonas aeruginosa* and *Pseudomonas putida* model biofilms. *FEMS Microbiol. Lett.* **261:**1–11.
67. **Ko, M., and C. Park.** 2000. Two novel flagellar components and H-NS are involved in the motor function of *Escherichia coli*. *J. Mol. Biol.* **303:**371–382.
68. **Lacour, S., and P. Landini.** 2004. σ^S-dependent gene expression at the onset of stationary phase in *Escherichia coli:* function of σ^S-dependent genes and identification of their promoter sequences. *J. Bacteriol.* **186:**7186–7195.
69. **Lane, M. C., A. N. Simms, and H. L. Mobley.** 2007. Complex interplay between type 1 fimbrial expression and flagellum-mediated motility of uropathogenic *Escherichia coli*. *J. Bacteriol.* **189:**5523–5533.
70. **Lange, R., D. Fischer, and R. Hengge-Aronis.** 1995. Identification of transcriptional start sites and the role of ppGpp in the expression of *rpoS,* the structural gene for the σ^S subunit of RNA-polymerase in *Escherichia coli*. *J. Bacteriol.* **177:** 4676–4680.
71. **Lange, R., and R. Hengge-Aronis.** 1994. The cellular concentration of the σ^S subunit of RNA-polymerase in *Escherichia*

coli is controlled at the levels of transcription, translation and protein stability. *Genes Dev.* **8**:1600–1612.

72. **Lange, R., and R. Hengge-Aronis.** 1991. Growth phase-regulated expression of *bolA* and morphology of stationary phase *Escherichia coli* cells is controlled by the novel sigma factor σ^S *(rpoS)*. *J. Bacteriol.* **173**:4474–4481.
73. **Lange, R., and R. Hengge-Aronis.** 1991. Identification of a central regulator of stationary-phase gene expression in *Escherichia coli*. *Mol. Microbiol.* **5**:49–59.
74. **Lee, J., A. Jayaraman, and T. K. Wood.** 2007. Indole is an inter-species biofilm signal mediated by SdiA. *BMC Microbiol.* **7**:42.
75. **Lee, J., R. Page, R. García-Contreras, J.-M. Palermino, X.-S. Zhang, O. Doshi, T. K. Wood, and W. Peti.** 2007. Structure and function of the *Escherichia coli* protein YmgB: a protein critical for biofilm formation and acid resistance. *J. Mol. Biol.* **373**:11–26.
76. **Lin, Z., L. C. Johnson, H. Weissbach, N. Brot, M. O. Livel, and W. T. Lowther.** 2007. Free methionine-(R)-sulfoxide reductase from *Escherichia coli* reveals a new GAF domain function. *Proc. Natl. Acad. Sci. USA* **104**:9597–9602.
77. **Maeda, H., N. Fujita, and A. Ishihama.** 2000. Competition among seven Escherichia coli sigma subunits: relative binding affinities to the core RNA polymerase. *Nucleic Acids Res.* **28**: 3497–3503.
78. **Majdalani, N., and S. Gottesman.** 2005. The Rcs phosphorelay: a complex signal transduction system. *Annu. Rev. Microbiol.* **599**:379–405.
79. **McCann, M. P., J. P. Kidwell, and A. Matin.** 1991. The putative sigma factor KatF has a central role in development of starvation-mediated general resistance in *Escherichia coli*. *J. Bacteriol.* **173**:4188–4194.
80. **Méndez-Ortiz, M. M., M. Hyodo, Y. Hayakawa, and J. Membrillo-Hernández.** 2006. Genome wide transcription profile of *Escherichia coli* in response to high levels of the second messenger c-diGMP. *J. Biol. Chem.* **281**:8090–8099.
81. **Merighi, M., V. T. Lee, M. Hyodo, Y. Hayakawa, and S. Lory.** 2007. The second messenger bis-(3′-5′)-cyclic-GMP and its PilZ domain-containing receptor Alg44 are required for alginate biosynthesis in *Pseudomonas aeruginosa*. *Mol. Microbiol.* **65**:876–895.
82. **Minasov, G., S. Padavattan, L. Shuvalova, J. S. Brunzelle, D. J. Miller, A. Baslé, C. Massa, F. R. Collart, T. Schirmer, and W. F. Anderson.** 2009. Crystal structures of YkuI and its complex with second messenger cyclic di-GMP suggest catalytic mechanism of phosphodiester bond cleavage by EAL domains. *J. Biol. Chem.* **284**:13174–13184.
83. **Nakasone, Y., T. A. Ono, A. Ishii, S. Masuda, and M. Terazima.** 2007. Transient dimerization and conformational change of a BLUF protein: YcgF. *J. Am. Chem. Soc.* **129**:7028–7035.
84. **Newell, P. D., R. D. Monds, and G. A. O'Toole.** 2009. LapD is a bis-(3′,5′)-cyclic dimeric GMP-binding protein that regulates surface attachment by *Pseudomonas fluorescens* Pf0-1. *Proc. Natl. Acad. Sci. USA* **106**:3461–3466.
85. **Nikolskaya, A. N., A. Y. Mulkidjanian, I. B. Beech, and M. Y. Galperin.** 2003. MASE1 and MASE2: two novel integral membrane sensory domains. *J. Mol. Microbiol. Biotechnol.* **5**: 11–16.
86. **Oglesby, L., S. Jain, and D. E. Ohman.** 2008. Membrane topology and roles of *Pseudomonas aeruginosa* Alg8 and Alg44 in alginate polymerization. *Microbiology* **154**:1605–1615.
87. **Olsén, A., A. Jonsson, and S. Normark.** 1989. Fibronectin binding mediated by a novel class of surface organelles on *Escherichia coli*. *Nature* **338**:652–655.
88. **Partridge, J. D., G. Sanguinetti, D. P. Dibden, R. E. Roberts, R. K. Poole, and J. Green.** 2007. Transition of *Escherichia coli* from aerobic to micro-aerobic conditions involves fast and slow reacting regulatory components. *J. Biol. Chem.* **282**: 11230–11237.
89. **Patten, C. L., M. G. Kirchhhof, M. R. Schertzberg, R. A. Morton, and H. E. Schellhorn.** 2004. Microarray analysis of RpoS-mediated gene expression in *Escherichia coli* K-12. *Mol. Genet. Genomics* **272**:580–591.
90. **Paul, R., S. Weiser, N. Amiot, C. Chan, T. Schirmer, B. Giese, and U. Jenal.** 2004. Cell cycle-dependent dynamic localization of a bacterial response regulator with a novel diguanylate cyclase output domain. *Genes Dev.* **18**:715–727.
91. **Pesavento, C., G. Becker, N. Sommerfeldt, A. Possling, N. Tschowri, A. Mehlis, and R. Hengge.** 2008. Inverse regulatory coordination of motility and curli-mediated adhesion in *Escherichia coli*. *Genes Dev.* **22**:2434–2446.
92. **Pesavento, C., and R. Hengge.** 2009. Bacterial nucleotide-based second messengers. *Curr. Opin. Microbiol.* **12**:170–176.
93. **Peters, J. E., T. E. Thate, and N. L. Craig.** 2003. Definition of the *Escherichia coli* MC4100 genome by use of a DNA array. *J. Bacteriol.* **185**:2017–2021.
94. **Potrykus, K., and M. Cashel.** 2008. (p)ppGpp: still magical? *Annu. Rev. Microbiol.* **62**:35–51.
95. **Pratt, L. A., and R. Kolter.** 1999. Genetic analyses of bacterial biofilm formation. *Curr. Opin. Microbiol.* **2**:598–603.
96. **Pratt, L. A., and T. J. Silhavy.** 1998. Crl stimulates RpoS activity during stationary phase. *Mol. Microbiol.* **29**:1225–1236.
97. **Prigent-Combaret, C., E. Brombacher, O. Vidal, A. Ambert, P. Lejeune, P. Landini, and C. Dorel.** 2001. Complex regulatory network controls initial adhesion and biofilm formation in *Escherichia coli* via regulation of the *csgD* gene. *J. Bacteriol.* **183**:7213–7223.
98. **Prigent-Combaret, C., G. Prensier, T. T. Le Thi, Q. Vidal, P. Lejeune, and C. Dorel.** 2000. Developmental pathway for biofilm formation in curli-producing *Escherichia coli* strains: role of flagella, curli and colanic acid. *Environ. Microbiol.* **2**: 450–464.
99. **Prigent-Combaret, C., O. Vidal, C. Dorel, and P. Lejeune.** 1999. Abiotic surface sensing and biofilm-dependent regulation of gene expression in *Escherichia coli*. *J. Bacteriol.* **181**: 5993–6002.
100. **Prüβ, B. M., C. Besemann, A. Denton, and A. J. Wolfe.** 2006. A complex transcription network controls the early stages of biofilm development in *Escherichia coli*. *J. Bacteriol.* **188**:3731–3739.
101. **Rajagopal, S., J. M. Key, E. B. Purcell, D. J. Boerema, and K. Moffat.** 2004. Purification and initial characterization of a putative blue light-regulated phosphodiesterase from *Escherichia coli*. *Photochem. Photobiol.* **80**:542–547.
102. **Rao, F., Y. Yang, Y. Qi, and Z. X. Liang.** 2008. Catalytic mechanism of c-di-GMP specific phosphodiesterase: a study of the EAL domain-containing RocR from *Pseudomonas aeruginosa*. *J. Bacteriol.* **190**:3622–3631.
103. **Regonesi, M. E., M. Del Favero, F. Basilico, F. Briani, L. Benazzi, P. Tortora, P. Mauri, and G. Dehò.** 2006. Analysis of the *Escherichia coli* RNA degradosome composition by a proteomic approach. *Biochimie* **88**:151–161.
104. **Römling, U., Z. Bian, M. Hammar, W. D. Sierralta, and S. Normark.** 1998. Curli fibers are highly conserved between *Salmonella typhimurium* and *Escherichia coli* with respect to operon structure and regulation. *J. Bacteriol.* **180**:722–731.
105. **Römling, U., M. Gomelsky, and M. Y. Galperin.** 2005. C-di-GMP: the dawning of a novel bacterial signalling system. *Mol. Microbiol.* **57**:629–639.

106. **Römling, U., M. Rohde, A. Olsén, S. Normark, and J. Reinköster.** 2000. AgfD, the checkpoint of multicellular and aggregative behaviour in *Salmonella typhimurium* regulates at least two independent pathways. *Mol. Microbiol.* **36:**10–23.
107. **Römling, U., W. D. Sierralta, K. Eriksson, and S. Normark.** 1998. Multicellular and aggregative behaviour of *Salmonella typhimurium* strains is controlled by mutations in the *agfD* promoter. *Mol. Microbiol.* **28:**249–264.
108. **Ross, P., H. Weinhouse, Y. Aloni, D. Michaeli, P. Weinberger-Ohana, R. Mayer, S. Braun, E. de Vroom, G. A. van der Marel, J. H. van Boom, and M. Benziman.** 1987. Regulation of cellulose synthesis in *Acetobacter xylinum* by cyclic diguanylate. *Nature* **325:**279–281.
109. **Ryan, R. P., Y. Fouhy, F. Lucey, and J. M. Dow.** 2006. Cyclic di-GMP signaling in bacteria: recent advances and new puzzles. *J. Bacteriol.* **188:**8327–8334.
110. **Ryan, R. P., Y. Fouhy, J. F. Lucey, L. C. Crossman, S. Spiro, Y.-W. He, L.-H. Zhang, S. Heeb, P. Williams, and J. M. Dow.** 2006. Cell-cell signaling in *Xanthomonas campestris* involves an HD-GYP domain protein that functions in cyclic di-GMP turnover. *Proc. Natl. Acad. Sci. USA* **103:**6712–6717.
111. **Rychlik, I., G. Martin, U. Methner, M. Lovell, L. Cardova, A. Sebkova, M. Sevcik, J. Damborsky, and P. A. Barrow.** 2002. Identification of *Salmonella enterica* serovar Typhimurium genes associated with growth suppression in stationary-phase nutrient broth cultures and in the chicken intestine. *Arch. Microbiol.* **178:**411–420.
112. **Ryjenkov, D. A., R. Simm, U. Römling, and M. Gomelsky.** 2006. The PilZ domain is a receptor for the second messenger c-di-GMP: the PilZ protein YcgR controls motility in enterobacteria. *J. Biol. Chem.* **281:**30310–30314.
113. **Ryjenkov, D. A., M. Tarutina, O. V. Moskvin, and M. Gomelsky.** 2005. Cyclic diguanylate is a ubiquitous signaling molecule in bacteria: insights into biochemistry of the GGDEF protein domain. *J. Bacteriol.* **187:**1792–1798.
114. **Schmidt, A. J., D. A. Ryjenkov, and M. Gomelsky.** 2005. The ubiquitous protein domain EAL is a cyclic diguanylate-specific phosphodiesterase: enzymatically active and inactive EAL domains. *J. Bacteriol.* **187:**4774–4781.
115. **Schroeder, C., K. Werner, H. Otten, S. Krätzig, H. Schwalbe, and L.-O. Essen.** 2008. Influence of a joining helix on the BLUF domain of the YcgF photoreceptor from *Escherichia coli*. *Chembiochem* **9:**2463–2473.
116. **Sezonov, G., D. Joseleau-Petit, and R. D'Ari.** 2007. *Escherichia coli* physiology in Luria-Bertani broth. *J. Bacteriol.* **189:** 8746–8749.
117. **Shenoy, A. R., K. Sivakumar, A. Krupa, N. Srinivasan, and S. S. Visweswariah.** 2004. A survey of nucleotide cyclases in actinobacteria: unique domain organisation and expansion of the class III cyclase family in *Mycobacterium tuberculosis*. *Comp. Funct. Genomics* **5:**17–38.
118. **Shenoy, A. R., and S. S. Visweswariah.** 2004. Class III nucleotide cyclases in bacteria and archaebacteria: lineage-specific expansion of adenylyl cyclases and a dearth of guanylyl cyclases. *FEBS Lett.* **561:**11–21.
119. **Shenoy, A. R., and S. S. Visweswariah.** 2006. New messages from old messengers: cAMP and mycobacteria. *Trends Microbiol.* **14:**543–550.
120. **Simm, R., A. Lusch, A. Kader, M. Andersson, and U. Römling.** 2007. Role of EAL-containing proteins in multicellular behavior of *Salmonella enterica* serovar Typhimurium. *J. Bacteriol.* **189:**3613–3623.
121. **Sommerfeldt, N., A. Possling, G. Becker, C. Pesavento, N. Tschowri, and R. Hengge.** 2009. Gene expression patterns and differential input into curli fimbriae regulation of all GGDEF/EAL domain proteins in *Escherichia coli*. *Microbiology* **155:**1318–1331.
122. **Soutourina, O., A. Kolb, E. Krin, C. Laurent-Winter, S. Rimsky, A. Danchin, and P. Bertin.** 1999. Multiple control of flagellum biosynthesis in *Escherichia coli:* role of H-NS protein and the cyclic AMP-catabolite activator protein complex in transcription of the *flhDC* master operon. *J. Bacteriol.* **181:** 7500–7508.
123. **Storz, G., and R. Hengge-Aronis (ed.).** 2000. *Bacterial Stress Responses.* ASM Press, Washington, DC.
124. **Sudarsan, N., E. R. Lee, Z. Weinberg, R. H. Moy, J. N. Kim, K. H. Link, and R. R. Breaker.** 2008. Riboswitches in eubacteria sense the second messenger cyclic di-GMP. *Science* **321:**411–413.
125. **Suzuki, K., P. Babitzke, S. R. Kushner, and T. Romeo.** 2006. Identification of a novel regulatory protein (CsrD) that targets the global regulatory RNAs CsrB and CsrC for degradation by RNase E. *Genes Dev.* **20:**2605–2617.
126. **Tal, R., H. C. Wong, R. Calhoon, D. Gelfand, A. L. Fear, G. Volman, R. Mayer, P. Ross, D. Amikam, H. Weinhouse, A. Cohen, S. Sapir, P. Ohana, and M. Benziman.** 1998. Three cdg operons control cellular turnover of cyclic di-GMP in *Acetobacter xylinum:* genetic organization and occurrence of conserved domains in isoenzymes. *J. Bacteriol.* **180:**4416–4425.
127. **Tamayo, R., A. D. Tischler, and A. Camilli.** 2005. The EAL domain protein VieA is a cyclic diguanylate phosphodiesterase. *J. Biol. Chem.* **280:**33324–33330.
128. **Taylor, B. L., and I. G. Zhulin.** 1999. PAS domains: internal sensors of oxygen, redox potential, and light. *Microbiol. Mol. Biol. Rev.* **63:**479–506.
129. **Tomoyasu, T., A. Takaya, E. Isogai, and T. Yamamoto.** 2003. Turnover of FlhD and FlhC, master regulator proteins for *Salmonella* flagellum biogenesis, by the ATP-dependent ClpXP protease. *Mol. Microbiol.* **48:**443–452.
130. **Tschowri, N., S. Busse, and R. Hengge.** 2009. The BLUF-EAL protein YcgF acts as a direct anti-repressor in a blue light response of *E. coli*. *Genes Dev.* **23:**522–534.
131. **Typas, A., C. Barembruch, and R. Hengge.** 2007. Stationary phase reorganisation of the *E. coli* transcription machinery by Crl protein, a fine-tuner of σ^S activity and levels. *EMBO J.* **26:**1569–1578.
132. **Vianney, A., G. Jubelin, S. Renault, C. Dorel, P. Lejeune, and J. C. Lazzaroni.** 2005. *Escherichia coli tol* and *rcs* genes participate in the complex network affecting curli synthesis. *Microbiology* **151:**2487–2497.
133. **Vidal, O., R. Longin, C. Prigent-Combaret, C. Dorel, M. Heooreman, and P. Lejeune.** 1998. Isolation of an *Escherichia coli* K-12 mutant strain able to form biofilms on inert surfaces: involvement of a new *ompR* allele that increases curli expression. *J. Bacteriol.* **180:**2442–2449.
134. **Wan, X., J. R. Tuckerman, J. A. Saito, T. A. K. Freitas, J. S. Newhouse, J. R. Denery, M. Y. Galperin, G. Gonzalez, M.-A. Gilles-Gonzalez, and M. Alam.** 2009. Globins synthesize the second messenger bis-(3′-5′)-cyclic diguanosine monophosphate in bacteria. *J. Mol. Biol.* **388:**262–270.
135. **Wang, X., A. K. Dubey, K. Suzuki, C. S. Baker, P. Babitzke, and T. Romeo.** 2005. CsrA post-transcriptionally represses *pgaABCD,* responsible for synthesis of a biofilm polysaccharide adhesin of *Escherichia coli*. *Mol. Microbiol.* **56:**1648–1663.
136. **Wassmann, P., C. Chan, R. Paul, A. Beck, H. Heerklotz, U. Jenal, and T. Schirmer.** 2007. Structure of BeF3-modified response regulator PleD: implications of diguanylate cyclase activation, catalysis, and feedback inhibition. *Structure* (Cambridge) **15:**915–927.

137. **Weber, H., C. Pesavento, A. Possling, G. Tischendorf, and R. Hengge.** 2006. Cyclic-di-GMP-mediated signaling within the σ^S network of *Escherichia coli*. *Mol. Microbiol.* **62:**1014–1034.

138. **Weber, H., T. Polen, J. Heuveling, V. Wendisch, and R. Hengge.** 2005. Genome-wide analysis of the general stress response network in *Escherichia coli*: σ^S-dependent genes, promoters and sigma factor selectivity. *J. Bacteriol.* **187:** 1591–1603.

139. **Weinhouse, H., S. Sapir, D. Amikam, Y. Shilo, G. Volman, P. Ohana, and M. Benziman.** 1997. c-di-GMP-binding protein, a new factor regulating cellulose synthesis in *Acetobacter xylinum*. *FEBS Lett.* **416:**207–211.

140. **Willoughby, D., and D. M. Cooper.** 2007. Organization and Ca^{2+} regulation of adenylyl cyclases in cAMP microdomains. *Physiol. Rev.* **87:**965–1010.

141. **Wolfe, A. J., and K. L. Visick.** 2008. Get the message out: cyclic-di-GMP regulates multiple levels of flagellum-based motility. *J. Bacteriol.* **190:**463–475.

142. **Wood, T. K., A. F. González-Barrios, M. Herzberg, and J. Lee.** 2006. Motility influences biofilm architecture in *Escherichia coli*. *Appl. Environ. Microbiol.* **72:**361–367.

143. **Wozniak, C. E., C. Lee, and K. T. Hughes.** 2009. T-POP array identifies EcnR and PefI-SrgD as novel regulators of flagellar gene expression. *J. Bacteriol.* **191:**1498–1508.

144. **Wu, Q., and K. Gardner.** 3 February 2009. Structure and insight into the blue light-induced changes in the BirP1 BLUF domain. *Biochemistry* **48:**2620–2629.

145. **Yamamoto, K., and A. Ishihama.** 2006. Characterization of copper-inducible promoters regulated by CpxA/CpxR in *Escherichia coli*. *Biosci. Biotechnol. Biochem.* **70:**1688–1695.

146. **Zhao, K., M. Liu, and R. R. Burgess.** 2007. Adaptation in bacterial flagellar and motility systems: from regulon members to "foraging"-like behavior in *E. coli*. *Nucleic Acids Res.* **35:**4441–4452.

147. **Zhou, X., X. Meng, and B. Sun.** 2008. An EAL domain protein and cyclic AMP contribute to the interaction between the two quorum sensing systems in *Escherichia coli*. *Cell Res.* **18:**937–948.

148. **Zogaj, X., M. Nimtz, M. Rohde, W. Bokranz, and U. Römling.** 2001. The multicellular morphotypes of *Salmonella typhimurium* and *Escherichia coli* produce cellulose as the second component of the extracellular matrix. *Mol. Microbiol.* **39:**1452–1463.

149. **Zoraghi, R., J. D. Corbin, and S. H. Francis.** 2004. Properties and functions of GAF domain in cyclic nucleotide phosphodiesterases and other proteins. *Mol. Pharmacol.* **65:**267–278.

The Second Messenger Cyclic Di-GMP
Edited by A. J. Wolfe and K. L. Visick

Chapter 17

Cyclic Di-GMP Signaling in *Vibrio cholerae*

SINEM BEYHAN AND FITNAT H. YILDIZ

Cyclic di-GMP (c-di-GMP) is a ubiquitous second messenger that is produced by diguanylate cyclases (DGCs), which contain GGDEF domains, and degraded by c-di-GMP phosphodiesterases (PDEs), which contain EAL or HD-GYP domains (42, 44, 46) (see chapters 4 and 5). Microbial genome sequencing projects have revealed that the proteins containing GGDEF, EAL, or HD-GYP domains (hereafter referred to as GGDEF/EAL/HD-GYP proteins) are present in most bacteria (14) (see also chapter 3). *Vibrio cholerae,* a facultative human pathogen and the causative agent of the disease cholera, has 62 genes (hereafter referred to as GGDEF/EAL/HD-GYP genes) predicted to encode GGDEF/EAL/HD-GYP proteins (Fig. 1). Specifically, *V. cholerae* contains 31 GGDEF, 12 EAL, 10 GGDEF-EAL, and 9 HD-GYP genes in its genome (14). The physiological role of c-di-GMP in *V. cholerae* has emerged from the analysis of phenotypes of mutants unable to produce GGDEF/EAL/HD-GYP proteins as well as from phenotypes of strains overproducing GGDEF/EAL/HD-GYP proteins. In *V. cholerae,* c-di-GMP is involved in the regulation of multiple cellular processes including biofilm formation, motility, and virulence. In this chapter, we will first introduce the processes regulated by c-di-GMP signaling, discuss what is currently known about the involvement of specific GGDEF/EAL/HD-GYP proteins in these processes, and then summarize our current understanding of regulation of GGDEF/EAL/HD-GYP genes.

c-di-GMP-REGULATED PROCESSES IN *V. CHOLERAE*

In *V. cholerae,* c-di-GMP is intimately linked to the regulation of biofilm formation, motility, and virulence. In the following section, we discuss each of these processes.

Biofilm Formation

c-di-GMP positively regulates biofilm formation in *V. cholerae.* Biofilms are surface-attached microbial communities composed of microorganisms and the extracellular polymeric substances they produce. The biofilm-forming capacity of *V. cholerae,* both in natural habitats and under laboratory conditions, is well documented (60). Biofilm formation facilitates *V. cholerae* growth, survival, and persistence in aquatic ecosystems and facilitates transmission of the disease (37). Production of mature biofilms requires extracellular matrix components. A major component of the *V. cholerae* biofilm matrix is VPS (Vibrio polysaccharide) exopolysaccharide (62). VPS production is essential for the development of three-dimensional (3-D) biofilm structures. The *vps* genes are clustered in two regions on the large chromosome of *V. cholerae* O1 El Tor (*vpsU,* VC0916; *vpsA-K,* VC0917–27 [*vps-I* cluster]; *vpsL-Q,* VC0934–9 [*vps-II* cluster]) (62). Biofilm matrix proteins RbmA (VC0928), RbmC (VC0930), and Bap1 (VC1888) are also required for wild-type biofilm formation (10, 12).

Different strains of *V. cholerae* and even phenotypic variants derived from a single strain can generate profoundly different biofilms (62). *V. cholerae* generates spontaneous phenotypic variants, termed smooth and rugose (Fig. 2A). Rugose variants form colonies with increased corrugation, which is largely due to increased levels of VPS (37). Rugose variants have an increased capacity to form biofilm, and patterns of biofilm formation by the smooth and rugose variants are quite distinct (Fig. 2B). Measurements of average thickness, surface roughness, substratum cov-

Sinem Beyhan and Fitnat H. Yildiz • Department of Microbiology and Environmental Toxicology, University of California, Santa Cruz, Santa Cruz, CA 95064.

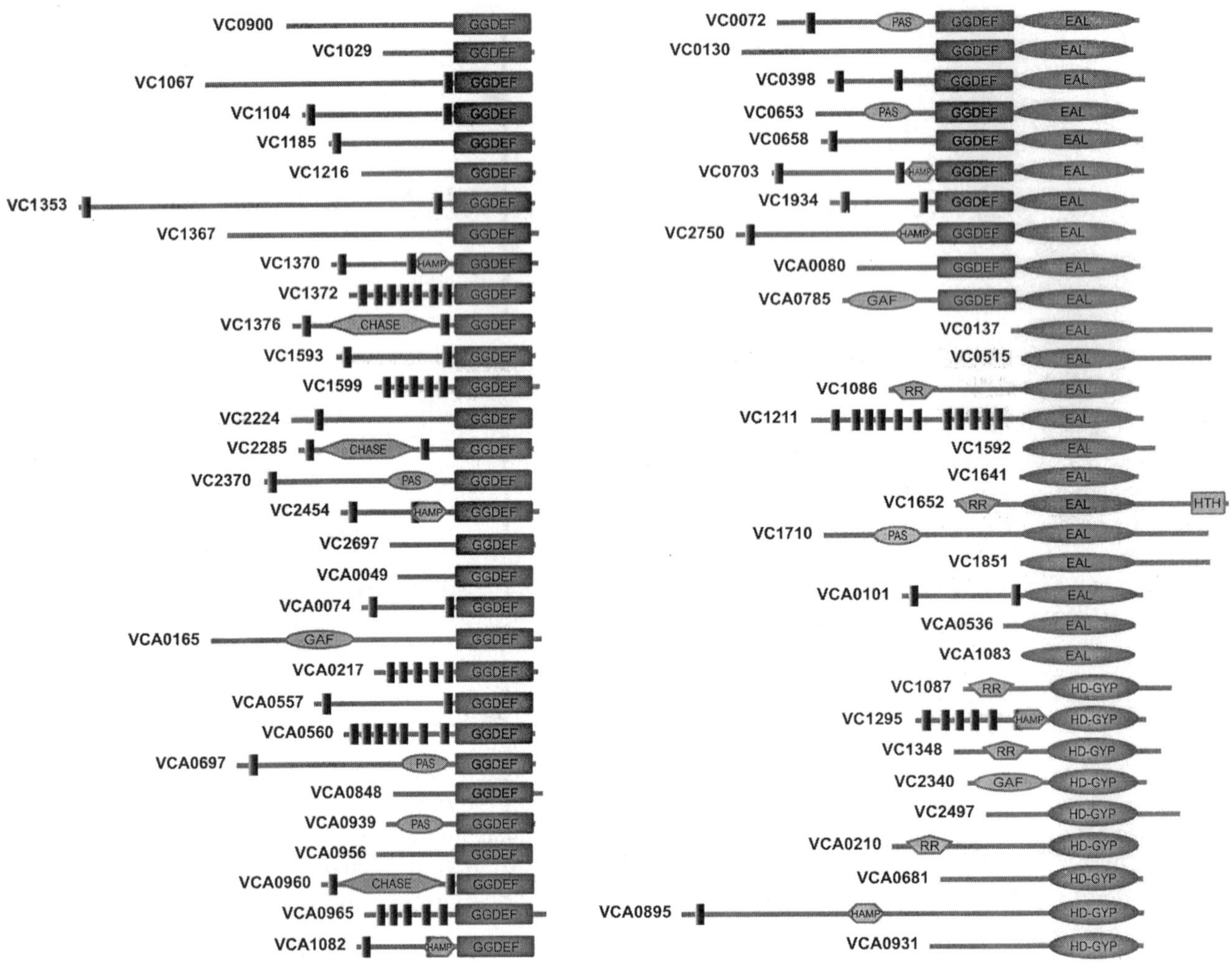

Figure 1. Proteins harboring GGDEF/EAL/HD-GYP domains in *V. cholerae*. There are 31 proteins with GGDEF domains, 10 proteins with GGDEF and EAL domains, 12 proteins with EAL domains, and 9 proteins with HD-GYP domains. Domain information and organization were obtained from The Institute of Genomic Research (TIGR) Comprehensive Microbial Resource. Locations and numbers of transmembrane helices (shown as black rectangles) were determined using the TMHMM 2.0 program. HTH, helix-turn-helix.

erage, and total biomass of rugose biofilms are markedly higher than those of the smooth variant (61). Rugose variants appear to be selected for under a variety of conditions including carbon limitation, growth in biofilms, and treatment with bacteriocidal agents (20, 35, 41, 62). Recent studies revealed that specific proteins of the c-di-GMP signaling system control rugosity and biofilm formation; these proteins will be discussed in this chapter.

Biofilm formation in *V. cholerae* is regulated by numerous positive and negative transcriptional regulators. The core biofilm regulatory network consists of two positive transcriptional regulators, VpsR and VpsT, and two negative transcriptional regulators, HapR and the cAMP-CRP (cAMP receptor protein) complex (9, 11, 17, 29, 59, 61, 63).

VpsR exhibits homology to the NtrC subclass of two-component response regulators. Disruption of *vpsR* prevents expression of the *vps* genes and production of VPS and abolishes formation of typical 3-D biofilm structure (59). VpsT is similar to proteins belonging to the UhpA (FixJ) family of transcriptional regulators. Disruption of *vpsT* reduces *vps* gene expression and biofilm-forming capacity (9). VpsT is homologous to CsgD of *Salmonella enterica* serovar Typhimurium. CsgD positively regulates production of extracellular matrix compounds curli pili and cellulose, processes controlled by c-di-GMP signaling in *Salmonella* serovar Typhimurium (see chapters 10 and 16). Characterization of Δ*vpsT,* Δ*vpsR,* and Δ*vpsT*Δ*vpsR* mutants with respect to biofilm formation and *vps* gene expression revealed that VpsR is

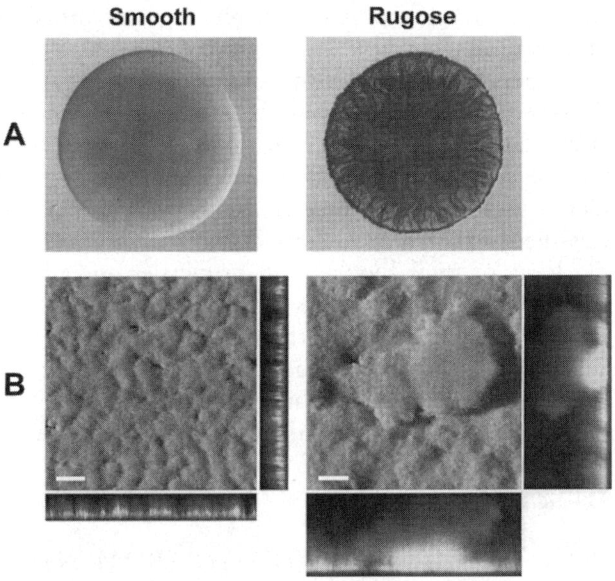

Figure 2. Phenotypic characterization of smooth and rugose variants of the *V. cholerae* O1 El Tor strain A1552. (A) Photos of colony morphologies of smooth and rugose variants were taken after strains were grown at 30°C for 2 days on LB agar plates. (B) 3-D biofilm structures of smooth and rugose variants that are formed 24 h postinoculation in flow cell chambers. Images were acquired with confocal scanning laser microscopy. Top-down (large panels) and orthogonal (side panels) views of biofilms are shown. Bars, 30 μm.

essential for VPS production and biofilm formation, while VpsT plays an accessory role (1).

Quorum sensing (QS), a bacterial cell-cell communication system, negatively regulates biofilm formation in *V. cholerae* (17, 61, 63). In QS, bacteria produce signaling molecules termed autoinducers (AIs) that are secreted and accumulate in the medium in proportion to population density. AIs are detected when they reach a critical concentration. *V. cholerae* produces two AIs, known as CAI-1 [(*S*)-3-hydroxytridecan-4-one] and AI-2 [2*S*,4*S*)-2-methyl-2,3,3,4-tetrahydroxytetrahydrofuran borate] which are produced by the CqsA and LuxS AI synthases, respectively (55). CAI-1 and AI-2 are detected via sensor histidine kinase receptors CqsS and LuxQ, respectively. At low cell density, when the concentrations of AIs are low, CqsS and LuxQ function as kinases and are autophosphorylated. Information from the sensors is then transduced through a phosphorelay, first to the LuxU phosphotransfer protein and then to LuxO, which is a σ^{54}-dependent response regulator. LuxO activity is defined by its phosphorylation state (13, 54). When phosphorylated, LuxO indirectly represses expression of the major QS regulator, HapR, by activating the expression of a set of small regulatory RNAs (sRNAs) termed Qrr1-4 (*q*uorum *r*egulatory *R*NA). The Qrr sRNAs, in conjunction with the RNA chaperone Hfq, destabilize the *hapR* mRNA and prevent production of HapR. In contrast, at high cell density, when the concentrations of AIs are high, CqsS and LuxQ recognize their cognate signal and function as phosphatases. Upon dephosphorylation, LuxO can no longer activate transcription of the Qrr sRNAs; consequently, *hapR* mRNA is stabilized and HapR is produced. HapR represses expression of a large set of genes, including genes required for biofilm formation and virulence factor production, and activates expression of a large set of genes, including the genes necessary for flagellum production (36, 61, 63).

Recently, the regulatory complex cAMP-CRP was identified as a negative regulator of biofilm formation in *V. cholerae* (11, 29). cAMP-CRP negatively regulates the transcription of genes required for the production of VPS and biofilm matrix proteins (11, 29). The effect of cAMP-CRP on *vps* genes can be indirect, as cAMP-CRP negatively regulates *vpsR* and *vpsT* expression and positively regulates *hapR* expression (11).

DGCs and PDEs regulating biofilm formation in *V. cholerae* have been identified. Furthermore, regulation of GGDEF/EAL/HD-GYP genes by VpsR, VpsT, HapR, and the cAMP-CRP complex has also been investigated. In this chapter, we will discuss our current understanding of regulation of biofilm formation through c-di-GMP signaling and transcriptional regulation of genes encoding GGDEF/EAL/HD-GYP proteins.

Motility

In *V. cholerae,* similar to other organisms, c-di-GMP controls a motile-to-sessile lifestyle switch and negatively regulates motility (see chapter 8). *V. cholerae* is motile by means of a single polar flagellum which consists of three major parts: a basal body, a hook, and a filament. A large set of proteins is required for flagellum biosynthesis, assembly, and function. Genes encoding flagellar components are located in six different regions and organized into 18 different transcriptional units on the large chromosome (34). Five genes, *flaA, flaB, flaC, flaD,* and *flaE,* encode flagellin subunits and are required to build the flagellin filament. FlaA is the major flagellin protein essential for motility, while other flagellin subunits play accessory roles. The flagellin is capped by a flagellar cap protein FliD. Flagellin filaments are connected to the cell through a hook and a basal body structure. The basal body is composed of rod proteins FlgB, FlgC, FlgF, FlgG, and FliE and ring proteins FlgH (L ring), FlgI (P ring), FliF (MS ring), FliG (C ring), MotX (T ring), and MotY (T ring), which span the cytoplasmic and outer membranes. The hook is composed of FlgE, and the hook junction is composed of FlgL and FlgK. The *V. cholerae* flagellum is powered by a flagellar motor that is driven by the sodium motive force. Functioning of the motor requires the general motor switch complex proteins FliG, FliM, and FliN as well as proteins MotX, MotY, PomA, and PomB, which form a motor channel specific for Na^+ (15, 24).

To coordinate the production of a flagellum, a hierarchical transcriptional arrangement controlling flagellar gene expression in *V. cholerae* has been proposed. According to this model, the flagellar gene transcriptional hierarchy is composed of four classes of flagellar genes (39). Class I is only composed of *flrA*, which encodes a σ^{54}-dependent activator. FlrA, together with σ^{54}, is required for the expression of class II genes that encode components of the MS ring-switch complex (Fli proteins [FliEFGHIJKLMNOPQR], FlhB, FlhAFG), components of the chemotaxis system (CheABWYZ), and the regulatory proteins FlrB, FlrC, and FliA (σ^{28}). Class III genes are dependent on FlrC and σ^{54} for their expression and encode the basal body and hook proteins (FlgBCDEFGHIJKL), core flagellin (FlaA), flagellar cap (FliD), and motor component (MotX). Class IV genes are dependent on alternative sigma factor σ^{28} and encode alternative flagellins (FlaBCDE) and motor components (FlgMN, PomAB, and MotY).

DGCs and PDEs regulating motility in *V. cholerae* have been identified. We will discuss these proteins and our current understanding of regulation of flagellar motility by c-di-GMP signaling.

Virulence

c-di-GMP regulates virulence in *V. cholerae*. The major virulence factors of *V. cholerae* are cholera toxin (CT) and the toxin coregulated pilus (TCP) colonization factor (33). A complex transcriptional regulatory cascade controls CT and TCP production, whereby transcription of *ctxAB* (the genes encoding CT) and the *tcp* operon is regulated by the regulatory proteins ToxR and TcpP. Two additional proteins, ToxS and TcpH, enhance transcriptional activity of ToxR and TcpP, respectively. These two regulatory units control the expression of *toxT*, encoding ToxT, the most downstream regulator of virulence factors in *V. cholerae*. In addition, expression of *tcpP* and *tcpH* is under the control of two other regulatory proteins, AphA and AphB (27, 47). The QS transcriptional regulator HapR acts as a negative regulator of *ctxAB* and *tcpA* expression by repressing expression of *aphA* (26).

There are over 200 different serogroups of *V. cholerae* based on a lipopolysaccharide O-antigen classification scheme (22). Of these serogroups, only the O1 and O139 serogroups have endemic and epidemic potentials. Strains of the O1 serogroup are classified further as classical or El Tor biotypes on the basis of their biochemical properties and phage susceptibility (22). The role of c-di-GMP in virulence appears to be different in the O1 classical and O1 El Tor strains, and we will briefly discuss their differences and similarities in this chapter. The role of c-di-GMP in *V. cholerae* virulence is discussed in detail in chapter 20.

ROLES OF GGDEF/EAL/HD-GYP PROTEINS IN *V. CHOLERAE* BIOLOGY

In this section, we review what is known about the involvement of the GGDEF/EAL/HD-GYP proteins in biofilm formation, motility, and virulence. We will first discuss the proteins whose absence leads to an observable phenotype.

Genes Encoding Proteins Harboring Both GGDEF and EAL Domains

In the *V. cholerae* genome, there are 10 genes encoding proteins with both a GGDEF and an EAL domain. Of these, MbaA, RocS, CdgC, and CdpA have been studied in detail. These proteins act as PDEs, and they affect biofilm formation, motility, and virulence (6, 30, 31, 40, 49) (Table 1).

MbaA

The importance of GGDEF/EAL/HD-GYP proteins in biofilm formation in *V. cholerae* was first indicated by Bomchil et al. This study showed that mutants of *mbaA* (VC0703, *m*aintenance of *b*iofilm *a*rchitecture), encoding a GGDEF-EAL domain protein, had increased capacity to produce extracellular matrix and consequently formed biofilms with increased thickness that lacked the typical biofilm architecture (6). MbaA is an inner membrane protein that has cytoplasmic GGDEF and EAL domains. The GGDEF domain lacks the conserved residues required for DGC activity, and MbaA is predicted to act as a PDE. MbaA is encoded by the second gene in a three-gene operon. The first gene, *nspS*, is predicted to encode a periplasmic polyamine-binding protein, while the third gene in the operon is predicted to encode a cytoplasmic protein of unknown

Table 1. Genes encoding proteins with GGDEF and EAL domains in *V. cholerae*

Gene identification[a]	Domains[b]	Phenotype[c]		
		Biofilm formation	Motility	Virulence
VC0072	GGDEF-EVL			
VC0130 *(cdpA)*	GVGEW-ECL	–		
VC0398	YEDVF-ELF			
VC0653 *(rocS)*	GGDEF-EAL	–	+	
VC0658 *(cdgI)*[d]	GGDEF-EAL			
VC0703 *(mbaA)*	SGDEF-EVL	–	+	
VC1934	GADEF-EAL			
VC2750	GGDEF-EAL			
VCA0080	AADEF-EIL			
VCA0785 *(cdgC)*	NSTNL-EAL	–	+	–/+

[a] Gene list was obtained from The Institute of Genomic Research Comprehensive Microbial Resource.
[b] Conserved amino acid sequences of each of the GGDEF/EAL/HD-GYP domains of the proteins were identified with multiple-sequence alignments using the ClustalW program.
[c] + and – indicate positive and negative regulation, respectively.
[d] Deletion of this gene resulted in observable phenotypes only in the Δ*crp* mutant (11).

function. The polyamine norspermidine enhances biofilm formation in *V. cholerae* in an NpsS-MbaA-dependent manner (23). It is proposed that MbaA activity is regulated by the interaction of NpsS or norspermidine-NpsS with the periplasmic domain of MbaA. Norspermidine-NpsS could reduce the ability of MbaA to inhibit *V. cholerae* biofilm formation by modulating the activity of the cytoplasmically localized GGDEF-EAL output domains, either by preventing formation of an MbaA dimer, which may be important for enzymatic activity of the protein, or by a conformational change that can alter the function of the protein (23).

Studies by Karatan et al. and Lim et al. identified a set of genes regulated by MbaA in the smooth variant of the O139 strain MO10 and the rugose variant of El Tor O1 strain A1552 using whole-genome expression profiling (23, 31). These studies showed that MbaA regulates gene expression; in particular, MbaA represses expression of *vps* genes in *V. cholerae*. However, the molecular mechanism by which this regulation takes place has yet to be determined.

RocS

RocS (VC0653, *r*egulation *o*f *c*ell *s*ignaling) was first identified as a regulator of the smooth-to-rugose switch in *V. cholerae* (40). It was shown that strains harboring a *rocS* mutation are defective in the smooth-to-rugose switch in *V. cholerae* O1 El Tor strain N16961 (40). RocS is a cytoplasmic protein that has a PAS sensory domain (a domain present in *P*er, period circadian protein; *A*rnt, Ah receptor nuclear translocator protein; and *S*im, single-minded protein) and tandem GGDEF-EAL domains that harbor the conserved residues required for DGC and PDE activities, suggesting that RocS is a bifunctional enzyme. Phenotypic analysis of Δ*rocS* mutants in *V. cholerae* O1 El Tor strains N16961 and A1552 showed that the Δ*rocS* mutant has reduced motility and enhanced VPS production and biofilm formation (30, 40). Thus, RocS appears to be acting as a PDE under the conditions utilized in these studies (growth in LB [Luria-Bertani medium] at either 30°C or 37°C). It appears that the EAL domain of RocS is functional only in the presence of its GGDEF domain (51). The mechanisms by which RocS regulates transcription, whether RocS is a bifunctional enzyme, and whether its N-terminal PAS domain is involved in modulating RocS enzymatic activity have yet to be determined.

CdgC

CdgC (VCA0785, *c*yclic *d*i*g*uanylate) was first identified as a gene differentially regulated between the smooth and rugose variants of *V. cholerae* O1 El Tor A1552 (30). Expression of *cdgC* is higher in the rugose than in the smooth variant (61). CdgC is a cytoplasmic protein that has an N-terminal GAF domain (domain present in cGMP-specific and -stimulated PDEs, *Anabaena a*denylate cyclases, and *E. coli F*hlA) and tandem GGDEF and EAL domains. The GGDEF domain lacks the conserved residues required for DGC activity, and in accordance, CdgC acts as a PDE based on phenotypic and mutational analyses. Tischler and Camilli also showed that the EAL domain of CdgC is active and that when overproduced it leads to a decrease in *vps* gene expression (51). Overexpression studies using CdgC and a CdgC allele harboring an E407A mutation (changing the EAL residues critical to PDE activity to AAL) showed that while parent strains harboring a wild-type copy of the *cdgC* exhibited a significant increase in motility, strains harboring the mutated version, *cdgC*-AAL, did not exhibit any significant change in motility (31). These results indicate that the EAL domain of CdgC is in an active form and is required for increased flagellar motility. This analysis also showed that the GGDEF domain of CdgC is not active, since it does not lead to any change in flagellar motility even when the EAL domain is mutated.

Consistent with an increase in c-di-GMP levels due to a loss of a key PDE, a Δ*cdgC* mutant of the rugose variant (RΔ*cdgC*, hereafter R represents rugose parent) forms colonies that are more opaque and wrinkled than the rugose parent strain (30). As expected, the biofilm-forming capacities and *vps* gene expression in the Δ*cdgC* mutants of both rugose and smooth variants are increased compared to the parent strains (30).

CdgC also regulates virulence in *V. cholerae.* Expression analysis and reporter gene assays revealed that during the stationary growth phase, the RΔ*cdgC* mutant exhibits an increase in the transcript abundance of *tcpA,* encoding the major pilin subunit of TCP pili, compared to the rugose parent strain (31). The expression of virulence factors among *V. cholerae* El Tor strains occurs under specific in vitro growth conditions, termed AKI growth conditions (21). Compared to the rugose parent, TcpA production was higher in the RΔ*cdgC* mutant grown under AKI conditions (31). Similarly, CT production was also greatly increased in the RΔ*cdgC* mutant, whereby culture supernatants of RΔ*cdgC* yielded a higher amount of CT than those of the rugose. It was further shown that CdgC represses *ctxA, tcpA, aphA,* and *tcpP* expression in the rugose variant (31). RΔ*cdgC* exhibited a slight increase in cellular c-di-GMP levels compared to the rugose parent, and it is possible that such an increase could induce the expression of virulence genes. Alternatively, a decrease in the *hapR* transcript, as observed in the RΔ*cdgC* mutant, could lead to an increase in virulence gene transcription, since HapR negatively regulates virulence gene expression by repressing *aphA* transcription in *V. cholerae* (26).

While expression of *ctx* genes (encoding CT) was similar between the smooth wild type and the smooth Δ*cdgC* mutant (SΔ*cdgC,* hereafter S represents smooth parent), expression of *tcpA* and *toxT* decreased in the SΔ*cdgC* mutant relative to the smooth wild-type strain (24). This study indicated that the regulation of virulence factors by CdgC differs between the *V. cholerae* rugose and smooth variants (31). It is critical to note that the rugose variant can cause cholera when given orally to human volunteers (35), and virulence factor production is significantly greater in the rugose variant than in the smooth variant (31). However, at present, little information is available on the virulence characteristics of rugose variants of *V. cholerae.*

CdpA

C-di-GMP regulates virulence gene expression in a biotype-specific manner. In the *V. cholerae* classical biotype, but not in the El Tor biotype, virulence is positively regulated by a PDE, VieA. To identify a PDE that could modulate in vivo virulence in El Tor biotype strains, Tamayo et al. studied a gene termed *cdpA* (VC0130, *c*yclic *d*iguanylate *p*hosphodiesterase), which is induced in vivo in mice and in humans and encodes a protein with GGDEF and EAL domains (49). CdpA has a degenerate GG(D/E)EF motif and divergent EAL (ECL) motif. Enzymatic activity and in vitro phenotypic analysis of the strains overproducing wild-type and mutated versions of CdpA revealed that CdpA has only a PDE activity and that the GGDEF domain is necessary for full PDE activity. Although phenotypic analyses showed that the Δ*cdpA* mutant has increased biofilm formation capacity (49), it exhibited motility and in vivo colonization phenotypes similar to that of the wild type. Thus, CdpA affects only a subset of c-di-GMP-regulated phenotypes.

In addition to *cdpA,* two other GGDEF/EAL/HD-GYP genes, VC2697 and VC1593 *(acgB),* were also identified as late-infection-induced genes of *V. cholerae* (45). Single-deletion mutants of these genes did not exhibit any colonization defect. However, each of the single-deletion mutants and the triple mutant exhibited a fitness defect relative to the wild type in stool and pond water (45), suggesting that *cdpA,* VC2697, and VC1593 *(acgB)* may be involved in the adaptation of *V. cholerae* to a nonhost environment during the transition from the host to an aquatic environment.

Genes Encoding Proteins Harboring an EAL Domain

There are 12 genes encoding proteins with an EAL domain in the *V. cholerae* genome (Table 2). Of these proteins, only VieA, which acts as a PDE, has been studied in detail (50–53). Below, we discuss the involvement of VieA in biofilm formation and motility.

VieA

The importance of c-di-GMP in *V. cholerae* biology has been conclusively shown in studies involving VieA (VC1652) (51, 52). The response regulator VieA, which is a component of the VieSAB three-component signal transduction system, was first identified in a genetic screen designed to identify in vivo positive regulators of *ctxAB* transcription (8). Subsequent studies on the Vie signal transduction system revealed that VieS is a sensor histidine kinase that can selectively phosphorylate VieA, a response regulator (32). The role of VieB, a second response regulator, in this signal transduction event has yet to be determined.

The VieA response regulator harbors an EAL domain and has PDE activity. VieA represses transcription of *vps* genes and induces transcription of flagel-

Table 2. Genes encoding proteins with an EAL domain in *V. cholerae*

Gene identification[a]	Domain[b]	Phenotype[c]		
		Biofilm formation	Motility	Virulence
VC0137	ELL			
VC0515	ELL			
VC1086	EAL			
VC1211	EAL			
VC1592 *(acgA)*[d]	ESL			
VC1641	EAL			
VC1652 *(vieA)*	EAL	−	+	+
VC1710	EAL			
VC1851	ELL			
VCA0101	EVL			
VCA0536	EAL			
VCA1083	EVF			

[a] Gene list was obtained from The Institute of Genomic Research Comprehensive Microbial Resource.
[b] Conserved amino acid sequences of each of the GGDEF/EAL/HD-GYP domains of the proteins were identified with multiple-sequence alignments using the ClustalW program.
[c] + and − indicate positive and negative regulation, respectively.
[d] Deletion of this gene in the wild-type genetic background did not yield observable phenotypes, and it was characterized through overexpression studies (Table 4).

lar biosynthesis genes in the *V. cholerae* O1 classical biotype in a defined M9 minimal medium supplemented with the amino acids asparagine, arginine, glutamate, and serine (M9-NRES) (3, 51). The EAL domain of VieA is capable of hydrolyzing c-di-GMP, and in the Δ*vieA* mutant, total cellular levels of c-di-GMP are increased compared to the wild type (50, 51). Increased c-di-GMP levels activate expression of *vpsR* and the *vps* genes through a mechanism yet to be identified. Interestingly, VieA does not play a role in *vps* gene regulation, biofilm formation, and motility in the El Tor biotype (3). VieA also regulates virulence in a biotype-specific manner. Studies on VieA revealed, for the first time, the importance of c-di-GMP in *V. cholerae* virulence (52, 53), which is discussed in detail in chapter 20.

Table 3. Genes encoding proteins with a GGDEF domain in *V. cholerae*

Gene identification[a]	Domain[b]	Phenotype[c]		
		Biofilm formation	Motility	Virulence
VC0900 *(cdgG)*	SGEEF	−	+	
VC1029 *(cdgB)*[d]	GGDEF			
VC1067 *(cdgH)*	GGDEF	+	−	
VC1104	GGEEF			
VC1185	GGEEF			
VC1216	GGDEF			
VC1353	GGEEF			
VC1367 *(cdgE)*[d]	GDDEF			
VC1370	GGDEF			
VC1372	GGEEF			
VC1376	GGDEF			
VC1593 *(acgB)*[e]	GGEEF			
VC1599	GGEEF			
VC2224	GGEEF			
VC2285	GGEEF			
VC2370	GGEEF			
VC2454 *(vpvC)*	GGDEF	+	−	
VC2697	GGEEF			
VCA0049	GGEEF			
VCA0074 *(cdgA)*	GGDEF	+		
VCA0165	GGEEF			
VCA0217	GGDEF			
VCA0557	GGEEF			
VCA0560	GGEEF			
VCA0697 *(cdgD)*	GGEEF	+	−	
VCA0848	GGEEF			
VCA0939	GGEEF			
VCA0956 *(cdgF)*[e]	GGEEF			
VCA0960	GGDEF			
VCA0965	AGDEF			
VCA1082	STYEF			

[a] Gene list was obtained from The Institute of Genomic Research Comprehensive Microbial Resource.
[b] Conserved amino acid sequences of each of the GGDEF/EAL/HD-GYP domains of the proteins were identified with multiple-sequence alignments using the ClustalW program.
[c] + and − indicate positive and negative regulation, respectively.
[d] These genes were deleted in the *V. cholerae* O1 El Tor A1552 strain; however, their deletion in the wild-type genetic background did not yield observable phenotypes (30).
[e] Deletion of these genes in the wild-type genetic background did not yield observable phenotypes, and they were characterized through overexpression studies (Table 4).

Genes Encoding Proteins Harboring a GGDEF Domain

There are 31 genes encoding proteins with a GGDEF domain in the *V. cholerae* genome (Table 3). Of these proteins, five of them, CdgA, VpvC, CdgH, CdgG and CdgD, have been studied in detail. These proteins, except CdgG, harbor the GG(D/E)EF residues essential for DGC activity and act as DGCs and regulate biofilm formation and motility (1, 2, 5, 30).

CdgA

CdgA (VCA0074, *c*yclic *d*i*g*uanylate) was first identified in studies designed to determine differences in whole-genome expression profiles of smooth and rugose variants of *V. cholerae* O1 El Tor A1552. Expression of *cdgA* was shown to be higher in rugose relative to that of the smooth variant (61). *cdgA* is predicted to be in a two-gene operon with VCA0075, which encodes a hypothetical protein of unknown function. CdgA is an inner membrane protein that has a cytoplasmic GGDEF domain harboring the conserved residues required for DGC activity; thus, CdgA is predicted to act as a DGC.

Studies by Lim et al. showed that mutations in *cdgA* in the rugose variant (RΔ*cdgA*) caused a decrease in colony corrugation compared to the parent strain, indicating a role for CdgA in maintaining rugosity (30). Subsequent studies showed that CdgA is required for HapR-mediated rugosity (1). The Δ*hapR* mutant forms rugose colonies and has increased *vps* transcription. Expression of *cdgA* is increased in the Δ*hapR* mutant relative to the wild type (36, 61, 63). Phenotypic analysis of Δ*cdgA* single and Δ*cdgA*Δ*hapR* mutants revealed that Δ*cdgA*Δ*hapR* mutants are smooth, as opposed to Δ*hapR* mutants, which are rugose (1). Furthermore, expression of the *vps* genes is decreased in the Δ*cdgA*Δ*hapR* mutant relative to that of the Δ*hapR* mutant. Through whole-genome expression profiling, it was also shown that CdgA positively regulates *vps* expression (1); however, the molecular mechanism by which this regulation occurs remains to be investigated.

VpvC

Levels of c-di-GMP are elevated in *V. cholerae* rugose variants (5, 30). An increase in c-di-GMP levels can be manifested through several different but specific pathways. In *V. cholerae* O1 El Tor A1552, the smooth-to-rugose switch can be mediated by a mutation in the coding sequence of the *vpvC* (VC2454, *Vibrio phase variation*) gene (5). *vpvC*, which is predicted to be the last gene in a three-gene operon (VC2456, named *vpvA*, and VC2455, named *vpvB*), encodes a protein with a GGDEF domain, while the first two genes encode hypothetical proteins. VpvC is predicted to be located on the membrane and carries a HAMP domain (domain present in *h*istidine kinases, *a*denylyl cyclases, *m*ethyl-accepting proteins, and *p*hosphatases) and a GGDEF domain with conserved residues.

Studies by Beyhan and Yildiz showed that *vpvC* is required for the increase in c-di-GMP levels in the prototypical rugose variant of *V. cholerae* O1 El Tor A1552 strain, as no detectable c-di-GMP accumulation was observed in the rugose Δ*vpvC* mutant (RΔ*vpvC*) (5). It was further shown that the DGC activity of VpvC is critical for rugosity and that disruption of *vpvC* in the rugose variant yields smooth colonies, decreases expression of the *vps* genes, and abolishes the formation of typical 3-D biofilms. The *vpvC* mutation can be complemented only with the *vpvC* coding region from the rugose variant, and introduction of this locus to the smooth strain converts the smooth variant to rugose (5). It was determined that VpvC of the smooth variant [VpvC(S)] and VpvC of the rugose variant [VpvC(R)] differ by a single point mutation converting a tryptophan in the smooth to an arginine in the rugose (W240R) (5). It is not yet known how this change alters the activity of VpvC, although it is possible that the rugose VpvC protein may mimic the activated state of the protein.

CdgH

CdgH (VC1067, *c*yclic *d*i*g*uanylate) was first identified in a study designed to identify GGDEF/EAL/HD-GYP proteins modulating rugosity in *V. cholerae* (2). CdgH has a GGDEF domain and is predicted to be located on the membrane. Using enzymatic analysis, it was shown that CdgH functions as a DGC (2). RΔ*cdgH* and SΔ*cdgH* mutants exhibited decreased *vps* gene expression and biofilm formation compared to that in the respective parent strains (2), indicating the importance of CdgH in VPS production. Further phenotypic analysis revealed that Δ*cdgH* mutants exhibited an increase in motility compared to the parent strains (2).

CdgG

CdgG (VC0900, *c*yclic *d*i*g*uanylate) was also identified in a study designed to identify GGDEF/EAL/HD-GYP proteins modulating rugosity in *V. cholerae* (2). CdgG is predicted to have a GGDEF domain. However, CdgG carries SGEEF residues [instead of the conserved GG(D/E)EF residues] in the active site of the enzyme, suggesting that it may not act as a DGC. Enzymatic activity of CdgG was tested by measuring intracellular c-di-GMP levels in the strains overproducing either a wild-type copy of CdgG or a version (CdgG-SGAAF) harboring point mutations which convert SGEEF residues to SGAAF. Overproduction of CdgG or CdgG-SGAAF did not lead to an accumulation of c-di-GMP (2). Thus, CdgG does not appear to function as a DGC.

The rugose Δ*cdgG* mutant (RΔ*cdgG*) formed super-rugose colonies with increased corrugation compared to the rugose parent (2). Interestingly, CdgG does not affect the transcription of *vps* genes, and the transcriptional profiles of the RΔ*cdgG* mutant and that of its rugose parent are similar, suggesting that CdgG acts at the posttranscriptional level to regulate rugosity. The RΔ*cdgG* mutant also exhibited a decrease in motility compared to the rugose parent (2). Increased biofilm formation capacity and decreased motility were also observed in the SΔ*cdgG* mutant compared to the smooth parent strain (2). The mechanism through which CdgG negatively regulates rugosity and biofilm formation has yet to be determined.

CdgD

cdgD (VCA0697, *c*yclic *d*i*g*uanylate) was identified as a gene whose expression is decreased in the rugose relative to that in the smooth variant (61). CdgD is predicted to have a GGDEF domain with conserved residues and a PAS domain and is predicted to be localized on the membrane. *cdgD* is in a two-gene operon with VCA0698, which encodes a hypothetical protein with no known domains and is predicted to be localized in the cytoplasm. Deletion of *cdgD* causes an increase in motility and alters biofilm development (30). The Δ*cdgD* mutant has a decrease in the surface coverage during the early stages of biofilm formation (30), suggesting that CdgD may play a key role during the initial stages of biofilm formation. Because *cdgD* mutation does not alter *vps* or flagellar gene expression, it is thought that, like CdgG, CdgD regulates biofilm formation and motility at the posttranscriptional level.

c-di-GMP OVEREXPRESSION STUDIES

A set of GGDEF/EAL/HD-GYP proteins produce a phenotype only when they are overproduced (Table 4). Below we discuss the results of such studies.

CdgF

cdgF (VCA0956, *c*yclic *d*iguanylate, also referred to as *vdcA* for *V*ibrio *d*iguanylate *c*yclase) (49) encodes a protein which is predicted to localize in the cytoplasm and have a GGDEF domain with conserved residues. Overexpression of *cdgF* increases intracellular c-di-GMP levels in the cell (51). CdgF was used in a functional genomics study designed to determine the physiological consequences of an elevated c-di-GMP level (4). The study revealed that 4.5% of *V. cholerae* genes are differentially expressed in response to an increased level of c-di-GMP (4). The main responses were an increase in the expression of *vps* genes, *eps* genes involved in the extracellular protein secretion system, and *msh* genes required for the mannose-sensitive hemagglutinin type IV pilus biogenesis as well as a decrease in the expression of *fla* genes required for flagellum biogenesis (4).

Studies in other organisms have indicated that c-di-GMP can regulate motility at least at three different levels: transcriptional, posttranscriptional, and functional (58). As discussed earlier, the genes required for the flagellum biogenesis and flagellar motility fall into four temporally distinct classes (I through IV). Transcriptional profiling studies revealed that the expression of several class III and IV genes, which encode many of the flagellins and flagellar basal body rod proteins, decreased in response to an increased level of c-di-GMP (4). Similarly, the expression level of the transcription factor σ^{28} *(fliA)*, which controls the transcription of class IV genes, was also lower (4). The effect of the increased intracellular c-di-GMP levels on motility was also further supported by a decreased motility phenotype. The strain overproducing the EAL domain of VieA exhibited increased motility, whereas a strain overexpressing *cdgF* exhibited a decrease in motility (4). In addition, the percentage of motile cells was dramatically decreased when intracellular c-di-GMP levels were increased. Taken together, these findings suggest that c-di-GMP can regulate motility by controlling the transcription of flagellar genes and also by modulating flagellum function.

cdgF overexpression also affected biofilm formation. Cells with increased c-di-GMP levels exhibited enhanced *vps*, *vpsT*, and *vpsR* expression and, consequently, formed thicker and more developed biofilms than those with wild-type levels of c-di-GMP (4).

The effect of an increase in cellular c-di-GMP on the ability of a *V. cholerae* El Tor strain to colonize the intestine was analyzed through ectopic production of the DGC CdgF (VdcA) (49). It was shown

Table 4. Effect of overexpression of the genes encoding GGDEF/EAL/HD-GYP proteins on biofilm formation, motility, and virulence phenotypes

Protein	Phenotype[a]		
	Biofilm formation	Motility	Virulence
pCdpA	Decreased (C6709)	Increased (C6709)	
pCdgF	Increased (A1552)	Decreased (A1552)	Decreased CT in vitro (O395) Decreased mouse colonization and *toxT* expression (C6709)
pAcgA	Decreased (C6706)	Increased (C6706)	
pAcgB	Increased (C6706)	Decreased (C6706)	
pVC1086	Decreased (C6706-*lux*OD47E)		
pVCA0681	Decreased (C6706)		

[a]Affected strains are indicated in parentheses.

that an increase in c-di-GMP levels during infection leads to a decrease in colonization, and this effect is due to reduced expression of *toxT* (49). Although this study showed that when overproduced in vivo, c-di-GMP decreases the virulence properties of El Tor strains, it is yet to be determined which PDE or DGC is involved in the regulation of in vivo virulence in El Tor strains.

Taken together, these studies revealed that significant changes to the transcriptome occur in response to an increased c-di-GMP level, which in turn alters cell physiology and leads to alterations in biofilm development, motility, and virulence.

AcgA and AcgB

acgA (VC1592, *a*cetoin *c*oregulated gene) and *acgB* (VC1593, *a*cetoin *c*oregulated *g*ene) encode EAL and GGDEF domain-containing proteins, respectively. These genes were originally identified as AphA-repressed genes in a whole-genome expression profiling study designed to identify AphA-regulated genes (25). In this study, AphA was found to regulate the expression of genes involved in acetoin production (25). Acetoin, a neutral metabolic by-product in cells grown in the presence of glucose, is predicted to prevent intracellular acidification and associated lethality. *acgA* and *acgB* are predicted to be a part of the acetoin biosynthesis operon but are not required for production of acetoin in *V. cholerae*. Biofilm formation and motility phenotypes of Δ*acgA* and Δ*acgB* mutants were similar to that of the wild-type strain (25). However, overproduction of AcgA increased motility and reduced the ability to form biofilms. Overproduction of AcgA and AcgB in the same strain impaired the ability of AcgA to increase motility, suggesting that AcgB may regulate the activity of AcgA; alternatively, AcgA and AcgB may be cognate pairs and thereby antagonize each other's activities.

REGULATION OF c-di-GMP SIGNALING PATHWAYS

c-di-GMP is important to *V. cholerae,* as suggested by the presence of a large number of predicted GGDEF/EAL/HD-GYP proteins. The abundance of these proteins, however, brings with it the complication of specificity and noise in c-di-GMP signaling. One way to ensure specificity is to synthesize DGCs and PDEs under specific conditions only when they are needed. This could be accomplished by transcriptional and posttranscriptional regulation (i.e., mRNA stability and protein stability) of GGDEF/EAL/HD-GYP genes/proteins. Our understanding of the transcriptional regulation of GGDEF/EAL/HD-GYP genes is increasing. In the following section, we will summarize what is known about the transcriptional regulation of these genes.

VpsR- and VpsT-Regulated Genes

A comparison of the transcriptional profile of the RΔ*vpsR* and RΔ*vpsT* mutants with that of the rugose parent during exponential-phase growth revealed that the expression of a set of GGDEF/EAL/HD-GYP genes was altered in the RΔ*vpsR* and RΔ*vpsT* mutants (Table 5) (1, 61). Of the VpsR- and/or VpsT-regulated genes, seven encode proteins containing a GGDEF domain (six have conserved residues and one has degenerate residues), one encodes a protein containing an EAL domain, and one encodes a protein containing GGDEF and EAL domains. The promoter regions of GGDEF/EAL/HD-GYP genes *cdgA, cdgC, cdgD,* and VCA0165 are predicted to have a VpsR-binding site, suggesting that these genes may be directly regulated by VpsR.

QS-Regulated Genes

Population density-dependent regulatory systems, known as QS systems, regulate the expression of genes encoding GGDEF/EAL/HD-GYP proteins in *V. cholerae.* HapR, the master regulator of the QS regulatory system, regulates the transcription of a set of GGDEF/EAL/HD-GYP genes. Initially, HapR-regulated genes were revealed through whole-genome expression profiling comparisons of the wild-type and Δ*hapR* mutant strains (1, 36, 63). More recently, GGDEF/EAL genes transcriptionally regulated by the QS system were systematically identified using *gfp* transcriptional fusions (56). Predicted upstream regulatory regions of 52 of the 53 GGDEF/EAL genes were used to construct *gfp* transcriptional fusions (56). Expression of each of the fusions was then determined in a *V. cholerae* strain harboring a mutated version of *luxO*(D47E) that constitutively mimics the low-cell-density state and compared to expression in the strain lacking *luxO* that constitutively mimics the high-cell-density state. The *gfp* expression evaluated at two different growth phases (prestationary and stationary) showed that 14 of the 52 promoters exhibited QS regulation (Table 5) (56). Of the QS-regulated genes, nine encode proteins containing a GGDEF domain (seven have conserved residues and two have degenerate residues), two encode proteins containing an EAL domain, and three encode proteins containing GGDEF and EAL domains. Of the QS-regulated promoters, HapR was shown to di-

Table 5. Regulation of GGDEF/EAL/HD-GYP genes in *V. cholerae*

Gene identification[b]	Result for[a]:					
	VpsR[c]	VpsT[d]	HapR[e]	QS[f]	cAMP-CRP[g]	VieA[h]
VC0072					–	
VC0130 *(cdpA)*				–		+
VC0137					–	
VC0653 *(rocS)*				+	–	
VC0658 *(cdgI)*					–	
VC0703 *(mbaA)*			+		–	
VC0900 *(cdgG)*	+		–	–		
VC1029 *(cdgB)*	+		–		–	–
VC1067 *(cdgH)*	–	–			–	
VC1086				+	+	
VC1087					+	
VC1185				–		
VC1211					–	
VC1216					+	
VC1348					+	
VC1353					+	–
VC1367 *(cdgE)*	–	–			+	
VC1370			+	+	+	
VC1376					–	
VC1599					+	
VC1641					–	
VC1710			+		+	
VC1851				+		+
VC1934			+		+	
VC2340			+	+	–	
VC2370				–		
VC2497					+	
VC2697			–			
VC2750	+		–		–	
VCA0049					+	
VCA0074 *(cdgA)*	+	+	–	–	–	
VCA0080				+		
VCA0165	–				+	–
VCA0210				+	–	
VCA0217				+	–	
VCA0536	–	–	+			+
VCA0560			–			
VCA0681				+		
VCA0697 *(cdgD)*	–	–	+		+	
VCA0785 *(cdgC)*					–	–
VCA0848				+		
VCA0895				+	+	
VCA0931					+	+
VCA0939				–		
VCA0956 *(cdgF)*				+	–	
VCA0960			–			
VCA0965				+	+	
VCA1082	+	+				

[a] + and – indicate positive and negative regulation, respectively.
[b] Gene list was obtained from The Institute of Genomic Research Comprehensive Microbial Resource.
[c] VpsR transcriptome data were obtained from previously published studies (1, 61).
[d] VpsT transcriptome data were obtained from a previously published study (1).
[e] HapR transcriptome data were obtained from previously published studies (1, 36, 61).
[f] QS transcriptome data were obtained from previously published studies (16, 18, 56).
[g] cAMP-CRP transcriptome data were obtained from a previously published study (11).
[h] VieA transcriptome data were obtained from a previously published study (3).

rectly regulate gene expression by directly binding to the promoter sequences of *cdgA, cdgG,* VCA0080, VC2370, VC1851, and VC1086 (56). Additional analyses showed that overexpression of VC1086, which encodes an EAL domain protein with the conserved residues, led to a decrease in biofilm formation in the *luxO(D47E)* mutant compared to the wild-type parent. Similarly, overexpression of VC1086 in the wild type resulted in decreased *vpsL* expression (56). Altogether, these results suggest that QS regulation of biofilm formation is mediated in part by the components of the c-di-GMP signaling system in *V. cholerae.*

A genetic screen designed to identify HapR-independent QS-regulated genes led to the identification of VCA0939, which is predicted to encode a GGDEF domain-containing protein (18). It was shown that expression of VCA0939 required LuxO, Qrr sRNAs, and Hfq but not HapR. Further analysis revealed that Qrr sRNAs directly base-pair with VCA0939 mRNA and disrupt an inhibitory stem-loop structure in the VCA0939 mRNA, leading to activation of VCA0939 expression (18). Deletion of VCA0939, however, did not lead to altered *vps* gene expression and biofilm formation (18).

In another study, QS regulation of the HD-GYP genes was evaluated using *lux* transcriptional fusions in QS-locked mutants as discussed above (16). In the *V. cholerae* genome, there are nine genes predicted to encode proteins containing an HD-GYP domain. Of these, seven (VCA0681, VCA0895, VC1295, VC1348, VC2340, VCA0210, and VCA0931) contain conserved HD and GYP residues and other residues shown to be critical for PDE activity in RpfG from *Xanthomonas campestris.* Of the seven fusions tested, expression of VCA0681, VCA0895, VC2340, and VCA0210 was found to be activated by QS (16). Further analysis showed that overexpression of VCA0681 reduces *vpsL* and *vpsT* expression and that the HD-GYP domain is required for this activity. However, the VCA0681 mutant did not exhibit any alteration in *vps* expression or biofilm formation (16). The absence of such phenotypes could be due to functional redundancy and/or growth conditions under which the phenotypes were tested.

cAMP-CRP-Regulated Genes

cAMP-CRP negatively regulates biofilm formation in *V. cholerae.* A recent study identified GGDEF/EAL/HD-GYP genes that are regulated through the cAMP-CRP regulatory complex (Table 5) (11). The genes whose transcription levels were higher in the Δ*crp* mutant and encoded proteins with a conserved

GGDEF domain were further studied in the wild type as well as in the Δ*crp* genetic background. This study showed that deletion of GGDEF genes in the wild type did not significantly alter biofilm formation phenotypes (11). However, when the genes were deleted in the enhanced biofilm-forming Δ*crp* genetic background, only the Δ*crp*Δ*cdgA* double-deletion mutant exhibited a reduced biofilm-forming capacity (11). This finding indicates that the increased biofilm-forming capacity of the Δ*crp* mutant is due to increased expression of *cdgA*. It was also observed that deletions of *rocS* and *cdgI* (VC0658, *c*yclic *d*iguanylate) in the Δ*crp* genetic background further enhanced biofilm formation (11). Both RocS and CdgI contain conserved GGDEF-EAL domains. Such proteins can function as either DGCs or PDEs. Although the enzymatic activities of these proteins have not yet been determined, based on biofilm-forming phenotypes, RocS and CdgI appear to function as PDEs under the experimental conditions that were utilized.

The effect of deletion of cAMP-CRP-regulated GGDEF genes on motility was also analyzed (11). Results revealed that, of the GGDEF single-deletion mutants tested, the Δ*rocS* mutant exhibited a decrease in motility, while the Δ*cdgH* mutant exhibited an increase in motility compared to the wild type. The Δ*crp* mutant exhibits a decrease in motility. Deletion of *crp* in individual GGDEF deletion mutants led to a further decrease in motility compared to the corresponding GGDEF single-deletion mutants (11). In summary, of the 10 GGDEF genes tested, deletions of *cdgA, rocS,* and *cdgI* in the Δ*crp* genetic background altered biofilm-forming and/or motility phenotypes.

Expression of EAL and HD-GYP genes was also differentially regulated in the Δ*cyaA* (*cyaA* encodes an adenylate cyclase, which produces cAMP) and/or Δ*crp* mutant (11). It is possible that a decrease in the expression of these genes encoding proteins predicted to have PDE activity may also contribute to an overall increase in the c-di-GMP level, thus giving rise to the increased biofilm-forming capacities observed for the Δ*cyaA* and Δ*crp* mutants.

Based on the biofilm phenotypes of the Δ*vpsR* and Δ*vpsT* mutants, one would expect that expression of the genes encoding a functional DGC would be decreased or expression of the genes encoding functional PDE would be increased in these mutants, leading to a decrease in cellular c-di-GMP levels and, in turn, biofilm formation. With the same logic, the converse would be expected in expression patterns of DGCs and PDEs in Δ*hapR* and Δ*crp* mutants. However, the above-discussed studies revealed that there is no obvious regulatory pattern, indicating that our understanding of the functions of DGC and PDE and the phenotypes regulated by them is far from complete.

VieA-Regulated Genes

VieA-regulated genes were identified by whole-genome level expression profiling under the conditions where virulence gene production is enhanced in the *V. cholerae* classical biotype (3). VieA regulates the expression of eight GGDEF/EAL/HD-GYP genes (Table 5). Of the VieA-regulated genes, *cdpA* and *cdgC* were shown to regulate virulence gene expression; however, it is yet to be determined how VieA regulates the expression of *cdpA* and *cdgC* or whether VieA uses any other GGDEF/EAL/HD-GYP proteins to regulate virulence gene expression in *V. cholerae.*

c-di-GMP-BINDING PROTEINS

PilZ Domain Proteins

c-di-GMP must be sensed to affect cellular processes. Recent studies have shown that a novel protein domain, designated PilZ, is involved in c-di-GMP binding (43). *V. cholerae* contains five predicted PilZ domain proteins: VC0697 (PlzA), VC1885 (PlzB), VC2344 (PlzC), VCA0042 (PlzD), and VCA0735 (PlzE). All of these proteins, except PlzB, contain the conserved residues required to bind c-di-GMP. PlzC and PlzD can bind c-di-GMP, while the others did not bind c-di-GMP (38). Pratt et al. deleted *plzB, plzC,* and *plzD* genes in both the wild-type (*V. cholerae* classical biotype strain) and *vieA(E170A)* (overproduces c-di-GMP due to defective EAL domain that lacks PDE activity) genetic backgrounds (38) and determined their roles in c-di-GMP-regulated cellular processes. Phenotypic analysis revealed that a *plzB* deletion resulted in reduced motility when tested in tryptone-based complex medium, whereas deletion of *plzC* or *plzD* had no effect on motility (38). However, when motility was assayed on minimal medium (M9-NRES), both Δ*plzB* and Δ*plzC* mutants exhibited a motility defect. Strains harboring *vieA(E170A)* exhibited a reduced motility; however, deletion of any of the *plz* genes did not restore motility (38). Taken together, these studies showed that PlzB and PlzC act as positive regulators of motility.

The role of Plz proteins in biofilm formation was also evaluated. A Δ*plzB* mutant had reduced biofilm formation compared to the wild type, and this phenotype was dependent on a functional PilZ domain, as a mutated version harboring a point mutation in the PilZ domain (PlzBR96A) had a similar reduction in biofilm formation (38). This finding indicates that PlzB positively regulates biofilm formation. Deletion of *plzB* or *plzC* in the c-di-GMP-overproducing *vieA(E170A)* strain resulted in a decrease in biofilm formation. This observation indicated that PlzB and PlzC function as positive regulators of biofilm formation when cellular c-di-GMP levels are increased. On the other hand, overproduction of each of the Plz proteins in both the wild-type and *vieA(E170A)* strains did not alter biofilm phenotypes (38).

The role of Plz proteins in intestinal colonization was analyzed using an infant mouse colonization assay. Strains harboring a *plzB* deletion or a mutated version of the PilZ domain exhibited a decrease in colonization (38). While single deletions of *plzC* or *plzD* had no effect on colonization, the double-deletion mutant had a decrease in colonization, which could be complemented by the addition of either gene in *trans*. Although Plz proteins were found to modulate intestinal colonization, PlzB, PlzC, and PlzD were not necessary for c-di-GMP-mediated regulation of cholera toxin production in the wild-type or *vieA(E170A)* strain (38). Taken together, this study showed that PilZ domain proteins can function as c-di-GMP receptors and regulate some of the c-di-GMP-dependent processes in *V. cholerae* under specific growth conditions.

In a recent study, the contribution of the PilZ proteins to *vps* gene expression and rugosity was analyzed in the rugose prototype strain of *V. cholerae* El Tor strain A1552 (2). In-frame deletion mutants of each of the five PilZ-encoding genes, as well as double-, triple-, quadruple-, and pentuple-deletion mutants were generated. Based on these colony morphologies, as well as biofilm-forming phenotypes, it was concluded that these PilZ domain proteins contribute minimally to rugosity and that there must be additional c-di-GMP-binding proteins (e.g., non-PilZ domain) in *V. cholerae* involved in regulating rugosity and biofilm formation.

Riboswitch

Riboswitches are structured mRNA domains that control gene expression by binding to small metabolites (7, 57). The analysis of bacterial genome sequences for the presence of a conserved RNA domain (GEMM RNA) with characteristics of riboswitches revealed that *V. cholerae* has two such sequences residing upstream of the *gbpA* (Vc1) and VC1722 (Vc2) genes (48). Biochemical analysis, via in-line probing with and without 100 μM c-di-GMP, showed changes in the spontaneous cleavage patterns, revealing that these sequences can function as aptamers for c-di-GMP. Vc2 had a dissociation constant (K_d) of 1 nM for c-di-GMP, suggesting that Vc2 has a very high affinity for c-di-GMP (48). Further genetic analysis revealed that the Vc2 aptamer functions as an on switch, as mutated versions of the aptamer, where conserved structure was abolished, were unable to induce gene expression. Interestingly, the transcription of Vc2 was higher in rugose variants, which has high c-di-GMP levels (5). It is suggested that in cells with high levels of c-di-GMP, the Vc2 on switch leads to higher gene expression. The role of riboswitches in c-di-GMP sensing is discussed in chapter 15.

Candidate c-di-GMP Receptor Proteins

A non-PilZ domain c-di-GMP receptor protein, PelD (from *Pseudomonas aeruginosa*), was recently identified (28). PelD harbors an RXXD motif similar to the ones found in the I-site (inhibition site) of DGCs. The RXXD motif of PelD is able to bind c-di-GMP (28). As discussed above, CdgG regulates biofilm formation. CdgG harbors the conserved RXXD motif in its GGDEF domain but does not have a DGC activity. Mutational and complementational analyses revealed that while SGEEF residues in the active site of CdgG are not required, RDSD residues at the I-site, which is predicted to bind c-di-GMP, are essential for the function of CdgG. It is yet to be determined whether CdgG is able to bind c-di-GMP.

Recently, FleQ, the master regulator of flagellar gene expression in *P. aeruginosa*, was shown to be a c-di-GMP-responsive transcriptional regulator (19). This study showed that FleQ binds to the *pel* promoter, and its binding is inhibited by c-di-GMP (19) (see chapter 11). *V. cholerae* has a FleQ homolog, FlrA; whether FlrA is also a c-di-GMP-responsive transcriptional regulator is yet to be determined.

CONCLUSIONS

There remains a significant information gap in our understanding of the molecular mechanisms by which c-di-GMP regulates biofilm formation, motility, and virulence in *V. cholerae*. However, the identity of GGDEF/EAL/HD-GYP proteins involved in these processes, as well as regulatory interaction points where c-di-GMP feeds into the previously identified regulatory networks controlling biofilm formation, motility, and virulence, is emerging. Our current understanding of the transcriptional regulation of biofilm formation, motility, and virulence by c-di-GMP in *V. cholerae* is summarized in Fig. 3.

The abundance of the GGDEF/EAL/HD-GYP proteins in *V. cholerae*, similar to other organisms, is

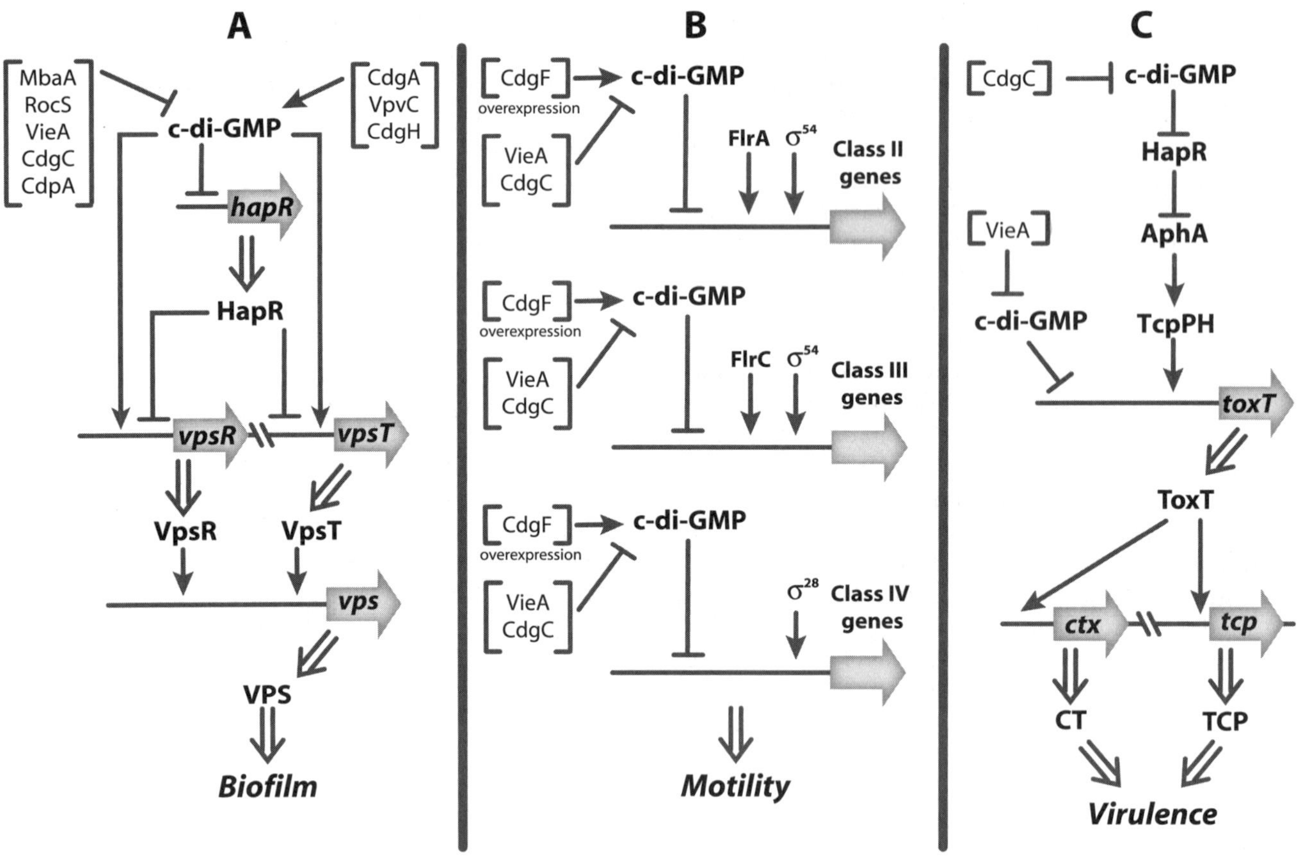

Figure 3. Model of transcriptional regulation by c-di-GMP in *V. cholerae*. (A) An increase in c-di-GMP level, either by increasing the transcription or activity of a key DGC (CdgA, VpvC, or CdgH) or by decreasing the transcription or activity of a key PDE (MbaA, RocS, VieA, CdgC, or CdpA), enhances biofilm formation. High c-di-GMP levels could decrease *hapR* expression and/or increase *vpsR* and *vpsT* expression, leading to an increase in biofilm formation. (B) Increased levels of c-di-GMP, either by overproduction of a DGC (CdgF) or deletions of PDEs (VieA or CdgC), repress the transcription of class III and class IV flagellar genes. Although the mechanism by which class III gene expression is repressed is not known, a decrease in the expression of class IV genes can in part be explained by a decrease in transcription of *fliA*, encoding σ^{28}, necessary for class IV gene expression. (C) In the classical biotype, the PDE VieA enhances virulence gene expression by decreasing the intracellular c-di-GMP level, which in turn activates *toxT* and *ctx* expression. In El Tor strains, the PDE CdgC modulates in vitro virulence factor production. CdgC positively regulates *hapR* expression; hence, expression of *aphA* and *tcpP* is increased in the Δ*cdgC* mutant, leading to increased expression of *ctx* and *tcp*. In El Tor strains producing low levels of c-di-GMP, mutation of *cdgC* does not affect *ctx* expression but causes a slight decrease in *tcp* gene expression.

expected to be a problem in achieving specificity in c-di-GMP signaling due to potential cross talk or noise in the system. It has been proposed that spatial sequestration of these proteins to create localized c-di-GMP pools in the cell could reduce cross talk. Interestingly, half of the GGDEF/EAL/HD-GYP proteins are predicted to have one or more transmembrane domains (Fig. 1). In particular, among 31 GGDEF proteins, 21 carry at least one transmembrane helix. It is likely that each of these proteins could form independent c-di-GMP signaling clusters in different regions of the cell together with their effector proteins (transcriptional regulators or proteins that can change the activity/function of transcriptional regulators) and target proteins. It is also likely that each of these proteins generates a different c-di-GMP pool that can be degraded by the cognate PDEs. c-di-GMP signaling complexes can be dynamic, and effector proteins may be released from these complexes and participate directly or indirectly in gene expression. Alternatively, some portion of c-di-GMP may be freely diffusible into the cytoplasm and c-di-GMP might interact with cytoplasmically localized c-di-GMP-binding transcriptional regulators, thereby regulating gene expression. Although the presence of parallel c-di-GMP signaling pathways regulating biofilm formation has been shown through epistasis analysis (2, 30), cognate c-di-GMP signaling pairs (DGC and PDE) have not been identified. Hence, the identification of structural and regulatory proteins interacting with key c-di-GMP-producing proteins and c-di-GMP-degrading proteins is necessary to gain a complete understanding of molecular mechanisms by which c-di-GMP controls such diverse cellular processes in *V. cholerae.*

Another critical question that remains unanswered relates to the regulation of activities of GGDEF/EAL/HD-GYP proteins. In *V. cholerae,* similar to other organisms, DGC and PDE enzymatic domains are linked to various sensory domains and regulatory domains. This suggests that the perception and cellular responses to various environmental and intracellular signals could be mediated through c-di-GMP signaling. Intriguingly, the *Vibrio* species contain much larger numbers of DGCs and PDEs than most other bacteria (14). The abundance of enzymes controlling the synthesis and degradation of c-di-GMP in vibrios indicates the importance of c-di-GMP signaling to the biology of vibrios. The life cycle of *V. cholerae* involves relatively short-term growth within the human intestine and prolonged survival in aquatic habitats. Thus, during growth in the human intestine or growth in aquatic ecosystems as well as during transitions between intestinal and aquatic lifestyles, the pathogen needs to sense, respond, and adapt to its environment. Studies on the importance of c-di-GMP signaling in such processes are an active area of investigation. There is much to be discovered about the mechanisms by which the c-di-GMP signaling system regulates biofilm formation, motility, and virulence in *V. cholerae.* A better understanding of mechanisms of c-di-GMP signaling in *V. cholerae* biology will not only be useful for the development of future strategies for predicting and controlling cholera epidemics but will also improve our understanding of the biology of a quintessential human pathogen.

Acknowledgments. c-di-GMP signaling work in the Yildiz laboratory is supported by a grant from the NIH (AI055987).

We thank Nicholas Shikuma, Jiunn Fong, Kivanc Bilecen, and Karen Ottemann for their valuable comments on the manuscript.

REFERENCES

1. **Beyhan, S., K. Bilecen, S. R. Salama, C. Casper-Lindley, and F. H. Yildiz.** 2007. Regulation of rugosity and biofilm formation in *Vibrio cholerae:* comparison of VpsT and VpsR regulons and epistasis analysis of *vpsT, vpsR,* and *hapR. J. Bacteriol.* **189:**388–402.
2. **Beyhan, S., L. S. Odell, and F. H. Yildiz.** 2008. Identification and characterization of cyclic diguanylate signaling systems controlling rugosity in *Vibrio cholerae. J. Bacteriol.* **190:**7392–7405.
3. **Beyhan, S., A. D. Tischler, A. Camilli, and F. H. Yildiz.** 2006. Differences in gene expression between the classical and El Tor biotypes of *Vibrio cholerae* O1. *Infect. Immun.* **74:**3633–3642.
4. **Beyhan, S., A. D. Tischler, A. Camilli, and F. H. Yildiz.** 2006. Transcriptome and phenotypic responses of *Vibrio cholerae* to increased cyclic di-GMP level. *J. Bacteriol.* **188:**3600–3613.
5. **Beyhan, S., and F. H. Yildiz.** 2007. Smooth to rugose phase variation in *Vibrio cholerae* can be mediated by a single nucleotide change that targets c-di-GMP signalling pathway. *Mol. Microbiol.* **63:**995–1007.
6. **Bomchil, N., P. Watnick, and R. Kolter.** 2003. Identification and characterization of a *Vibrio cholerae* gene, *mbaA,* involved in maintenance of biofilm architecture. *J. Bacteriol.* **185:**1384–1390.
7. **Breaker, R. R.** 2008. Complex riboswitches. *Science* **319:** 1795–1797.
8. **Camilli, A., and J. J. Mekalanos.** 1995. Use of recombinase gene fusions to identify *Vibrio cholerae* genes induced during infection. *Mol. Microbiol.* **18:**671–683.
9. **Casper-Lindley, C., and F. H. Yildiz.** 2004. VpsT is a transcriptional regulator required for expression of *vps* biosynthesis genes and the development of rugose colonial morphology in *Vibrio cholerae* O1 El Tor. *J. Bacteriol.* **186:**1574–1578.
10. **Fong, J. C., K. Karplus, G. K. Schoolnik, and F. H. Yildiz.** 2006. Identification and characterization of RbmA, a novel protein required for the development of rugose colony morphology and biofilm structure in *Vibrio cholerae. J. Bacteriol.* **188:**1049–1059.
11. **Fong, J. C., and F. H. Yildiz.** 2008. Interplay between cyclic AMP-cyclic AMP receptor protein and cyclic di-GMP signaling in *Vibrio cholerae* biofilm formation. *J. Bacteriol.* **190:**6646–6659.
12. **Fong, J. C., and F. H. Yildiz.** 2007. The *rbmBCDEF* gene cluster modulates development of rugose colony morphology and biofilm formation in *Vibrio cholerae. J. Bacteriol.* **189:** 2319–2330.
13. **Freeman, J. A., and B. L. Bassler.** 1999. A genetic analysis of the function of LuxO, a two-component response regulator involved in quorum sensing in *Vibrio harveyi. Mol. Microbiol.* **31:**665–677.
14. **Galperin, M. Y., A. N. Nikolskaya, and E. V. Koonin.** 2001. Novel domains of the prokaryotic two-component signal transduction systems. *FEMS Microbiol. Lett.* **203:**11–21.
15. **Gosink, K. K., and C. C. Hase.** 2000. Requirements for conversion of the Na^+-driven flagellar motor of *Vibrio cholerae* to the H^+-driven motor of *Escherichia coli. J. Bacteriol.* **182:** 4234–4240.
16. **Hammer, B. K., and B. L. Bassler.** 2009. Distinct sensory pathways in *Vibrio cholerae* El Tor and classical biotypes modulate cyclic dimeric GMP levels to control biofilm formation. *J. Bacteriol.* **191:**169–177.
17. **Hammer, B. K., and B. L. Bassler.** 2003. Quorum sensing controls biofilm formation in *Vibrio cholerae. Mol. Microbiol.* **50:**101–104.
18. **Hammer, B. K., and B. L. Bassler.** 2007. Regulatory small RNAs circumvent the conventional quorum sensing pathway in pandemic *Vibrio cholerae. Proc. Natl. Acad. Sci. USA* **104:** 11145–11149.
19. **Hickman, J. W., and C. S. Harwood.** 2008. Identification of FleQ from *Pseudomonas aeruginosa* as a c-di-GMP-responsive transcription factor. *Mol. Microbiol.* **69:**376–389.
20. **Islam, M. S., M. I. Jahid, M. M. Rahman, M. Z. Rahman, M. S. Islam, M. S. Kabir, D. A. Sack, and G. K. Schoolnik.** 2007. Biofilm acts as a microenvironment for plankton-associated *Vibrio cholerae* in the aquatic environment of Bangladesh. *Microbiol. Immunol.* **51:**369–379.
21. **Iwanaga, M., K. Yamamoto, N. Higa, Y. Ichinose, N. Nakasone, and M. Tanabe.** 1986. Culture conditions for stimu-

lating cholera toxin production by *Vibrio cholerae* O1 El Tor. *Microbiol. Immunol.* **30:**1075–1083.
22. **Kaper, J. B., J. G. Morris, Jr., and M. M. Levine.** 1995. Cholera. *Clin. Microbiol. Rev.* **8:**48–86.
23. **Karatan, E., T. R. Duncan, and P. I. Watnick.** 2005. NspS, a predicted polyamine sensor, mediates activation of *Vibrio cholerae* biofilm formation by norspermidine. *J. Bacteriol.* **187:** 7434–7443.
24. **Kojima, S., K. Yamamoto, I. Kawagishi, and M. Homma.** 1999. The polar flagellar motor of *Vibrio cholerae* is driven by an Na^+ motive force. *J. Bacteriol.* **181:**1927–1930.
25. **Kovacikova, G., W. Lin, and K. Skorupski.** 2005. Dual regulation of genes involved in acetoin biosynthesis and motility/biofilm formation by the virulence activator AphA and the acetate-responsive LysR-type regulator AlsR in *Vibrio cholerae. Mol. Microbiol.* **57:**420–433.
26. **Kovacikova, G., and K. Skorupski.** 2002. Regulation of virulence gene expression in *Vibrio cholerae* by quorum sensing: HapR functions at the *aphA* promoter. *Mol. Microbiol.* **46:** 1135–1147.
27. **Kovacikova, G., and K. Skorupski.** 1999. A *Vibrio cholerae* LysR homolog, AphB, cooperates with AphA at the *tcpPH* promoter to activate expression of the ToxR virulence cascade. *J. Bacteriol.* **181:**4250–4256.
28. **Lee, V. T., J. M. Matewish, J. L. Kessler, M. Hyodo, Y. Hayakawa, and S. Lory.** 2007. A cyclic-di-GMP receptor required for bacterial exopolysaccharide production. *Mol. Microbiol.* **65:**1474–1484.
29. **Liang, W., A. J. Silva, and J. A. Benitez.** 2007. The cyclic AMP receptor protein modulates colonial morphology in *Vibrio cholerae. Appl. Environ. Microbiol.* **73:**7482–7487.
30. **Lim, B., S. Beyhan, J. Meir, and F. H. Yildiz.** 2006. Cyclic-diGMP signal transduction systems in *Vibrio cholerae:* modulation of rugosity and biofilm formation. *Mol. Microbiol.* **60:** 331–348.
31. **Lim, B., S. Beyhan, and F. H. Yildiz.** 2007. Regulation of *Vibrio* polysaccharide synthesis and virulence factor production by CdgC, a GGDEF-EAL domain protein, in *Vibrio cholerae. J. Bacteriol.* **189:**717–729.
32. **Martinez-Wilson, H. F., R. Tamayo, A. D. Tischler, D. W. Lazinski, and A. Camilli.** 2008. The *Vibrio cholerae* hybrid sensor kinase VieS contributes to motility and biofilm regulation by altering the cyclic diguanylate level. *J. Bacteriol.* **190:** 6439–6447.
33. **Matson, J. S., J. H. Withey, and V. J. DiRita.** 2007. Regulatory networks controlling *Vibrio cholerae* virulence gene expression. *Infect. Immun.* **75:**5542–5549.
34. **McCarter, L. L.** 2001. Polar flagellar motility of the *Vibrionaceae. Microbiol. Mol. Biol. Rev.* **65:**445–462.
35. **Morris, J. G., Jr., M. B. Sztein, E. W. Rice, J. P. Nataro, G. A. Losonsky, P. Panigrahi, C. O. Tacket, and J. A. Johnson.** 1996. *Vibrio cholerae* O1 can assume a chlorine-resistant rugose survival form that is virulent for humans. *J. Infect. Dis.* **174:**1364–1368.
36. **Nielsen, A. T., N. A. Dolganov, G. Otto, M. C. Miller, C. Y. Wu, and G. K. Schoolnik.** 2006. RpoS controls the *Vibrio cholerae* mucosal escape response. *PLoS Pathog.* **2:**e109.
37. **Pasmore, M., and J. W. Costerton.** 2003. Biofilms, bacterial signaling, and their ties to marine biology. *J. Ind. Microbiol. Biotechnol.* **30:**407–413.
38. **Pratt, J. T., R. Tamayo, A. D. Tischler, and A. Camilli.** 2007. PilZ domain proteins bind cyclic diguanylate and regulate diverse processes in *Vibrio cholerae. J. Biol. Chem.* **282:**12860–12870.
39. **Prouty, M. G., N. E. Correa, and K. E. Klose.** 2001. The novel sigma54- and sigma28-dependent flagellar gene transcription hierarchy of *Vibrio cholerae. Mol. Microbiol.* **39:**1595–1609.
40. **Rashid, M. H., C. Rajanna, A. Ali, and D. K. Karaolis.** 2003. Identification of genes involved in the switch between the smooth and rugose phenotypes of *Vibrio cholerae. FEMS Microbiol. Lett.* **227:**113–119.
41. **Rice, E. W., C. J. Johnson, R. M. Clark, K. R. Fox, D. J. Reasoner, M. E. Dunnigan, P. Panigrahi, J. A. Johnson, and J. G. Morris, Jr.** 1992. Chlorine and survival of "rugose" *Vibrio cholerae. Lancet* **340:**740.
42. **Ryan, R. P., Y. Fouhy, J. F. Lucey, L. C. Crossman, S. Spiro, Y. W. He, L. H. Zhang, S. Heeb, M. Camara, P. Williams, and J. M. Dow.** 2006. Cell-cell signaling in *Xanthomonas campestris* involves an HD-GYP domain protein that functions in cyclic di-GMP turnover. *Proc. Natl. Acad. Sci. USA* **103:**6712–6717.
43. **Ryjenkov, D. A., R. Simm, U. Romling, and M. Gomelsky.** 2006. The PilZ domain is a receptor for the second messenger c-di-GMP: the PilZ domain protein YcgR controls motility in enterobacteria. *J. Biol. Chem.* **281:**30310–30314.
44. **Ryjenkov, D. A., M. Tarutina, O. V. Moskvin, and M. Gomelsky.** 2005. Cyclic diguanylate is a ubiquitous signaling molecule in bacteria: insights into biochemistry of the GGDEF protein domain. *J. Bacteriol.* **187:**1792–1798.
45. **Schild, S., R. Tamayo, E. J. Nelson, F. Qadri, S. B. Calderwood, and A. Camilli.** 2007. Genes induced late in infection increase fitness of *Vibrio cholerae* after release into the environment. *Cell Host Microbe* **2:**264–277.
46. **Schmidt, A. J., D. A. Ryjenkov, and M. Gomelsky.** 2005. The ubiquitous protein domain EAL is a cyclic diguanylate-specific phosphodiesterase: enzymatically active and inactive EAL domains. *J. Bacteriol.* **187:**4774–4781.
47. **Skorupski, K., and R. K. Taylor.** 1999. A new level in the *Vibrio cholerae* ToxR virulence cascade: AphA is required for transcriptional activation of the *tcpPH* operon. *Mol. Microbiol.* **31:**763–771.
48. **Sudarsan, N., E. R. Lee, Z. Weinberg, R. H. Moy, J. N. Kim, K. H. Link, and R. R. Breaker.** 2008. Riboswitches in eubacteria sense the second messenger cyclic di-GMP. *Science* **321:** 411–413.
49. **Tamayo, R., S. Schild, J. T. Pratt, and A. Camilli.** 2008. Role of cyclic di-GMP during El Tor biotype *Vibrio cholerae* infection: characterization of the in vivo-induced cyclic di-GMP phosphodiesterase CdpA. *Infect. Immun.* **76:**1617–1627.
50. **Tamayo, R., A. D. Tischler, and A. Camilli.** 2005. The EAL domain protein VieA is a cyclic diguanylate phosphodiesterase. *J. Biol. Chem.* **280:**33324–33330.
51. **Tischler, A. D., and A. Camilli.** 2004. Cyclic diguanylate (c-di-GMP) regulates *Vibrio cholerae* biofilm formation. *Mol. Microbiol.* **53:**857–869.
52. **Tischler, A. D., and A. Camilli.** 2005. Cyclic diguanylate regulates *Vibrio cholerae* virulence gene expression. *Infect. Immun.* **73:**5873–5882.
53. **Tischler, A. D., S. H. Lee, and A. Camilli.** 2002. The *Vibrio cholerae vieSAB* locus encodes a pathway contributing to cholera toxin production. *J. Bacteriol.* **184:**4104–4113.
54. **Vance, R. E., J. Zhu, and J. J. Mekalanos.** 2003. A constitutively active variant of the quorum-sensing regulator LuxO affects protease production and biofilm formation in *Vibrio cholerae. Infect. Immun.* **71:**2571–2576.
55. **Waters, C. M., and B. L. Bassler.** 2005. Quorum sensing: cell-to-cell communication in bacteria. *Annu. Rev. Cell Dev. Biol.* **21:**319–346.
56. **Waters, C. M., W. Lu, J. D. Rabinowitz, and B. L. Bassler.** 2008. Quorum sensing controls biofilm formation in *Vibrio cholerae* through modulation of cyclic di-GMP levels and repression of *vpsT. J. Bacteriol.* **190:**2527–2536.

57. **Winkler, W. C., and R. R. Breaker.** 2005. Regulation of bacterial gene expression by riboswitches. *Annu. Rev. Microbiol.* **59:**487–517.
58. **Wolfe, A. J., and K. L. Visick.** 2008. Get the message out: cyclic-Di-GMP regulates multiple levels of flagellum-based motility. *J. Bacteriol.* **190:**463–475.
59. **Yildiz, F. H., N. A. Dolganov, and G. K. Schoolnik.** 2001. VpsR, a member of the response regulators of the two-component regulatory systems, is required for expression of *vps* biosynthesis genes and EPS^{ETr}-associated phenotypes in *Vibrio cholerae* O1 El Tor. *J. Bacteriol.* **183:**1716–1726.
60. **Yildiz, F. H., and R. Kolter.** 2008. Genetics and microbiology of biofilm formation by *Vibrio cholerae,* p. 123–139. *In* S. M. Faruque and G. B. Nair (ed.), Vibrio cholerae: *Genomics and Molecular Biology.* Caister Academic Press, Norfolk, United Kingdom.
61. **Yildiz, F. H., X. S. Liu, A. Heydorn, and G. K. Schoolnik.** 2004. Molecular analysis of rugosity in a *Vibrio cholerae* O1 El Tor phase variant. *Mol. Microbiol.* **53:**497–515.
62. **Yildiz, F. H., and G. K. Schoolnik.** 1999. *Vibrio cholerae* O1 El Tor: identification of a gene cluster required for the rugose colony type, exopolysaccharide production, chlorine resistance, and biofilm formation. *Proc. Natl. Acad. Sci. USA* **96:** 4028–4033.
63. **Zhu, J., and J. J. Mekalanos.** 2003. Quorum sensing-dependent biofilms enhance colonization in *Vibrio cholerae. Dev. Cell* **5:**647–656.

The Second Messenger Cyclic Di-GMP
Edited by A. J. Wolfe and K. L. Visick

Chapter 18

Role of Cyclic Di-GMP in Biofilm Development and Signaling in *Yersinia pestis*

ROBERT D. PERRY AND ALEXANDER G. BOBROV

Yersinia pestis, a gram-negative bacterium and member of the *Enterobacteriaceae* family, is the causative agent of bubonic, septicemic, and pneumonic plague. Humans are accidental hosts that are highly susceptible to plague. In humans, bubonic plague from a flea bite can develop into a lung infection (secondary pneumonic plague); this can lead to respiratory droplet transmission (primary pneumonic plague) (Fig. 1). While *Y. pestis* is infamous for causing three pandemics (Justinian plague, 541 to 544 A.D.; the second pandemic, 8th to 14th centuries; and the third pandemic, 1855 to the present), it remains a serious problem with ~1,800 WHO-reported human plague cases occuring annually worldwide. The number of yearly cases is highly dependent upon epidemic outbreaks, and the WHO-reported cases are likely only a fraction of the actual cases. Epidemic outbreaks with hundreds of human plague cases still occur frequently in 14 countries (23, 52, 54, 55, 82).

Plague is a zoonotic disease primarily affecting rodents and their associated fleas; the organism has an obligate life cycle, alternating between growth in mammals and fleas (Fig. 1). After infecting a flea via a bacteremic/septicemic blood meal, the bacterium grows in the midgut (stomach) of fleas and does not spread beyond the flea alimentary tract. The bacteria colonize the proventriculus (a spiny valve between the stomach and esophagus) by growing as a biofilm mass that partially or fully blocks that organ (flea blockage) in many but not all fleas. In the oriental rat flea *(Xenopsylla cheopis),* a classic plague vector, blockage can occur as early as 5 days after taking an infected blood meal. These blocked fleas are unable to pump fresh blood into the midgut and increase their feeding attempts, which dislodge bacteria from the biofilm, contaminating blood that flows from the flea back into the mammal. However, early-phase transmission by unblocked fleas is also important in infection of mammals. After transmission of plague bacilli from fleas to mammals, the bacteria likely revert to a planktonic state to invade and disseminate within the mammalian host and cause acute disease. *Y. pestis* cells spread from the site of the infected flea bite to the regional lymph nodes, grow to high numbers, causing formation of a bubo, and spill into the bloodstream where bacteria are removed in the liver and spleen. Growth continues in the liver and spleen, spreads to other organs, and causes septicemia. Fleas feeding on septicemic/bacteremic animals complete the life cycle (Fig. 1). Maintenance of plague in nature appears to depend, at least partially, on the ability of *Y. pestis* to switch between biofilm and planktonic states (23, 52, 54, 55). The global second messenger signaling molecule cyclic di-GMP (c-di-GMP) plays a critical role in controlling this *Y. pestis* lifestyle switch (Fig. 2) (8, 36, 40).

c-di-GMP governs the transition between biofilm and planktonic states for a number of other bacteria, including *Vibrio cholerae, Salmonella enterica* serovar Typhimurium, *Pseudomonas aeruginosa, Pseudomonas fluorescens, Escherichia coli,* and *Shewanella oneidensis.* In general, increased levels of c-di-GMP inside bacterial cells correlate with biofilm formation and expression of exopolysaccharide (EPS) and adherence factors, while decreased intracellular levels stimulate motility and a planktonic lifestyle (see chapter 8). c-di-GMP homeostasis is carried out by GGDEF domain proteins, cyclic diguanylate cyclases (DGCs), which synthesize c-di-GMP, as well as EAL

Robert D. Perry and Alexander G. Bobrov • Department of Microbiology, Immunology, and Molecular Genetics, University of Kentucky, Lexington, KY 40536-0298.

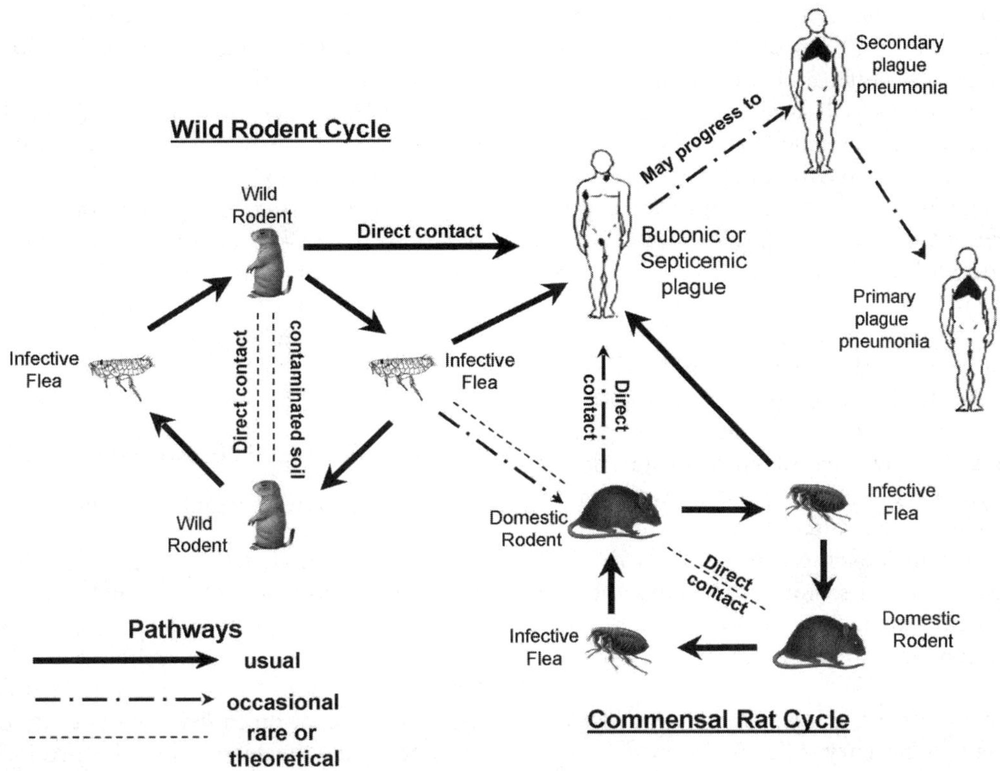

Figure 1. Transmission routes of plague. For simplicity, some occasional transmission routes are not shown. Transmission via ingestion of infected rodents by wild or domesticated carnivores and subsequent interaction with humans is not shown. This figure was slightly modified from a figure kindly provided by Ken Gage, CDC.

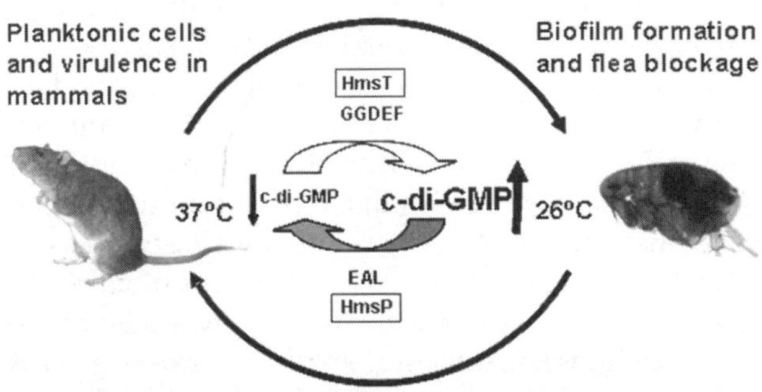

Figure 2. Proposed model for life cycle regulation of *Y. pestis* by c-di-GMP. We hypothesize that c-di-GMP reciprocally regulates biofilm and virulence factors in *Y. pestis*. High levels of c-di-GMP mediated by the GGDEF domain protein HmsT have been proven to be required for biofilm formation in the flea. During growth in mammals, decreases in c-di-GMP in *Y. pestis* cells due to the EAL domain protein HmsP lead to a transition to a planktonic state. We hypothesize that decreased c-di-GMP levels may also enhance expression of mammalian virulence factors. The image of the blocked flea was reproduced from Fig. 1B from Hinnebusch et al. (29) with permission of the American Association for the Advancement of Science.

and HD-GYP domain proteins, c-di-GMP phosphodiesterases (PDEs), which degrade this nucleotide (12, 35, 58, 69, 81) (see also chapters 4 and 5).

Like many other bacteria, the *Y. pestis* genome encodes multiple open reading frames (ORFs) with DGC/GGDEF and/or PDE/EAL/HD-GYP domains. To date, *Y. pestis* c-di-GMP signaling has been implicated only in biofilm regulation and development.

BIOFILM FORMATION IN *Y. PESTIS*

The *hmsHFRS* operon, predicted to encode enzymes required for the synthesis of polymeric β-1,6-*N*-acetyl-D-glucosamine (poly-β-1,6-GlcNAc), is essential for biofilm development in vitro, in fleas for colonization and eventual blockage of the proventriculus, and on the mouthparts of *Caenorhabditis elegans*. *Y. pestis* was reported to express an Hms-dependent biofilm EPS, based on electron microscopy, reaction with antisera against poly-β-1,6-GlcNAc, and on the ability of cells to bind Congo red (CR), calcofluor white, and ruthenium red, each shown to stain various polysaccharides (13, 33, 34, 36, 40, 53). Orthologues of the *hmsHFRS* genes have been shown to be necessary for adherence to surfaces and cell-cell attachment during biofilm formation in *Staphylococcus epidermidis, Staphylococcus aureus, E. coli,* and *Actinobacillus pleuropneumoniae* (6, 32, 46, 75). All four HmsHFRS proteins have 50 to 83% similarity to products encoded by the *E. coli pga-ABCD* operon and are required for biofilm development in vitro (45). On the basis of these similarities among Hms, staphylococcal Ica, and *E. coli* Pga proteins as well as the presence of known catalytic domains and various experimental studies, we have developed a model of the function of individual Hms proteins in biofilm development (Fig. 3A). One unique feature of *Y. pestis* biofilm development is its temperature regulation. Biofilm is formed at ambient but not mammalian temperatures (Fig. 2 and 3) (33, 68).

HmsR, a putative GT2 family glycosyltransferase, likely plays a primary role in the synthesis of poly-β-1,6-GlcNAc-like molecules as part of the EPS. Single alanine substitutions in a conserved D, D, D35QXRW glycosyltransferase motif of HmsR cause loss of biofilm formation and CR binding in *Y. pestis* (21). The D, D, D35QXRW motif is essential for enzymatic activity of glycosyltransferases from several bacteria (10, 65). Like several other GT2 glycosyltransferases (11, 28), inner membrane (IM) protein HmsR has this domain in the cytoplasm (6, 51). Recently, using immunodot blot analysis with monospecific antisera against polysaccharide intercellular adhesion of *S. epidermidis* (previously used to identify poly-β-1,6-GlcNAc in *S. epidermidis, E. coli,* and *Actinobacillus* species) (25, 37, 47), we detected a plague poly-β-1,6-GlcNAc-like EPS that requires *hmsHFRS* gene products for its synthesis (Fig. 4) (6).

There is evidence that HmsR interacts with the IM protein HmsS whose function is unknown (6, 21, 45, 51). PgaD, an HmsS orthologue in *E. coli,* is essential for the production of poly-β-1,6-GlcNAc (31). We hypothesize that HmsS assists HmsR in the assembly of polysaccharide repeating units (Fig. 3A).

HmsF is an outer membrane protein with a putative polysaccharide deacetylase domain. Single alanine substitutions in conserved residues of the deacetylase domain of HmsF completely abrogate biofilm formation (21, 45, 51). Some of these functional residues have been shown to be essential for enzymatic activity of the peptidoglycan deacetylase SpPgdA from *Streptococcus pneumoniae* (5). Similar to its homologue IcaB in *S. epidermidis,* HmsF is likely involved in the modification of poly-β-1,6-GlcNAc (Fig. 3A), which is important for *S. epidermidis* cell-cell interaction (74).

HmsH is a proven outer membrane protein predicted to have a β-barrel structure with a large periplasmic domain (51). Our hypothesis is that HmsH is the porin through which the polysaccharide chain passes to reach the bacterial surface (Fig. 3A). PgaA, a homologue of HmsH in *E. coli,* has been recently shown to be important for export of poly-β-1,6-GlcNAc to the bacterial surface (31).

c-di-GMP AND THE BIOFILM PHENOTYPE

HmsT and HmsP Regulate Biofilm Development

To our knowledge, *Y. pestis* is the third organism for which a biological phenotype for a c-di-GMP-metabolizing enzyme, HmsT, a DGC, was described (27, 36, 40, 66). It is a positive regulator of CR and hemin binding—an indicator of biofilm formation in *Y. pestis* (36). A mutation in *hmsT* causes loss of CR binding and a drastic decrease in the production of the β-1,6-GlcNAc-like EPS (Fig. 4) and biofilm formation (Fig. 5), while overexpression of HmsT results in overproduction of poly-β-1,6-GlcNAc (Fig. 4) and biofilm at ambient temperatures (Fig. 5) as well as loss of temperature regulation of biofilm development [Hms(Con)] (6, 27, 36, 40). HmsT is an IM protein with four transmembrane (TM) domains and N- and C-terminal cytoplasmic domains (Fig. 6). The C-terminal region contains the GGDEF domain associated with DGC activity. Indeed, the overexpression of HmsT in *Y. pestis* enhances c-di-GMP production, indicating that HmsT has DGC activity (66). An alanine substitution in each of the GGEE residues of the GGDEF domain of HmsT prevented complementation of the chromosomal *hmsT* mutation. An alanine substitution in residue F of this motif resulted in an unstable protein (40). A *Salmonella enterica* serovar Typhimurium DGC, AdrA, restores biofilm formation in the *Y. pestis hmsT* mutant, while *hmsT* restores cellulose production in the *S. enterica adrA* mutant (66). Thus, the DGC activity of HmsT is an essential component of the positive regulation of plague biofilm development. However, HmsT can be replaced by overexpression of foreign DGCs (66).

In addition to *hmsT,* genomic analysis of *Y. pestis* KIM10+ (15) identified five ORFs with GGDEF domains. In three ORFs *(hmsP, y0203,* and *y3389)* encoding both GGDEF and EAL domains, the GGDEF domains have substitutions in amino acid residues shown to be essential for DGC activity. Thus, the GGDEF-EAL proteins likely possess no

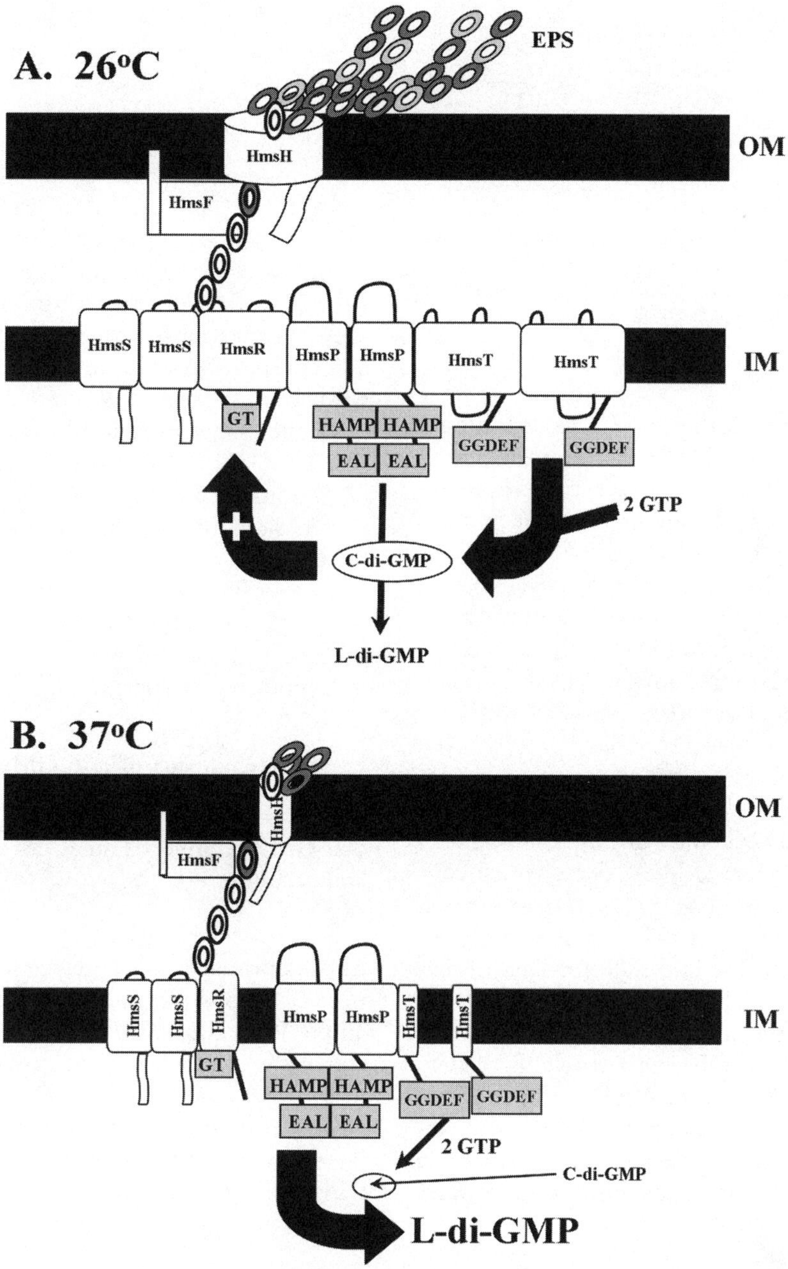

Figure 3. Hms-dependent biofilm model of protein interactions and temperature regulation. Chains of circles represent linked monomers of the EPS component of the biofilm (different shades represent acetylated and deacetylated monomers). Predicted or hypothesized protein activities: HmsH, porin for polysaccharide export; HmsF, polysaccharide deacetylase; HmsR, glycosyl transferase; HmsS, assists HmsR in EPS synthesis. HmsT is a proven DGC (GGDEF domain), while HmsP is a proven c-di-GMP PDE (EAL domain). HmsP and HmsT form homodimers with HmsP interacting with HmsR and HmsT. HmsR also interacts with HmsS. HmsS is shown as a homodimer due to preliminary results that have not yet been confirmed. Cytoplasmic domains and periplasmic loops of IM proteins are shown as well as the putative lipoporotein linkage of HmsF and the periplasmic domain of HmsH. For simplicity, the defective GGDEF domain of HmsP is not shown. The plus sign indicates likely stimulation of HmsR enzymatic activity by c-di-GMP. At 26 to 34°C, Hms proteins are highly expressed (A), while at 37°C, the levels of Hms proteins are reduced to varying degrees (B). HmsT is degraded, thus reducing c-di-GMP levels. This, along with low levels of HmsR and HmsH, prevents significant biofilm formation. L-di-GMP, linear di-GMP; OM, outer membrane; GT, glycosyltransferase.

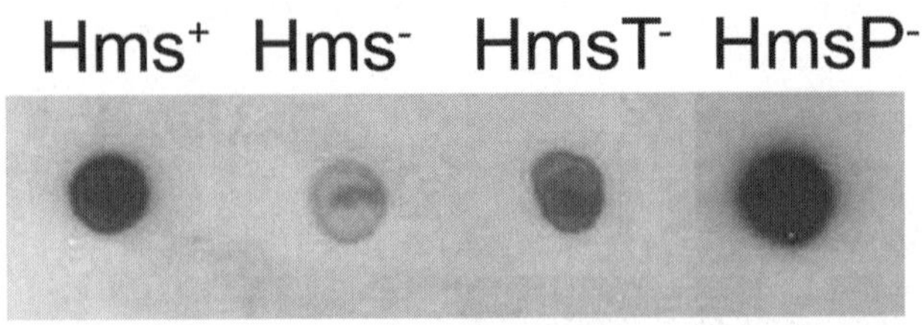

Figure 4. Dot blot analysis of poly-β-1,6-GlcNAc-like EPS production by *Y. pestis* strains. *Y. pestis* cells were grown at 26°C overnight in PMH2 with 0.05% D-glucosamine, and crude polysaccharide extracts were spotted onto a nitrocellulose membrane. Blots were developed with antisera against purified polysaccharide intercellular adhesion of *S. epidermidis* (47).

DGC activity. In contrast, the two ORFs encoding a GGDEF domain but no EAL domain, *y2559* and *y3730,* appear to have all the conserved residues in the GGDEF domain (15). Nucleotide sequence comparison of *y2559* with its *Yersinia pseudotuberculosis* strain IP 32953 orthologue *(yptb1628)* suggests that *y2559* has an N-terminal truncation due to a frameshift mutation. This computational analysis suggests that HmsT and Y3730 may be the only functional DGCs in *Y. pestis* KIM6+ (the parental strain of KIM10+).

IM protein HmsP is a negative regulator of biofilm development. The C terminus of HmsP is in the cytoplasm and has HAMP and EAL domains as well as a GGDEF domain that is likely defective since it lacks the GGDEF motif (Fig. 6). A *Y. pestis hmsP* mutant produces biofilm at 37°C and shows enhanced poly-β-1,6-GlcNAc-like EPS production and biofilm formation at ambient temperatures (Fig. 4 and 5). The EAL domain alone of HmsP complements this phenotype and has in vitro PDE activity against the artificial substrate bis(*p*-nitrophenyl) phosphate. In addition, overexpression of *hmsP* abrogates EPS production and biofilm development in *Y. pestis* (6, 7, 40). Alanine substitutions in residues critical for PDE activity result in loss of PDE activity and an inability to complement the *hmsP* mutant, indicating that PDE activity is central to the negative regulation of biofilm development by HmsP (7, 8, 40).

In *Y. pestis* KIM10+ (15), there are five additional ORFs with EAL domains and one with an HD-GYP domain, another bacterial c-di-GMP PDE domain (62). Alignment of *Y. pestis* EAL domains with functional PDEs revealed a number of substitutions in highly conserved residues in two ORFs, *y0203* and *y1612,* while the other three had no or only a few such substitutions (15). However, sequence analysis and comparison of these three EAL-containing ORFs with their *Y. pseudotuberculosis* strain IP 32953 orthologues suggests that all three are pseudogenes.

In *Y. pestis* KIM10+, Y2472 has the only HD-GYP domain. However, this domain lacks two conserved signature residues shown to be critical for c-di-GMP PDE activity in the HD-GYP protein, RpfG, from *Xanthomonas campestris* (62). Consequently, it is possible that HmsP is the only functional c-di-GMP PDE in *Y. pestis* KIM6+ and plays the primary role in biofilm regulation.

Mechanisms of c-di-GMP Regulation of Biofilm Formation

c-di-GMP regulates HmsR enzymatic activity and stability

In other organisms, c-di-GMP affects various behaviors primarily through regulation at the transcriptional and posttranslational levels. In one classic example, *Gluconacetobacter xylinus,* cellulose synthase activity is allosterically controlled by c-di-GMP (59, 61).

A *Y. pestis hmsP* mutant overproduces biofilm but shows no increase in the levels of HmsHFRS proteins. Indeed, the level of HmsR is drastically lower

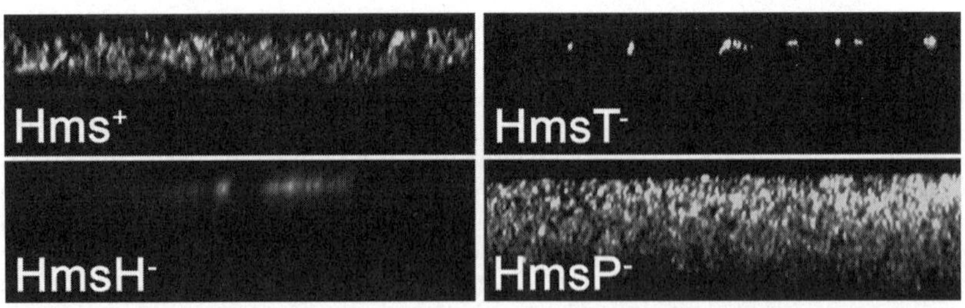

Figure 5. Confocal laser scanning microscopy images of green fluorescent protein-expressing *Y. pestis* cells. Cultures were grown at 26°C overnight in PMH2 with 0.2% glucose in a conical tube with a coverslip. Cells attached to the glass coverslip (at the air-liquid interface) were visualized with a Leica TCS laser scanning confocal microscope. The xzy plane (a cross-sectional view) is shown. Hms+ and HmsH− strains are positive and negative controls, respectively. The figure uses black and white unpublished images from the experiment depicted in Fig. 6 from Kirillina et al. (40).

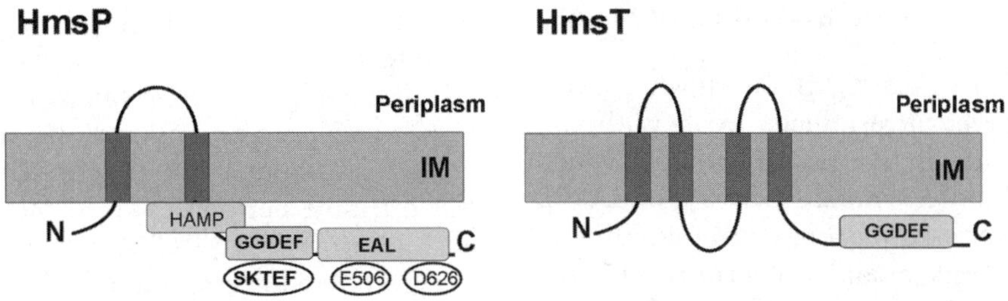

Figure 6. IM topology models of HmsP and HmsT. Dark grey boxes represent TM domains. Periplasmic and cytoplasmic domains are based on analysis of *phoA* and *lacZ* fusions (6). HAMP, EAL, and GGDEF domains are shown as light grey boxes. For HmsP, the GGDEF domain likely lacks DGC activity, since residues essential for this activity have been altered to SKTEF (shown in oval). E506 and D626 residues as well as the EAL motif within the EAL domain of HmsP are critical for PDE activity while the GGDEF residues of HmsT are essential for biofilm development (8, 36, 40). Modified and reproduced from reference 6 with permission of Blackwell Publishing.

in this mutant which should have higher levels of c-di-GMP (Fig. 7) (6, 40). Overproduction of biofilm with reduced levels of HmsR is indirect evidence that, like the *G. xylinus* cellulose synthase, the *Y. pestis* glycosyltransferase HmsR is allosterically activated by c-di-GMP. However, conclusive proof of this hypothesis awaits in vitro analysis of the enzymatic activity of HmsR.

The level of HmsS (but not HmsH or HmsF) was also significantly reduced in an *hmsP* mutant compared to the parental strain (Fig. 7). The effect on HmsS levels may be indirect since HmsR stabilizes HmsS. The levels of HmsR and HmsS were increased in the *hmsP* mutant by complementation with *hmsP* (40). While studies elucidating whether this regulation is transcriptional, translational, or posttranslational have not been published, we suggest that c-di-GMP either directly or indirectly affects HmsR levels. Overproduction of EPS is likely harmful to *Y. pestis* cells. Thus, the effect of c-di-GMP on HmsR stability likely represents a feedback mechanism for controlling biofilm formation by reducing the protein levels of HmsR when its enzymatic activity is high.

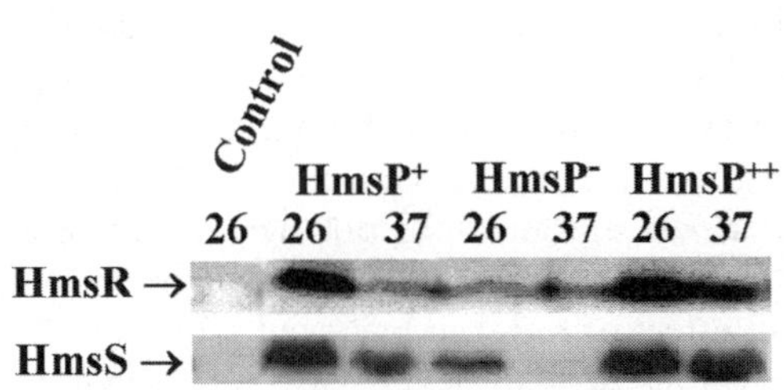

Figure 7. Western blot analysis of HmsR and HmsS protein levels in *Y. pestis* strains after growth at 26°C or 37°C. Equal concentrations of whole-cell lysates were separated by sodium dodecyl sulfate-polyacrylamide gel electrophoresis; immunoblots were reacted with antiserum against the indicated Hms proteins. The negative control lane has a whole-cell lysate from a Δ*pgm Y. pestis* mutant that lacks 102 kb of chromosomal DNA that includes the *hmsHFRS* operon. Modified and reproduced from Kirillina et al. (40) with permission of Blackwell Publishing.

c-di-GMP levels in the Hms enzyme complex

All other potential DGCs and PDEs encoded by the *Y. pestis* KIM6+ genome do not compensate for *hmsT* or *hmsP* mutations (36, 40), and our analysis (see above) suggests that HmsT and Y3730 are the only two functional DGCs and that HmsP is the only functional PDE in *Y. pestis.* While *y3730* gene expression may account for its lack of influence on biofilm development in vitro, we hypothesize that the local level of c-di-GMP is more important than the overall cellular c-di-GMP level to promote EPS biosynthesis in *Y. pestis*. HmsT overexpression only modestly increases c-di-GMP cellular levels, while *Salmonella* AdrA expression increases c-di-GMP levels >100-fold. Despite this, HmsT expression yields equivalent or higher biofilm levels than AdrA expression in *Y. pestis* (66). While the lower levels of c-di-GMP producted by HmsT may be sufficient for maximal biofilm production, we hypothesize that c-di-GMP produced by HmsT may be compartmentalized to an Hms protein complex for use in EPS synthesis. Recently, using a number of methods, we obtained evidence for an IM Hms EPS biosynthetic complex. HmsR interacts with HmsS, while HmsP interacts with HmsR and HmsT. In addition, the glycosyltransferase domain of HmsR, the DGC domain of HmsT, and the PDE domain of HmsP are localized to the cytoplasm (Fig. 3A and 6), suggesting an interplay between c-di-GMP metabolism and polysaccharide biosynthesis. Our current model proposes that HmsR together with HmsS is essential for the synthesis of the β-1,6-GlcNAc-like polysaccharide while HmsT and HmsP regulate the glycosyltransferase activity and stability of HmsR by controlling the local c-di-GMP level in this protein complex (6).

Mechanism of HmsR activation—c-di-GMP binding

While the effect of c-di-GMP on HmsR enzyme activity remains to be determined, we hypothesize that c-di-GMP binds to the HmsR-HmsS protein complex. The cellulose synthase activity of BcsAB from *G. xylinus* is strongly promoted by binding c-di-GMP (60, 77). Amikam and Galperin identified a PilZ domain in the C-terminal portion of BscA (1), and in vitro biochemical analyses have confirmed that PilZ domains from BcsA, *E. coli* YcgR, *P. aeruginosa* PA4608 and Alg44, and *V. cholerae* PlzC and PlzD bind c-di-GMP. Only two conserved motifs (RXXXR and D/NXSXGG) were identified among diverse PilZ domains, and a mutation in one conserved R residue of the RXXXR motif abrogated c-di-GMP binding by YcgR and PlzD (47, 56, 57, 64, 66). While neither HmsR nor HmsS possesses a PilZ domain, arginine residues with similar spacing to the RXXXR motif occur in both proteins and we propose that regulation of EPS biosynthesis may occur by direct binding of c-di-GMP to HmsR and/or HmsS.

HmsS interacts with HmsR (the putative glycosyl transferase) and is essential for biofilm formation, but its function is undetermined. HmsS has two RXXXR motifs—one in the N terminus (RllpR; residues 11 to 15) just before the first TM domain and the other (RRghR; residues 92 to 96) after the second TM domain. Both motifs reside in the cytoplasm (6). So it is possible that HmsS serves as a c-di-GMP binding protein to activate HmsR. Alternatively, HmsR may directly bind c-di-GMP to enhance glycosyltransferase activity. A search for RXXXR motifs within cytoplasmic regions of HmsR identified three such sequences—two within the glycosyltransferase domain and the other within the KRKRAR sequence in the cytoplasmic C terminus (I396 to S457) of HmsR. Previously, we found that a spontaneous deletion of six *hmsR* nucleotides (removing one set of KR residues in this region) caused a loss of CR binding (45). Whether this KRKRAR sequence serves to bind c-di-GMP or some other essential function remains to be determined.

REGULATION OF HmsT AND HmsP

Regulation by Temperature

One of the first striking observations of plague hemin/CR binding (i.e., biofilm formation) was temperature regulation with biofilm forming at ambient but not mammalian temperatures (33, 68). Insight into how temperature affects biofilm development came with the discovery that the levels of three key Hms proteins (HmsH, HmsR, and HmsT) were significantly lower at 37°C (Fig. 3B). Within 4 hours of a shift from 26°C to 37°C, the levels of HmsH, HmsF, HmsR, and HmsS fall to 11, 44, 32, and 68%, respectively, of the amounts observed during growth at 26°C. Although HmsT was relatively stable over this 4-h time period, it was virtually undetectable by 11 h of growth at 37°C. Results with chromosomally integrated reporters and mRNA dot blots indicate that the temperature effects do not involve transcriptional regulation or significant changes in mRNA stability (51). Degradation of HmsT appears to play a key role, since overexpression of HmsT at 37°C abrogates normal temperature regulation of CR binding. We have found that a mutation in *clpB* negates temperature regulation and that HmsT degradation at 37°C does not occur in a Δ*lon clpPX::kan* triple mutant. Together, these genes encode two different proteases. Temperature regulation is also overcome by increased copies of the *hmsHFRS* operon and by a mutation in *hmsP* (7, 36, 40, 51). Thus, we believe that the primary mechanisms for temperature regulation of biofilm development are twofold: (i) lower HmsH, HmsR, and HmsT protein levels at 37°C, which reduce the porin and enzymes responsible for c-di-GMP as well as polysaccharide synthesis; and (ii) a lower enzymatic activity for HmsR due to lower c-di-GMP levels at 37°C as a result of lower HmsT protein levels and higher HmsP PDE activity (Fig. 3B) (7, 40, 51).

Polyamine Requirement for Biofilm Development

An unexpected finding was that polyamines are required for *Y. pestis* biofilm formation. Mutations in *speC* (encoding ornithine decarboxylase), *speA* (encoding arginine decarboxylase), or both genes cause progressively lower intracellular levels of putrescine and a corresponding loss of crystal violet staining as a measure of adherence. EPS levels as measured by CR binding were negligible in the *speA speC* double mutant (50). Biofilm development is restored by complementation with *speA* or by addition of putrescine to the growth medium. Unlike *E. coli*, the *Y. pestis* Δ*speA* Δ*speC* mutant does not have a severe growth defect. Thus, the loss of biofilm formation is not an indirect effect of reduced bacterial growth (50). Analysis of Hms protein levels at 26°C in our Δ*speA* Δ*speC* mutant revealed normal levels of HmsH, HmsF, and HmsP, a <50% reduction in HmsS, and greatly reduced levels of HmsR and HmsT. Increased levels of HmsT alone were unable to complement the defect in biofilm formation in this mutant. However, expression of both HmsT and HmsR in the Δ*speA* Δ*speC* mutant did restore biofilm formation. In addition, exogenous putrescine added to cultures of the

Δ*speA* Δ*speC* mutant increased the protein levels of all three Hms proteins and restored biofilm formation. Analysis of *hmsH::lacZ* and *hmsT::lacZ* transcriptional reporters indicated that putrescine did not affect transcription from these promoters. Quantitative real-time reverse transcription-PCR showed no difference in mRNA levels for *hmsH, hmsF, hmsR, hmsS,* and *hmsT,* indicating that mRNA stability is not responsible for the reduced levels of Hms proteins in the polyamine mutant. β-Galactosidase fusions to *hmsT, hmsR,* and *hmsS* but not to *hmsP* showed increased translation when the Δ*speA* Δ*speC* mutant was grown in the presence of putrescine. The mechanism by which putrescine affects translation of *hmsT* and *hmsR* is undetermined. It is also likely that polyamine deficiency has additional effects on protein stability (B. W. Wortham, M. A. Oliveira, J. A. Fetherston, and R. D. Perry, submitted for publication).

PhoP Regulation of Biofilm Development through HmsT

PhoP is the response regulator for the PhoQ-PhoP two-component system that adapts bacterial cells to low-magnesium environments. Recently, Sun et al. found that a *Y. pestis phoP* mutant produced a more extensive biofilm under in vitro conditions than its PhoP$^+$ parent. A *Y. pseudotuberculosis phoP* mutant strain showed a reduced level of HmsT protein and biofilm formation when transformed with a PhoP-expressing plasmid. Thus, PhoP negatively regulates *Y. pestis* biofilm development (67). Whether PhoP affects HmsT expression in *Y. pestis* and the mechanism of regulation of HmsT in *Y. pestis* and/or *Y. pseudotuberculosis* remain to be determined. It has been demonstrated that PhoPQ-PmrAB regulatory mechanisms differ in these two closely related bacteria (48, 80)

ROLE(S) FOR c-di-GMP IN PLAGUE MAINTENANCE, FLEA TRANSMISSION, AND MAMMALIAN PATHOGENESIS

Virulence and biofilm factors have been shown to be reciprocally expressed in a number of pathogens (12, 22, 26, 69). For *V. cholerae, S. enterica* serovar Typhimurium, *P. aeruginosa, Bordetella pertussis, Brucella melitensis,* and *Xanthomonas campestris,* inactivation of some c-di-GMP PDEs inhibits virulence in animals or plants (12, 30, 41, 43, 49, 63, 69–72). Since maintenance of *Y. pestis* in nature depends on circulation between rodents and fleas, there are vastly different hosts in which c-di-GMP may affect the survival and growth of *Y. pestis* (Fig. 1 and 2).

Transmission from Fleas to Mammals and Maintenance in Nature

Since the early observations by Bacot (2, 3) of autoaggregative growth of *Y. pestis* in the flea midgut and subsequent blockage of the proventriculus, the blocked flea model has been a paradigm for transmission of plague from fleas to mammals. It is clear that the macrophenomenon of blockage requires biofilm development and leads to greatly enhanced transmission of bacteria 5 to 14 days after the infectious blood meal (29, 34, 55). Kutyrev et al. (41a) found that an Hms$^-$ strain of *Y. pestis* was cleared more rapidly by the vole flea, *Nosopsyllus laeviceps,* than an Hms$^+$ strain. Similarly, infection of *X. cheopis* fleas with Hms$^-$ mutants was more likely to be cleared than infection with the Hms$^+$ parent (29). Thus, biofilm formation may increase the infection rate or prolong the infection in fleas. The *Y. pestis* aggregates in fleas have an extracellular matrix whose production is Hms dependent. Since synthesis of poly-β-1,6-GlcNAc is dependent upon HmsT in vitro, c-di-GMP is certain to play a positive regulatory role for in vivo biofilm development and flea blockage.

While flea blockage may have positive elements for flea infection and transmission to mammals, it also leads to flea mortality, reducing the period of blockage-dependent transmission. Indeed, blockage-dependent transmission alone cannot account for plague epizootics among prairie dogs (76). Plague transmission by unblocked fleas was described in the 1940s (9). More recently, it has been demonstrated that fleas transmit plague efficiently shortly after taking an infectious blood meal—long before blockage could develop. For *Oropsylla montana,* which infests California ground squirrels and rock squirrels, and *Oropsylla hirsute* (the prairie dog flea), effective transmission of plague from these fleas to mice was detected as early as 3 h after taking an infected blood meal, with a decline in transmission at 72 h postingestion. Early-phase transmission of plague from a variety of fleas, including *X. cheopis,* has now been documented and is thought to be a critical parameter that helps explain the rapid spread of *Y. pestis* during plague epizootics and rat-associated human bubonic plague outbreaks (16–19, 76, 78, 79).

The relative importance of blocked-flea transmission versus early-phase transmission for long-term maintenance in nature and in epidemic outbreaks is in flux. Biofilm development may moderately increase flea infection rates and extend the transmission period by blockage development. In addition, biofilm formation (without blockage) could either enhance or inhibit early-phase transmission. While blockage of the flea is not essential for transmission during epidemic outbreaks, it has been hypothesized for over 50 years that blockage may be critical for maintenance of plague enzootic animal reservoirs. Transmission by low populations of infected fleas may require blockage to extend the transmission period and to increase attempts to take a blood meal; both would increase transmission. This might serve to maintain a low level of infected mammals within a population (4, 9, 14, 20, 23, 24, 38, 39, 73).

Bubonic and Pneumonic Plague in Mammals

The temperature regulation of biofilm development suggests that high c-di-GMP levels and biofilm development are unlikely to be factors in pneumonic plague or the late stages of bubonic plague. However, either of these factors might play a role early in infection from a flea bite. Although Lillard et al. showed that an Hms^+ strain was more resistant to some reactive oxygen species than its Hms^- mutant, this did not translate into increased resistance to killing by neutrophils. In addition, an *hmsR* mutant was fully virulent via subcutaneous injection (44).

However, c-di-GMP could still play a role in mammalian disease, independent of its regulatory role in biofilm development. As noted above, a number of pathogens show reciprocal expression of virulence and biofilm factors linked to c-di-GMP metabolism. In one study, signature-tagged mutagenesis of *Y. pestis* CO92 identified an *hmsP::*Tn*5* mutant with reduced invasion of human type 1 pneumocyte WI26 cells. However, the cause of this invasion defect was not established (42). While direct studies on the virulence of various c-di-GMP mutants in *Y. pestis* remain to be performed, we hypothesize that lower c-di-GMP levels enhance expression of *Y. pestis* virulence factors (Fig. 2).

CONCLUSIONS AND FUTURE DIRECTIONS

Our investigation of the mechanism for temperature regulation of biofilm development in *Y. pestis* uncovered c-di-GMP as a major regulator, with HmsT and HmsP serving as positive and negative regulators, respectively, by modulating levels of this second messenger. c-di-GMP produced by HmsT likely affects both the level and enzymatic activity of HmsR, a glycosyltransferase required for EPS production. HmsP, a PDE that serves as a negative regulator of biofilm development, also prevents excessive biofilm matrix synthesis (6, 7, 36, 40, 66).

Environmental cues such as temperature and polyamine levels affect biofilm development by affecting the protein levels of HmsT. At a mammalian temperature, HmsT is degraded to lower c-di-GMP production (Fig. 2 and 3B). While HmsR and HmsH are also degraded, normal temperature regulation can be abrogated by overexpression of HmsT or another DGC and by mutation of *hmsP*. A *lon clpPX* mutant also exhibits temperature-independent biofilm formation and increased levels of HmsT at 37°C. Levels of HmsR and HmsH are still reduced, indicating a separate degradative mechanism for these two proteins (40, 50, 51). Putrescine deficiency reduces translation of both *hmsT* and *hmsR* mRNA but may also affect biofilm development by another mechanism (Wortham et al., submitted). PhoP has recently been identified as another negative regulator of plague biofilm (67). Whether the observation that HmsT levels are affected by PhoP in *Y. pseudotuberculosis* holds for *Y. pestis* remains to be determined.

Many bacteria have multiple ORFs with GGDEF and/or EAL domains; a number of alternative, but not mutually exclusive, mechanisms have been proposed for the apparent lack of redundancy for these ORFs. Our investigation of the biofilm system and computational analysis of GGDEF and EAL domain ORFs in *Y. pestis* supports two mechanisms. First, HmsP is likely the only functional c-di-GMP PDE, while HmsT and Y3730 may be the only two functional DGCs in *Y. pestis*. Under our in vitro conditions, HmsT is the DGC essential for biofilm development. The remaining ORFs appear to have undergone frameshift mutations or amino acid residue changes that should render them enzymatically inactive. This suggests that *Y. pestis* has reduced the biological functions controlled by this second messenger or the enzymes capable of modulating c-di-GMP levels.

Second, the local c-di-GMP pool in an enzyme complex may cause specificity for biofilm formation. We have begun characterization of an IM Hms complex of c-di-GMP-metabolizing enzymes and EPS biosynthetic enzymes (6). The relatively low levels of c-di-GMP produced by HmsT (66) suggest that the local c-di-GMP pool in this complex rather than total cellular levels is key to EPS production. HmsR interacts with HmsP, which interacts with HmsT (Fig. 3A), physically linking both EPS biosynthesis and c-di-GMP metabolism. Similar complexes in other organisms may cause the observed specificity of mutations in genes encoding c-di-GMP homeostasis enzymes for different biological functions. Of course, it remains possible that the observed low total cellular levels of c-di-GMP due to HmsT in *Y. pestis* are sufficient for normal biofilm development.

There remains a myriad of unresolved issues surrounding c-di-GMP signaling in *Y. pestis*. With regard to biofilm development, any effect of c-di-GMP on the glycosyltransferase activity of HmsR remains to be investigated. Whether c-di-GMP binds to HmsS, HmsR, or an unknown protein is also unanswered.

Currently, only biofilm development is known to be controlled by c-di-GMP signaling in *Y. pestis*. Identification of other genes or proteins affected by c-di-GMP would yield valuable clues as to additional cellular functions that may be regulated by this mechanism. Several environmental cues modulate HmsT levels; identification of additional signals controlling HmsT, HmsP, and Y3730 will also provide valuable information on biofilm regulation and other cellular functions controlled by c-di-GMP.

Further study is needed to determine whether c-di-GMP signaling plays a role in mammalian virulence and/or survival and growth in the flea. Such roles could be independent of biofilm formation. It appears that investigation of c-di-GMP signaling in *Y. pestis* will be a fruitful line of research for the foreseeable future.

Acknowledgments. We thank M. Oliveira, J. Fetherston, and B. Wortham for submitted results cited in this review. We also thank J. Fetherston and O. Kirillina for thoughtful discussions and suggestions.

The results of our studies cited in this review were supported by Public Health Service grant AI25098 from the National Institutes of Health.

We thank Ken Gage (CDC) for providing Fig. 1, which we have modified slightly for this review.

REFERENCES

1. **Amikam, D., and M. Y. Galperin.** 2006. PilZ domain is part of the bacterial c-di-GMP binding protein. *Bioinformatics* **22:** 3–6.
2. **Bacot, A. W.** 1915. LXXXI. Further notes on the mechanism of the transmission of plague by fleas. *J. Hyg.* **14**(Plague Suppl. 4):774–776.
3. **Bacot, A. W., and C. J. Martin.** 1914. LXVII. Observations on the mechanism of the transmission of plague by fleas. *J. Hyg.* **13**(Plague Suppl. 3):423–439.
4. **Bazanova, L. P., I. Zhovtyi, M. P. Maevskii, V. T. Klimov, and A. F. Popkov.** 1991. The seasonal dynamics of blocking in the flea *Citellophorus tesquorum altaicus* from the Tuva natural plague focus. *Med. Parazitol.* (Moscow) **Jan-Feb:**24–26.
5. **Blair, D. E., A. W. Schüttelkopf, J. I. MacRae, and D. M. F. van Aalten.** 2005. Structure and metal-dependent mechanism of peptidoglycan deacetylase, a streptococcal virulence factor. *Proc. Natl. Acad. Sci. USA* **102:**15429–15434.
6. **Bobrov, A. G., O. Kirillina, S. Forman, D. Mack, and R. D. Perry.** 2008. Insights into *Yersinia pestis* biofilm development: topology and co-interaction of Hms inner membrane proteins involved in biofilm exopolysaccharide production. *Environ. Microbiol.* **10:**1419–1432.
7. **Bobrov, A. G., O. Kirillina, and R. D. Perry.** 2005. The phosphodiesterase activity of the HmsP EAL domain is required for negative regulation of biofilm formation in *Yersinia pestis*. *FEMS Microbiol. Lett.* **247:**123–130.
8. **Bobrov, A. G., O. Kirillina, and R. D. Perry.** 2007. Regulation of biofilm formation in *Yersinia pestis*. *Adv. Exp. Med. Biol.* **603:**201–210.
9. **Burroughs, A. L.** 1947. Sylvatic plague studies. The vector efficiency of nine species of fleas compared with *Xenopsylla cheopis*. *J. Hyg.* **45:**371–396.
10. **Ciocchini, A. E., M. S. Roset, G. Briones, N. I. de Iannino, and R. A. Ugalde.** 2006. Identification of active site residues of the inverting glycosyltransferase Cgs required for the synthesis of cyclic β-1,2-glucan, a *Brucella abortus* virulence factor. *Glycobiology* **16:**679–691.
11. **Ciocchini, A. E., M. S. Roset, N. Inon de Iannino, and R. A. Ugalde.** 2004. Membrane topology analysis of cyclic glucan synthase, a virulence determinant of *Brucella abortus*. *J. Bacteriol.* **186:**7205–7213.
12. **Cotter, P. A., and S. Stibitz.** 2007. c-di-GMP-mediated regulation of virulence and biofilm formation. *Curr. Opin. Microbiol.* **10:**17–23.
13. **Darby, C., J. W. Hsu, N. Ghori, and S. Falkow.** 2002. *Caenorhabditis elegans:* plague bacteria biofilm blocks food intake. *Nature* **417:**243–244.
14. **Degtiareva, L. V., N. F. Labunets, S. P. Osipova, and V. I. Shchedrin.** 1990. The ability of flea species on the common vole from mountainous Dagestan to transmit and preserve the causative agent of plague. *Parazitologiia* **24:**106–112. (In Russian.)
15. **Deng, W., V. Burland, G. Plunkett III, A. Boutin, G. F. Mayhew, P. Liss, N. T. Perna, D. J. Rose, B. Mau, S. Zhou, D. C. Schwartz, J. D. Fetherston, L. E. Lindler, R. R. Brubaker, G. V. Plano, S. C. Straley, K. A. McDonough, M. L. Nilles, J. S. Matson, F. R. Blattner, and R. D. Perry.** 2002. Genome sequence of *Yersinia pestis* KIM. *J. Bacteriol.* **184:**4601–4611.
16. **Eisen, R., J. Borchert, J. Holmes, G. Amatre, K. Van Wyk, R. Enscore, N. Babi, L. A. Atiku, A. Wilder, S. Vetter, S. Bearden, J. Montenieri, and K. Gage.** 2008. Early-phase transmission of *Yersinia pestis* by cat fleas *(Ctenocephalides felis)* and their potential role as vectors in a plague endemic region of Uganda. *Am. J. Trop. Med. Hyg.* **78:**949–956.
17. **Eisen, R. J., S. W. Bearden, A. P. Wilder, J. A. Montenieri, M. F. Antolin, and K. L. Gage.** 2006. Early-phase transmission of *Yersinia pestis* by unblocked fleas as a mechanism explaining rapidly spreading plague epizootics. *Proc. Natl. Acad. Sci. USA* **103:**15380–15385.
18. **Eisen, R. J., J. L. Lowell, J. A. Montenieri, S. W. Bearden, and K. L. Gage.** 2007. Temporal dynamics of early-phase transmission of *Yersinia pestis* by unblocked fleas: secondary infectious feeds prolong efficient transmission by *Oropsylla montana* (Siphonaptera: Ceratophyllidae). *J. Med. Entomol.* **44:**672–677.
19. **Eisen, R. J., A. P. Wilder, S. W. Bearden, J. A. Montenieri, and K. L. Gage.** 2007. Early-phase transmission of *Yersinia pestis* by unblocked *Xenopsylla cheopis* (Siphonaptera: Pulicidae) is as efficient as transmission by blocked fleas. *J. Med. Entomol.* **44:**678–682.
20. **Engelthaler, D. M., B. J. Hinnebusch, C. M. Rittner, and K. L. Gage.** 2000. Quantitative competitive PCR as a technique for exploring flea-*Yersinia pestis* dynamics. *Am. J. Trop. Med. Hyg.* **62:**552–560.
21. **Forman, S., A. G. Bobrov, O. Kirillina, S. K. Craig, J. Abney, J. D. Fetherston, and R. D. Perry.** 2006. Identification of critical amino acid residues in the plague biofilm Hms proteins. *Microbiology* **152:**3399–3410.
22. **Furukawa, S., S. L. Kuchma, and G. A. O'Toole.** 2006. Keeping their options open: acute versus persistent infections. *J. Bacteriol.* **188:**1211–1217.

23. **Gage, K. L., and M. Y. Kosoy.** 2005. Natural history of plague: perspectives from more than a century of research. *Annu. Rev. Entomol.* **50:**505–528.
24. **Gan, N. V., G. A. Voronova, L. N. Iuzvik, and V. A. Beliaeva.** 1990. The efficiency of *Rhadinopsylla rothschildi* and *R. dahurica* fleas as vectors of the causative agent of plague in a Transbaikal natural focus. *Parazitologiia* **24:**151–154. (In Russian.)
25. **Goller, C., X. Wang, Y. Itoh, and T. Romeo.** 2006. The cation-responsive protein NhaR of *Escherichia coli* activates *pgaABCD* transcription, required for production of the biofilm adhesin poly-β-1,6-*N*-acetyl-D-glucosamine. *J. Bacteriol.* **188:** 8022–8032.
26. **Goodman, A. L., B. Kulasekara, A. Rietsch, D. Boyd, R. S. Smith, and S. Lory.** 2004. A signaling network reciprocally regulates genes associated with acute infection and chronic persistence in *Pseudomonas aeruginosa*. *Dev. Cell* **7:**745–754.
27. **Hare, J. M., and K. A. McDonough.** 1999. High-frequency RecA-dependent and -independent mechanisms of Congo red binding mutations in *Yersinia pestis*. *J. Bacteriol.* **181:**4896–4904.
28. **Heldermon, C., P. L. DeAngelis, and P. H. Weigel.** 2001. Topological organization of the hyaluronan synthase from *Streptococcus pyogenes*. *J. Biol. Chem.* **276:**2037–2046.
29. **Hinnebusch, B. J., R. D. Perry, and T. G. Schwan.** 1996. Role of the *Yersinia pestis* hemin storage *(hms)* locus in the transmission of plague by fleas. *Science* **273:**367–370.
30. **Hisert, K. B., M. MacCoss, M. U. Shiloh, K. H. Darwin, S. Singh, R. A. Jones, S. Ehrt, Z. Zhang, B. Gaffney, S. Gandotra, D. W. Holden, D. Murray, and C. Nathan.** 2005. A glutamate-alanine-leucine (EAL) domain protein of *Salmonella* controls bacterial survival in mice, antioxidant defence and killing of macrophages: role of cyclic diGMP. *Mol. Microbiol.* **56:**1234–1245.
31. **Itoh, Y., J. D. Rice, C. Goller, A. Pannuri, J. Taylor, J. Meisner, T. J. Beveridge, J. F. Preston III, and T. Romeo.** 2008. Roles of *pgaABCD* genes in synthesis, modification, and export of the *Escherichia coli* biofilm adhesin poly-β-1,6-*N*-acetyl-D-glucosamine. *J. Bacteriol.* **190:**3670–3680.
32. **Izano, E. A., I. Sadovskaya, E. Vinogradov, M. H. Mulks, K. Velliyagounder, C. Ragunath, W. B. Kher, N. Ramasubbu, S. Jabbouri, M. B. Perry, and J. B. Kaplan.** 2007. Poly-N-acetylglucosamine mediates biofilm formation and antibiotic resistance in *Actinobacillus pleuropneumoniae*. *Microb. Pathog.* **43:**1–9.
33. **Jackson, S., and T. W. Burrows.** 1956. The pigmentation of *Pasteurella pestis* on a defined medium containing haemin. *Br. J. Exp. Pathol.* **37:**570–576.
34. **Jarrett, C. O., E. Deak, K. E. Isherwood, P. C. Oyston, E. R. Fischer, A. R. Whitney, S. D. Kobayashi, F. R. DeLeo, and B. J. Hinnebusch.** 2004. Transmission of *Yersinia pestis* from an infectious biofilm in the flea vector. *J. Infect. Dis.* **190:**783–792.
35. **Jenal, U., and J. Malone.** 2006. Mechanisms of cyclic-di-GMP signaling in bacteria. *Annu. Rev. Genet.* **40:**385–407.
36. **Jones, H. A., J. W. Lillard, Jr., and R. D. Perry.** 1999. HmsT, a protein essential for expression of the haemin storage (Hms^+) phenotype of *Yersinia pestis*. *Microbiology* **145:**2117–2128.
37. **Kaplan, J. B., K. Velliyagounder, C. Ragunath, H. Rohde, D. Mack, J. K. M. Knobloch, and N. Ramasubbu.** 2004. Genes involved in the synthesis and degradation of matrix polysaccharide in *Actinobacillus actinomycetemcomitans* and *Actinobacillus pleuropneumoniae* biofilms. *J. Bacteriol.* **186:**8213–8220.
38. **Kartman, L., and F. M. Prince.** 1956. Studies on *Pasteurella pestis* in fleas. V. The experimental plague-vector efficiency of wild rodent fleas compared with *Xenopsylla cheopis,* together with observations on the influence of temperature. *Am. J. Trop. Med. Hyg.* **5:**1058–1070.
39. **Kartman, L., F. M. Prince, S. F. Quan, and H. E. Stark.** 1958. New knowledge on the ecology of sylvatic plague. *Ann. N. Y. Acad. Sci.* **70:**668–711.
40. **Kirillina, O., J. D. Fetherston, A. G. Bobrov, J. Abney, and R. D. Perry.** 2004. HmsP, a putative phosphodiesterase, and HmsT, a putative diguanylate cyclase, control Hms-dependent biofilm formation in *Yersinia pestis*. *Mol. Microbiol.* **54:**75–88.
41. **Kulasakara, H., V. Lee, A. Brencic, N. Liberati, J. Urbach, S. Miyata, D. G. Lee, A. N. Neely, M. Hyodo, Y. Hayakawa, F. M. Ausubel, and S. Lory.** 2006. Analysis of *Pseudomonas aeruginosa* diguanylate cyclases and phosphodiesterases reveals a role for bis-(3′-5′)-cyclic-GMP in virulence. *Proc. Natl. Acad. Sci. USA* **103:**2839–2844.
41a. **Kutyrev, V. V., A. A. Filippov, O. S. Oparina, and O. A. Protsenko.** 1992. Analysis of *Yersinia pestis* chromosomal determinants Pgm^+ and Pst^s associated with virulence. *Microb. Pathog.* **12:**177–186.
42. **Leigh, S. A., S. Forman, R. D. Perry, and S. C. Straley.** 2005. Unexpected results from the application of signature-tagged mutagenesis to identify *Yersinia pestis* genes required for adherence and invasion. *Microb. Pathog.* **38:**259–266.
43. **Lestrate, P., A. Dricot, R.-M. Delrue, C. Lambert, V. Martinelli, X. De Bolle, J.-J. Letesson, and A. Tibor.** 2003. Attenuated signature-tagged mutagenesis mutants of *Brucella melitensis* identified during the acute phase of infection in mice. *Infect. Immun.* **71:**7053–7060.
44. **Lillard, J. W., Jr., S. W. Bearden, J. D. Fetherston, and R. D. Perry.** 1999. The haemin storage (Hms^+) phenotype of *Yersinia pestis* is not essential for the pathogenesis of bubonic plague in mammals. *Microbiology* **145:**197–209.
45. **Lillard, J. W., Jr., J. D. Fetherston, L. Pedersen, M. L. Pendrak, and R. D. Perry.** 1997. Sequence and genetic analysis of the hemin storage *(hms)* system of *Yersinia pestis*. *Gene* **193:** 13–21.
46. **Mack, D., W. Fischer, A. Krokotsch, K. Leopold, R. Hartmann, H. Egge, and R. Laufs.** 1996. The intercellular adhesin involved in biofilm accumulation of *Staphylococcus epidermidis* is a linear β-1,6-linked glucosaminoglycan: purification and structural analysis. *J. Bacteriol.* **178:**175–183.
47. **Mack, D., J. Riedewald, H. Rohde, T. Magnus, H. H. Feucht, H.-A. Elsner, R. Laufs, and M. E. Rupp.** 1999. Essential functional role of the polysaccharide intercellular adhesin of *Staphylococcus epidermidis* in hemagglutination. *Infect. Immun.* **67:** 1004–1008.
48. **Marceau, M., F. Sebbane, F. Ewann, F. Collyn, B. Lindner, M. A. Campos, J.-A. Bengoechea, and M. Simonet.** 2004. The *pmrF* polymyxin-resistance operon of *Yersinia pseudotuberculosis* is upregulated by the PhoP-PhoQ two-component system but not by PmrA-PmrB, and is not required for virulence. *Microbiology* **150:**3947–3957.
49. **Merkel, T. J., S. Stibitz, J. M. Keith, M. Leef, and R. Shahin.** 1998. Contribution of regulation by the *bvg* locus to respiratory infection of mice by *Bordetella pertussis*. *Infect. Immun.* **66:**4367–4373.
50. **Patel, C. N., B. W. Wortham, J. L. Lines, J. D. Fetherston, R. D. Perry, and M. A. Oliveira.** 2006. Polyamines are essential for the formation of plague biofilm. *J. Bacteriol.* **188:** 2355–2363.
51. **Perry, R. D., A. G. Bobrov, O. Kirillina, H. A. Jones, L. L. Pedersen, J. Abney, and J. D. Fetherston.** 2004. Temperature

regulation of the hemin storage (Hms+) phenotype of *Yersinia pestis* is posttranscriptional. *J. Bacteriol.* **186:**1638–1647.

52. **Perry, R. D., and J. D. Fetherston.** 1997. *Yersinia pestis*—etiologic agent of plague. *Clin. Microbiol. Rev.* **10:**35–66.
53. **Perry, R. D., M. L. Pendrak, and P. Schuetze.** 1990. Identification and cloning of a hemin storage locus involved in the pigmentation phenotype of *Yersinia pestis. J. Bacteriol.* **172:** 5929–5937.
54. **Poland, J. D., T. J. Quan, and A. M. Barnes.** 1994. Plague, p. 93–112. *In* G. W. Beran (ed.), *CRC Handbook Series in Zoonoses,* 2nd ed. *Section A: Bacterial, Rickettsial, and Mycotic Diseases.* CRC Press, Inc., Ann Arbor, MI.
55. **Pollitzer, R.** 1954. Plague. *Monogr. Ser. World Health Organ.* **22:**1–698.
56. **Pratt, J. T., R. Tamayo, A. D. Tischler, and A. Camilli.** 2007. PilZ domain proteins bind cyclic diguanylate and regulate diverse processes in *Vibrio cholerae. J. Biol. Chem.* **282:**12860–12870.
57. **Ramelot, T. A., A. Yee, J. R. Cort, A. Semesi, C. H. Arrowsmith, and M. A. Kennedy.** 2007. NMR structure and binding studies confirm that PA4608 from *Pseudomonas aeruginosa* is a PilZ domain and a c-di-GMP binding protein. *Proteins* **66:** 266–271.
58. **Römling, U., M. Gomelsky, and M. Y. Galperin.** 2005. c-di-GMP: the dawning of a novel bacterial signalling system. *Mol. Microbiol.* **57:**629–639.
59. **Ross, P., Y. Aloni, H. Weinhouse, D. Michaeli, P. Weinberger-Ohana, R. Mayer, and M. Benziman.** 1986. Control of cellulose synthesis in *Acetobacter xylinum.* A unique guanyl oligonucleotide is the immediate activator of the cellulose synthase. *Carbohydr. Res.* **149:**101–117.
60. **Ross, P., R. Mayer, H. Weinhouse, D. Amikam, Y. Huggirat, M. Benziman, E. de Vroom, A. Fidder, P. de Paus, L. A. J. M. Sliedregt, G. A. van der Marel, and J. H. van Boom.** 1990. The cyclic diguanylic acid regulatory system of cellulose synthesis in *Acetobacter xylinum.* Chemical synthesis and biological activity of cyclic nucleotide dimer, trimer, and phosphothioate derivatives. *J. Biol. Chem.* **265:**18933–18943.
61. **Ross, P., H. Weinhouse, Y. Aloni, D. Michaeli, P. Weinberger-Ohana, R. Mayer, S. Braun, E. de Vroom, G. A. van der Marel, J. H. van Boom, and M. Benziman.** 1987. Regulation of cellulose synthesis in *Acetobacter xylinum* by cyclic diguanylic acid. *Nature* **325:**279–281.
62. **Ryan, R. P., Y. Fouhy, J. F. Lucey, L. C. Crossman, S. Spiro, Y.-W. He, L.-H. Zhang, S. Heeb, M. Cámara, P. Williams, and J. M. Dow.** 2006. Cell-cell signaling in *Xanthomonas campestris* involves an HD-GYP domain protein that functions in cyclic di-GMP turnover. *Proc. Natl. Acad. Sci. USA* **103:**6712–6717.
63. **Ryan, R. P., Y. Fouhy, J. F. Lucey, B.-L. Jiang, Y.-Q. He, J.-X. Feng, J.-L. Tang, and J. M. Dow.** 2007. Cyclic di-GMP signalling in the virulence and environmental adaptation of *Xanthomonas campestris. Mol. Microbiol.* **63:**429–442.
64. **Ryjenkov, D. A., R. Simm, U. Römling, and M. Gomelsky.** 2006. The PilZ domain is a receptor for the second messenger c-di-GMP. The PilZ domain protein YcgR controls motility in enterobacteria. *J. Biol. Chem.* **281:**30310–30314.
65. **Saxena, I. M., J. Brown, R. Malcolm, and T. Dandekar.** 2001. Structure-function characterization of cellulose synthase: relationship to other glycosyltransferases. *Phytochemistry* **57:** 1135–1148.
66. **Simm, R., J. D. Fetherston, A. Kader, U. Römling, and R. D. Perry.** 2005. Phenotypic convergence mediated by GGDEF-domain-containing proteins. *J. Bacteriol.* **187:**6816–6823.
67. **Sun, Y.-C., A. Koumoutsi, and C. Darby.** 2009. The response regulator PhoP negatively regulates *Yersinia pseudotuberculosis* and *Yersinia pestis* biofilms. *FEMS Microbiol. Lett.* **290:** 85–90.
68. **Surgalla, M. J., and E. D. Beesley.** 1969. Congo red-agar plating medium for detecting pigmentation in *Pasteurella pestis. Appl. Microbiol.* **18:**834–837.
69. **Tamayo, R., J. T. Pratt, and A. Camilli.** 2007. Roles of cyclic diguanylate in the regulation of bacterial pathogenesis. *Annu. Rev. Microbiol.* **61:**131–148.
70. **Tamayo, R., S. Schild, J. T. Pratt, and A. Camilli.** 2008. Role of cyclic di-GMP during El Tor biotype *Vibrio cholerae* infection: characterization of the in vivo-induced cyclic di-GMP phosphodiesterase CdpA. *Infect. Immun.* **76:**1617–1627.
71. **Tischler, A. D., and A. Camilli.** 2005. Cyclic diguanylate regulates *Vibrio cholerae* virulence gene expression. *Infect. Immun.* **73:**5873–5882.
72. **Tischler, A. D., S. H. Lee, and A. Camilli.** 2002. The *Vibrio cholerae vieSAB* locus encodes a pathway contributing to cholera toxin production. *J. Bacteriol.* **184:**4104–4113.
73. **Voronova, G. A.** 1989. Enhanced blocking capacity of the plague microbe in the body of the flea. *Parazitologiia* **23:**427–429. (In Russian.)
74. **Vuong, C., S. Kocianova, J. M. Voyich, Y. Yao, E. R. Fischer, F. R. DeLeo, and M. Otto.** 2004. A crucial role for exopolysaccharide modification in bacterial biofilm formation, immune evasion, and virulence. *J. Biol. Chem.* **279:**54881–54886.
75. **Wang, X., J. F. Preston III, and T. Romeo.** 2004. The *pgaABCD* locus of *Escherichia coli* promotes the synthesis of a polysaccharide adhesin required for biofilm formation. *J. Bacteriol.* **186:**2724–2734.
76. **Webb, C. T., C. P. Brooks, K. L. Gage, and M. F. Antolin.** 2006. Classic flea-borne transmission does not drive plague epizootics in prairie dogs. *Proc. Natl. Acad. Sci. USA* **103:** 6236–6241.
77. **Weinhouse, H., S. Sapir, D. Amikam, Y. Shilo, G. Volman, P. Ohana, and M. Benziman.** 1997. c-di-GMP-binding protein, a new factor regulating cellulose synthesis in *Acetobacter xylinum. FEBS Lett.* **416:**207–211.
78. **Wilder, A. P., R. J. Eisen, S. W. Bearden, J. A. Montenieri, K. L. Gage, and M. F. Antolin.** 2008. *Oropsylla hirsuta* (Siphonaptera: Ceratophyllidae) can support plague epizootics in black-tailed prairie dogs *(Cynomys ludovicianus)* by early-phase transmission of *Yersinia pestis. Vector-Borne Zoonotic Dis.* **8:**359–368.
79. **Wilder, A. P., R. J. Eisen, S. W. Bearden, J. A. Montenieri, D. W. Tripp, R. J. Brinkerhoff, K. L. Gage, and M. F. Antolin.** 2008. Transmission efficiency of two flea species *(Oropsylla tuberculata cynomuris* and *Oropsylla hirsuta)* involved in plague epizootics among prairie dogs. *EcoHealth* **5:**205–212.
80. **Winfield, M. D., T. Latifi, and E. A. Groisman.** 2005. Transcriptional regulation of the 4-amino-4-deoxy-L-arabinose biosynthetic genes in *Yersinia pestis. J. Biol. Chem.* **280:** 14765–14772.
81. **Wolfe, A. J., and K. L. Visick.** 2008. Get the message out: cyclic-di-GMP regulates multiple levels of flagellum-based motility. *J. Bacteriol.* **190:**463–475.
82. **World Health Organization.** 2000, posting date. WHO report on global surveillance of epidemic-prone infectious diseases. World Health Organization, Geneva, Switzerland. http://www.who.int/csr/resources/publications/surveillance/WHO_CDS_CSR_ISR_2000_1/en/.

The Second Messenger Cyclic Di-GMP
Edited by A. J. Wolfe and K. L. Visick

Chapter 19

Environmental Control of Cyclic Di-GMP Signaling in *Pseudomonas fluorescens:* from Signal to Output

PETER D. NEWELL AND GEORGE A. O'TOOLE

Pseudomonas fluorescens is a ubiquitous bacterium found in soil and water and in association with plant roots where it can protect against plant pathogens and promote plant growth (7, 14). Research into the genetic basis of biofilm formation has utilized this model organism because growth on surfaces is a typical lifestyle for *P. fluorescens* in nature and one with important ecological consequences in the rhizosphere (7, 8). Therefore, exploring the environmental cues that control the transition between biofilm and planktonic growth of this microbe may provide general insight into how environmental bacteria regulate the switch between motile and sessile lifestyles.

Inorganic phosphate (P_i) is an essential nutrient that is found at very low concentrations in the aqueous phase in most natural environments. In soil, most P_i is adsorbed to mineral surfaces or associated with plant roots (32), two niches colonized by fluorescent pseudomonads. Recently it was shown that P_i concentration regulates biofilm formation by *P. fluorescens* Pf0-1 through a cyclic di-GMP (c-di-GMP) signaling pathway (25). Study of this response has produced one of the more complete pictures of how c-di-GMP signaling can link an environmental signal to a complex biological output: from conditional regulation of a c-di-GMP phosphodiesterase (PDE) to the control of attachment via a c-di-GMP effector protein. This research supports the idea that c-di-GMP is a conserved modality in biofilm regulation, with distinct outputs in different organisms, and provides a paradigm for conditional, transcriptional control of c-di-GMP signaling pathways.

REGULATION OF BIOFILM FORMATION BY ENVIRONMENTAL SIGNALS

Growth in a surface-attached community, or biofilm, is considered to be a common lifestyle for bacteria in nature (see chapter 8). Choosing this lifestyle under the appropriate conditions likely provides an important adaptive advantage, given that bacteria in biofilms are known to be more resistant to a number of environmental stressors (1, 8). Despite the widely held belief that specific signals control the transition from free swimming to biofilm growth, few such signals are known and understood at the level of molecular cause and effect. There are many instances of environmental or nutritional conditions that affect formation or morphology of *Pseudomonas* biofilms, such as iron concentration (3, 31) and carbon source (29, 30, 38), but the cellular systems that sense and respond to these signals and/or the mechanism(s) by which these systems control biofilm formation are largely unknown. Some examples of signals that have been well characterized include control of biofilm formation by catabolite repression in *Escherichia coli* (19, 20, 42) and regulation of *P. fluorescens* biofilm formation by P_i (25, 28). In both of these cases, genetic dissection of the signaling pathways involved was essential to proving that signal transduction, and not indirect physiological effects, is responsible for the observed change in biofilm formation.

Inorganic phosphate is an essential micronutrient for all life. There are typically low levels of dissolved P_i in soil, where it is often growth-limiting for

Peter D. Newell and George A. O'Toole • Department of Microbiology and Immunology, Dartmouth Medical School, Hanover, NH 03755.

plants and microorganisms. Many bacteria respond to low environmental P_i concentrations through the induction of specific genetic pathways that adapt the cell for growth under these conditions. The Pho regulon, first described in *E. coli,* is perhaps the best-studied example (43). Consistent with the *E. coli* paradigm, *P. fluorescens* senses extracellular P_i with the two-component regulatory system PhoRB. Studies with *E. coli* indicate that low P_i conditions activate the sensor histidine kinase PhoR, which phosphorylates the response regulator PhoB. Phosphorylation of PhoB leads to its dimerization and makes PhoB competent for DNA binding (44). Expression of the Pho regulon, defined as genes directly regulated by PhoB~P, enables increased P_i scavenging and assimilation (e.g., through the expression of phosphatases) but can also control other cellular functions. The Pho regulon has been shown to regulate the production of secondary metabolites (24), cell morphogenesis (12), and virulence in a number of bacteria (21). In *P. fluorescens* Pf0-1, the Pho regulon also exerts control over biofilm formation (25).

The *pst* genes of *E. coli* encode a high-affinity phosphate transporter that is required for repression of PhoR's kinase activity, and thus the Pho regulon, in phosphate-replete conditions (43). Mutation of the *pst* genes in *E. coli* and *P. fluorescens* leads to constitutive expression of the Pho regulon (26, 43). Correlation of phenotypes associated with P_i starvation with those caused by mutational activation of the P_i response in the *pst* mutant has been instrumental in showing that this regulatory pathway, and not simply the physiological effects of P_i starvation, forms the basis for P_i control of biofilm formation by *P. fluorescens*.

REGULATION OF *P. FLUORESCENS* BIOFILM FORMATION BY THE Pho REGULON

P. fluorescens does not form a biofilm under P_i-limiting conditions. This conditional defect is dependent on the activation of the Pho regulon, as deletion of the response regulator PhoB fully restores biofilm formation in low P_i (25). Constitutive activation of the Pho regulon via deletion of the *pst* genes blocks biofilm formation under P_i-replete conditions, and deletion of *phoB* is epistatic to the *pst* mutant, enabling the *pst* mutant to form a biofilm (25). Together, these results indicate that biofilm formation is regulated by P_i concentration at the level of PhoB-dependent transcription.

The core machinery required for surface attachment by some strains of *P. fluorescens* consists of a large adhesin, LapA, and an ABC transporter encoded by *lapEBC* that transports LapA to the cell surface (Fig. 1). *lap* mutants maintain only transient, reversible contact with surfaces and do not transition to the stable, irreversible attachment required for biofilm formation (17). This phenotype, a defect in the transition to irreversible attachment, is identical to that displayed by wild-type (WT) cells grown in P_i limiting conditions (25). Therefore, the *lap* genes were prime candidates for transcriptional regulation of biofilm formation by the Pho regulon. Interestingly, constitutive activation of Pho leads to only a minor decrease in expression of the *lapEBC* genes and no change in *lapA* gene expression. Several lines of evidence demonstrated that transcriptional regulation of *lapEBC* is not the basis for Pho regulation of biofilm formation (25). While P_i does not significantly impact the Lap system at the level of transcription, it does control Lap function via a c-di-GMP signaling mechanism described below.

ACTIVATION OF A c-di-GMP PDE BY THE Pho REGULON

The PhoB dimer binds to a conserved DNA sequence consisting of two 7-bp direct repeats separated by a 4-bp linker sequence, referred to as a Pho box. The Pho box sequence was shown to be functionally conserved in *P. fluorescens* Pf0-1 (26). In an effort to identify putative Pho genes that might play a role in biofilm formation, the Pf0-1 genome was searched for the consensus Pho box sequence from *E. coli*. This candidate approach identified *rapA* (PfO1_1678), a gene encoding a putative EAL/GGDEF domain protein. *rapA* is preceded by two well-conserved, overlapping Pho boxes and is specifically induced by the Pho regulon (Fig. 1) (25). Consistent with this conditional regulation, mutation of *rapA* has no effect on biofilm formation in P_i-replete conditions. Instead, the *rapA* mutant is rescued for biofilm formation in low-P_i conditions (25). This rescue is partial (~70% of WT), however, when compared to the effect of inactivating the entire Pho regulon in the *phoB* mutant. This simply suggests that other, Pho-regulated genes contribute to repression of biofilm formation by low P_i. Consistent with these data, mutation of *rapA* in the *pst* mutant restores biofilm formation to ~70% of that of the WT (25).

The phenotype of the *rapA* mutant and the presence of a predicted EAL domain in RapA suggested a model in which RapA would act as a c-di-GMP PDE, suppressing adherence by decreasing cellular c-di-GMP levels. In support of this model, purified RapA had PDE activity in vitro, cleaving c-di-GMP to the linear pGpG (25). Despite the presence of a

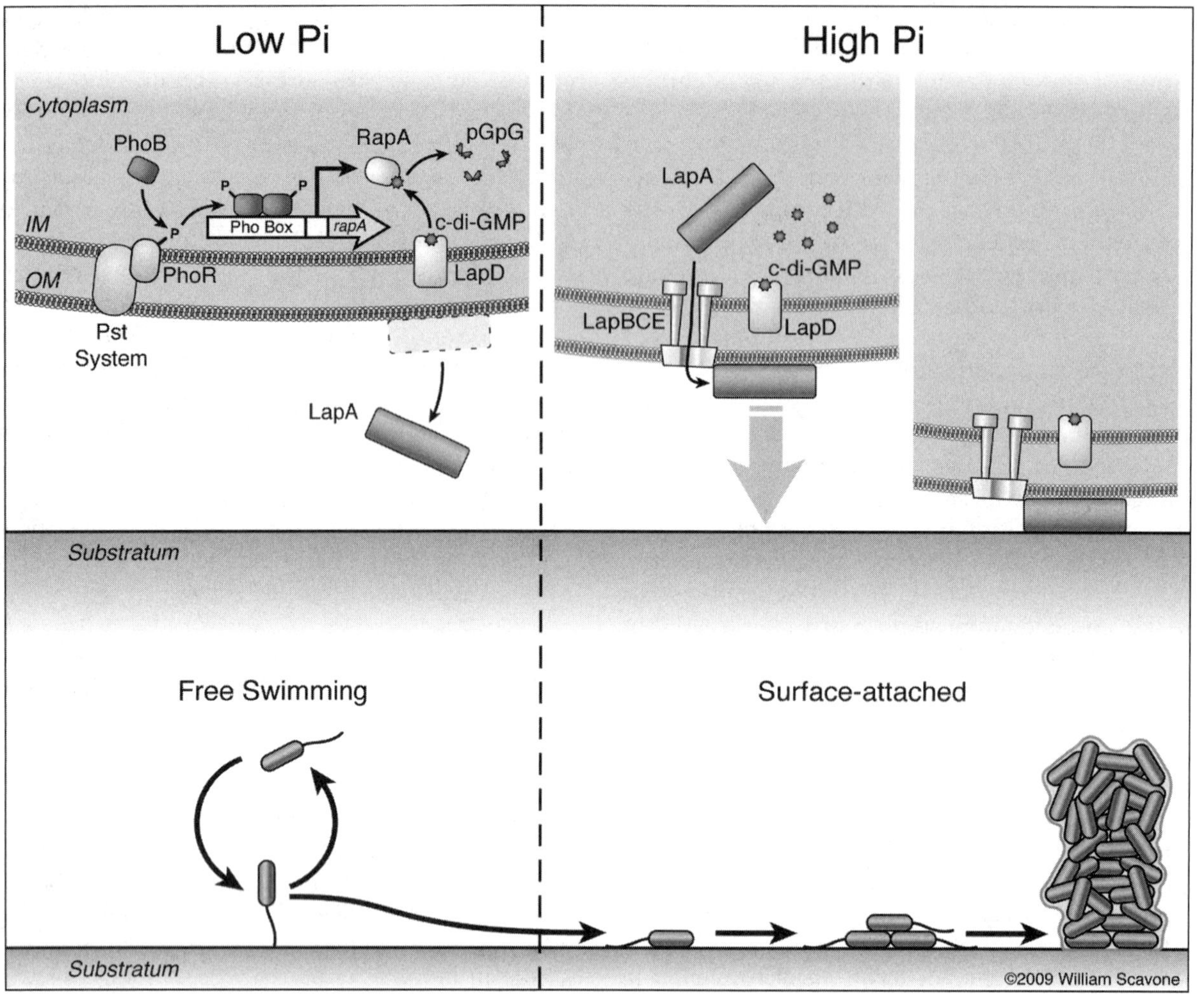

Figure 1. A schematic summarizing the current model for how phosphate concentration controls biofilm formation by *P. fluorescens* is shown at the molecular (top) and microscopic (bottom) levels. Left: Low extracellar P_i is sensed by the PhoR/Pst system complex, and this leads to activation of the PhoR kinase and phosphorylation of PhoB. PhoB~P forms a dimer and binds to the Pho Box sequence upstream of *rapA*, activating its transcription. The RapA protein cleaves c-di-GMP to form pGpG through its PDE activity, depleting cellular c-di-GMP pools. Decreased cellular c-di-GMP leads to dissociation of the nucleotide from the c-di-GMP effector LapD, and this signal promotes the egress of LapA to the culture supernatant. Cells cannot maintain stable surface attachments in low P_i. Right: Under high-P_i conditions RapA is not expressed, and c-di-GMP accumulates in the cell. LapD binds c-di-GMP and sends a signal promoting the maintenance of LapA on the cell surface. Cells can form irreversible attachments to the substratum and can go on to form a biofilm. IM, inner membrane; OM, outer membrane.

GGDEF domain, RapA did not show DGC activity in vitro (25). Mutation of the glutamic acid of the EAL motif to alanine inactivated RapA's PDE activity and rendered the gene unable to complement the *rapA* mutant, showing that PDE activity is essential to RapA's function in vivo. Finally, the cellular c-di-GMP levels in WT and the *rapA* mutant were compared under P_i starvation conditions. Here, the *rapA* mutant shows a 2.2-fold increase in c-di-GMP over the WT in low P_i (25). Thus, activation of the Pho regulon by P_i starvation leads to depletion of cellular c-di-GMP levels and loss of biofilm formation through the induction of the PDE RapA.

REGULATION AT THE LEVEL OF ADHESIN SECRETION AND LOCALIZATION

What is the effect of c-di-GMP depletion by RapA on the *P. fluorescens* cell? Typical outputs of c-di-GMP regulation, including exopolysaccharide (EPS) production and motility, do not appear to be affected (25). To answer this question, the core machinery for *P. fluorescens* Pf0-1 surface attachment, LapA, was investigated. The LapA adhesin can be detected inside cells, in the culture supernatant, and also loosely associated with the cell surface where it can be released by vortexing in a small volume of buffer (referred to as the cell-associated fraction). The abundance of LapA in the cell-associated protein fraction is a strong indicator of a strain's propensity for attachment and biofilm formation (18, 25, 28). Under low-P_i conditions, a distinct localization pattern for LapA was observed as compared to high P_i: there was a loss of LapA from the cell-associated fraction and an increase in LapA both inside cells and in the culture supernatant (25). This result suggested that the secretion and localization of the adhesin to the cell surface is a point of control for *P. fluorescens* attachment in response to P_i limitation.

To discern the role of the Pho regulon in LapA secretion and localization, the LapA localization profiles of the WT, *pst, pstrapA,* and *pstphoB* strains were compared. Consistent with a P_i-regulated change in LapA secretion, the *pst* mutant had decreased cell-associated LapA and an increased cellular pool. This change was dependent on Pho regulon activation, as the *pstphoB* mutant had a WT LapA localization pattern. Similar to its effects on biofilm formation, mutation of *rapA* in the *pst* background restored LapA localization to the cell surface to ~65% of WT levels (25). These data demonstrate that Pho regulon activation inhibits the secretion of the LapA adhesin and its localization to the cell surface, and the c-di-GMP PDE RapA is largely responsible for this effect.

LapD, A c-di-GMP EFFECTOR PROTEIN, REGULATES ATTACHMENT VIA LapA

RapA's role in regulating adhesin localization to the cell surface is the first instance of protein secretion or localization being regulated by c-di-GMP. While diguanylate cyclases (DGCs) and PDEs have been shown to control a variety of phenotypes, the mechanisms by which the c-di-GMP signal is received and translated into phenotypic outputs are unknown in most cases. The search for a candidate effector protein that could link c-di-GMP and LapA function in *P. fluorescens* Pf0-1 was aided by the identification of a strain carrying a mutation in another gene in the *lap* locus. Null mutants for the *lapD* gene were unable to form a biofilm or transition from reversible to irreversible attachment (18) and were specifically defective in localization of LapA to the cell surface (28). While LapA was secreted by *lapD* mutants, it was not retained in the cell-associated protein fraction and was lost to the culture supernatant (28). This phenomenon is notably similar to what happens to LapA in these fractions when the WT is grown in low P_i (25), prompting investigation of LapD's role in Pho regulon control of LapA localization, discussed below.

The predicted domain organization of the LapD protein appears much like that of a typical transmembrane signal transducer. The periplasmic domain of this inner membrane protein is connected to the cytoplasm via a HAMP domain, a key component in many signaling proteins (41). The cytoplasmic domains of LapD include a GGDEF and an EAL domain, both of which are degenerate and were shown to be catalytically inactive (28). When characterized in vitro, LapD specifically bound c-di-GMP via its degenerate EAL domain, and mutations in LapD that blocked c-di-GMP binding rendered it unable to promote biofilm formation (28), suggesting that it acts as a c-di-GMP effector protein.

Initial structure/function analyses revealed that LapD acts as an inside-out signaling protein with a cytoplasmic input and periplasmic output. Similar to what has been shown for other signaling proteins, mutations in LapD's HAMP domain modulated its output, the ability to promote biofilm formation (28). While some of these mutations could uncouple the necessity of LapD's input (c-di-GMP binding) from biofilm-promoting output, the periplasmic domain was shown to be required for LapD's effects on biofilm formation (28). In fact, expression of the periplasmic domain alone was able to complement the *lapD* mutant, demonstrating that the periplasmic domain is both necessary and sufficient for LapD's output. Thus, c-di-GMP binding in the cytoplasm is linked to a periplasmic output through the HAMP domain. The precise nature of LapD's output is not yet known, but what is clear is that mutations in LapD that cause increases or decreases in biofilm formation lead to concomitant and proportional effects on the levels of LapA on the surface of cells (28). This observation, together with the phenotype of the *lapD* mutant, suggests an essential role for LapD in regulating LapA localization to the cell surface in response to cytoplasmic c-di-GMP levels.

LapD's ROLE IN THE P_i-DEPENDENT c-di-GMP SIGNALING PATHWAY

There are several lines of evidence that suggest that LapD is the c-di-GMP receptor that is responsive to RapA expression in low P_i conditions. First, LapD and RapA both affect LapA localization to the cell surface, and *lapD* was required for biofilm formation by the *pstrapA* mutant (28). Further evidence came from studying a constitutively active *lapD* allele that promotes biofilm formation in the absence of c-di-GMP binding. This allele (dubbed ΔH1) was epistatic to the *pst* mutation and caused the *pst* mutant to form prodigious biofilms (28). The ΔH1 mutant was also insensitive to physiological P_i starvation, forming biofilms when the WT did not. Finally, ectopic induction of *rapA* expression in high-P_i conditions caused a 63% drop in attachment by the WT but had no effect on the ΔH1 mutant (28). Collectively, these data show that LapD's ability to sense c-di-GMP concentrations is required for repression of biofilm formation by Pho-regulated RapA and support a model in which LapD acts as the receptor in this P_i-regulated c-di-GMP pathway.

SURFACE-SENSING MODEL

As mentioned above, dissolved P_i concentrations are typically low in soils (<2 μM), where most P_i is found adsorbed to surfaces (32). This provides a compelling rationale for why *P. fluorescens* has evolved to regulate adherence to surfaces in response to extracellular P_i levels. Monds et al. synthesized the data for P_i regulation of biofilm formation and propose that expression of the Pho regulon confers a surface-sensing mode on *P. fluorescens* (25). In this mode, a cell can reversibly attach to a surface but not transition to irreversible attachment unless the local P_i concentration is sufficient to repress the Pho regulon. If suitable P_i sources are present, repression of RapA expression will allow c-di-GMP to accumulate, directing LapA localization to the cell surface and promoting irreversible attachment. In the absence of sufficient P_i, depletion of c-di-GMP by RapA will suppress stable attachment and promote dispersal of the bacteria to seek a more nutrient-rich environment (25). While the surface-sensing model has not yet been tested, it provides a framework for considering how local nutrient concentrations may regulate surface attachment by bacteria in general and has particular relevance to nutrients that are poorly soluble or adsorbed to surfaces in the natural environment.

REMAINING QUESTIONS

Even a generous appraisal of the literature to date suggests that we have a limited understanding of how c-di-GMP signaling networks function in a living cell. In the case of the system described here, general questions include what is the basal level of c-di-GMP in the cell and which enzymes (DGC and PDE) set that basal level under a given condition? Preliminary genetic screens did not identify a DGC required for *P. fluorescens* WCS365 biofilm formation, perhaps because there are redundant genes that fulfill this function under the conditions tested (30). Identifying the DGC(s) that generate the pool of c-di-GMP acted upon by RapA will bring us closer to understanding what signals promote biofilm formation in *P. fluorescens* Pf0-1.

Depletion of c-di-GMP by RapA has at least two effects on LapA biology. First, it causes a loss of cell-associated LapA to the culture supernatant, through inhibition of LapD signaling (Fig. 1) (25). Inhibition of LapA secretion is a second effect observed in low P_i and in the *pst* mutant background, which is rescued by deletion of *rapA* (25) but not accounted for by LapD. Future study could assess whether another c-di-GMP receptor plays a role in promoting LapA secretion or if the LapEBC transporter itself is responsive to c-di-GMP levels.

In addition to RapA, other PDEs could potentially impact attachment via LapD, and this possibility remains to be explored. Finally, how does LapD control LapA localization? LapD could impact LapA maintenance on the cell surface through a direct interaction with the LapA protein or an associated protein complex. Alternatively, LapD's effect on LapA could be indirect, through the activity of another protein(s) in the periplasm. The output of LapD signaling and the molecular mechanism that regulates LapA localization to the cell surface are still under investigation (28).

INTEGRATING c-di-GMP SIGNALING SYSTEMS

The description above represents the current extent of our knowledge about c-di-GMP signaling in *P. fluorescens* strain Pf0-1. Other signaling systems have, however, been described in closely related strains and species of *Pseudomonas*. It is interesting to consider the conservation of c-di-GMP systems among bacteria and if and how they may interact in the same organism. The brief analysis presented here suggests that while c-di-GMP regulatory components may be conserved in different organisms, the outputs they affect are likely to be different. In light of these observations, one must be cautious when making predictions about the roles that conserved systems might play, even in different strains of the same species.

Wsp Chemosensory System

The Wsp chemosensory pathway regulates adherence and EPS production by *P. fluorescens* strain

SBW25 and *Pseudomonas aeruginosa* PA01 (4, 16); Wsp of *P. aeruginosa* is discussed in chapter 11 of this volume. Much of the information about Wsp function has been uncovered through the characterization of the wrinkly spreader (WS) phenotype of *P. fluorescens* SBW25 by Rainey and colleagues. The WS phenotype arises in experimental populations of *P. fluorescens* through adaptive mutations that result in the activation of the Wsp pathway. Wsp activation results in overproduction of EPS, among other effects, and allows WS cells to colonize the air-liquid interface in static cultures, i.e., form a pellicle (4, 13, 39, 40). These studies have revealed Wsp to be a chemosensory system based on a predicted methyl-accepting chemotaxis protein and associated regulatory proteins, which is likely to function much like the canonical Che system of *E. coli* (4).

The output of signaling through Wsp is phosphorylation of the response regulator WspR and activation of WspR's DGC activity (23). c-di-GMP synthesis by WspR in *P. fluorescens* SBW25 stimulates the production of a cellulose polymer by transcriptionally and posttranscriptionally activating the cellulose synthase encoded by the *wss* operon (10, 13, 39). WspR also activates EPS synthesis in *P. aeruginosa,* but from two different EPS biosynthetic loci, both of which produce distinct noncellulose polysaccharides (28, 37). WspR of *P. aeruginosa* activates EPS production at the level of transcription via *psl* and *pel* operons (15, 16) and posttranslationally through the c-di-GMP binding protein PelD (22). This is a prime example of a conserved c-di-GMP regulatory apparatus directing distinct but analogous outputs. Is the Wsp system likely to play a similar role in other pseudomonads? The *wsp* operon appears to be well conserved in *P. fluorescens* strains Pf0-1 and Pf-5 as well as *Pseudomonas putida* KT2440 (Table 1). These loci include the following genes with their predicted proteins/function: *wspA* (methyl-accepting chemotaxis signal transducer), *wspB* (CheW domain protein), *wspC* (methyl transferase), *wspD* (CheW domain protein), *wspE* (sensor histidine kinase/response regulator fusion protein), *wspF* (CheB-like methyl esterase), and *wspR* (response regulator/DGC). In contrast to the conservation of the *wsp* signaling components, putative targets of *wspR* are not universally conserved among the strains examined. The *wss* genes, which code for the cellulose synthase of *P. fluorescens* SBW25, are absent from *P. fluorescens* Pf0-1 and Pf-5 and *P. putida*. Interestingly, homologues of the *pel* and *psl* operons can be found in Pf-5 but not in Pf0-1 or *P. putida*. Therefore, the potential target(s) of *wsp* regulation in the latter organisms remains an open question. To date, no EPS biosynthetic loci are known to play a role in surface attachment by *P. fluorescens* Pf0-1 or Pf-5 or *P. putida*. The published genome annotation for *P. putida* KT2440 does contain putative polysaccharide synthesis genes (27), but they have not yet been investigated. Nothing is currently known about EPS biosynthesis or composition in *P. fluorescens* Pf0-1. The Pf0-1 genome annotation has not been pub-

Table 1. Conservation of c-di-GMP signaling components and attachment factors among a few pseudomonads[a]

Gene	*P. fluorescens* Pf0-1	*P. fluorescens* Pf-5	*P. putida* KT2440	*P. fluorescens* SBW25	*P. syringae* DC3000	*P. aeruginosa* PA01
wspR[b]	89/96	91/96	82/90	100	84/92	74/85
wssE[c]	NA	NA	NA	100	51/67	NA
pelA[d]	NA	64/78	NA	NA	NA	100
pslA[d]	NA	65/80	NA	69/84	66/79	100
lapA[e]	100	73/81	70/80	NA	NA	NA
lapD	100	89/95	75/85	NA	NA	61/75
fleQ[f]	100	97/98	92/96	96/98	92/96	83/90
rapA	100	92/97	67/78	87/94	81/90	75/86

[a] Predicted protein sequences of the genes listed were subjected to a BLASTP search. Values reported in the table are the percent amino acid identity/similarity between the query sequence (listed as 100) and the subject. Not applicable (NA) is reported if no match was found with an identity ≥25% or similarity ≥40%.

[b] In characterized systems, *wspR* is in a multigene operon encoding all the Wsp chemosensory proteins. In all organisms above, *wspR* homologs occur with other Wsp homologs.

[c] The *wssE* gene is one gene in the *wss* operon; it encodes the C subunit of the cellulose synthase. This gene was shown to be specifically induced by *P. fluorescens* SBW25 during growth in the rhizosphere (31).

[d] The *pelA* and *pslA* genes are each the first gene in multigene operons encoding EPS synthesis functions. In all cases where a homolog of *pelA* or *pslA* was identified, a subsequent operon with conserved synteny and predicted functions was present.

[e] For ease of analysis, only the N-terminal 235 amino acids of LapA were used as a query. This includes all of the protein sequence prior to the highly repetitive region of LapA. The homologs identified also meet our cutoff criteria if their entire protein sequences are compared.

[f] *fleQ* is also known as *adnA* in *P. fluorescens* Pf0-1.

lished, and thorough bioinformatic searches for EPS genes in this organism have not been undertaken. Whether or not Wsp plays a role in biofilm formation in *P. fluorescens* Pf0-1 via c-di-GMP signaling through LapD or through enzymatic activation of a yet uncharacterized polysaccharide synthesis system awaits further investigation.

Regulation of Flagellar Synthesis and Adhesion by FleQ/AdnA

Another interesting point of comparison is the potential for a conserved role for the flagellar synthesis regulator FleQ in c-di-GMP-dependent regulation of biofilm formation. In multiple strains of *P. fluorescens* and also in *P. aeruginosa,* FleQ (also known as AdnA in *P. fluorescens*) is known to be an activator of flagellar gene transcription and is well conserved among these organisms (2, 5, 6) (Table 1). In both *P. fluorescens* SBW25 and *P. aeruginosa,* FleQ has also been shown to inhibit biofilm formation through repression of EPS gene transcription (10, 15). For *P. fluorescens* SBW25, repression of the *wss* genes is thought to be indirect, through the control of other regulators (10). For *P. aeruginosa,* however, FleQ directly binds the promoter of the *pel* operon (15). Repression of *pel* gene expression and in vitro DNA binding is inhibited by c-di-GMP through binding of the cyclic dinucleotide to FleQ (15) (see also chapter 11). Whether or not FleQ of *P. fluorescens* SBW25 is responsive to c-di-GMP remains to be tested, but the data for *P. aeruginosa* suggest a plausible mechanism by which c-di-GMP production by WspR could alleviate FleQ repression of EPS gene transcription.

The apparent conservation of FleQ's role in repression of attachment and activation of motility in *P. fluorescens* SBW25 and *P. aeruginosa* is intriguing and consistent with the theme of shared regulatory components acting on distinct outputs. This regulatory role does not, however, appear to be conserved across all fluorescent pseudomonads. For *P. fluorescens* Pf0-1, AdnA/FleQ is an activator of attachment, not a repressor, and is independently required for flagellum production and biofilm formation (36). Additionally, FleQ is required for biofilm formation by *P. putida* (45). How AdnA/FleQ promotes attachment in *P. fluorescens* Pf0-1 and *P. putida* is not yet known, but this distinct role is an instance of significant differences in the role played by the same regulator between closely related pseudomonads.

Conservation of *lap* Genes

Some of the largest genes yet discovered in bacterial genomes encode predicted adhesins that have highly repetitive amino acid sequences (35). As functional and structural data are still lacking for most proteins in this group, LapA is one of its best-characterized members. The outer membrane-associated adhesin encoded by *lapA* is essential for attachment and initiation of biofilm formation by *P. fluorescens* Pf0-1, *P. fluorescens* WCS356, and *P. putida* (9, 17, 25, 28). LapA of *P. putida* is required for adhesion to both biotic and abiotic surfaces (45), as is the *P. putida* homologue of *lapD,* PP0165 (11), suggesting that c-di-GMP regulation of adhesin localization may be important for plant colonization by this and other species. Indeed, a role for *lapD*-like PP0165 in regulating LapA function is likely, as mutants in PP0165 were not adhesive and could not transition from reversible to irreversible attachment (11).

Comparing the small group of organisms in Table 1, it is clear that a plethora of attachment strategies have been selected for in the *Pseudomonas* genus. The apparent differences in the occurrence of c-di-GMP-regulated adherence factors among closely related bacteria highlight both the diversity within the genus and the difficulty of making generalizations about the mechanisms they utilize for biofilm formation. Determining which components are required for attachment and how they are regulated by c-di-GMP in a given species would require a thorough experimental investigation of that particular system.

Conservation of *rapA* and Implications for Pho Control of c-di-GMP Signaling

As mentioned above, P_i is essential to life, and many bacteria sense and respond to limiting concentrations of P_i through induction of the Pho regulon. This signaling pathway has also been shown to regulate other traits, including biofilm formation in *P. fluorescens* (25) and virulence in pathogenic bacteria (21). Despite the considerable differences among the pseudomonads described above, is it possible that P_i-dependent modulation of c-di-GMP levels is conserved? In the initial characterization of RapA, Monds et al. noted that the *rapA* gene was well conserved among the pseudomonads and that overlapping Pho boxes with good matches to the *E. coli* consensus sequence could be located upstream of each *rapA* homolog (25) (Table 1). This observation provides some initial indication that P_i regulation of c-di-GMP signaling and biofilm formation may be conserved among these organisms even if the specific outputs of that signal are different.

CONCLUDING REMARKS

The presence of many predicted GGDEF and EAL domain proteins within individual bacterial genomes (*P. fluorescens* Pf0-1 has 43) (25) raises the possibility that c-di-GMP signaling networks are extraordinarily complex. Transcriptional regulation of RapA in *P. fluorescens* Pf0-1 seems a relatively straightforward strategy for modulating c-di-GMP levels when compared to other mechanisms shown to regulate DGC and PDE activities, including allosteric activation (33), subcellular localization (34), and mRNA stability (20). Whether this strategy is commonly employed or if additional layers of regulation act on RapA remains to be seen. The challenge of understanding any signaling system, simple or complex, requires tracing input to output under physiologically relevant conditions.

RapA was the first PDE discovered whose transcriptional expression is directly regulated by an environmental signal, P_i (25). By coupling PDE expression to the Pho regulon, *P. fluorescens* has evolved a mechanism to rapidly control attachment to surfaces in response to the availability of a key nutrient. Control is achieved, in part, through the action of the c-di-GMP receptor LapD, which directs the localization of the LapA adhesin to the cell surface (28). While much remains to be understood about how this system functions and how it may interact with other c-di-GMP signaling systems in the cell, it provides a foundational example of environmental control of c-di-GMP signaling.

REFERENCES

1. **Anderson, G. G., and G. A. O'Toole.** 2008. Innate and induced resistance mechanisms of bacterial biofilms. *Curr. Top. Microbiol. Immunol.* **322:**85–105.
2. **Arora, S. K., B. W. Ritchings, E. C. Almira, S. Lory, and R. Ramphal.** 1997. A transcriptional activator, FleQ, regulates mucin adhesion and flagellar gene expression in *Pseudomonas aeruginosa* in a cascade manner. *J. Bacteriol.* **179:**5574–5581.
3. **Banin, E., M. L. Vasil, and E. P. Greenberg.** 2005. Iron and *Pseudomonas aeruginosa* biofilm formation. *Proc. Natl. Acad. Sci. USA* **102:**11076–11081.
4. **Bantinaki, E., R. Kassen, C. G. Knight, Z. Robinson, A. J. Spiers, and P. B. Rainey.** 2007. Adaptive divergence in experimental populations of *Pseudomonas fluorescens*. III. Mutational origins of wrinkly spreader diversity. *Genetics* **176:**441–453.
5. **Capdevila, S., F. M. Martinez-Granero, M. Sanchez-Contreras, R. Rivilla, and M. Martin.** 2004. Analysis of *Pseudomonas fluorescens* F113 genes implicated in flagellar filament synthesis and their role in competitive root colonization. *Microbiology* **150:**3889–3897.
6. **Casaz, P., A. Happel, J. Keithan, D. L. Read, S. R. Strain, and S. B. Levy.** 2001. The *Pseudomonas fluorescens* transcription activator AdnA is required for adhesion and motility. *Microbiology* **147:**355–361.
7. **Compant, S., B. Duffy, J. Nowak, C. Clement, and E. A. Barka.** 2005. Use of plant growth-promoting bacteria for biocontrol of plant diseases: principles, mechanisms of action, and future prospects. *Appl. Environ. Microbiol.* **71:**4951–4959.
8. **Davey, M. E., and G. A. O'Toole.** 2000. Microbial biofilms: from ecology to molecular genetics. *Microbiol. Mol. Biol. Rev.* **64:**847–867.
9. **Espinosa-Urgel, M., A. Salido, and J. L. Ramos.** 2000. Genetic analysis of functions involved in adhesion of *Pseudomonas putida* to seeds. *J. Bacteriol.* **182:**2363–2369.
10. **Giddens, S. R., R. W. Jackson, C. D. Moon, M. A. Jacobs, X. X. Zhang, S. M. Gehrig, and P. B. Rainey.** 2007. Mutational activation of niche-specific genes provides insight into regulatory networks and bacterial function in a complex environment. *Proc. Natl. Acad. Sci. USA* **104:**18247–18252.
11. **Gjermansen, M., P. Ragas, C. Sternberg, S. Molin, and T. Tolker-Nielsen.** 2005. Characterization of starvation-induced dispersion in *Pseudomonas putida* biofilms. *Environ. Microbiol.* **7:**894–906.
12. **Gonin, M., E. M. Quardokus, D. O'Donnol, J. Maddock, and Y. V. Brun.** 2000. Regulation of stalk elongation by phosphate in *Caulobacter crescentus*. *J. Bacteriol.* **182:**337–347.
13. **Goymer, P., S. G. Kahn, J. G. Malone, S. M. Gehrig, A. J. Spiers, and P. B. Rainey.** 2006. Adaptive divergence in experimental populations of *Pseudomonas fluorescens*. II. Role of the GGDEF regulator WspR in evolution and development of the wrinkly spreader phenotype. *Genetics* **173:**515–526.
14. **Haas, D., and G. Defago.** 2005. Biological control of soil-borne pathogens by fluorescent pseudomonads. *Nat. Rev. Microbiol.* **3:**307–319.
15. **Hickman, J. W., and C. S. Harwood.** 2008. Identification of FleQ from *Pseudomonas aeruginosa* as a c-di-GMP-responsive transcription factor. *Mol. Microbiol.* **69:**376–389.
16. **Hickman, J. W., D. F. Tifrea, and C. S. Harwood.** 2005. A chemosensory system that regulates biofilm formation through modulation of cyclic diguanylate levels. *Proc. Natl. Acad. Sci. USA* **102:**14422–14427.
17. **Hinsa, S. M., M. Espinosa-Urgel, J. L. Ramos, and G. A. O'Toole.** 2003. Transition from reversible to irreversible attachment during biofilm formation by *Pseudomonas fluorescens* WCS365 requires an ABC transporter and a large secreted protein. *Mol. Microbiol.* **49:**905–918.
18. **Hinsa, S. M., and G. A. O'Toole.** 2006. Biofilm formation by *Pseudomonas fluorescens* WCS365: a role for LapD. *Microbiology* **152:**1375–1383.
19. **Jackson, D. W., J. W. Simecka, and T. Romeo.** 2002. Catabolite repression of *Escherichia coli* biofilm formation. *J. Bacteriol.* **184:**3406–3410.
20. **Jonas, K., A. N. Edwards, R. Simm, T. Romeo, U. Romling, and O. Melefors.** 2008. The RNA binding protein CsrA controls cyclic di-GMP metabolism by directly regulating the expression of GGDEF proteins. *Mol. Microbiol.* **70:**236–257.
21. **Lamarche, M. G., B. L. Wanner, S. Crepin, and J. Harel.** 2008. The phosphate regulon and bacterial virulence: a regulatory network connecting phosphate homeostasis and pathogenesis. *FEMS Microbiol. Rev.* **32:**461–473.
22. **Lee, V. T., J. M. Matewish, J. L. Kessler, M. Hyodo, Y. Hayakawa, and S. Lory.** 2007. A cyclic-di-GMP receptor required for bacterial exopolysaccharide production. *Mol. Microbiol.* **65:**1474–1488.
23. **Malone, J. G., R. Williams, M. Christen, U. Jenal, A. J. Spiers, and P. B. Rainey.** 2007. The structure-function relationship of WspR, a *Pseudomonas fluorescens* response regulator with a GGDEF output domain. *Microbiology* **153:**980–994.
24. **Martin, J. F.** 2004. Phosphate control of the biosynthesis of antibiotics and other secondary metabolites is mediated by the PhoR-PhoP system: an unfinished story. *J. Bacteriol.* **186:** 5197–5201.

25. **Monds, R. D., P. D. Newell, R. H. Gross, and G. A. O'Toole.** 2007. Phosphate-dependent modulation of c-di-GMP levels regulates *Pseudomonas fluorescens* Pf0-1 biofilm formation by controlling secretion of the adhesin LapA. *Mol. Microbiol.* **63:** 659–679.
26. **Monds, R. D., P. D. Newell, J. A. Schwartzman, and G. A. O'Toole.** 2006. Conservation of the Pho regulon in *Pseudomonas fluorescens* Pf0-1. *Appl. Environ. Microbiol.* **72:**1910–1924.
27. **Nelson, K. E., C. Weinel, I. T. Paulsen, R. J. Dodson, H. Hilbert, V. A. Martins dos Santos, D. E. Fouts, S. R. Gill, M. Pop, M. Holmes, L. Brinkac, M. Beanan, R. T. DeBoy, S. Daugherty, J. Kolonay, R. Madupu, W. Nelson, O. White, J. Peterson, H. Khouri, I. Hance, P. C. Lee, E. Holtzapple, D. Scanlan, K. Tran, A. Moazzez, T. Utterback, M. Rizzo, K. Lee, D. Kosack, D. Moestl, H. Wedler, J. Lauber, D. Stjepandic, J. Hoheisel, M. Straetz, S. Heim, C. Kiewitz, J. A. Eisen, K. N. Timmis, A. Düsterhöft, B. Tümmler, and C. M. Fraser.** 2002. Complete genome sequence and comparative analysis of the metabolically versatile *Pseudomonas putida* KT2440. *Environ. Microbiol.* **4:**799–808.
28. **Newell, P. D., R. D. Monds, and G. A. O'Toole.** 2009. LapD is a bis-(3′,5′)-cyclic dimeric GMP binding protein that regulates surface attachment by *Pseudomonas fluorescens* Pf0-1. *Proc. Natl. Acad. Sci. USA* **106:**3461–3466.
29. **O'Toole, G. A., K. A. Gibbs, P. W. Hager, P. V. Phibbs, Jr., and R. Kolter.** 2000. The global carbon metabolism regulator Crc is a component of a signal transduction pathway required for biofilm development by *Pseudomonas aeruginosa*. *J. Bacteriol.* **182:**425–431.
30. **O'Toole, G. A., and R. Kolter.** 1998. Initiation of biofilm formation in *Pseudomonas fluorescens* WCS365 proceeds via multiple, convergent signalling pathways: a genetic analysis. *Mol. Microbiol.* **28:**449–461.
31. **Patriquin, G. M., E. Banin, C. Gilmour, R. Tuchman, E. P. Greenberg, and K. Poole.** 2008. Influence of quorum sensing and iron on twitching motility and biofilm formation in *Pseudomonas aeruginosa*. *J. Bacteriol.* **190:**662–671.
32. **Paul, E. A., and F. E. Clark.** 1989. Soil as a habitat for organisms and their reactions, p. 13–30. *In* E. A. Paul and F. E. Clark (ed.), *Soil Microbiology and Biochemistry,* vol. 2. Academic Press, San Diego, CA.
33. **Paul, R., S. Abel, P. Wassmann, A. Beck, H. Heerklotz, and U. Jenal.** 2007. Activation of the diguanylate cyclase PleD by phosphorylation-mediated dimerization. *J. Biol. Chem.* **282:** 29170–29177.
34. **Paul, R., S. Weiser, N. C. Amiot, C. Chan, T. Schirmer, B. Giese, and U. Jenal.** 2004. Cell cycle-dependent dynamic localization of a bacterial response regulator with a novel diguanylate cyclase output domain. *Genes Dev.* **18:**715–727.
35. **Reva, O., and B. Tummler.** 2008. Think big—giant genes in bacteria. *Environ. Microbiol.* **10:**768–777.
36. **Robleto, E. A., I. Lopez-Hernandez, M. W. Silby, and S. B. Levy.** 2003. Genetic analysis of the AdnA regulon in *Pseudomonas fluorescens:* nonessential role of flagella in adhesion to sand and biofilm formation. *J. Bacteriol.* **185:**453–460.
37. **Ryder, C., M. Byrd, and D. J. Wozniak.** 2007. Role of polysaccharides in *Pseudomonas aeruginosa* biofilm development. *Curr. Opin. Microbiol.* **10:**644–648.
38. **Shrout, J., D. Chopp, C. Just, M. Hentzer, M. Givskov, and M. Parsek.** 2006. The impact of quorum sensing and swarming motility on *Pseudomonas aeruginosa* biofilm formation is nutritionally conditional. *Mol. Microbiol.* **62:**1264–1277.
39. **Spiers, A. J., J. Bohannon, S. M. Gehrig, and P. B. Rainey.** 2003. Biofilm formation at the air-liquid interface by the *Pseudomonas fluorescens* SBW25 wrinkly spreader requires an acetylated form of cellulose. *Mol. Microbiol.* **50:**15–27.
40. **Spiers, A. J., S. G. Kahn, J. Bohannon, M. Travisano, and P. B. Rainey.** 2002. Adaptive divergence in experimental populations of *Pseudomonas fluorescens*. I. Genetic and phenotypic bases of wrinkly spreader fitness. *Genetics* **161:**33–46.
41. **Taylor, B. L.** 2007. Aer on the inside looking out: paradigm for a PAS-HAMP role in sensing oxygen, redox and energy. *Mol. Microbiol.* **65:**1415–1424.
42. **Wang, X., A. K. Dubey, K. Suzuki, C. S. Baker, P. Babitzke, and T. Romeo.** 2005. CsrA post-transcriptionally represses pgaABCD, responsible for synthesis of a biofilm polysaccharide adhesin of *Escherichia coli. Mol. Microbiol.* **56:**1648–1663.
43. **Wanner, B. L.** 1996. Phosphorus assimilation and control of the phosphate regulon, p. 1371. *In* F. C. Neidhart, R. Curtiss III, J. L. Ingraham, E. C. C. Lin, K. B. Low, B. Magasanik, et al. (ed.), Escherichia coli *and* Salmonella typhimurium: *Cellular and Molecular Biology*. ASM Press, Washington, DC.
44. **Wanner, B. L.** 1996. Signal transduction in the control of phosphate-regulated genes of *Escherichia coli*. *Kidney Int.* **49:** 964–967.
45. **Yousef-Coronado, F., M. L. Travieso, and M. Espinosa-Urgel.** 2008. Different, overlapping mechanisms for colonization of abiotic and plant surfaces by *Pseudomonas putida*. *FEMS Microbiol. Lett.* **288:**118–124.

IV. CYCLIC DI-GMP, PATHOGENESIS, AND HOST IMMUNITY

The Second Messenger Cyclic Di-GMP
Edited by A. J. Wolfe and K. L. Visick

Chapter 20

Role of Cyclic Di-GMP in *Vibrio cholerae* Virulence

Jason T. Pratt, Rita Tamayo, and Andrew Camilli

Cholera is a severe secretory diarrheal disease caused by the motile gram-negative facultative pathogen *Vibrio cholerae*. Cholera is an acute infection; the incubation period can be as short as 12 hours, and patients of low body weight can die within hours of the onset of symptoms due to hypotensive shock (7). Patients suffering from cholera can lose many liters of fluid per day in the form of rice-water stool (3). Although rehydration therapy is generally an effective treatment, the disease remains a large social and economic burden in the developing world where access to clean water is limited and poor sanitation practices prevail. The World Health Organization (WHO) reports that there were 236,896 cases of cholera resulting in 6,311 deaths worldwide in 2006 (82). However, this is believed to be a gross underestimate of the total disease burden due to underreporting.

Cholera is a disease largely associated with epidemic spread; there have been seven cholera pandemics since 1817. Of the more than 200 serogroups of *V. cholerae* identified in environmental reservoirs, only the O1 and O139 serogroups have been associated with epidemic disease (4, 54). In fact, all seven pandemics have been attributed to *V. cholerae* O1. The O1 serogroup is subdivided into two biotypes: classical, which caused the first six pandemics, and El Tor, the causative agent of the ongoing seventh pandemic (4).

V. cholerae lives naturally in temperate waters around the world, including salt, brackish, and some fresh waters (32). There is evidence that *V. cholerae* persists in these environments in biofilms formed on biotic surfaces, such as the chitinous exoskeleton of zooplankton and phytoplankton (30, 76, 81). Cholera is thus a water-borne disease, and infection occurs upon ingestion of contaminated water or food. Upon ingestion of *V. cholerae,* the bacteria pass through the gastric acid barrier of the stomach and colonize the surface of the small intestine without tissue invasion. As the bacterium transitions from its natural aquatic environment to that of the small intestine, it undergoes a shift from environmental to virulence gene expression (27, 43, 44, 53, 77).

The ability of *V. cholerae* to efficiently adapt to the two very disparate environments of the host intestine and natural aquatic environments is paramount to its lifestyle and overall fitness. During transitions between each environment, the bacteria are exposed to a variety of stresses (e.g., changes in pH, osmolarity, and nutrient availability) that they must withstand and adapt to in order to survive. It stands to reason that shifts from one environment to another require coordinate regulation of a number of genes required in one condition but not the other. Indeed, *V. cholerae* undergoes large changes in gene transcription upon entry and exit from the host (41, 50, 55, 56, 68). Therefore, there is a necessity for regulatory/messenger systems to sense changes in the external environment and signal downstream for alterations in gene transcription. The bacterial allosteric regulator and secondary messenger cyclic di-GMP (c-di-GMP) plays an integral role in the life cycle of pathogenic *V. cholerae* by helping to mediate the environment-to-host transition (see also chapter 17).

The intracellular concentration of c-di-GMP is regulated by the opposing activities of diguanylate cyclase (DGC) and phosphodiesterase (PDE) enzymes (2, 73). DGC activity has been attributed to proteins containing the GGDEF domain, while PDE activity has been attributed to the EAL and HD-GYP domains, each named for conserved amino acid residues (11, 17, 57, 64, 65, 67, 74) (see also chapters 2 to 5). The regulation of the intracellular concentration

Jason T. Pratt and Andrew Camilli • Howard Hughes Medical Institute and Department of Molecular Biology and Microbiology, Tufts University School of Medicine, Boston, MA 02111. **Rita Tamayo** • Department of Microbiology and Immunology, The University of North Carolina at Chapel Hill, Chapel Hill, NC 27599.

of c-di-GMP in *V. cholerae* is complex, as there is an abundance of genes coding for DGCs and PDEs. In total, *V. cholerae* encodes 30 GGDEF, 12 EAL, 9 HD-GYP, and 10 GGDEF-EAL proteins (23), but at the present time, only a small number have been studied. Examination of the genome content of pathogenic and nonpathogenic *V. cholerae* does not reveal any difference in the repertoire of c-di-GMP metabolic enzymes encoded despite their differing lifestyles. This suggests that c-di-GMP was originally used for environmental persistence and that pathogenic *V. cholerae* strains have assimilated this system to regulate virulence gene expression as well.

Experiments have shown that c-di-GMP inversely regulates the expression of biofilm- and virulence-specific genes in *V. cholerae* (8, 9, 78, 79). Specifically, when the intracellular concentration of c-di-GMP is high, transcription of biofilm genes increases and transcription of virulence genes decreases. When the concentration of c-di-GMP is low, the opposite is true. In addition, c-di-GMP has been shown to negatively regulate motility in *V. cholerae*. It is largely because of this inverse relationship between biofilm and virulence gene expression that it is proposed that c-di-GMP mediates the transition of the bacterium between its natural environment, where biofilms are advantageous, and the host small intestine, where expression of virulence factors is essential. This suggests a model whereby c-di-GMP concentration fluctuates during the life cycle of pathogenic *V. cholerae* in response to environmental cues. Furthermore, the bacterium uses these changes in c-di-GMP concentration as a signal to modify its behavior to adapt to changing conditions. Specifically, the c-di-GMP concentration in *V. cholerae* is predicted to be high during biofilm formation in the aquatic environment, which promotes persistence, and low in the host environment, which promotes virulence gene expression. In this chapter we will focus on the role of c-di-GMP in the regulation of *V. cholerae* virulence mechanisms during colonization and dissemination. The role of c-di-GMP in *V. cholerae* biofilm formation is discussed in chapter 17.

CONVERGING PATHWAYS REGULATE VIRULENCE GENE EXPRESSION IN *V. CHOLERAE*

The two most important virulence factors expressed by *V. cholerae* are cholera toxin (CT) and the toxin-coregulated pilus (TCP). TCP is composed of a single structural subunit, TcpA, and its biosynthetic machinery is encoded by the *tcp* operon located within a 40-kb *Vibrio* pathogenicity island. TCP has been shown to be essential for colonization of the small intestine in both human and animal studies, although its exact function during colonization is unknown (27, 77). It appears to mediate interbacterial aggregation, thus facilitating microcolony formation on the surface of the intestinal epithelium; however, it does not seem to mediate direct interaction with host tissue (35). The role of CT during infection is better understood. The toxin is encoded by the *ctxAB* operon located within the genome of the temperate filamentous bacteriophage CTXφ. Interestingly, TCP serves as the receptor for CTXφ infection of *V. cholerae* and therefore denotes dependence of CTXφ horizontal transmission on the presence of TCP (80). CT is the prototype bipartite A-B toxin composed of one A subunit and five B subunits. The B subunit pentamer binds directly to intestinal epithelial cells through interaction with GM_1 gangliosides on their surfaces (31, 34, 59). Upon binding, the A subunit is translocated into the host cell cytosol where it ADP-ribosylates the host G protein, G_S, leading to activation of adenylate cyclase and an increase in intracellular cAMP concentration (24, 49). Increased cAMP level signals for a reduction of NaCl uptake and an increase of secretion of Cl^- ions, resulting in an osmotic imbalance, which leads to secretion of large amounts of water from host cells into the lumen of the intestine manifested as severe diarrhea (21).

Regulation of CT and TCP expression occurs primarily at the level of transcription. Two converging pathways, the ToxR regulon and quorum sensing (QS), mediate this regulation. The ToxR regulon, so named for the first regulator identified in this pathway, is a regulatory cascade consisting of a number of transcription factors that end in regulation of ToxT (20, 52). The core components of this pathway are illustrated in Fig. 1. ToxT, an AraC family transcription factor encoded within the *tcp* operon, is regulated at both the transcriptional and posttranscriptional levels (20, 26, 29, 69). It has been shown that ToxT activity can be regulated by specific environmental signals, including temperature and bile. Additionally, it has been proposed that the activity of ToxT is regulated by c-di-GMP. Regulation of ToxT expression is coordinated by two transmembrane transcription factors, ToxR and TcpP. Both of these proteins bind to a *toxT* proximal promoter in a cooperative manner to activate transcription (18, 40). ToxT then acts in an autoregulatory manner, binding at a distal promoter at the beginning of the *tcp* operon, to fully activate the operon (84).

The regulation of ToxR and TcpP is complex and not fully understood. Both are coexpressed with additional membrane-bound binding partners, ToxS and TcpH, respectively. While the exact role of ToxS

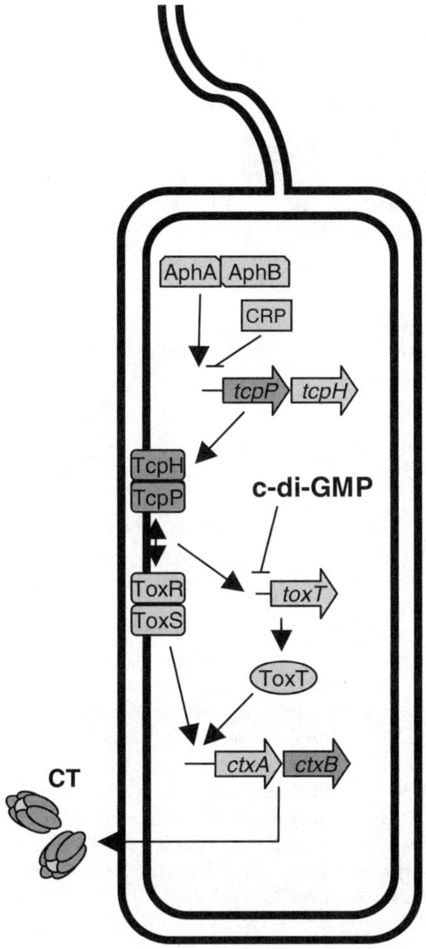

Figure 1. The ToxR regulon in *V. cholerae* is repressed by c-di-GMP. Arrows indicate activation of genes or proteins; parallel lines in place of an arrowhead represent repression. The predicted interaction of c-di-GMP with the regulon is shown. High c-di-GMP concentration results in decreased *toxT* transcription, consequently causing reduced *ctxAB* expression and CT production. c-di-GMP may also regulate ToxT activity, because *tcp* gene expression is unaffected by high c-di-GMP. CRP, cAMP receptor protein.

is unknown, it has been shown to be required for maximal activity of ToxR (19, 58). TcpH is required to maintain the stability of TcpP, which is degraded by the protease YaeL in the absence of TcpH (5, 48). In addition, transcription from the *tcpPH* promoter is tightly regulated by AphA, AphB, and cAMP receptor protein (CRP)-cAMP (37, 38, 70). AphA and AphB act as positive regulators at the *tcpPH* promoter, while CRP-cAMP negatively regulates expression by competing for binding sites with AphA. Therefore, carbon source availability is one of the environmental signals that regulates virulence gene expression in *V. cholerae*. Additional extracellular signals regulate *tcpPH* expression indirectly, as expression of *aphA* is negatively regulated by HapR, a transcription factor controlled by QS signals (39, 51, 85). When cell density is high and QS molecules accumulate, HapR is expressed and acts as a direct repressor of *aphA,* thus repressing virulence gene expression. Upon entry into the host, the bacteria and quorum signals are potentially diluted, leading to repression of HapR and thus expression of virulence genes.

In summary, AphA is induced in response to environmental stimuli, including reduction of QS signals, and in conjunction with AphB, activates transcription of *tcpPH*. TcpP and the constitutively expressed ToxR activate expression of ToxT, which then goes on to induce expression of TCP and CT.

REGULATION OF *V. CHOLERAE* VIRULENCE BY c-di-GMP

In vitro expression studies have shown that c-di-GMP has a negative effect on virulence gene expression, specifically CT. Western blot analysis and GM_1 ganglioside-based enzyme-linked immunosorbent assay have shown that CT expression is reduced approximately fivefold by increased intracellular c-di-GMP as a result of mutation of the PDE VieA in the classical biotype or ectopic expression of the DGC VdcA (VCA0956, CdgF [9]) (79). Transcriptional profiling experiments have shown a similar repression of *ctxAB* expression with high c-di-GMP concentration, and this defect appears to be the result of decreased expression of *toxT* (8, 9) (Fig. 1). Surprisingly, this effect does not extend to expression of the *tcp* operon, which appears to be unaffected by c-di-GMP. Though the basis for this difference is unknown, it is possible that c-di-GMP affects the expression and/or activity of ToxT in a manner that allows for discrimination between the *ctxAB* and *tcp* promoters. Alternatively, other c-di-GMP regulated factors may moderate this difference in virulence gene expression.

As mentioned above, ToxT is autoregulatory via read-through transcription from the upstream *tcp* operon promoter. That expression of TCP is not regulated by c-di-GMP suggests that c-di-GMP-mediated effects on *toxT* transcription occur at the *toxT* proximal promoter within the *tcp* operon. Alterations in c-di-GMP concentration did not affect the activity of ToxR or the expression of TcpP; therefore, the mechanism of c-di-GMP-mediated repression of ToxT remains unknown (79). Like other members of the AraC family, ToxT has an N-terminal domain that may function as a binding site for a small molecule effector. If this effector were c-di-GMP or a c-di-GMP-regulated molecule, then the resulting alterations in ToxT activity in the presence of c-di-GMP could mediate the observed pattern of regulation.

The model in which c-di-GMP inversely regulates biofilm and virulence genes suggests that c-di-GMP concentrations are high during life in a biofilm within aquatic environments. Consequently, it is proposed that the c-di-GMP concentration needs to be decreased upon entry into humans to permit virulence gene expression for establishment of an infection (or vice versa, when disseminating from the host into the aquatic environment the c-di-GMP concentration will need to increase). Experiments have shown that *V. cholerae* does have the capacity to change the intracellular concentration of c-di-GMP in response to a changing environment in the manner predicted by this model.

Studies using recombination-based in vivo expression technology (RIVET) to examine the expression of *V. cholerae* genes during infection of the infant mouse model host demonstrated that c-di-GMP concentration is decreased upon infection to allow maximal virulence gene expression (15, 42). Briefly, RIVET utilizes the function of the resolvase enzyme (TnpR) as a reporter to measure the expression of specific genes (Fig. 2). Chromosomal transcriptional fusions are made between *tnpR* and a gene of interest in a *V. cholerae* strain containing a reporter cassette (i.e., antibiotic resistance gene) located at a neutral site in the genome. The reporter cassette is flanked by *res* sites, which are recognized by TnpR. When the gene of interest is expressed, so too is TnpR, which functions to excise DNA between the *res* sites, removing it from the genome. Thus, loss of the reporter cassette serves as a proxy to measure expression of the gene in question in a given condition (71).

This technique was used to identify the *vieSAB* operon as induced in classical biotype *V. cholerae* during infection of the infant mouse (15). While the function of VieB remains unknown, VieSA have been shown to comprise a two-component system, where VieS is a membrane-bound sensor histidine kinase and VieA is a soluble response regulator with PDE function (47, 74). Insertional mutation of *vieS* was shown to negatively regulate expression of CT during infection (in vivo) (43). Moreover, inactivation of *vieA* either by clean deletion (Δ*vieA*) or by mutation of a conserved glutamate residue required for PDE function *[vieA(E170A)]*, led to a 10-fold defect in colonization of the infant mouse in competition with the wild type (79). Surprisingly, the same mutations of *vieA* in the El Tor biotype do not result in the same attenuation, suggesting that the two biotypes have developed different regulatory mechanisms for modulating the intracellular concentration of c-di-GMP upon entry into the host. This opens the possibility

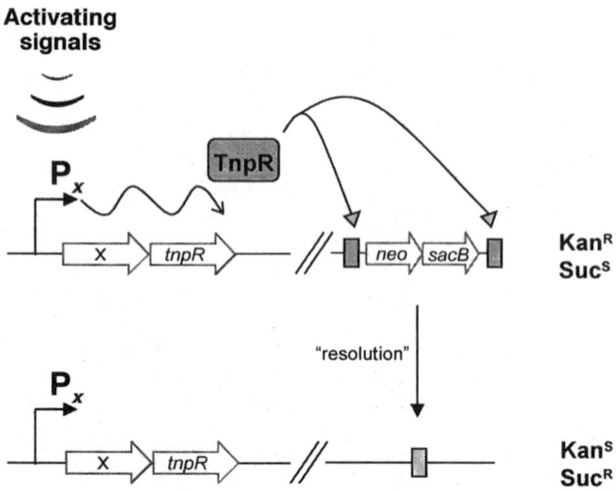

Figure 2. RIVET. RIVET was used to examine the expression of *V. cholerae* genes during infection of the infant mouse. Transcriptional fusions of *tnpR,* which encodes a resolvase, to promoters of interest (here, P_x) are made. When a promoter is activated in response to undefined signals present in the host, TnpR is produced and targets *res* sites (indicated by grey rectangles). The *res* sites flank *neo* (kanamycin resistance) and *sacB* (sucrose sensitivity) genes, and recognition by TnpR results in excision, or resolution, of *neo-sacB.* Thus, loss of kanamycin resistance and sucrose sensitivity represents expression of the gene of interest in a given condition (71). RIVET was used to show that c-di-GMP concentration is decreased upon infection to allow maximal virulence gene expression (15, 42). In addition, a modified RIVET was used to identify genes expressed late during infection, including three GGDEF genes and one GGDEF-EAL hybrid.

that there are more differences in the regulation of c-di-GMP concentration between the biotypes given the large number of c-di-GMP metabolic genes present in the *V. cholerae* genome. The recent completion of a classical biotype strain genome should facilitate identification of these differences using comparative genomics as a starting point.

To confirm the role of c-di-GMP during colonization of El Tor, the DGC VdcA was constitutively expressed from a plasmid vector and, similar to *vieA* mutations in the classical biotype, attenuation of colonization was observed compared to the wild type (75). This attenuation was specific to the enzymatic activity of VdcA as ectopic expression of a noncatalytic mutant of VdcA [VdcA(E258A)] exhibited wild-type colonization.

Due to the observed c-di-GMP-mediated virulence gene regulation in vitro and attenuation associated with high c-di-GMP concentration in vivo, one would hypothesize that c-di-GMP regulates virulence gene expression in vivo. Recently, experiments using RIVET to monitor the expression of *toxT* confirmed this hypothesis (75). As anticipated, *V. cholerae* constitutively expressing VdcA showed significantly less induction of *toxT* during infection than cells expressing the catalytic site mutant VdcA(E258A). Similar experiments showed that *ctxA* expression is regulated in response to c-di-GMP concentration in vivo. These in vivo experiments confirm that c-di-GMP does regulate *V. cholerae* virulence gene expression during infection.

The observations that high c-di-GMP concentration inhibits pathogenesis of *V. cholerae* and that c-di-GMP concentration is decreased during the transition from aquatic environments to the host suggest that a concomitant increase in c-di-GMP concentration must occur during the opposite transition from host to aquatic environment. Data suggest that high c-di-GMP-mediated biofilm formation is important for survival in aquatic environments; therefore, an increase in intracellular c-di-GMP during dissemination seems likely (33, 46, 78). Experiments using a modified RIVET system designed to specifically identify genes induced during late stages of infection (>7 hours postinoculation) revealed that three DGC-encoding genes, VC1593, VC2370, and VC2697, are induced by El Tor *V. cholerae* late during infection of the infant mouse (66). Similar regulation of each DGC gene was observed using quantitative reverse transcription-PCR. While these genes were shown to be dispensable during intestinal colonization, experiments modeling dissemination, which directly test the ability of *V. cholerae* to survive the transition from the infant mouse small intestine to an aquatic environment (in this case, Bangladeshi pond water), suggest they play a role in survival in the pond environment. In addition, some of these genes were required for survival in rice-water stool, suggesting a role for these genes during passage out of the host in the secretory diarrhea. Due to their redundant function in synthesizing c-di-GMP, single deletion of any one of these three genes does not greatly alter pond survival compared to the wild type; it is only mutation of all three DGC genes that leads to attenuated survival. Consistent with this notion of overlapping function, the defect in pond water survival could be complemented by expressing one of the DGCs in *trans* from a plasmid.

Surprisingly, this modified RIVET study also identified a GGDEF-EAL hybrid protein, CdpA, as induced at late stages of infection. CdpA has since been shown to function solely as a PDE both in vitro and in vivo (75). The naturally inactive DGC domain is divergent from the consensus sequence and appears to regulate the activity of the PDE domain. The induction of enzymes with opposing activities late in infection presents a paradox. It may be that induction of *cdpA* in combination with the three DGC-encoding genes serves to temper the increase in c-di-GMP late during infection. Alternatively, their activities may be spatially separated within the cell. Interestingly, *cdpA* is also induced during *V. cholerae* growth in a biofilm, another result that is unexpected given the importance of high c-di-GMP concentration in mediating biofilm formation (75). However, this also supports the hypothesis that CdpA may function to fine-tune the intracellular c-di-CMP concentration in either a global or spatially restricted manner.

Therefore, *V. cholerae* appears to control the intracellular concentration of c-di-GMP in response to (or in anticipation of) transition to new environments. The data collected suggest that PDEs, VieA in the classical biotype and, potentially, other(s) in the El Tor biotype, are active during early stages of colonization to lower the intracellular concentration of c-di-GMP to allow virulence gene expression. It is hypothesized that once the bacteria have colonized and the infection has progressed toward the late stages, the bacteria begin to prepare for the next step in their life cycle, which will include shedding from their host and a return to their natural aquatic environment. It is proposed that the bacteria respond with increases in cellular c-di-GMP concentration to survive dissemination into the aquatic environment. It appears that *V. cholerae* has evolved to prepare for this transition, perhaps by sensing changes in the intestinal environment late during infection by inducing expression of DGC genes. This induction is important to combat the stresses (e.g., osmolarity changes and nutrient limitation) encountered upon passage out of the host and during the transition to an aquatic environment.

Data obtained thus far have focused on transcription of DGC and PDE genes; however, it remains possible that their enzyme activities could be controlled by changes in environment. Further study will be required to fully understand the function and regulation of c-di-GMP metabolic enzymes during environmental transitions.

c-di-GMP REGULATION OF *V. CHOLERAE* MOTILITY

V. cholerae is a motile bacterium with a single sheathed polar flagellum. Although *V. cholerae* motility has been well studied, the role of motility in pathogenesis remains unclear. Studies suggest that motility is an important virulence determinant and that the flagellum may facilitate interactions with the intestinal epithelium. Nonmotile mutants lacking flagella are attenuated for colonization in both the rabbit and infant mouse models of infection (63). Nonmotile mutants have also been shown to be unable to penetrate the intestinal crypts in infant mice (25). Mutants lacking flagella or proper flagellar function are defective for interaction with homogenized small intestine tissue or epithelial cells in vitro, respectively (36, 60). While motility per se is clearly important for colonization, some studies have shown that motile but nonchemotactic mutants are hyperinfectious compared to the wild type in the infant mouse model (13, 14). Contrasting this, others have shown that a functional chemotaxis system is required for efficient penetration of intestinal mucus layers in primary tissue culture (22). A model reconciling these seemingly contradictory results has been proposed whereby motile *V. cholerae* cells shed in secretory diarrhea have transiently turned off chemotaxis to enhance initial colonization but can turn chemotaxis back on later in infection, if and when needed (14). Finally, the presence of the flagellum appears to regulate virulence gene expression. It was observed that *V. cholerae* mutants lacking the major flagellin, FlaA, did not express CT in vitro or in vivo (43). These data taken together suggest that motility, chemotaxis, and the flagellum play multifaceted roles in the virulence of *V. cholerae,* and a clear model encompassing all of these roles remains to be constructed.

Studies have shown that c-di-GMP is a key regulator of motility in *V. cholerae*. Overexpression of the PDE VieA leads to an increase in motility, while null mutations in *vieA* or overexpression of the DGC VdcA leads to a decrease in motility in *V. cholerae* (8, 9). These data clearly indicate that c-di-GMP acts as a negative regulator of *V. cholerae* motility and may at least partially explain the colonization defect of a *vieA* mutant.

In *V. cholerae,* there is a hierarchical structure to the regulation of genes required for motility, with genes falling into four differentially regulated classes (I to IV) (62). The master regulator FlrA (class I) sits atop the hierarchy and, in conjunction with sigma-54, positively activates the transcription of class II genes, which include the genes encoding the MS ring switch, the chemotaxis system, the two-component system FlrBC, and the alternative sigma factor sigma-28. Phosphorylated FlrC, also a sigma-54-dependent transcription factor, regulates the expression of class III genes, encoding FlaA and components of the basal body hook. Sigma-28 is responsible for regulation of class IV genes, which encode minor flagellins (FlaB, FlaC, FlaD, and FlaE) in addition to components of the motor (Fig. 3).

Transcriptional profile studies on *V. cholerae* strains with artificially increased intracellular c-di-GMP, either by ectopic expression of DGC VdcA or mutation of the PDE VieA, have shown a considerable repression of genes involved in motility (8, 9). A clear repression of class III and IV genes was observed, with additional repression observed in some but not all class II genes. Included in the repressed class II genes is sigma-28, which explains the decrease in transcription of class IV genes (Fig. 3). However, transcript of *flrBC* and the master regulator *flrA* remained unchanged. Therefore, although c-di-GMP-mediated repression of motility in *V. cholerae* occurs at the transcriptional level, there is no obvious mechanism to explain the pattern of gene repression ob-

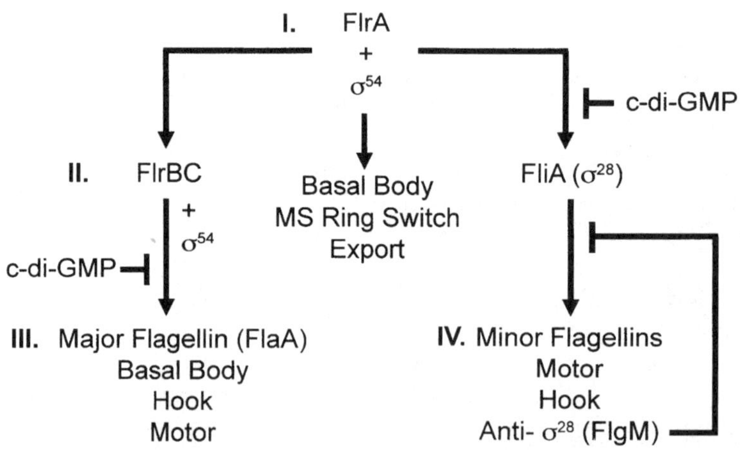

Figure 3. c-di-GMP regulates motility gene expression in *V. cholerae*. Schematic of *V. cholerae* motility gene hierarchy, represented by the function of genes within each class. Classes are denoted by roman numerals I to IV. Gene regulators or proteins of note are named. Arrows indicate activation of genes, and parallel lines in place of an arrowhead indicate gene repression. Transcriptional profiling studies have shown that c-di-GMP represses the expression of all class III and IV genes and some class II genes, including FliA, which is required for expression of class IV genes. Despite this information, there is no model to account for the effects of c-di-GMP on *V. cholerae* motility at this time.

served. More in-depth studies are needed to fully explain this phenomenon.

MECHANISMS OF c-di-GMP REGULATION

There is an abundance of information regarding physiological behaviors regulated by c-di-GMP, but relatively little is known about the mechanism(s) of this regulation. We understand that the intracellular concentration of c-di-GMP is modulated by the enzymatic activities of DGCs and PDEs, but we do not know how these changes in concentration signal for the vast alterations in bacterial behavior observed. Therefore, elucidating the mechanistic details of the c-di-GMP regulatory system beyond alterations in signal concentration will be the next frontier and will provide greater understanding of how bacteria utilize this secondary messenger and allosteric regulator. To date, two c-di-GMP-binding protein domains and one c-di-GMP-binding riboswitch have been identified in bacteria including *V. cholerae*. These are the autoregulatory I-site, the PilZ domain, and the RNA domain GEMM, respectively (16, 61, 72). These domains shed some light on potential regulatory mechanisms and may play a role in pathogenesis. The I-site has been shown to serve as an allosteric feedback inhibition site controlling the enzymatic activity of some DGCs.

PilZ Domain

As detailed in chapter 4, the PilZ protein domain has been identified as a c-di-GMP-binding domain. Based on homology, five genes in *V. cholerae* are predicted to encode PilZ domain-containing proteins, each with unknown function, subsequently named PlzA to PlzE (1, 61). Two of the five PilZ proteins (PlzC and PlzD) have been purified and shown to bind c-di-GMP in vitro. Additionally, mutational analysis of PlzD revealed a number of conserved PilZ domain residues required for c-di-GMP binding. The role of these conserved residues in contacting c-di-GMP was corroborated by the solution of the PlzD crystal structure bound to c-di-GMP (6). While in vitro binding was not observed for all *V. cholerae* PilZ proteins, based on the conservation of the PilZ domain sequence, it is assumed that most, if not all, will bind c-di-GMP in vivo.

Mutational analysis showed that some PilZ proteins play a role in colonization of the infant mouse small intestine (61). Deletion of *plzB* leads to a 10-fold attenuation in vivo compared to the wild type; this attenuation may be due to a defect in motility. Although PlzB exhibits the poorest homology to the consensus PilZ motif among the five PilZ proteins in *V. cholerae*, a site-specific mutation in a highly conserved residue recapitulated this attenuation. While individual mutation of *plzC* or *plzD* has no effect on colonization, deletion of both leads to a threefold colonization defect compared to the wild type, suggesting a potential redundant role for these proteins. This defect could be complemented by reintroduction of either gene in *trans*. Also, mutation of conserved PilZ domain residues required for c-di-GMP binding activity led to a similar decrease in colonization. Mutation of PlzC causes a minor reduction of motility, again suggesting a role for this family of proteins in regulating motility. However, none of the PilZ proteins regulates CT expression in vitro, which suggests that PilZ proteins regulate other pathways, such as motility, that contribute to pathogenesis. Thus the five PilZ domain proteins do not appear to be the main link between c-di-GMP concentration and virulence gene regulation.

GEMM Riboswitch Domain

Riboswitches are mRNA domains capable of binding specific ligands, and they typically modulate translation in response to changing ligand concentrations (12). A conserved RNA domain, named GEMM, was identified upstream of a number of open reading frames predicted to encode DGCs, PDEs, or genes regulated by c-di-GMP (83). GEMM was identified in a variety of bacteria and was proposed to function as a c-di-GMP sensing riboswitch (72) (see chapter 15). *V. cholerae* was predicted to encode two GEMM sequences, Vc1 and Vc2, which are located upstream of *gbpA* and VC1722, respectively. Both Vc1 and Vc2 were demonstrated to form alternate structures in the presence and absence of c-di-GMP. Moreover, in-depth study of Vc2 showed that the riboswitch binds c-di-GMP tightly and specifically and that binding of c-di-GMP regulated the expression of downstream coding sequences. Vc2 was cloned upstream of the *lacZ* gene and, using β-galactosidase activity as a reporter, it was shown that high intracellular c-di-GMP levels block expression of the reporter gene while low c-di-GMP levels increase expression. Although similar studies to outline the response of Vc1 to c-di-GMP were not done, the observation of structural changes in the presence of c-di-GMP suggests that this riboswitch is also responsive to c-di-GMP.

Although a function has not been ascribed to VC1722, GbpA was previously identified as a surface adhesin required for chitin binding, a process likely to be critical for the persistence of *V. cholerae* in aquatic environments. In addition, GbpA was shown to be involved in *V. cholerae* host cell attachment (36). Deletion of *gbpA* caused reduced adhesion to cultured intestinal epithelial cells (HT-29) in vitro and attenuated colonization of the infant mouse small intestine. GbpA binding to chitin and to host cell surfaces appears to be dependent on the presence of GlcNAc residues on both surfaces. GbpA is a unique protein because it represents an adhesin important for survival in both aquatic and host environments.

GbpA has been shown to interact with mucin, which is the first surface encountered by *V. cholerae* in the small intestine, indicating that GbpA is involved in the initial steps of colonization (10). In fact, GbpA and mucin seem to have a cooperative relationship whereby the presence of one increases expression of the other. This appears to be a mechanism used by the bacterium to augment adherence to the host epithelium because increases in *gbpA* and mucin expression lead to enhanced cell interactions.

Although the exact effect of c-di-GMP on the function of Vc1 has yet to be studied, it is tempting to speculate that expression of *gbpA* is regulated by c-di-GMP through Vc1. This would serve as a way for the bacterium to control expression of this adhesin during transitions into and out of the host. For example, initial colonization aided by GbpA binding to intestinal mucus, coupled with the subsequent decrease in c-di-GMP concentration early during colonization, may alter expression of *gbpA* and allow other adhesins to mediate subsequent steps in adherence and colonization. Late in infection, and upon dissemination into the environment, rising c-di-GMP levels could signal an increase in GbpA expression to aid in attachment to chitinous surfaces. Further study of Vc1 will be required to confirm a potential regulatory role for Vc1 during *V. cholerae* infection and in aquatic environments.

While PilZ and GEMM domains have been informative regarding potential c-di-GMP regulatory mechanisms, they do not answer key questions regarding the effects of c-di-GMP on *V. cholerae* pathogenesis and virulence gene regulation. For example, these systems do not account for the observed c-di-GMP-mediated regulation of core virulence (*toxT* and CT) or motility genes, which suggests that there are most likely additional mechanisms left to be discovered. The identification of additional c-di-GMP-binding protein domains in *Pseudomonas aeruginosa* with limited similarity to known binding domains (PilZ and I-site) opens the door to the possibility that additional binding domains, protein or RNA, exist in *V. cholerae* (28, 45). The discovery of new c-di-GMP-binding domains is an exciting prospect and may hold the key to a more complete understanding of the c-di-GMP regulatory circuit.

CONCLUSION

c-di-GMP plays an integral role in the life cycle of pathogenic *V. cholerae* by regulating gene expression during environmental transitions. Experiments have shown that c-di-GMP inversely regulates behaviors specific to the aquatic and host environments; more precisely, c-di-GMP inversely regulates genes mediating biofilm formation and virulence. Additionally, *V. cholerae* regulates the concentration of c-di-GMP upon transition to new environments by inducing the expression of PDEs at the initial stages of infection and DGCs late during infection. This suggests that *V. cholerae* coordinates gene expression by modulating c-di-GMP concentration. It also suggests that c-di-GMP concentration fluctuates throughout the *V. cholerae* life cycle, leaving us with the model that c-di-GMP concentration is high during life in aquatic reservoirs, but upon entry into the host small intestine, PDEs are induced, lowering the c-di-GMP concentration and allowing for virulence gene expression. Once the infection is fully developed and bacteria are to be shed back into the aquatic environment in secretory diarrhea, we propose that *V. cholerae* begins to increase the c-di-GMP concentration by inducing expression of DGCs and perhaps by decreasing PDE expression. This results in increased c-di-GMP concentration and promotes survival in aquatic environments by inducing biofilm genes among others (Fig. 4). Additionally, PDE and DGC enzymatic activities could be regulated by external signals upon environmental transitions. This could account for swift changes in intracellular c-di-GMP concentration in response to rapidly changing environments. It is very possible that a combination of transcriptional and posttranscriptional regulation contributes to regulation of intracellular c-di-GMP concentration.

Despite the importance of c-di-GMP signaling to the survival of *V. cholerae* (and many other bacteria), there is still relatively little known about c-di-GMP regulatory circuits. *V. cholerae* encodes 61 predicted c-di-GMP metabolic enzymes, and only a few have been studied. With this large number of enzymes committed to controlling the concentration of c-di-GMP, one might predict that there is interplay, redundancy, and specialization among these 61 proteins. In the *V. cholerae* classical biotype, the PDE VieA has been shown to be important during early stages of infection for reducing c-di-GMP concentration and allowing virulence gene expression. In addition, experiments in the El Tor biotype have identified specific DGCs, induced late during infection of infant mice, dispensable for colonization but required for survival in aquatic environments postdissemina-

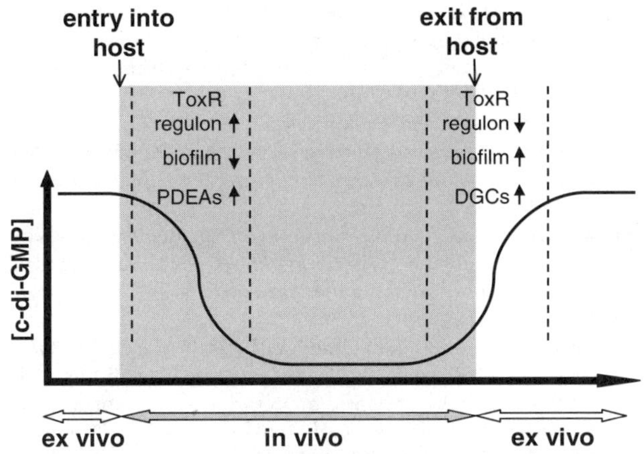

Figure 4. The intracellular concentration of c-di-GMP fluctuates throughout the life cycle of *V. cholerae*. The solid line indicates the relative level of intracellular c-di-GMP, which is predicted to oscillate as *V. cholerae* shifts from aquatic reservoirs (ex vivo) to the host small intestine (in vivo) and back again; dashed lines demarcate these transition points. c-di-GMP is predicted to be high in the aquatic environment and in the biofilm state. The induction of PDEs upon entry into the host and the negative effect of c-di-GMP, in vitro and in vivo, on virulence gene expression suggest that c-di-GMP must be lowered upon infection. When *V. cholerae* is disseminated from the host, c-di-GMP must be elevated again to aid survival in the environment. Evidence suggests that elevation of c-di-GMP by *V. cholerae* begins prior to exiting the host during advanced cholera, possibly in response to changing conditions in the intestine.

tion from the host. This leaves 56 other enzymes unaccounted for during these critical stages of the life cycle of pathogenic *V. cholerae*. Certainly, not all are involved in every *V. cholerae* behavior, but it is very likely that other enzymes play at least a supporting role in these transitions.

It has been shown that alterations in c-di-GMP concentration lead to numerous changes in gene transcription throughout the *V. cholerae* genome. However, the signaling mechanisms connecting fluctuations in c-di-GMP concentration with transcriptional changes remain largely unknown. As mentioned above, two regulatory mechanisms, one protein based and one riboswitch based, have been identified in *V. cholerae,* but they do not fully explain the observed large-scale gene regulation by c-di-GMP or the concomitant changes in *V. cholerae* behavior. Therefore, there must be alternative regulatory mechanisms that have yet to be discovered. Identification of new regulatory pathways will provide new insight into the c-di-GMP regulatory circuit and will help to explain how this secondary messenger molecule is so important to *V. cholerae* pathogenesis and environmental fitness.

REFERENCES

1. **Amikam, D., and M. Y. Galperin.** 2006. PilZ domain is part of the bacterial c-di-GMP binding protein. *Bioinformatics* **22:** 3–6.
2. **Ausmees, N., R. Mayer, H. Weinhouse, G. Volman, D. Amikam, M. Benziman, and M. Lindberg.** 2001. Genetic data indicate that proteins containing the GGDEF domain possess diguanylate cyclase activity. *FEMS Microbiol. Lett.* **204:** 163–167.
3. **Banwell, J. G., N. F. Pierce, R. C. Mitra, K. L. Brigham, G. J. Caranasos, R. I. Keimowitz, D. S. Fedson, J. Thomas, S. L. Gorbach, R. B. Sack, and A. J. Mondal.** 1970. Intestinal fluid and electrolyte transport in human cholera. *J. Clin Investig.* **49:**183–195.
4. **Barua, D.** 1991. History of cholera, p. 1–35. *In* D. Barua and W. B. Greenough III (ed.), *Cholera.* Plenum Medical Book Company, New York, NY.
5. **Beck, N. A., E. S. Krukonis, and V. J. DiRita.** 2004. TcpH influences virulence gene expression in *Vibrio cholerae* by inhibiting degradation of the transcription activator TcpP. *J. Bacteriol.* **186:**8309–8316.
6. **Benach, J., S. S. Swaminathan, R. Tamayo, S. K. Handelman, E. Folta-Stogniew, J. E. Ramos, F. Forouhar, H. Neely, J. Seetharaman, A. Camilli, and J. F. Hunt.** 2007. The structural basis of cyclic diguanylate signal transduction by PilZ domains. *EMBO J.* **26:**5153–5166.
7. **Bennish, M. L.** 1994. Cholera: pathophysiology, clinical features, and treatment, p. 3–25. *In* K. Wachsmut, P. A. Blake, and O. Olsvik (ed.), Vibrio cholerae *and Cholera: Molecular to Global Perspectives.* ASM Press, Washington, DC.
8. **Beyhan, S., A. D. Tischler, A. Camilli, and F. H. Yildiz.** 2006. Differences in gene expression between the classical and El Tor biotypes of *Vibrio cholerae* O1. *Infect. Immun.* **74:** 3633–3642.
9. **Beyhan, S., A. D. Tischler, A. Camilli, and F. H. Yildiz.** 2006. Transcriptome and phenotypic responses of *Vibrio cholerae* to increased cyclic di-GMP level. *J. Bacteriol.* **188:**3600–3613.
10. **Bhowmick, R., A. Ghosal, B. Das, H. Koley, D. R. Saha, S. Ganguly, R. K. Nandy, R. K. Bhadra, and N. S. Chatterjee.** 2008. Intestinal adherence of *Vibrio cholerae* involves a coordinated interaction between colonization factor GbpA and mucin. *Infect. Immun.* **76:**4968–4977.
11. **Bobrov, A. G., O. Kirillina, and R. D. Perry.** 2005. The phosphodiesterase activity of the HmsP EAL domain is required for negative regulation of biofilm formation in *Yersinia pestis. FEMS Microbiol. Lett.* **247:**123–130.
12. **Breaker, R. R.** 2008. Complex riboswitches. *Science* **319:** 1795–1797.
13. **Butler, S. M., and A. Camilli.** 2004. Both chemotaxis and net motility greatly influence the infectivity of *Vibrio cholerae. Proc. Natl. Acad. Sci. USA* **101:**5018–5023.
14. **Butler, S. M., E. J. Nelson, N. Chowdhury, S. M. Faruque, S. B. Calderwood, and A. Camilli.** 2006. Cholera stool bacteria repress chemotaxis to increase infectivity. *Mol. Microbiol.* **60:**417–426.
15. **Camilli, A., and J. J. Mekalanos.** 1995. Use of recombinase gene fusions to identify *Vibrio cholerae* genes induced during infection. *Mol. Microbiol.* **18:**671–683.
16. **Christen, B., M. Christen, R. Paul, F. Schmid, M. Folcher, P. Jenoe, M. Meuwly, and U. Jenal.** 2006. Allosteric control of cyclic di-GMP signaling. *J. Biol. Chem.* **281:**32015–32024.
17. **Christen, M., B. Christen, M. Folcher, A. Schauerte, and U. Jenal.** 2005. Identification and characterization of a cyclic di-

GMP-specific phosphodiesterase and its allosteric control by GTP. *J. Biol. Chem.* **280:**30829–30837.

18. **Crawford, J. A., E. S. Krukonis, and V. J. DiRita.** 2003. Membrane localization of the ToxR winged-helix domain is required for TcpP-mediated virulence gene activation in *Vibrio cholerae*. *Mol. Microbiol.* **47:**1459–1473.
19. **DiRita, V. J., and J. J. Mekalanos.** 1991. Periplasmic interaction between two membrane regulatory proteins, ToxR and ToxS, results in signal transduction and transcriptional activation. *Cell* **64:**29–37.
20. **DiRita, V. J., C. Parsot, G. Jander, and J. J. Mekalanos.** 1991. Regulatory cascade controls virulence in *Vibrio cholerae*. *Proc. Natl. Acad. Sci. USA* **88:**5403–5407.
21. **Field, M.** 1979. Mechanism of action of cholera and *Escherichia coli* toxins. *Am. J. Clin. Nutr.* **32:**189–196.
22. **Freter, R., P. C. O'Brien, and M. S. Macsai.** 1981. Role of chemotaxis in the association of motile bacteria with intestinal mucosa: in vivo studies. *Infect. Immun.* **34:**234–240.
23. **Galperin, M. Y., A. N. Nikolskaya, and E. V. Koonin.** 2001. Novel domains of the prokaryotic two-component signal transduction systems. *FEMS Microbiol. Lett.* **203:**11–21.
24. **Gill, D. M., and C. A. King.** 1975. The mechanism of action of cholera toxin in pigeon erythrocyte lysates. *J. Biol. Chem.* **250:**6424–6432.
25. **Guentzel, M. N., L. H. Field, E. R. Eubanks, and L. J. Berry.** 1977. Use of fluorescent antibody in studies of immunity to cholera in infant mice. *Infect. Immun.* **15:**539–548.
26. **Gupta, S., and R. Chowdhury.** 1997. Bile affects production of virulence factors and motility of *Vibrio cholerae*. *Infect. Immun.* **65:**1131–1134.
27. **Herrington, D. A., R. H. Hall, G. Losonsky, J. J. Mekalanos, R. K. Taylor, and M. M. Levine.** 1988. Toxin, toxin-coregulated pili, and the toxR regulon are essential for *Vibrio cholerae* pathogenesis in humans. *J. Exp. Med.* **168:**1487–1492.
28. **Hickman, J. W., and C. S. Harwood.** 2008. Identification of FleQ from *Pseudomonas aeruginosa* as a c-di-GMP-responsive transcription factor. *Mol. Microbiol.* **69:**376–389.
29. **Higgins, D. E., E. Nazareno, and V. J. DiRita.** 1992. The virulence gene activator ToxT from *Vibrio cholerae* is a member of the AraC family of transcriptional activators. *J. Bacteriol.* **174:**6974–6980.
30. **Huq, A., E. B. Small, P. A. West, M. I. Huq, R. Rahman, and R. R. Colwell.** 1983. Ecological relationships between *Vibrio cholerae* and planktonic crustacean copepods. *Appl. Environ. Microbiol.* **45:**275–283.
31. **Kaper, J. B., A. Rasano, and M. Truskis.** 1994. Toxins of *Vibrio cholerae*, p. 145–176. *In* K. Wachsmut, P. A. Blake, and O. Olsvik (ed.), Vibrio cholerae *and Cholera: Molecular to Global Perspectives*. ASM Press, Washington, DC.
32. **Kaysner, C. A., and W. E. Hill.** 1994. Toxigenic *Vibrio cholerae* 01 in food and water, p. 27–40. *In* K. Wachsmut, P. A. Blake, and O. Olsvik (ed.), Vibrio cholerae *and Cholera: Molecular to Global Perspectives*. ASM Press, Washington, DC.
33. **Kierek, K., and P. I. Watnick.** 2003. Environmental determinants of *Vibrio cholerae* biofilm development. *Appl. Environ. Microbiol.* **69:**5079–5088.
34. **King, C. A., and W. A. van Heyningen.** 1973. Deactivation of cholera toxin by a sialidase-resistant monosialosylganglioside. *J. Infect. Dis.* **127:**639–647.
35. **Kirn, T. J., M. J. Lafferty, C. M. Sandoe, and R. K. Taylor.** 2000. Delineation of pilin domains required for bacterial association into microcolonies and intestinal colonization by *Vibrio cholerae*. *Mol. Microbiol.* **35:**896–910.
36. **Kirn, T. J., B. A. Jude, and R. K. Taylor.** 2005. A colonization factor links *Vibrio cholerae* environmental survival and human infection. *Nature* **438:**863–866.
37. **Kovacikova, G., and K. Skorupski.** 1999. A *Vibrio cholerae* LysR homolog, AphB, cooperates with AphA at the *tcpPH* promoter to activate expression of the ToxR virulence cascade. *J. Bacteriol.* **181:**4250–4256.
38. **Kovacikova, G., and K. Skorupski.** 2001. Overlapping binding sites for the virulence gene regulators AphA, AphB and cAMP-CRP at the *Vibrio cholerae tcpPH* promoter. *Mol. Microbiol.* **41:**393–407.
39. **Kovacikova, G., and K. Skorupski.** 2002. Regulation of virulence gene expression in *Vibrio cholerae* by quorum sensing: HapR functions at the *aphA* promoter. *Mol. Microbiol.* **46:**1135–1147.
40. **Krukonis, E. S., R. R. Yu, and V. J. Dirita.** 2000. The *Vibrio cholerae* ToxR/TcpP/ToxT virulence cascade: distinct roles for two membrane-localized transcriptional activators on a single promoter. *Mol. Microbiol.* **38:**67–84.
41. **Larocque, R. C., J. B. Harris, M. Dziejman, X. Li, A. I. Khan, A. S. Faruque, S. M. Faruque, G. B. Nair, E. T. Ryan, F. Qadri, J. J. Mekalanos, and S. B. Calderwood.** 2005. Transcriptional profiling of *Vibrio cholerae* recovered directly from patient specimens during early and late stages of human infection. *Infect. Immun.* **73:**4488–4493.
42. **Lee, S. H., M. J. Angelichio, J. J. Mekalanos, and A. Camilli.** 1998. Nucleotide sequence and spatiotemporal expression of the *Vibrio cholerae vieSAB* genes during infection. *J. Bacteriol.* **180:**2298–2305.
43. **Lee, S. H., S. M. Butler, and A. Camilli.** 2001. Selection for in vivo regulators of bacterial virulence. *Proc. Natl. Acad. Sci. USA* **98:**6889–6894.
44. **Lee, S. H., D. L. Hava, M. K. Waldor, and A. Camilli.** 1999. Regulation and temporal expression patterns of *Vibrio cholerae* virulence genes during infection. *Cell* **99:**625–634.
45. **Lee, V. T., J. M. Matewish, J. L. Kessler, M. Hyodo, Y. Hayakawa, and S. Lory.** 2007. A cyclic-di-GMP receptor required for bacterial exopolysaccharide production. *Mol. Microbiol.* **65:**1474–1484.
46. **Lim, B., S. Beyhan, J. Meir, and F. H. Yildiz.** 2006. Cyclic-diGMP signal transduction systems in *Vibrio cholerae:* modulation of rugosity and biofilm formation. *Mol. Microbiol.* **60:**331–348.
47. **Martinez-Wilson, H. F., R. Tamayo, A. D. Tischler, D. W. Lazinski, and A. Camilli.** 2008. The *Vibrio cholerae* hybrid sensor kinase VieS contributes to motility and biofilm regulation by altering the cyclic diguanylate level. *J. Bacteriol.* **190:**6439–6447.
48. **Matson, J. S., and V. J. DiRita.** 2005. Degradation of the membrane-localized virulence activator TcpP by the YaeL protease in *Vibrio cholerae*. *Proc. Natl. Acad. Sci. USA* **102:**16403–16408.
49. **Mekalanos, J. J., R. J. Collier, and W. R. Romig.** 1979. Enzymic activity of cholera toxin. I. New method of assay and the mechanism of ADP-ribosyl transfer. *J. Biol. Chem.* **254:**5849–5854.
50. **Merrell, D. S., S. M. Butler, F. Qadri, N. A. Dolganov, A. Alam, M. B. Cohen, S. B. Calderwood, G. K. Schoolnik, and A. Camilli.** 2002. Host-induced epidemic spread of the cholera bacterium. *Nature* **417:**642–645.
51. **Miller, M. B., K. Skorupski, D. H. Lenz, R. K. Taylor, and B. L. Bassler.** 2002. Parallel quorum sensing systems converge to regulate virulence in *Vibrio cholerae*. *Cell* **110:**303–314.
52. **Miller, V. L., and J. J. Mekalanos.** 1984. Synthesis of cholera toxin is positively regulated at the transcriptional level by *toxR*. *Proc. Natl. Acad. Sci. USA* **81:**3471–3475.
53. **Miller, V. L., and J. J. Mekalanos.** 1985. Genetic analysis of the cholera toxin-positive regulatory gene *toxR*. *J. Bacteriol.* **163:**580–585.

54. **Morris, J. G., Jr., and the Cholera Laboratory Task Force.** 1994. *Vibrio cholerae* 0139 Bengal, p. 95–102. *In* K. Wachsmut, P. A. Blake, and O. Olsvik (ed.), Vibrio cholerae *and Cholera: Molecular to Global Perspectives.* ASM Press, Washington, DC.

55. **Nelson, E. J., A. Chowdhury, J. Flynn, S. Schild, L. Bourassa, Y. Shao, R. C. LaRocque, S. B. Calderwood, F. Qadri, and A. Camilli.** 2008. Transmission of *Vibrio cholerae* is antagonized by lytic phage and entry into the aquatic environment. *PLoS Pathog.* **4:**e1000187.

56. **Nielsen, A. T., N. A. Dolganov, G. Otto, M. C. Miller, C. Y. Wu, and G. K. Schoolnik.** 2006. RpoS controls the *Vibrio cholerae* mucosal escape response. *PLoS Pathog.* **2:**e109.

57. **Paul, R., S. Weiser, N. C. Amoit, C. Chan, T. Schirmer, B. Giese, and U. Jenal.** 2004. Cell cycle-dependent dynamic localization of a bacterial response regulator with a novel di-guanylate cyclase output domain. *Genes Dev.* **18:**715–727.

58. **Pfau, J. D., and R. K. Taylor.** 1998. Mutations in *toxR* and *toxS* that separate transcriptional activation from DNA binding at the cholera toxin gene promoter. *J. Bacteriol.* **180:** 4724–4733.

59. **Pierce, N. F.** 1973. Differential inhibitory effects of cholera toxoids and ganglioside on the enterotoxin of *Vibrio cholerae* and *Escherichia coli. J. Exp. Med.* **137:**1009–1023.

60. **Postnova, T., O. G. Gómez-Duarte, and K. Richardson.** 1996. Motility mutants of *Vibrio cholerae* O1 have reduced adherence in vitro to human small intestinal epithelial cells as demonstrated by ELISA. *Microbiology* **142:**2767–2776.

61. **Pratt, J. T., R. Tamayo, A. D. Tischler, and A. Camilli.** 2007. PilZ domain proteins bind cyclic diguanylate and regulate diverse processes in *Vibrio cholerae. J. Biol. Chem.* **282:** 12860–12870.

62. **Prouty, M. G., N. E. Correa, and K. E. Klose.** 2001. The novel sigma54- and sigma28-dependent flagellar gene transcription hierarchy of *Vibrio cholerae. Mol. Microbiol.* **39:**1595–1609.

63. **Richardson, K.** 1991. Roles of motility and flagellar structure in pathogenicity of *Vibrio cholerae:* analysis of motility mutants in three animal models. *Infect. Immun.* **59:**1384–1386.

64. **Ryan, R., Y. Fouhy, J. Lucey, L. Crossman, S. Spiro, Y. He, L. Zhang, S. Heeb, M. Camara, P. Williams, and J. Dow.** 2006. Cell-cell signaling in *Xanthomonas campestris* involves an HD-GYP domain protein that functions in cyclic di-GMP turnover. *Proc. Natl. Acad. Sci. USA* **103:**6712–6717.

65. **Ryjenkov, D. A., M. Tarutina, O. V. Moskvin, and M. Gomelsky.** 2005. Cyclic diguanylate is a ubiquitous signaling molecule in bacteria: insights into biochemistry of the GGDEF protein domain. *J. Bacteriol.* **187:**1792–1793.

66. **Schild, S., R. Tamayo, E. J. Nelson, F. Qadri, S. B. Calderwood, and A. Camilli.** 2007. Genes induced late in infection increase fitness of *Vibrio cholerae* after release into the environment. *Cell Host Microbe* **2:**264–277.

67. **Schmidt, A. J., D. A. Ryjenkov, and M. Gomelsky.** 2005. The ubiquitous protein domain EAL is a cyclic diguanylate-specific phosphodiesterase: enzymatically active and inactive EAL domains. *J. Bacteriol.* **187:**4774–4781.

68. **Schoolnik, G. K., M. I. Voskuil, D. Schnappinger, F. H. Yildiz, K. Meibom, N. A. Dolganov, M. A. Wilson, and K. H. Chong.** 2001. Whole genome DNA microarray expression analysis of biofilm development by *Vibrio cholerae* O1 E1 Tor. *Methods Enzymol.* **336:**3–18.

69. **Schuhmacher, D. A., and K. E. Klose.** 1999. Environmental signals modulate ToxT-dependent virulence factor expression in *Vibrio cholerae. J. Bacteriol.* **181:**1508–1514.

70. **Skorupski, K., and R. K. Taylor.** 1999. A new level in the *Vibrio cholerae* ToxR virulence cascade: AphA is required for transcriptional activation of the *tcpPH* operon. *Mol. Microbiol.* **31:**763–771.

71. **Slauch, J. M., and A. Camilli.** 2000. IVET and RIVET: use of gene fusions to identify bacterial virulence factors specifically induced in host tissues. *Methods Enzymol.* **326:**73–96.

72. **Sudarsan, N., E. R. Lee, Z. Weinberg, R. H. Moy, J. N. Kim, K. H. Link, and R. R. Breaker.** 2008. Riboswitches in eubacteria sense the second messenger cyclic di-GMP. *Science* **321:** 411–413.

73. **Tal, R., H. C. Wong, R. Calhoon, D. Gelfand, A. L. Fear, G. Volman, R. Mayer, P. Ross, D. Amikam, H. Weinhouse, A. Cohen, S. Sapir, P. Ohana, and M. Benziman.** 1998. Three *cdg* operons control cellular turnover of cyclic di-GMP in *Acetobacter xylinum:* genetic organization and occurrence of conserved domains in isoenzymes. *J. Bacteriol.* **180:**4416–4425.

74. **Tamayo, R., A. D. Tischler, and A. Camilli.** 2005. The EAL domain protein VieA is a cyclic diguanylate phosphodiesterase. *J. Biol. Chem.* **280:**33324–33330.

75. **Tamayo, R., S. Schild, J. T. Pratt, and A. Camilli.** 2008. Role of cyclic di-GMP during El Tor biotype *Vibrio cholerae* infection: characterization of the in vivo-induced cyclic di-GMP phosphodiesterase CdpA. *Infect. Immun.* **76:**1617–1627.

76. **Tamplin, M. L., A. L. Gauzens, A. Huq, D. A. Sack, and R. R. Colwell.** 1990. Attachment of *Vibrio cholerae* serogroup O1 to zooplankton and phytoplankton of Bangladesh waters. *Appl. Environ. Microbiol.* **56:**1977–1980.

77. **Taylor, R. K., V. L. Miller, D. B. Furlong, and J. J. Mekalanos.** 1987. Use of *phoA* gene fusions to identify a pilus colonization factor coordinately regulated with cholera toxin. *Proc. Natl. Acad. Sci. USA* **84:**2833–2837.

78. **Tischler, A. D., and A. Camilli.** 2004. Cyclic diguanylate (c-di-GMP) regulates *Vibrio cholerae* biofilm formation. *Mol. Microbiol.* **53:**857–869.

79. **Tischler, A. D., and A. Camilli.** 2005. Cyclic diguanylate regulates *Vibrio cholerae* virulence gene expression. *Infect. Immun.* **73:**5873–5882.

80. **Waldor, M. K., and J. J. Mekalanos.** 1996. Lysogenic conversion by a filamentous phage encoding cholera toxin. *Science* **272:**1910–1914.

81. **Watnick, P. I., and R. Kolter.** 1999. Steps in the development of a *Vibrio cholerae* El Tor biofilm. *Mol. Microbiol.* **34:** 586–595.

82. **Weekly Epidemiological Record.** 2007. Cholera, 2006. *Wkly. Epidemiol. Rec.* **82:**273–284.

83. **Weinberg, Z., J. E. Barrick, Z. Yao, A. Roth, J. N. Kim, J. Gore, J. X. Wang, E. R. Lee, K. F. Block, N. Sudarsan, S. Neph, M. Tompa, W. L. Ruzzo, and R. R. Breaker.** 2007. Identification of 22 candidate structured RNAs in bacteria using the CMfinder comparative genomics pipeline. *Nucleic Acids Res.* **35:**4809–4819.

84. **Yu, R. R., and V. J. DiRita.** 1999. Analysis of an autoregulatory loop controlling ToxT, cholera toxin, and toxin-coregulated pilus production in *Vibrio cholerae. J. Bacteriol.* **181:**2584–2592.

85. **Zhu, J., M. B. Miller, R. E. Vance, M. Dziejman, B. L. Bassler, and J. J. Mekalanos.** 2002. Quorum-sensing regulators control virulence gene expression in *Vibrio cholerae. Proc. Natl. Acad. Sci. USA* **99:**3129–3134.

The Second Messenger Cyclic Di-GMP
Edited by A. J. Wolfe and K. L. Visick

Chapter 21

Cyclic Di-GMP Signaling and Host Immunity

David K. R. Karaolis and Theodore J. Standiford

3′,5′-Cyclic diguanylic acid, also known as cyclic di-GMP (c-di-GMP) (Fig. 1), is an intracellular signaling molecule that was initially identified by the Benziman laboratory in 1985 in the bacterium *Acetobacter xylinum* (renamed *Gluconacetobacter xylinus*) and shown to regulate the production of cellulose by this microbe (1, 28–30) (see also chapter 2). This cyclic dinucleotide molecule has been subsequently identified in multiple bacterial species but not in higher eukaryotes (7, 9, 16, 23, 25, 26). Importantly, c-di-GMP is now recognized to control many complex biological processes in bacteria, including survival, adhesion, colonization, biofilm formation, and virulence (6, 12, 16, 20, 23, 27) (see also chapter 8). Many laboratories are now performing basic research studying the intracellular signaling role of c-di-GMP in bacteria and how it regulates gene expression in bacteria.

There is now also compelling data to indicate that this bacterial product also regulates fundamental eukaryotic and mammalian host cellular and immune processes that should be exploited for the development of new and effective clinical applications and platform technologies to prevent and treat various infections and diseases. In support of its effects in eukaryotic systems, c-di-GMP was found to regulate basal and growth factor-induced proliferation of murine and human tumor cells (2, 17, 31), leading Karaolis et al. to first suggest that cyclic dinucleotides, such as c-di-GMP, represent an attractive approach to preventing or treating cancer (17). More recently, numerous in vitro and in vivo animal model studies using synthesized c-di-GMP (11, 13, 14) have demonstrated that c-di-GMP has potent immunomodulatory (immunostimulatory and adjuvant) effects on cellular components of both the innate and adaptive or acquired arms of host immunity (4, 18, 19, 22).

EFFECTS OF c-di-GMP ON HOST ANTIMICROBIAL IMMUNITY

In 2003, based on the observation in *Vibrio cholerae* that a gene encoding a novel regulatory protein they termed RocS (regulation of cell signaling) containing GGDEF and EAL domains known to be associated with c-di-GMP levels was important for the rugose variant phenotype, exopolysaccharide production, biofilm formation, and motility, Rashid et al. suggested a clinical use of c-di-GMP in which they proposed that “modulation of cyclic dinucleotide (e.g., c-di-GMP) levels might have application in regulating various phenotypes of prokaryotes” (23).

Testing the hypothesis that c-di-GMP can be used clinically to inhibit infection, synthetic exogenous c-di-GMP was found not to be bactericidal or bacteriostatic but could significantly inhibit *Staphylococcus aureus* biofilm formation in vitro (20). Using a murine model of infection, Brouillette et al. showed that in vivo administration of synthetic c-di-GMP resulted in markedly reduced numbers of *S. aureus* in mammary tissue and that it can promote resolution of *Staphylococcus aureus*-induced mastitis in this murine model of mastitis (4). Given that c-di-GMP lacked direct bactericidal or bacteriostatic effects on *S. aureus* but promoted resolution of infection, this observation suggested the possibility that c-di-GMP might modulate or regulate host protective antimicrobial responses and be an immune enhancer.

Subsequent studies have clearly shown that c-di-GMP exerts important immunomodulatory effects on

David K. R. Karaolis • Intragenics Research Institute, Havre de Grace, MD 21078, and Karagen Pharmaceuticals, Baltimore, MD 21210. **Theodore J. Standiford** • Department of Internal Medicine, Division of Pulmonary and Critical Care Medicine, 6301 MSRB III, University of Michigan Medical Center, Ann Arbor, MI 48109.

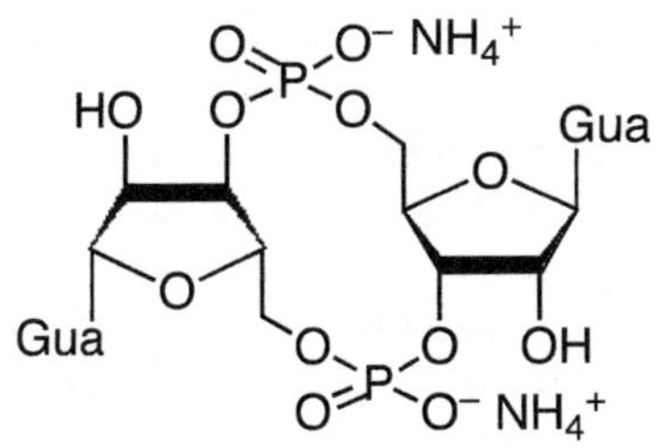

Figure 1. Molecular structure of c-di-GMP. Gua, guanine.

host immune cells that participate in antibacterial immunity. Notably, intramammary treatment of mice with c-di-GMP 12 and 6 h before challenge with the gram-positive pathogen *S. aureus* gave a protective effect and a 10,000-fold reduction in CFUs in tissues ($P < 0.001$) (18). Studies have also shown that the intranasal (i.n.) or subcutaneous administration of c-di-GMP delivered 24 and/or 48 h prior to intratracheal (i.t.) challenge with the gram-negative bacterial pathogen *Klebsiella pneumoniae* enhanced survival in this lethal pneumonia model, and protective effects were attributable to improved lung bacterial clearance and reduced bacterial dissemination (19). Similar protective effects of c-di-GMP have been observed in a murine *Streptococcus pneumoniae* (pneumococcal) pneumonia model, again due to more effective clearance of the organism from the lung (22). In the mastitis model and both bacterial pneumonia models, exogenous administration of c-di-GMP in vivo significantly reduced the numbers of organisms in tissue without exerting direct effects on *K. pneumoniae* or *S. pneumoniae* viability or replication, indicating its ability to modulate the host antibacterial immunity.

EFFECTS OF c-di-GMP ON HOST INNATE IMMUNITY

Tissue macrophages, exudate macrophages, and neutrophils serve as the first line of defense against invading pathogens. Key antimicrobial responses of these phagocytes are regulated by c-di-GMP. For example, the intraperitoneal (i.p.) administration of nanomolar concentrations of c-di-GMP to mice resulted in the influx of polymorphonuclear leukocytes (PMN) and newly arriving blood monocytes (referred to as exudate macrophages), comparable in magnitude to that observed after i.p. lipopolysaccharide administration (18). In addition, i.n. pretreatment of animals with c-di-GMP but not cGMP resulted in increased accumulation of PMN in the lung 48 hours after i.t. *K. pneumoniae* administration (19). In contrast, pretreatment of mice with c-di-GMP i.n. prior to *S. pneumoniae* administration did not substantially alter PMN influx but did enhance the accumulation of newly arriving exudate macrophages (identified by the cell surface expression of CD11b) (22). Interestingly, cyclic dinucleotide analogs (including *tert*-butyldimethylsilyl–c-di-GMP, c-GpAp, cGpIp, and cGpsGp) have been found to promote monocyte and granulocyte recruitment in vivo using murine models (D. K. R. Karaolis, unpublished data). The direct effects of c-di-GMP on either PMN or monocyte/macrophage chemotactic responses have not been addressed. However, c-di-GMP has been shown to induce the expression of neutrophil and monocyte active chemokines from dendritic cells (DC) in vitro and in animals i.t. challenged with *K. pneumoniae,* suggesting that the chemotactic effects of this molecule may be indirect (18, 19).

In addition to facilitating the influx of leukocytes, c-di-GMP may also serve as an important activator of macrophage effector responses (18). Incubation of human monocyte-derived macrophages with c-di-GMP induces a modest and transient activation of the mitogen-activated protein kinase (MAPK) extracellular signal-regulated kinase (ERK) 1/2, but no activation of p38 and Jun N-terminal protein kinase. Likewise, c-di-GMP failed to induce the expression of inflammatory cytokines and chemokines from primary murine lung macrophages in culture (T. Standiford, unpublished data). However, pretreatment of mice with c-di-GMP i.n. 24 and 48 h prior to i.t. administration of *K. pneumoniae* resulted in substantial induction of spontaneous inducible nitric oxide synthase mRNA expression and nitric oxide production by lung macrophages cultured ex vivo compared to cells isolated from control cGMP-treated mice (19). Nitric oxide has previously been shown to be a required component of effective lung innate immunity in gram-negative bacterial pneumonia, and nitric oxide production is partially dependent on activation of ERK (32, 34). The mechanism of enhanced macrophage activation in vivo has not been defined but may be attributable to increased early expression of gamma interferon (IFN-γ) in the lungs of c-di-GMP-treated mice.

Natural killer (NK) cells are active participants in innate immunity against bacterial, fungal, and viral pathogens, in part by elaborating activating cytokines, including IFN-γ, in response to pathogen-associated molecular patterns (8). The production of IFN-γ by NK cells is further amplified in the presence of DC or macrophage-derived interleukin-12 (IL-12) with or without IL-18. It has been demonstrated that i.n. administration of c-di-GMP prior to i.t. bacterial challenge induced an increase in the accumulation of activated NK cells (as indicated by the expression of the activational marker CD69), which was associated with increases in whole-lung IFN-γ levels in c-di-GMP-treated mice (19). The mechanism accounting for NK cell accumulation/activation in these mice is uncertain, but the i.n. administration of c-di-GMP results in a substantial pulmonary induction of CXCL10/IP-10, a chemokine with potent NK cell chemotactic and activating effects (19, 24, 34).

Structural cells, including epithelial cells and interstitial cells, are active participants in innate responses to microbial pathogens (10). These cells are rich cellular sources of chemokines, activating and suppressive cytokines, and antimicrobial peptides. Cyclic dinucleotides have been shown to modulate proliferative responses in human tumor cells of epithelial origin. However, specific effects of c-di-GMP on mucosal innate responses involving structural cells have not been defined but may represent an important future line of investigation.

EFFECTS OF c-di-GMP ON HOST ACQUIRED IMMUNITY

The generation of acquired immunity is necessary for effective cellular and humoral responses to invading pathogens. DC are the most potent antigen-presenting cells and are integrated within mucosal surfaces, where they are well positioned to survey antigens (3, 33). Upon exposure to antigen, DC mature and then migrate from lung peripheral sites to regional lymph nodes, where they present antigen to naive T cells in the context of costimulation and DC-derived cytokines, resulting in antigen-specific T-cell proliferation and expression of subset-specific cytokines (15, 33). Th1 T cells secrete IFN-γ, which activates macrophage effector functions and is required for clearance of most intracellular pathogens. This cytokine is also involved in B-cell activation and subsequent isotype switching to the immunoglobulin G2a (IgG2a) subclass (Fig. 2), immunoglobulins that facilitate phagocytosis of microbes, complement fixation, and antibody-dependent cellular cytotoxicity required for tumor lysis (5). Th2 T cells secrete IL-4, IL-5, and IL-13 and participate in immunity against parasites. Th2 cytokines also induce isotype switching of immunoglobulins to IgG subtypes 1 (mice) and 4 (humans), IgA, and IgE production.

There is considerable evidence to suggest that c-di-GMP acts as a strong activator of DC function and drives the development of type 1 immunity. Incubation of murine splenic DC with c-di-GMP promoted DC maturation, as manifested by enhanced expression of the costimulatory molecules CD80 and CD86 (18). Similarly, treatment of c-di-GMP induced the cell surface expression of CD80, CD86, CD83, and major histocompatibility complex class II in human monocyte-derived DC (18), indicative of DC maturation and improved antigen-presenting function. Treatment with c-di-GMP also enhanced CCR7 expression, which mediates homing of DC to regional lymph nodes. In addition, c-di-GMP induced monocyte-derived DC expression of inflammatory cytokines (tumor necrosis factor alpha, IL-1, CXCL8/IL-8, CCL5/RANTES, monocyte chemoattractant protein 1/CCR2, and MIP-α/CCL3) and type 1 promoting cytokines (IL-12, MIG/CXCL9, and IP-10/CXCL10) (18). Moreover, c-di-GMP-treated DC dis-

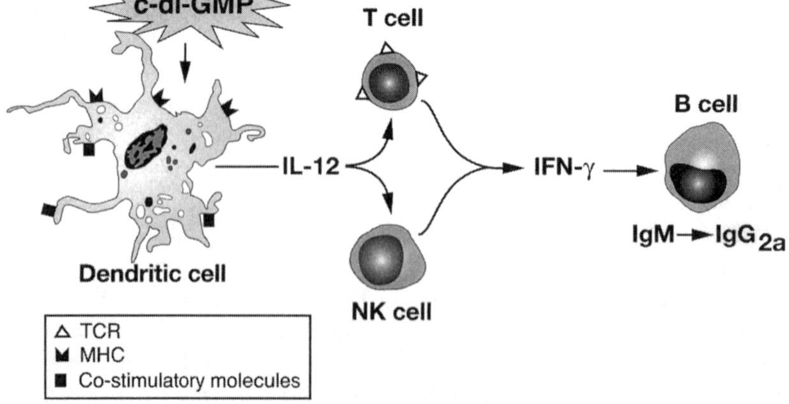

Figure 2. Schematic depicting effects of c-di-GMP on type 1 cytokine production and antibody class switching. c-di-GMP promotes DC IL-12 production and antigen presentation, resulting in enhanced production of type 1 cytokines (e.g., IFN-γ) by T cells and NK cells. In the presence of IFN-γ, B cells switch from predominantly IgM production to IgG2a production. TCR, T-cell receptor; MHC, major histocompatibility complex class II.

played markedly enhanced antigen-presenting function, as these cells induced a four- to sevenfold increase in allospecific T-cell proliferation compared to that induced by control DC (18). Signaling pathways involved in stimulation of DC function by c-di-GMP have not been completely defined but have been associated with selective activation of p38 but not ERK 1/2 or Jun N-terminal protein kinase MAPK activation (18). The stimulatory effects of c-di-GMP also appeared to target myeloid, but not plasmacytoid, DC, as treatment of human plasmacytoid DC with this molecule failed to induce the production of IFN-α by these cells (18).

Studies aimed at defining the immunostimulatory effects of c-di-GMP on DC function and T-cell responses have primarily been performed in vitro using cells in isolation or coculture systems. Some insights into effects on DC and T-cell responses have been provided by observations made in murine models of bacterial infection. For instance, pretreatment of mice with c-di-GMP i.n. prior to *S. pneumoniae* administration enhanced CD40 and CD86 expression by lung DC and CD86 expression by DC isolated in draining lymph node (22). In a murine *K. pneumoniae* model, c-di-GMP increased the expression of type 1 promoting cytokines (e.g., IL-12) and chemokines (IP-10/CXCL10) as well as an accumulation of activated ($CD69^+$) $\alpha\beta$ T cells in the lung (19). Finally, coadministration of c-di-GMP with either staphylococcal or streptococcal surface antigens strongly promoted the production of IgG2a humoral responses, suggestive of a vigorous Th1, rather than Th2, response (18, 22). Collectively, these in vivo studies are consistent with observations made in cell culture systems and point to a possible role for this cyclic dinucleotide in driving type 1 immunity (depicted in Fig. 2). However, the effect of c-di-GMP in the setting of infections that are dependent on robust cell-mediated and humoral immune responses requires further study.

c-di-GMP AS AN ADJUVANT IN VACCINE DEVELOPMENT AND DISEASE PREVENTION

The ability of cyclic dinucleotides such as c-di-GMP to be used as an adjuvant and immune enhancer in vaccine development was first proposed in 2006 by Karaolis et al. in a study in which it was shown that intramuscular vaccination of mice with two injections of c-di-GMP coinjected with *S. aureus* ClfA antigen produced serum with significantly higher anti-ClfA IgG antibody titers ($P < 0.001$) than vaccination with ClfA alone (18, 21).

A later vaccine and challenge study showed that i.p. coadministration of c-di-GMP with PdB or PspA before challenge with virulent *Streptococcus pneumoniae* resulted in significantly higher antigen-specific antibody titers and increased survival of mice than that obtained with the traditionally used alum adjuvant (22). The higher antibody titer produced suggests that c-di-GMP (and its analogs) can be used instead of alum as an immune modulator or vaccine adjuvant. Additional studies examining the role of c-di-GMP in vaccine development and supporting its clinical development are also reviewed in chapter 22. These findings demonstrate that local or systemic c-di-GMP administration stimulates innate and adaptive immunity against invasive and systemic disease and that c-di-GMP can be used as an effective broad-spectrum immunomodulator and vaccine adjuvant to prevent infectious diseases.

There remains an obvious and urgent need for improvements in our prevention and treatment regimens for many types of infections and diseases. Strategies such as immunomodulation that go beyond simple antibiotic treatment are needed. Cyclic dinucleotide molecules such as c-di-GMP and its analogs which may augment targeted aspects of the innate immune response (better than current adjuvants) could be used as immunoprophylactic agents or immune enhancers for prevention of infections and disease in high-risk patients in combination with traditional antibiotics and antimicrobials for prevention or treatment of infections and as a vaccine adjuvant to inhibit infections and disease. Further studies to test the ability of c-di-GMP and analogs either alone, as a vaccine adjuvant, or in combination with traditional antimicrobials and other drugs are under way in various animal models of infection and disease (Table 1).

c-di-GMP-INDUCED CELL ACTIVATION

Incubation of cells with c-di-GMP results in intracellular accumulation of this molecule. The mech-

Table 1. Potential clinical applications of cyclic dinucleotides

Clinical applications
Immunomodulator
Immune enhancer
Vaccine adjuvant
Inhibition of infection and infectious disease
Prevention and treatment of cancer, including metastatic tumors
General immune enhancer to stimulate the host immune system
Suppression of the allergic response

anism of internalization is uncertain, but passive diffusion, rather than active transport, is thought to be the most likely route. Exposure to c-di-GMP results in rapid activation of the MAPK cascade and downstream expression of relevant effector molecules (18). This molecule does not appear to directly activate either the toll-like receptor or the NOD 1/2 signaling pathway (18). Interaction with G-protein-coupled receptors is the most plausible, yet unproven, possibility.

c-di-GMP IN VIVO STABILITY AND SAFETY

Detailed high-pressure liquid chromatography analysis has shown that c-di-GMP is stable in varied physiological conditions (18). For instance, this molecule retains its size and two-dimensional structure in pooled human serum for 24 hours at 37°C. Moreover, c-di-GMP has remained biologically active in tissues, including lung, subcutaneous, and mammary tissues. However, it is worth noting that c-di-GMP is most effective as an immunostimulant when administered prophylactically (before infectious challenge) rather than as active therapy (administered during infection), raising the possibility that the dinucleotide may lose biological activity within the context of an inflammatory milieu. This remains to be defined experimentally.

In genotoxicity tests, c-di-GMP has been found to be negative for carcinogenicity and mutagenicity in the mini-Ames assay and negative for the induction of structural and numerical chromosome aberrations in CHO cells using the mini-CAB assay (Karaolis, unpublished). Toxicity studies have also revealed that exposure of normal rat kidney cells and human neuroblastoma cells to c-di-GMP at biologically relevant doses showed no lethal cytotoxicity (17).

REGULATION OF TUMOR CELL GROWTH AND REPLICATION

c-di-GMP has been shown to potently modulate DNA synthesis and cell replication in eukaryotic systems (2). Specifically, incubation of the lymphoblast cell line Molt 4 with c-di-GMP in micromolar concentrations resulted in a marked increase in DNA synthesis while inhibiting cell division. Cell cycle analysis revealed a block in S phase, resulting in failure of cells to progress to the G_2/M phase of the cell cycle. Extracellular c-di-GMP accumulated within Molt 4 cells and mediated cell cycle arrest by interacting with the GTP-binding $p21^{ras}$ protein. This same group found similar antiproliferative effects of c-di-GMP on Jurkat cells, a CD4-bearing lymphoblastoid cell line. In addition to induction of S-phase cell cycle arrest, c-di-GMP also enhanced the cell surface expression of CD4 by these cells (31).

To test the hypothesis that c-di-GMP can be used clinically to prevent or treat cancer, it was shown that c-di-GMP can exert antiproliferative effects on human colon cancer (H508) cells (17). Treatment of H508 colon cancer cells with c-di-GMP resulted in dose-dependent inhibition of basal and growth factor-induced cell proliferation. The mechanism of growth inhibition was not identified, but the antiproliferative effects occurred in the absence of cell death.

Collectively, these observations suggest that c-di-GMP can selectively regulate important cell cycle responses in cancer cells. The biological relevance of these effects and the specificity of antiproliferative effects in malignant and nonmalignant cells require further investigation. These data together with the dramatic effects on host immunity justify that in vivo testing using animal models for the development and clinical application of using c-di-GMP either alone or as an adjuvant in combination with other compounds (e.g., as a cancer vaccine) for tumor prevention, tumor progression, and tumor metastasis requires further study. Studies to test the ability of c-di-GMP alone and as part of a cancer vaccine are under way in animal models.

CONCLUSION

Research over the past decade has identified c-di-GMP as an important signaling molecule regulating multiple aspects of eukaryotic cell biology. While the functional significance of naturally bacterium-derived c-di-GMP in regulating immunological response is largely untested, based on data from numerous in vivo and in vitro models, it is clear that c-di-GMP has drug-like properties and can affect host immunity. Synthetic exogenously delivered c-di-GMP can serve as a potent immunomodulator and adjuvant in the promotion of innate and acquired host immunity and as a key regulator of cell cycle progression in replicating cells. The biological effects of c-di-GMP on both myeloid and structural/parenchymal cells are depicted in Fig. 3. Based on these biological activities, cyclic dinucleotides such as c-di-GMP and its analogs used alone or in combination with other compounds such as antimicrobials, vaccines, and drugs may serve as very promising clinically relevant and well-tolerated general immunomodulators, immune enhancers, and adjuvant thera-

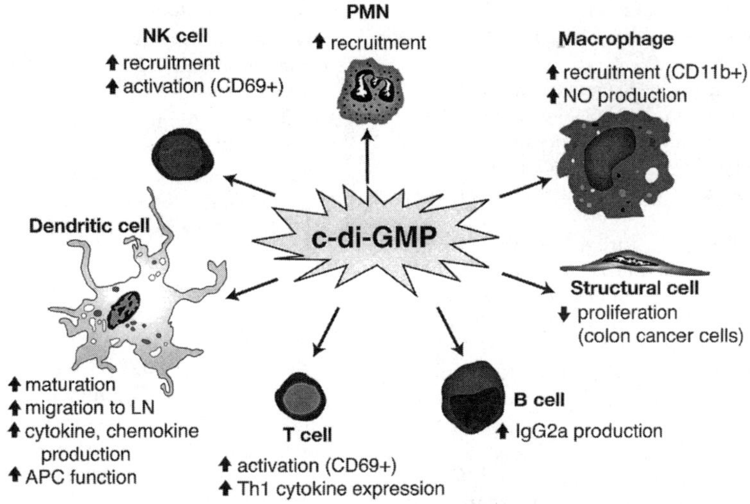

Figure 3. Schematic summarizing immunomodulatory effects of c-di-GMP on myeloid and structural cells (e.g., dendritic cells, PMN, macrophages, NK cells, T and B cells, and cancer cells). APC, antigen-presenting cell.

pies in the prevention or treatment of diseases and disorders ranging from infection to malignancy.

REFERENCES

1. **Amikam, D., and M. Benziman.** 1989. Cyclic diguanylic acid and cellulose synthesis in *Agrobacterium tumefaciens*. *J. Bacteriol.* **171:**6649–6655.
2. **Amikam, D., O. Steinberger, T. Shkolnik, and Z. Ben-Ishai.** 1995. The novel cyclic dinucleotide 3′-5′ cyclic diguanylic acid binds to p21ras and enhances DNA synthesis but not cell replication in the Molt 4 cell line. *Biochem. J.* **311**(Pt 3):921–927.
3. **Banchereau, J., and R. M. Steinman.** 1998. Dendritic cells and control of immunity. *Nature* **392:**245–252.
4. **Brouillette, E., M. Hyodo, Y. Hayakawa, D. K. R. Karaolis, and F. Malouin.** 2005. 3′,5′-cyclic diguanylic acid reduces the virulence of biofilm-forming *Staphylococcus aureus* strains in a mouse model of mastitis infection. *Antimicrob. Agents Chemother.* **49:**3109–3113.
5. **Coffman, R. L., H. F. Savelkoul, and D. A. Lebman.** 1989. Cytokine regulation of immunoglobulin isotype switching and expression. *Semin. Immunol.* **1:**55–63.
6. **D'Argenio, D. A., M. W. Calfee, P. B. Rainey, and E. C. Pesci.** 2002. Autolysis and autoaggregation in *Pseudomonas aeruginosa* colony morphology mutants. *J. Bacteriol.* **184:**6481–6489.
7. **D'Argenio, D. A., and S. I. Miller.** 2004. Cyclic di-GMP as a bacterial second messenger. *Microbiology* **150:**2497–2502.
8. **Ferlazzo, G., B. Morandi, A. D'Agostino, R. Meazza, G. Melioli, A. Moretta, and L. Moretta.** 2003. The interaction between NK cells and dendritic cells in bacterial infections results in rapid induction of NK cell activation and in the lysis of uninfected dendritic cells. *Euro. J. Immunol.* **33:**306–313.
9. **Galperin, M. Y., A. N. Nikolskaya, and E. V. Koonin.** 2001. Novel domains of the prokaryotic two-component signal transduction systems. *FEMS Microbiol. Lett.* **203:**11–21.
10. **Hajjar, A. M., H. Harowicz, H. D. Liggitt, P. J. Fink, C. B. Wilson, and S. J. Skerrett.** 2005. An essential role for non-bone marrow-derived cells in control of *Pseudomonas aeruginosa* pneumonia. *Am. J. Respir. Cell Mol. Biol.* **33:**470–475.
11. **Hayakawa, Y., R. Nagata, A. Hirata, M. Hyodo, and R. Kawai.** 2003. A facile synthesis of cyclic bis(3′-5′) diguanylic acid. *Tetrahedron* **59:**6465–6471.
12. **Hecht, G. B., and A. Newton.** 1995. Identification of a novel response regulator required for the swarmer-to-stalked-cell transition in *Caulobacter crescentus*. *J. Bacteriol.* **177:**6223–6229.
13. **Hyodo, M., and Y. Hayakawa.** 2004. An improved method for synthesizing cyclic bis(3′-5′)diguanylic acid (c-di-GMP). *Bull. Chem. Soc. Jpn.* **77:**2089–2093.
14. **Hyodo, M., Y. Hayakawa, and D. K. R. Karaolis.** 2006. Organic synthesis, chemical properties, and biological activities of cyclic bis(3′-5′)diguanylic acid (c-di-GMP) and its analogs. *J. Synth. Org. Chem. Jpn.* **64:**359–370.
15. **Jahnsen, F. L., D. H. Strickland, J. A. Thomas, I. T. Tobagus, S. Napoli, G. R. Zosky, D. J. Turner, P. D. Sly, P. A. Stumbles, and P. G. Holt.** 2006. Accelerated antigen sampling and transport by airway mucosal dendritic cells following inhalation of a bacterial stimulus. *J. Immunol.* **177:**5861–5867.
16. **Jones, H. A., J. W. Lillard, Jr., and R. D. Perry.** 1999. HmsT, a protein essential for expression of the haemin storage (Hms+) phenotype of *Yersinia pestis*. *Microbiology* **145**(Pt 8): 2117–2128.
17. **Karaolis, D. K. R., K. Cheng, M. Lipsky, A. Elnabawi, J. Catalano, M. Hyodo, Y. Hayakawa, and J.-P. Raufman.** 2005. 3′,5′-Cyclic diguanylic acid (c-di-GMP) inhibits basal and growth factor-stimulated human colon cancer cell proliferation. *Biochem. Biophys. Res. Commun.* **329:**40–45.
18. **Karaolis, D. K. R., T. K. Means, D. Yang, M. Takahashi, T. Yoshimura, E. Muraille, D. Philpott, J. T. Schroeder, M. Hyodo, Y. Hayakawa, B. G. Talbot, E. Brouillette, and F. Malouin.** 2007. Bacterial c-di-GMP is an immunostimulatory molecule. *J. Immunol.* **178:**2171–2181.
19. **Karaolis, D. K. R., M. W. Newstead, X. Zeng, M. Hyodo, Y. Hayakawa, U. Bhan, H. Liang, and T. J. Standiford.** 2007. c-di-GMP stimulates protective innate immunity in bacterial pneumonia. *Infect. Immun.* **75:**4942–4950.
20. **Karaolis, D. K. R., M. H. Rashid, C. Rajanna, W. Luo, M. Hyodo, and Y. Hayakawa.** 2005. c-di-GMP (3′-5′-cyclic di-

guanylic acid) inhibits *Staphylococcus aureus* cell-cell interactions and biofilm formation. *Antimicrob. Agents Chemother.* **49**:1029–1038.

21. **Karaolis, D. K. R., T. K. Means, E. Brouillette, B. G. Talbot, D. Yang, E. Muraille, M. Hyodo, Y. Hayakawa, and F. Malouin.** 2006. c-di-GMP is an immunostimulatory molecule with prophylactic and adjuvant activity, abstr. E-060, p. 235. *Abstr. 106th Gen. Meet. Am. Soc. Microbiol.*, Orlando, FL, 21 to 25 May 2006.
22. **Ogunniyi, A. D., J. C. Paton, A. C. Kirby, J. A. McCullers, J. Cook, M. Hyodo, Y. Hayakawa, and D. K. Karaolis.** 2008. c-di-GMP is an effective immunomodulator and vaccine adjuvant against pneumococcal infection. *Vaccine* **26**:4676–4685.
23. **Rashid, M. H., C. Rajanna, A. Ali, and D. K. R. Karaolis.** 2003. Identification of genes involved in the switch between the smooth and rugose phenotypes of *Vibrio cholerae*. *FEMS Microbiol. Lett.* **227**:113–119.
24. **Romagnani, P., F. Annunziato, E. Lazzeri, L. Cosmi, C. Beltrame, L. Lasagni, G. Galli, M. Francalanci, R. Manetti, F. Marra, V. Vanini, E. Maggi, and S. Romagnani.** 2001. Interferon-inducible protein 10, monokine induced by interferon gamma, and interferon-inducible T-cell alpha chemoattractant are produced by thymic epithelial cells and attract T-cell receptor (TCR) alphabeta+ CD8+ single-positive T cells, TCRgammadelta+ T cells, and natural killer-type cells in human thymus. *Blood* **97**:601–607.
25. **Römling, U., and D. Amikam.** 2006. Cyclic di-GMP as a second messenger. *Curr. Opin. Microbiol.* **2**:218–228.
26. **Römling, U., M. Gomelsky, and M. Y. Galperin.** 2005. C-di-GMP: the dawning of a novel bacterial signalling system. *Mol. Microbiol.* **57**:629–639.
27. **Römling, U., M. Rohde, A. Olsen, S. Normark, and J. Reinkoster.** 2000. AgfD, the checkpoint of multicellular and aggregative behaviour in *Salmonella typhimurium* regulates at least two independent pathways. *Mol. Microbiol.* **36**:10–23.
28. **Ross, P., Y. Aloni, C. Weinhouse, D. Michaeli, P. Weinberger-Ohana, R. Meyer, and M. Benziman.** 1985. An unusual guanyl oligonucleotide regulates cellulose synthesis in *Acetobacter xylinum*. *FEBS Lett.* **186**:191–196.
29. **Ross, P., Y. Aloni, H. Weinhouse, D. Michaeli, P. Weinberger-Ohana, R. Mayer, and M. Benziman.** 1986. Control of cellulose synthesis in *Acetobacter xylinum:* a unique guanyl oligonucleotide is the immediate activator of the cellulose synthase. *Carbohydr. Res.* **149**:101–117.
30. **Ross, P., H. Weinhouse, Y. Aloni, D. Michaeli, P. Weinberger-Ohana, R. Mayer, S. Braun, E. de Vroom, G. A. van der Marel, J. H. van Boom, and M. Benziman.** 1987. Regulation of cellulose synthesis in *Acetobacter xylinum* by cyclic diguanylic acid. *Nature* **325**:279–281.
31. **Steinberger, O., Z. Lapidot, Z. Ben-Ishai, and D. Amikam.** 1999. Elevated expression of the CD4 receptor and cell cycle arrest are induced in Jurkat cells by treatment with the novel cyclic dinucleotide 3′,5′-cyclic diguanylic acid. *FEBS Lett.* **444**:125–129.
32. **Tsai, W. C., R. Strieter, D. A. Zisman, J. M. Wilkowski, K. A. Bucknell, G.-H. Chen, and T. J. Standiford.** 1997. Nitric oxide is required for effective innate immunity against *Klebsiella pneumoniae*. *Infect. Immun.* **65**:1870–1875.
33. **von Garnier, C., L. Filgueira, M. Wikstrom, M. Smith, J. A. Thomas, D. H. Strickland, P. G. Holt, and P. A. Stumbles.** 2005. Anatomical location determines the distribution and function of dendritic cells and other APCs in the respiratory tract. *J. Immunol.* **175**:1609–1618.
34. **Zeng, X., T. A. Moore, M. W. Newstead, J. C. Deng, S. L. Kunkel, A. D. Luster, and T. J. Standiford.** 2005. Interferon-inducible protein 10, but not monokine induced by gamma interferon, promotes protective type 1 immunity in murine *Klebsiella pneumoniae* pneumonia. *Infect. Immun.* **73**:8226–8236.

The Second Messenger Cyclic Di-GMP
Edited by A. J. Wolfe and K. L. Visick

Chapter 22

Bis-(3′,5′)-Cyclic Di-GMP: Promising Adjuvant for Vaccine Design

THOMAS EBENSEN, RIMMA LIBANOVA, AND CARLOS A. GUZMÁN

Infectious diseases have a tremendous impact on human health, being directly responsible for at least 26% of all deaths occurring every year worldwide. In addition, infectious agents are directly responsible for several types of cancer, and are involved in the pathogenesis of many chronic noninfectious diseases. Thus, even healthy economies cannot cope with the exploding costs associated with infectious diseases and their sequels. Therefore, one of the most pressing issues in public health is the development of prophylactic interventions against infection.

Vaccines represent the most cost-efficient intervention against infections, and their therapeutic use against infections, cancer, and chronic inflammatory diseases raised considerable interest. Traditional vaccines were based on the use of killed or attenuated microorganisms or relatively crude antigenic extracts. These rather crude preparations were often associated with local and/or systemic side effects. From a historical perspective, the development of vaccines has largely relied on empiric approaches. In contrast, the development of modern vaccines is underscored by an in-depth understanding of the immunological paradigms developed over the past 30 years, which enables a rational and cost-effective selection of the most promising vaccination strategies. A better understanding of microbial pathogenesis together with the identification of antigens able to evoke protective immune responses led to a new generation of subunit vaccines, which are based on a discrete number of well-defined native or recombinant antigens and exhibit improved safety profiles. However, formulations based on highly purified antigens are considerably less immunogenic than those containing whole cells or semicrude extracts, making essential their coadministration with an adjuvant to trigger robust immune responses.

We have also considerably expanded our knowledge on the effector mechanisms responsible for microbial clearance during natural infections. This is particularly important, since it is not sufficient to stimulate a strong immune response following vaccination. In fact, the elicitation of the wrong type of immune response may even result in more severe forms of disease after exposure (48). This would be possible by exploiting adjuvants which are able to polarize the elicited responses, according to the specific clinical needs. Therefore, the development of new adjuvants for human and veterinary use which are able to promote strong, broad, and sustained immune responses still remains an important challenge in vaccinology.

Adjuvants can exert their biological activity through different mechanisms, such as (i) affecting the intrinsic properties of a given antigen, (ii) creating a depot which results in slow antigen release, and (iii) direct targeting and activation of key components of the innate immune system. Their effector functions on innate immune cells are particularly important, since they represent the bridge for the elicitation of strong adaptive responses. More specifically, several adjuvants are able to target professional antigen-presenting cells (APC), such as dendritic cells (DC) or macrophages, thereby promoting their activation and maturation. This in turn leads to improved antigen processing and presentation. Furthermore, different adjuvants can promote the secretion of different soluble mediators (e.g., cytokines and chemokines) by bystander cells. This results in cellular migration and recruitment, creates a microenvironment conducive to antigen processing, and contributes to the modulation and polarization of the subsequent adaptive response. Thus, adjuvants not only promote the

Thomas Ebensen, Rimma Libanova, and Carlos A. Guzmán • Department of Vaccinology and Applied Microbiology, Helmholtz Centre for Infection Research, 38124 Braunschweig, Germany.

induction of strong immune responses, thereby reducing the amount of antigen needed (i.e., antigen sparing), but also enable us to fine-tune the elicited responses through their immune modulatory properties (Fig. 1). The stimulation of the designated right type of immune response is a key aspect of achieving protection and avoiding immune pathological reactions.

Adjuvants have been routinely used to increase the immunogenicity of vaccines for more than 70 years. However, only a few compounds have been approved and licensed for human use (e.g., aluminum salts, the microfluidized squalene-in-water emulsion MF59, and monophosphoryl lipid A). Until recently, the discovery of novel adjuvants has not been an extremely successful process, which was mainly driven by the rules of serendipity. However, recent advances in our understanding of the immune system, particularly with respect to early proinflammatory signals, have led to the identification of new potential targets for immune modulation (3, 4, 6, 10, 25, 42). The establishment of improved techniques for the total synthesis of complex chemical entities has also allowed the production of well-defined molecules. This is expected to lead to the generation of new vaccines exhibiting optimal safety and efficacy profiles (23, 38, 39).

However, there is no vaccine formulation containing a mucosal adjuvant approved for human use in the market (40). This is particularly important considering that mucosal vaccines promote both systemic and local mucosal responses, which can protect against both disease (e.g., symptoms) and infections (e.g., colonization). In addition, they are more accepted by the public and associated with considerably simple administration logistics. Thus, there is still an urgent need for new candidate adjuvants, particularly those active by the mucosal route.

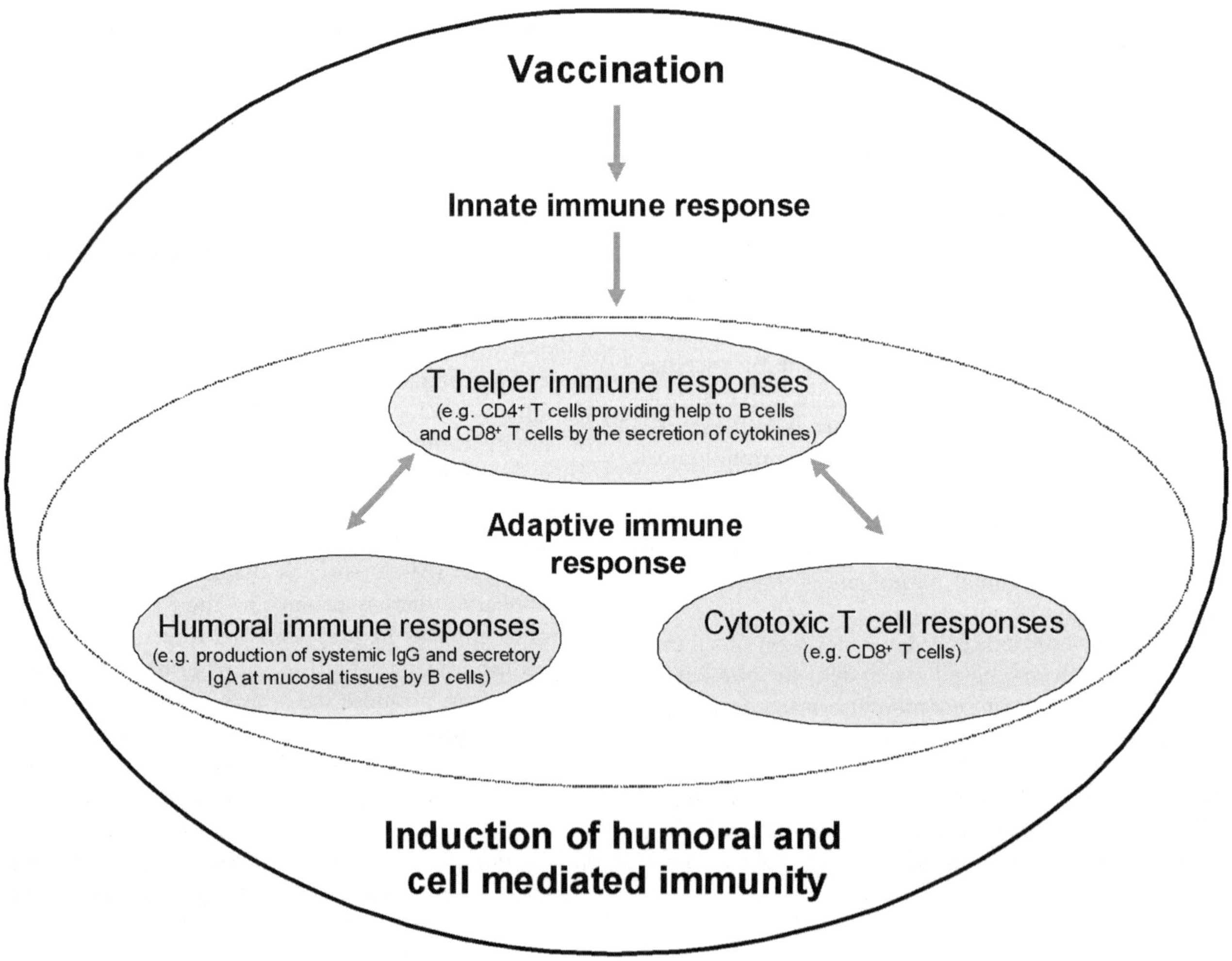

Figure 1. Vaccination with an antigen stimulates the innate and adaptive immune responses which afford humoral and cellular immunity.

A-B moiety toxins (e.g., cholera toxin and heat-labile toxin from *Escherichia coli*) and their derivatives are probably the class of compounds which has been most frequently studied as mucosal adjuvants (7). Nevertheless, there are some concerns regarding their safety profile, particularly when administered by intranasal (i.n.) route (35, 45). In addition, their intrinsic immunogenicity suggests that preexisting immunity in response to previous (i) infection by agents expressing these molecules, (ii) contact with cross-reacting moieties, and/or (iii) vaccination with formulations containing these adjuvants might affect the elicited responses. In this regard, it would be particularly appealing to identify nonprotein compounds exhibiting activity as mucosal adjuvants, which are poorly immunogenic and amenable for incorporation in different vaccine formulations.

Cyclic di-GMP (c-di-GMP) was first identified in *Gluconacetobacter xylinus,* where it regulates the production of cellulose through the modulation of the cellulose synthase activity (34) (see also chapter 2). It was also shown that increased levels of c-di-GMP correlate with enhanced bacterial aggregation (36, 37, 43, 44). The reduction or degradation of the intracellular concentration of c-di-GMP by several proteins and phosphodiesterases resulted in decreased exopolysaccharide production and biofilm formation in different bacteria (11, 14, 20, 31, 43) (see also chapter 4), thereby playing an important role in the regulation of biofilm formation (see chapter 8 and reviews in references 1, 5, 33, and 41). It was also shown that c-di-GMP can affect eukaryotic cells (17) (see also chapter 21).

Further studies demonstrated the capacity of c-di-GMP to act as an adjuvant for model- or disease-related antigens when given by either the systemic or the mucosal route (Fig. 2). Coadministration of c-di-GMP with model antigens by the subcutaneous (s.c.) or intramuscular (i.m.) route resulted in the stimulation of strong humoral and cellular immune responses in mice, which were characterized by a balanced T helper type 1 (Th1)/T helper type 2 (Th2) cytokine pattern, thereby supporting both antigen-specific humoral and cellular immune responses (8, 30).

Additional work showed that c-di-GMP stimulates the maturation of human DC (18). Interestingly, antigen coadministration with c-di-GMP by the i.n. route also resulted in strong humoral responses at the systemic level as well as at local and distant mucosal territories (9). Moreover, T helper and cytotoxic-T-cell responses were also stimulated (8, 9). This body of experimental evidence suggests that the immune modulatory properties of c-di-GMP can be exploited to develop interventions against infections and cancer (17, 19, 20, 30).

IMMUNE MODULATORY PROPERTIES OF c-di-GMP

In vitro experiments showed that c-di-GMP is able to stimulate the proliferation of naive murine splenocytes and to promote the maturation and activation of both murine and human DC. In the presence of c-di-GMP, these cells showed an increased expression of major histocompatibility complex (MHC) class I, MHC class II, and costimulatory molecules (e.g., CD80 and CD86), as well as maturation markers (CD83) in the human system. Furthermore, c-di-GMP promoted the expression of different cytokines, chemokines, and chemokine receptors (e.g., interleukin-12 [IL-12], gamma interferon [IFN-γ], IL-8, monocyte chemoattractant protein 1, IFN-γ-inducible protein 10, RANTES, CCR1, CCR7, and CXCR4), which play an important role in cell-cell communication, cell maturation and differentiation, and cellular migration and homing. Nevertheless, the in vitro activity was rather modest compared with other compounds (e.g., toll-like receptor agonists). It is of interest that the application of c-di-GMP promotes the activation of p38 mitogen-activated protein kinase (MAPK) and extracellular signal-regulated kinase phosphorylation (ERK) in APC (8, 18). These studies tried to elucidate the signal transduction events triggered by c-di-GMP. However, its receptor has not yet been identified.

The initial in vivo studies performed with mice demonstrated that coadministration of antigens with c-di-GMP by the s.c. route resulted in the stimulation of significantly increased antigen-specific serum immunoglobulin G (IgG) titers in comparison to animals receiving the antigen alone (8, 18). Cellular immune responses were also stronger in mice immunized with the antigen coadministered with c-di-GMP. The analysis of the antigen-specific IgG subclasses present in the sera of vaccinated animals together with the pattern of cytokines secreted by spleen cells suggested that both Th1 and Th2 cells were stimulated. The results obtained after in vitro restimulation of splenocytes recovered from vaccinated animals showed that not only IFN-γ, IL-10, and IL-2 were increased but also proinflammatory mediators, such as granulocyte-macrophage colony-stimulating factor (GM-CSF), IL-17, and macrophage inflammatory protein 1 alpha (MIP-1α).

IL-17 is involved in the recruitment of neutrophils and eosinophils and promotes the expression of Th2 cytokines (15, 21–24, 46). This in turn stimu-

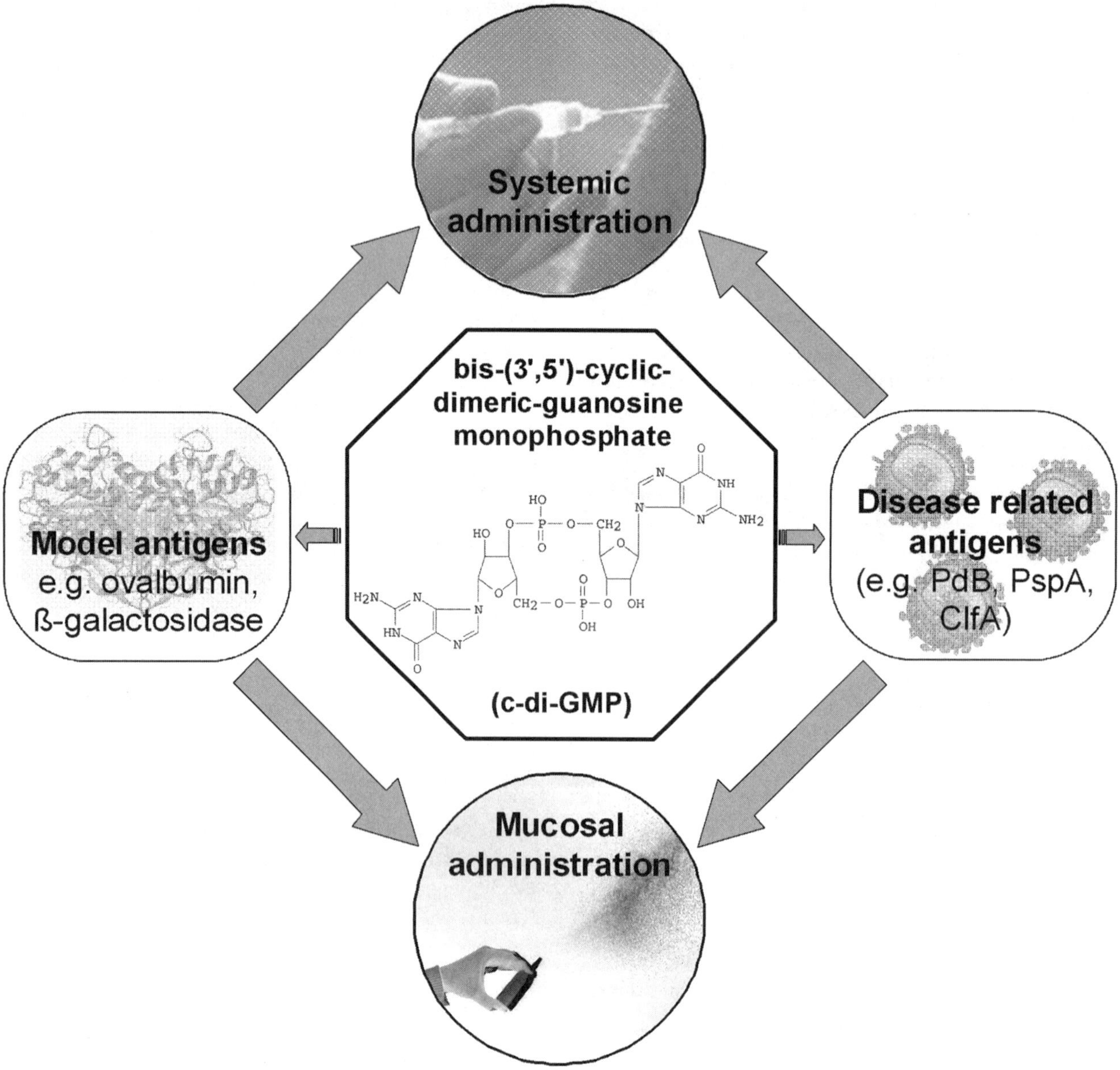

Figure 2. Coadministration of c-di-GMP as a parenteral or mucosal adjuvant with model- or disease-related antigens resulted in efficient humoral and cellular immune responses in the host.

lates DC maturation and enhancement of their antigen presentation functions (49). In addition, IL-17 is able to induce GM-CSF via the Raf I-MEK-ERK pathway (22), which in turn promotes Th1 responses. The stimulation of MIP-1α, which affects the magnitude and polarity of T-cell responses, may also favor IL-12 and IFN-γ expression by an alternative Th1/Th2 polarization pathway (26). Taken together, these data contribute to explain, at least in part, the mixed Th1/Th2 responses observed in mice immunized by using c-di-GMP as an adjuvant (Fig. 3).

The observed Th response pattern suggests that the use of c-di-GMP results in the stimulation of T cells, which can provide help for both antibody production and stimulation of cytotoxic T-cell responses. In fact, besides elevated antibody titers, a clear increment in the number of IFN-γ secreting cells was also observed when splenocytes from mice vaccinated with ovalbumin were stimulated with a peptide encompassing an MHC class I-restricted epitope. The same was true when an in-depth analysis of the influence of c-di-GMP on the elicitation of lymphocyte-mediated cytotoxic responses was performed by measuring the in vivo cytotoxic T lymphocyte (CTL) activity against ovalbumin in C57BL/6 mice (9, 13). This suggests that c-di-GMP is indeed a promising adjuvant for vaccines administered by a systemic route (e.g., s.c. or i.m.).

c-di-GMP AS MUCOSAL ADJUVANT

Pathogenic microorganisms are an extremely diverse group of living entities which are adapted to different environments, including their potential hosts. Despite their intrinsic diversity, they share common gateways into the human body (i.e., the mucosal surfaces). Thus, the mucosal immune system serves as the first line of defense against bacteria and viruses. Therefore, significant efforts have been invested in the development of mucosal vaccination strategies which are amenable to oral, i.n., or pulmonary administration. All of these approaches share the capacity to promote efficient immune responses at both the systemic and mucosal levels.

The implementation of a mucosal vaccination strategy can not only result in disease prevention but also block early infection events, thereby reducing the likelihood of horizontal transmission to susceptible hosts. In addition, mucosal vaccination reduces or makes negligible the risk of cross contamination. However, as mentioned above, most antigens are extremely poorly immunogenic when administered by this route. This is in part due to enzymatic degradation, structural modification resulting from extreme pH, and/or mechanical removal. Mucosal adjuvants can be exploited to overcome this bottleneck.

Immunization with model antigens coadministered with c-di-GMP by the i.n. route resulted in a significantly improved stimulation of antigen-specific immune responses at both the systemic and mucosal levels in comparison to mice immunized by the parenteral route. More specifically, high titers of antigen-specific IgG were detected in the sera of vaccinated mice, even after a single boost. Furthermore, the production of antigen-specific secretory IgA was efficiently stimulated not only locally in the lungs but also at distant mucosal territories, such as in vaginal lavage fluid (9). Cellular responses were also stronger in mice receiving the antigen with c-di-GMP than in animals receiving the antigen alone. Similar results were obtained with C57BL/6 mice, thereby demonstrating that the adjuvant activity of c-di-GMP is not restricted to a particular mouse haplotype.

Of remark, the use of c-di-GMP as a mucosal adjuvant also promoted MHC class I-restricted immune responses, as shown by the analysis of CTL responses in vivo. The i.n. vaccination strategy promoted a shift from a Th2 to a more balanced Th1/Th2 response pattern, as demonstrated by the improved secretion of IFN-γ. A more in-depth analysis showed that cells derived from mice vaccinated with antigen coadministered (i.e., add-mixed) with c-di-GMP have an enhanced expression of cytokines and chemokines, which play a role as attractants for naive and effector T cells as well as a role in T-cell differentiation (e.g., IL-17 and GM-CSF). In contrast, splenocytes recovered from control mice immunized with the antigen alone showed no or only marginal secretion of these soluble mediators. In conclusion, coadministration of antigens with c-di-GMP by the i.n. route resulted in the stimulation of strong humoral and cellular immune responses at both the systemic and mucosal levels, which were comparable in strength and breadth to the responses obtained following vaccination by the systemic route.

USE OF c-di-GMP FOR IMMUNE INTERVENTIONS AGAINST INFECTIONS

Most microbial pathogens cause infections that are restricted to the mucosa or should transit across them to invade, disseminate, and reach their target organs. Thus, the respiratory, intestinal, and genitourinary tracts are niches where the main battles against infectious agents are fought. In this context, components of the innate immune system have a central role in the rapid and unspecific early response, which usually determines whether a symptomatic disease will develop in a host. Therefore, compounds which are able to either reduce the virulence of microorganisms with a mucosal portal of entry or stimulate the innate immune system can have an important role in defense. It would also be possible to exploit these moieties as adjuvants for vaccine formulations, thereby bridging the innate and adaptive immune systems.

Previous studies demonstrated that c-di-GMP exhibits antimicrobial properties. When *Staphylococcus aureus* cells were treated with c-di-GMP, they showed impaired attachment to HeLa cells as well as a reduced capacity to form biofilms in vitro (2, 20). In contrast, intramammary injections of c-di-GMP decreased the in vivo virulence and colonization of *S. aureus* and allowed in higher doses the clearance of the bacteria below the detection limit compared to the untreated control groups. The speeding up of the recovery from experimental infections allows the development of novel intervention strategies for the prevention of biofilms and the treatment of infections (2, 20). In another experimental animal infection

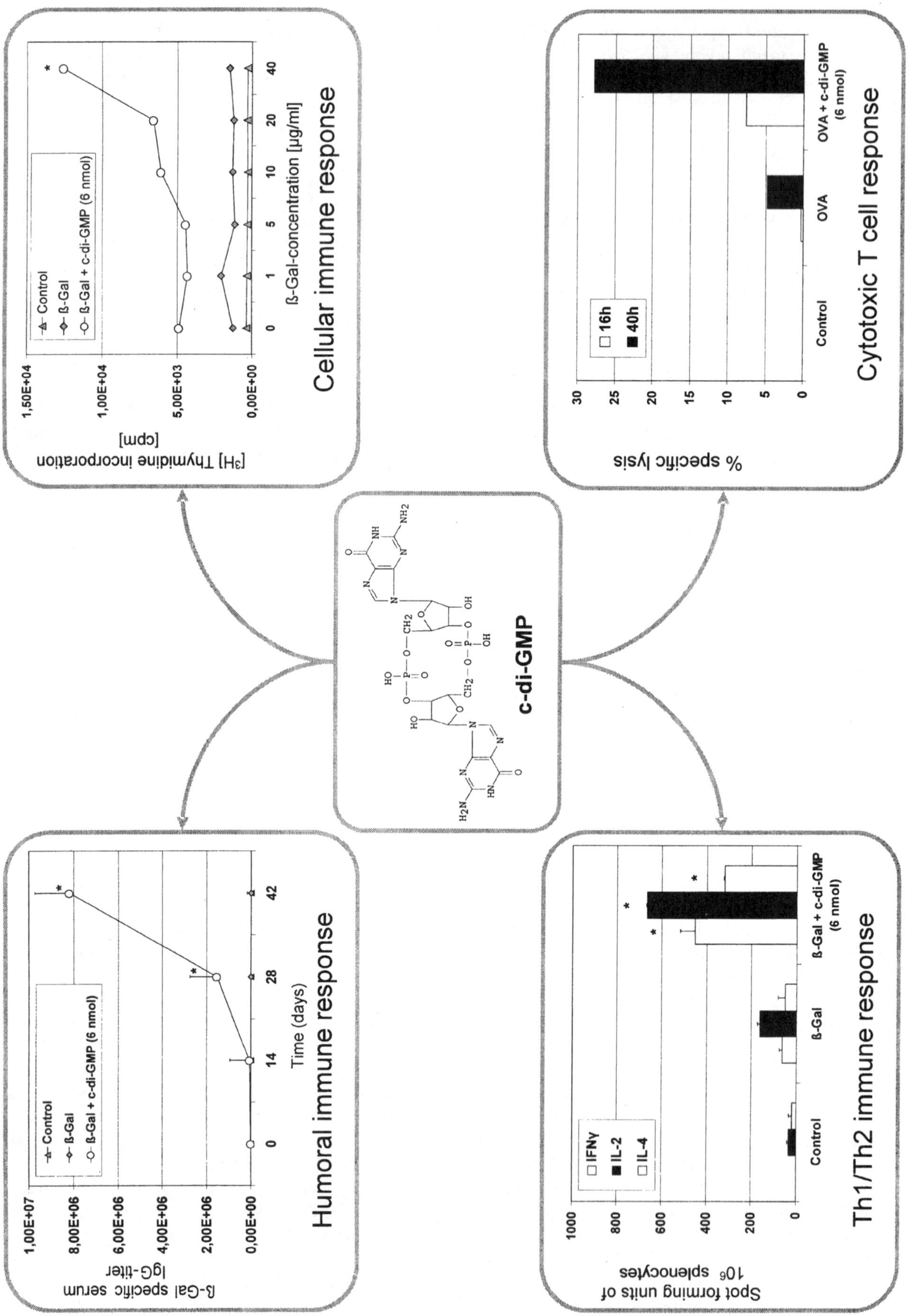
Cellular immune response
[3H] Thymidine incorporation [cpm]
ß-Gal-concentration [µg/ml]
Control
ß-Gal
ß-Gal + c-di-GMP (6 nmol)
Cytotoxic T cell response
% specific lysis
16h
40h
Control
OVA
OVA + c-di-GMP (6 nmol)
c-di-GMP
Humoral immune response
ß-Gal specific serum IgG-titer
Time (days)
Th1/Th2 immune response
Spot forming units of 10^6 splenocytes
IFNγ
IL-2
IL-4

model, the i.n. or s.c. administration of c-di-GMP prior to intratracheal challenge with *Klebsiella pneumoniae* resulted in a significantly increased survival (19).

The i.n. coadministration of c-di-GMP with either pneumolysin toxoid (PdB) or the pneumococcal surface protein A (PspA) also resulted in a decreased bacterial load in the blood and/or lungs after challenge with pneumococcal strains from serotypes 2, 3, and 4. When mice were vaccinated with these antigens coadministered with c-di-GMP by the intraperitoneal route before challenge, significantly higher antigen-specific antibody titers and increased survival were observed, as compared to animals immunized with antigens and alum as the gold standard (30). c-di-GMP was also successfully evaluated as an adjuvant with other disease-relevant antigens. For example, i.m. vaccination of mice with c-di-GMP together with the *S. aureus* clumping factor A resulted in significantly higher antigen-specific antibody titers than in the controls (18). Similar results were obtained after i.n. vaccination with antigens derived from the influenza (including the H5N1 strain), hepatitis B, and human immunodeficiency viruses (C. A. Guzmán and T. Ebensen, unpublished data).

NOVEL MEMBERS OF THE CYCLIC DINUCLEOTIDE FAMILY

Over the last years, members of the cyclic dinucleotide and nucleotide families, such as bis-(3′,5′)-c-di-AMP, cAMP, and cGMP have become well recognized as low-molecular-weight signaling molecules in prokaryotes. c-di-AMP is a second messenger that signals DNA integrity in *Bacillus subtilis* during sporulation (32, 47). This compound is widespread in bacteria and archaea, thereby suggesting that it might act as a second messenger in response to other signals besides branched DNA (12, 32). This is further supported by the presence of a diadenylate cyclase domain in many proteins from different prokaryotes (47). In addition, cAMP has a role in prokaryotes by alleviating glucose catabolite repression (16, 27), whereas cGMP has been shown to act as a signaling molecule in *Synechocystis* cyanobacteria (28, 29). The cyclic nucleotide analogs cGMP and cAMP have a lesser inhibitory effect on biofilms, whereas 5′-GMP exhibits no major effect (20). Studies from Karaolis et al. suggest that cGMP and cAMP have adjuvant capacities, but to a far lesser extent than c-di-GMP (20).

c-di-AMP has a cyclic structure which, although intrinsically distinct, is still reminiscent at the conformational level of c-di-GMP. This prompted us to evaluate the potential of c-di-AMP as an immune modulator. The results obtained demonstrated that the novel compound c-di-AMP exhibits an improved capacity to activate in vitro bone marrow-derived murine DC with respect to c-di-GMP (T. Ebensen and C. A. Guzmán, unpublished data). The i.n. coadministration of different antigens with c-di-AMP or its derivative bis-(3′,5′)-c-di-IMP resulted in the stimulation of strong humoral and cellular immune responses at both the systemic and local levels (T. Ebensen, R. Libanova, and C. A. Guzmán, unpublished data).

In contrast to traditional adjuvants, such as alum, the cyclic dinucleotides offer the opportunity to support the development of vaccines which can be administered by different mucosal routes (e.g., i.n., oral, or intravaginal). These cyclic compounds are also able to induce vigorous CTL responses in addition to enhanced humoral immune responses.

CONCLUSIONS

The results obtained using different in vitro and in vivo experimental models demonstrated that c-di-GMP, c-di-AMP, and c-di-IMP exhibit strong immune modulatory properties, being able to stimulate the innate and adaptive immune systems. c-di-AMP, c-di-IMP, and, to a lesser extent, c-di-GMP are able to promote the activation and maturation of professional APC. They also exhibit powerful adjuvant activities when coadministered with a broad range of antigens from bacteria or viruses by either the systemic or the mucosal route. In the particular case of c-di-GMP, pretreatment or coadministration to animals receiving a bacterial challenge resulted in enhanced clearance. Besides this antimicrobial approach to attenuating the virulence and pathogenesis of *S. aureus,* c-di-GMP has striking properties regarding the inhibition of cancer cell proliferation (17). Thus,

Figure 3. Coadministration of c-di-GMP stimulates a broad spectrum of antigen-specific immune responses irrespective of the administration route (8, 9, 18). Animals were immunized with phosphate-buffered saline alone (control), model antigens alone, such as ovalbumin (OVA) or β-galactosidase (β-Gal), or antigen coadministered with c-di-GMP on days 0, 14, and 28. Mice showed strong antigen-specific humoral and cellular immune responses in comparison to mice vaccinated with the antigen alone.

the cyclic dinucleotides represent a promising new class of molecules for the establishment of immune interventions against infectious diseases and cancer.

REFERENCES

1. **Bobrov, A. G., O. Kirillina, and R. D. Perry.** 2007. Regulation of biofilm formation in Yersinia pestis. *Adv. Exp. Med. Biol.* **603:**201–210.
2. **Brouillette, E., M. Hyodo, Y. Hayakawa, D. K. Karaolis, and F. Malouin.** 2005. 3′,5′-Cyclic diguanylic acid reduces the virulence of biofilm-forming *Staphylococcus aureus* strains in a mouse model of mastitis infection. *Antimicrob. Agents Chemother.* **49:**3109–3113.
3. **Burdin, N., B. Guy, and P. Moingeon.** 2004. Immunological foundations to the quest for new vaccine adjuvants. *BioDrugs* **18:**79–93.
4. **Cavallo, F., A. Astolfi, M. Iezzi, F. Cordero, P. L. Lollini, G. Forni, and R. Calogero.** 2005. An integrated approach of immunogenomics and bioinformatics to identify new tumor associated antigens (TAA) for mammary cancer immunological prevention. *BMC Bioinformatics* **6**(Suppl. 4):S7.
5. **Cotter, P. A., and S. Stibitz.** 2007. c-di-GMP-mediated regulation of virulence and biofilm formation. *Curr. Opin. Microbiol.* **10:**17–23.
6. **Di Guilmi, A. M., and A. Dessen.** 2002. New approaches towards the identification of antibiotic and vaccine targets in Streptococcus pneumoniae. *EMBO Rep.* **3:**728–734.
7. **Douce, G., M. Fontana, M. Pizza, R. Rappuoli, and G. Dougan.** 1997. Intranasal immunogenicity and adjuvanticity of site-directed mutant derivatives of cholera toxin. *Infect. Immun.* **65:**2821–2828.
8. **Ebensen, T., K. Schulze, P. Riese, C. Link, M. Morr, and C. A. Guzman.** 2007. The bacterial second messenger cyclic diGMP exhibits potent adjuvant properties. *Vaccine* **25:**1464–1469.
9. **Ebensen, T., K. Schulze, P. Riese, M. Morr, and C. A. Guzman.** 2007. The bacterial second messenger cdiGMP exhibits promising activity as a mucosal adjuvant. *Clin. Vaccine Immunol.* **14:**952–958.
10. **Guy, B., and N. Burdin.** 2005. New adjuvants for parenteral and mucosal vaccines. *Therapie* **60:**235–241.
11. **Hammer, B. K., and B. L. Bassler.** 2009. Distinct sensory pathways in *Vibrio cholerae* El Tor and classical biotypes modulate cyclic dimeric GMP levels to control biofilm formation. *J. Bacteriol.* **191:**169–177.
12. **Harwood, C.** 2008. A microbiologist hopes to disrupt bacterial 'decisions'. *Nature* **455:**5.
13. **Hermans, I. F., J. D. Silk, J. Yang, M. J. Palmowski, U. Gileadi, C. McCarthy, M. Salio, F. Ronchese, and V. Cerundolo.** 2004. The VITAL assay: a versatile fluorometric technique for assessing CTL- and NKT-mediated cytotoxicity against multiple targets in vitro and in vivo. *J. Immunol. Methods* **285:** 25–40.
14. **Hickman, J. W., D. F. Tifrea, and C. S. Harwood.** 2005. A chemosensory system that regulates biofilm formation through modulation of cyclic diguanylate levels. *Proc. Natl. Acad. Sci. USA* **102:**14422–14427.
15. **Huang, S. H., S. Frydas, D. Kempuraj, R. C. Barbacane, A. Grilli, W. Boucher, R. Letourneau, B. Madhappan, N. Papadopoulou, N. Verna, M. A. De Lutiis, T. Iezzi, G. Riccioni, T. C. Theoharides, and P. Conti.** 2004. Interleukin-17 and the interleukin-17 family member network. *Allergy Asthma Proc.* **25:**17–21.
16. **Jackson, D. W., J. W. Simecka, and T. Romeo.** 2002. Catabolite repression of *Escherichia coli* biofilm formation. *J. Bacteriol.* **184:**3406–3410.
17. **Karaolis, D. K., K. Cheng, M. Lipsky, A. Elnabawi, J. Catalano, M. Hyodo, Y. Hayakawa, and J. P. Raufman.** 2005. 3′,5′-Cyclic diguanylic acid (c-di-GMP) inhibits basal and growth factor-stimulated human colon cancer cell proliferation. *Biochem. Biophys. Res. Commun.* **329:**40–45.
18. **Karaolis, D. K., T. K. Means, D. Yang, M. Takahashi, T. Yoshimura, E. Muraille, D. Philpott, J. T. Schroeder, M. Hyodo, Y. Hayakawa, B. G. Talbot, E. Brouillette, and F. Malouin.** 2007. Bacterial c-di-GMP is an immunostimulatory molecule. *J. Immunol.* **178:**2171–2181.
19. **Karaolis, D. K., M. W. Newstead, X. Zeng, M. Hyodo, Y. Hayakawa, U. Bhan, H. Liang, and T. J. Standiford.** 2007. Cyclic di-GMP stimulates protective innate immunity in bacterial pneumonia. *Infect. Immun.* **75:**4942–4950.
20. **Karaolis, D. K., M. H. Rashid, R. Chythanya, W. Luo, M. Hyodo, and Y. Hayakawa.** 2005. c-di-GMP (3′-5′-cyclic diguanylic acid) inhibits *Staphylococcus aureus* cell-cell interactions and biofilm formation. *Antimicrob. Agents Chemother.* **49:**1029–1038.
21. **Kawaguchi, M., M. Adachi, N. Oda, F. Kokubu, and S. K. Huang.** 2004. IL-17 cytokine family. *J. Allergy Clin. Immunol.* **114:**1265–1273.
22. **Kawaguchi, M., F. Kokubu, M. Odaka, S. Watanabe, S. Suzuki, K. Ieki, S. Matsukura, M. Kurokawa, M. Adachi, and S. K. Huang.** 2004. Induction of granulocyte-macrophage colony-stimulating factor by a new cytokine, ML-1 (IL-17F), via Raf I-MEK-ERK pathway. *J. Allergy Clin. Immunol.* **114:** 444–450.
23. **Kelly, M. N., J. K. Kolls, K. Happel, J. D. Schwartzman, P. Schwarzenberger, C. Combe, M. Moretto, and I. A. Khan.** 2005. Interleukin-17/interleukin-17 receptor-mediated signaling is important for generation of an optimal polymorphonuclear response against *Toxoplasma gondii* infection. *Infect. Immun.* **73:**617–621.
24. **Kolls, J. K., and A. Linden.** 2004. Interleukin-17 family members and inflammation. *Immunity* **21:**467–476.
25. **Lizotte-Waniewski, M., W. Tawe, D. B. Guiliano, W. Lu, J. Liu, S. A. Williams, and S. Lustigman.** 2000. Identification of potential vaccine and drug target candidates by expressed sequence tag analysis and immunoscreening of *Onchocerca volvulus* larval cDNA libraries. *Infect. Immun.* **68:**3491–3501.
26. **Luther, S. A., and J. G. Cyster.** 2001. Chemokines as regulators of T cell differentiation. *Nat. Immunol.* **2:**102–107.
27. **Notley-McRobb, L., A. Death, and T. Ferenci.** 1997. The relationship between external glucose concentration and cAMP levels inside Escherichia coli: implications for models of phosphotransferase-mediated regulation of adenylate cyclase. *Microbiology* **143**(Pt 6):1909–1918.
28. **Ochoa de Alda, J. A., G. Ajlani, and J. Houmard.** 2000. *Synechocystis* strain PCC 6803 *cya2,* a prokaryotic gene that encodes a guanylyl cyclase. *J. Bacteriol.* **182:**3839–3842.
29. **Ochoa de Alda, J. A., and J. Houmard.** 2000. Genomic survey of cAMP and cGMP signalling components in the cyanobacterium Synechocystis PCC 6803. *Microbiology* **146**(Pt 12): 3183–3194.
30. **Ogunniyi, A. D., J. C. Paton, A. C. Kirby, J. A. McCullers, J. Cook, M. Hyodo, Y. Hayakawa, and D. K. Karaolis.** 2008. c-di-GMP is an effective immunomodulator and vaccine adjuvant against pneumococcal infection. *Vaccine* **26:** 4676–4685.
31. **Rashid, M. H., C. Rajanna, A. Ali, and D. K. Karaolis.** 2003. Identification of genes involved in the switch between the

smooth and rugose phenotypes of Vibrio cholerae. *FEMS Microbiol. Lett.* **227**:113–119.

32. **Romling, U.** 2008. Great times for small molecules: c-di-AMP, a second messenger candidate in Bacteria and Archaea. *Sci. Signal.* **1**:pe39.
33. **Romling, U., and D. Amikam.** 2006. Cyclic di-GMP as a second messenger. *Curr. Opin. Microbiol.* **9**:218–228.
34. **Ross, P., R. Mayer, and M. Benziman.** 1991. Cellulose biosynthesis and function in bacteria. *Microbiol. Rev.* **55**:35–58.
35. **Schulze, K., E. Medina, G. S. Chhatwal, and C. A. Guzman.** 2003. Stimulation of long-lasting protection against Streptococcus pyogenes after intranasal vaccination with non adjuvanted fibronectin-binding domain of the SfbI protein. *Vaccine* **21**:1958–1964.
36. **Simm, R., J. D. Fetherston, A. Kader, U. Romling, and R. D. Perry.** 2005. Phenotypic convergence mediated by GGDEF-domain-containing proteins. *J. Bacteriol.* **187**:6816–6823.
37. **Simm, R., M. Morr, A. Kader, M. Nimtz, and U. Romling.** 2004. GGDEF and EAL domains inversely regulate cyclic di-GMP levels and transition from sessility to motility. *Mol. Microbiol.* **53**:1123–1134.
38. **Stanic, A. K., A. D. De Silva, J. J. Park, V. Sriram, S. Ichikawa, Y. Hirabyashi, K. Hayakawa, L. Van Kaer, R. R. Brutkiewicz, and S. Joyce.** 2003. Defective presentation of the CD1d1-restricted natural Va14Ja18 NKT lymphocyte antigen caused by beta-D-glucosylceramide synthase deficiency. *Proc. Natl. Acad. Sci. USA* **100**:1849–1854.
39. **Stevceva, L., and M. G. Ferrari.** 2005. Mucosal adjuvants. *Curr. Pharm. Des.* **11**:801–811.
40. **Tagliabue, A., and R. Rappuoli.** 2008. Vaccine adjuvants: the dream becomes real. *Hum. Vaccin.* **4**:347–349.
41. **Tamayo, R., J. T. Pratt, and A. Camilli.** 2007. Roles of cyclic diguanylate in the regulation of bacterial pathogenesis. *Annu. Rev. Microbiol.* **61**:131–148.
42. **Teicher, B. A.** 2000. Molecular targets and cancer therapeutics: discovery, development and clinical validation. *Drug Resist. Updat.* **3**:67–73.
43. **Tischler, A. D., and A. Camilli.** 2004. Cyclic diguanylate (c-di-GMP) regulates Vibrio cholerae biofilm formation. *Mol. Microbiol.* **53**:857–869.
44. **Tischler, A. D., and A. Camilli.** 2005. Cyclic diguanylate regulates *Vibrio cholerae* virulence gene expression. *Infect. Immun.* **73**:5873–5882.
45. **van Ginkel, F. W., R. J. Jackson, Y. Yuki, and J. R. McGhee.** 2000. Cutting edge: the mucosal adjuvant cholera toxin redirects vaccine proteins into olfactory tissues. *J. Immunol.* **165**:4778–4782.
46. **Witowski, J., K. Ksiazek, and A. Jorres.** 2004. Interleukin-17: a mediator of inflammatory responses. *Cell. Mol. Life Sci.* **61**:567–579.
47. **Witte, G., S. Hartung, K. Buttner, and K. P. Hopfner.** 2008. Structural biochemistry of a bacterial checkpoint protein reveals diadenylate cyclase activity regulated by DNA recombination intermediates. *Mol. Cell* **30**:167–178.
48. **Zareba, G.** 2006. A new combination vaccine for measles, mumps, rubella and varicella. *Drugs Today* (Barcelona) **42**:321–329.
49. **Zou, G. M., and Y. K. Tam.** 2002. Cytokines in the generation and maturation of dendritic cells: recent advances. *Eur. Cytokine Netw.* **13**:186–199.

V. CONCLUSIONS

The Second Messenger Cyclic Di-GMP
Edited by A. J. Wolfe and K. L. Visick

Chapter 23

Cyclic Di-GMP: Using the Past To Peer into the Future

Karen L. Visick and Alan J. Wolfe

Just over 20 years have elapsed since cyclic di-GMP (c-di-GMP) was discovered as an allosteric effector of cellulose biosynthesis (chapter 2). While still in its early adolescence, the c-di-GMP field is growing at an amazing rate, due in large part to the maturation of bioinformatics, the near ubiquity of this molecule in bacteria, and the role that it plays in key life-style choices. In relatively short order, the field has established and discarded numerous paradigms for the roles of c-di-GMP, its metabolic enzymes, and receptor/effector proteins. This book describes the established paradigms and reveals new and interesting exceptions to the rules. It also discusses conceptual problems associated with the relative abundance of genes for c-di-GMP metabolism in numerous bacterial genomes. In this concluding chapter, we will dispel myths concerning current paradigms and highlight key questions currently under investigation.

PARADIGM 1: c-di-GMP IS SYNTHESIZED BY ENZYMES THAT CONTAIN A GGDEF DOMAIN AND DEGRADED BY ENZYMES THAT CONTAIN AN EAL DOMAIN

Bioinformatics, and the prevalence of this molecule, provided the necessary boost to lift the study of c-di-GMP from a single laboratory to a community of researchers. As described in detail in chapter 3, the GGDEF domain was first identified as an output domain in the *Caulobacter crescentus* protein PleD (22). That PleD was not alone in containing the GGDEF domain became apparent with the availability of fully sequenced bacterial genomes. Finally, the connection made by Benziman and colleagues between the GGDEF and EAL domains and the enzymes involved in c-di-GMP metabolism provided the key to understanding the function of these domains (chapter 2). Subsequent work narrowed down the possibilities for these domains between c-di-GMP synthesis and degradation, provided proof of their enzymatic activities, and established the importance of the signature amino acids in the function of these proteins (chapter 4).

At the same time, however, it was becoming increasingly clear that the situation was complicated. For example, many genomes containing GGDEF domains lacked corresponding EAL domain genes; this glaring absence led to the discovery of the HD-GYP domain, which also functions in c-di-GMP degradation (chapters 3 and 5). Furthermore, many proteins contained both GGDEF and EAL domains and, thus, naively might be predicted to exhibit both diguanylate cyclase (DGC) and phosphodiesterase (PDE) activities. The facts, however, confound the simple interpretation. Most often, these dual-domain proteins possess one highly conserved domain and one poorly conserved domain and exhibit a single activity: the one from the more well conserved domain. For example, the dual-domain protein CC3396 of *C. crescentus* functions as a PDE (13), while DgcA1 of *Gluconacetobacter xylinus* acts as a DGC (57). In some cases, enzymatically inactive domains can bind the substrate: GTP by the inactive GGDEF domain for PDEs and c-di-GMP by the inactive EAL domain for DGCs (chapter 4). Less often, dual-domain proteins possess both activities. Currently, only three such bifunctional dual-domain proteins (ScrC *[Vibrio parahaemolyticus]*, MSDGC-1 *[Mycobacterium smegmatis]*, and BphG1 *[Rhodobacter sphaeroides]*) have been shown to possess the capacity to function both as a DGC and as a PDE (chapters 4 and 12) (31).

Some dual-domain proteins have lost both enzymatic activities. For example, both domains of

Karen L. Visick and Alan J. Wolfe • Department of Microbiology and Immunology, Loyola University Chicago, Maywood, IL 60153.

CsrD from *Escherichia coli* and *Salmonella enterica* and both domains of LapD from *Pseudomonas fluorescens* deviate substantially from the consensus and neither CsrD nor LapD exhibits DGC or PDE activities. Whereas LapD retains the ability to bind c-di-GMP through its degenerate EAL domain, CsrD might not: it does not seem to possess any of the obvious c-di-GMP binding signatures (see below). Instead, CsrD binds small Csr RNAs and mediates their degradation (chapters 10, 14, 16, and 19). Similarly, single-domain proteins can evolve away from their association with c-di-GMP. For example, the EAL domain protein YcgF and the GGDEF domain protein GdpS, neither of which carries out enzymatic activity, perform functions that do not include c-di-GMP or proteins normally thought to be associated with c-di-GMP metabolism of signaling (chapters 4 and 16).

The existence of composite proteins containing domains associated with both synthetic and degradative activities presents a dilemma to researchers, who have often used bioinformatic analyses to develop the hypothesis that their gene of interest likely encodes a DGC or a PDE due to the predicted existence of a GGDEF or EAL (or HY-GYP) domain, respectively, and have proceeded accordingly. Indeed, these are not unreasonable assumptions, especially given the previous successful characterization of a number of such proteins. However, the paradigm weakens as the conservation within the domain decreases, especially within the signature amino acid motif. Thus, caution should be used in assigning and interpreting protein annotations for c-di-GMP-associated proteins. Enzymatic activity will not always follow a conserved domain sequence. Although it is clear that certain amino acids within the signature motif are critical to the function of these proteins, additional work is necessary to fully understand the contribution of various residues within and outside these motifs, as protein-specific changes may lead to enzymatically active or inactive proteins. Furthermore, domains that appear to be poorly conserved may, in fact, retain function; ultimately, it is the three-dimensional structure that is key to function rather than the primary amino acid sequence (chapter 3). Our understanding will be deepened through additional work in this area, including structure-function and crystallography studies of additional conserved and poorly conserved proteins.

PARADIGM 2: PilZ DOMAINS SERVE AS c-di-GMP-BINDING PROTEINS

To regulate cellular processes, c-di-GMP must be sensed. Thus, effectors must exist that bind c-di-GMP and, in response, impact some downstream target. The search for such an effector began early, with Benziman and coworkers determining that c-di-GMP activated the *G. xylinus* cellulose synthase complex and showing that c-di-GMP bound tightly with this complex (chapter 2). However, they were unable to identify the effector (reviewed in reference 50). Success was achieved when a bioinformatic analysis, applied to essential subunits of the cellulose synthase complex, identified a separate protein domain located at the C terminus of BcsA that also resided within some proteins associated with c-di-GMP metabolism (chapter 3). This led to the proposal that this so-called PilZ domain was the elusive c-di-GMP effector (1). Rapid progress followed. One study (52) demonstrated that some isolated PilZ domains specifically bind c-di-GMP, that a full-length PilZ domain protein has a physiologically relevant dissociation constant, that conserved residues are required for c-di-GMP binding in vitro, and that these residues are in close proximity within the tertiary structure of the domain and, thus, likely comprise the c-di-GMP binding site (chapter 4). Several reports followed in rapid succession that supported these observations but made it clear that only a subset of PilZ domains bind c-di-GMP, i.e., those that possess the so-called c-di-GMP loop (3, 12, 34, 37, 46, 47). The focus now has shifted toward the effort to learn how PilZ domain proteins that bind c-di-GMP exert their influence on their downstream targets (chapters 3, 4, and 7).

It quickly became clear, however, that the newly identified PilZ domain could not be the whole answer. First, in many bacteria, the total number of GGDEF, EAL, and/or HD-GYP domains far exceeds the number of PilZ domains. Second, some bacteria with GGDEF, EAL, and/or HD-GYP domains lack PilZ domains altogether. Third, in *Vibrio cholerae,* deletion of all PilZ domain-encoding genes does not abolish the effect of c-di-GMP on colony rugosity (5) (chapter 17). Taken together, these observations provided compelling evidence that non-PilZ effectors must exist (see chapters 3, 4, 17, and 20). The void quickly began to fill. The first non-PilZ domain c-di-GMP receptors to be identified were degenerate GGDEF domains. These enzymatically inactive domains had retained the ability to bind c-di-GMP (chapters 3, 4, 9, and 11). For example, *C. crescentus* PopA contains a GGDEF domain that possesses no DGC activity, presumably because the GGDEF signature motif is degenerate. Indeed, only two residues within the signature amino acid motif are conserved, ATEVF, and at least one of these, E, is not even strictly required for activity of the protein (17). Intriguingly, however, the GGDEF domain is required for the ability of PopA to control the cell cycle regulator CtrA. Additional work revealed a role for a sequence known as the I-site, identified through crystallographic analysis of the *C. crescentus* protein PleD and conserved in PopA (chapter 7). In PleD, the I-site was assigned an inhibitory role on the basis of its ability to bind c-di-GMP and, while bound, to provide negative feedback control on enzymatic activity. In the case of PopA, it appears that c-di-GMP binding at the conserved I-site within the GGDEF domain is responsible for the activity of the protein in mediating degradation of CtrA. An intriguing variation on the I-site theme arose with the discovery that the *Pseudomonas aeruginosa* protein PelD bound c-di-GMP (33). While a GGDEF domain is not apparent in PelD, this protein contains sequences similar to the I-site and the predicted secondary structure in that region is similar to GGDEF domains (chapter 4). Because the I-site is present in numerous GGDEF proteins that do not possess a conserved signature motif, these degenerated GGDEF proteins might, with caution, be predicted to function as c-di-GMP effectors. However, the PelD example demonstrates that other variations are possible.

Similarly, degenerate EAL domain proteins were found to function as c-di-GMP effectors, as the degenerate and enzymatically inactive EAL domains of LapD (41) and YkuI from *Bacillus subtilis* (38) were each determined to bind c-di-GMP via their EAL domain. It is not hard to imagine that an enzymatically inactive HD-GYP domain could retain its ability to bind c-di-GMP and thus also serve as a c-di-GMP effector; evidence of such an effector may just be a matter of time (chapter 4).

A surprising new class of c-di-GMP effectors is the *Pseudomonas* transcription factor FleQ/AdnA (chapters 4, 11, and 19). That this protein does not fall into any of the previously identified effector classes suggests the potential for the existence of several unpredictable new kinds of c-di-GMP effectors. Given that c-di-GMP binding to the characterized c-di-GMP-binding effectors (i.e., PilZ domains, I-site-containing domains, and EAL domains) apparently requires a small number of strategically located amino acid residues, one can easily imagine the existence of several different types of c-di-GMP effectors. The hunt for these effector classes should be illuminating (chapter 4).

Like with PilZ, the discovery of a c-di-GMP-specific riboswitch owes much to bioinformatic analysis. The GEMM class of riboswitches had been previously identified, but its ligand remained unknown until it was recognized that GEMM was often located upstream of c-di-GMP-associated genes. These genes either encoded DGCs or PDEs or were predicted to be regulated by c-di-GMP. On the basis of these observations, it was hypothesized that GEMM RNAs might serve as a c-di-GMP effector. Subsequent studies both supported this hypothesis and demonstrated the ubiquity of this class of c-di-GMP effectors (chapter 15).

Thus, to date, several classes of c-di-GMP effectors have been identified: the PilZ domain, the I-site, the enzymatically inactive EAL (and perhaps HD-GYP) domain, FleQ, and the GEMM riboswitch. Yet our understanding of these effectors lags well behind our knowledge of the processes by which c-di-GMP is synthesized and hydrolyzed. The diversity of these effectors (from PilZ-containing proteins and catalytically inactive GGDEF and EAL domain proteins to riboswitches) is impressive and suggests that a bioinformatics-based approach will have to lag behind other experimental approaches until the range of distinct effectors becomes more defined. Furthermore, for the most part, we do not know how c-di-GMP alters the function of these effectors and we know precious little about how those effectors regulate a set of extremely diverse downstream processes. Considerable energy and resources should be applied toward understanding these processes.

PARADIGM 3: c-di-GMP CONTROLS THE "STICK-OR-SWIM" DECISION

It is now universally accepted that c-di-GMP contributes to the decision to transit between the motile planktonic and the sessile biofilm lifestyles (chapter 8) (23, 26, 29, 58). Indeed, the evidence overwhelming supports this view. Accumulation of c-di-GMP generally favors cellular processes that contribute to biofilm formation (e.g., extracellular polysaccharide biosynthesis), while inhibiting processes that favor the planktonic lifestyle (e.g., motility) (72) (chapter 8).

Although this paradigm appears valid in *Proteobacteria* (for example, the *Enterobacteriaceae,* the pseudomonads, and the vibrios), it remains unclear whether c-di-GMP plays that role in other phylogenetic branches. Data from those branches are scarce, but reports on *M. smegmatis (Actinobacteria)* (31), *Anabaena* sp. *(Cyanobacteria)* (40), *Borrelia burgdorferi (Spirochetes)* (48), and *Anaplasma phagocytophilum (Rickettsiales)* (32) suggest that c-di-GMP may regulate processes that are unrelated to the stick-or-swim decision (chapter 4).

Even in proteobacteria, evidence is building that c-di-GMP plays other roles. For example, genetic evidence supports the hypothesis that c-di-GMP regulates protein secretion and controls bacterial virulence via pathways that do not involve motility or polysaccharide biosynthesis (5, 36, 46; see also chapters 4, 5, 18, and 20). The observation that c-di-GMP impacts the host immune system supports the hypothesis that this second messenger plays a role in virulence (chapters 10, 21, and 22). Furthermore, it seems unlikely that the dozens of experimentally determined and predicted c-di-GMP-associated genes found in many bacterial species could all be involved in this single transition. Undoubtedly, c-di-GMP plays an important role in the stick-or-swim lifestyle decision, but it is reasonable to suggest the hypothesis that this is not its only role. A rigorous test of this hypothesis will require the completion of more thorough genetic/phenotypic analyses of several of the model organisms already under study and the investigation of several species from several distinct phylogenetic branches of bacteria. The information derived from such studies almost certainly will permit the construction of a new and improved paradigm.

BEYOND THE PARADIGMS: OPEN QUESTIONS

Beyond the clear need for a more complete understanding of the domains involved in c-di-GMP synthesis and degradation, the identification of additional c-di-GMP binding domains, and a more complete survey of the phenotypes controlled by c-di-GMP, additional concerns remain. For example, how do the many diverse sensory domains associated with c-di-GMP modules influence their activity? Signals controlling the activity of c-di-GMP-associated proteins have been identified or implicated in a few cases. For example, norspermidine enhances *V. cholerae* biofilm development in a manner that depends on both the periplasmic sensor protein NspS and the putative PDE MbaA (28), light stimulates the phytochrome-coupled DGC activity of BphG1 from *Rhodobacter sphaeroides* (60), blue light activates the PDE BlrP1 from *Klebsiella pneumoniae* (2, 73, 74), and oxygen regulates two heme-PAS domain PDEs (PdeA1 from *G. xylinus* [63] and Dos from *E. coli* [16, 56]) and a globin-coupled DGC from *Bordetella pertussis* (67). These and other studies (e.g., chapter 19) represent an important new direction for c-di-GMP research.

Below, we will focus on two questions under active investigation: how does structure impact function and how is specificity achieved? The answers to these questions will drive the field forward.

How Does Structure Impact Function?

Crystallography has provided great insights into the mechanisms of action of DGCs, c-di-GMP PDEs, and at least one PilZ domain protein (chapter 7). The structure of the DGC PleD, for example, revealed the presence of the I-site involved in feedback inhibition (11; see also chapter 7), while the structure of a second DGC, WspR from *P. aeruginosa*, highlighted the conservation of core regulatory mechanisms but also suggested differences between the two proteins that perhaps arise from differences in their domain architecture and structural motifs (11). Each DGC contains both a catalytically active GGDEF domain and a signaling-proficient REC domain, and PleD (but not WspR) contains an additional REC domain, called REC′. In PleD, it is the first REC domain, rather than the second REC′ domain, that participates in the phosphorylation-dependent signal transduction events that lead to catalytic activation (chapters 7 and 9). The PleD structure revealed not only binding of dimeric c-di-GMP to the I-site but also binding of c-di-GMP to arginine residues within both the GGDEF and REC′ domains (10). These results led to the hypothesis that cross-linking could inactivate the protein by preventing the active half-sites from coming together. In the activated state of PleD (and of WspR), c-di-GMP-mediated cross-linking involves a second arginine residue on an adjacent GGDEF domain. For unphosphorylated PleD, an alternative mechanism was envisioned in which c-di-GMP bridges the REC′ and GGDEF domains. Both mechanisms may be relevant at different stages in the regulatory cycle of PleD (11, 69). While the intramolecular cross-linkage between the GGDEF and REC′ domains may be limited to PleD and similarly constructed proteins, GGDEF domain dimerization is likely to be a more prevalent inhibitory mechanism. WspR is subject to a similar negative feedback control, although in this case a c-di-GMP dimer binds to the WspR I-site and to a specific arginine residue located within a second GGDEF domain, stabilizing an interaction that occurs between dimers rather than within dimers (15).

Thus, from structural studies of even just two DGC proteins, relatively similar ones at that, potential mechanistic differences have emerged. While it is reasonable to expect that GGDEF domains will fold into a common three-dimensional structure, it may be unwise to anticipate such commonality for the full-length proteins. Given the great diversity of domains found in association with GGDEF domains (chapter 3), it is unlikely that all DGC activity will be regulated similarly. These differences should be reflected in structure. The same should be true of the other c-di-GMP-associated domains, i.e., EAL, HD-GYP, and PilZ.

Indeed, the recently solved structure of the first enzymatically active EAL domain PDE revealed that two metal ions are required for c-di-GMP hydrolysis, something that had been overlooked in earlier structures of EAL domain proteins (chapter 7). This structure also offered a definitive explanation for the functions of conserved residues in the EAL domains (2) and explained why some EAL domains are not enzymatically active. It also showed how a delicate positioning of a catalytic water molecule in the active site of the PDE is influenced by an external stimulus, light.

The X-ray and nuclear magnetic resonance structures of the PilZ domain proteins revealed that most structural elements of a globular PilZ domain are not involved in c-di-GMP signaling (chapter 7). Instead, the critical feature is a specific c-di-GMP loop that extends from the PilZ domain (3). Can the c-di-GMP loop be implanted onto a different domain and engage that domain in c-di-GMP reception? The determination of structure and its relationship to function will be an exciting and fruitful one to pursue.

How Is Signaling Specificity Achieved from a Readily Diffusible Signal?

The multiplicity of DGCs and PDEs within a single organism raises the question of signaling specificity. It is clear that specificity exists: mutants that lack individual GGDEF, EAL, or HD-GYP proteins are most often defective in only a subset of c-di-GMP-associated traits (see, for example, chapters 10, 13, and 16). Yet a common pool of c-di-GMP also must exist: c-di-GMP is freely diffusible and thus heterogeneity should, without some counteracting mechanism, equilibrate to a common homogeneous concentration. Empirical evidence supports this contention: repeatedly, researchers have found that overexpression of DGCs or PDEs tends to cause global behavioral changes (72).

How can these two apparently opposing observations be reconciled? Several simple solutions are apparent. First, c-di-GMP modules might be expressed differentially: certain combinations of modules might not be expressed at the same time. Second, c-di-GMP modules might be activated differentially: modules might be activated in response to distinctly different stimuli that rarely exist simultaneously. Third, c-di-GMP modules and/or their targets might possess distinctly different affinities for c-di-GMP or the precursor GTP. Fourth, a DGC or PDE or both might colocalize with a c-di-GMP receptor (chapters 1, 4, 9, 16, and 17).

The first three solutions could work even in the context of a homogeneous common pool. In contrast, the freely diffusible nature of c-di-GMP and the lack of obvious compartmentalization in most bacteria challenge the simplicity of the last proposal. How might a c-di-GMP microenvironment be maintained against the forces that push for homogeneity? To answer this question, we might look to models proposed by the mammalian cyclic nucleotide signaling community (4, 8, 65, 70; see also chapters 1 and 4). Such models attempt to explain how cells can exert specificity within large networks of multiple modules that each produce the same end product: a cyclic nucleotide like c-di-GMP. As described in chapter 4, one (or perhaps a few) general purpose DGC and/or PDE might set the intracellular c-di-GMP concentration. For a given external environment, different general purpose enzymes with different activities might be expressed or activated, thereby adjusting the level of the common c-di-GMP pool. Upon this background concentration, a specific DGC might be expressed, activated, or localized to generate a local c-di-GMP cloud that a neighboring c-di-GMP receptor can sense and, in response, produce a specific outcome. A colocalized PDE could control the amplitude, duration, and diameter of the cloud. Alternatively, a PDE colocalized with a c-di-GMP receptor could deplete the local vicinity of c-di-GMP and thus ensure that the receptor does not become activated. As was the case with mammalian cAMP signaling (see chapter 1), the development of probes that can monitor concentrations of c-di-GMP in live cells would go a long way toward understanding how cells attain specificity and avoid cross-talk (chapter 6).

LESSONS FROM THE BUGS: MODULARITY, LOCALIZATION, AND NETWORKS

Ongoing studies have begun to elucidate some mechanisms bacteria have evolved to make good use of c-di-GMP. Here, we will discuss just three of the many examples described in great detail throughout this book. The first example describes how cells might sense subtle environmental changes and, in response, alter the local concentration of c-di-GMP. The second example describes how cells might use dynamic localization of various c-di-GMP signal components to ensure that key cellular processes proceed properly. The third example describes how cells might utilize multiple c-di-GMP modules for diverse purposes while maintaining specificity.

In *P. aeruginosa,* the activity of one biofilm-associated c-di-GMP module is controlled by the Wsp pathway, a set of proteins highly homologous to those that control bacterial chemotaxis (chapter 11, Fig. 2). In bacterial chemotaxis, a chemoreceptor senses a temporal or spatial gradient. In response, it simultaneously initiates a rapid excitation signal that leads to a change in behavior and a slower adaptation signal that inhibits generation of the excitatory signal. At the core of each signal pathway is a two-component signal transduction pathway composed of a common sensor kinase and either an excitatory or an adaptation response regulator. The outcome in both cases is phosphorylation of the REC domain and enhancement of the response regulator's fundamental activity (21). On the basis of homology and substantial experimental evidence, it is reasonable to propose that the Wsp chemoreceptor senses some environmental gradient and, in response, initiates phosphorylation of WspR's REC domain, which leads to enhancement of WspR's inherent DGC activity. Simultaneous initiation of a slower adaptation signal would eventually lead to cessation of the excitation signal. Intriguingly, WspR appears to form discrete clusters upon phosphorylation (chapters 11 and 19). Clustering may reflect an inherent property of phospho-WspR to oligomerize and/or its affinity for binding to a Wsp aggregate composed of clusters of Wsp chemoreceptors and their associated Wsp signaling components. Aggregate formation is a well-documented characteristic of chemoreceptors and their associated signaling machinery (21). In either case, the anticipated result would be a burst of highly localized c-di-GMP synthesis that somehow favors the transition from motile individual to sessile member of a biofilm community (see also reference 29). Intriguingly, the SadC/BifA c-di-GMP module of *P. aeruginosa* sits upstream of another chemotaxis-like pathway, which controls the transition between flagellum-based swarming and the initiation of biofilm development (reviewed in reference 72). Thus, this single bacterium has used the same two fundamental modules (c-di-GMP and chemotaxis) to evolve two separate circuits to perform two quite different functions to control two distinct aspects of the same lifestyle transition. More examples of the use of such modularity to build pathways with very particular specifications are certain to be revealed.

While studies with *P. aeruginosa* and other organisms provide evidence that c-di-GMP modules might be confined to specific cellular locations, the paradigm for coordinated spatial and temporal regulation comes from investigations of the *C. crescentus* life cycle. This life cycle requires a flagellated, motile swarmer cell that cannot replicate its chromosome or undergo cell division to differentiate into a sessile nonmotile stalked cell that can both replicate its DNA and divide into both cell types (chapter 9, Fig. 2). At distinct stages throughout this cycle, multiple c-di-GMP-associated proteins localize to specific cell poles, where they coordinate discrete events in the process of cell differentiation. For example, the DGC activity of PleD controls flagellar ejection, holdfast biogenesis, and stalk formation, events that must occur in the proper order to enable the swarmer-to-stalk cell transition. To perform this function, PleD must transiently localize to the flagellated pole of the swarmer cell. This localization requires that PleD dimerize, an interaction favored by phosphorylation of the REC domain. The phosphorylation status of PleD and, hence, its ability to localize depend on the action of a highly integrated circuit comprised of three signal proteins that control each others' phosphorylation state, interactions, and/or ability to localize to a specific pole (see chapter 9, Fig. 3 for details). Other c-di-GMP components, DGCs, PDEs, and effectors do likewise. In an elaborate dance, these proteins move from pole to pole, synthesizing, degrading, or binding c-di-GMP and interacting with downstream targets to ensure that the cell cycle proceeds in an orderly fashion (chapter 9, see Fig. 5B and 7). The concerted effort to uncover the details of this ballet of proteins and second messenger will undoubtedly propel the field toward a greater understanding of mechanism.

Like *P. aeruginosa* and many other bacteria, the accumulation of c-di-GMP enhances the motile-to-sessile transition of the *Enterobacteriaceae* family members *S. enterica* serovar Typhimurium and *E. coli*. This transition requires that the cell assembles the organelle of motility (the flagellum) only when it does not synthesize the machinery for biofilm development (curli fibers and cellulose) and vice versa. Such inverse regulation utilizes multiple c-di-GMP modules that appear to be organized into highly integrated and hierarchical signaling networks that operate at multiple levels. In *E. coli*, FliZ has been shown to play a central role in this inverse regulation. A member of the flagellar regulon, FliZ inhibits σ^S. This alternative sigma factor inhibits flagellar function indirectly through two DGCs and the c-di-GMP-binding PilZ domain effector YcgR while activating curli assembly and cellulose synthesis via several other c-di-GMP modules (chapters 4, 10, and 16). In *S. enterica* serovar Typhimurium, eight distinct GGDEF, EAL, and GGDEF-EAL composite proteins have been shown to directly or indirectly and positively or negatively regulate transcription of *csgD*. The product of this gene, the transcription factor CsgD, activates transcription of the DGC AdrA, which in turn activates cellulose synthesis and thus permits biofilm formation. Such networks are dynamic: components are differentially expressed and/or differentially activated. For example, in *S. enterica* serovar Typhimurium, the CsgD-dependent AdrA is required to form biofilms in a nutrient-rich environment, while the CsgD-independent DGC GcpA (YedQ) is required to construct a biofilm in a nutrient-poor environment. Finally, within these networks, specificity is maintained. For example, in *E. coli,* it has been shown that the YegE/YhjH DGC/PDE module controls motility and modulates transcription of *csgD,* but the YdaM/YciR module exclusively controls *csgD* transcription. Similar dynamic, integrated, and hierarchical networks must operate in other organisms that possess large numbers of c-di-GMP modules (chapters 10 and 16) (see also reference 29).

MEDICAL APPLICATIONS

It is widely accepted that biofilm formation is a major virulence factor for many pathogenic bacteria (44, 45). It is estimated that about 80% of human microbial infections occur within the context of a biofilm (9, 14; http://grants.nih.gov/grants/guide/pa-files/PA-07-288.html). Importantly, biofilm cells exhibit significantly increased resistance to antibiotics and other stresses (54). Because numerous organisms use c-di-GMP to control the transition between planktonic and biofilm lifestyles (chapter 8), there is clear potential to control infections by modulating the ability of bacteria to form biofilms in the context of an infection.

A key question for feasibility is whether mature biofilms can be disrupted, and the answer appears to be yes. *Shewanella oneidensis,* for example, forms biofilms that depend upon a GGDEF domain protein (61). Loss of this protein prevented biofilm formation, which could be restored by the overexpression of a heterologous DGC from *V. cholerae*. Importantly, heterologous induction of the *E. coli* PDE YhjH caused rapid detachment. Similarly, biofilms formed by *Pseudomonas putida* quickly dispersed upon heterologous overexpression of YhjH (20).

Numerous studies have focused on the role of c-di-GMP in virulence (see, e.g., references 7, 27, 30, 32, 51, 53, 59, and 62). Importantly for possible therapeutic applications, a few of these have evaluated the effects of exogenous c-di-GMP in controlling infection. For example, the addition of c-di-GMP impairs the ability of *Staphylococcus aureus* to form biofilms in culture, attach to host cells, and colonize in a mouse model (7, 27). In *Streptococcus mutans,* extracellular c-di-GMP decreased both biofilm formation and adherence to tooth surfaces (75). These studies reported that c-di-GMP exerted minimal to no effect on the growth of these bacteria. These effects are opposite to the paradigm: c-di-GMP tends to enhance biofilm formation, and thus, it might be expected to enhance attachment. Relatively little is known about the role of c-di-GMP in either of these organisms, both of which are gram-positive bacteria, other than that they appear to contain very few c-di-GMP-associated genes (two and one GGDEF domain proteins for *S. aureus* and *S. mutans,* respectively). However, these studies are exciting because they suggest that c-di-GMP and anti-c-di-GMP therapeutics can be developed to control bacterial infections. In further support of this idea, a recent report revealed that infection by the gram-negative microbe *Anaplasma phagocytophilum* can be inhibited by a hydrophobic c-di-GMP analog, 2′-O-di(*tert*-butyldimethylsilyl)-c-di-GMP (32).

These studies are just the beginning, and many questions remain. For example, does c-di-GMP enter bacterial cells (both gram-positive and gram-negative) to exert its effect? If so, are there specific c-di-GMP transporters? Does c-di-GMP exist outside bacterial cells in natural/environmental settings? If so, do some bacteria specifically export it? Is there competition for this molecule? And/or are natural analogs produced in nature? There is evidence that c-di-GMP itself is fairly stable under a variety of conditions (27), suggesting that if it gets exported (or inadvertently released through cell lysis), it could survive well enough to impact another bacterium. These questions parallel those that have been addressed in the quorum-sensing field, in which acyl homoserine lactone molecules serve as signals for determining cell density (18, 71). It is clear from work in that field that competition exists, that cheater bacteria can take up signal molecules that they themselves do not synthesize, and that organisms exist that can degrade these external signals (reviewed in references 64, 66, and 68). Finally, the quorum-sensing field has identified both agonists and antagonists for signaling in specific bacteria. While much remains to be learned about the same in the c-di-GMP field, these early studies are very encouraging.

Beyond controlling bacterial infections, c-di-GMP may also be useful for stimulating the immune system. As described in greater detail in chapters 21 and 22, c-di-GMP induces specific production of cytokines by host cells. These studies support the possibility that c-di-GMP could be used in vaccines as an adjuvant in combination with antigen to induce a greater immune response. This is a promising area of current research.

CONCLUDING REMARKS: LESSONS FROM TWO-COMPONENT SIGNAL TRANSDUCTION

The evolution of the c-di-GMP field is not unlike that of the two-component signal transduction field.

In the early 1980s, many laboratories were actively dissecting diverse signal transduction pathways, among them initiation of sporulation, chemotaxis, the regulation of nitrogen and phosphate assimilation, osmoregulation, and nitrogen fixation. For the most part, each was considered to be a completely separate field. All of that changed in 1987 when an early bioinformatic analysis revealed that these and other signaling pathways possessed at their core a common module predicted to include two distinct sequences, HisKA and REC (49). Following as it did on the heels of a critical report that showed that activation of the nitrogen assimilation signal pathway required phosphotransfer from a HisKA-containing protein to a REC domain protein (43), this study led to the identification of the so-called two-component module composed of histidine sensor kinases and response regulators and thus generated the seeds that led to a new paradigm. Indeed, by the end of the 1980s, several groups had demonstrated that their own two-component signal transduction pathways involved a similar phosphotransfer event, had begun to identify the amino acid residues required for the transfer, and had begun structural studies. For reviews, see references 6, 25, 35, 39, 42, and 55.

For the next few years, the main focus was to discover the fundamental structure-function relationships of what was perceived as the two-component central module, and each new report seemed to more firmly establish the mechanism by which that central module worked. But reports of strange exceptions also leaked out; for example, the presence of extra REC domains, the ability of some sensor kinases to function as phosphatases, and the lack of conserved phosphoacceptor residues within certain REC domains. During this period, the issue of cross talk and specificity came to the fore along with discourse concerning orphan response regulators and small phosphodonors. And throughout, it was clearly obvious that each central module was associated with distinct sensory input and effector domains. For reviews, see reference 24.

Eventually, it became clear that one could indeed define the typical two-component module but that many variations on that theme had evolved. This was quite apparent from studies of one of the two components, the response regulator. In most cases, the REC domain within the response regulator is linked to an effector domain. In the majority of these cases, this effector domain functions in the binding of the response regulator to DNA. Despite these commonalities, comparisons have shown many differences. Indeed, a comparison of the structures of the inactive and active forms of multiple members of one subset of the response regulator family (the PhoB/OmpR subfamily) led to the conclusion "...that, despite common structures and mechanisms of function within individual domains, a range of interactions between receiver and effector domains confer great diversity in regulatory strategies..." (19). Presumably, these differences are necessary to fulfill the specific contextual requirements of each two-component system in each organism.

The parallels with the nascent c-di-GMP field are readily apparent. In each case, bioinformatics provided the critical push, thrusting groups of individual laboratories working diligently to understand their own systems into a field. In the time between the establishment of the two fields, bioinformatics itself matured to the extent that it played a much more prominent and fast-paced role in the development of the c-di-GMP field. For both fields, there also was a rush to identify the common themes and the inevitable discovery of the exceptions. Maturation of the c-di-GMP field from early adolescence into adulthood will be smoothed if we pay attention to the lessons learned in the two-component field: exceptions exist and there is danger in prematurely establishing dogma. While some paradigms are likely true, others will not stand the test of time.

Acknowledgments. We thank Mark Gomelsky, Holger Sondermann, Regine Hengge, and Urs Jenal for helpful comments.

Work in our laboratories is funded by NIH R01 grants GM066130 to A.J.W. and GM59690 to K.L.V. We also thank the Loyola University Medical Center Research Funding Council for funding our work on the role of c-di-GMP in the motility of *V. fischeri*.

REFERENCES

1. **Amikam, D., and M. Y. Galperin.** 2006. PilZ domain is part of the bacterial c-di-GMP binding protein. *Bioinformatics* **22:** 3–6.
2. **Barends, T. R., E. Hartmann, J. J. Griese, T. Beitlich, N. V. Kirienko, D. A. Ryjenkov, J. Reinstein, R. L. Shoeman, M. Gomelsky, and I. Schlichting.** 2009. Structure and mechanism of a bacterial light-regulated cyclic nucleotide phosphodiesterase. *Nature* **459:**1015–1018.
3. **Benach, J., S. S. Swaminathan, R. Tamayo, S. K. Handelman, E. Folta-Stogniew, J. E. Ramos, F. Forouhar, H. Neely, J. Seetharaman, A. Camilli, and J. F. Hunt.** 2007. The structural basis of cyclic diguanylate signal transduction by PilZ domains. *EMBO J.* **26:**5153–5166.
4. **Bender, A. T., and J. A. Beavo.** 2006. Cyclic nucleotide phosphodiesterases: molecular regulation to clinical use. *Pharmacol. Rev.* **58:**488–520.
5. **Beyhan, S., L. S. Odell, and F. H. Yildiz.** 2008. Identification and characterization of cyclic diguanylate signaling systems controlling rugosity in *Vibrio cholerae*. *J. Bacteriol.* **190:**7392–7405.
6. **Bourret, R. B., J. F. Hess, K. A. Borkovich, A. A. Pakula, and M. I. Simon.** 1989. Protein phosphorylation in chemotaxis and two-component regulatory systems of bacteria. *J. Biol. Chem.* **264:**7085–7088.

7. **Brouillette, E., M. Hyodo, Y. Hayakawa, D. K. Karaolis, and F. Malouin.** 2005. 3′,5′-Cyclic diguanylic acid reduces the virulence of biofilm-forming *Staphylococcus aureus* strains in a mouse model of mastitis infection. *Antimicrob. Agents Chemother.* **49:**3109–3113.
8. **Brunton, L. L.** 2003. PDE4: arrested at the border. *Sci. STKE* **2003:**PE44.
9. **Bryers, J. D.** 2008. Medical biofilms. *Biotechnol. Bioeng.* **100:** 1–18.
10. **Chan, C., R. Paul, D. Samoray, N. C. Amiot, B. Giese, U. Jenal, and T. Schirmer.** 2004. Structural basis of activity and allosteric control of diguanylate cyclase. *Proc. Natl. Acad. Sci. USA* **101:**17084–17089.
11. **Christen, B., M. Christen, R. Paul, F. Schmid, M. Folcher, P. Jenoe, M. Meuwly, and U. Jenal.** 2006. Allosteric control of cyclic di-GMP signaling. *J. Biol. Chem.* **281:**32015–32024.
12. **Christen, M., B. Christen, M. G. Allan, M. Folcher, P. Jeno, S. Grzesiek, and U. Jenal.** 2007. DgrA is a member of a new family of cyclic diguanosine monophosphate receptors and controls flagellar motor function in *Caulobacter crescentus. Proc. Natl. Acad. Sci. USA* **104:**4112–4117.
13. **Christen, M., B. Christen, M. Folcher, A. Schauerte, and U. Jenal.** 2005. Identification and characterization of a cyclic di-GMP-specific phosphodiesterase and its allosteric control by GTP. *J. Biol. Chem.* **280:**30829–30837.
14. **Costerton, J. W., P. S. Stewart, and E. P. Greenberg.** 1999. Bacterial biofilms: a common cause of persistent infections. *Science* **284:**1318–1322.
15. **De, N., M. Pirruccello, P. V. Krasteva, N. Bae, R. V. Raghavan, and H. Sondermann.** 2008. Phosphorylation-independent regulation of the diguanylate cyclase WspR. *PLoS Biol.* **6:**e67.
16. **Delgado-Nixon, V. M., G. Gonzalez, and M. A. Gilles-Gonzalez.** 2000. Dos, a heme-binding PAS protein from *Escherichia coli*, is a direct oxygen sensor. *Biochemistry* **39:** 2685–2691.
17. **Duerig, A., S. Abel, M. Folcher, M. Nicollier, T. Schwede, N. Amiot, B. Giese, and U. Jenal.** 2009. Second messenger-mediated spatiotemporal control of protein degradation regulates bacterial cell cycle progression. *Genes Dev.* **23:**93–104.
18. **Dunny, G. M., and S. C. Winans (ed.).** 1999. *Cell-Cell Signaling in Bacteria.* American Society for Microbiology Press, Washington, DC.
19. **Gao, R., T. R. Mack, and A. M. Stock.** 2007. Bacterial response regulators: versatile regulatory strategies from common domains. *Trends Biochem. Sci.* **32:**225–234.
20. **Gjermansen, M., P. Ragas, and T. Tolker-Nielsen.** 2006. Proteins with GGDEF and EAL domains regulate *Pseudomonas putida* biofilm formation and dispersal. *FEMS Microbiol. Lett.* **265:**215–224.
21. **Hazelbauer, G. L., J. J. Falke, and J. S. Parkinson.** 2008. Bacterial chemoreceptors: high-performance signaling in networked arrays. *Trends Biochem. Sci.* **33:**9–19.
22. **Hecht, G., and A. Newton.** 1995. Identification of a novel response regulator required for the swarmer-to-stalked-cell transition in *Caulobacter crescentus. J. Bacteriol.* **177:**6223–6229.
23. **Hengge, R.** 2009. Principles of c-di-GMP signalling in bacteria. *Nat. Rev. Microbiol.* **7:**263–273.
24. **Hoch, J. A., and T. J. Silhavy (ed.).** 1995. *Two-Component Signal Transduction.* ASM Press, Washington, DC.
25. **Igo, M. M., J. M. Slauch, and T. J. Silhavy.** 1990. Signal transduction in bacteria: kinases that control gene expression. *New Biol.* **2:**5–9.
26. **Jenal, U., and J. Malone.** 2006. Mechanisms of cyclic-di-GMP signaling in bacteria. *Annu. Rev. Genet.* **40:**385–407.
27. **Karaolis, D. K., M. H. Rashid, R. Chythanya, W. Luo, M. Hyodo, and Y. Hayakawa.** 2005. c-di-GMP (3′-5′-cyclic diguanylic acid) inhibits *Staphylococcus aureus* cell-cell interactions and biofilm formation. *Antimicrob. Agents Chemother.* **49:**1029–1038.
28. **Karatan, E., T. R. Duncan, and P. I. Watnick.** 2005. NspS, a predicted polyamine sensor, mediates activation of *Vibrio cholerae* biofilm formation by norspermidine. *J. Bacteriol.* **187:** 7434–7443.
29. **Karatan, E., and P. Watnick.** 2009. Signals, regulatory networks, and materials that build and break bacterial biofilms. *Microbiol. Mol. Biol. Rev.* **73:**310–347.
30. **Kulasakara, H., V. Lee, A. Brencic, N. Liberati, J. Urbach, S. Miyata, D. G. Lee, A. N. Neely, M. Hyodo, Y. Hayakawa, F. M. Ausubel, and S. Lory.** 2006. Analysis of *Pseudomonas aeruginosa* diguanylate cyclases and phosphodiesterases reveals a role for bis-(3′-5′)-cyclic-GMP in virulence. *Proc. Natl. Acad. Sci. USA* **103:**2839–2844.
31. **Kumar, M., and D. Chatterji.** 2008. Cyclic di-GMP: a second messenger required for long-term survival, but not for biofilm formation, in *Mycobacterium smegmatis. Microbiology* **154:** 2942–2955.
32. **Lai, T. H., Y. Kumagai, M. Hyodo, Y. Hayakawa, and Y. Rikihisa.** 2009. The *Anaplasma phagocytophilum* PleC histidine kinase and PleD diguanylate cyclase two-component system and role of cyclic di-GMP in host cell infection. *J. Bacteriol.* **191:**693–700.
33. **Lee, V. T., J. M. Matewish, J. L. Kessler, M. Hyodo, Y. Hayakawa, and S. Lory.** 2007. A cyclic-di-GMP receptor required for bacterial exopolysaccharide production. *Mol. Microbiol.* **65:**1474–1484.
34. **Li, T. N., K. H. Chin, J. H. Liu, A. H. Wang, and S. H. Chou.** 2009. XC1028 from *Xanthomonas campestris* adopts a PilZ domain-like structure without a c-di-GMP switch. *Proteins* **75:** 282–288.
35. **Magasanik, B.** 1989. Regulation of transcription of the *glnALG* operon of *Escherichia coli* by protein phosphorylation. *Biochimie* **71:**1005–1012.
36. **McCarthy, Y., R. P. Ryan, K. O'Donovan, Y. Q. He, B. L. Jiang, J. X. Feng, J. L. Tang, and J. M. Dow.** 2008. The role of PilZ domain proteins in the virulence of *Xanthomonas campestris* pv. *campestris. Mol. Plant Pathol.* **9:**819–824.
37. **Merighi, M., V. T. Lee, M. Hyodo, Y. Hayakawa, and S. Lory.** 2007. The second messenger bis-(3′-5′)-cyclic-GMP and its PilZ domain-containing receptor Alg44 are required for alginate biosynthesis in *Pseudomonas aeruginosa. Mol. Microbiol.* **65:**876–895.
38. **Minasov, G., S. Padavattan, L. Shuvalova, J. S. Brunzelle, D. J. Miller, A. Basle, C. Massa, F. R. Collart, T. Schirmer, and W. F. Anderson.** 2009. Crystal structures of YkuI and its complex with second messenger cyclic di-GMP suggest catalytic mechanism of phosphodiester bond cleavage by EAL domains. *J. Biol. Chem.* **284:**13174–13184.
39. **Mizuno, T., and S. Mizushima.** 1990. Signal transduction and gene regulation through the phosphorylation of two regulatory components: the molecular basis for the osmotic regulation of the porin genes. *Mol. Microbiol.* **4:**1077–1082.
40. **Neunuebel, M. R., and J. W. Golden.** 2008. The *Anabaena* sp. strain PCC 7120 gene all2874 encodes a diguanylate cyclase and is required for normal heterocyst development under high-light growth conditions. *J. Bacteriol.* **190:**6829–6836.
41. **Newell, P. D., R. D. Monds, and G. A. O'Toole.** 2009. LapD is a bis-(3′,5′)-cyclic dimeric GMP-binding protein that regulates surface attachment by *Pseudomonas fluorescens* Pf0-1. *Proc. Natl. Acad. Sci. USA* **106:**3461–3466.

42. **Ninfa, A. J.** 1991. Protein phosphorylation and the regulation of cellular processes by the homologous two-component regulatory systems of bacteria. *Genet. Eng.* (New York) **13:**39–72.
43. **Ninfa, A. J., and B. Magasanik.** 1986. Covalent modification of the *glnG* product, NRI, by the *glnL* product, NRII, regulates the transcription of the *glnALG* operon in *Escherichia coli*. *Proc. Natl. Acad. Sci. USA* **83:**5909–5913.
44. **Palmer, R. J., Jr., and P. Stoodley.** 2007. Biofilms 2007: broadened horizons and new emphases. *J. Bacteriol.* **189:**7948–7960.
45. **Parsek, M. R., and C. Fuqua.** 2004. Biofilms 2003: emerging themes and challenges in studies of surface-associated microbial life. *J. Bacteriol.* **186:**4427–4440.
46. **Pratt, J. T., R. Tamayo, A. D. Tischler, and A. Camilli.** 2007. PilZ domain proteins bind cyclic diguanylate and regulate diverse processes in *Vibrio cholerae*. *J. Biol. Chem.* **282:**12860–12870.
47. **Ramelot, T. A., A. Yee, J. R. Cort, A. Semesi, C. H. Arrowsmith, and M. A. Kennedy.** 2007. NMR structure and binding studies confirm that PA4608 from *Pseudomonas aeruginosa* is a PilZ domain and a c-di-GMP binding protein. *Proteins* **66:**266–271.
48. **Rogers, E. A., D. Terekhova, H. M. Zhang, K. M. Hovis, I. Schwartz, and R. T. Marconi.** 2009. Rrp1, a cyclic-di-GMP-producing response regulator, is an important regulator of *Borrelia burgdorferi* core cellular functions. *Mol. Microbiol.* **71:**1551–1573.
49. **Ronson, C. W., B. T. Nixon, and F. M. Ausubel.** 1987. Conserved domains in bacterial regulatory proteins that respond to environmental stimuli. *Cell* **49:**579–581.
50. **Ross, P., H. Weinhouse, Y. Aloni, D. Michaeli, P. Weinberger-Ohana, R. Mayer, S. Braun, E. de Vroom, G. A. van der Marel, J. H. van Boom, and M. Benziman.** 1987. Regulation of cellulose synthesis in *Acetobacter xylinum* by cyclic diguanylic acid. *Nature* **325:**279–281.
51. **Ryan, R. P., J. Lucey, K. O'Donovan, Y. McCarthy, L. Yang, T. Tolker-Nielsen, and J. M. Dow.** 2009. HD-GYP domain proteins regulate biofilm formation and virulence in *Pseudomonas aeruginosa*. *Environ. Microbiol.* **11:**1126–1136.
52. **Ryjenkov, D. A., R. Simm, U. Romling, and M. Gomelsky.** 2006. The PilZ domain is a receptor for the second messenger c-di-GMP. The PilZ domain protein YcgR controls motility in enterobacteria. *J. Biol. Chem.* **281:**30310–30314.
53. **Solano, C., B. Garcia, C. Latasa, A. Toledo-Arana, V. Zorraquino, J. Valle, J. Casals, E. Pedroso, and I. Lasa.** 2009. Genetic reductionist approach for dissecting individual roles of GGDEF proteins within the c-di-GMP signaling network in *Salmonella*. *Proc. Natl. Acad. Sci. USA* **106:**7997–8002.
54. **Stewart, P. S., and J. W. Costerton.** 2001. Antibiotic resistance of bacteria in biofilms. *Lancet* **358:**135–138.
55. **Stock, J. B., A. J. Ninfa, and A. M. Stock.** 1989. Protein phosphorylation and regulation of adaptive responses in bacteria. *Microbiol. Rev.* **53:**450–490.
56. **Takahashi, H., and T. Shimizu.** 2006. Phosphodiesterase activity of Ec DOS, a heme-regulated enzyme from *Escherichia coli*, toward 3′,5′-cyclic diguanylic acid is obviously enhanced by O2 and CO binding. *Chem. Lett.* **35:**970–971.
57. **Tal, R., H. C. Wong, R. Calhoon, D. Gelfand, A. L. Fear, G. Volman, R. Mayer, P. Ross, D. Amikam, H. Weinhouse, A. Cohen, S. Sapir, P. Ohana, and M. Benziman.** 1998. Three *cdg* operons control cellular turnover of cyclic di-GMP in *Acetobacter xylinum:* genetic organization and occurrence of conserved domains in isoenzymes. *J. Bacteriol.* **180:**4416–4425.
58. **Tamayo, R., J. T. Pratt, and A. Camilli.** 2007. Roles of cyclic diguanylate in the regulation of bacterial pathogenesis. *Annu. Rev. Microbiol.* **61:**131–148.
59. **Tamayo, R., S. Schild, J. T. Pratt, and A. Camilli.** 2008. Role of cyclic di-GMP during El Tor biotype *Vibrio cholerae* infection: characterization of the in vivo-induced cyclic di-GMP phosphodiesterase CdpA. *Infect. Immun.* **76:**1617–1627.
60. **Tarutina, M., D. A. Ryjenkov, and M. Gomelsky.** 2006. An unorthodox bacteriophytochrome from *Rhodobacter sphaeroides* involved in turnover of the second messenger c-di-GMP. *J. Biol. Chem.* **281:**34751–34758.
61. **Thormann, K. M., S. Duttler, R. M. Saville, M. Hyodo, S. Shukla, Y. Hayakawa, and A. M. Spormann.** 2006. Control of formation and cellular detachment from *Shewanella oneidensis* MR-1 biofilms by cyclic di-GMP. *J. Bacteriol.* **188:**2681–2691.
62. **Tischler, A. D., and A. Camilli.** 2005. Cyclic diguanylate regulates *Vibrio cholerae* virulence gene expression. *Infect. Immun.* **73:**5873–5882.
63. **Tomita, T., G. Gonzalez, A. L. Chang, M. Ikeda-Saito, and M. A. Gilles-Gonzalez.** 2002. A comparative resonance Raman analysis of heme-binding PAS domains: heme iron coordination structures of the BjFixL, AxPDEA1, EcDos, and MtDos proteins. *Biochemistry* **41:**4819–4826.
64. **Uroz, S., Y. Dessaux, and P. Oger.** 2009. Quorum sensing and quorum quenching: the yin and yang of bacterial communication. *Chembiochem* **10:**205–216.
65. **Vandecasteele, G., F. Rochais, A. Abi-Gerges, and R. Fischmeister.** 2006. Functional localization of cAMP signalling in cardiac myocytes. *Biochem. Soc. Trans.* **34:**484–488.
66. **von Bodman, S. B., J. M. Willey, and S. P. Diggle.** 2008. Cell-cell communication in bacteria: united we stand. *J. Bacteriol.* **190:**4377–4391.
67. **Wan, X., J. R. Tuckerman, J. A. Saito, T. A. Freitas, J. S. Newhouse, J. R. Denery, M. Y. Galperin, G. Gonzalez, M. A. Gilles-Gonzalez, and M. Alam.** 2009. Globins synthesize the second messenger bis-(3′-5′)-cyclic diguanosine monophosphate in bacteria. *J. Mol. Biol.* **388:**262–270.
68. **Wang, Y. J., J. J. Huang, and J. R. Leadbetter.** 2007. Acyl-HSL signal decay: intrinsic to bacterial cell-cell communications. *Adv. Appl. Microbiol.* **61:**27–58.
69. **Wassmann, P., C. Chan, R. Paul, A. Beck, H. Heerklotz, U. Jenal, and T. Schirmer.** 2007. Structure of BeF(3)(-)-modified response regulator PleD: implications for diguanylate cyclase activation, catalysis, and feedback inhibition. *Structure* **15:**915–927.
70. **Willoughby, D., and D. M. Cooper.** 2007. Organization and Ca^{2+} regulation of adenylyl cyclases in cAMP microdomains. *Physiol. Rev.* **87:**965–1010.
71. **Winans, S. C., and B. L. Bassler.** 2008. *Chemical Communication among Bacteria*. ASM Press, Washington, DC.
72. **Wolfe, A. J., and K. L. Visick.** 2008. Get the message out: cyclic-di-GMP regulates multiple levels of flagellum-based motility. *J. Bacteriol.* **190:**463–475.
73. **Wu, Q., and K. Gardner.** 3 February 2009. Structure and insight into blue light-induced changes in the BlrP1 BLUF domain. *Biochemistry* [Epub ahead of print.]
74. **Wu, Q., W. H. Ko, and K. H. Gardner.** 2008. Structural requirements for key residues and auxiliary portions of a BLUF domain. *Biochemistry* **47:**10271–10280.
75. **Yan, W., T. Qu, H. Zhao, L. Su, Q. Yu, J. Gao, and B. Wu.** 30 January 2009. The effect of c-di-GMP (3′-5′-cyclic diguanylic acid) on the biofilm formation and adherence of *Streptococcus mutans*. *Microbiol. Res.* [Epub ahead of print.]

INDEX